ASTRONOMICAL DATA ANALYSIS SOFTWARE AND SYSTEMS XIV

COVER ILLUSTRATION:

The Grist project is developing a library of compute-based web services to enable astronomers and the public to interact with massive, complex datasets in the Virtual Observatory. This Figure has been extracted from the paper by Jacob et al. entitled, Grist: Grid-based Data Mining for Astronomy, p. 306.

ASTRONOMICAL SOCIETY OF THE PACIFIC
CONFERENCE SERIES

A SERIES OF BOOKS ON RECENT DEVELOPMENTS IN
ASTRONOMY AND ASTROPHYSICS

Volume 347

Beginning 2004, published volumes may be found as e-books with color images at:
http://www.aspbooks.org

A listing of ASP Conference Series and IAU volumes
published by the ASP during the past three years may be found at the back of this volume.

ASTRONOMICAL SOCIETY OF THE PACIFIC
CONFERENCE SERIES

Volume 347

ASTRONOMICAL DATA ANALYSIS SOFTWARE AND SYSTEMS XIV

Proceedings of a conference held in
Pasadena, California, USA
24-27 October 2004

Edited by

Patrick L. Shopbell
California Institute of Technology, Pasadena, California, USA

Matthew C. Britton
California Institute of Technology, Pasadena, California, USA

and

Rick Ebert
Infrared Processing & Analysis Center, California Institute of Technology
Pasadena, California, USA

SAN FRANCISCO

ASTRONOMICAL SOCIETY OF THE PACIFIC

390 Ashton Avenue
San Francisco, California, 94112-1722, USA
Phone: 415-337-1100
Fax: 415-337-5205
E-mail: service@astrosociety.org
Web Site: www.astrosociety.org
E-books: www.aspbooks.org

ASP Conference Series - First Edition

ISBN: 1-58381-215-6
ISSN: 1080-7926

Library of Congress Cataloging in Publication Data
Main entry under title

Printed in United States of America by Sheridan Books, Ann Arbor, Michigan

Contents

Part 4. Algorithms and Data Analysis

Part 6. The Virtual Observatory: Grid

Part 8. Great Space Observatories

Part 10. Data Pipelines

Part 11. Data Archives

Preface

This volume contain papers presented at the fourteenth annual conference for Astronomical Data Analysis Software and Systems (ADASS XIV). The meeting was held during October 24-27, 2004 at the Pasadena Sheraton Hotel in Pasadena, California. The meeting was hosted by the Infrared Processing and Analysis Center (IPAC) and the California Institute of Technology (Caltech).

1. Conference Overview

Pasadena, California is the home of the famous Rose Parade and Rose Bowl, the well-known Huntington Gardens, the California Institute of Technology (Caltech), and the Jet Propulsion Laboratory (JPL). Pasadena and the greater Los Angeles area provided the conference participants with ample distractions and pleasant weather, typical of southern California in October.

ADASS XIV was attended by 309 registered participants from 19 countries. Over one third of the attendees were visiting from outside of the United States. The program comprised 15 invited speakers and 22 contributed oral presentations, 157 posters, 12 exhibit demo booths, 5 Birds of a Feather (BoF) sessions, and 13 "focus" demo sessions (a new addition with this conference).

Presentations were solicited by the Program Organizing Committee (POC) on key topics, including "Algorithms", "Great Space Observatories", "Ground-based Observatories", and the "Virtual Observatory". Easily a third of the conference (and three sections of this volume) were dedicated to the last of these topics, as the first "real world" Virtual Observatory (VO) applications started to supplant the past emphasis on VO protocols and standards.

The BoF sessions took place on Sunday evening and Monday afternoon. The topics and organizers were: "FITS" (William Pence [GSFC]), "Sky Indexation, Pixelization and the VO" (William O'Mullane [JHU]), "Future Astronomical Data Analysis Environments" (Preben Grosbøl [ESO] and Doug Tody [NRA0]), "Python: A Better Development Language for Astronomical Data Analysis" (Lee Rottler [IPAC], Paul Barrett [STScI], and Perry Greenfield [STScI]), and "Customising and Extending DS9: Community Discussion" (De Clarke [UCO/Lick], Bill Joye [SAO], and Steve Allen [UCO/Lick]). Thanks goes to these volunteers for making the BoF sessions possible.

A new feature this year were the "Focus Demonstration" sessions, in which dedicated meeting rooms were provided to developers as an opportunity to demonstrate their application to a small, focused audience. These sessions were well received and seem likely to continue as an aspect of ADASS at future meetings.

2. Organizing Committees and Sponsors

The semi-permanent ADASS Program Organizing Committee (POC) at the time of ADASS XIV consisted of: Tim Cornwell (NRAO), Dick Crutcher (UIUC/NCSA), Daniel Durand (CADC), Daniel Egret (CDS), Carlos Gabriel (ESA-VILSPA), Tom Handley (IPAC/Caltech/JPL), Richard Hook (ST-ECF), Jim Lewis (IoA), Glenn Miller (STScI), Koh-Ichiro Morita (NAOJ), François Ochsenbein (CDS), Michele Peron (ESO), Arnold Rots (SAO; POC Chair), Dick Shaw (NOAO), Betty Stobie (Univ. of Arizona), and Christian Veillet (CFHT).

The Local Organizing Committee (LOC) was chaired by Tom Handley, and the members (all at IPAC except where noted) were: Mary Ellen Barba, Rick Ebert, Bill Green, Eloise Kennedy, Mike Khuong, Jake Llamas, Steven Lo, Rosanne Scholey, Patrick Shopbell (Caltech), and Carl Smay. We also thank all of the volunteers from IPAC and Caltech who helped ensure the success of the conference.

ADASS XIV is pleased to express our deep appreciation for the support of the following organizations and companies: the Infrared Processing and Analysis Center (IPAC), the California Institute of Technology (Caltech), the European Southern Observatory (ESO), the National Optical Astronomy Observatories (NOAO), the National Radio Astronomy Observatory (NRAO), the Canada-France-Hawaii Telescope (CFH), the Smithsonian Astrophysical Observatory (SAO), Centre de Donnes astronomiques de Strasbourg (CDS), the Space Telescope Science Institute (STScI), the National Science Foundation (NSF), the National Aeronautics and Space Administration (NASA), and the Jet Propulsion Laboratory (JPL).

3. ADASS Information

The ADASS XV conference will be held in San Lorenzo de El Escorial, Spain, October 2-5, 2005. It will be hosted by the European Space Agency (ESA). Details concerning all of the ADASS meetings, as well as electronic versions of the proceedings, may be found on the ADASS web site: www.adass.org.

The ADASS XIV Editors

Patrick L. Shopbell
California Institute of Technology

Matthew C. Britton
California Institute of Technology

Rick Ebert
Infrared Processing and Analysis Center,
California Institute of Technology

August 2005

Participants

Ghaleb Abdulla, LLNL, Lawrence Livermore National Lab, P.O. Box 808, L-560, Livermore, CA 94551, U.S.A. ⟨ abdulla1@llnl.gov ⟩

Rudolf Albrecht, ST-ECF, Karl-Schwarzschild-Str. 2, Garching, BY D-85748, Germany ⟨ ralbrech@eso.org ⟩

Anastasia Alexov, IPAC/Caltech, 770 S. Wilson Ave., Mail Code 100-22, Pasadena, CA 91125, U.S.A. ⟨ aalexov@ipac.caltech.edu ⟩

Alasdair Allan, University of Exeter, Astrophysics Group, School of Physics, Stocker Road, Exeter, EX4 4QL, United Kingdom ⟨ aa@astro.ex.ac.uk ⟩

Steven Allen, UCO/Lick Observatory, UCO/Lick Observatory ISB, University of California, Santa Cruz, CA 95064, U.S.A. ⟨ sla@ucolick.org ⟩

Terence Allen, Johns Hopkins University, Dept of Physics & Astronomy , JHU, 3400 N. Charles St, Baltimore, MD 21218, U.S.A. ⟨ terrya@pha.jhu.edu ⟩

Narsipur Amarnath, University of Maryland, Department of Astronomy, University of Maryland, College Park, MD 20723, U.S.A. ⟨ amar@astro.umd.edu ⟩

Kenneth Anderson, Johns Hopkins University, Dept of Physics & Astronomy, JHU, 3400 N. Charles St, Baltimore, MD 21218, U.S.A. ⟨ anderson@pha.jhu.edu ⟩

Salim Ansari, European Space Agency/ESTEC, ESA - ESTEC, Keplerlaan 1, Noordwijk, ZH 2200 AG, Netherlands ⟨ Salim.Ansari@esa.int ⟩

Cosimo Antonio Volpicelli, INAF/OATo, str. Osservatorio 20, Pino Torinese, TO 10025, Italy ⟨ volpicelli@to.astro.it ⟩

Raymond Arvidson, Washington University, One Brookings Dr, Campus Box 1169, St. Louis, MO 63130, U.S.A. ⟨ arvidson@wunder.wustl.edu ⟩

Christophe Aviset, ESA, PO Box 50727, Madrid 28080, Spain ⟨ Christophe.Arviset@esa.int ⟩

Charles Backus, NASA/JPL, 3400 Ellington Villa Drive, Altadena, CA 91001, U.S.A. ⟨ charles.backus@jpl.nasa.gov ⟩

Klaus Banse, E.S.O., Karl-Schwarzschild-str. 2, Garching 85748, Germany ⟨ kbanse@eso.org ⟩

Mary Ellen Barba, IPAC/Caltech, 770 South Wilson Ave., MS 100-22, Pasadena, CA 91125, U.S.A. ⟨ meb@ipac.caltech.edu ⟩

Paul Barrett, STScI, 3700 San Martin Drive, Baltimore, MD 21218, U.S.A. ⟨barrett@stsci.edu⟩

Andrew Beard, CARMA, 100 Leighton Ln, Big Pine, CA 93513, U.S.A. ⟨andy@ovro.caltech.edu⟩

Ugo Becciani, INAF - Astrophysical Observatory of Catania, INAF - Osservatorio Astrofisico di Catania, Via S. Sofia 78, Catania I-95123, Italy ⟨ube@ct.astro.it⟩

Mathias Beck, INTEGRAL Science Data Centre, Chemin d'Ecogia 16, Versoix 1290, Switzerland ⟨Mathias.Beck@obs.unige.ch⟩

Kevin Benson, MSSL/UCL, Mullard Space Science Laboratory, Holmbury House, Holmbury St Mary, Dorking, Surrey RH5 6NT, United Kingdom ⟨kmb@mssl.ucl.ac.uk⟩

Bruce Berriman, IPAC/Caltech, 770 S. Wilson Ave., Mail Code 100-22, Pasadena, CA 91125, U.S.A. ⟨gbb@ipac.caltech.edu⟩

Emmanuel Bertin, IAP/Obs. Paris, IAP, 98bis boulevard Arago, Paris F-75014, France ⟨bertin@iap.fr⟩

Sanjay Bhatnagar, NRAO Socorro, 1003 Lopezville Road, Socorro, NM 87801, U.S.A. ⟨sbhatnag@aoc.nrao.edu⟩

Sarah Blecksmith, SAO, 60 Garden St, MS 67, Cambridge MA 02138, U.S.A. ⟨sarah@head.cfa.harvard.edu⟩

David Bohlender, Canadian Astronomy Data Centre, National Research Council Canada, 5071 West Saanich Road, Victoria, BC V9E 2E7, Canada ⟨david.bohlender@nrc-cnrc.gc.ca⟩

Bruce Bon, Subaru Telescope, 650 N. A'ohoku Place, Hilo, HI 96720, U.S.A. ⟨Bruce.Bon@SubaruTelescope.org⟩

Kirk Borne, George Mason University, 602 Stamford Rd., Baltimore, MD 21229, U.S.A. ⟨kborne@gmu.edu⟩

Todd Boroson, NOAO, PO Box 26732, Tucson, AZ 85726-6732, U.S.A. ⟨tyb@noao.edu⟩

Rychard Bouwens, UCO Lick Observatory, Astronomy Department, 1156 High Street, Santa Cruz, CA 95060, U.S.A. ⟨bouwens@ucolick.org⟩

Heidi Brandenburg, IPAC/SSC, MS 220-6, 1200 E California Blvd, Pasadena, CA 91125, U.S.A. ⟨heidi@ipac.caltech.edu⟩

Paul Bristow, ST-ECF, Karl-Schwarzschild-Str. 2, Garching, BY 85748, Germany ⟨bristowp@eso.org⟩

Matthew Britton, Caltech, MC 105-24, 1200 E. California Blvd., Pasadena, CA 91125, U.S.A. ⟨mbritton@astro.caltech.edu⟩

Marianne Brouty, IPAC/Caltech/NED, IPAC MS 100-22, Pasadena, CA 91125, U.S.A. ⟨marianne@ipac.caltech.edu⟩

Peter Bunclark, Institute of Astronomy, Madingley Road, Cambridge, CB3 0HA, United Kingdom ⟨psb@ast.cam.ac.uk⟩

Douglas Burke, Harvard-Smithsonian Center for Astrophysics, 60 Garden St., MS-83, Cambridge, MA 02138, U.S.A. ⟨dburke@cfa.harvard.edu⟩

HOWARD BUSHHOUSE, STScI, 3700 San Martin Drive, Baltimore, MD 21218, U.S.A. ⟨ bushouse@stsci.edu ⟩

TOM CALDERWOOD, SAO/CXC, 60 Garden St, MS-81, Cambridge, MA 02138, U.S.A. ⟨ tcalderw@head-cfa.harvard.edu ⟩

WILLIAM CARITHERS, Lawrence Berkeley Lab, LBNL, MS 50R-6048, 1 Cyclotron Rd, Berkeley, CA 94720, U.S.A. ⟨ wccarithers@lbl.gov ⟩

BRAD CAVANAGH, Joint Astronomy Centre, 660 N. A'ohoku Place, University Park, Hilo, HI 96720, U.S.A. ⟨ b.cavanagh@jach.hawaii.edu ⟩

TONY CHAN, UCLA, The UCLA College - L & S, 2300 Murphy Hall, Box 951438, Los Angeles, CA 90095-1438, U.S.A. ⟨ tonyc@college.ucla.edu ⟩

ANIL CHANDRA, CSIRO, CSIRO - ATNF, PO Box 76, Epping, NSW 1710, Australia ⟨ anil.chandra@csiro.au ⟩

JOE CHAVEZ, Caltech, Mail Code 220-6, 1200 E. California Blvd, Pasadena, CA 91125, U.S.A. ⟨ jchavez@ipac.caltech.edu ⟩

JAMES CHIANG, GSFC/UMBC, SLAC MS 43A, 2575 Sand Hill Road, Menlo Park, CA 94025, U.S.A. ⟨ jchiang@slac.stanford.edu ⟩

IGOR CHILINGARIAN, SAI MSU, Apt. 76, Leninsky prospect 73/8, Moscow 119296, Russia ⟨ chil@sai.msu.su ⟩

NIAN-MING CHIU, Caltech/IPAC, 770 S. Wilson Ave., Mail Code 100-22, Pasadena, CA 91125, U.S.A. ⟨ nmc@ipac.caltech.edu ⟩

DE CLARKE, UCO/Lick Observatory, Lick Obs ISB UCSC, 1156 High Street, Santa Cruz, CA 95964, U.S.A. ⟨ de@ucolick.org ⟩

GHELLER CLAUDIO, CINECA, Via Magnanelli 6/3, Casalecchio di Reno, Bologna I-40033, Italy ⟨ c.gheller@cineca.it ⟩

JOERG COLBERG, University of Pittsburgh, 3941 O'Hara Street, 100 Allen Hall, Pittsburgh, PA 15260, U.S.A. ⟨ astro@jmcolberg.com ⟩

ALBERT CONRAD, W. M. Keck Observtory, 65-1120 Mamalahoa Hwy, Kamuela, HI 96743, U.S.A. ⟨ aconrad@keck.hawaii.edu ⟩

ALBERTO CONTI, Space Telescope Science Institute, 3700 San Martin Dr, Baltimore, MD 21218, U.S.A. ⟨ aconti@stsci.edu ⟩

ANDREW COOKE, NOAO/CTIO/AURA, 950 N Cherry Av, Tucson, AZ 85719-4933, U.S.A. ⟨ acooke@noao.edu ⟩

TIM CORNWELL, NRAO, PO Box 0, Socorro, NM 87801, U.S.A. ⟨ tcornwel@nrao.edu ⟩

ALESSANDRO COSTA, INAF Astrophysical Observatory of Catania, Via S. Sofia N. 78, Catania, I-95123, Italy ⟨ acosta@ct.astro.it ⟩

MARK CRESITELLO-DITTMAR, SAO, 60 Garden St., MS 81, Cambridge, MA 02138, U.S.A. ⟨ mdittmar@cfa.harvard.edu ⟩

RICHARD CRUTCHER, University of Illinois, 1002 W. Green St., Urbana, IL 61801, U.S.A. ⟨ crutcher@uiuc.edu ⟩

MALCOLM CURRIE, Starlink, R68 Rutherford Appleton Laboratory, Chilton , Didco, Oxfordshire OX11 0QX, United Kingdom ⟨ mjc@star.rl.ac.uk ⟩

Roc Cutri, IPAC/Caltech, IPAC, Caltech MS 100-22, Pasadena, CA 91125, U.S.A. ⟨roc@ipac.caltech.edu⟩

Philip Daly, NOAO, 950 N. Cherry Avenue, Tucson, AZ 85719, U.S.A. ⟨pnd@noao.edu⟩

Adrian Damian, HIA/NRC, 5071 West Saanich Road, Victoria, BC V9E 2E7, Canada ⟨Adrian.Damian@nrc.ca⟩

John Davis, MIT, 70 Vassar Street, NE80-6019, Cambridge, MA 02155, U.S.A. ⟨davis@space.mit.edu⟩

David DeBonis, NRAO, PO Box O, Socorro, NM 87801, U.S.A. ⟨ddebonis@nrao.edu⟩

Jean-Pierre De Cuyper, Royal Observatory of Belgium, Ringlaan 3, Ukkel 1180, Belgium ⟨Jean-Pierre.DeCuyper@oma.be⟩

Michele De La, LBTO, 8230 E. Ridgewood Drive, Tucson, AZ 85750, U.S.A. ⟨mdelapena@as.arizona.edu⟩

Wim De Meester, Institute of Astronomy - K.U.Leuven - Belgium, Celestijnenlaan 200B, Leuven 3001, Belgium ⟨wim.demeester@ster.kuleuven.ac.be⟩

David De Young, NOAO, 950 N. Cherry Ave, Tucson AZ 85719, U.S.A. ⟨deyoung@noao.edu⟩

Ewa Deelman, USC/ISI, 4676 Admiralty Way Suite 1001, Marina Del Rey, CA 90292, U.S.A. ⟨deelman@isi.edu⟩

Drew Devereux, CSIRO Australia, CSIRO ICT Centre, Private Bag No. 5, Wembley, WA 6163, Australia ⟨drew.devereux@csiro.au⟩

Shaun DeWitt, Joint Astronomy Centre, 660 N. A'ohoku Place, University Park, Hilo, HI 96720, U.S.A. ⟨s.dewitt@jach.hawaii.edu⟩

Adam Dobrzycki, ESO, European Southern Observatory, Karl-Schwarzschild-Strasse 2, Garching bei Muenchen, Bavaria 85748, Germany ⟨adam.dobrzycki@eso.org⟩

Sheperd Doeleman, MIT Haystack Observatory, MIT Haystack Observatory, Off Route 40, Westford, MA 01886, U.S.A. ⟨dole@haystack.mit.edu⟩

Peter W. Draper, Starlink, Dept. of Physics, University of Durham, South Road, Durham DH1 3LE, United Kingdom ⟨p.w.draper@durham.ac.uk⟩

Richard Dubois, SLAC, 2575 Sand Hill Road, Menlo Park, CA 94025, U.S.A. ⟨richard@slac.stanford.edu⟩

Daniel Durand, National Research Council Canada, 5071 W. Saanich Rd, Victoria, BC V9E 2E7, Canada ⟨daniel.durand@nrc.ca⟩

Rick Ebert, IPAC/Caltech, MS 314-6, Pasadena, CA 91125, U.S.A. ⟨rick@ipac.caltech.edu⟩

Ian Evans, SAO/CXC, 60 Garden Street MS-81, Cambridge, MA 02138, U.S.A. ⟨ievans@cfa.harvard.edu⟩

Janet Evans, SAO/CXC, 60 Garden St. MS 80, Cambridge MA 02138, U.S.A. ⟨janet@cfa.harvard.edu⟩

Hajime Ezawa, National Astronomical Observatory of Japan, 2-21-1 Osawa, Mitaka, Tokyo 181-8588, Japan ⟨h.ezawa@nao.ac.jp⟩

Tony Farrell, Anglo-Australian Observatory, 167 Vimiera rd., Eastwood, NSW 1710, Australia ⟨tjf@aaoepp.aao.gov.au⟩

Thomas Fenouillet, LAM-OAMP, Traverse du Siphon, Les trois Lucs, Marseille BP8-13376, France ⟨thomas.fenouillet@oamp.fr⟩

Mauro Fiorentino, ST-ECF, Karl-Schwarzschild-Str. 2, Garching BY D-85748, Germany ⟨mfioren@eso.org⟩

Michael Fitzpatrick, NOAO, 950 N Cherry Ave, Tucson, AZ 85719, U.S.A. ⟨fitz@noao.edu⟩

Daniel Flath, SLAC, 2575 Sand Hill Road, Menlo Park, CA 94025, U.S.A. ⟨dflath@slac.stanford.edu⟩

John Fowler, IPAC/Caltech/SSC, Caltech/IPAC MS 314-6, 1200 E. California Blvd., Pasadena, CA 91125, U.S.A. ⟨jwf@ipac.caltech.edu⟩

jay frogel, AURA Inc, 8726 Hickory Bend Trailq, Potomac, MD 20854, U.S.A. ⟨jfrogel@aura-astronomy.org⟩

Carlos Gabriel, ESA/ESAC, Villafranca del Castillo, PO BOX 50727, Madrid 28080, Spain ⟨cgabriel@xmm.vilspa.esa.es⟩

Elizabeth Galle, Harvard-Smithsonian Center for Astrophysics, 60 Garden Street, MS-21, Cambridge, MA 02138, U.S.A. ⟨egalle@head.cfa.harvard.edu⟩

Jorge Garcia, Gemini Observatory, Colina el Pino s/n, Casilla 603, La Serena, Coquimbo 00000, Chile ⟨jgarcia@gemini.edu⟩

Jeffrey Gardner, Pittsburgh Supercomputing Center, 4400 Fifth Ave, MI-216, Pittsburgh, PA 15213, U.S.A. ⟨jenets@psc.edu⟩

David Gasson, National Optical Astronomy Observatory, Rm 90 950 N. Cherry Ave., Tucson, AZ 85719, U.S.A. ⟨dgasson@noao.edu⟩

Severin Gaudet, National Research Council Canada, Herzberg Institute of Astrophysics, 5071 West Saanich Road, Victoria, BC V9E 2E7, Canada ⟨Severin.Gaudet@nrc-cnrc.gc.ca⟩

David Giaretta, Starlink, Rutherford Appleton Lab, Chilton, Didcot, Oxon OX11 0QX, United Kingdom ⟨d.l.giaretta@rl.ac.uk⟩

Andy Gibb, University of British Columbia, Department of Physics & Astronomy, University of British Columbia, Vancouver, BC V6T 1Z1, Canada ⟨agg@astro.ubc.ca⟩

Brian Glendenning, NRAO, P.O. Box O, Socorro, NM 87801, U.S.A. ⟨bglenden@nrao.edu⟩

Kenny Glotfelty, SAO/CXC, 60 Garden St, MS 81, Cambridge, MA 02138, U.S.A. ⟨kjg@head.cfa.harvard.edu⟩

Alexey Goldin, Michelson Science Center, Caltech, 1200 E. California Blvd., Pasadena, CA 91125, U.S.A. ⟨alexey@ipac.caltech.edu⟩

Juan Carlos Gomez, INAOE, Luis Enrique Erro #1, Tonantzintla, Puebla, Pue 72000, Mexico ⟨jcgc@inaoep.mx⟩

John Good, IPAC/Caltech, 770 S. Wilson Ave., Mail Code 100-22, Pasadena, CA 91125, U.S.A. ⟨jcg@ipac.caltech.edu⟩

Matthew Graham, Caltech CACR, 1200 E. California Blvd, MC 158-79, Pasadena, CA 91125, U.S.A. ⟨mjg@cacr.caltech.edu⟩

Arno Granados, AURA/NOAO, 950 N Cherry Ave, Tucson, AZ 85719, U.S.A. ⟨agranados@noao.edu⟩

Alexander Gray, Carnegie Mellon University, 1100 Liberty Ave., #1005, Pittsburgh, PA 15222, U.S.A. ⟨agray@cs.cmu.edu⟩

Norman Gray, Starlink/Glasgow University, Dept of Physics & Astronomy, Uni Glasgow, University Avenue, Glasgow G12 8QQ, United Kingdom ⟨norman@astro.gla.ac.uk⟩

William Green, IPAC/SSC (Ret.), 11786 Encino Ave., Granada Hills, CA 91344, U.S.A. ⟨billg55@earthlink.net⟩

Perry Greenfield, STScI, 3700 San Martin Drive, Baltimore, MD 21218, U.S.A. ⟨perry@stsci.edu⟩

Eric W. Greisen, NRAO, P.O. Box O, Socorro, NM 87801, U.S.A. ⟨egreisen@nrao.edu⟩

Preben Grosbøl, ESO, Karl-Schwarzschild-Str 2, Garching D 85748, Germany ⟨pgrosbol@eso.org⟩

David Grumm, STScI, 3700 San Martin Drive, Baltimore, MD 21218, U.S.A. ⟨grumm@stsci.edu⟩

Chul Gwon, University of Maryland, University of Maryland, Department of Astronomy, College Park, MD 20742, U.S.A. ⟨cgwon@astro.umd.edu⟩

Jonas Haase, ST-ECF, Karl-Schwarzschild-Str. 2, Garching BY D-85748 Germany ⟨jhaase@eso.org⟩

Roger Hain, Harvard-Smithsonian Center for Astrophysics, MS 81, 60 Garden St., Cambridge, MA 02138, U.S.A. ⟨rhain@cfa.harvard.edu⟩

Tom Handley, IPAC/Caltech, MC 100-22, 770 S. Wilson Ave, Pasadena, CA 91125, U.S.A. ⟨thh@ipac.caltech.edu⟩

Robert Hanisch, STScI, Space Telescope Science Institute, 3700 San Martin Drive, Baltimore, MD 21218, U.S.A. ⟨hanisch@stsci.edu⟩

Vivek Haridas, Johns Hopkins University, Department of Physics and Astronomy, Baltimore, MD 21218, U.S.A. ⟨haridas@pha.jhu.edu⟩

Paul Harrison, Astrogrid, Jodrell Bank, Lower Withington, Macclesfield SK8 4RR, United Kingdom ⟨pah@jb.man.ac.uk⟩

Jeffrey Hayes, NASA HQ, 300 E St SW, SMD/Universe Division, Washington, DC 20456, U.S.A. ⟨jhayes@nasa.gov⟩

Helen He, SAO, 60 Garden Street, MS 81, Cambridge, MA 02138, U.S.A. ⟨hhe@cfa.harvard.edu⟩

Martin Heemskerk, Astronomical Institute Anton Pannekoek, Kruislaan 403, Amsterdam 1098SJ, Netherlands ⟨martin@science.uva.nl⟩

David Henderson, IPAC/Caltech, Mail Code 220-6, 1200 E California Blvd, Pasadena, CA 91125 U.S.A. ⟨davidh@ipac.caltech.edu⟩

Ted Hesselroth, IPAC/Caltech, Mail Code 314-6, 1200 E California Blvd, Pasadena CA 91125, U.S.A. ⟨tdh@ipac.caltech.edu⟩

Martin Hill, AstroGrid @ ROE, Royal Observatory Edinburgh, Blackford Hill, Edinburgh EH9 3HJ, United Kingdom ⟨mch@roe.ac.uk⟩

Rafael Hiriart, NOAO, 950 North Cherry Avenue, Tucson, AZ 85719, U.S.A. ⟨rhiriart@noao.edu⟩

Phil Hodge, STScI, 3700 San Martin Drive, Baltimore, MD 21218, U.S.A. ⟨hodge@stsci.edu⟩

Satoshi Honda, National Astronomical Observatory of Japan, 2-21-1 Osawa Mitaka Tokyo, Tokyo 181-8588, Japan ⟨honda.satoshi@nao.ac.jp⟩

Richard Hook, ST-ECF at STScI, 3700 San Martin Drive, Baltimore, MD 21218, U.S.A. ⟨hook@stsci.edu⟩

Allan Hornstrup, Danish Space Research Institute, Juliane Maries Vej 30, Copenhagen 2100, Denmark ⟨allan@dsri.dk⟩

Jin-chung Hsu, STScI, 3700 San Martin Drive, Baltimore, MD 21218, U.S.A. ⟨hsu@stsci.edu⟩

Carlo Izzo, European Southern Observatory, Karl-Schwarzschild-Str. 2, Garching bei Munchen, Bavaria 85748, Germany ⟨cizzo@eso.org⟩

Joseph Jacob, Jet Propulsion Laboratory, 4800 Oak Grove Drive, Mail Stop 126-104, Pasadena, CA 91109-8099, U.S.A. ⟨Joseph.Jacob@jpl.nasa.gov⟩

Mark Jarnyk, ANU, RSAA Mount Stromlo Observatory, Cotter Road, Weston Creek, ACT 2611, Australia ⟨jarnyk@mso.anu.edu.au⟩

Robert Jedrzejewski, STScI, 3700 San Martin Drive, Baltimore, MD 21218, U.S.A. ⟨rij@stsci.edu⟩

Isabelle Jegouzo, Observatoire de Paris - Migale, Observatoire de Meudon - bat 18, 5 place Jules Janssen, Meudon 92 92195, France ⟨Isabelle.Jegouzo@obspm.fr⟩

Tim Jenness, Joint Astronomy Centre, 660 N. A'ohoku Place, University Park, Hilo, HI 96720, U.S.A. ⟨t.jenness@jach.hawaii.edu⟩

Vicki Johnson, Interconnect Technologies Corp, POB 788, Claremont, CA 91711, U.S.A. ⟨vlj@interconnect.com⟩

William Joye, SAO, 60 Garden Street, Cambridge, MA 02138, U.S.A. ⟨wjoye@cfa.harvard.edu⟩

nick Kaiser, U. Hawaii Institute for Astronomy, 2680 Woodlawn Drive, Honolulu, HI 96822, U.S.A. ⟨kaiser@ifa.hawaii.edu⟩

Inga Kamp, STScI, 3700 San Martin Drive, Baltimore, MD 21204, U.S.A. ⟨kamp@stsci.edu⟩

Heather Kelly, GSFC/SSAI, 14652 Billy Lane, Poway, CA 92064, U.S.A. ⟨heather@lheapop.gsfc.nasa.gov⟩

Athol Kemball, NCSA/UIUC, 605 E. Springfield Ave, Champaign, IL 61801, U.S.A. ⟨akemball@uiuc.edu⟩

Eloise Kennedy, IPAC, 770 S. Wilson, Pasadena, CA 91125, U.S.A. ⟨esk@ipac.caltech.edu⟩

Jeff Kern, NRAO, 1003 Lopezville Rd, Socorro, NM 87801, U.S.A. ⟨jkern@nrao.edu⟩

Mike Khuong, IPAC, Mailcode 220-6, 1200 E. California Blvd., Pasadena, CA 91125, U.S.A. ⟨mkhuong@ipac.caltech.edu⟩

David L. King, NRAO, P.O. Box O, Socorro, NM 87801, U.S.A. ⟨dking@nrao.edu⟩

Anton Koekemoer, STScI, 3700 San Martin Drive, Baltimore, MD 21218, U.S.A. ⟨koekemoe@stsci.edu⟩

Mitsuhiro Kohama, RIKEN, Cosmic radiation lab, Hirosawa 2-1 , Wako Saitama 351-0198, Japan ⟨kohama@crab.riken.jp⟩

Mihseh Kong, IPAC/Caltech, 770 S. Wilson Ave., Mail Code 100-22, Pasadena, CA 91125, U.S.A. ⟨mihseh@ipac.caltech.edu⟩

Tevfik Kosar, University of Wisconsin, Computer Science Dept., 1210 W. Dayton St., Madison, WI 53706, U.S.A. ⟨kosart@cs.wisc.edu⟩

Karl Krughoff, University of Pittsburgh, Dept. of Physics and Astronomy, 3941 O'Hara St., Pittsburgh, PA 15260, U.S.A. ⟨simon@phyast.pitt.edu⟩

Sergey Kudryavtsev, Sternberg Astronomical Institute of Moscow State University, Universitetsky Pr. 13, Moscow 119992, Russia ⟨ksm@sai.msu.ru⟩

Martin Kuemmel, ST-ECF, Karl-Schwarzschild-Str. 2, Garching BY 85748, Germany ⟨mkuemmel@eso.org⟩

Russ Laher, IPAC/Caltech/SSC, 1200 E. California Blvd., M/C 314-6, Pasadena, CA 91125, U.S.A. ⟨laher@ipac.caltech.edu⟩

Anastasia Laity, IPAC/Caltech, 770 S. Wilson Ave., Mail Code 100-22, Pasadena, CA 91125, U.S.A. ⟨laity@ipac.caltech.edu⟩

Uwe Lammers, European Space Agency, ESA/ESTEC/SCI-SD, Noordwijk NL 2200 AG, Netherlands ⟨ulammers@rssd.esa.int⟩

Matt Langston, SLAC, 2575 Sand Hill Road, Menlo Park, CA 94025, U.S.A. ⟨langston@slac.stanford.edu⟩

Wen P. Lee, IPAC/Caltech, Mail code 314-6 SSC, 1200 E. California Blvd., Pasadena, CA 91125, U.S.A. ⟨wplee@ipac.caltech.edu⟩

Jamie Leech, Joint Astronomy Centre, 660 N. A'ohoku Place, University Park, Hilo, HI 96720, U.S.A. ⟨j.leech@jach.hawaii.edu⟩

Marco Leoni, ESO, KarlSchwarzschild Str. 2, Garching bei Muenchen 85748, Germany ⟨marco.leoni@eso.org⟩

James Lewis, University of Cambridge, Institute of Astronomy, Madingley Road, Cambridge CB3 0HA, United Kingdom ⟨jrl@ast.cam.ac.uk⟩

Nolan Li, Johns Hopkins University, 74 E. Padonia #103, TImonium, MD 21093, U.S.A. ⟨nli@pha.jhu.edu⟩

Tony Linde, AstroGrid University of Leicester, University Road, Leicester LE1 7RH, United Kingdom ⟨ael@star.le.ac.uk⟩

Jacob Llamas, IPAC/Caltech, 770 S. Wilson, Mail Code 100-22, Pasadena, CA 91016, U.S.A. ⟨jllamas@ipac.caltech.edu⟩

Steven Lo, IPAC/Caltech, 770 S. Wilson Ave., Mail Code 100-22, Pasadena, CA 91125 U.S.A. ⟨slo@ipac.caltech.edu⟩

Colin Lonsdale, MIT, MIT Haystack Observatory, Westford, MA 01886, U.S.A. ⟨clonsdale@haystack.mit.edu⟩

Jeff Lusted, University of Leicester, Dept of Physics and Astronomy, University Road, Leicester, Leicestershire LE1 7RH, United Kingdom ⟨jl99@star.le.ac.uk⟩

Daniel Magee, UCO/Lick Observatory, 1156 High St., Santa Cruz, CA 95064, U.S.A. ⟨magee@ucolick.org⟩

Justin Maki, NASA/JPL, M/S 264-621, Pasadena, CA 91103, U.S.A. ⟨Justin.N.Maki-104410@jpl.nasa.gov⟩

David Makovoz, IPAC/Caltech, MC 220-6, Pasadena, CA 91125, U.S.A. ⟨davidm@ipac.caltech.edu⟩

Robert Mann, University of Edinburgh, Royal Observatory, Blackford Hill, Edinburgh EH93HJ, United Kingdom ⟨rgm@roe.ac.uk⟩

Frank Masci, SSC/Caltech, 1200 East California Blvd., Pasadena, CA 91125, U.S.A. ⟨fmasci@ipac.caltech.edu⟩

Keiichi Matsuzaki, JAXA/ISAS, 3-1-1, Yoshinodai, Sagamihara, Kanagawa 229-8510, Japan ⟨matuzaki@plain.isas.jaxa.jp⟩

Steve mazuk, Aerospace Corporation, PO Box 92957 , Mail Code M2-266, Los Angeles, CA 90009-2957, U.S.A. ⟨steve.mazuk@aero.org⟩

Joe Mazzarella, IPAC, MS 100-22, 770 S. Wilson Ave, Pasadena, CA 91125, U.S.A. ⟨ned@ipac.caltech.edu⟩

William McCann, The Johns Hopkins University, Dept Physics & Astronomy, 3400 N. Charles St, Baltimore, MD 21218, U.S.A. ⟨mccann@jhu.edu⟩

Michael McCollough, Harvard-Smithsonian Center for Astrophysics, 60 Garden St., MS 67, Cambridge, MA 02138-1516, U.S.A. ⟨mmccollough@head.cfa.harvard.edu⟩

Sabine McConnell, Queen's University, 61 Casterton Avenue, Kingston, Ontario K7M 1R7, Canada ⟨mcconnell@cs.queensu.ca⟩

Jonathan McDowell, SAO, 60 Garden St, Cambridge, MA 02138, U.S.A. ⟨jcm@cfa.harvard.edu⟩

David Mehringer, UIUC/NCSA, 605 E. Springfield Ave., Champaign, IL 61820, U.S.A. ⟨dmehring@beren.ncsa.uiuc.edu⟩

Laurent Michel, Observatoire de Strasbourg, 11 Rue de l'Universite, Strasbourg - 67000, France ⟨laurent.michel@astro.u-strasbg.fr⟩

Alberto Micol, ST-ECF, Karl-Schwarzschild-Str. 2, Garching BY D-85748, Germany ⟨amicol@eso.org⟩

Manfred Miebach, ESA, 3700 San Martin Drive, Baltimore, MD 21218, U.S.A. ⟨miebach@stsci.edu⟩

Chris Miller, NOAO/CTIO, NOAO/CTIO, 950 N. Cherry Ave., Tucson, AZ 85719, U.S.A. ⟨cmiller@noao.edu⟩

Glenn Miller, STScI, 3700 San Martin Drive, Baltimore, MD 21218, U.S.A. ⟨miller@stsci.edu⟩

Jessica Miller, IPAC/Caltech, 1618 Pilgram Way, Monrovia, CA 91016, U.S.A. ⟨jllamas@ipac.caltech.edu⟩

Joseph Miller, SAO, 360 May St, Worcester, MA 01602, U.S.A. ⟨jmiller@cfa.harvard.edu⟩

Todd Miller, STScI, 3700 San Martin Drive, Baltimore, MD 21218, U.S.A. ⟨jmiller@stsci.edu⟩

W. Marcus Miller, LLNL, 7000 East Ave, Livermore, CA 94550, U.S.A. ⟨marcusm@llnl.gov⟩

Douglas Mink, Harvard-Smithsonian Center for Astrophysics, 60 Garden Street, MS 20, Cambridge, MA 02138, U.S.A. ⟨dmink@cfa.harvard.edu⟩

Palle Moller, ESO, Karl-Schwarzschild-Str. 2, Garching B 85748, Germany ⟨pmoller@eso.org⟩

Serge Monkewitz, IPAC/Caltech, 770 S. Wilson Ave., Mail Code 100-22, Pasadena, CA 91125, U.S.A. ⟨smm@ipac.caltech.edu⟩

Thomas Morgan, NRAO, Array Operations Center, 1003 Lopezville Rd, Socorro, NM 87801, U.S.A. ⟨tmorgan@aoc.nrao.edu⟩

Mehrdad Moshir, IPAC/Caltech/SSC, Caltech, MC314-6, Pasadena, CA 91125, U.S.A. ⟨mmm@ipac.caltech.edu⟩

Robert Narron, IPAC/Caltech/SSC, M/S 220-6, 1200 E. California Blvd., Pasadena, CA 91125, U.S.A. ⟨bob@ipac.caltech.edu⟩

Ngoc Hoan Nguyen, Observatoire Astronomique Strasbourg, 11 Rue de l'Universite, Strasbourg, Alsace 67000, France ⟨nguyen@astro.u-strasbg.fr⟩

Maria Nieto-Santisteban, Johns Hopkins University, 3400 N. Charles St., Phyisics & Astronomy, Baltimore, MD 21218, U.S.A. ⟨nieto@pha.jhu.edu⟩

Michael Noble, MIT Center For Space Research, 70 Vassar Street, Room NE80-6081, Cambridge, MA 02139, U.S.A. ⟨mnoble@space.mit.edu⟩

Keith Noddle, University Of Leicester, Dept of Physics & Astronomy, Leicester, Leicestershire LE1 7RH, United Kingdom ⟨ktn@star.le.ac.uk⟩

Jan Noordam, ASTRON, P.O.Box 2, Dwingeloo, Drenthe 7990AA, Netherlands ⟨jnoordam@astron.nl⟩

Francois Ochsenbein, CDS Strasbourg, Observatoire Astronomique, 11 rue de l'Universite, Strasbourg 67000, France ⟨francois@astro.u-strasbg.fr⟩

Ryusuke Ogasawara, Subaru Telescope NAOJ, 650 North Aohoku Place, Hilo, HI 96720, U.S.A. ⟨ryu@naoj.org⟩

Masatoshi Ohishi, NAOJ, 2-21-1 Osawa, Mitaka, Tokyo 181-8588, Japan ⟨ masatoshi.ohishi@nao.ac.jp ⟩

Eduardo Ojero, European Space Astronomy Centre - European Space Agency, Apdo. Correos 50727, Madrid 28080, Spain ⟨ eduardo.ojero@esa.int ⟩

William O'Mullane, JHU, JHU Bloomberg Center, 3400 N. Charles Street, Baltimore, MD 21218, U.S.A. ⟨ womullan@jhu.edu ⟩

Inaki Ortiz, ESA, PO Box 50727, Madrid, Madrid 28080, Spain ⟨ Inaki.Ortiz@sciops.esa.int ⟩

Pedro Osuna, ESA/ESAC, Villafranca del Castillo, PO BOX 50727, Madrid 28080, Spain ⟨ Pedro.Osuna@esa.int ⟩

Seathrun Ó Tuairisg, National University of Ireland Galway, Dept of Information Technology, Galway City, Ireland ⟨ seathrun@it.nuigalway.ie ⟩

Paolo Padovani, ST-ECF/ESO, Karl-Schwarzschild-Str. 2, Garching bei Munchen D 85748, Germany ⟨ Paolo.Padovani@eso.org ⟩

Ralf Palsa, European Southern Observatory, Karl-Schwarzschild-Str. 2, Garching bei Munchen, Bavaria 85748, Germany ⟨ rpalsa@eso.org ⟩

Fabio Pasian, INAF - O.A.Trieste, Osservatorio Astronomico, Via G.B.Tiepolo 11, Trieste, I 34131, Italy ⟨ pasian@ts.astro.it ⟩

Mikhail Pavlov, INAF O.A.Capodimonte, Salita Moiariello 16, Napoli 80131, Italy ⟨ mike@na.astro.it ⟩

William Pence, NASA/GSFC, 8377 Glenmar Road, Ellicott City, MD 21043, U.S.A. ⟨ pence@milkyway.gsfc.nasa.gov ⟩

Michele Peron, ESO, Karl-Schwarzschild-Str. 2, Garching 85748, Germany ⟨ mperon@eso.org ⟩

Meyer Pesenson, IPAC/Caltech, Mail Code 220-6, Pasadena, CA 91125, U.S.A. ⟨ misha@ipac.caltech.edu ⟩

Olga Pevunova, IPAC/Caltech/NED, Olga Pevunova, 429 Lincoln Avenue, Apt #1, Glendale, CA 91205, U.S.A. ⟨ olga@ipac.caltech.edu ⟩

Matt Phelps, Harvard - Smithsonian Center for Astrophysics, 60 Garden St, MS-39, Cambridge, MA 02138, U.S.A. ⟨ mphelps@cfa.harvard.edu ⟩

Francesco Pierfederici, NOAO, 950 N. Cherry Ave, Tucson, AZ 85719, U.S.A. ⟨ fpierfed@noao.edu ⟩

Raymond Plante, NCSA, 152 CAB, 605 E. Springfield Ave., Champaign, IL 61820, U.S.A. ⟨ rplante@ncsa.uiuc.edu ⟩

Moshe Pniel, Caltech/IPAC, Mail Code 100-22, Pasadena, CA 91125, U.S.A. ⟨ mp@ipac.caltech.edu ⟩

Adrian Pope, Johns Hopkins University, Dept. of Physics & Astronomy, 3701 San Martin Dr., Baltimore, MD 21218, U.S.A. ⟨ pope@pha.jhu.edu ⟩

Marc Pound, University of Maryland, Astronomy Department, College Park, MD 20742, U.S.A. ⟨ mpound@astro.umd.edu ⟩

FRANCIS A. PRIMINI, SAO, 60 Garden St., Mail Stop 3, Cambridge, MA 02138,, U.S.A. ⟨fap@head.cfa.harvard.edu⟩

SASKIA PRINS, Nordic Optical Telescope, Apartado 474, Santa Cruz de La Palma (TF) E-38700, Spain ⟨stp@not.iac.es⟩

THOMAS QUINN, University of Washington, Astronomy, Box 351580, Seattle, WA 98195, U.S.A. ⟨trq@astro.washington.edu⟩

JORDAN RADDICK, Johns Hopkins University, 3701 San Martin Dr., Baltimore, MD 21218, U.S.A. ⟨raddick@pha.jhu.edu⟩

STEPHEN RANKIN, CCLRC RAL STARLINK, Rutherford Appleton Laboratory, Chilton, DIDCOT, Oxon OX11 0QX, United Kingdom ⟨ser@star.rl.ac.uk⟩

KEVIN RAUCH, University of Maryland, Dept. of Astronomy, College Park, MD 20742, U.S.A. ⟨rauch@astro.umd.edu⟩

MICHAEL READ, IfA Edinburgh, IfA Royal Observatory, Blackford Hill, Edinburgh, Lothian EH93HJ, United Kingdom ⟨mar@roe.ac.uk⟩

GUY RIXON, AstroGrid, Institute of Astronomy, Madingley Road, Cambridge, Cambs. CB3 0HA, United Kingdom ⟨gtr@ast.cam.ac.uk⟩

WILLIAM ROBY, Caltech, Mail Code 100-22, Pasadena, CA 91125, U.S.A. ⟨roby@caltech.edu⟩

NATHALIE ROSSAT, ESO, European Southern Observatory, Karl-Schwarzschild-Strasse 2, Garching bei Munchen D-85748, Germany ⟨nrossat@eso.org⟩

ARNOLD ROTS, CfA, 60 Garden Street - MS 67, Cambridge, MA 02138, U.S.A. ⟨arots@cfa.harvard.edu⟩

LEE ROTTLER, IPAC/Caltech, 2385 Oneida St, Pasadena, CA 91107, U.S.A. ⟨rottler@ipac.caltech.edu⟩

NELSON SAAVEDRA, NOAO/DPP, 950 N. Cherry Ave., Tucson, AZ 85719, U.S.A. ⟨nsaavedra@ctio.noao.edu⟩

GRAN SANDELL, SOFIA-USRA, NASA ARC, MS 144-2, Moffett Field, CA 94035, U.S.A. ⟨gsandell@mail.arc.nasa.gov⟩

RENAUD SAVALLE, CFHT, 65-1238 Mamalahoa Hwy, Kamuela, HI 96743, U.S.A. ⟨savalle@cfht.hawaii.edu⟩

TSUYOSHI SAWADA, Nobeyama Radio Observatory, 462-2 Nobeyama, Minamimaki, Minamisaku, Nagano 384-1305, Japan ⟨sawada@nro.nao.ac.jp⟩

PHILIP SCHALLER, Las Campanas Observatory, Casilla 601, La Serena - -, Chile ⟨skip@lco.cl⟩

MARION SCHMITZ, IPAC/Caltech/NED, 770 S. Wilson Ave., Pasadena, CA 91125, U.S.A. ⟨mschmitz@ipac.caltech.edu⟩

ROSANNE SCHOLEY, IPAC/Caltech, 770 S. Wilson Ave., Pasadena, CA 91125, U.S.A. ⟨rosanne@ipac.caltech.edu⟩

Isabelle Scholl, International Space University, 1 rue JD Cassini, Parc d'Innovation, Illkirch-Graffenstaden n/a F-67400, France ⟨scholl@isu.isunet.edu⟩

Bernhard Schulz, IPAC/Caltech, Mail Code 100-22, Pasadena, CA 91125, U.S.A. ⟨bschulz@ipac.caltech.edu⟩

Arnold Schwartz, IPAC/Caltech/MSC, Mail Code 100-22, Pasadena, CA 91125, U.S.A. ⟨arnold@ipac.caltech.edu⟩

Greg Schwarz, AAS Journal Editoral Scientist, Steward Observatory, University of Arizona, 933 N. Cherry Avenue, Tucson, AZ 85721, U.S.A. ⟨gschwarz@as.arizona.edu⟩

Stephen Scott, Caltech/CARMA, OVRO, POB 968, Big Pine, CA 93514, U.S.A. ⟨scott@ovro.caltech.edu⟩

Robert Seaman, NOAO, 950 N. Cherry Ave., Tucson, AZ 85719, U.S.A. ⟨seaman@noao.edu⟩

Carey Sean, IPAC/Caltech/SSC, MS 220-6, Pasadena, CA 91125, U.S.A. ⟨carey@ipac.caltech.edu⟩

Eric Sessoms, NRAO, 2515 Efland-Cedar Grove Rd., Efland, NC 27243, U.S.A. ⟨esessoms@nrao.edu⟩

Diego Sforna, ST-ECF, Karl-Schwarzschild-Str. 2, Garching BY 85748, Germany ⟨dsforna@eso.org⟩

Julie Shaw, Harvard-Smithsonian Center for Astrophysics, 60 Garden St., MS-39, Cambridge, MA 02138, U.S.A. ⟨jshaw@cfa.harvard.edu⟩

Richard Shaw, NOAO, 950 N. Cherry Avenue, Tucson, AZ 85719, U.S.A. ⟨shaw@noao.edu⟩

Ed Shaya, University of Maryland, Astronomy Dept, College Park, MD 20742, U.S.A. ⟨shaya@gsfc.nasa.gov⟩

Andrew Shearer, NUI Galway, IT Dept., NUI Galway, Galway, Galway 12345, Ireland ⟨andy.shearer@nuigalway.ie⟩

Yuji Shirasaki, National Astronomical Observatory of Japan, Osawa 2-21-1, Mitaka, Tokyo 181-8588, Japan ⟨yuji.shirasaki@nao.ac.jp⟩

Patrick L. Shopbell, Caltech, MC 105-24, 1200 E. California Blvd., Pasadena, CA 91125, U.S.A. ⟨pls@astro.caltech.edu⟩

Sergiy Shumko, Big Bear Solar Observatory, 40386 North Shore Ln., Big Bear City, CA 92314, U.S.A. ⟨shoumko@bbso.njit.edu⟩

David Shupe, IPAC/Caltech/SSC, 1200 E. California Blvd, MS 314-6, Pasadena, CA 91125, U.S.A. ⟨shupe@ipac.caltech.edu⟩

Petr Skoda, Astr. Inst. Academy of Sciences, Fricova 298, Ondrejov, CR 251 65, Czech Republic ⟨skoda@pleione.asu.cas.cz⟩

Riccardo Smareglia, INAF- OATs, Via G.B. Tipolo 11, Trieste, Italy 34131, Italy ⟨smareglia@ts.astro.it⟩

Oleg Smirnov, ASTRON, Postbus 2, Dwingeloo D 7990 AA, Netherlands ⟨smirnov@astron.nl⟩

Robert Smith, Liverpool John Moores University, ARI Twelve Quays House, Egerton Wharf, Birkenhead, Wirral CH41 1LD, United Kingdom ⟨ rjs@astro.livjm.ac.uk ⟩

Robert Smith, NOAO/DPP, NOAO/CTIO, 950 N. Cherry Ave., Tucson, AZ 85719, U.S.A. ⟨ csmith@ctio.noao.edu ⟩

Alvaro Soto, Pontificia Universidad Catolica de Chile, Vicuna Mackenna 4860 (143), Santiago, Sa 22, Chile ⟨ asoto@ing.puc.cl ⟩

Elizabeth Stobie, University of Arizona, Steward Observatory, 933 N. Cherry Avenue, Tucson, AZ 85721, U.S.A. ⟨ bstobie@as.arizona.edu ⟩

Jeff Stoner, , Suite 204 PMB 205, 700 Ken Pratt Blvd, Longmont, CO 80501, U.S.A. ⟨ jls2@bearclan.com ⟩

Christian Surace, LAM/OAMP, LAM, TRaverse du siphon BP 8, Marseille cedex 12, 13376, France ⟨ christian.surace@oamp.fr ⟩

Giuliano Taffoni, INAF - Trieste, via Tiepolo 2, Trieste ITL 34122, Italy ⟨ taffoni@ts.astro.it ⟩

Naveed Tahir-Kheli, Caltech/IPAC, 770 S. Wilson Ave., Mail Code 100-22, Pasadena, CA 91125, U.S.A. ⟨ naveed@ipac.caltech.edu ⟩

Francoise Tajahmady, Observatoire de Paris, CAI-GEPI-Observatoire de Paris, Batiment Perrault 77 Avenue Denfert Rochereau, Paris 75014, France ⟨ francoise.tajahmady@obspm.fr ⟩

Masahiro Tanaka, NAOJ, 2-21-1 Osawa, Mitaka, Tokyo 181-8588, Japan ⟨ masahiro.tanaka@nao.ac.jp ⟩

Mark Taylor, Starlink (Bristol UK), H H Wills Physics Laboratory, Tyndall Avenue, Bristol, Avon BS8 1TL, United Kingdom ⟨ m.b.taylor@bristol.ac.uk ⟩

Peter Teuben, University of Maryland, Astronomy Department, College Park, MD 20742, U.S.A. ⟨ teuben@astro.umd.edu ⟩

Ani Thakar, JHU, Ctr for Astrophysical Sciences, 3701 San Martin, Baltimore, MD 21218, U.S.A. ⟨ thakar@pha.jhu.edu ⟩

Brian Thomas, University of Maryland, Department of Astronomy, College Park Md 20742, U.S.A. ⟨ thomas@astro.umd.edu ⟩

Randall Thompson, STScI/CSC, 304 King James Court, Upper Marlboro, MD 20774, U.S.A. ⟨ rthomp@stsci.edu ⟩

Douglas Tody, NRAO, PO Box O, 1003 Lopezville Rd, Socorro NM 87801, U.S.A. ⟨ dtody@nrao.edu ⟩

Gabriella Turek, IPAC/Caltech, Mail Code 314-6, Pasadena, CA 91125, U.S.A. ⟨ gaby@ipac.caltech.edu ⟩

Francisco Valdes, NOAO, P.O. Box 26732, Tucson, AZ 85726, U.S.A. ⟨ valdes@noao.edu ⟩

Julio J. Valdes, National Research Council of Canada, M50 1200 Montreal Rd., Ottawa, ON K1A 0R6, Canada ⟨ Julio.Valdes@nrc.ca ⟩

Robert Vallance, University of Birmingham, School of Physics and Astronomy, Birmingham, W Midlands B15 2TT, United Kingdom ⟨ rjv@star.sr.bham.ac.uk ⟩

David van Dyk, University of California, Department of Statistics, Irvine, CA 92697-1250, U.S.A. ⟨ dvd@ics.uci.edu ⟩

Gustaaf van Moorsel, NRAO, 1003 Lopezville, Socorro, NM 87801, U.S.A. ⟨ gvanmoor@nrao.edu ⟩

Kari Vasko, CSC - Scientific Computing Ltd., Box 405, Espoo, FIN-02101, Finland ⟨ vasko@csc.fi ⟩

Christian Veillet, CFHT, 65-1238 Mamalahoa Hwy, Kamuela, HI 96743, U.S.A. ⟨ veillet@cfht.hawaii.edu ⟩

Urvashi Rao Venkata, NRAO/NMT, 607 Coulson Avenue, Socorro, NM 87801, U.S.A. ⟨ rurvashi@aoc.nrao.edu ⟩

Catherine Venturini, The Aerospace Corporation, 2350 E. El Segundo Blvd, M2-266, El Segundo, CA 90245, U.S.A. ⟨ catherine.c.venturini@aero.org ⟩

Harro Verkouter, JIVE, Oude Hoogeveensedijk 4, Dwingeloo 7991PD, Netherlands ⟨ verkouter@jive.nl ⟩

Bruno Voisin, CAL NUI Galway, IT Department, NUI Galway, Galway Co., Galway 1 , Ireland ⟨ bruno.voisin@nuigalway.ie ⟩

Alessandra Volpato, INAF Astronomical Observatory of Padova, Viale Bacchiglione 3, Vicenza, VI 36100, Italy ⟨ ale.volpato@ieee.org ⟩

Claudio Vuerli, INAF-O.A.Trieste, Osservatorio Astronomico di Trieste, Via G.B.Tiepolo 11, Trieste, TS I-34131, Italy ⟨ vuerli@ts.astro.it ⟩

Kiri Wagstaff, Jet Propulsion Laboratory, 1792 New York Drive, Altadena, CA 91001, U.S.A. ⟨ kiri.wagstaff@jpl.nasa.gov ⟩

Patrick Wallace, RAL UK, CCLRC/RAL, Chilton, Didcot Oxon OX11 0QX, United Kingdom ⟨ ptw@star.rl.ac.uk ⟩

Roland Walter, ISDC, ISDC, Chemin d'Ecogia 16, Versoix 1290, Switzerland ⟨ Roland.Walter@obs.unige.ch ⟩

Nicholas Walton, Institute of Astronomy, University of Cambridge, Madingley Road, Cambridge CB3 0HA, United Kingdom ⟨ naw@ast.cam.ac.uk ⟩

Rein Warmels, European Southern Observatory, Karl-Schwarzschild-Strasse 2, Garching bei Munich D 85748, Germany ⟨ rwarmels@eso.org ⟩

Phillip Warner, NOAO, 950 North Cherry Ave, Tucson, AZ 85719, U.S.A. ⟨ pwarner@noao.edu ⟩

Saille Warner-Norton, IPAC/Caltech, 770 S. Wilson Ave., Mail Code 100-22, Pasadena, CA 91125, U.S.A. ⟨ saille@ipac.caltech.edu ⟩

Boyd Waters, NRAO, PO Box 0, Socorro, NM 87801, U.S.A. ⟨ bwaters@nrao.edu ⟩

Marc Wenger, CDS - Strasbourg, 11 rue de l'Universite, Strasbourg - 67000, France ⟨ wenger@astro.u-strasbg.fr ⟩

Michael Werner, NASA/JPL, M/S 264-767, JPL, Pasadena, CA 91107, U.S.A. ⟨ mwerner@sirtfweb.jpl.nasa.gov ⟩

Richard White, STScI, 3700 San Martin Dr, Baltimore, MD 21218, U.S.A. ⟨ rlw@stsci.edu ⟩

Ekkehard Wieprecht, MPE Garching, Max Planck Institut fuer extraterrestrische Physik, Giessenbachstrasse, Garching 85741, Germany ⟨ ewieprec@mpe.mpg.de ⟩

Roy Williams, Caltech CACR, 1200 E. California Blvd., MC 158-79, Pasadena, CA 91125, U.S.A. ⟨ roy@cacr.caltech.edu ⟩

Greg Wilson, Australian National University, Cotter Road Weston Creek, Canberra, ACT 2611, Australia ⟨ gmw@mso.anu.edu.au ⟩

Justin Kanoa Withington, CFHT, 65-1506 Kawaihae Rd, Kamuela, HI 96743, U.S.A. ⟨ kanoa@cfht.hawaii.edu ⟩

Xiuqin Wu, IPAC/Caltech/SSC, Mail Code 220-6, Pasadena, CA 91125, U.S.A. ⟨ xiuqin@ipac.caltech.edu ⟩

Shilong Stanley Yao, NOAO/AURA Inc., 950 N. Cherry Avenue, NOAO/AURA Inc., Tucson, AZ 85719, U.S.A. ⟨ yao@noao.edu ⟩

Jason Ye, Brown University, Box 4087, Providence, RI 02912, U.S.A. ⟨ jye@brown.edu ⟩

Nelson Zarate, NOAO, 950 N. Cherry Ave, Tucson, AZ 85719, U.S.A. ⟨ zarate@noao.edu ⟩

Angela Zhang, IPAC/Caltech, 770 S. Wilson Ave., Mail Code 100-22, Pasadena, CA 91125, U.S.A. ⟨ azhang@ipac.caltech.edu ⟩

Lijun Zhang, IPAC/Caltech, 770 S. Wilson Ave., Mail Code 220-22, Pasadena, CA 91125, U.S.A. ⟨ zhang@ipac.caltech.edu ⟩

Olga Zhelenkova, SAO RAS, b.2 app.64, Niznij Arkhyz, KChR 369167, Russia ⟨ zhe@sao.ru ⟩

Ivan Zolotukhin, Sternberg Astronomical Institute of Moscow State University, Rokotova 4-2-17, Moscow, Moscow 117593, Russia ⟨ iz@sai.msu.ru ⟩

Part I
Birds of a Feather Sessions

ASTRONOMICAL DATA ANALYSIS SOFTWARE AND SYSTEMS XIV
ASP Conference Series, Vol. 347, 2005
P. L. Shopbell, M. C. Britton, and R. Ebert, eds.

Sky Indexation, Pixelization, and the VO

William O'Mullane
The Johns Hopkins University

Kris Gorski
Jet Propulsion Laboratories

Abstract. This is the short report from the Birds of a Feather session on Sky Indexing. The purpose of the session was to bring participants up to speed on a few of the ideas in the area of sky indexing and spatial querying and to provide a forum for discussion for those who need one of the presented schemes or similar. The format was a series of short presentations followed by questions.

1. Presentations

A brief overview of HTM (Hierarchical Triangular Mesh) was given by W. O'Mullane. A brief overview of HEALPix (Hierarchical Equal Area isoLatitude Pixelization) was given by K. Gorski. The different goals of these schemes were pointed out i.e., for HTM complex geometric queries while for HEALPix the ability to do spherical FFT. Both schemes are hierarchical; HTM uses recursion while HEALPix does not. In SDSS, the main driver for HTM, HEALPix is now also beginning to be used.

A common language for expressing geometric shapes would be useful. Effectively the manipulation and simplification of shapes is a separate issue from that of tessellation, a problem which A. Szalay and J. Gray are currently looking into. The JPL and JHU teams continue to work together on these issues despite being turned down in our individual and joint proposals for funding for these topics this year.

A. Rots then presented the Space Time Coordinate model which may provide such a language.

Finally, M. Nieto-Santiesteban presented the simple zone model which works well in RDBMS for certain problems.

There was a short discussion on balanced trees, in the data access sense.

The session was attended by around twenty people and is probably worth repeating.

Part II
Demonstration Sessions

ASTRONOMICAL DATA ANALYSIS SOFTWARE AND SYSTEMS XIV
ASP Conference Series, Vol. 347, 2005
P. L. Shopbell, M. C. Britton, and R. Ebert, eds.

The NASA/IPAC Infrared Science Archive (IRSA): The Demo

Anastasia Alexov, G. Bruce Berriman, Nian-Ming Chiu, John C. Good, Thomas H. Jarrett, Mih-seh Kong, Anastasia Clower Laity, Serge Michel Monkewitz, Naveed D. Tahir-Kheli, Saille Warner Norton, and Anzhen Zhang

Infrared Processing and Analysis Center, California Institute of Technology, Mail Code 100-22, CA 91125

Abstract. This paper describes the services available at the NASA/IPAC Infrared Science Archive (IRSA).

Currently there are nearly 250,000 data requests a month, taking advantage of IRSA's data repository which includes 660 million sources (60 catalogs), 10 million images (22 image sets; 10.4 TB) and over 30,000 spectra (7 spectroscopic datasets). These data are the science products of: The Two Micron All Sky Survey (2MASS), The Infrared Astronomical Satellite (IRAS), The Mid-course Space Experiment (MSX), The Submillimeter Wave Astronomy Satellite (SWAS), The Infrared Space Observatory (ISO), The Infrared Telescope in Space (IRTS), The Spitzer First Look Survey (FLS), Spitzer Legacy & Ancillary data, Spitzer Reserved Observations (ROC) and the Spitzer Space Telescope data.

IRSA is also seamlessly interoperable with ten remote archives and services: GOODS, ISO, MAST, VizieR, DSS, NVSS, FIRST, HEASARC, NED and JPL, which help expand the available dataset wavelength range from X-ray to radio. The majority of IRSA's image collections are Simple Image Access (SIA) compliant and are available through the Virtual Observatory (VO) data mining tools.

The IRSA demo includes IRSA's five main services: inventory service RADAR, catalog query service Gator, data fusion service OASIS, general search service for complex data collections Atlas, and IRSA's 2MASS Image data access services.

IRSA's website is **http://irsa.ipac.caltech.edu**.

1. IRSA Overview

Over the past four years, the NASA/IPAC Infrared Science Archive (IRSA) has transformed itself from the curator of the 2MASS datasets to a multi-mission archive for NASA's infrared and sub-millimeter (IR/SM) astronomy data. Today, it serves 60 source catalogs, 22 image datasets, and 7 spectroscopic datasets from seven projects and missions, and is seamlessly interoperable with ten remote archives and services. Astronomers have taken advantage of the expansion of the archive. The average number of requests to archive services has risen nearly thirty-fold since 2000, to nearly 250,000 data requests per month, and there have been $\sim$ 2000 scientific publications based on IRSA-served datasets. The growth is due partly to an increasing number of customers accessing the data through web browsers, and partly due to remote services, such as observation planning tools for the Spitzer Space Telescope, taking advantage of seamless access to IRSA's data. The archive has recently released a data inventory and data mining service, called RADAR, that allows rapid access to all IRSA holdings (see section 4.1). The archive is built on the foundation of a robust architecture whose costs have been substantially reduced since 2000, allowing IRSA to serve new datasets in a highly efficient fashion.

2. IRSA's Scientific Data Holdings

Table 1 lists the major IRSA missions and data collections and their associated wavelength ranges; Tables 2 and 3 summarize the source catalogs, images, and spectroscopic datasets in IRSA as of March 2004. Totals for May 2000 highlight the expansion of the archive and show large increases in the number of datasets and their volume. A breakdown of the holdings is at: **http://irsa.ipac.caltech.edu/holdings.html**.

IRSA hosts data from seven missions and projects. Three of them are major infrared sky surveys. The Two Micron All Sky Survey (2MASS) is the first multi-terabyte, all-sky, fully digital image and catalog dataset to be released in astronomy. The Infrared Astronomical Satellite (IRAS) performed the first wide-area sky survey in the thermal infrared. Its successor, the Midcourse Space Experiment (MSX), provided higher spatial resolution coverage in the Galactic Plane, and surveyed those areas missed by IRAS. The Sub-millimeter Wave Astronomical Satellite (SWAS) and the Infrared Telescope in Space (IRTS) conducted new infrared spectroscopic observations, and, in the case of IRTS, far-infrared photometry. IRSA supports the Spitzer Space Telescope by serving ground-based ancillary data on behalf of the Spitzer First Look Survey and the Spitzer Legacy Science teams, consisting of source catalogs, images, and spectra covering targeted regions of the sky.

IRSA preserves the technical content of the datasets by assimilating and serving product and mission documentation at the end of the mission's lifetime. Archive staff work closely with data providers to ensure that datasets are delivered with all necessary documentation and metadata to allow proper and maximal use of the data by archive users. As part of this effort, IRSA delivered a major upgrade to the IRAS documentation in 2003, and is now serving on-line the IRAS Explanatory Supplements for the source catalogs, image atlas, small scale structure, and serendipitous catalogues.

3. Interoperability with Remote Datasets

Table 4 lists the remote datasets and services now accessible from IRSA; all of these program interfaces have been deployed since 2000. IRSA is interoperable with seven major astronomical datasets, including the Infrared Space Observatory (ISO), and all catalogs accessible through the VizieR interface at the Centre de Donnes astronomiques de Strasbourg (CDS).

Table 1. IRSA Missions Summary

Abbreviation	Dataset/Mission	Wavelength
2MASS	Two Micron All Sky Survey	J, H, K band
IRAS	Infrared Astronomical Satellite	12 μm, 25 μm, 60 μm, 100 μm
MSX	Midcourse Space Experiment	8.28 μm, 12.13 μm, 14.65 μm, 21.3 μm
SWAS	Submillimeter Wave Astronomy Sat.	1.4-2.8 GHz
ISO	Infrared Space Observatory	2.5 μm - 240 μm
Spitzer	Spitzer Space Telescope	3-180 μm; ancillary data: X-ray to Radio
IRTS	Infrared Telescope in Space	1 - 1000 μm

Table 2. Scientific Catalog Holdings at IRSA, March 2004

Dataset/Mission	Wavelength	Num of Catalogs	Num of Sources
IRAS	12 μm, 25 μm, 60 μm, 100 μm	9	1,717,014
2MASS	J, H, K band	13	656,106,982
MSX	8.28 μm, 12.13 μm, 14.65 μm, 21.3 μm	1	331,117
IRTS	1 - 700 μm	1	14,294
Spitzer	3-180 μm; ancillary data: X-ray to Radio	36	1,270,042
	Total, March 2004:	60	659,439,819
	Total, May 2000:	17	185,053,997

Table 3. Scientific Image and Spectra Data Holdings at IRSA, March 2004

Dataset/Mission	Wavelength	Volume (GB)	Number
Image Data			
IRAS	12 μm, 25 μm, 60 μm, 100 μm	20.91	37,460
2MASS	J, H, K band	10,358.00	9,572,125
MSX	8.28 μm, 12.13 μm, 14.65 μm, 21.3 μm	38.77	7,316
IRTS	1 - 700 μm	0.18	1,067
Spitzer	3-180 μm; ancillary data: X-ray to Radio	14.69	155
	Total, March 2004:	10,432.55	9,618,123
	Total, May 2000:	77.00	807,380
Spectra			
IRTS	1 - 700 μm	0.12	536
SWAS	1.4-2.8 GHz	5.90	27,026
ISO	2.5 μm - 240 μm	9.40	2,524
	Total, March 2004:	15.42	30,086
	Total, May 2000:	0.00	0

Table 4. Interoperable Datasets and Services

Dataset/Archive	Location	Wavelength	Data
Great Observatories Deep Survey (GOODS)	STScI	1150-11,200Å	280 FITS files
Infrared Space Observatory (ISO)	Vilspa	2.4-240 μm	22,966 observations
MAST Scrapbook	STScI	70-25,00Å	7,383 observations
VizieR	CDS	X-ray to Radio	4,014 catalogs
Digital Sky Survey (DSS)	STScI	0.35-1 μm	7.2 TB
New Very Large Array Sky Survey (NVSS)	NRAO	1.4 GHz	2,326 images 1,800,000 sources
Faint Images of the Radio Sky at 20 cm	LLNL	20 cm	811,117 sources
HEASARC	Goddard	X-ray to Radio	40+ sky surveys
NASA Extragalactic Database (NED)	IPAC	X-ray to Radio	7.5 million objects 1.9 million images
JPL Horizons & Small Body Ephemeris	JPL	N/A	$\geq$16,300 Objects

4. Major Archive Services

IRSA uses a component-based architecture, in which new applications are built by linking existing services through a simple executive program. In this model, the user interfaces are simply "thin" front ends that accept input from users. The following subsections describe five of the most widely used applications.

4.1. RADAR

http://irsa.ipac.caltech.edu/applications/Radar

RADAR is an inventory and data access service for navigating and exploring the infrared sky (Zhang et al. 2005). For a target or region, RADAR quickly generates an inventory of IRSA's science products. RADAR supports one-click retrieval of subsets of IRSA science products and provides links to dataset-specific query and download services.

4.2. Gator

http://irsa.ipac.caltech.edu/applications/Gator

Gator is a completely general query building service that allows users to extract data from any of the catalogs held at IRSA. Gator supports positional queries as well as queries by parameter. It also supports upload of a table of positions for cross-comparison with catalogs.

4.3. OASIS

http://irsa.ipac.caltech.edu/applications/Oasis

OASIS is IRSA's data integration tool (Good et al. 2003), which provides visualization of and access to data from many providers. OASIS has access to several image archives, IRSA's catalogs, and all CDS VizieR catalog holdings; it can perform contour map generation and region statistics; and it has extensive overlay capabilities for image footprints and spectra/catalog source locations. Most of IRSA's data search results will be available to the user for download as well as for viewing in OASIS, by clicking on the blue "OASIS" button.

4.4. Atlas

http://irsa.ipac.caltech.edu/applications/Atlas

Many of the datasets served by IRSA consist of images, source tables, spectra, or light curves from multiple missions, often covering a particular region of the sky. **Atlas** is a single CGI program which can be used to search any such data collection in a general and uniform manner (Alexov & Good 2004). The Atlas front page has a listing of all of the data collections currently available; the user must choose the collection they are interested in to get to its front page, which contains further information and a search form. Atlas currently serves the following datasets: Great Observatories Origins Deep Survey (**GOODS**), 2MASS 6X Lockman Hole (**LH**), IRAS Galaxy Atlas (**IGA**), Extended IRAS Galaxy Atlas (**EIGA**), Mid-Infrared Galaxy Atlas (**MIGA**), IRAS Sky Survey Atlas (**ISSA**), 2MASS Large Galaxy Atlas (**LGA**), spectra from the ISO

Short Wavelength Spectrometer (**SWS**), NASA/IPAC Extragalactic Database Image Data Atlas (**NED**), Midcourse Space Experiment (**MSX**), Spitzer Reserved Observation Catalog Atlas (**ROC**), Spitzer First Look Survey (FLS) – Ancillary VLA Data (**FLS VLA**), Spitzer First Look Survey (FLS) – NOAO Extragalactic – R (**FLS MAIN R**), Spitzer First Look Survey (FLS) – NOAO ELAIS N1 – R (**FLS ELAISN1 R**), Infrared Telescope in Space (**IRTS**), The MAST Image/Spectra Scrapbook (**MAST Scrapbook**), Spitzer Wide-area InfraRed Extragalactic Survey (**SWIRE**), Cosmic Evolution Survey with HST (**COSMOS**) and Spitzer Space Telescope Data Atlas (**SPITZER**).

4.5. 2MASS Image Services

http://irsa.ipac.caltech.edu/applications/2MASS/IM

The **2MASS Image Services** provide access to the Two Micron All Sky Survey (2MASS) Atlas and Quicklook Images.

1. **Interactive Image Service**: Enables rapid interactive viewing and retrieval of single 2MASS image sets (J, H, and K) covering a specified location or object. Indicates image artifacts. This is useful for fast examination of individual sources and/or small fields.
2. **Image Inventory Service**: Provides a listing of all 2MASS images and their metadata that cover a point or region on the sky. It is useful for determining if a position is covered by more than one image, as is the case in tile overlap regions, and for selectively finding and retrieving images that cover a broader region.
3. **Batch Image Service**: Retrieves sets of all 2MASS images that cover a specified region, sets of images that cover an input list of sky positions, or sets of images that match an input list of image descriptors. This is useful for retrieving large numbers of images to construct large area image mosaics, and for generating finding charts for large numbers of targets.

4.6. Custom services

IRSA has many custom services for other datasets. These can be found on the IRSA **Home** page, or through the full list of IRSA **Services**: **http://irsa.ipac.caltech.edu/services**.

References

Alexov, A., & Good, G. C. 2004, in ASP Conf. Ser., Vol. 314, ADASS XIII, ed. F. Ochsenbein, M. Allen, & D. Egret (San Francisco: ASP), 205

Good, J. C. , Kong, M. , Berriman, G. B. 2003, in ASP Conf. Ser., Vol. 295, ADASS XII, ed. H. E. Payne, R. I. Jedrzejewski, & R. N. Hook (San Francisco: ASP), 89

Zhang, A., et al. 2005, in ASP Conf. Ser., Vol. 347, ADASS XIV, ed. P. L. Shopbell, M. C. Britton, & R. Ebert (San Francisco: ASP), 418

ASTRONOMICAL DATA ANALYSIS SOFTWARE AND SYSTEMS XIV
ASP Conference Series, Vol. 347, 2005
P. L. Shopbell, M. C. Britton, and R. Ebert, eds.

ESA VO : Archives and VO Services

Christophe Arviset, Isa Barbarisi, Inaki Ortiz, Pedro Osuna, and Jesus Salgado

European Space Astronomy Centre, ESA, Villafranca del Castillo, 28080 Madrid, Spain

Abstract. ESA has recently decided to put more effort in the development of the Virtual Observatory (VO). This effort will be based at ESAC where most of ESA's current astronomy missions archives are already located. There was created the ESA VO Project, which is a natural extension of the Archive and VO work already carried out at ESAC. Tasks included making the ISO and XMM-Newton archives VO complian through SIAP and SSAP VO protocols as well as developing some VO tools, like VOSpec, which handles spectra coming from VO SSAP compliant servers.

1. ESA VO Project

1.1. ESA VO Presentation

The ESA centre in Spain near Madrid, previously called VILSPA, has now been renamed ESAC for European Space Astronomy Centre. ESAC will host all ESA astronomy missions' Science Operations Centres, as well as all ESA astronomy and planetary missions' archives (Rosetta, Mars Express, Smart-1, and Huygens for the time being). That currently includes ISO, XMM-Newton, Integral, and the ESA's Planetary missions archives. In the future, Herschel, Planck, and GAIA will also have their archives located at ESAC.

Since 2002, a dedicated Science Archive Group has been created at ESAC. This team provides support to various archive projects. Each member of the group is not dedicated to one project in particular, but is working on a specific subsystem across all the archive projects. There are some Database engineers, some experts on User Interface and Middle-Tier layer, and others on Data Storage, Data Distribution, and Interoperability aspects.

As part of the development of activities at ESAC, a specific budget has now been setup to allow increased VO involvement by the ESAC Science Archive Group. From 2005 onwards, there will be the equivalent of five people working on VO issues. These people will be fully part of the Science Archive Group and will be developing archives to make sure that ESAC archives are developed in a way that can be easily integrated into the VO.

The ESA VO project main goals are the following:

1. Ensure that all ESAC archives are VO compliant.
2. Develop VO applications, in particular an ESA VO Portal and ESA multi-mission archive using VO protocols, and VOSpec to handle spectra from SSAP compatible servers.

3. A natural link to all new ESA astronomy space-based projects are their science operations centres, also located at ESAC.
4. Build an ESA VO registry.
5. Increase overall participation in VO working groups (both in Europe and worldwide).

1.2. Existing Archives at ESAC

The ISO Data Archive (IDA[1]; Figure 1) has been open to the community since December 1998 and offers a state-of-the art archive facility to access ISO data products and auxiliary files. It contains all of the ISO raw and fully processed, science and calibration data, as well as all ancillary data (engineering, uplink and downlink data) for a total of about 400 GBytes stored on magnetic disks. Through a powerful and user friendly Java User Interface, over 1200 registered users have already downloaded the archive content more than eight times since it was opened to the public.

The XMM-Newton Science Archive (XSA[2]) has been available to the scientific community since mid-April 2002. It contains all of the XMM-Newton raw and fully processed, science and calibration data, as well as some ancillary data for an expected total of about 2 TBytes stored on magnetic disks. It also contains the XMM-Newton source catalogue, with powerful search criteria. Based on the IDA architecture and code, it offers a similar data query and retrieval (FTP and CDROM) facility through a user friendly User Interface. Proprietary data is accessible to the observation owner only while public data has unrestricted access. Furthermore, the XSA offers an on-the-fly reprocessing facility to allow the users to calibrate the data with the latest version of the pipeline processing software.

The INTEGRAL SOC Data Archive (ISDA) is for the time being an internal archive for INTEGRAL Science Operations purposes. It has been available since December 2004. It contains all of the INTEGRAL raw and fully processed data for a current size of 2 TBytes stored on magnetic disks. Based on the XSA, it offers the same types of functionalities, including various level of searches as per the INTEGRAL data organization (observations, observations data sets, science window, etc.).

The Planetary Science Archive (PSA[3]) is planned to hold all ESA Planetary missions datasets. The first version was released in March 2004 with GIOTTO data and some ROSETTA calibration data. In the course of 2004, Mars Express, Rosetta, and Smart-1 data will be included, later followed by Huygens. All future ESA planetary missions (Venus Express, BepiColombo) data will also be part of the PSA. Also based on the XSA architecture, the PSA is the central place for all ESA planetary missions and offers similar powerful search, view, and retrieving capabilities. Furthermore, a PSA Validation Tool has been developed to allow the Mission Instruments Teams to validate their Datasets against the PDS standards before they are ingested into the PSA.

---

[1] `http://www.iso.vilspa.esa.es/ida/`

[2] `http://xmm.vilspa.esa.es/xsa/`

[3] `http://www.rssd.esa.int/PSA`

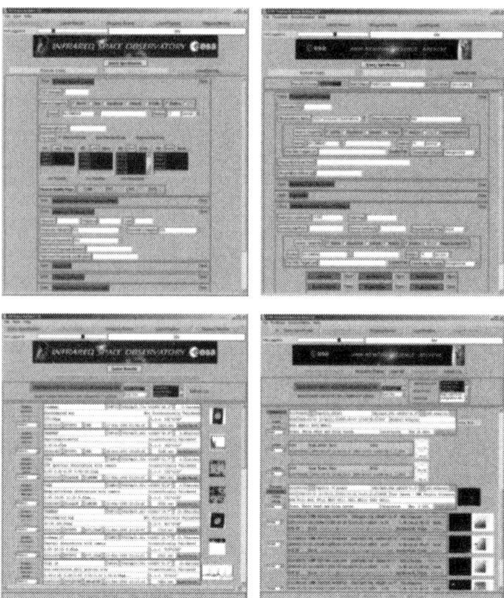

Figure 1. ESA RSSD Archives Snapshots

1.3. VO Compliant Architecture

The Virtual Observatory is a world-wide initiative in astronomy which aims is to allow astronomers to perform new science by providing them with a "federation of astronomical archives and databases around the world, together with analysis tools and computational services, all linked into an integrated facility". Some Inter-Operability standards have been defined, but these protocols are still under definition and evolution is to be expected. Therefore, there was a need to keep our Archives Inter-Operability (AIO) system flexible so we could adapt easily with existing and future VO standards.

By building "translation layers" using XML transformation rules, we can easily adapt our existing AIO systems to these new VO standards (see Figure 2). VO standard metadata query and data requests are translated into our Archive Inter Operability System metadata and data requests to be passed to our business logic. Similarly, metadata results and data products are translated from our system towards the VO standards.

Full AIO documentation can be found at:

- XMM-Newton : `http://xsa.vilspa.esa.es:8080/aio/doc/`
- ISO : `http://pma.iso.vilspa.esa.es:8080/aio/doc/`

2. XMM-Newton and ISO SIAP and SSAP Services

The IVOA (International Virtual Observatory Alliance) has defined various Data Access Layer protocols, in particular to access simple image or simple spectra:

- SIAP for Simple Image Access Protocol
- SSAP for Simple Spectra Access Protocol

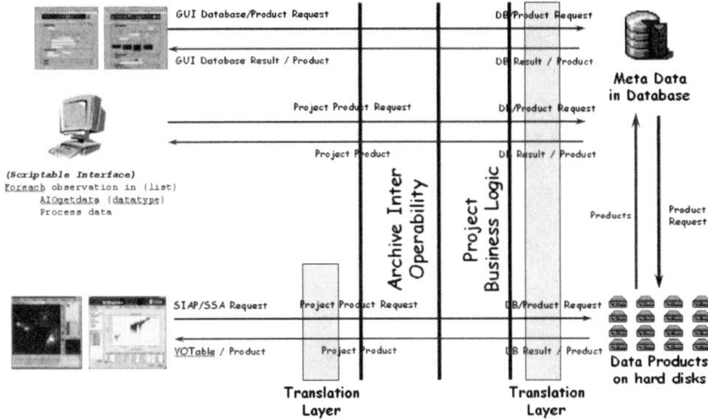

Figure 2. VO Compliant Architecture

Each protocol defines a query method (to query the content of the remote server and return a VOTable containing links to the data and their relevant metadata information) and a data access method (to actually download the data - Image or Spectrum).

Both the XMM-Newton Science Archive and the ISO Data Archive have been one of the very first archives worldwide to implement SIAP and SSAP around the end of 2003. Since then, they have been used in various VO applications, such as the VO-compliant Aladin, the NVO (US National VO) Data Inventory Service, the VOSpec, etc.

The XMM-Newton SIAP service (e.g., cone search) can be reached by
`http://xsa.vilspa.esa.es:8080/aio/jsp/siap.jsp?POS=10.5,41.2`
The ISO SIAP service (e.g., target search) can be reached by
`http://pma.iso.vilspa.esa.es:8080/aio/jsp/siap.jsp?ID_TARGET=m51`
The ISO SSAP service (e.g., target search) can be reached by
`http://pma.iso.vilspa.esa.es:8080/aio/jsp/siap.jsp?`
`ID_TARGET=m51&imagetype=spectrum`

3. VOSpec : A Tool to Handle Spectra from VO SSA Servers

VOSpec is a tool that has been developed at ESAC which allows one to display and superimpose spectra coming from various VO compliant servers, including automatic unit conversion (Figure 3).

Designed and developed with VO in mind, VOSpec already makes use of various VO standards:

- Access the Registry to get information on available SSAP servers for the region of interest.
- Use available Web Services when needed (e.g., the SIMBAD name resolver for the target name resolution).
- Display VOTable information (the standard for VO data.)
- Get Spectra from SSAP servers.

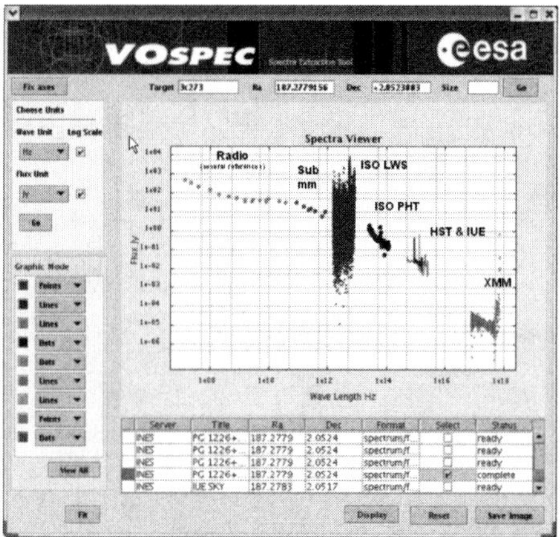

Figure 3. VOSpec

Furthermore, by making clever use of dimensional analysis, the spectra units are described with specific metadata parameters which then allow various functionalities within the tool, in particular:

- Display and superimpose spectra.
- Accept spectra in both VOTable and FITS binary table formats.
- Handle units automatically through the dimensional analysis proposal above.
- SED display and multi-wavelength analysis.
- Polynomial, blackbody, and Gaussian fitting.

More details of VOSpec and the dimensional analysis method for units conversion can be found in Osuna et al. (2005).

4. Conclusions

The existing archives at ESAC (in particular, ISO and XMM-Newton) have been at the front-edge of the VO, as they have implemented already the agreed VO interface protocols standards (SIAP, SSAP). By having a dedicated VO budget, ESAC involvement in the VO is expected to grow, with the goal of the ESA VO Project becoming the VO node for all European Space-based astronomy data.

Acknowledgments. The authors would like to acknowledge the Archive Development Team at ESAC who participated to the development of all of the archive systems which have been the base of the interoperability and VO services. In particular, thanks to John Dowson, Jose Hernandez, Guillermo San Miguel, and Aurele Venet.

References

Osuna, et al. 2005, in ASP Conf. Ser., Vol. 347, ADASS XIV, ed. P. L. Shopbell, M. C. Britton, & R. Ebert (San Francisco: ASP), 198

ASTRONOMICAL DATA ANALYSIS SOFTWARE AND SYSTEMS XIV
ASP Conference Series, Vol. 347, 2005
P. L. Shopbell, M. C. Britton, and R. Ebert, eds.

Extending the Capabilities of *CIAO* with *S-Lang*-based Tools

F. Primini,[1] M. Noble,[2] and the CXC Science Data Systems Group

Chandra X-ray Center, 60 Garden Street, Cambridge, MA 02138

Abstract. We demonstrate a number of tools that extend the capabilities of the *Chandra* X-Ray Center's data analysis software system *CIAO*[3]. These tools were developed using *S-Lang*[4], an interpreted language and multi-platform programmer's library that may be easily embedded into other applications. The *S-Lang* interpreter's C-like syntax and sophisticated vector and array manipulation capabilities allow rapid development by scientists of programs that meet their individual analysis needs.

1. Introduction

S-Lang is an open-source interpreted language, bundled with Linux and installed on millions of machines world-wide. It provides most of the usual benefits one expects from a scripting language: extensibility, rapid prototyping, loose data-typing, robust memory management, etc., and is especially well-suited to scientific and engineering tasks due to its powerful, native multi-dimensional numerical capabilities. For example, it supports complex numbers, as well as highly optimized and transparently vectorized operators and functions. This fosters the use of concise analytic expressions such as $c = sin(a) + b/10.0$ without regard to whether a, and b are scalars, vectors, or n-dimensional arrays, achieving performance and capability on a par with compiled code and commercial analysis packages.

These features are built in to *S-Lang* and distinguish it from more widely known languages like *Perl*, *Python*, or *Tcl*, which lack native high-performance multi-dimensional numerical capability. Native *S-Lang* operators can out-perform these languages by two orders of magnitude in representative calculations such as matrix multiplication[5] and *S-Lang* arithmetic functions can perform up to twice as fast as *Python* extensions specifically crafted to optimize numerical work. In function and syntax, *S-Lang* bears a strong resemblance to what is arguably the most popular scripting language in Astronomy, *IDL*, and our colleagues routinely convert *IDL* scripts to *S-Lang* with little effort. Mapping such scripts to non-vectorized languages is considerably more difficult.

---

[1] Harvard-Smithsonian Center for Astrophysics

[2] MIT Center for Space Research

[3] `http://cxc.harvard.edu/ciao`

[4] `http://www.s-lang.org`

[5] `http://dada.perl.it/shootout/matrix.html`

2. *S-Lang* in *CIAO*

S-Lang has been part of the *CIAO* distribution[6] since release 2.0 in 2001. A number of *CIAO* tools are *S-Lang* scripts. Other scripts are used in analysis threads[7] or in the Sherpa spectral analysis package (Burke 2005). The *CIAO* graphics tool *CHIPS*, spectral analysis tools *Sherpa* and *ISIS*, and the *slsh* shell all embed the *S-Lang* interpreter and can launch user-developed *S-Lang* scripts.

The *CIAO* distribution also includes a number of *S-Lang modules*[8], shared objects that can be dynamically linked to *S-Lang* applications during runtime. These provide access to low-level *CIAO* libraries such as coordinate conversion (pixlib), position filtering (regions), and file I/O (ASCII & FITS). Other general modules[9], compatible with *CIAO*, are also available. Using these, scientists can rapidly code algorithms, scripts, and even applications, all without confronting the details of the *CIAO* system architecture.

In the following sections, we present three examples that illustrate the use of *S-Lang* in *CIAO*, adopting *ISIS* as our *S-Lang* platform.

3. Example 1: Where Are My Data

The *WHERE* function is an intrinsic *S-Lang* function that determines indices of array elements that fulfill arbitrary boolean expressions. It can be a powerful tool for rapidly selecting subsets of arrays meeting complex criteria. Here we use it to filter a *Chandra* event list, producing the result shown in Figure 1. A more sophisticated GUI-based application is presented in Noble (2005).

```
% Read Event Data into Arrays
(x,y,s) = fits_read_col("acisf01575_001N001_evt1.fits","x","y","status");

% These coordinates correspond to an x-ray source
x0 = 4138.54; y0 = 4013.76;

% Find events within 30 pixels from source center
l  = where(hypot(x-x0,y-y0)<30);
dx = x[l] - x0; dy = y[l] - y0;

% Make a scatter plot of events in region
pl = open_plot("/xwin");
label("dx [pix]","dy [pix]", "Source Events Profile (red=afterglow)");
plot(dx,dy);

% Now find afterglow events (status bits 16-19) and overplot in red
ag_bits = 0xf shl 16;
lg = where(s[l]&ag_bits);
oplot(dx[lg],dy[lg],red);
```

---

[6] http://cxc.harvard.edu/ciao/why/slang.html

[7] http://cxc.harvard.edu/ciao/threads/slang.html

[8] http://cxc.harvard.edu/ciao/slang/functions/modules.html

[9] http://space.mit.edu/CXC/software/slang/modules/

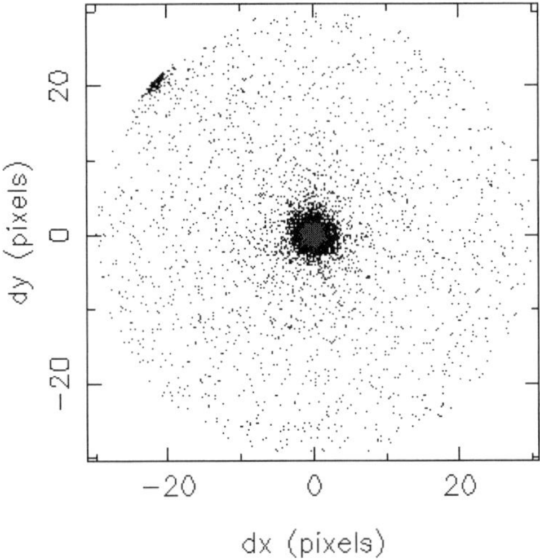

Figure 1. The *S-Lang WHERE* function determines arbitrary array subsets, such as ACIS events (in red) flagged as afterglow events.

4. Example 2: Interactive Aperture Photometry with *DS9*

The image display program *DS9* can communicate with other processes via the X Public Access (XPA) mechanism. The *S-Lang ds9* package and *xpa* module[10] allow *S-Lang* scripts to retrieve image data and metadata from a running *DS9* process, analyse them, and send updated information to the *DS9* display. The *APPHOT* package[11] may be launched from *DS9's Analysis Menu* (see Figure 2) and uses the *xpa* module to

- retrieve source locations from *DS9* and compute off-axis angles;
- compute standard (e.g., 90 & 95%) Enclosed Count Fraction (ECF) aperture sizes using ECF calibration data[12] appropriate to those off-axis angles;
- compute net counts for each source, using the *CIAO* tool *DMEXTRACT*;
- correct net counts for ECFs in source and background apertures.

5. Example 3: Using *DS9* and the GNU Scientific Library

The *GSL* module[13] provides access to a number of numerical analysis routines in the GNU Scientific Library[14]. Here we use it to smooth a column extracted

---

[10] `http://space.mit.edu/cxc/software/slang/modules/xpa/`

[11] `http://hea-www.harvard.edu/~fap/APPHOT.html`

[12] `http://cxc.harvard.edu/cal/Hrma/psf/index.html`

[13] `http://space.mit.edu/CXC/software/slang/modules/gsl/`

[14] `http://sources.redhat.com/gsl/`

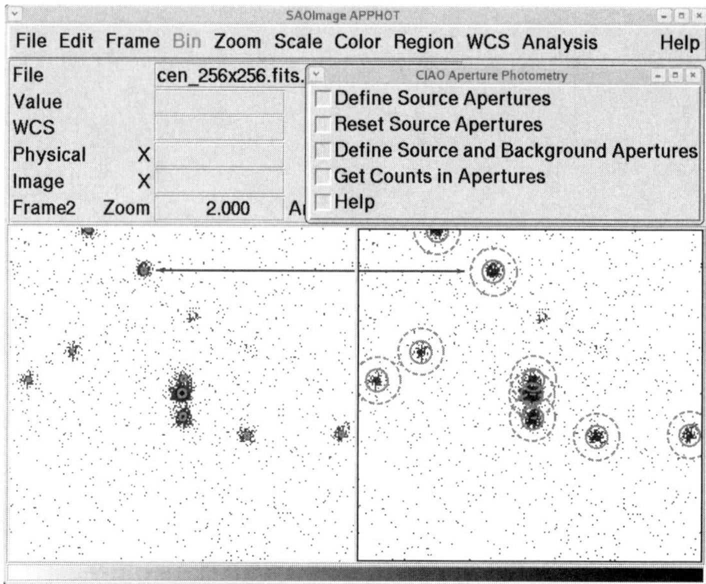

Figure 2. *APPHOT* uses the *xpa* module to retrieve *DS9* image data and metadata, ECF calibration data to determine appropriate aperture sizes, and *DMEXTRACT* to compute net counts, corrected for ECF.

from the x-ray image displayed in the previous example. The results are shown in Figure 3.

```
% Import the GSL module
require("gsl");

% Connect to DS9 process
require("ds9");
xpah=ds9_launch("APPHOT");

% Read data in ds9 display frame and get an interesting slice
m31=ds9_get_array(xpah); m31column=m31[*,129];

% Plot it
ipl=open_plot("/xw");
xrange(100,150);
label("Y [pix]","Counts","");
hplot([0:255],[1:256],m31column);

% Use every fourth point as an interpolation point
xa=[96:160:4]; ya=m31column[[96:160:4]];

% Use GSL's cubic spline interpolation function to to smooth the data
yinterp=interp_cspline([96:160],xa,ya);

% Overplot the smoothed data on the original curve
oplot([96:160],yinterp);
```

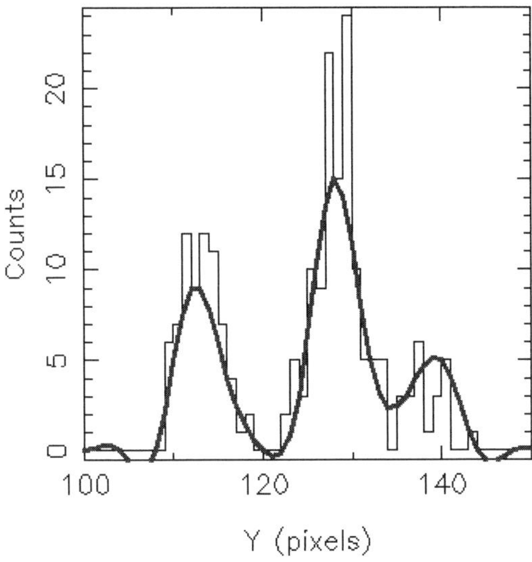

Figure 3. A number of *GSL* numerical analysis routines, such as cubic spline interpolation, can be imported to analyse, for example, *Chandra* data extracted from *DS9* displays.

6. Summary

S-Lang provides scientists wishing to analyse *Chandra* data with powerful array and numeric capabilities in a scripting language already distributed with the *CIAO* x-ray analysis package. By using *S-Lang* shared object modules specific to *CIAO*, as well as other more general modules such as *GSL*, scientists can rapidly develop new applications that build on the existing *CIAO* architecture to address their individual scientific needs.

Acknowledgments. This work is supported by NASA under contract NAS8-03060 to the Smithsonian Astrophysical Observatory for operation of the *Chandra* X-ray Center. We would also like to acknowledge the work of D. Jerius and M. Karovska in developing the ECF calibration files used in Example 2.

References

Burke, D. 2005, in ASP Conf. Ser., Vol. 347, ADASS XIV, ed. P. L. Shopbell, M. C. Britton, & R. Ebert (San Francisco: ASP), 105

Noble, M. 2005, in ASP Conf. Ser., Vol. 347, ADASS XIV, ed. P. L. Shopbell, M. C. Britton, & R. Ebert (San Francisco: ASP), 237

ASTRONOMICAL DATA ANALYSIS SOFTWARE AND SYSTEMS XIV
ASP Conference Series, Vol. 347, 2005
P. L. Shopbell, M. C. Britton, and R. Ebert, eds.

Starlink Software Developments

Peter W. Draper

Department of Physics, University of Durham, South Road, Durham, DH1 3LE, UK

Alasdair Allan

University of Exeter, UK

David S. Berry

University of Central Lancashire, UK

Malcolm J. Currie, David Giaretta, and Steve Rankin

Rutherford Appleton Laboratory, UK

Norman Gray

University of Glasgow, UK

Mark B. Taylor

University of Bristol, UK

Abstract. Various recent changes to the software produced by Starlink are demonstrated. These cover areas such as table handling, time-series analysis, pipeline processing, astrometric calibration, spectral and cube visualisation, and ports to the Mac OS X and Cygwin environments. Particular emphasis was given to the applicability to the Virtual Observatory.

1. Changes to the Classic Starlink Software Collection

The Starlink Software Collection has very recently been converted to use a build system based on GNU Autotools, moved to a single CVS repository, and re-licensed under the GPL. These are seen as key developments in the move to a more open strategy for future support, and so that the whole system can be more easily ported to Unix-like operating systems (in particular POSIX systems which include an X server). The details of this work are described in Gray et al. (2005). Figure 1 shows some of our applications running under Mac OS X (similar screen shots of Cygwin/Windows, Solaris. and various flavours of Linux have been omitted for space reasons).

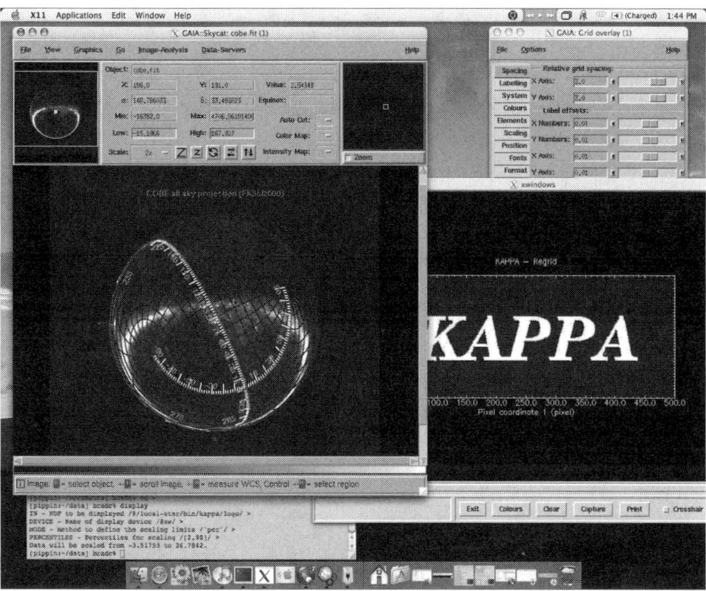

Figure 1. GAIA and KAPPA running under Mac OS X, a development recently made possible by the introduction of a GNU Autotools-based build system.

2. TOPCAT & STIL, New Features

Starlink TOPCAT is a user-friendly graphical program for viewing, analysing and editing tables in many different formats. It is based on STIL – the Starlink Tables Infrastructure Library – a pure-Java class library for accessing and manipulating astronomical tables. Recent changes to TOPCAT include:

- New ‘activation actions’, such as causing the display of a region around an object from an image cutout service. These are extensible.
- Many plotting improvements – faster, tidier, flippable axes.
- Graphical selection of arbitrary shaped subsets.
- Faster and improved matching.
- Support for more table formats (VOTable 1.1, CSV, FITS-plus).

For more details of TOPCAT and STIL see Taylor (2005).

3. FROG, New Application

Starlink FROG is a new time-series analysis and display application written in Java. It has been designed to provide a simple user interface for astronomers wanting to do time domain astrophysics, but still have access to powerful features like those found in PERIOD. A full description of FROG can be found in Allan (2005).

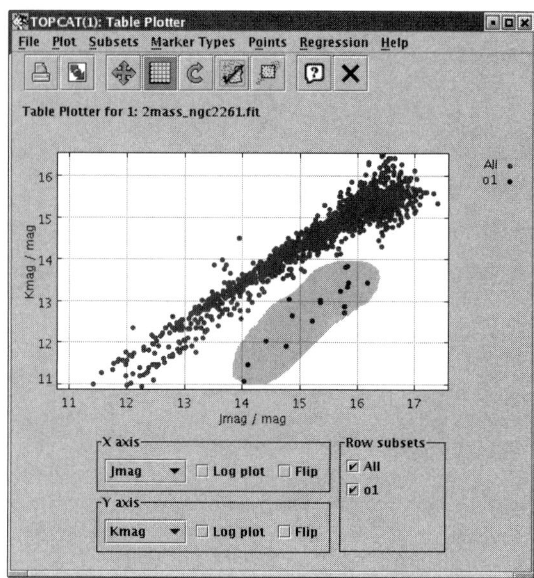

Figure 2. TOPCAT shown selecting an arbitrary region to create a subset

4. ORAC-DR Pipelines

ORAC-DR is a generic pipeline system written by the Joint Astronomy Centre, Hawaii (Cavanagh et al. 2003), which uses Starlink applications to process data. As part of its support for UK astronomy, Starlink has been developing new ORAC-DR pipelines and recipes for some unsupported, or partly supported instruments, most recently SOFI (ESO) and NIRI (Gemini). Currie (2005) compares one such recipe for the ISAAC instrument with the developing ESO equivalent.

5. GAIA, New Features

GAIA[1] is a classic Starlink application (Draper 2004a), based on the ESO Skycat tool, for the visualisation and analysis of images and catalogues. It offers many proto-VO features, such as remote catalogue and image downloads, which are widely used. Recent changes (beyond its arrival on Mac OS X and Cygwin, along with the other Starlink applications) are the ability to display slices from cubes and to perform automated astrometric calibrations.

[1] `http://www.starlink.ac.uk/gaia`

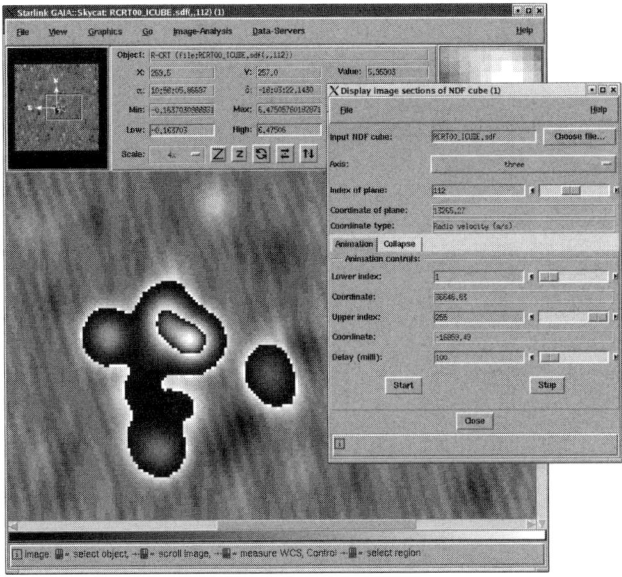

Figure 3. GAIA displaying a slice, along the third axis of an NDF cube. It is also possible to step automatically through a range of slices, and collapse a range into an image. Displayed images can be analysed like any other image.

6. SPLAT, New Features

SPLAT is a Java based tool for visualising and analysing spectra, which is described in Draper (2004b). Some of its more recent enhancements are:

- Tables access, including VOTable as provided by the STIL library (Taylor 2005).
- Graphic annotations and log axes.
- Interpolated backgrounds using hand-drawn splines.
- Hierarchical data browsing and access.
- Spectral line deblending (from the command-line only at present).

7. Virtual Observatory Tie-ins

Several of the applications described here process VOTables and offer Web-service access to their functionality. They are also written in Java, so are ideally placed to quickly adapt and interact with VO services as they become available (importantly much of this is also from a desktop point of view). This stance has recently been demonstrated by the extension of SPLAT (now SPLAT-VO) to include a Simple Spectral Access Protocol UI. The Starlink classic code base is now much more portable, and freely available, which makes it more attractive for processing data, and it comes with a well proven pipeline system, as well as also being wrappable in Java so that it can act as a ready supply of web services.

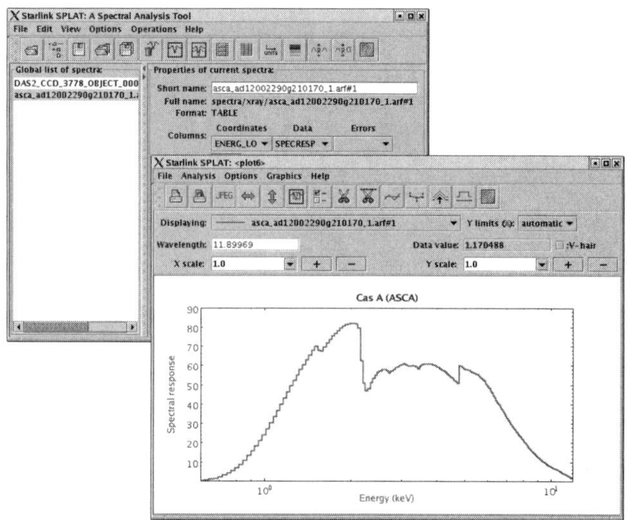

Figure 4. SPLAT reading a FITS binary table and displaying the spectrum formed from the two selected columns. Note the use of log axes.

8. Finding Out More About Starlink Applications

Information about Starlink applications and how to download them is available at `http://www.starlink.ac.uk`.

References

Allan, A. 2005, in ASP Conf. Ser., Vol. 347, ADASS XIV, ed. P. L. Shopbell, M. C. Britton, & R. Ebert (San Francisco: ASP), 370

Cavanagh, B., Hirst, P., Jenness, T., Economou, F., Currie, M. J., Todd, S., & Ryder, S. D. 2003, in ASP Conf. Ser., Vol. 295, ADASS XII, ed. H. E. Payne, R. I. Jedrzejewski, & R. N. Hook (San Francisco: ASP), 237

Currie, M. J. 2005, in ASP Conf. Ser., Vol. 347, ADASS XIV, ed. P. L. Shopbell, M. C. Britton, & R. Ebert (San Francisco: ASP), 580

Draper, P. W. 2004a, Starlink User Note 243, Starlink Project CCLRC

Draper, P. W. 2004b, Starlink User Note 214, Starlink Project CCLRC

Gray, N., Jenness, T., Allan, A., Berry, S. B., Currie, M. J., Draper, P. W., Taylor, M. B., Cavanagh, B. C. 2005, in ASP Conf. Ser., Vol. 347, ADASS XIV, ed. P. L. Shopbell, M. C. Britton, & R. Ebert (San Francisco: ASP), 119

Taylor, M. B. 2005, in ASP Conf. Ser., Vol. 347, ADASS XIV, ed. P. L. Shopbell, M. C. Britton, & R. Ebert (San Francisco: ASP), 29

Part III
Focus Sessions

ASTRONOMICAL DATA ANALYSIS SOFTWARE AND SYSTEMS XIV
ASP Conference Series, Vol. 347, 2005
P. L. Shopbell, M. C. Britton, and R. Ebert, eds.

TOPCAT & STIL: Starlink Table/VOTable Processing Software

Mark B. Taylor

H. H. Wills Physics Laboratory, Bristol University, Tyndall Avenue, Bristol, UK

Abstract. The Starlink Tables Infrastructure Library (STIL) is a pure-Java, open source library for I/O and manipulation of tabular data such as astronomical catalogs. It is designed to be high-performance and to cope with large tables. The core library is format-neutral, with the work of serialization and deserialization performed by pluggable format-specific I/O handlers. This means that the programmer seees a high-level abstraction of a table which is easy to work with, and also that support for new data formats can be added easily. Supplied handlers provide support for VOTables, FITS table extensions, relational databases via SQL and plain text tables, amongst others. The VOTable handler is believed to be the only existing library capable of reading or writing all the defined VOTable encoding formats (TABLEDATA, FITS, BINARY).

TOPCAT, based on STIL, is a user-friendly graphical program for viewing, analysis and editing of tables. It has facilities for plotting, cross matching, row selection, sorting and manipulation of data and metadata. Synthetic columns can be created and row selections made using a powerful and extensible algebraic expression language.

1. Introduction

Tables are common in astronomy, and are a prominent feature of the data produced, transmitted and consumed by human and software elements of the Virtual Observatory. The fact that the VOTable format is one of the first standards to become an IVOA recommendation[1] bears witness to this observation. A common example of astronomical tabular data is an object catalogue, but other examples, such as event lists, are important too. This paper describes STIL, a library for generic I/O and processing of tables, and TOPCAT, a graphical user application built on top of STIL. Both of these products are open source (released under the GNU Public License) and pure Java (J2SE1.4), which makes them highly portable and easy to deploy. They have been developed for the Starlink Project.

[1] `http://www.ivoa.net/Documents/latest/VOT.html`

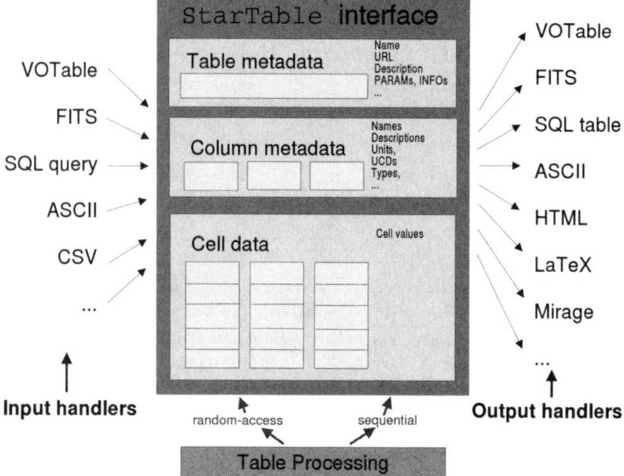

Figure 1. Schematic of STIL architecture.

2. STIL

The Starlink Tables Infrastructure Library (STIL)[2] is a generic I/O and processing library for tabular data. Central to STIL is a relatively simple model of what a table is, namely a data structure which has some per-table metadata, per-column metadata and the cell data themselves. The `StarTable` Java language interface embodies this data model, and it is `StarTable` objects that the programmer deals with when manipulating tables within STIL.

STIL has no native external data format, but a number of pluggable input and output handlers. Input handlers perform the job of deserializing tables from external storage to turn them into StarTables, while output handlers perform the opposite task of taking a StarTable and serializing it to external storage. Since these I/O handlers are separate from the core of the library, it is easy to change their implementation and add new ones without affecting application code. It is possible to install a new handler at run time by specifying its class name to an application. This design resembles the way that Java's JDBC database communication layer works. A schematic diagram of the architecture is given in Figure 1.

Among the I/O handlers supplied are ones which can read/write VOTable documents, FITS BINTABLE and TABLE extensions, relational databases (using SQL), and a number of text-based file formats including Comma-Separated Values. A feature of this design is that conversion between any of the supported formats is trivial, and STIL comes with a simple command-line utility `tablecopy` which performs this task.

As well as I/O, STIL provides a number of facilities for manipulation of tables including ways to add, remove and rearrange columns and rows, join and split

[2] `http://www.starlink.ac.uk/stil/`

tables, and modify data and metadata. Flexible and efficient facilities for cross-matching are also distributed with the library, although these are currently not fully documented and somewhat experimental.

Care has been taken to make the I/O and processing facilities scalable throughout STIL; the aim in particular is that it should be possible to process a table with an unlimited number of rows in a limited amount of memory. The `StarTable` interface provides both sequential and random methods of data access for different processing requirements; in the former case limited memory use can be achieved by streaming the data a row at a time, and in the latter by caching the cell data in a temporary disk file, for which the library provides facilities.

STIL is fully documented; the public classes have comprehensive javadocs, and a user document (SUN/252), which contains examples and overview documentation, is available in HTML and PDF formats.

2.1. STIL for VOTables

As noted above, one of the formats supported by STIL is the XML-based VOTable format. Since this is of particular importance in the Virtual Observatory era, and because it has some notable features, there follow a few comments on this handler in particular.

The STIL VOTable parser is at time of writing, as far as we know, the only one which fully supports the VOTable standard. Although other available parsers can read the pure-XML TABLEDATA variant of the format, no others can read the FITS and BINARY variants. For reasons of bandwidth and CPU efficiency, TABLEDATA is inappropriate for encoding very large amounts of data, so this ability is of considerable significance. As well as reading, STIL also makes it easy to write VOTables in any of the three variants.

STIL can also provide to the programmer a hierarchical in-memory view (DOM) of the structure of a VOTable document. By use of custom stream-based processing (SAX) it is able to do this using modest amounts of memory even when the tables contained are large.

3. TOPCAT

TOPCAT[3] is a graphical user application for viewing, analysis and editing of tables. Being based on STIL, it can read and write tables in many formats, and it is extensible to new ones in the same way. It is not in practice able to cope with tables of unlimited size, but is happy to manipulate fairly large ones; on a normal desktop machine tables of order 10^6 rows $\times$ 10^2 columns can be processed easily. The program can be deployed in a number of ways, including as a WebStart application or from a single local jar file. Comprehensive user documentation (SUN/253) is available either in PDF or HTML form, or from within the program's context-sensitive help browser.

The program offers many ways to view and manipulate the data and metadata of tables. Some of the actions it permits are:

---

[3]`http://www.starlink.ac.uk/topcat/`

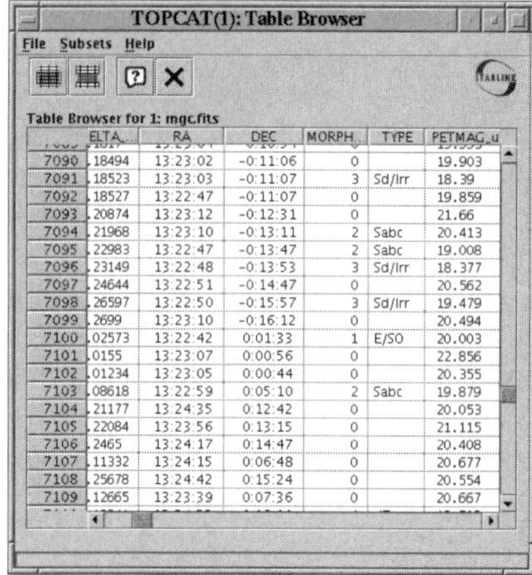

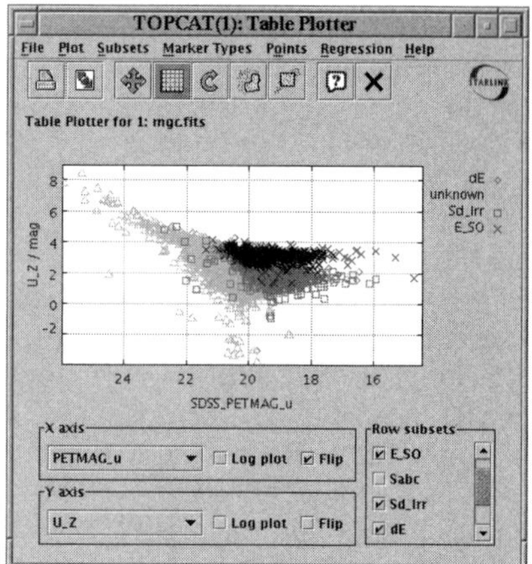

Figure 2. Example windows from TOPCAT: the table browser window and the plot window

- View/edit table data in a scrollable browser
- View/edit table and column metadata
- Re-order, hide and unhide existing columns
- Insert algebraically-defined "synthetic" columns
- Sort the rows
- Define row subsets (selections of the rows) in various ways
- Plot columns against each other, distinguishing different row subsets
- Calculate statistics on each column for some or all rows
- Perform cross-matching between tables or within a table
- Create a new table by concatenating the rows of two existing ones

Two of TOPCAT's many windows are shown by way of example in Figure 2. Some, though by no means all, of the program's key capabilities are described in more detail in the remaining sections.

3.1. Plotting

The Plot Window displays a scatter-plot of values from one of a table's columns against those from another. The columns can be selected interactively, and points may be plotted on logarithmic or reversed axes if desired. It is easy to zoom in and out to focus on regions of interest in the plot. If multiple row subsets have been defined, they will be plotted with different symbols, and the user can select which subsets are displayed and control the kinds of plotting symbols that are used. New subsets can be defined from the plot by indicating a rectangular region or by drawing an arbitrary shape or shapes with the mouse. The plot currently in view can be exported at any time to Postscript or GIF format.

3.2. Joining Tables

TOPCAT provides flexible and efficient facilities for cross-matching, either between two or more tables, or internally to a single one. The most common match criterion is a maximum angular separation between two (RA,Dec) points on the celestial sphere, but a range of other criteria are available including ones based on separation in isotropic or anisotropic Cartesian space of one, two or more dimensions, value equivalence, and combinations of these. In most cases, the speed of the match algorithm scales as $O(N \log N)$, where N is the total number of rows involved.

It is also possible to concatenate two tables "head-to-toe" by specifying the correspondence of columns between them.

3.3. Algebraic Expressions

A powerful feature of TOPCAT is the ability to create new columns or specify row subsets algebraically. In both cases the user enters a textual expression in which the names or identifiers of table columns serve as variable names; this expression can be evaluated for each row with each column identifier evaluating to that column's entry in the row. The expression thus defines a new "synthetic" column, and in the case of a boolean-valued expression it can be taken to define a row subset (if the expression evaluates true for a given row, that row is taken to be included in the subset).

The expression syntax is powerful and extensible; expressions are actually written in the Java language and compiled to bytecode prior to evaluation. This means that the full power of a programming language can be used to define operations. A number of functions such as arithmetic, trigonometric and string manipulations are initially available, but the user can augment these by supplying Java classes which define new functions and making them known to the program at run time. Despite this flexibility, it is in many cases easy for the non-specialist to specify algebraic expressions; for instance to create a new column which contains the average of columns named RMAG and BMAG, it is only necessary to enter the expression "`(RMAG+BMAG)*0.5`".

3.4. Activation Actions

Facilities exist in TOPCAT to focus on a particular row of a table, which can be useful for instance if it is an outlier in some sense. If you click on a row in the table browser, the corresponding point becomes highlighted in the plot window, and vice versa. It is also possible to cause other actions to take place when a row is "activated" in this way; one possibility is to display in SoG or SPLAT (Giaretta et al. 2005) an image or spectrum related to the selected row. There are pre-packaged facilities to display an image of the sky region surrounding a row obtained from certain cutout servers (DSS, 2MASS quick-look and SDSS), but activation can be configured to trigger almost any action, defined by user-supplied classes.

References

Giaretta, D, et al. 2005, in ASP Conf. Ser., Vol. 347, ADASS XIV, ed. P. L. Shopbell, M. C. Britton, & R. Ebert (San Francisco: ASP), 22

ASTRONOMICAL DATA ANALYSIS SOFTWARE AND SYSTEMS XIV
ASP Conference Series, Vol. 347, 2005
P. L. Shopbell, M. C. Britton, and R. Ebert, eds.

Montage: An Astronomical Image Mosaic Service for the NVO

Anastasia Clower Laity, Nate Anagnostou, G. Bruce Berriman, and John C. Good

Infrared Processing and Analysis Center, California Institute of Technology

Joseph C. Jacob and Daniel S. Katz

Jet Propulsion Laboratory, California Institute of Technology

Thomas Prince

California Institute of Technology

Abstract. Montage is a software system for generating astronomical image mosaics according to user-specified size, rotation, WCS-compliant projection and coordinate system, with background modeling and rectification capabilities. Its architecture has been described in the proceedings of ADASS XII and XIII (Berriman et al. 2003, 2004). It has been designed as a toolkit, with independent modules for image reprojection, background rectification and co-addition, and will run on workstations, clusters and grids. The primary limitation of Montage thus far has been in the projection algorithm. It uses a spherical trigonometry approach that is general at the expense of speed. The reprojection algorithm has now been made 30 times faster for commonly used tangent plane to tangent plane reprojections that cover up to several square degrees, through modification of a custom algorithm first derived for the Spitzer Space Telescope. This focus session will describe this algorithm, demonstrate the generation of mosaics in real time, and describe applications of the software. In particular, we will highlight one case study which shows how Montage is supporting the generation of science-grade mosaics of images measured with the Infrared Array Camera aboard the Spitzer Space Telescope.

1. Introduction

Montage builds science-grade image mosaics from any input WCS-compliant data. Separate modules perform reprojection of input images to a common projection, background modeling and rectification, and image co-addition. Montage is portable and highly parallelizable, and can be run in multi-processor or grid environments as well as on a desktop computer. The modular design provides great flexibility to end users. One example is provided by the Spitzer Wide-area Infrared Extragalactic Survey (SWIRE), which uses Montage in its processing pipeline. SWIRE has a wide range of ancillary observations taken using ground- and space-based telescopes, with a variety of image parameters (rotation, projection type, pixel scale, etc.). The SWIRE team uses Montage to re-project and mosaic ancillary data into the same tiling scheme and pixel scale as the mosaics

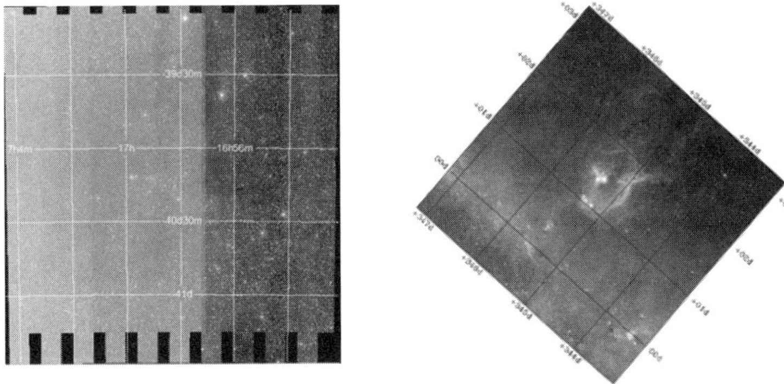

Figure 1. Original 2MASS and MSX images

created by the Spitzer pipeline. This allows optical ancillary data to be easily compared to Spitzer data, facilitating multi-wavelength source extraction and quick recognition of high-redshift objects.

2. Enhancements to Montage Version 2.2: Fast Reprojections and Arbitrary Image Sizes

The latest release of Montage, version 2.2, contains several significant enhancements upon the initial public release, v1.7, and can be downloaded from: http://montage.ipac.caltech.edu

Version 1.7 supported photometrically and astrometrically accurate image reprojections, at the expense of processing speed. The reprojection algorithm, while general, was slow because it projected both input and output pixels onto the celestial sphere in order to calculate the overlap of these pixels on the sky. Version 2.2 of Montage includes a new fast reprojection module for tangent-plane reprojections. mProjectPP has been developed in collaboration with the Spitzer Science Center, and is based on the Mopex algorithm (http://ssc.spitzer.caltech.edu/postbcd/doc/mosaicker.pdf), which uses plane-to-plane solutions to calculate the overlap entirely in pixel space, instead of projecting pixels onto the celestial sphere. This produces a significant speed-up; for 2MASS Atlas images, the improvement is a factor of 20.

Over a small enough area (up to several degrees), many non-tangent plane projections can be approximated by a TAN header with distortion parameters. Montage v2.2 analyzes a FITS header in any projection and determines the tolerances on an equivalent, distorted-TAN projection. It outputs a header template which can be used in conjunction with the fast reprojection module to speed up transformations to or from non-TAN projections.

3. Application of Montage: Creating a 2MASS/MSX Mosaic

This section walks the user through the creation of a 3-color mosaic using data from the 2MASS and MSX missions. On the left of Figure 1 is a mosaic of 170 2MASS Atlas images, with no background rectification, covering a total area of about 3 square degrees. On the right is a MSX A-band (8 μm) image retrieved using IRSA's MSX image server, covering about 2.4 square degrees and in a Cartesian projection. In the following walk-through, each Montage command is preceded with a ">", and is followed by the output of the module. Each Montage module prints a structured output message to stdout, of the form [struct stat="*status*", key1="*val1*", ... , keyn="*valn*"].

Step One: Setup metadata tables and header templates. First, create tables of image metadata (WCS information, FITS geometry, etc.) for the Montage modules to read, instead of having to repeatedly open and close FITS files:

```
>mImgtbl raw_K raw_K.tbl
[struct stat="OK", count=170, failed=0, nooverlap=0]
```

Call mMakeHdr to create a header template that completely encloses all the 2MASS images:

```
>mMakeHdr raw_K.tbl template.hdr
[struct stat="OK", count=170, clon=254.587292, clat=-40.25175
3, lonsize=2.353611, latsize=2.450000, posang=359.891421, lon
1=256.154189, lat1=-41.468162, lon2=253.014309, lat2=-41.4636
21, lon3=253.076184, lat3=-39.014964, lon4=256.104469, lat4=
-39.019343]
```

The 2MASS images are in a tangent-plane projection, and so can use the fast reprojection module. To speed up the MSX reprojection, however, we need to create a distorted-TAN header for the MSX data:

```
>mGetHdr raw_MSX/msx_4deg.fits msx.hdr
[struct stat="OK", ncard=23]
>mTANHdr -c eq msx.hdr msxtan.hdr
[struct stat="OK", fwdxerr=0.00351429, fwdyerr=0.0054
6297, fwditer=51, revxerr=0.00335636, revyerr=0.03825
81, reviter=9]
```

Step Two: Image reprojection. Launch the 2MASS image reprojections by calling the reprojection executable, mProjExec, and using the "-f" flag to instruct it to use fast reprojection:

```
>mProjExec -f -p raw_K raw_K.tbl template.hdr proj_K
stats_K.tbl
[struct stat="OK", count=170, failed=0, nooverlap=0]
```

Next, call the fast reprojection module directly on the MSX data, instructing it to use the distorted-TAN template header instead of reading the native MSX FITS header:

```
>mProjectPP -i msxtan.hdr raw_MSX/msx_4deg.fits
final_MSX.fits template.hdr
[struct stat="OK", time=6082]
```

Once the reprojections are complete, it is necessary to regenerate the image metadata tables, as the FITS geometry has changed:

```
>mImgtbl proj_K proj_K.tbl
[struct stat="OK", count=170, badfits=0]
```

Step Three: Background rectification. The background rectification process matches each image's background to its surrounding images, globally minimizing the inter-image differences. First, create a table which lists the files that overlap each other and gives each overlap pair a unique identifier:

```
>mOverlaps proj_K.tbl diff_K.tbl
[struct stat="OK", count=454]
```

Second, create "difference" images for each overlap region by subtracting FITS files from each other:

```
>mDiffExec -p proj_K diff_K.tbl template.hdr diff_K
[struct stat="OK", count=454, failed=0]
```

Third, call mFitplane on each difference image to find the plane that best fits each one:

```
>mFitExec diff_K.tbl fits_K.tbl diff_K
[struct stat="OK", count=454, failed=0, warning=0,
missing=0]
```

Fourth, using the information found by mFitplane, calculate what planes need to be removed from each image in order to globally minimize the background differences:

```
>mBgModel proj_K.tbl fits_K.tbl corrections_K.tbl
[struct stat="OK"]
```

Fifth and finally, call mBackground on each projected image to subtract the plane calculated by mBgModel:

```
>mBgExec -p proj_K proj_K.tbl corrections_K.tbl corr_K
[struct stat="OK", count=170, nocorrection=0, failed=0]
```

Step Four: Coaddition.
The last step for the two bands of 2MASS data is to coadd them into mosaics:

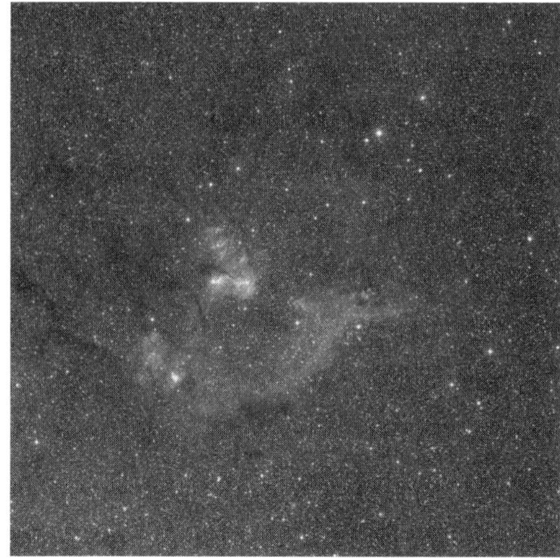

Figure 2. Final 3-color mosaic: MSX A-band is red, 2MASS K-band is green, and 2MASS J-band is blue.

```
>mAdd -e -p corr_K proj_K.tbl template.hdr final_K.fits
[struct stat="OK", time=144]
```

Step Five: Creating a 3-color JPEG image.
After cropping the edges out of each mosaic, call mJPEG (another new module; not included in the public release, but available on request) to create a 3-color JPEG image from the 3 FITS files. The user assigns a color to each image and inputs the desired color-stretch, which can be found using a visualization tool such as IRSA's OASIS. The output of mJPEG is shown in Figure 2.

```
>mJPEG -red final_MSX_crop_4.fits 0% 99.95% 2
-green final_K_crop_4.fits 0% 99.3% 2
-blue final_J_crop_4.fits 0% 99.4% 2
-out jpeg/r99.95_g99.3_b99.4_crop_4.jpg
[struct stat=OK]
```

Acknowledgments. Montage is funded by NASA's Earth Science Technology Office, Computational Technnologies Project, under Cooperative Agreement Number NCC5-626 between NASA and the California Institute of Technology.

References

Berriman, G. B., et al. 2003, in ASP Conf. Ser., Vol. 295, ADASS XII, ed. H. E. Payne, R. I. Jedrzejewski, & R. N. Hook (San Francisco: ASP), 343

Berriman, G. B., et al. 2004, in ASP Conf. Ser., Vol. 314, ADASS XIII, ed. F. Ochsenbein, M. Allen, & D. Egret (San Francisco: ASP), 593

ASTRONOMICAL DATA ANALYSIS SOFTWARE AND SYSTEMS XIV
ASP Conference Series, Vol. 347, 2005
P. L. Shopbell, M. C. Britton, and R. Ebert, eds.

New Software for Ensemble Creation in the Spitzer Space Telescope Operations Database

Russ Laher and John Rector
Spitzer Science Center, MS 314-6, California Institute of Technology, Pasadena, CA 91125; laher@ipac.caltech.edu, jar@ipac.caltech.edu

Abstract. Some of the computer pipelines used to process digital astronomical images from NASA's Spitzer Space Telescope require multiple input images, in order to generate high-level science and calibration products. The images are grouped into ensembles according to well-documented ensemble-creation rules by making explicit associations in the operations Informix database at the Spitzer Science Center (SSC). The advantage of this approach is that a simple database query can retrieve the required ensemble of pipeline input images. New and improved software for ensemble creation has been developed. The new software is much faster than the existing software because it uses pre-compiled database stored procedures written in Informix SPL (SQL programming language). The new software is also more flexible because the ensemble-creation rules are now stored in and read from newly defined database tables. This table-driven approach was implemented so that ensemble rules can be inserted, updated, or deleted without modifying software.

1. Introduction

Roughly one third of the 230 downlink-data pipelines for computer processing Spitzer Space Telescope DCEs (data-collection events or raw images) rely on explicit groupings, or ensembles, of images made *a priori* in the Spitzer Science Center (SSC) operations database for *en masse* data processing of images for creating calibration files, co-added-image mosaics, band-merged source lists, and other high-level science products (e.g., Fang et al. 2003). Information about how the DCEs are grouped together as ensembles is stored in the SSC operations database so that a simple database query can retrieve the required pipeline input images. The SSC pipeline operator developed a script to perform this task, and it has been used successfully in operations since Spitzer's launch (Aug. 2003). It became apparent that the script would be more convenient if it were faster (some observing campaigns took as long as twelve hours to assemble and store the ensemble relationship information). Furthermore, because the rules for ensemble creation were evolving for a long time and the script had to be modified each time the rules changed, a table-driven design for specifying the ensemble-creation rules was identified as a desideratum. This report briefly describes our next version of ensemble-creation software with these improvements. The design concepts are general and can be applied to other projects that require ensemble processing.

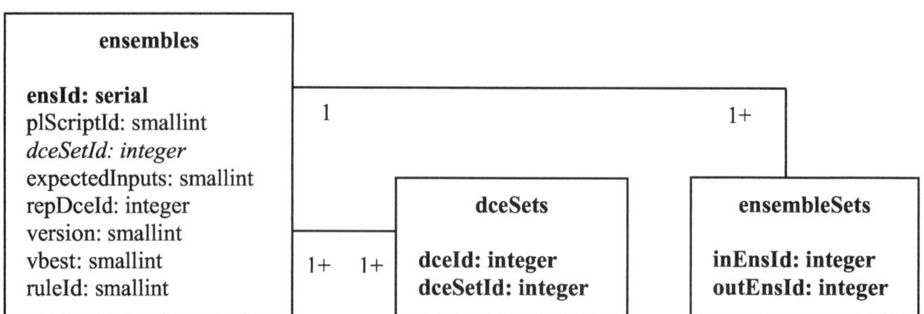

Figure 1. Schema for the ensembles, dceSets, and ensembleSets database tables.

2. Database Storage of Ensembles

The SSC operations database has three tables for storing information about how ensembles of images are defined and how they are to be processed (see Figure 1). The dceSets database table associates a unique index (dceSetId) with all images (dceId) in an ensemble. The ensembles database table associates ensembles (dceSetId) with pipelines (plScriptId). The ensembleSets database table associates multiple records in the ensembles database table, which is useful for pipeline processing that requires multiple input ensembles, such as band-merging astronomical point-source lists from multiple instrument channels.

The ensembles database table is referenced by the ensembleProducts database table (not shown), in which the products of pipeline ensemble processing are registered, thus enabling traceback to their pipeline input images.

3. Ensemble-Creation Rules

Our design calls for storing the ensemble-creation rules in two new database tables (see Figure 2). Records in the ensRules database table specify the rules, i.e., how various attributes of the DCEs are used both to query the database for specific DCEs and as discriminants for grouping the returned DCEs into one or more DCE sets. Each ensRules record has a unique index (ruleId). Records in the ensPlScripts database table specify the ensemble-processing pipelines that are to be associated with a given rule.

The table-driven feature of the design is encapsulated in the "sql" field of the ensRules database table, which stores the actual SQL query for the DCEs and their unique sets of discriminants for creating ensembles. Specifying an ensemble-creation rule requires expert knowledge of SQL and familiarity with the SSC operations database schema. However, this design allows powerful control over ensemble creation.

The "instrument" field associates the rule with the relevant Spitzer science instrument (IRAC, MIPS, or IRSX).

The "minInputs" field specifies the minimum number of images required in the ensemble for a given ensemble-creation rule. Ensembles not satisfying this criterion will not be created and stored in the database.

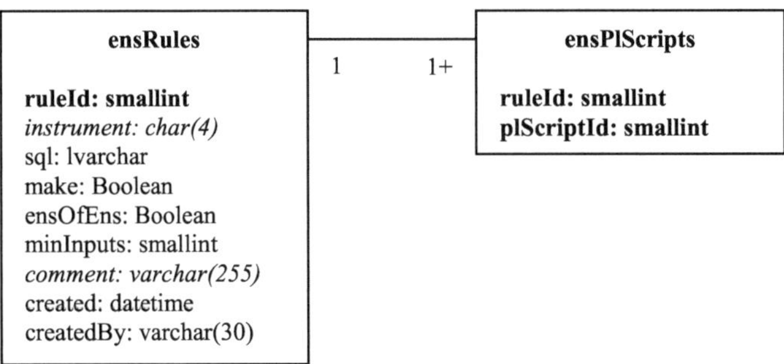

Figure 2. Schema for the ensRules and ensPlScripts database tables.

The "make" field gives the capability of moving an ensemble-creation rule into abeyance without having to delete the rule from the database.

The 12 known discriminants of the 56 ensemble-creation rules specified thus far at the SSC are hard wired into the new and improved software, in order to satisfy the faster ensemble-creation requirement. These rules are limited to using the known discriminants. Modifications to the software, database schema, and database stored procedures are required if additional discriminants are needed.

As part of our design, a new ruleId field has been added to the ensembles database table, in order to associate each ensemble with the rule that was applied to create it, for tracking purposes (see Figure 1). This is a distinct improvement that facilitates *post facto* analysis of how a given ensemble was created.

4. Ensemble-Creation Software

Figure 3 gives a flow chart of the software. The top-level command is implemented in perl. Although it works on one observation request at a time, it can be easily scripted to loop over all requests in a campaign. For a given request, only those ensemble-creation rules that are applicable to the associated science instrument are executed. The perl script calls our custom database stored procedures, which use several temporary database tables to do the bulk of the work. Only the freely formulated SQL for the rules cannot be executed via database stored procedures and therefore are executed with system calls to Informix's dbaccess command. There is a test mode that just loads the temporary database tables (not the dceSets, ensembles, and ensembleSets database tables) and outputs a very readable table, all of which provide useful diagnostics for determining whether the rule as specified is giving the expected ensembles. Finally, the software has options to 1) just give a listing of the ensemble-creation rules currently stored in the database for either a given one or all three Spitzer science instruments; and 2) print out a tutorial on the software's usage.

The software does not support multiple simultaneous users. Only one instance of the software should be executed at a time.

Acknowledgments. This work was performed at the Spitzer Science Center as part of a mission/project managed by Jet Propulsion Laboratory, Cali-

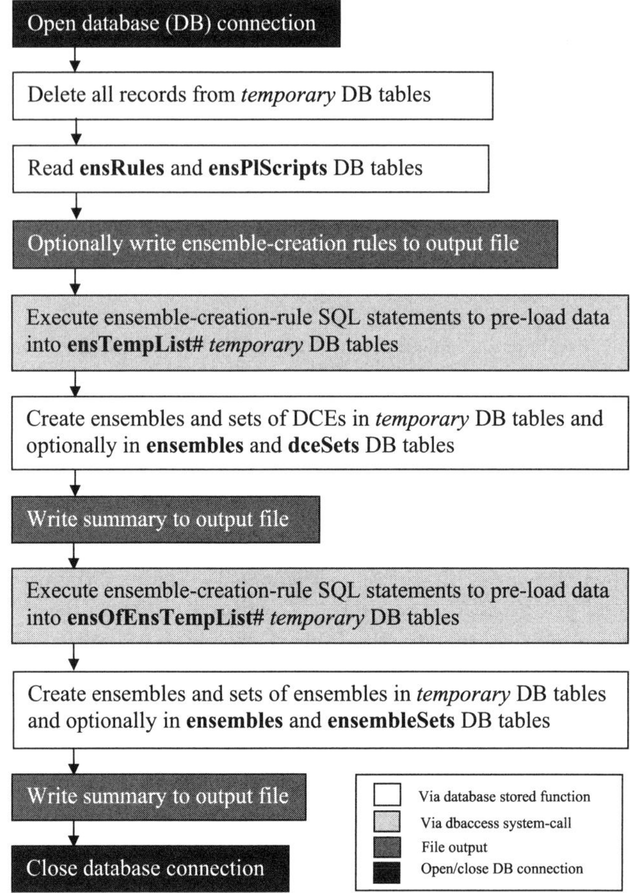

Figure 3. Flow chart for the perl script createEnsembles.pl.

fornia Institute of Technology, under a contract with the National Aeronautics and Space Administration.

References

Fang, Fan, Li, J., Narron, B., Waterson, C., Khan, I., Lee, W. P., Fowler, J., Laher, R., & Moshir, M. 2003, in ASP Conf. Ser., Vol. 295, ADASS XII, ed. H. E. Payne, R. I. Jedrzejewski, & R. N. Hook (San Francisco: ASP), 253

ASTRONOMICAL DATA ANALYSIS SOFTWARE AND SYSTEMS XIV
ASP Conference Series, Vol. 347, 2005
P. L. Shopbell, M. C. Britton, and R. Ebert, eds.

RedShift 5 - Virtual Planetarium for Amateurs and Professionals

Nick Maris, Sergey M. Kudryavtsev,[1] Oleg Kalitsev, and Oleg Margorin

Maris Technologies Ltd, Cams Hall, Cams Hill, Fareham, Hants PO16 5AB, UK

Kurt Tomaszewski and Michael Fleissner

United Soft Media Verlag GmbH, Thomas-Wimmer-Ring 11, 80539, München, Germany

Claus Vester

cc-live, Ferdinand-Miller-Platz 12, 80335 München, Germany

Abstract. A description of the product, database, and main features of RedShift 5, the virtual desktop planetarium, are presented.

1. Introduction

RedShift is a desktop planetarium being developed by Maris Technologies Ltd. and its licensors since 1992. The software visualizes the sky view in any direction as seen from an arbitrary point in the Solar system covering a time span of more than 15 thousand years, from 4713 BC to 9999 AD. The high visualization accuracy is ensured by use of the latest catalogs and motion theories of all types of celestial objects and precision calculation algorithms.

2. RedShift 5 database

The latest, 5th edition of the program includes:

- The complete Hipparcos (ESA 1997) and Tycho-2 (Hög et al. 2000) stellar catalogues (2.5 million stars) as well as some 15 million stars of up to the 14th - 15th mag from the Hubble Guide Star Catalog, v.2.2;
- The 4th edition of the General Catalogue of Variable Stars and New Catalogue of Suspected Variable Stars (Kukarkin et al. 1982);
- Catalogue of Principal Galaxies (Paturel et al. 1989) (70,000+ objects) and latest catalogues of optically visible open clusters and nebulae, radio and X-ray sources, quasi-stellar objects and BL lacertae;
- updated catalogue of more than 50,000 asteroids from The Asteroid Orbital Elements Database and 1,500 comets from the 13th edition of the

[1]Sternberg Astronomical Institute, Universitetsky Pr.13, Moscow 119992, Russia

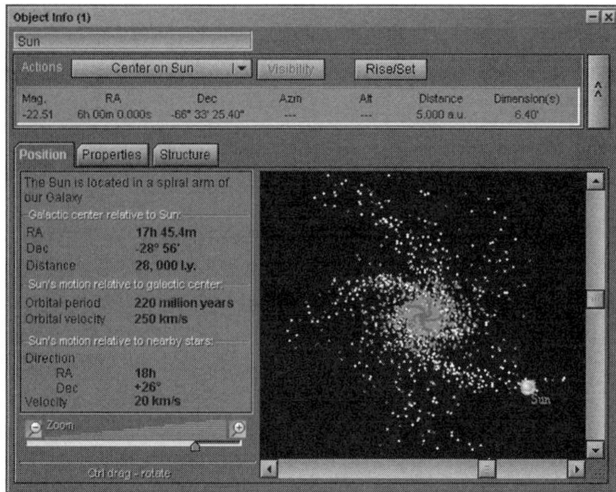

Figure 1. A sample "Object Info" screen

Catalogue of Cometary Orbits and Cometary Notes of Bureau des Longitudes (I.M.C.C.E.), Paris;
- expanded collection of images of galactic and extragalactic objects from Messier and Caldwell catalogues;
- motion theory of major planets VSOP87 (Bretagnon & Francou 1988);
- all natural moons of planets including recently discovered moons of Jupiter, Saturn, Uranus and Neptune.

The database of asteroids, comets and spacecraft can be easily updated by importing the latest orbital elements from NORAD, MPC and other Internet resources. The next step in RedShift development is to provide users with access to other on-line resources from the world-wide Virtual Observatories.

Every object included in the database has an exhaustive "Object Info" display onscreen, giving extensive positional and photometric data about the object at any given date (Figure 1).

3. RedShift 5 main features

- Macroplayer for playing the guided tours - specially prepared scenarios using the Redshift software engine to illustrate many fascinating aspects of astronomy;
- Revised high-resolution maps of planets obtained from the latest space missions (Figure 2);
- Downloading and instant visualization of new objects (Figure 3);
- Custom magnification of planets and their moons;
- Add Image tool - user can take any image of a Deep Sky object from an external source and put it to the right place on the sky in RedShift;
- Multi-window mode that enables users to visualize various astronomical events from different locations on the Earth and in the sky at the same time (Figure 4);

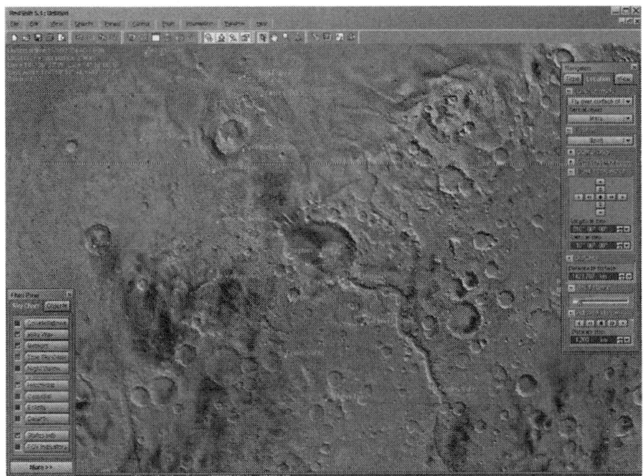

Figure 2. An image of the Martian surface

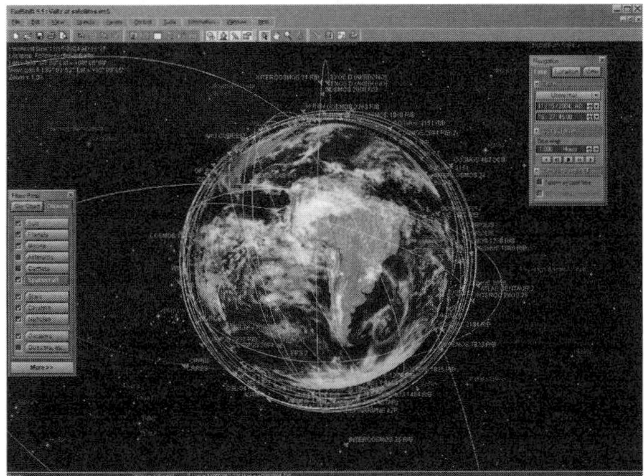

Figure 3. A dynamically updated visualization of Earth spacecraft

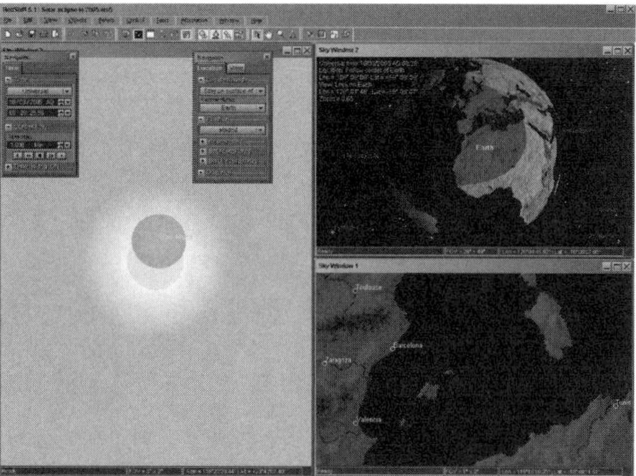

Figure 4. Multi-window mode

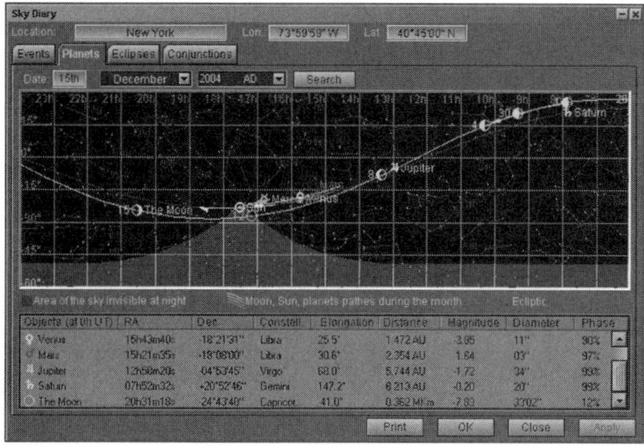

Figure 5. Sky Diary

- Numerical integration engine precisely models the motion of comets and asteroids in relation to all planetary gravitation forces. User can even simulate the famous impact of Shoemaker-Levi 9 Comet to Jupiter in July 1994;
- Full screen mode and Night Vision option - dimming and reddening of the screen to reduce its intensity during night time observation;
- Sky Diary - a powerful tool for calculation of various celestial events at any date or time period (eclipses, conjunctions, planets configurations, etc.) (Figure 5);
- Solar eclipse path and other major lines/points on a vector map of the Earth (Figure 6);
- Quick set-up of the sky view with newly integrated Open GL technologies;

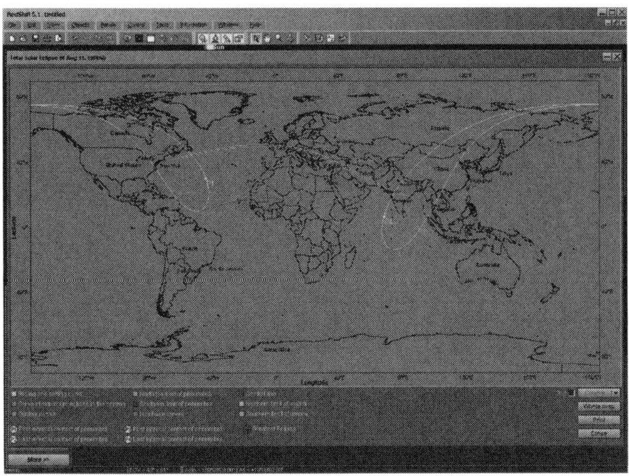

Figure 6. Solar eclipse path

The RedShift family will be expanded by release of a special version for beginners (RedShift Planetarium) at the end of 2004 and a version for teachers (RedShift 5 College) in 2005.

Acknowledgments. We would like to acknowledge the excellent work of a great number of programmers, designers and astronomers who have worked on RedShift over more than 12 years of development.

References

Bretagnon, P., & Francou, G. 1988, A&A, 202, 309
ESA 1997, The Hipparcos Catalogue, ESA SP-1200
Hög, E., et al. 2000, A&A, 355, L27
Kukarkin, B., et al. 1982, Moscow, Nauka Publ. House
Paturel, G., et al. 1989, ApJS, 80, 299

Part IV
Algorithms and Data Analysis

ASTRONOMICAL DATA ANALYSIS SOFTWARE AND SYSTEMS XIV
ASP Conference Series, Vol. 347, 2005
P. L. Shopbell, M. C. Britton, and R. Ebert, eds.

Visual Data Mining of Astronomic Data with Virtual Reality Spaces: Understanding the Underlying Structure of Large Data Sets

Julio J. Valdés
National Research Council Canada
Institute for Information Technology
1200 Montreal Rd.
Ottawa, ON K1A 0R6
julio.valdes@nrc.ca

Abstract. The information explosion in astronomy requires the development of data mining procedures that speed up the process of scientific discovery, and the in-depth understanding of the internal structure of the data. This is crucial for the identification of valid, novel, potentially useful, and understandable patterns (regularities, oddities, etc.).

A Virtual Reality (VR) approach for large heterogeneous, incomplete and imprecise information is introduced for the problem of visualizing and analyzing astronomic data. The method is based on mappings between one heterogeneous space representing the data, and a homogeneous virtual reality space. This VR-based visual data mining technique allows the incorporation of the unmatched geometric capabilities of the human brain into the knowledge discovery process, and helps in understanding data structure and patterns. This approach has been applied successfully to a wide variety of real-world domains, and it has a large potential in astronomy. Examples are presented from the domain of galaxy research.

1. Introduction

The science of astronomy has experienced unprecedented progress in the last years. In particular, the advances in computer, communication, and observation technologies have increased in many orders of magnitude the quantity and quality of astronomic data. This information explosion requires the development of data mining procedures that speed up the process of scientific discovery, and the in-depth understanding of the internal structure of the data. This is crucial for the non-trivial process of identifying valid, novel, potentially useful, and ultimately *understandable patterns* in data; that is, for knowledge discovery (Fayyad et al 1996).

The information explosion requires analytic and interpretation procedures which enable users to *understand* their data rapidly and with greater ease. Further, the increasing complexity of the data analysis procedures makes it more difficult for the user to extract useful information out of the results given by the various techniques applied. Visual techniques are, therefore, very appealing. In general, objects under study are described in terms of collections of *heterogeneous* properties. For example, an astronomic source can be characterized by a set of properties represented by nominal, ordinal or real-valued variables (scalar), as well as by other of a more complex nature like images (in the visible wavelength region, infrared, and others), time-series (e.g., spectra), etc. In addition, the information comes with different degrees of precision, uncertainty and completion (missing data is quite common). Classical data mining and analysis methods are sometimes difficult to use, the output of many procedures may be large and time consuming to analyze, and often their interpretation requires special expertise. Moreover, some methods are based on assumptions about the data which limit their application, specially for the purpose of exploration, comparison, hypothesis formation, etc, typical of the first stages of scientific investigation.

This makes graphical representation directly appealing. Humans perceive most of the information through vision, in large quantities and at very high input rates. The human brain is extremely well qualified for the fast understanding of complex visual patterns, and still outperforms the computer. Several reasons make Virtual Reality (VR) a suitable paradigm: Virtual Reality is *flexible*, as it allows the choice of different representation models to better suit different human perception preferences. It allows the construction of different virtual worlds representing *the same* underlying information, but with different look and feel. Thus, the user can choose the most appealing representation. VR allows *immersion.* The user can navigate inside the data, and interact with the objects in the world. VR creates a *living* experience. The user is not merely a passive observer or an outsider, but an actor in the world, in fact, part of the information itself. VR is *broad and deep.* The user may see the VR world as a whole, and/or concentrate the focus of attention on specific details or portions of the world. Of no less importance is the fact that in order to interact with a Virtual World only minimal skills are required.

In this paper a *Virtual Reality* approach for understanding large heterogeneous, incomplete and imprecise data (Valdés 2002, 2002b; Valdés & Bonham-Carter 2003; Valdés 2003; Valdés 2004), is introduced in the domain of astronomy. In this approach, the notion of data is not restricted to databases, but also includes logical relations and other forms of structured knowledge.

2. The Heterogeneous Space

Consider an *information system* $S =< U, A >$ where U and A are non-empty finite sets, called the *universe* and the set of *attributes* respectively, such that each $a \in A$ has a domain V_a and an evaluation function f_a assigns to each $u \in U$ an element $f_a(u) \in V_a$ (i.e. $f_a(u) : U \to V_a$) (here the V_a are not required to be finite). An example is shown in Figure 1. There are attributes with domains of different kinds (nominal, ordinal, ratio, fuzzy, images, time-series and graphs), and also containing missing values (represented as ?).

Heterogeneous and incomplete information systems will be considered as follows. Let ? be a special symbol having two basic properties: *i*) if $? \in \Omega$ (Ω being an arbitrary set) and f is any unary function defined on Ω, then $f(?) = ?$, and *ii*) ? is an incomparable element w.r.t any ordering relation defined on Ω. A heterogeneous domain is defined as a Cartesian product of a collection of *source*

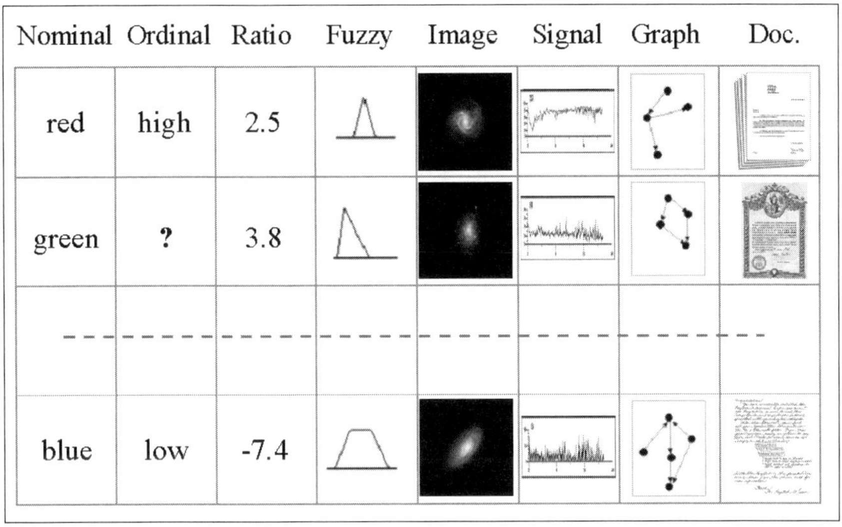

Nominal	Ordinal	Ratio	Fuzzy	Image	Signal	Graph	Doc.
red	high	2.5					
green	?	3.8					
blue	low	-7.4					

Figure 1. An example of a heterogeneous database.

sets (Ψ_i): $\hat{\mathcal{H}}^n = \Psi_1 \times \ldots \times \Psi_n$, where $n > 0$ is the number of *information sources* to consider.

As an example, consider the case of a heterogeneous domain where objects are characterized by attributes given by continuous crisp quantities, discrete features, fuzzy features, graphs and digital images. Let $\mathbf{R}$ be the reals with the usual ordering, and $\mathcal{R} \subseteq \mathbf{R}$. Now define $\hat{\mathcal{R}} = \mathcal{R} \cup \{?\}$ to be a source set and extend the ordering relation to a partial order accordingly ($\hat{\mathcal{R}}$ may model scalar measurements, with missing values). Now let $\mathbf{N}$ be the set of natural numbers and consider a family of n_r sets ($n_r \in \mathbf{N}^+ = \mathbf{N} - \{0\}$) given by $\hat{\mathcal{R}}^{n_r} = \hat{\mathcal{R}}_1 \times \ldots \times \hat{\mathcal{R}}_{n_r}$ (n_r times) where each $\hat{\mathcal{R}}_j$ ($0 \leq j \leq n_r$) is constructed as $\hat{\mathcal{R}}$, and define $\hat{\mathcal{R}}^0 = \phi$ (the empty set). Now let $\mathcal{O}_j$, $1 \leq j \leq n_o \in \mathbf{N}^+$ be a family of finite sets with cardinalities k_j^o respectively, composed by arbitrary elements, such that each set has a fully ordering relation $\leq_{\mathcal{O}_j}$. Construct the sets $\hat{\mathcal{O}}_j = \mathcal{O}_j \cup \{?\}$, and for each of them define a partial ordering $\hat{\leq}_{\mathcal{O}_j}$ by extending $\leq_{\mathcal{O}_j}$ according to the definition of ?. Analogously construct the set $\hat{\mathcal{O}}^{n_o} = \hat{\mathcal{O}}_1 \times \ldots \times \hat{\mathcal{O}}_{n_o}$ (n_o times and $\hat{\mathcal{O}}^0 = \phi$). For the special case of nominal variables, let $\mathcal{N}_j$, $1 \leq j \leq n_m$ ($n_m \in \mathbf{N}^+$) be a family of finite sets with cardinalities $k_j^m \in \mathbf{N}^+$ composed by arbitrary elements but such that no ordering relation is defined on any of the $\mathcal{N}_j$ sets. Now construct the sets $\hat{\mathcal{N}}_j = \mathcal{N}_j \cup \{?\}$, and define $\hat{\mathcal{N}}^{n_m} = \hat{\mathcal{N}}_1 \times \ldots \times \hat{\mathcal{N}}_{n_m}$, ($n_m$ times and $\hat{\mathcal{N}}^0 = \phi$). Sets $\hat{\mathcal{O}}^{n_o}$, $\hat{\mathcal{N}}^{n_m}$ may represent the case of n_o ordinal variables and n_m nominal variables respectively. Similarly, a collection of n_f extended fuzzy sets $\hat{\mathcal{F}}_j$ ($1 \leq j \leq n_f$) , n_g extended graphs $\hat{\mathcal{G}}_j$ ($1 \leq j \leq n_g$) and n_i extended digital images $\hat{\mathcal{I}}_j$ ($1 \leq j \leq n_i$), can be used for constructing the corresponding cartesian products given by $\hat{\mathcal{F}}^{n_f}$, $\hat{\mathcal{G}}^{n_g}$ and $\hat{\mathcal{I}}^{n_i}$.

The heterogeneous domain is given by $\hat{\mathcal{H}}^n = \hat{\mathcal{R}}^{n_r} \times \hat{\mathcal{O}}^{n_o} \times \hat{\mathcal{N}}^{n_m} \times \hat{\mathcal{F}}^{n_f} \times \hat{\mathcal{G}}^{n_g} \times \hat{\mathcal{I}}^{n_i}$. Elements of this domain will be objects $o \in \hat{\mathcal{H}}^n$ given by tuples of length $n = n_r + n_o + n_m + n_f + n_g + n_i$, with $n > 0$ (the empty set is excluded). Other kinds of heterogeneous domains can be constructed in a similar manner, using the appropriate source sets. More general information systems are those in which the universe is endowed with a set of relations of different arities. Let $t =< t_1, \ldots, t_p >$ be a sequence of p natural integers, called *type*, and $\underline{Y} =< Y, \gamma_1, \ldots, \gamma_p >$ a relational structure as defined in (Hajek & Havranek 1978), where Y is a non-empty domain of objects and the $\Gamma = \{\gamma_i\}$ $(i = 1, \ldots, p)$ are different relations of various arities defined on Y (according to t). The extended information system will be $\hat{S} =< U, A, \Gamma >$, endowed with the relational system $\underline{U} =< U, \Gamma >$.

3. The Virtual Reality Space

A *virtual reality space* is a structure composed of different sets and functions defined as $\Upsilon =< \underline{O}, G, B, \Re^m, g_o, l, g_r, b, r >$. $\underline{O}$ is a relational structure defined as above ($\underline{O} =< O, \Gamma^v >$, $\Gamma^v =< \gamma_1^v, \ldots, \gamma_q^v >$, $q \in \mathbf{N}^+$ and the $o \in O$ are objects), G is a non-empty set of *geometries* representing the different objects and relations (the *empty* or *invisible* geometry is a possible one). B is a non-empty set of *behaviors* (i.e. ways in which the objects from the virtual world will express themselves: movement, response to stimulus, etc.). $\Re^m \subset \mathbf{R}^m$ is a *metric space* of dimension m (euclidean or not) which will be the actual virtual reality geometric space. The other elements are mappings: $g_o : O \rightarrow G$, $l : O \rightarrow \Re^m$, $g_r : \Gamma^v \rightarrow G$, $b : O \rightarrow B$, r is a collection of characteristic functions for Γ^v, $(r_1, \ldots, r_q)$ s.t. $r_i : \gamma_i^{v t_i} \rightarrow \{0, 1\}$, according to the type t associated with Γ^v.

The representation of an extended information system $\hat{S}$ in a virtual world implies the construction of another $\hat{S}^v =< O, A^v, \Gamma^v >, \underline{O}$ in Υ, which requires the specification of several sets and a collection of extra mappings (w.r.t. those required for Υ). A desideratum for $\hat{S}^v$ is to keep as many properties from $\hat{S}$ as possible. Thus, a natural requirement is that U and O are in one-to-one correspondence (with a mapping $\xi : U \rightarrow O$). The structural link is given by a mapping $f : \hat{\mathcal{H}}^n \rightarrow \Re^m$. If $u =< f_{a_1}(u), \ldots, f_{a_n}(u) >$ and $\xi(u) = o$, then $l(o) = f(\xi(< f_{a_1}(u), \ldots, f_{a_n}(u) >)) =< f_{a_1^v}(o), \ldots, f_{a_m^v}(o) >$ ($f_{a_i^v}$ are the evaluation functions of A^v). This gives *semantics* to the pair $< g_o(o), l(o) >$ (it determines important properties like geometry, visibility and location).
It is natural to require that $\Gamma^v \subseteq \Gamma$ (possibly empty), thus having a virtual world portraying selected relations from the information system, represented according to the choices made for G and g_r.

4. The Problem of Large Datasets

Regardless of the criteria followed when computing a virtual reality space, complex optimization procedures are applied involving the estimation of the image of the data objects. The objective function surface becomes more complex and convoluted with the increase of the dimensionality of the parameter space, and local extrema entrapment is typical. Even if all of the difficulties related with the amount of memory and the numeric computation involved are put aside (note that a dissimilarity matrix grows quadratically with the number of objects), the graphical representation of millions or possibly billions of objects in a screen with the current computer technologies, is neither feasible, nor practical. On the other hand, assuming that it would be possible, the amount of information presented to the user will be overwhelming, and will obscure, rather than clarify, the presence of meaningful or interesting patterns. The approach followed here is to study the properties of the dataset ($\mathbf{X}$), possibly huge, in order to extract a subset of a sufficiently smaller cardinality which will either retain as much structural information as possible, or guarantee its preservation up to a predefined threshold. In this approach only the non-redundant objects up to a predefined degree are preserved, thus producing a kernel or core representation of the original dataset. If a similarity measure S is chosen as a redundancy criterium, and a similarity threshold T_s is set forth as a parameter, it is possible to construct a set $\mathbf{L} \subseteq \mathbf{X}$, such that $\forall x \in \mathbf{X}, \exists\, l \in \mathbf{L}, S(x,l) \geq T_s$ (Figure 2). There are efficient algorithms which can generate $\mathbf{L}$-sets at different T_s-levels, and this parameter will determine both the cardinality of the resulting $\mathbf{L}$-set, as well as its semantics. According to this approach, a VR representation of a large or huge dataset is obtained by first extracting a $\mathbf{L}$-set according to a suitable similarity threshold, and then computing its VR space. Since each of the data objects is represented by a sufficiently similar l-object (lower bounded by T_s), the VR space is compliant with the similarity structure of the whole dataset $\mathbf{X}$ at that level.

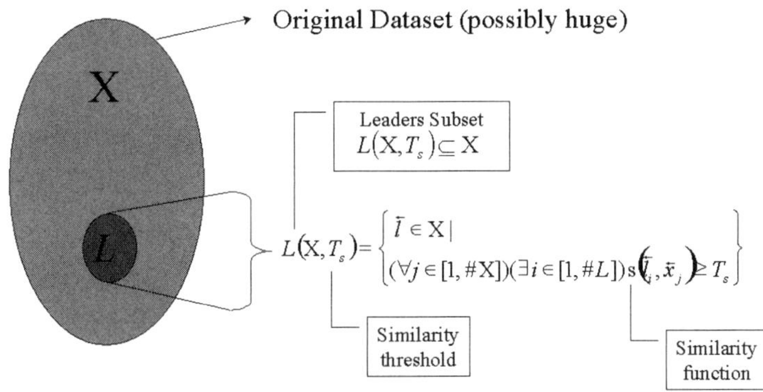

Figure 2. Relation between a dataset $\mathbf{X}$ and its corresponding $\mathbf{L}$-subset at the T_s-similarity level (# denotes set cardinality).

4.1. The Direct and Inverse Transforms

As mentioned, f plays an important role in giving semantics to the virtual world, and there are many ways in which such a mapping can be defined. In a great extent it depends on which features from the original data need to be highlighted. In particular, adjacency relationships between the objects O in Υ should give an indication about the *similarity relationships* (Chandon, Pinson

1981) between the objects in the original heterogeneous space $\hat{\mathcal{H}}^n$ (Valdés 2002b). Other interpretations about internal structure are related with the linear/non-linear separability of class membership relations defined on the data (Jianchang & Jain 1995). In this sense, f can be constructed according to several criteria: *i)* to maximize some metric/non-metric structure preservation criteria as has been done for decades in multidimensional scaling (Kruskal 1964, Borg & Lingoes 1987), *ii)* minimize some error measure of information loss, *iii)* maximize some measure of class separability (in a supervised case), or *iv)* satisfy several criteria simultaneously. For example, in the case of *i)*, if δ_{ij} is a dissimilarity measure between any two $i, j \in U$ $(i, j \in [1, N]$, where n is the number of objects), and $\zeta_{i^v j^v}$ is another dissimilarity measure defined on objects $i^v, j^v \in O$ from Υ $(i^v = \xi(i), j^v = \xi(j)$, they are in one-to-one correspondence), two examples of error measures frequently used are:

$$S\ stress = \sqrt{\frac{\sum_{i<j}(\delta_{ij}^2 - \zeta_{ij}^2)^2}{\sum_{i<j}\delta_{ij}^4}} \tag{1}$$

$$Sammon\ error = \frac{1}{\sum_{i<j}\delta_{ij}} \frac{\sum_{i<j}(\delta_{ij} - \zeta_{ij})^2}{\delta_{ij}} \tag{2}$$

The f mappings obtained using approaches of this kind are only implicit, as no functional representations are found, and its usefulness depends the final errors obtained in the optimization process. Explicit mappings can be obtained from these solutions using neural network, genetic programming, and other techniques. An explicit f is useful for both practical and theoretical reasons. On one hand, in dynamic data sets (e.g. systems being monitored or databases formed incrementally from continuous processes) an explicit direct transform f will speed up the incremental update of the VR information system S^v. On the other hand, it can give semantics to its attributes, thus acting as a dimensionality reducer or as a generator of new attributes.

The possibilities derived from this approach are practically unlimited, since the number of different similarity, dissimilarity and distance functions definable is immense. Moreover, similarities and distances can be transformed into dissimilarities according to a wide variety of schemes. This provides a rich framework where one can find appropriate measures better suited to both the internal structure of the data, and external criteria.

The existence of an *inverse transformation* f^{-1} from Υ back to $\hat{\mathcal{H}}^n$ is, in many cases, worth considering. If a sense is made of patterns of objects in Υ in terms of abstract concepts, and new conjectured objects or relations are conceived, it is natural to ask what kind of previously unseen or undiscovered objects or relations they would correspond to in $\hat{\mathcal{H}}^n$. Several approaches for finding the inverse transformation can be followed, and neural networks are among the obvious choices (Valdés 2002b).

5. An Astronomic Example

In order to illustrate the possibilities of the proposed approach, a dataset containing information about 33055 galaxies was used. This information is part of the Canada-France-Hawaii Legacy Survey (the CFH telescope), and the observational conditions, and preprocessing related with the dataset were the following:

- I-band (red filter) exposure time is 46740 seconds
- Total 5-band exposure time is 77180 seconds (u,g,r,i,z filters were applied, and the i-Band was used to get the morphologies)
- Image reduction, with photometry, and photometric redshifts (courtesy of Stephen Gwyn from the University of Victoria)
- The seeing was 0.9 arcseconds to 1.1 arcseconds (moderate).
- Morphologic analysis by David Schade (Herzberg Institute for Astrophysics, National Research Council Canada)

Each galaxy was characterized by a collection of 11 attributes: *1)* The I-band (red) magnitude, *2-6)* five variables describing the color of the galaxy (derived from the values obtained by the u,g,r,i,z filters, *7)* the half-light radius of the galaxy image, *8)* the half-light radius, as a measure of the size of the galaxy, *9)* the exponential index of the slope of the light profile, *10)* the axial ratio (longer half-axis/smaller half-axis of an ellipse), and *11)* the Photometric redshift.

When presenting the VR spaces corresponding to the experiments, it must be taken into account that it is impossible to illustrate appropriately the look, feel and immersion of a virtual reality, color, 3D environment within the limits imposed by printed paper. Thus, grey level screen snapshots from the examples are presented only to give a rough idea. The design of the virtual reality spaces was kept simple in terms of the geometries used, (in particular, behaviors were excluded). The snapshots were simplified w.r.t the information included in the corresponding Υs to avoid information overload. The criterium for computing the VR space was to preserve the similarity structure, and the direct transform between the original space and Υ was found by minimizing Sammon error, with ζ_{ij} given by the euclidean distance in Υ and $\delta_{ij} = (1 - \hat{s}_{ij})/\hat{s}_{ij}$, where $\hat{s}_{ij}$ is Gower's similarity (Gower 1973). In all cases, the computed VR space corresponds to **L**-sets extracted from the database containing all of the galaxies.

5.1. Experiment 1: All Galaxies

For this first experiment, all of the variables were used as descriptor attributes, and the **L**-sets were computed at a similarity threshold of 0.85. In addition, the values of the Photometric redshift were used as a classification criterium, and the galaxies were divided into three classes: < 0.5, in the $[0.5, 1)$ interval, and ≥ 1. Accordingly, additional objects were included in the space, namely, transparent membranes wrapping the classes induced by the previously defined partition. The resulting space is shown in Figure 3. In the left hand side, each element of the **L**-set is represented as a sphere with a radius proportional to the number of objects of the original database represented by the corresponding **L**-object, thus giving an idea of the relative distribution of the elements of the whole database in the VR space. In the right hand side, the elements of the same space are wrapped with semitransparent surfaces corresponding to the classes induced by the partition derived from the Photometric redshift. This variable is related with the distance to a given galaxy, and the differential concentration of the galaxies within each class. The clear distinction of the wrapping surfaces, indicates that their intrinsic properties have a dependency w.r.t. their Photometric redshift.

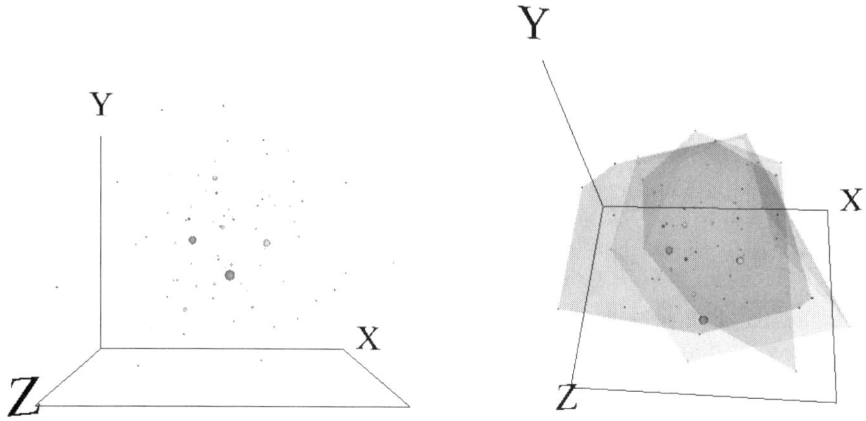

Figure 3. Virtual Reality Space corresponding to the 33055 galaxies database. Left: **L**-set computed with a similarity threshold of 0.85. Right: the same set but with transparent membranes wrapping subsets having specific ranges of the Photometric redshift attribute (see text).

5.2. Experiment 2: Three Groups of Galaxies According to the Photometric Redshift

In this case, the dataset was partitioned into three separate subsets according to the value of the Photometric redshift as described in the previous experiment. Then, a four-fold set of VR-spaces was computed (for the whole dataset, and for the three subsets). In all cases the Photometric redshift was excluded as a descriptor attribute in order not to bias the computation of the **L**-sets and their corresponding VR-spaces, hence, each galaxy was described by a set of 10 attributes. The similarity threshold used for computing the **L**-sets in all cases was 0.75, and the results are shown in Figure 4.

The shapes and structure of the VR-spaces corresponding to the galaxy subgroups in comparison with the whole are different. This provides an indication of the influence of distance to the galaxy (expressed by the Photometric redshift), on its nature and properties.

6. Conclusion

The construction of virtual reality spaces for astronomic databases allows the visualization and the understanding of the underlying structure of datasets, possibly large. As illustrated by examples from the domain of galaxy research, this tool is potentially useful in knowledge discovery and data mining in astronomy.

Acknowledgments. The author would like to thank David Schade from the Herzberg Institute of Astrophysics (for providing the data, and for his suggestions when constructing the VR-spaces), as well as to Alan Barton from the Institute for Information Technology. Both institutes belong to the National Research Council Canada.

References

Borg, I., & Lingoes, J. 1987, Multidimensional Similarity Structure Analysis (Springer-Verlag)

Chandon, J.L., & Pinson, S. 1981, Analyse Typologique. Thorie et Applications (Masson, Paris)

Fayyad, U., Piatesky-Shapiro, G., & Smyth, P. 1996, From Data Mining to Knowledge Discovery. In U.M. Fayyad et al. (eds.) Advances in Knowledge Discovery and Data Mining (AAAI Press), 1

Gower, J.C. 1973, A General Coefficient of Similarity and Some of its Properties, Biometrics, 1, 27, 857

Hajek, P., & Havranek, T. 1978, Mechanizing Hypothesis Formation (Springer Verlag)

Jianchang, M., & Jain, A. 1995, Artificial Neural Networks for Feature Extraction and Multivariate Data Projection, IEEE Trans. On Neural Networks, 6, 2, 296

Kruskal, J. 1964, Multidimensional Scaling by Optimizing Goodness of Fit to a Non-metric Hypothesis, Psichometrika, 29, 1

Valdés, J.J. 2002, Virtual Reality Representation of Relational Systems and Decision Rules: An exploratory Tool for understanding Data Structure. In Theory and Application of Relational Structures as Knowledge Instruments. Meeting of the COST Action 274 (P. Hajek. Ed). Prague, November 14-16

Valdés, J.J. 2002b, Similarity-based Heterogeneous Neurons in the Context of General Observational Models, Neural Network World, 12, 5, 499

Valdés, J.J., & Bonham-Carter, G.F. 2003, Virtual Reality Representation of Geoscience Databases and Decision Making Knowledge. Proc. of the 2003 Anual Conference of the Int. Assoc. for Mathematical Geology, Portsmouth, UK, September 7-12

Valdés, J.J. 2003, Virtual Reality Representation of Information Systems and Decision Rules: An Exploratory Tool for Understanding Data and Knowledge. Lecture Notes in Artificial Intelligence LNAI 2639 (Springer-Verlag), 615

Valdés, J.J. 2004, Interpreting fuzzy clustering results with virtual reality-based visual data mining: application to microarray gene expression data. Proc. NAFIPS04 Int. Conf. of the North American Fuzzy Information Processing Society, 302

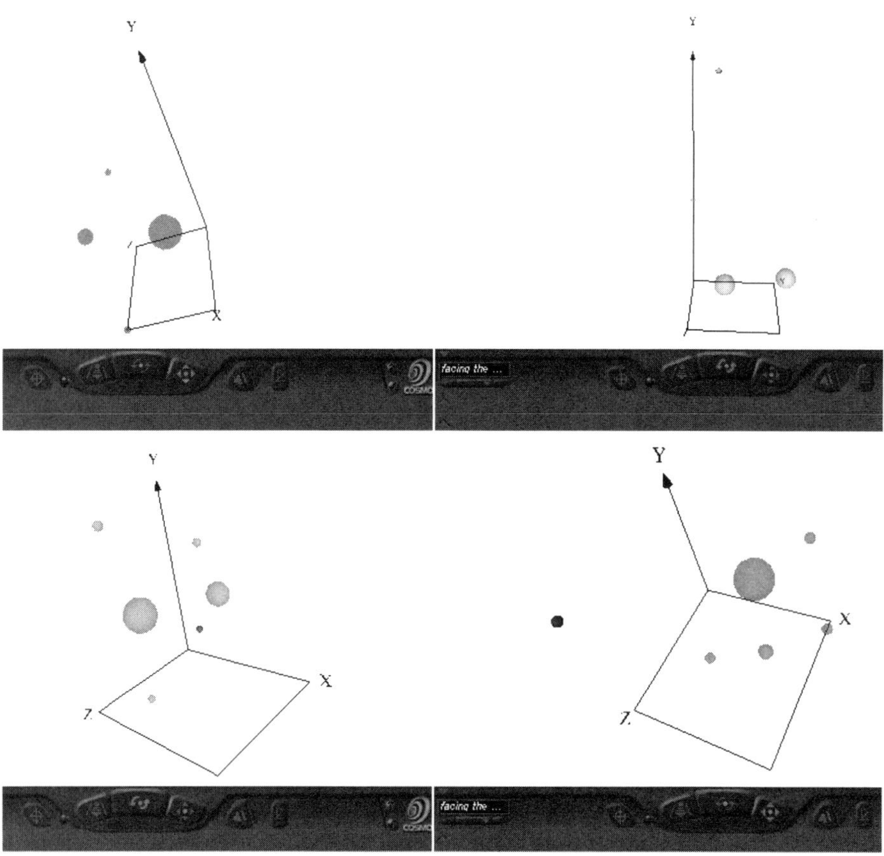

Figure 4. Virtual Reality Space corresponding to the 33055 galaxies database according to the values of the Photometric redshift. The **L**-sets were computed with a similarity threshold of 0.75. Upper left: All of the galaxies. Upper right: Galaxies with Photometric redshift < 0.5. Lower left: Galaxies with Photometric redshift in the $[0.5, 1)$ interval. Lower right: Galaxies with Photometric redshift ≥ 1. The toolbar at the bottom of each representation corresponds to the navigation controls of the virtual reality browser.

ASTRONOMICAL DATA ANALYSIS SOFTWARE AND SYSTEMS XIV
ASP Conference Series, Vol. 347, 2005
P. L. Shopbell, M. C. Britton, and R. Ebert, eds.

Hires: Super-resolution for the Spitzer Space Telescope

C. Backus, T. Velusamy, T. Thompson, and J. Arballo

Jet Propulsion Laboratory, Pasadena, CA 91109-8099, Email: backus@jpl.nasa.gov

Abstract. We present a description of Hires, a super-resolution program based upon the Richardson-Lucy algorithm, generalized to the case of redundant coverage, with higher order optical image distortion, implemented for the Spitzer Space Telescope.

1. Introduction

Spitzer is approximately critically sampled and has good SNR, but the telescope is relatively small (85cm). To compensate for the effect of aperture size on angular resolution there are observing strategies which employ redundant coverage. Hires was implemented to reassemble the information from overlapping observations into a single image with refined resolution, justified by the increased spatial frequency coverage provided by the redundancy, in conjunction with reliable knowledge of the point spread functions.

2. The Hires Algorithm

The algorithm is based upon the Maximum Correlation Method (MCM) developed at the Caltech Infrared Analysis Center (IPAC) (Aumann et al. 1990) for use with IRAS data. MCM is a generalization of the Richardson-Lucy (R-L) algorithm (Richardson 1974; Lucy 1972) to redundant coverage. We reformulated the MCM calculations so that convolutions could be evaluated using the Convolution Theorem, which enables the use of FFT's.

Let

D:	Acquired image
P:	Point spread function (psf)
$\tilde{P}$:	Reflected psf, $\tilde{P}(v) = P(-v)$
f^n:	Current trial image estimate.

Then the single coverage R-L algorithm can be described as a series of iterations shown in vectorized form as

$$f^{n+1} = f^n \frac{D}{f^n * P} * \tilde{P}, \tag{1}$$

a procedure which consists of three main steps: First, simulate acquisition of the data by convolving the current image with the psf and create an image of correction factors by dividing the simulated data into the acquired data pixel

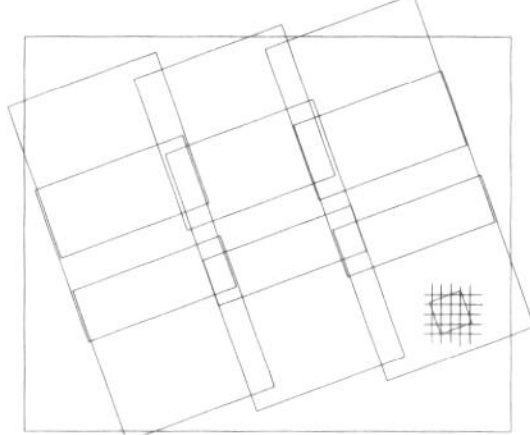

Figure 1. The trial image with data coverage, or correction factors.

by pixel. Second, convolve the correction factor image with the reflected psf. Finally, apply the smoothed correction factors to the current image to produce the next estimate.

The initial image estimate is customarily uniformly set to unity, a Bayesian prior commonly referred to as "uniform ignorance." *A priori* knowledge can be introduced by placing delta functions at the locations of known point sources[1], in hopes of solving the point source removal problem connected with stellar debris disks and galactic nuclei. But, this is difficult because the $\tilde{P}$ convolutions cause a spike amplitude to adjust slowly.

Extension to the multiple coverage case involves what is essentially weighted averaging.

$$f^{n+1} = f^n \frac{\sum_j \left[\frac{D_j}{f^n * P_j}\right] * \tilde{P}_j}{\sum_j U_j * \tilde{P}_j} = f^n \frac{\sum_{j,k} c_{jk}(u_{jk} * \tilde{P}_j)}{\sum_{j,k} u_{jk} * \tilde{P}_j} \tag{2}$$

where c_{jk} is the kth pixel of the jth data correction factor image, and U_j is a unit 2D pulse function coaligned with D_j with (unit data) pixels u_{jk}. The right hand side of this equation is quite similar to the formulation given in the appendix of Aumann et al. (1990), because the $u_{jk} * \tilde{P}_j$ are response functions.

The extension retains the key attributes of the R-L method: no negative flux, conservation of flux[2], and increasing likelihood[3]. The above expression of the R-L algorithm in Hires depends upon two important assumptions:

1. Psf Isotropy. The psf does not vary over the detector array. The D_j could be decomposed into subarrays small enough to assume needed stationarity with a different psf on each one, but Hires presently does not handle separate psf's.
2. Noise Isotropy. Hires assumes that the rms measurement error of each pixel in each observation is constant. Noise weighting each pixel is feasible, as in Aumann et al. (1990), by dividing each c_{jk} and each u_{jk} by a σ^2_{jk}.

---

[1] John Fowler, IPAC.

[2] Provable in closed form by Yu Cao, in informal communication with IPAC, but with our edge effect treatment, conservation is no longer exact near the edges.

[3] The proof is difficult. Lucy's (1974) proof is close. An informal communication Lucy referred us to Dempster et al. (1977), very deep.

2.1. Implementation of the Formalism

Figure 1 shows the overall setting for the above calculations, using the middle member of eq (2). The trial image is sampled at a level appropriate to the amount of redundant coverage. The data images are in general not aligned with the trial image, and they are not necessarily coaligned with each other, although for a given Spitzer observation set taken on the same day they would be co-oriented to within one degree of rotation. In this case it is important to note that equal (or nearly equal) $\tilde{P}_j$ can be factored out of partial sums, greatly reducing the number of FFT's that need to be performed. The same is true of the simulation convolutions: resampling to the detector image resolution and division into each member of a coaligned subset of the D_j can be performed using the trial image convolved once with one of the P_j oriented with the subset.

2.2. Services Performed by Hires

The iterative procedure itself is a relatively simple implementation. Hires performs various calculations to prepare for it:

1. Resampling. Striping is mitigated by correcting accumulations with ratios computed from pixel areas and inclusion counts.
2. Edge effects. Abrupt flux transitions at the image boundary can cause high frequency ringing artifacts. We mitigate this by adopting the included psf solid angle weighting technique discussed in Aumann et al. (1990).
3. Projection calculation. Some observation scenarios can cover as much as 4 degrees of sky, mandating an exact calculation to perform tangent plane transfers. This is done with 3D homogeneous linear coordinate transformations followed by renormalization to the destination tangent plane.
4. Distortion. Spitzer optical distortion is recorded in the form of FITS keywords giving coefficients for quadratic or cubic bivariate Taylor series. These polynomials need to be evaluated for each trial image pixel for each observation image to determine pixel inclusions for resampling. We use the recursive Horner algorithm[4], which we generalized to the multivariate case with a second recursion on the coefficients, which in this conception are themselves polynomials, with the number of variables reduced by one. The Jacobian of the distortion transformation is needed because pixel areas in Spitzer images can vary by as much as 10%.

For FFT's a suitable convolution pad is added to the trial image. Then dimensions are further increased slightly, up to the nearest higher numbers which are products of powers of the first few primes. A substantial percentage of the execution time is spent evaluating the Fourier transforms, which are RAM and CPU intensive. We use as much memory as we can and use an FFT package that supports threading.

---

[4]Horner published the technique in 1819. It involves successive factorizations of the independent variable to reach a form with nested multiplications, familiar to most students of computer science.

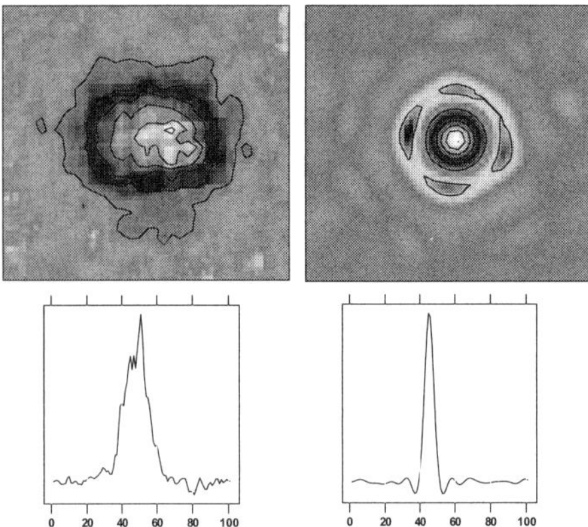

Figure 2. Response to Isolated 24μm Point Source. Left is coadd, right is 100 Hires iterations with 10x redundancy.

3. Invocation of Hires

All of the image files used are in FITS format. The program is invoked with a script run in background mode, and information about progress is piped to a log file. Various switches set output image registration, resolution, dimensions, and orientation, the latter being important in cases where coverage is long and narrow, and diagonal to the local celestial frame. The trial images and correction factor images from successive iterations can be examined with a FITS viewer during execution. Correction factor images are useful for diagnosis of bad input pixels and poorly matched psf's. At convergence they should uniformly be within a percent or so of all 1's. Our largest runs, 4096x4096 pixel trial image and 300 256 x 256 input images, take 5 minutes per iteration on a modern 2 CPU PC with 2 GB of RAM running linux.

The number of iterations to run is an imposition of *a priori* knowledge on the part of the user. There is a switch to determine the iteration count, but the user can examine intermediate output and decide which image to use.[5]

4. Examples

Figure 2 shows the 3:1 reduction in the half-power width of the point source response which is typical for many Spitzer observations. The ringing is due to the high spatial frequency limitation inherent to the process. It can be mitigated by running fewer iterations , or, for cosmetic purposes, close background removal.

Figure 3 shows the effect of 50 Hires iterations on a Spitzer 70μm observation of the debris disk around Fomalhaut, where a single bulge of ice and dust

[5]The literature is extensive on stop rules for R-L. Cf Prasad (2002).

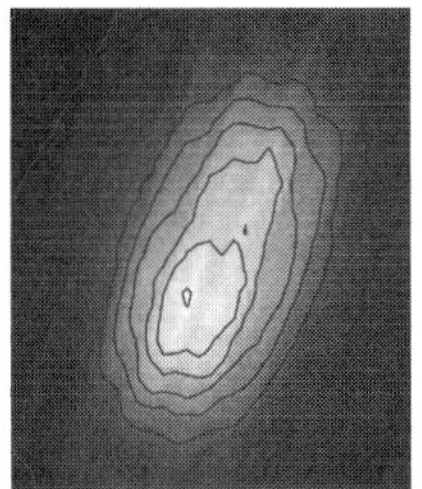 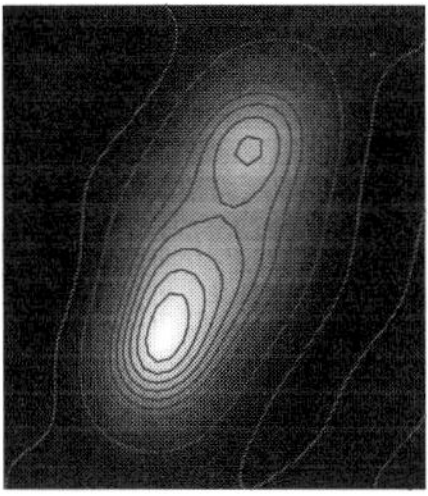

Figure 3. Fomalhaut Debris Disk. Coadd on left, 50 Hires iterations on right. The central source, not visible at 70μm, is between the peaks.

emission is resolved into a nearly edge on view of a toroidal shape, with an asymmetry possibly due to an unseen planet (Stapelfeldt et al. 2004).

Acknowledgments. We are grateful to John Fowler, David Shupe, and David Makovoz at IPAC, Jim Cadien and Jane Morrison at the University of Arizona. Lucas Kamp at JPL wrote the routine to rotate and resample the psf. This research was carried out with the support of the Center for Long Wavelength Astrophysics at JPL, California Institute of Technology, under a contract with NASA.

References

Aumann, H., Fowler, J., & Melnick, M. 1990, ApJ, 99, 5

Dempster, A. P., Laird, N. M., & Rubin, D. B. 1977, J. Roy. Stat. Soc., B, 39, 1

Lucy, L.B. 1974, AJ, 79, 6

Prasad, S. 2002, J. Opt. Soc. Am., 19, 7

Richardson, W.H. 1972, J. Opt. Soc. Am., 62, 1

Stapelfeldt, K., et al. 2004, ApJ, 154, 458

ASTRONOMICAL DATA ANALYSIS SOFTWARE AND SYSTEMS XIV
ASP Conference Series, Vol. 347, 2005
P. L. Shopbell, M. C. Britton, and R. Ebert, eds.

Detection of Rare Objects in Massive Astronomical Datasets Using Innovative Knowledge Discovery Technology

A. Soto, A. Cansado, and F. Zavala

Department of Computer Science, Pontificia Universidad Catolica de Chile, Santiago 22, Email: asoto@ing.puc.cl

Abstract. Our work presents an application of knowledge discovery technology aimed to help scientists in the detection of rare types of astrophysical objects. Our main idea is that while computer have the power to search huge amounts of data, an expert has the domain knowledge to efficiently lead this search. Our system builds upon two main components: a probabilistic model able to scale to large datasets and a set of modules to interact with the scientist. Here, we focus on the probabilistic model used to represent the joint uncertainty among the attributes of the objects registered in a sky survey catalog. This model consists of a combination of a Bayesian network and a set of Gaussian mixture models (GMMs) trained with an accelerated version of the expectation maximization (EM) algorithm. The model is currently being tested using data from the release 1 of the Sloan Digital Sky Survey. The results indicate that the system is able to accurately detect a set of simulated rare objects, but it also provides a large number of false positives.

1. Introduction

Today's Astronomy is living an information revolution, critically needing novel technologies that help in the analysis of new vast sources of data. Several recent projects to massively survey the sky are starting to generate many Terabytes of data, with billions of sources detected and tens or hundreds of parameters measured for each of them. Up to date, researchers have mainly handled these types of surveys manually. However, this approach is becoming no longer viable. As Astronomy is expanding its frontiers, automated knowledge discovery in databases (KDD) is emerging as a key technology to take advantage of the new data available.

There are several interesting applications where KDD technology may increase the efficiency of the analysis of massive astronomical datasets, such as cataloging the information of huge amounts of sky images, automating measurements of the clustering of galaxies, and developing better interactive visualization tools, among others. In particular, one of the most prominent new applications is the detection of unusual objects.

The discovery of rare or new types of astrophysical objects plays a key role in Astronomy (high redshift quasars, L and T stars, brown dwarfs, etc.; Djorgovski et al. 2001). With the next generation of massive sky surveys, the search for new phenomena in our cosmos will intensify. There will be wider frequency ranges

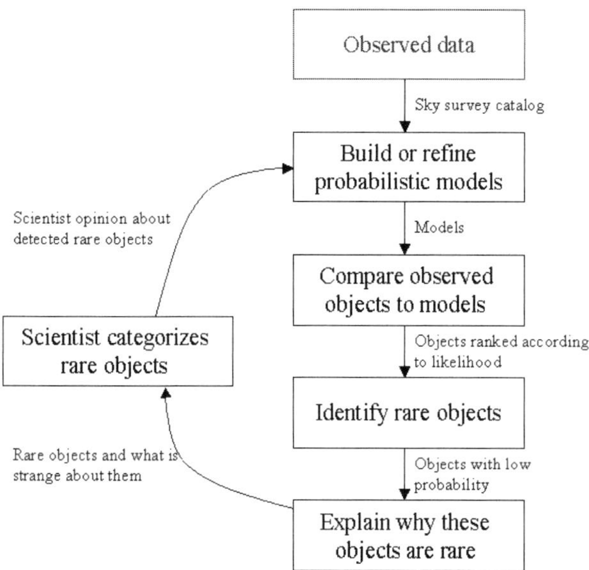

Figure 1. Diagram of the intended application.

available and significantly more sources and resolution, increasing the chance of discovering and studying new or unusual objects.

In terms of KDD technology, the detection of unusual objects in a sky survey catalog is understood as the detection of outliers in a database. Several KDD techniques can be used to solve this problem (Kou et al. 2004), however, their application to high dimensional and massive datasets faces devastating computational and statistical difficulties. Problems such as the curse of dimensionality, lack of labeled data, and an extremely low number of outliers prevent the use of off-the-shelf technology.

In this work, we propose the creation of a computing tool to aid the scientist in the discovery of rare types of astronomical objects. While computers have the power to search huge amounts of data, an expert has the domain knowledge to lead the search. This task requires the development of new algorithms to robustly identify anomalies in huge data sources. It also requires the presence of active learning, to intelligently decide which record to display to the human scientist. Both of these steps must be scalable to interactive speed on billion-record data.

Figure 1 sketches the main features of our approach. Using observations from a sky survey catalog the system is able to learn a probabilistic model of the objects in the database. Using the model, the system sorts the objects in the catalog according to their likelihood. Highly common objects like typical stars or galaxies will show a high likelihood, but rare objects will be poorly explained by the model, thus, they will show a low likelihood.

Using previous intuition the system selects the low likelihood points as a set of possible rare objects. Next, the system analyzes the attributes of the rare objects aiming to find an explanation about what is strange about them. Finally, the system sequentially presents to the scientist the most prominent rare objects, together with an explanation of what makes them unusual. At the same time the system uses the scientist's feedback to interactively improve its probabilistic model for the detection of rare objects. This online improvement of the model is a form of active learning that allows an efficient use of the scientist's feedback, helping to focus the efforts on key steps of the algorithm and avoiding reprocessing irrelevant objects.

In this paper we focus on describing the main steps of the method used to build the joint probability distribution that models the data. This model consists of a combination of a Bayesian network (Pearl 1988) and a set of GMMs (Duda & Hart 1973) trained with an accelerated version of the EM algorithm (Dempster et al. 1977).

2. Probabilistic Model

Our probabilistic model considers the joint probability distribution (JPD) of the attributes of the objects in the sky survey catalog. Using a Bayesian net, we take advantage of conditional independence relations among the attributes to obtain an efficient factorization and graphical representation of the JPD. These independence relations divide the JPD into simpler local conditional probabilistic distributions whose reduced dimensionality simplifies the estimation process.

The novelty of our approach resides in the incorporation of special techniques that provide the efficiency required to scale the approach to huge sky survey catalogs. This is especially critical since a slow KDD analysis severely limits the interaction between the scientist and the data. The probabilistic model provides three main features to our algorithm: scalability to large datasets, efficient detection of rare objects, and an explanation of what makes an object unusual. As for scalability and detection, Sections 2.1. and 2.2. describe the two main steps involved in finding an appropriate Bayesian net: learning the structure of the network and learning the conditional probability distributions that relate the nodes in the network. Next Section 2.3. explains our approach to find what makes an object unusual.

2.1. Learning the structure of the Bayesian Network

To learn the structure of the network, we use a variant of the Sparse Candidate algorithm (Friedman et al. 1999) that shares the same general basic steps: Restrict and Maximize. The first selects candidate parents for each variable, and the latter uses Greedy Hill Climbing and BIC scoring criteria to find the network structure with maximum score according to the current constraints. GMMs trained with EM are used to model continuous variables and to sample from the net. All calls to the EM algorithm are cached for later use. In the estimation of mutual information in the Restrict-step, the variables are discretized using an adaptive approach. Further details of the algorithm are given in a forthcoming paper.

2.2. Learning the local conditional probability distributions

To learn the local conditional probability distributions, we train the GMMs with an accelerated version of the EM algorithm. This version is based on the use of condensed representations (Moore 1999). These representations exploit the internal operation of EM to pre-computed statistical information that summarizes parts of the database. The intuition is to replace a set of similar points by a representative point (Bradley et al. 1998). In this way, at each iteration the EM algorithm does not need to visit each data point, but only isolated points and the statistics of the clustered points. At each iteration, we cluster or condense points using the membership factors between the points and the gaussians that form each GMM.

One of the limitations of training the GMMs with EM is the selection of the number of components that form the mixture. We tackle this problem using a similar approach to the KD-Clust algorithm (Sand & Moore 2001), however, we do not use KD-trees because of their limitations to operate in high dimensional datasets. We also extend our algorithm to search for convergence points using operators to add, merge, and delete clusters.

2.3. Finding an explanation of what makes an object unusual

One advantage of using Bayesian nets is the straightforward evaluation of the likelihood of each data point, which otherwise can be the bottleneck for the detection of rare objects. Furthermore, the local structure and conditional probability distributions embedded in the net provide key information about which groups of attributes are related to other groups, to what extent they are related, and under what circumstances. In this sense, the relative values of the local conditional probability distributions for a given object provide key information to detect the attributes that make the object unusual. We exploit this fact by comparing the attributes of each of the rare objects with the attributes of the most typical clusters in the data. These most typical clusters are given by local relations that explain a large part of the probability distribution.

3. Results

We tested the performance of an initial version of the probabilistic model using a subset of data from release 1 of the Sloan Digital Sky Survey (MySkyServer from SDSS DR1). We first eliminated a small subset of records with missing values, ending up with a database consisting of 166,000 records and 194 attributes for each of them. We treated these attributes as continuous variables, except in the calculation of the mutual information, where we discretized the variables to four labels. To learn the structure of the network we limited to five the maximum number of parents for each node.

In order to test the performance of the model to detect rare objects, we inserted in the database 1000 simulated rare objects with random values in some of their attributes. The results indicate that the system successfully identifies all of the inserted points as anomalies. The test also classified as rare 10% of the objects in the original database.

4. Conclusions and Future Work

We present an algorithm to aid the scientist in the detection of unusual objects in large databases generated by new massive sky surveys. Considering the simulated objects inserted in the database, the results indicate good capabilities to detect rare objects. However, the 10% of objects considered as unusual in the original database suggests that the system detects a large number of false positives, since the number of unusual objects in the original database should be less than 0.1%. We are currently working on fine tuning the numerical precision of the probabilistic model, since it is unable to handle the low values associated with the likelihood of most of the false positives. We are also working on including feedback from the scientist, which will be a key factor to focus the search on the real unusual objects.

In addition, we are testing a modified version of the distance-weighted K-nearest neighbor algorithm to explore the set of rare objects, with the goal of optimizing the use of time and feedback from the scientist. We are also modifying the GMM-EM based algorithm, in order to accommodate data with measurement errors. We expect to test the final system on the complete dataset from SDSS DR3.

Acknowledgments. This work is partially funded by FONDECYT grant 1030336.

References

Bradley, P., Fayyad, U., & Reina, C. 1998, MSR-TR-98-35, Microsoft Research

Dempster, A., Laird, N., & Rubin, D. 1977, Journal of the Royal Society, B 39, 1

Djorgovski, S., Carvalho, R., & Odewahn, S. 2001, in ASP conference series, 255

Duda, R., & Hart, P. 1973, John Wiley and Sons

Friedman, N., Nachman, I., & Peer, D. 1999, UAI, 206

Kou, Y., Lu, C.T., Sirwongwattana, S., & Huang, Y.P. 2004, in Proc. of Int. Conf. on Networking, Sensing, and Control, 749

Moore, A. 1999, in Advances in Neural Information Processing Systems 11

Pearl, J. 1988, Morgan Kaufmann

Sand, P., & Moore, A. 2001, in Proc. of Int. Conf. on Machine Learning,

ASTRONOMICAL DATA ANALYSIS SOFTWARE AND SYSTEMS XIV
ASP Conference Series, Vol. 347, 2005
P. L. Shopbell, M. C. Britton, and R. Ebert, eds.

SAADA:Astronomical Databases Made Easier

Laurent Michel, Hoan Ngoc Nguyen, and Christian Motch

Observatoire Astronomique, 11 rue de l'Université, 67000 Strasbourg, France

Abstract. Many astronomers wish to share datasets with their community but have not enough manpower to develop databases having the functionalities required for high-level scientific applications. The SAADA project aims at automatizing the creation and deployment process of such databases. A generic but scientifically relevant data model has been designed which allows one to build databases by providing only a limited number of product mapping rules. Databases created by SAADA rely on a relational database supporting JDBC and covered by a Java layer including a lot of generated code. Such databases can simultaneously host spectra, images, source lists and plots. Data are grouped in user defined collections whose content can be seen as one unique set per data type even if their formats differ. Datasets can be correlated one with each other using qualified links. These links help, for example, to handle the nature of a cross-identification (e.g., a distance or a likelihood) or to describe their scientific content (e.g., by associating a spectrum to a catalog entry). The SAADA query engine is based on a language well suited to the data model which can handle constraints on linked data, in addition to classical astronomical queries. These constraints can be applied on the linked objects (number, class and attributes) and/or on the link qualifier values. Databases created by SAADA are accessed through a rich WEB interface or a Java API. We are currently developing an inter-operability module implanting VO protocols.

1. Introduction

The amount and the quality of astronomical data available online is growing so much that public data centers are becoming real observational means. The worldwide Virtual Observatory project aims to federate such data centers into a single and global resource (McDowell 2004) including a number of powerful tools very helpful for scientific work. This high integration level is nevertheless still out of reach for data coming from scientific teams having no coding manpower. Building a database remains a difficult job and a number of small or medium teams still share data by down/uploading product files. There is a real need for a tool to help make individual data accessible through high level database front-ends, allowing users to select data by content and not necessarily by product files. The SAADA project proposes a standalone solution for this demand. Contrary to projects offering storage areas on remote nodes (O'Mullane 2003; Davenhall et al. 2003), we prefer to provide a tool to help to build databases locally because we think that this solution avoids large data transfers and gives the finest control over data diffusion. The price to pay for that is that system operations are also local. The present paper presents an overview of the process of automatic database creation with SAADA. It aims to highlight the scientific value provided

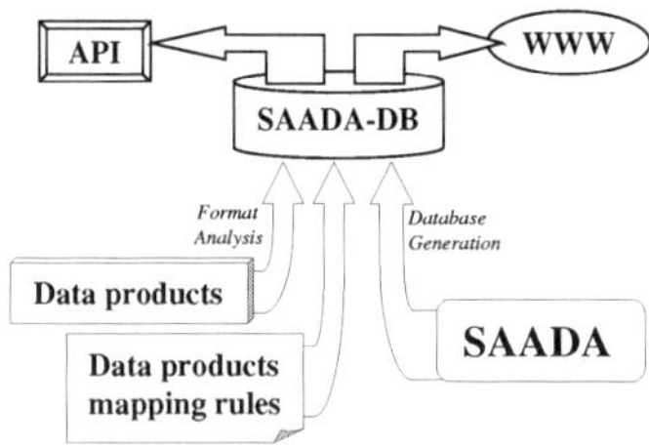

Figure 1. SAADA at a glance.

by SAADA rather than to develop technical issues. More information can be found at http://amwdb.u-strasbg.fr/saada.

2. What is SAADA

The purpose of SAADA is to provide astronomers with data repositories for individual astronomical products (FITS data files). These repositories, created by SAADA, are named SAADA-DBs. SAADA-DBs are dedicated to astronomical data. They rely on a built-in data model based on sky positions, WCS attributes and HTM indexing (Kunszt et al. 1999). SAADA-DBs are designed to be operated by astronomers, not by developers. The database creation process is meant to be very simple and fast as shown in Figure 1. It is achieved by an automatic analysis of input products driven by rules given by the operator (see Section 5.3.). No code has to be written, and computer system operations remain quite simple. A SAADA-DB is a full featured database allowing one to select, browse, and download data. It can simultaneously host products of different natures: images, spectra, source lists and list of vectors (e.g., time series). The same query can, for instance, be used to get images or spectra related to a given field. Data stored in SAADA-DBs can be retrieved independently of the products files they come from (location or format) and only from their scientific content. They can be queried with constraints on global attributes (e.g. sky position) or on specific attributes (FITS keywords). The content of a SAADA-DB can furthermore be enriched with qualified relationships set-up between data (see Section 6.).

3. A Common Data Model

All SAADA-DBs are based on the same common data model (see Figure 2): Data are stored in containers named collections. Collections are supposed to contain products having a common scientific interest. They can for instance hold data coming from a given mission or resulting from a given astrophysical model. One collection can host a single set of products of each kind (image, ...). The scientist building the SAADA_DB (the operator) is in charge of defining the collections. Products are modeled by SAADA super-classes allowing one to handle heterogeneous datasets. These super-classes have general attributes such as sky position or name. Business classes modeling individual products are

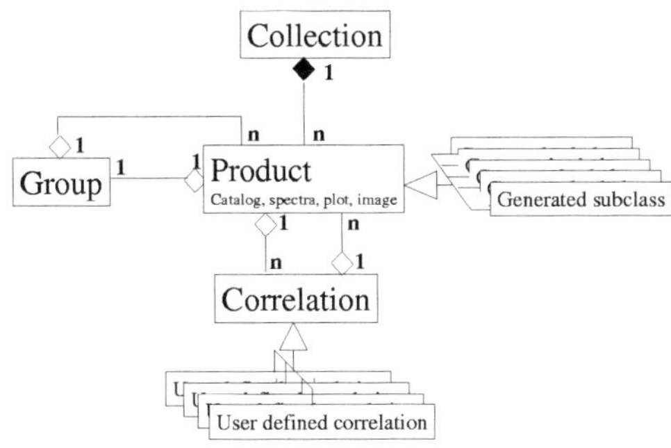

Figure 2. SAADA-DB data model.

automatically created by the data loader. The operator can also define groups. A group is a list of references of products spread out in different collections. Groups can be used for instance to easily retrieve all data coming from one observation run. Groups are automatically populated at data loading time. Data can also be linked together by persistent relationships defined by the operator (see Section 6.).

4. What Does a SAADA-DB Look Like?

All SAADA-DBs are built from the same template (see Figure 3). The storage layer can be in principle any SQL database supporting JDBC[1] (Flanagan et al1999). The structure of this SQL database is hidden to users. Tables are automatically created at configuration time or after some database modification. On top of it is a Java layer. Its kernel is common to all SAADA-DBs but classes modeling real data or relationships are automatically created and compiled. Java class behavior can be modified by hand by using inheritance. The purpose of this Java layer is to give users a data view based on collections and product classes rather than on SQL tables. The object cache is in charge of building Java instances from the object identifiers returned by the query engine (see Section 7.). The Web interface is based on Java servlets (Flanagan et al. 1999). The API allow users to build Java applications handling SAADA-DB data. This API is read-only.

5. Building a Database in 5 Steps

We assume here that the SQL database (PSQL currently) and the required JAVA packages are properly installed. SAADA uses XML configuration files.

[1]Java Data Base Connectivity: Java package giving access to SQL databases for Java applications.

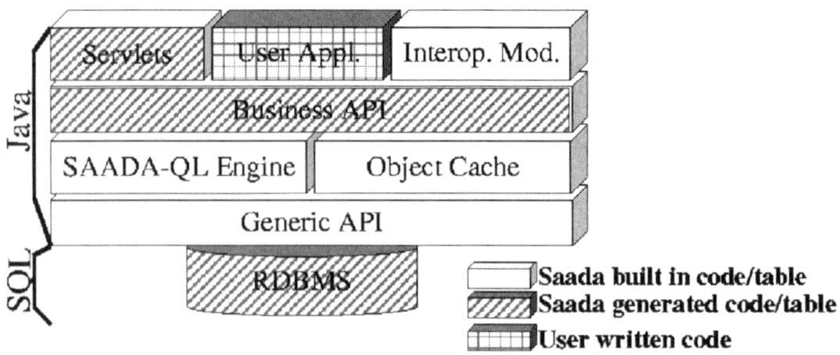

Figure 3. Structure of a SAADA-DB.

The following subsections are illustrated with XML code actually edited with a graphical tool.

5.1. Step 1: SAADA-DB Creation

The first step consists of running the SAADA command building a SAADA-DB. The SAADA package can be downloaded at `http://amwdb.u-strasbg.fr/saada`. It is just used to build generic SAADA-DBs. The code below shows the XML description of a SAADA-DB. Note that a coordinate system must be defined for the whole database.

```
<saada_db>
  <database>
    <name> MySAADADB </name>
    <administrator> develop </administrator>
    <password> "" </password>
    <sql_driver> org.postgresql.Driver </sql_driver>
    <jdbc_url> jdbc:postgresql:MySAADADB</jdbc_url>
    <db_root_dir> /raid50/saada/saada-db </db_root_dir>
    <repository_root_dir> /raid50/saada/repository </repository_root_dir>
  </database>
  <interface_web>
    <url_root> amwbd.u-strasbg.fr </url_root>
    <context> MySAADADB </context>
  </interface_web>
  <coordinate>
    <system> FK5 </system>
    <equinox> 2000 </equinox>
  </coordinate>
</saada_db>
```

After this operation there is no longer a link between the SAADA-DB and the SAADA package from which it comes.

5.2. Step 2: Defining Collections

Collections are just defined by names used by the command building them into the generic SAADA-DB.

```
<collection>
  <name> OBSERVED_DATA  </name>
  <name> SIMULATED_DATA  </name>
</collection>
```

5.3. Step 3: Product Classification

In this step, the operator must state for each input product the collection in which it will be stored. Obviously, this operation is not required for each individual product. Classification rules are applied to product sets. A classification rule gives the class mapping mode and the target collection for the specified product set. A product set can be defined by a directory, by a regular expression applied on filenames, or by a header keyword value. Product sets are used by the operator to load data but are not known to the users. All products of a given nature belonging to one collection are seen as a single set in order to allow queries onto the whole collection. This set is divided into classes in order to allow queries on similar products with constraints on specific attributes. This is a classical method to implement inheritance in SQL tables. Three class mapping modes are available. SAADA can either build one product class for each input file or one unique class for all products or finally process an automatic mapping where files having the same format will be grouped in a unique class. In the example below, the image set `MyExposureMaps` will gather all products having the header keyword `CONTENT='EPIC EXPOSURE MAP'`. These products will be stored into the collection `OBSERVED_DATA` as instances of a unique class named `ExpMapClass` . The instance names will be the merge of the keywords `OBS_ID` and `INSTRUME` and the keywords `CONTINUE` and `HISTORY` will not be stored.

```
<image2D name= "MyExposureMaps">
    <identification>
        <content>
            <attribute> CONTENT </attribute>
            <value>EPIC EXPOSURE MAP</value>
        </content>
    </identification>
    <class_name>ExpMapClass</class_name>
    <coord_system>
        <system> FK5 </system>
        <equinox> 2000 </equinox>
    </coord_system>
    <collection>OBSERVED_DATA</collection>
    <mapping>
        <instance_name> OBS_ID,INSTRUME </instance_name>
        <ignored_att> CONTINUE,HISTORY   </ignored_att>
    </mapping>
</image2D>
```

5.4. Step 4: Product Loading

The data loader facility loads all products matching the set given as a parameter and found in any subdirectory of the base directory also given as parameter. Before populating the SAADA-DB the data loader reads first all product files in order to built relevant Java classes and SQL tables. The first release of SAADA will not support database updates but only one shot data loading.

5.5. Step 5: Web Deployment

This operation, not always clear for non-J2EE[2] experts, is encapsulated in a simple script. It consists of compiling the servlet code including generated methods, building the archive files (ear, war), and invoking the J2EE server to deploy them.

6. Stored Relationships

We know that the trend of modern query engines is rather to compute correlations on the fly, but we estimate that good reasons do exist to propose a system storing correlation links and other relationships. A relationship between two datasets can be quite more valuable than a simple correlation by position. It can correlate simulated data and observed data bearing on some critical parameter of the model (Lemson 2004). Links can be qualified by values which are computed but also measured or refined by hand. Relationships can also be set between different kinds of data. For instance, sources can be correlated with the images they come from. Such links could be qualified with instrumental values (e.g., detection pixels) in order to easily include experimental considerations in scientific queries. In short, relationships seen as valuable information must be stored within the database. Examples are given in the following table:

Primary to	Secondary	Qualifiers	Interest
Sources	Image	Pixel Values	Discard sources detected on bad pixels
Spectra	Image	Flux	Complete spectral coverage with flux

Another reason to store relationships is a performance issue. If the algorithm computing the correlations is not a built-in function of the engine, join optimization might become very difficult especially for queries including considerations of the number of links involving the searched objects. SAADA takes advantage of persistent relationships to build indexes which speed up very complex queries including constraints on patterns drawn by relation vectors. A SAADA relationship links all data of a given type in a given collection to the data of another (or the same) type in another (or the same) collection. It is a set of links possibly qualified by values defined by the operator. Primary records can be correlated with several secondary records (1-N relationship). Relationships can be set-up between data of different natures (e.g., entries to images or entries to spectra). The example below shows the definition of a relationship modeling a correlation between sources and the images in which they have been detected. Links are qualified by pixel values.

```
<relation>
    ....
    <N_M >
           <relation_name> PixDetectCorr  </relation_name>
           <coll_primary
```

[2]J2EE (Java 2 Enterprise Edition):A Java-based runtime platform created by Sun Microsystems used for developing, deploying, and managing multi-tier server-centric applications on an enterprise-wide scale.

```
                name= "OBSERVED_DATA"
                type= "ENTRY" />
            <coll_secondary
                        name="OBSERVED_DATA"
                        type="IMAGE" />
            <algorithm>
                        <class_name> PixDetectAlgo </class_name>
                        <qualifier name= "PixX" type="int" />
                        <qualifier name= "PixY" type="int" />
            </algorithm>
    </N_M>
    ...
</relation>
```

PixDetectCorr is the name of the relationship which has to be used for queries involving it. Links will be modeled by a generated Java class with the same name, PixDetectCorr. Once the relation is declared, a piece of code must be written to compute the relation. The code below shows a simplified example of such algorithm. This class is invoked by the SAADA iterator processing the relationships.

```
public   class  PixDetectAlgo extends AlgorithmCorrelation {
  public   RelationSaada Algorithm(EntrySaada o1, Image2DSaada o2)  {
    /*
     * Compute (pix_x, pix_y) o1 position on o2
     * in pixels by using WCS attributes
     */
    int pix_x = ...;
    int pix_y = ...;
    if( /* pix_x and pix_y are into the image */ ) {
      PixDetectCorr pdc = new PixDetectCorr(o1,o2);
      pdc.setPixX(pix_x);
      pdc.setPixX(pix_y);
      return  pdc;
      }
    else {
      return null;
      }
    }
  }
```

7. SAADA-QL

It is not worth having a fancy data model if we have no query language to match it. It is important that user queries refer not to SQL tables but to collections, classes and relationships. For this reason a query language (SAADA-QL) has been developed. The structure of a complete SAADA-QL query is given below.

```
Select obj_type From Classes In  Collections
  WherePosition{...}
  WhereAttributeSaada{...}
  WhereAttributeClass{...}
  WhereRelation{...}
```

The Select parameter gives the nature of the objects to retrieve (entries, tables, images, spectra or vector lists). SAADA-QL queries always return SAADA object identifiers (OIDs), names and positions. The full content of queried data must be read by using the API. The `In` parameter gives the collection names of searched products. Queries can be apply to one collection (`In coll1`) or to some collections (`In coll1, coll2, ...`) or to any collection (`In *`). The `From` parameter gives the class names of searched products. Classes can only be given in queries on one collection. `From *` is otherwise required. Separate `Where` clauses are defined for each kind of constraint.

- `WherePosition` contains a list of positions defined by the operator (`isInBox` , `isInCone` or `isNearObject`)
- `WhereAttributeSaada` contains a SQL `where` statement bearing on the SAADA attributes (OID, name, position, HTM, WCS,..)
- `WhereAttributeClass` contains a SQL `where` statement bearing on the class attributes. Only queries applied on one class support this clause.
- `WhereRelation` contains a list of patterns on relationships. Each pattern is handled by an `matchPattern{...}` operator. The example below shows a pattern used to select data having more than 2 counterparts in the relation `My_Relation`. Each of these counterparts must be of the class `Other_Class` and must have a `Mag_V` (`Mag_V` is supposed to be a class attribute) greater than 20).

```
matchPattern{ "My_Relation"
            , Cardinality(">", 2, 0)
            , ObjAssClass("Other_Class")
            , ObjClassAtt("Mag_V > 20")
            }
```

SAADA-QL queries are split to three separate sub-queries (position, relation and attributes) processed separately. The details of the SAADA-QL syntax can be found on our Web site.

8. GUI

SAADA_DB web interfaces are based on Java servlets. They can run on any Web server supporting the J2EE architecture. In order to make the deployment as simple as possible, it has been decided not to use a middleware but to design servlets directly accessing the database.

Web pages displayed by a SAADA-DB result from Java method invocations. Most of this code is automatically generated at configuration time. For this reason we can not expect to have sophisticated graphical user interfaces. This interface, though not sophisticated, must allow users to make any query they want without any limitation due to the graphical tool. For this reason the main page (see Figure 4) is rather a query string editor than a database front-end. The query submission returns the list of selected objects with some relevant information (position, name, collection,..) and the detail of the objects can of course be seen by clicking on the appropriate anchors. The query editor is based on a text area where the query string is written thanks to pop-up menus containing relevant keywords (class names, attributes, ...). The consistency of these menus with the current query is handled by specific JavaScript procedures. The advantage of this method is that simple queries (e.g., search by position) can be run with a few clicks on HTML widgets as usual but more and more complex queries can also be set up by adding new constraints into the text area.

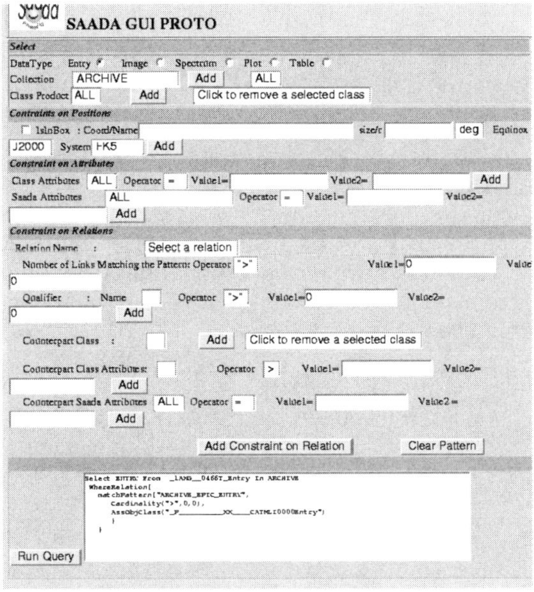

Figure 4. SAADA-DB Main Page.

The current query can also be modified by hand which is quite convenient for just adapting search parameters.

9. Software and Hardware Requirements

Saada is totally based on freeware. All code is written in Java and any JDBC DBMS can be used. The Web interface can be deployed on any server compliant with J2EE. The first release will however be limited to Linux/Solaris platforms due to some shell scripts. The data loader uses the HEASARC Java fitsio package (McGlynn 2004).

10. SAADA and the VO

The first purpose of SAADA is to help astronomers as much as possible to build personal databases for their individual data. To achieve this we focused on making product mapping as simple as possible. It refers only to basic astronomical attributes (position) and to collections and classes of the SAADA-DB. VO base concepts (UCD, DM features) have not been taken into account to design the heart of SAADA. A SAADA-DB can live without reference to the VO. It is nevertheless essential (and great) to allow users to publish their SAADA-DB into the VO. This functionality will be handled by an external module. This module will be able to produce VO Registry resources consistent with the real database content. It will also be able to deploy web services processing SIAP/SSAP and VOQL queries. A new VOQL operator will be proposed to process constraints on stored correlation patterns with the SAADA style. Collections and classes will be translated in VOQL by pseudo attributes as it has been done for the SIAP/VOQL translation (Shirasaki 2004). This development is just starting now.

11. Status and Prospect

The first release of SAADA will be delivered in December 2004. It aims to provide necessary functions to build operational databases but with a limited data model (no groups, spectra, or vector lists). A more complete release is scheduled for spring 2005.

	First release (12/2004)	Second release (Spring 2005)
data model	tables + images	spectra + groups + vector lists
data loader	one shot	append supported
VO	-	VOQL/SIAP/SSAP + registry pub.
platform	Unix/PSQL	Window/any JDBC

Acknowledgments. Part of this work is supported by the SSC of XMM-Newton and funded by the *Région Alsace* and by the *CNES* (French space agency)

References

Davenhall, A.C., et al. 2003, in ASP Conf. Ser., Vol. 314, ADASS XIII, ed. F. Ochsenbein, M. Allen, & D. Egret (San Francisco: ASP), 330

Flanagan , D., et al. 1999, in Java Enterprise in a Nutshell , ed. O'Reilly Cambridge, 15, 123

Kunszt, P. Z., Szalay, A. S., Csabai, I., & Thakar, A. R. 1999, in ASP Conf. Ser., Vol. 216, ADASS IX, ed. N. Manset, C. Veillet, & D. Crabtree (San Francisco: ASP), 141

Lemson, G., & Colberg, J. 2004, Theory in the VO at http://www.ivoa.net/pub/papers/TheoryInTheVO.pdf

McDowell, J.C. 2004, at http://www.spectrum.ieee.org/WEBONLY/publicfeature/aug04/0804vobs.html

McGlynn, T. 2004, at http://heasarc.gsfc.nasa.gov/docs/heasarc/fits/java/v0.9/JavaFits.doc

O'Mullane, W. 2003, in ASP Conf. Ser., Vol. 314, ADASS XIII, ed. F. Ochsenbein, M. Allen, & D. Egret (San Francisco: ASP), 372

Shirasaki, Y. 2004 , Unification of ADQL, SIAP, SSAP, SXAP, at http://jvo.nao.ac.jp/Documents/VOQL-WG-yshirasa.ppt

ASTRONOMICAL DATA ANALYSIS SOFTWARE AND SYSTEMS XIV
ASP Conference Series, Vol. 347, 2005
P. L. Shopbell, M. C. Britton, and R. Ebert, eds.

Mosaicking with MOPEX

David Makovoz and Iffat Khan
Spitzer Science Center, California Institute of Technology, Pasadena, CA 91125

Abstract. We present MOPEX - a software package for image mosaicking and point source extraction. MOPEX has been developed for the *Spitzer Space Telescope*. This paper concentrates on the mosaicking aspects of the package. MOPEX features the use of several interpolation techniques, coaddition schemes, and robust and flexible outlier detection based on spatial and temporal filtering. A number of original algorithms have been designed and implemented in MOPEX. Among them is direct plane-to-plane coordinate transformation, which allows at least an order of magnitude speed up in performing coordinate transformation by bypassing the sky coordinates. The dual outlier detection makes possible outlier detection in the areas of even minimal redundancy. Image segmentation based on adaptive thresholding is used for object detection, which is part of outlier detection. Efficient use of computer memory allows mosaicking of data sets of very deep coverage of thousands of images per pointing, as well as areas of sky covering many square degrees. Although designed for *Spitzer* data, MOPEX does not require any *Spitzer*-specific fits header keywords to run, and can be applied to other data, that have standard header information on the image geometry and pointing. The package is available for distribution at `http://ssc.spitzer.caltech.edu/postbcd/`.

1. Introduction

MOPEX is an astronomical image processing package developed at the *Spitzer* Science Center. The name MOPEX stands for MOsaicking and Point source EXtraction. The software for the package was originally developed for the automated pipeline to process *Spitzer Space Telescope* data. Up to date MOPEX has been used for processing *Spitzer* data by various groups, such as SSC, IRAC Instrument Team, GLIMPSE, SWIRE, C2D Legacy teams, and others. MOPEX has several major tasks - mosaicking with outlier detection, background matching, point source extraction, pointing refinement - and some other ancillary tasks. The full description of MOPEX capability can be found in the documentation on the *Spitzer* web site `http://ssc.spitzer.caltech.edu/postbcd/`. The pointing refinement is described in Masci, Makovoz, & Moshir (2004). This paper is devoted to the mosaicking capabilities of MOPEX.

MOPEX features several interpolation schemes which are described in Section 3. Image reprojection is performed using a new direct plane-to-plane coordinate transformation, the detailed description of which can be found in Makovoz (2004). As part of interpolation MOPEX performs a *Spitzer* distortion correction. Robust multiframe and dual outlier detection allows removing cosmic ray hits and other artifacts in the areas of minimal coverage. Efficient memory management using tiles allows performing outlier rejection in the areas of practically unlimited depth of coverage. The users have at their disposal various coaddition options which are described in Section 5. The user specifies the mosaic size, orientation, and pixel size. The only limitations on the mosaic size is the hard drive size.

2. Input Data

The minimal input requirements to make a mosaic with MOPEX are a set of FITS images with the following set of standard WCS keywords: $BITPIX$, $NAXIS$, $NAXIS1$, $NAXIS2$, $CRVAL1$, $CRVAL2$, $CRPIX1$, $CRPIX2$, $CTYPE1$, $CTYPE2$, and the CD-matrix elements, Alternatively, instead of the CD-matrix the $CDELT1$, $CDELT2$, $CROTA2$ keywords can be used.

Optionally, each input image can be accompanied with an uncertainty image, where each pixel in the uncertainty image give the measurement and processing uncertainty of the corresponding input image pixel. Also, for each input image the user can specify a number of mask images to mark pixels that are not usable for various reasons, such as dead pixels, saturated pixels, cosmic ray affected pixels, hot pixels, latent pixels, non-linearizable pixels, etc.

3. Image Interpolation

The main interpolation technique used by the Spitzer mosaicker MOPEX is the area overlap interpolation scheme. The interpolated pixel value $I(k)$ is equal to the sum of the values $J(l)$ of the input pixels l overlapping the interpolated pixel k and weighted with the area a_{kl} of the pixel overlap.

$$I(k) = \sum J(l)a_{kl} / \sum a_{kl} \tag{1}$$

This interpolation scheme accurately performs flux transfer between the input and output pixels, but is slow, since it requires computing the overlap areas.

In MOPEX we implemented a new fast interpolation technique. We call it the *grid* interpolation technique. A grid of points is placed on the input frame. Each grid point is assigned the value of the pixel it belongs to. Each grid point is projected onto the output frame, and the flux associated with the grid point is added to the output image pixel into which the grid point was projected. This scheme is equivalent to approximating the value of the overlap area with the number of grid points n_{kl} in the overlap area:

$$I(k) = \sum J(l)n_{kl} / \sum n_{kl} \tag{2}$$

The number of the grid points in an input pixel is equal to G^2, where G is the grid ratio. In the limit of $G \Rightarrow \infty$ this scheme is equivalent to the area overlap interpolation. But even for small values of $G \sim 1 - 2$ the resulting mosaics are comparable in quality to those produced by the area overlap interpolation scheme, but produced in a fraction of the time necessary for the area overlap method. The grid method is very useful for quick-look mosaics and for a background matching algorithm

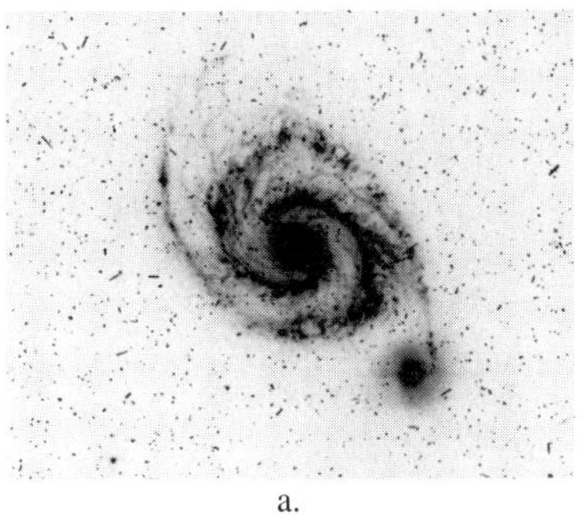

a.

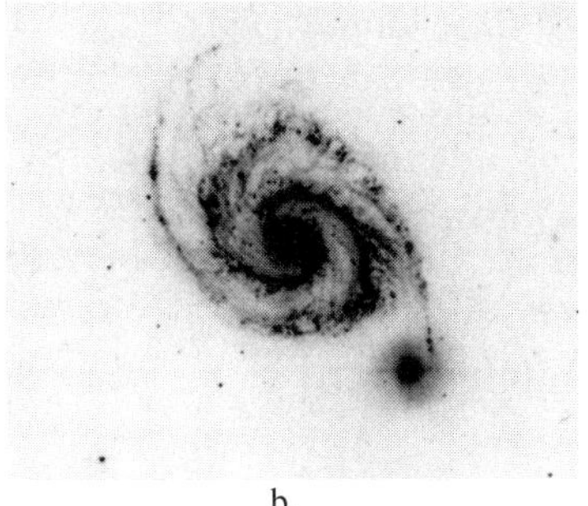

b.

Figure 1. Mosaic without (a.) and with (b.) outlier detection

MOPEX also provides an option of using the drizzle interpolation described, e.g., in Fruchter & Hook (2002).

Whatever interpolation scheme is used for the input images, the uncertainty images are interpolated in an identical fashion. A coverage map is created for each interpolated image. For each output pixel the coverage map gives the fraction of the pixel covered by the input image. It reflects the presence of any input bad pixels or pixels on the edge of the input image.

4. Outlier Detection

MOPEX is designed to perform effective detection of cosmic ray hits in the input images. One single frame and two multiframe algorithms for outlier detection are implemented. The single frame outlier detection represents spatial filtering of input images and is used to detect "spiky" objects.

The second outlier detection method represents temporal filtering of the interpolated images. Interpolated images are stacked up. In each stack the mean and standard deviation σ are found. Pixels outside of an $n - \sigma$ envelope are declared outliers, where n is defined by the user. This approach breaks down in cases of shallow coverage.

The third approach - dual outlier detection - consists of two-stage filtering. At the first stage, all spatial pixel outliers are detected and saved as detection maps. These detection maps include point sources and cosmic ray hits. Detection maps are interpolated to a common grid. Then for each spatial location, the pixels in the interpolated detection maps are matched. The unmatched pixels are declared outliers. This method effectively supplements the conventional multiframe outlier detection and is the only way of outlier detection in the areas of double coverage. In Figure 1 we show a portion of a mosaic of an IRAC channel 3 mosaic made without (a.) and with (b.) outlier detection. The average coverage of this observation is ~ 4 images per pointing. Both multiframe and dual outlier rejection were used here. Two examples of MOPEX outlier detection in action are given in Patten et al. (2004).

5. Image Coaddition

After the input images are interpolated to a common grid, they can be combined into a single mosaic image. Three weighting schemes are available in MOPEX. The interpolated images can be combined using straight averaging, they can be weighted with the interpolated uncertainty images, with the exposure time, or with both the interpolated uncertainty images and with the exposure time. The most general expression for the value O of a coadded pixel is

$$O = (\sum_j (c_j T_j I_j)/\sigma_j^2)/(CU^{-2}) \tag{3}$$

where I_j, σ_j, c_j, T_j are the values of the corresponding pixel in the $j-th$ interpolated image, interpolated uncertainty image, coverage map, and the exposure time for the $j-th$ image. C is the value of the coadded coverage map pixel

$$C = \sum_j c_j T_j, \tag{4}$$

and U is the value of the coadded uncertainty image pixel

$$U^2 = \frac{\sum_j c_j T_j / \sum_j c_j}{\sum_j (c_j T_j)/\sigma_j^2}. \tag{5}$$

The mosaic uncertainty image and coverage map are saved as separate products.

MOPEX also produces the median mosaic and the corresponding uncertainty images. It is created by using the median value in the stack of interpolated pixels for the mosaic pixel value and the 68 percentile width around the median for its uncertainty. It provides a quick way of creating clean (no cosmic ray hits) mosaic images without having to run outlier rejection. However, the flux is not conserved in the median mosaics.

6. Mosaic Mask

A new kind of mosaicking implemented in MOPEX is mosaicking of mask images. The mask image interpolation has to preserve the bit information in the input mask images. Regular interpolation destroys the bit information. There is no single "correct" way to set the value of the interpolated mask pixel. We implemented the following: The value of each output interpolated pixel is set to the result of the logical OR of the values of the input mask pixels overlapping the interpolated pixel.

The coaddition has also been modified to preserve the bit information. MOPEX makes two products: The first product is a mosaic fits data cube. Each plane in such a data cube corresponds to a single bit in the input masks. The pixel value in the plane equal to the number of interpolated masks that have that bit set in that pixel. The second product is single plane mosaic image. It has a bit to bit correspondence to the input masks. A bit in the pixel value of the single plane mosaic image is set if the number of the interpolated masks with that bit set in that pixel exceeds the user specified threshold.

7. Using MOPEX

MOPEX is implemented as a set of perl scripts each running a separate task. Each script runs a number of individual modules written in C/C++. The complete package consists of the perl scripts, binary executable files, sample data with configuration files, and documentation. The binaries are available for Solaris and Linux operating systems. MOPEX can be downloaded from `http://ssc.spitzer.caltech.edu/postbcd/download-mopex.html`. The perl scripts are run on the command line. Configuration files (aka namelists) are also given on the command line. They are used to turn on and of the individual modules run by the script, give the location of the input data, set up the output directories and also include separate namelist blocks for the individual modules.

Benchmarking of MOPEX is not trivial. The only step that clearly scales with the number of image and image sizes is interpolation. Interpolation of a 256x256 image without resampling on a 1 GHz Sparc 4GB RAM SunBlade takes 1.6 seconds using area overlap interpolation and 0.2 seconds using grid interpolation with the grid ratio $G = 1$. For the other steps the timing varies widely with the density of the cosmic ray hits and the coverage depth.

8. MOPEX Future

As a part of the JPL R&TD High Capability Computing in Engineering and Science Strategic Initiative a project has been funded to develop a parallelized implementation of MOPEX to run on JPL's 1,024-node Xeon computer. By conforming to the standard Message Passing Interface (MPI), the parallelized MOPEX software can be exported for use on other computing systems and will be of great value for other *Spitzer* surveys.

Some other current developments include mosaicking of moving objects, implementing the bicubic interpolation scheme, porting MOPEX to MAC OS X, and a graphical user interface (GUI). We also plan to extend the usage of MOPEX for other than *Spitzer* images.

Acknowledgments. This work was carried out at the *Spitzer* Science Center, with funding from NASA under contract 1407 to the California Institute of Technology and the Jet Propulsion Laboratory.

References

Fruchter & Hook 2002, PASP, 114, 144

Makovoz, D. 2004, PASP, 116, 971

Masci, F. J., Makovoz, D., & Moshir, M. 2004, PASP, 116, 842

Patten, B.M., Hora, J.L., Fazio, G.G., Barmby, P., Wang, Z., & Makovoz, D. 2004, in Proc SPIE 5487, eds. Marija S. Scholl and Bjorn F. Andresen, 223

ASTRONOMICAL DATA ANALYSIS SOFTWARE AND SYSTEMS XIV
ASP Conference Series, Vol. 347, 2005
P. L. Shopbell, M. C. Britton, and R. Ebert, eds.

W Projection: A New Algorithm for Wide Field Imaging with Radio Synthesis Arrays

T.J. Cornwell, K. Golap, and S. Bhatnagar

National Radio Astronomy Observatory[1], *PO Box 0, Socorro, NM, 87801*

Abstract. We present a novel interpretation of the non-coplanar baselines effect in synthesis radio telescopes as being due to differential Fresnel diffraction in the neighborhood of the array antennas, and a new algorithm to deal with this effect. Our new algorithm, which we call "w projection", has markedly superior performance compared to existing algorithms. At roughly equivalent levels of accuracy, w-projection can be up to an order of magnitude faster than the corresponding facet-based algorithms.

1. Does W trouble you?

The response of a narrow-band phase-tracking interferometer to spatially incoherent radiation from the far field can be expressed by the following relation between the spatial coherence, or 'visibility', $V(u, v, w)$, and the spectral intensity, or brightness, $I(\ell, m)$;

$$V(u, v, w) = \int \frac{I(\ell, m)}{\sqrt{1-\ell^2-m^2}} e^{-2\pi i[u\ell+vm+w(\sqrt{1-\ell^2-m^2}-1)]} d\ell dm \quad (1)$$

In this equation, the baseline coordinates, (u, v, w), and direction cosines, (ℓ, m) have their usual definitions. (See for example, Thompson, Moran, & Swenson, 2001.)

When the term $2\pi w(\sqrt{1-\ell^2-m^2}-1)$ is much less than unity, it may be ignored, and a two-dimensional Fourier relationship results. When this term is comparable to or exceeds unity, a two-dimensional Fourier transform cannot be used. The value of this term is roughly:

$$\frac{B\lambda}{D^2} = \left(\frac{r_F}{D}\right)^2 \quad (2)$$

where B is the maximum baseline length, D is the antenna diameter, and λ is the observing wavelength. The parameter r_F is the Fresnel zone diameter for a distance B. It is useful to work with the ratio of these two quantities:

$$N_F = \frac{D^2}{B\lambda} \quad (3)$$

---

[1]Associated Universities Inc. operates the National Radio Astronomy Observatory under cooperative agreement with the National Science Foundation.

Wide-field imaging is affected by this *non-coplanar baselines* effect when the Fresnel number N_F is less than unity: this occurs for small apertures, long baselines, or long wavelengths. In optics terminology, the effect is a *vignetting*: a limitation of the field of view due to the optical system.

Since the problem largely originates with the w part of (u, v, w), it is worth asking if there is any way to project w out of the problem, thus allowing a two dimensional Fourier transform to a single image to be used. Frater and Doherty (1980) noted that projection from a single plane w to $w = 0$ is possible, and they proposed using Clean to solve the resulting convolution relationship. Although they consider only the (unusual) case of all measurements occurring on a single plane, it is clear that their equation allows reprojection to and from any position in (u, v, w) space from and to the $w = 0$ plane. To derive this result, we must rewrite equation (1) as a convolution between the Fourier transform of the sky brightness and the Fourier transform of an image plane phase term parametrized by w.

$$V(u,v,w) = \int \frac{I(\ell,m)}{\sqrt{1-\ell^2-m^2}} G(\ell,m,w)\; e^{-2\pi i[u\ell+vm]} d\ell dm \tag{4}$$

$$G(\ell,m,w) = e^{-2\pi i[w(\sqrt{1-\ell^2-m^2}-1)]} \tag{5}$$

Applying the Fourier convolution theorem, we find that:

$$V(u,v,w) = \tilde{G}(u,v,w) * V(u,v,w=0) \tag{6}$$

To understand the form of $\tilde{G}(u,v,w)$, we can use a small angle approximation:

$$G(\ell,m,w) = e^{\pi i[w(l^2+m^2)]} \tag{7}$$

$$\tilde{G}(u,v,w) = \frac{i}{w} e^{-\pi i[\frac{(u^2+v^2)}{w}]} \tag{8}$$

This has remarkable and felicitous algorithmic implications: the visibility for non-zero w can be calculated from the visibility for zero w by convolution with the known function $\tilde{G}(u,v,w)$. Thus the three-dimensional function $V(u,v,w)$ is determined purely from the two-dimensional function $V(u,v,w=0)$. This holographic result arises because the original brightness is confined to a two-dimensional surface (the celestial sphere).

2. In which we use projection to get rid of W

Therefore to evaluate the visibility predicted for a pixellated model I of the sky brightness, we do the following:

1. Multiply the sky brightness model I by the taper function T.
2. Perform a two dimensional Fourier transform (real to complex) of the tapered $T.I$.
3. Evaluate the convolution (equation 4) for each sample point to obtain the predicted visibility.

To evaluate the dirty image:

1. For each sample visibility, evaluate the convolution for a gridded $(u, v, w = 0)$ plane.
2. Perform a two dimensional inverse Fourier transform (complex to real) to obtain the tapered $T.I^D$.
3. Divide out the image plane tapering function T to obtain the dirty image.

Most of the computational load lies in the gridding/degridding step which is directly determined by the number of visibility samples and the size of the convolution kernel. Taking the field of view to be λ/D, we find that the number of pixels in the support region along each of the u and v axes goes as $B\lambda/D^2$.

The costs for the standard facet based gridding (Cornwell and Perley 1992) go as the total number of facets (i.e., the product of the number along each of the two spatial axes). We then have that the costs per sample go as:

$$t_{facets} = N^2_{facets}.N^2_{GCFpoints}.t_{single} \tag{9}$$

$$t_{wproject} = 2(N^2_{wproject} + N^2_{GCFpoints}).t_{single} \tag{10}$$

In this equation, $N_{GCFpoints}$ is the support of the normal gridding convolution function in one axis (typically 9), and t_{single} is the time to grid a single sample to a single grid point. N_{facets} (the number of facets in one axis) and $N_{wproject}$ (the typical size of the $\tilde{G^T}$ gridding function) are both proportional to $B/(\lambda D^2)$ but with different proportionality constants. We have assumed that the sizes of the normal gridding convolution function and $\tilde{G}$ add in quadrature. Note also that $\tilde{G^T}$ is necessarily complex. If we take N_{facets} and $N_{wproject}$ to be roughly equal, then for large fields of view, the ratio of these times is roughly the total number of points in the *normal* gridding convolution function. Since the normal gridding convolution function is typically 7 by 7 or 9 by 9, the *asymptotic* speedup is between 25 and 50. Allowing for different proportionality constants, *we could conservatively expect at least an order of magnitude speed advantage for w projection.*

More details of the W projection algorithm are to be found in Cornwell, Golap, and Bhatnagar (2003, 2005).

3. A synthetic example

To demonstrate the w projection algorithm, we simulated a low frequency observation of a typically full field: a 74 MHz VLA C-configuration full synthesis on a field generated by taking the sixty-six WENSS sources brighter than 2Jy within 12 degrees of the specified center. The sources were scaled to 74 MHz by a spectral index of -0.7, and then multiplied by a simple model of the 74 MHz antenna primary beam. The data corresponding to these scaled sources were calculated using analytical transforms, and should thus be fully accurate to machine precision. The brightest source has strength 47.8 Jy and has been chosen to be at the field center.

The speed difference is apparent : 5419 seconds for w projection compared to 30488 seconds for the facet based algorithm. If the algorithms are configured for similar dynamic range, the difference in speed is usually about an order of magnitude.

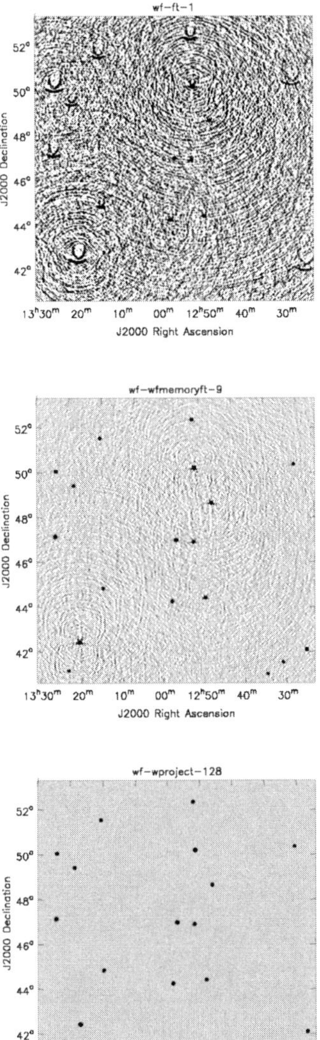

Figure 1. Clean images for a VLA 74 MHz simulation. Top: standard Fourier transform, Middle: facet-based algorithm (9 x 9), Bottom: w projection (128). The brightness range is -5 to +50 mJy/beam, and the peak brightness should be 47.2 Jy. The peak sidelobes around the brightest sources in the uvw-space facets image are about 0.3%. Calculation of these images took 784s, 30488s, and 5419s respectively.

References

Cornwell, T. J. & Perley, R. 1992, A&A, 261, 353

Cornwell, T. J., Golap, K., & Bhatnagar, S. 2003, EVLA memo 67

Cornwell, T. J., Golap, K., & Bhatnagar, S. 2005, A&A, (submitted)

Frater, R. & Docherty, I. 1980, A&A, 84, 75

Thompson, A., Moran, J., & Swenson, G. 2001, Interferometry and synthesis in radio astronomy (Wiley, New York)

ASTRONOMICAL DATA ANALYSIS SOFTWARE AND SYSTEMS XIV
ASP Conference Series, Vol. 347, 2005
P. L. Shopbell, M. C. Britton, and R. Ebert, eds.

matplotlib – A Portable Python Plotting Package

Paul Barrett
Space Telescope Science Institute

John Hunter
University of Chicago

J. Todd Miller, Jin-Chung Hsu, and Perry Greenfield
Space Telescope Science Institute

Abstract. matplotlib is a portable 2D plotting and imaging package aimed primarily at visualization of scientific, engineering, and financial data. matplotlib can be used interactively from the Python shell, called from python scripts, or embedded in a GUI application (GTK, Wx, Tk, Windows). Many popular hardcopy outputs are supported including JPEG, PNG, PostScript and SVG. Features include the creation of multiple axes and figures per page, interactive navigation, many predefined line styles and symbols, images, antialiasing, alpha blending, date and financial plots, W3C compliant font management and FreeType2 support, legends and tables, pseudocolor plots, mathematical text and more. It works with both numarray and Numeric. The goals of the package, basic architecture, current features (illustrated with examples), and planned enhancements will be described.

1. Introduction

matplotlib is designed with the philosophy that you should be able to create simple plots with just a few commands, or just one! If you want to see a histogram of your data, you shouldn't need to instantiate objects, call methods, set properties, etc; it should just work.

The initial goals of matplotlib were:

- Plots should be publication quality; particularly the text (antialiased, rotated, etc.).
- PostScript output for inclusion with TEX documents.
- Embeddable in a graphical user interface for application development.
- Code should be understandable.
- Making plots should be easy.
- The software is Open Source, so it can be downloaded, used, and distributed freely.

matplotlib can be used in a variety of settings. Most users are familiar with the command-line for interactively creating plots and images. This interface provides a simple pop-up window for displaying and manipulating the data. However, the true power of matplotlib is the underlying plotting library, which

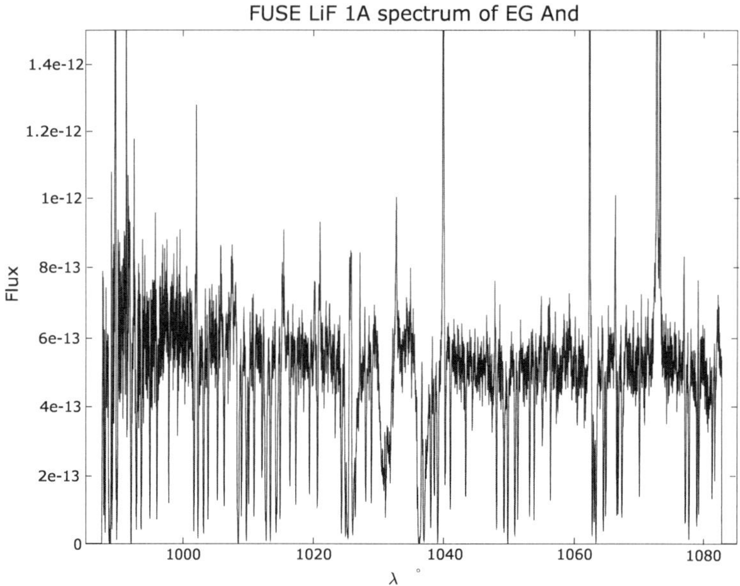

Figure 1. A line plot of FITS binary table data containing 10k points.

is operating system independent and graphical user interface (GUI) agnostic. It can be used without a GUI as part of a web server to create plots and images in a variety of hardcopy outputs; or can be embedded in a larger application using one of several GUIs (e.g. GTK, Tk, or WXwindows) running on one of several OSs (e.g. Windows, OS X, Solaris, and Linux).

2. Architecture

The matplotlib code is conceptually divided into three parts:

- The matlab interface is the set of functions that allow a user to create plots from the command line.
- The frontend or matplotlib API is the set of classes that do the heavy lifting by creating and managing figures, text, lines, plots, etc. This is the abstract interface that knows nothing about output.
- The backends are device dependent drawing devices or renderers that transform the frontend representation to hardcopy (JPEG, PNG, PDF, PS, SVG, Paint, GD) or a display device (Agg, GTK/GTKAgg, TkAgg, WX/WXAgg). Much of the critical rendering code is written in C/C++ and therefore provides very good performance.

Agg is the Anti-Grain Graphics library that enables writing vector graphics to a buffer, which can then be block transfered (or BLTed) to the display device. This means that all interactive implementations based on Agg avoid the

Figure 2. A financial plot that uses the daily high, low, and closing values of a stock price.

graphical limitations of the GUI and render identical graphics regardless of the GUI interface.

3. Plotting

The following Python session uses the matlab interface to create a quicklook spectrum of FUSE data (see Figure 1).

```
> python
Python 2.3.3 (#1, Jan  5 2004, 16:22:13)}
[GCC 2.96 20000731 (Red Hat Linux 7.3 2.96-113)] on linux2
Type "help", "copyright", "credits" or "license"
for more information.
>>> import pyfits
>>> fits = pyfits.open('fuse.fits')
>>> wave = fits[1].data.field('wave')
>>> flux = fits[1].data.field('flux')
>>> from matplotlib.matlab import *
>>> ylim(0, 1.5e-12)
>>> xlim(985, 1085)
>>> xlabel(r'$\lambda\ (\angstrom)$')
<matplotlib.text.Text instance at 0x41118f8c>
>>> ylabel(r'Flux')
```

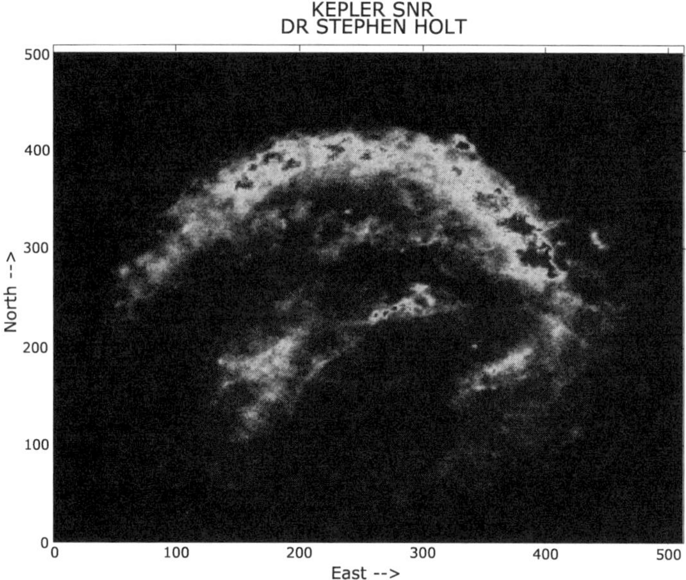

Figure 3. A graphic of FITS image data. Note the labeled axes and title.

```
<matplotlib.text.Text instance at 0x4112108c>
>>> title('FUSE LiF 1A spectrum of EG And')
<matplotlib.text.Text instance at 0x4112420c>
>>> plot(wave, flux)
```

Other available plot types are: 2-D vector plots, high-low-close plots (see Figure 2), histogram plots, log plots, pie charts and bar charts.

4. Images

The matlab interface has two functions for displaying image data: figimage, which will preserve the size and shape of the image; and imshow, which will resample the image to fit the size of the figure (see Figure 3). Images can be enhanced by annotations or graphical overlays.

5. Features

Key features that make matplotlib easy to use are:

- Integrated support for numarray or Numeric – the Python multi-dimensional array libraries.
- The plot window contains a simple interactive GUI with support for pan-and-zoom, history recall, and saving to hardcopy.

- The command line interface is modeled after the easy to use MatLab interface.
- Support for multiple plots and images per page.
- TrueType/FreeType fonts are available in the GD, Agg, Paint, and PostScript backends. SVG support is coming soon.
- Mathematical text ala TeX math mode is available whenever TrueType fonts are available.
- Images are automatically resampled to the size of the figure.
- A fully object-oriented design to ease programming and development.

6. Enchancements

Enhancements to matplotlib that are expected in the near future are:

- Contour plots which can be used for image overlays.
- The ability to handle general 2-D transforms, which are useful for map projections and world coordinate systems.

To learn more about matplotlib and to download the latest version, go to `http://matplotlib.sourceforge.net`.

ASTRONOMICAL DATA ANALYSIS SOFTWARE AND SYSTEMS XIV
ASP Conference Series, Vol. 347, 2005
P. L. Shopbell, M. C. Britton, and R. Ebert, eds.

Mosaicing with Interferometers: An Efficient Algorithm for Imaging and Image Plane Corrections

S. Bhatnagar, K. Golap, and T. J. Cornwell

National Radio Astronomy Observatory, Socorro, NM 87801. E-mails: sbhatnag@aoc.nrao.edu, kgolap@aoc.nrao.edu, tcornwel@aoc.nrao.edu

Abstract. The imaging dynamic range of an aperture synthesis telescope for mosaicing and for fields with significant flux throughout the antenna primary beams can be limited by the knowledge of the individual primary beams projected on the sky. For high dynamic range imaging of such fields, one requires an accurate measurement of the shape of the primary beams and the pointing offsets as a function of time. The effect of antenna pointing errors remain separable in the visibility domain. With at least two, well separated sources along the RA and Dec axis each to constrain the solutions, it is possible to solve for these errors in an antenna based fashion in the visibility domain.

Here we describe an algorithm to solve for these errors using a model for the sky brightness distribution. For a typical L-band eVLA simulation with typical pointing errors for the VLA antennas, the RMS noise can be reduced by a factor of $\sim$ 10 using this algorithm. The improvement in the image fidelity is even larger.

1. Introduction

In the absence of pointing offsets, the image made with an ideal aperture synthesis telescope with identical antennas is the true sky multiplied with the ideal primary beam pattern of the antennas. The observed visibilities are then expressed as $E_{ij} \star V^{True}$, where $E_{ij} = E_i \star E_j^*$ and E_i is the antenna aperture illumination pattern. With no pointing offsets E_{ij} is identical for all the baselines which can be used as the gridding convolution function for computing model visibilities during the deconvolution major cycle. In the presence of antenna pointing errors, E_{ij} is potentially different for each baseline, resulting in a systematic error pattern in the residual image which limits the dynamic range of images where there is significant flux throughout the field of view.

The multiplicative antenna based complex gain, constant across the field of view, can be measured from the observation of even a single unresolved source using the SelfCal algorithm. Antenna pointing errors on the other hand result in a gain which changes across the field of view. With at least two well separated sources along the RA and the Dec axis and a good model image, it should be possible to derive the antenna based pointing offsets by minimizing the differences between the observed and the model visibilities. In this paper we show that such errors can be solved for in an antenna based fashion.

2. The algorithm

The measurement equation for a single pointing observation with antenna based pointing offsets is:

$$V_{ij}^{\circ} = E_{ij} \star V_{ij}^{True} \tag{1}$$

where E_i is the complex aperture illumination pattern, $E_{ij} = E_i \star E_j^*$, V_{ij}° and V_{ij}^{True} are the observed and the true visibility from the baseline $i - j$, and $\star$ is the convolution operator.

For a test case of a collection of point sources of amplitude I_k located at l_k (model image $I^M(l_k) = I_k \delta(l - l_k)$), the corrupted visibilities are given by:

$$V_{ij}^{C} = E_{ij} \star V_{ij}^{M} = \int I^M(l) P_i(l).P_j(l).e^{2\pi\iota u_{ij} l}\, dl \tag{2}$$

and $\nabla V_{ij}^{R} = -2\sigma^2 \left[V_{ij}^{C'} - l_i V_{ij}^{C} \right]$ where $V_{ij}^{C'} = \int I^M(l) l P_i(l) P_j(l) e^{2\pi\iota u_{ij} l}\, dl$. The update equation for the estimated pointing offset at the nth iteration (l_i^n) is then given by:

$$l_i^n = l_i^{n-1} + \alpha \left. \nabla \chi^2 \right|_{l_i^{n-1}} \tag{3}$$

where $\alpha < 1.0$ is the step size. The iterations are stopped when $|\chi_n^2 - \chi_{n-1}^2| \leq \epsilon$.

For a general case where the model image is not a collection of just point sources, E_{ij} can be computed and used for predicting the visibilities for baseline $i - j$ using the Eq. 1.

3. Simulation and Results

The algorithm was tested using a VLA C-array multiple snapshot simulated observation at 1.4 GHz. The sky was simulated using 49 NVSS sources within the field of view with flux density in the range $\sim 2 - 200$mJy. The primary beams of the antennas were assumed to be a gaussian and similar except for the random pointing offsets. The mean of pointing offsets of various antennas was uniformly distributed between $\pm 20''$ about the phase center with an RMS of $\sim 5''$ (see Fig. 1). A baseline based gaussian random RMS noise corresponding to image plane RMS noise of $\sim 1\mu$Jy was added to simulate the thermal noise. With such a pointing error model in the above simulations, the pointing noise was $\sim 15\mu$Jy and the peak residual of $\sim 250\mu$Jy. The noise after pointing calibration was consistent with the thermal noise $\sim 1\mu$Jy and the peak residual was $\sim 5\mu$Jy. The residual images before and after pointing selfcal are shown in Fig. 2. The typical antenna based pointing offsets and the residual pointing offsets after pointing calibration as a function of time are shown in Fig. 1.

4. Discussion

At the advertised sensitivities of the upcoming new telescopes like the EVLA, ALMA and the LOFAR and other planned telescopes like the SKA, direction dependent (image plane) calibration is crucial. In solving for any direction dependent instrumental/atmospheric effect for interferometric observations, the

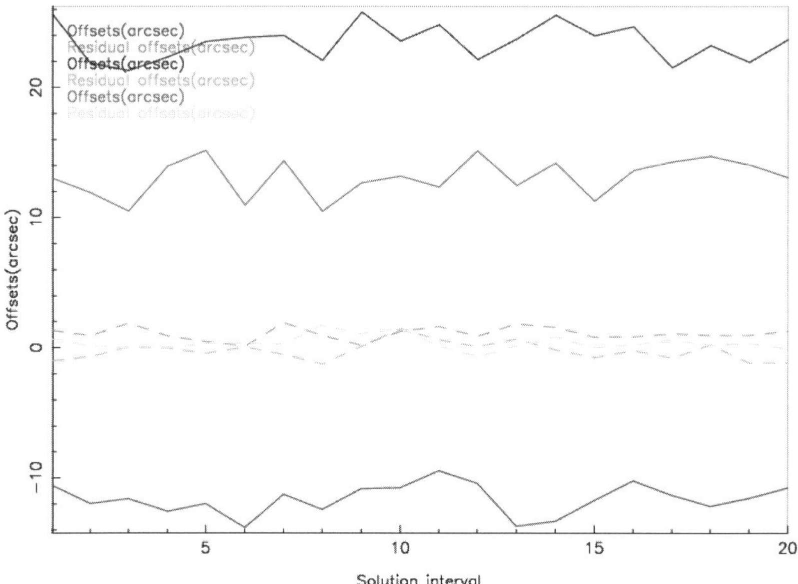

Figure 1. Figure shows the typical antenna based pointing offsets as a function of time used for the images shown in Fig. 2. The continuous lines show the pointing offsets used in the simulation and the dashed lines show the residual pointing offsets after pointing calibration. The RMS residual pointing errors are $\sim 1''$.

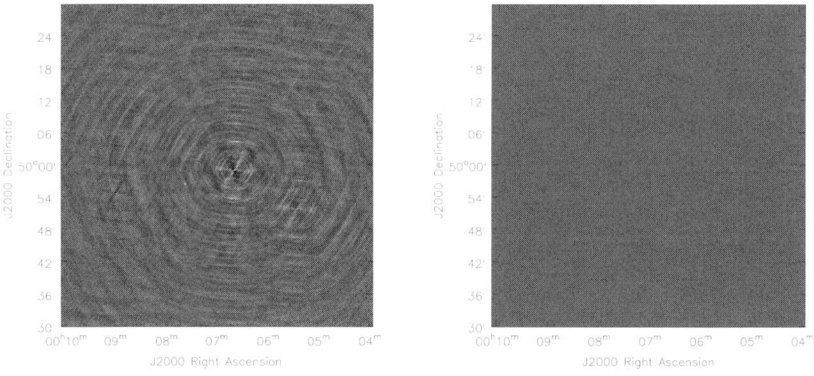

Figure 2. The image on the left is the residual image with the mean antenna pointing offsets between $\pm 25''$ with an RMS of $2''$ (see Fig. 1). The image on the right is the residual image after applying the pointing offset correction. The images have an RMS noise of $\sim 15\mu Jy$ and $\sim 1\mu Jy$ and peaks of $\sim 250\mu Jy$ and $\sim 5\mu Jy$ respectively.

transform between the image and visibility domains which includes the effects one is solving for is the most compute intensive step. Typically the data is re-sampled on a regular grid using a gridding convolution function (GCF) so that the FFT algorithm can be used for this transform. The algorithm presented here for solving for the antenna based pointing offsets was inspired by the w-projection algorithm (Cornwell et al, 2003) which corrects of the w-term in the visibility domain. It uses a modified GCF, which includes the effects of pointing offsets. The GCF however becomes a function of the two antennas making a baseline. An analytical approximation is therefore used to compute it on the fly. Since the re-sampling process is part of the imaging process, correcting for the pointing offsets can be done with minimal overhead.

Multiplicative image place based effects become a convolutional effect in the visibility domain. Various other direction dependent errors, like the non-isoplanatic ionospheric/atmospheric phase correction, polarization squint correction, correction for the asymmetric primary beams changing as a function of time, etc. can be similarly applied. Fast computation of these effects is crucial for algorithms which also solve for such effects. The algorithm present here forms the first step towards development of a more sophisticated solver for direction dependent calibration. Work for solvers for other image plane effects which might limit the achievable dynamic range and image fidelity with the new more sensitive interferometric telescopes is in progress.

Acknowledgments. We thank the AIPS++ group and the members of the Friday discussion group at Socorro for various fruitful discussions.

References

Cornwell, T. J., Golap, K., and Bhatnagar, S. 2003, Tech. Rep., EVLA Memo 67

ASTRONOMICAL DATA ANALYSIS SOFTWARE AND SYSTEMS XIV
ASP Conference Series, Vol. 347, 2005
P. L. Shopbell, M. C. Britton, and R. Ebert, eds.

BUCS: Automating Sample Selection, Volume Density Determinations, and Projection onto Different Image Sets and Redshift Regimes

Rychard J. Bouwens, Garth D. Illingworth, and Daniel K. Magee

Astronomy Department, University of California, Santa Cruz, CA 95064, Email: bouwens@ucolick.org

Abstract.
Interpreting today's high-resolution multiwavelength data on galaxies can be a challenging task. Not only do galaxies have complex wavelength-dependent morphologies, but attempts to measure these morphologies with different imaging data are subject to a wide-range of observational issues like beam smearing and differing signal-to-noise. Properly measuring the evolution of galaxy properties across cosmic time demands that these things be treated. It is to this end that we have developed the BUCS software suite (as it came to be known: Bouwens' Universe Construction Set). With this package, it is possible to select complete samples of galaxies off different image sets and then project these objects to higher redshift accounting for pixel k-corrections, cosmic surface brightness dimming, and image specific noise and PSF. The end result of these simulations are no-evolution images that can be directly compared against the observations. This suite is now available to do science and has been used in a large number of refereed publications (including recent studies on $z \sim 6-8$ samples from the HUDF). Work is also under way to port these capabilities to the web.

1. Introduction

One of the main goals of galaxy evolution studies is to quantify the evolution of galaxy properties across cosmic time. While much progress has been made towards this goal, making sense of this progress in terms of the details has been much more troublesome. The field is literally a mishmash of different samples, selection and analysis techniques, and it can often be difficult to interpret the differences that occur between studies. This wouldn't be a large problem if there were a simple means of transforming results from one study to the selection and measurement frame of another. Unfortunately, this cannot typically be done without making additional assumptions since most studies only tabulate derived quantities, like the luminosity or size function.

The principal difficulty is that generic studies of galaxy evolution *do not preserve the total information content* present in the original samples and therefore transformations to other selection or measurement frames can only be approximately done. Real galaxy samples consist of disparate sets of individual galaxies, each with their own pixel-by-pixel multiwavelength morphologies, volume densities, and interobject spacing. Choosing to represent these samples exclusively with select numbers of derived quantities (e.g., the LF, size functions) results

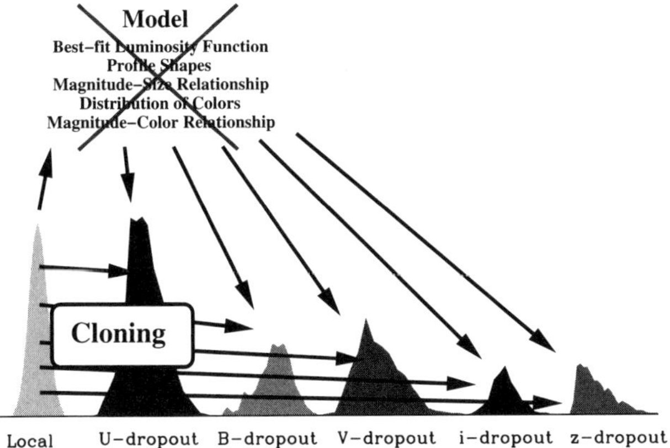

Figure 1. The BUCS software automates the *pixel-by-pixel* projection ("cloning") of individual objects from lower redshift samples to higher redshift. This procedure more effectively preserves the total information content than is possible through the use of simple models which attempt to represent these samples by e.g., their luminosity function and size distribution.

in a loss of information and can compromise the usefulness of these samples to other scientists.

At the same time, the shear quantity of multiwavelength data available on samples of high-redshift galaxies continues to increase. Eight years ago, essentially the only field with multiwavelength, high-resolution data was the HDF North. Now, with the availability of the NICMOS and ACS on HST, Chandra, Spitzer, GALEX, and $8-10$-m ground-based telescopes, the data volume has increased at least a hundred fold. While there have also been noticeable advances in computing speed, processing these data by hand to produce galaxy samples continues to be more and more tedious and time consuming. Telescopes like the LSST can only make these matters worse.

Foreseeing these trends, several years ago we set about to develop some software to automate the process. This software eventually became known as BUCS (Bouwens' Universe Construction Set) and was centered around the concept of a galaxy sample and the extraction of samples from deep galaxy images. The reason we decided to focus on galaxy samples is that these samples really act as our stepping stones for evaluating evolution across cosmic time, and it seemed obvious that procedures would be needed to quickly and easily convert observational data into these more scientifically meaningful units. It was also evident that such procedures would also provide a convenient way of preserving the total information content in such samples (i.e., multiwavelength pixel-by-pixel fluxes on all individual sample objects) so that they could be readily disseminated for comparison with newer samples being acquired.

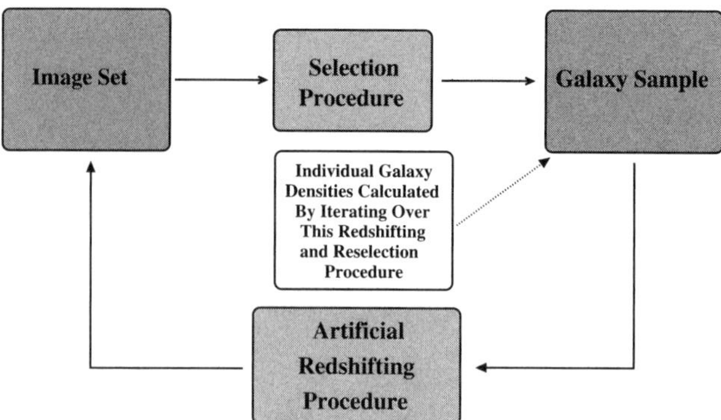

Figure 2. Schematic illustrating the general flow of the BUCS software library. The software automates the conversion of image sets (e.g., deep galaxy fields like the HUDF) into galaxy samples. Galaxy samples can then in turn be laid down over a specific redshift interval and thus used to generate image sets. Volume density of individual galaxies can be determined by artificially redshifting each galaxy template to all redshifts (Figure 3) and reselecting it using the selection criteria of the sample. Image sets are essentially what they sound like: coaligned sets of images in different filters, with PSFs, zero points, and noise properties. Samples can be anything from a standard two-color B-dropout sample selected from the HUDF to a $z \sim 0.3$ spectroscopic sample selected from GOODS.

2. Overview

The BUCS software suite is centered around the abstraction of a galaxy sample. Galaxy samples are defined by an image set, selection criteria, and a procedure for estimating redshifts. With this software, deep galaxy image sets can be seamlessly converted into galaxy samples. Conversely, galaxy samples can be used to simulate sections of the universe and hence produce image sets again. A simple illustration of the general flow is provided in Figure 2.

The software is designed to manipulate samples, image sets, and individual galaxies in much the same way that other software libraries manipulate the abstract objects which are the focus of those libraries (e.g., as CFITSIO allows one to manipulate FITS files). It should therefore be no surprise that basic calculations like the determination of sample selection functions or the projection of one sample onto another are automated in the BUCS software suite (Figure 1). This latter function naturally allows for tests of the null-evolution hypothesis and hence some measure of the actual evolution.

After years of development, this software is now quite mature. One testament to this is the large number of refereed publications which have already made heavy use of it (Bouwens et al. 1998a,b; Bouwens et al. 2003a,b; Bouwens et al. 2004a,b,c). Dissemination of the software and some much needed documentation should be available in the not too distant future.

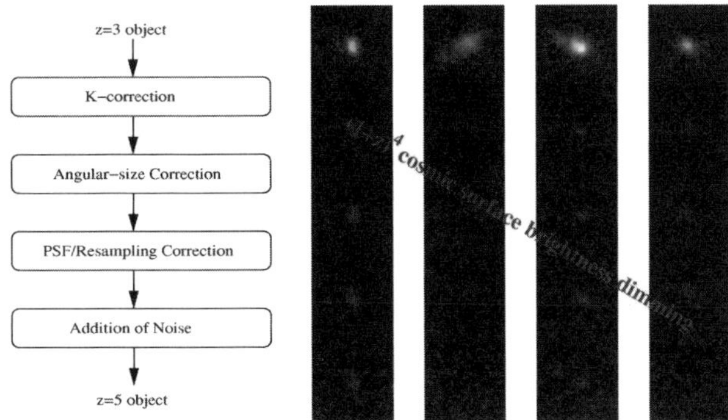

Figure 3. Schematic showing our artificial redshifting procedure (k-correction + angular size correction + PSF/resampling correction + addition of noise) applied to four $z \sim 3$ U-dropouts from the HDF-North. This figure also provides a good illustration of the dramatic effect that $(1+z)^4$ cosmic surface brightness dimming has on the overall appearance of galaxies, demonstrating the importance of this procedure.

3. WBUCS: A Web Simulator for Deep Galaxy Fields

By automating the procedure for quantifying the volume density of individual galaxies in different samples and modeling the pixel-by-pixel SEDs of each object, it is straightforward to use this information to simulate deep galaxy fields and to make this available to the community at large. In our implementation of this (Magee, Bouwens, & Illingworth 2005), we utilize galaxy samples at all redshifts, ranging from the Sloan Digital Sky Survey at $z \sim 0$ (Blanton et al. 2004) to a low-redshift GOODS sample at $z \sim 0.2 - 0.4$ (Bouwens et al. 2004e) to a wide variety of different dropout samples from $z \sim 2$ to $z \sim 6$ (Bouwens et al. 2004d). This service should be useful both at the proposal stage and for comparing with existing data.

Acknowledgments. We are grateful to Narciso Benítez, Emmanuel Bertin, Tom Broadhurst, Fred Courbin, Michele de la Pena, Harry Ferguson, Marijn Franx, Andy Fruchter, Jon McCann, and Gerhardt Meurer for input important for the development of this software. We also gladly acknowledge fruitful collaborations with other members of the ACS GTO team and support from NASA grant NAG5-7697.

References

Blanton, M.R., et al. 2004, AJ, submitted, astro-ph/0410166.

Bouwens, R.J., Broadhurst, T.J., & Silk, J. 1998a, ApJ, 506, 557.

Bouwens, R.J., Broadhurst, T.J., & Silk, J. 1998b, ApJ, 506, 579.

Bouwens, R.J., et al. 2003a, ApJ, 595, 589.

Bouwens, R.J., Broadhurst, T.J., & Illingworth, G.D. 2003b, ApJ, 593, 640.

Bouwens, R.J., et al. 2004a, ApJ, 606, L25.

Bouwens, R.J., et al. 2004b, ApJ, 611, L1.

Bouwens, R.J., et al. 2004c, ApJ, 616, L79.
Bouwens, R.J., et al. 2004d, ApJ, submitted.
Bouwens, R.J., et al. 2004e, ApJ, in preparation.
Magee, D.K., Bouwens, R.J., & Illingworth, G.D. 2005, in ASP Conf. Ser., Vol. 347, ADASS XIV, ed. P. L. Shopbell, M. C. Britton, & R. Ebert (San Francisco: ASP), 399.

ASTRONOMICAL DATA ANALYSIS SOFTWARE AND SYSTEMS XIV
ASP Conference Series, Vol. 347, 2005
P. L. Shopbell, M. C. Britton, and R. Ebert, eds.

Extending and Enhancing the Capabilities of the Sherpa Modeling and Fitting Tool with S-Lang

D. J. Burke, A. Siemiginowska, S. Doe, and D. Nguyen

Harvard-Smithsonian Center for Astrophysics, 60 Garden Street, Cambridge, MA 02138

Abstract. *Sherpa* is the fitting and modeling tool of the Chandra Interactive Analysis of Observations (CIAO[1]) package. It is designed to allow users to fit the 1D (e.g., spectra, luminosity functions) and 2D datasets (e.g., images) obtained from the Chandra X-ray Observatory[2] and other telescopes. The S-Lang programming language has been embedded into CIAO, which allows users the ability to access numerical data calculated by *Sherpa*. We present two examples of how S-Lang has been used to enhance and extend *Sherpa's* capabilities.

1. Introduction

The S-Lang library - which was designed to be easily embeddable within a program - provides an interpreted language with a C-like syntax which is used to enhance and extend the application. It is primarily used in the CIAO environment to:

- provide an alternative interface to the main command-line applications in CIAO: *Sherpa* and ChIPS,
- allow users to access the functionality of the CIAO code base by the use of S-Lang modules,
- as a means of developing algorithms useful for data reduction and analysis (e.g., Noble 2005, Primini & Noble 2005).

The *Sherpa* program (Freeman et al. 2001) provides a simple command-line environment for modeling and fitting 1D and 2D datasets, such as: spectra, luminosity functions, lightcurves, and images. The S-Lang interface to Sherpa (Doe et al. 2001, 2004) permits the use of in-memory datasets, which, coupled with the strong complement of elementary mathematical and array-manipulation capabilities of S-Lang, opens up the possibility of automating common data manipulation and filtering tasks. These routines and techniques enhance and extend *Sherpa's* command-line capabilities as well as allow *Sherpa* to be used in non-interactive applications, such as data-reduction pipelines. Since they do not require compilation they are easy to install, use, and modify by the CIAO user

[1]`http://cxc.harvard.edu/ciao/`

[2]`http://chandra.harvard.edu/`

community. In this presentation we highlight two of these additions. Further details and examples are available from the *Sherpa* web site[3].

2. Calculating Radial Profiles of 2D fits

Sherpa allows users to fit source models to their imaging data. Due to the nature of X-ray imaging data, these fits often require the use of more advanced fitting statistics than the χ^2 statistic which do not provide a measure of the goodness of fit. To help users evaluate how well the model represents the data, we provide a S-Lang routine that automatically calculates and displays the radial profile of the data and current model.

```
% calculate the radial profile of the data and model about the
% best-fit location of a 2D fit
%
define plot_rprof(dataset,rmin,rmax,deltar) {

  % find the center of the model
  % (look for the xpos and ypos parameters)
  %
  variable model_name = get_source_expr(dataset);
  variable xcen = get_par(model_name+".xpos").value;
  variable ycen = get_par(model_name+".ypos").value;

  % Get the pixel values of the image and model datasets
  % (as 1D S-Lang arrays)
  %
  variable image = get_data(dataset);
  variable model = get_mcounts(dataset);

  % create a 1D array which gives the distance from the
  % best-fit location of each valid pixel in the data
  % (use the center of the pixel in the calculation)
  %
  variable axes  = get_axes(dataset);
  variable xmid  = 0.5*(axes[0].lo + axes[0].hi);
  variable ymid  = 0.5*(axes[1].lo + axes[1].hi);
  variable rdist = sqrt( (xmid-xcen)*(xmid-xcen) +
                         (ymid-ycen)*(ymid-ycen) );

  % Calculate the minimum (rlo) and maximum (rhi) values
  % of each radial bin
  %
  variable nbins = int((rmax-rmin)/deltar);
  variable r = rmin + [0:nbins]*deltar;
  variable rlo = r[[0:-2]];
```

[3]http://cxc.harvard.edu/sherpa/threads/slang/

```
  variable rhi = r[[1:-1]];

  % Create the radial histogram by looping through each bin
  % and finding all the pixels that lie in that range
  %
  variable hist_data  = Double_Type [nbins];
  variable hist_model = Double_Type [nbins];
  variable hist_npix  = Int_Type [nbins];
  _for ( 0, nbins-1, 1 ) {
    variable i = ();
    variable index = where(rdist>=rlo[i] and rdist<rhi[i]);
    variable nindex = length(index);

    % if no pixels match then skip this bin
    if (0==nindex) continue;

    hist_data[i]  = sum(image[index]);
    hist_model[i] = sum(model[index]);
    hist_npix[i]  = nindex;
  }

  % remove those bins with no data and normalize by the
  % number of pixels
  %
  variable good = where(hist_npix>0);
  hist_npix  = hist_npix[good];
  hist_data  = hist_data[good]/hist_npix;
  hist_model = hist_model[good]/hist_npix;
  variable rmid = 0.5*(rlo[good]+rhi[good]);

  % Plot the radial profile of the image and model
  %
  chips_clear();
  () = curve(rmid, hist_data);
  chips.curvecolor = _chips->red;
  () = curve(rmid, hist_model);
}
```

An example of the output from this function is shown in Figure 1.

3. Using other CIAO modules

The following routine can be used from *Sherpa* to change the grouping scheme of an X-ray spectrum on-the-fly, without having to create temporary disk files. It shows how multiple S-Lang modules can be combined to enhance the analysis environment; here S-Lang routines in *Sherpa* are used to read and write the data and grouping columns of a file, while the `grpNumCounts()` routine from the `group` module is used to calculate the new grouping scheme.

```
require("group"); % ensure the group module is available

define groupByCounts(dataset,numCounts) {

  % Has the background been subtracted?
  %
  variable bgnd_flag = is_subtracted(dataset);

  % reset the state so that we have the original, ungrouped
  % data including the background contribution
  %
  if (bgnd_flag) () = set_unsubtract(dataset);
  variable groups = get_groups(dataset);
  if (sum(groups) != length(groups))
    () = sherpa_eval("ungroup " + string(dataset));

  % Use the grpNumCounts() routine from the CIAO group module to
  % calculate the new grouping scheme, given the users requested
  % number of counts per bin (the numCounts parameter)
  %
  variable data, grouping, quality;
  data = get_data(dataset);
  (grouping, quality) = grpNumCounts(data, numCounts);

  % Apply the new grouping scheme
  %
  () = set_groups(dataset,grouping);
  () = set_quality(dataset,quality);

  % Subtract the background (if necessary)
  if (bgnd_flag)
    () = set_subtract(dataset);
}
```

Acknowledgments. This work was supported by the Chandra X-ray Center under NASA grant NAS8-03060.

References

Doe, S., Freeman, P., Smith, R., Burke, D., & Siemiginowska, A. 2004, in ASP Conf. Ser., Vol. 314, ADASS XIII, ed. F. Ochsenbein, M. Allen, & D. Egret (San Francisco: ASP), 404

Doe, S., Noble, M., & Smith, R. 2001, in ASP Conf. Ser., Vol. 238, ADASS X, ed. F. R. Harnden, Jr., F. A. Primini, & H. E. Payne (San Francisco: ASP), 310

Freeman, P., Doe, S., & Siemiginowska, A. 2001, SPIE Proceedings, 4477, 76

Noble, M. 2005, in ASP Conf. Ser., Vol. 347, ADASS XIV, ed. P. L. Shopbell, M. C. Britton, & R. Ebert (San Francisco: ASP), 237

Primini, F., Noble, M. 2005, in ASP Conf. Ser., Vol. 347, ADASS XIV, ed. P. L. Shopbell, M. C. Britton, & R. Ebert (San Francisco: ASP), 17

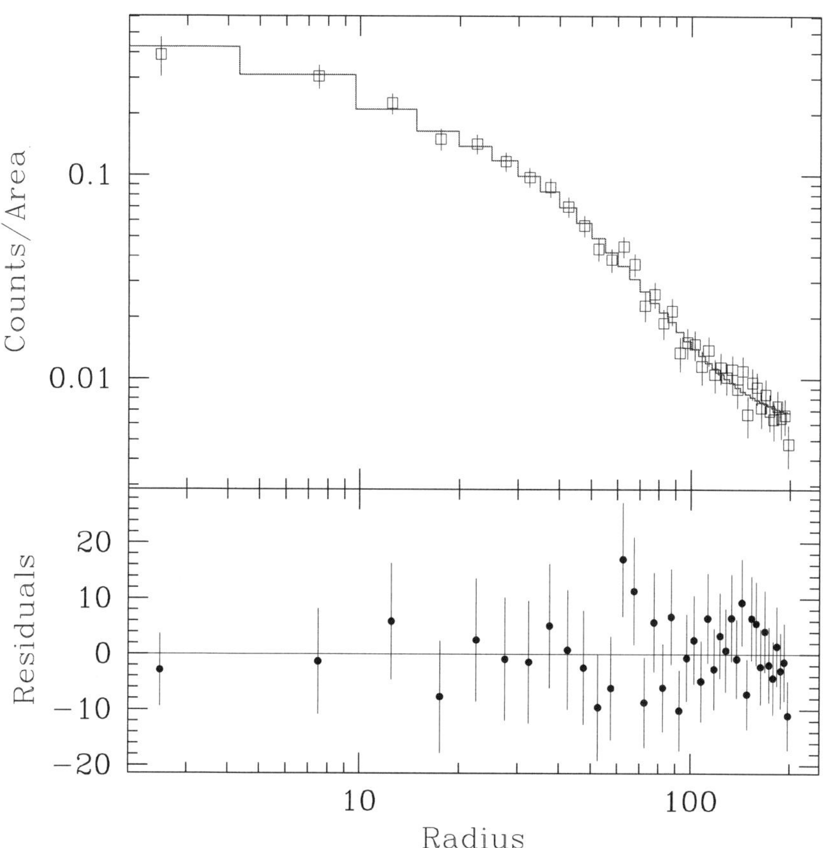

Figure 1. Plot showing the radial profile of the data (square points with error bars) and best-fit model (solid line) calculated by the `plot_rprof()` routine. The calculation of the error bars and residual points is not shown in the code presented here but is available in the full version of the routine, which is discussed at `http://cxc.harvard.edu/sherpa/threads/spatial/`.

ASTRONOMICAL DATA ANALYSIS SOFTWARE AND SYSTEMS XIV
ASP Conference Series, Vol. 347, 2005
P. L. Shopbell, M. C. Britton, and R. Ebert, eds.

The Development of SAOImage DS9: Lessons Learned from a Small but Successful Software Project

W. A. Joye and E. Mandel

Smithsonian Astrophysical Observatory, 60 Garden St., Cambridge, MA 02138

Abstract. We present a discussion of lessons learned during the development of SAOImage DS9. Starting with general observations on scientific software development, we discuss our design and implementation cycle, allocation of effort and resources, keys to our success, overall strategies that have worked well (and those that have not), and future challenges. We hope our experience will be of use to other small software development projects.

1. Introduction

SAOImage DS9 is a FITS visualization application whose development effort started more than six years ago as a research project with funding from NASA's Applied Information Systems Research Program. In the five years since its first public release, more than 35,000 copies of DS9 binaries and source code have been downloaded by 10,000 sites from all over the world. DS9 has grown into a collaborative effort of the astrophysical community, being composed of more than 20 Tcl/Tk, C, C++ software packages. In addition to the primary authors at SAO, code contributions come from Starlink, HEASARC, ATNF and NOAO. Furthermore, our list of collaborators and beta testers include individuals from UCO/Lick, ESA, MIT, and UMD. DS9 currently receives funding from the Chandra X-ray Center and NASA's High Energy Astrophysics Science Archive Center.

2. Observations on scientific software development

Five observations on scientific software development have influenced the evolution of DS9:

- User requirements evolve dynamically. There is little or no need to think up new requirements, since the user community has many more needs than can be satisfied at any given time. More importantly, these needs expand and change over time as new functionality gives rise to new ideas about how to improve data analysis techniques. Not only must the architecture of DS9 reflect this inherent dynamism, but we also have tried to maintain a flexible attitude toward the currently implemented code. We assume that requirements will change and we try not to get caught in a mind-set that resists changing the existing code to meet these new needs.
- In the face of constantly changing requirements, it is essential to maintain what Fred Brooks calls "conceptual integrity." We agree with Brooks' as-

sessment that "conceptual integrity is *the* most important consideration in [software] design ... Simplicity and straightforwardness proceed from conceptual integrity ... Ease of use, then, dictates unity of design, conceptual integrity." (Brooks 1982)

- Existing software degrades with time. Hardware, compilers, operating systems, and libraries change over time. For example, changes in gcc's support of C++ templates has required us to modify our code several times, as the template standard has become finalized. Thus, software requires continued care to support existing features, and it can take a surprisingly large amount of resources to maintain existing code.
- Software chases hardware. By definition, today's software can only be implemented on today's hardware. But the development time required for major advances (i.e. RGB support) can be a year or more, and hardware performance can increase significantly during that time. As a result, it is important to choose the timing of development projects carefully, and to design code so that it has the possibility of making use of hardware improvements.
- More bodies do not equal better results. Experience leads us to concur with Fred Brooks that "the design must proceed from one mind, or from a very small number of agreeing resonant minds." (Brooks 1982) We would go further in this assessment, believing that a small team of expert implementors, working together closely, can produce a better product in less time. Such an expert team has a good chance of maintaining conceptual integrity throughout the inevitable evolution of the program's requirements.

Of the above observations, the central one driving DS9 design and implementation is the first, i.e., that the user community provides a continual stream of evolving requirements. Long experience chasing this moving target has lead us to treat DS9 as one continuous prototyping project. Rather than try to tie static requirements to a "final" implementation, we provide DS9 to the astronomical community in a series of "beta releases" incorporating both bug fixes and functional improvements, occasionally punctuated by an official "public release." In reality, the only difference between one of the large 'beta releases" and a "public release" is that we ensure that the latter works on all supported platforms (beta releases often lag in support for the troublesome Windows release) when we publicly announce its availability.

3. A simple key to DS9's success

Following our second observation, we have religiously followed the golden rule: keep it simple. In particular, the DS9 graphical user interface must be simple enough to allow users to perform important functions without the aid of documentation. Furthermore, the installation process must be simple enough so that users are able to install and update the software without the aid of a system administrator. We also have adopted a secondary dictum: do no harm. We only add new capabilities that benefit the community at large and hide the complexity that only benefits the few. This is in keeping with a valuable set of design principles stated in the *The X Window System*, including:

- Do not add functionality unless an implementor cannot complete a real application without it.
- The only thing worse than generalizing from one example is generalizing from no examples at all. (Scheiffler & Gettys 1992)

4. With help from the community

By a combination of insight and luck, we have enjoyed continued support for the technology that underlies DS9. Tcl/Tk and other support packages are now open source, licensed under GPL, hosted at www.sourceforge.net, and have continuing support from the open source community. In fact, the open software model has worked very well for us. It allows the developers of packages used by DS9 to have total creative control over their products and encourages new development and continuing support. We thus benefit from the work of many others without expending valuable resources. We have also benefited from the large improvement in hardware performance. Following Moore's Law, hardware performance has increased over 400% since the first release of DS9. This has enabled us to implement algorithms that just a few years ago would have been too slow and unusable for an interactive program.

5. A weakness in the development model?

We believe that conceptual integrity requires a small team of developers, but success puts an expansion pressure on such a team. In particular, an increasing amount of our time is spent answering email and supporting users (20%) and updating and maintaining existing capabilities (30%). Configuration and porting issues consume a considerable amount of time as well. Thus, more than half of our time goes to maintenance issues. Additionally, we devote another 30% of our time to enhancing existing capabilities, leaving less than 20% of our time for adding new functionally. The problem of a maturing product requiring more and more maintenance resources is an unresolved issue with DS9 development.

6. Future Challenges

Our most pressing challenge is to continue development of an increasingly complex program to meet the ever-changing needs of the astronomical community, without sacrificing conceptual integrity. We have to find a better way to balance the increasing costs of maintenance and incremental evolution against the need for new initiatives. The inward pressure to keep the circle of developers small and tightly coordinated must be better balanced against the outward pressure to expand the developer group to meet increasing maintenance costs.

On the technical front, a major challenge we foresee is the explosion of FITS file sizes. Users routinely require the ability to display images that have a combined size of more than 2Gb and are now limited by maximum address space of 32-bit operating systems. In addition to adding file system support for file sizes over 2Gb, applications must be ported to 64-bit operating systems.

Another challenge will involve support for new capabilities provided by "dual-core" processors. As clocking speeds meet an upper limit due to heat dissipation problems, hardware designers are starting to implement multiple processors on one chip. It seems likely that code will have to be threaded to take advantage of this new trend. Sections of code must be analyzed for the possibility of parallelization, and prime candidates rewritten, bench-marked, and verified.

7. Summary

DS9 development is governed by the idea that community-based requirements inevitably change with time, even as the hardware and operating system environments also evolve. As a result, we consider DS9 to be a continually evolving prototype. This implementation model allows us to spend the requisite amount of time maintaining and updating existing capabilities while we also attempt to satisfy new needs. Throughout these efforts, we have been able to utilize the open source software model to our advantage.

8. Availability

DS9 is available on the Web at `http://hea-www.harvard.edu/saord/ds9` or via anonymous FTP from `ftp://sao-ftp.harvard.edu/pub/rd/ds9`.

Acknowledgments. This work was performed in large part with support from the Chandra X-ray Science Center (NAS8-39073) and from the High Energy Astrophysics Science Archive Center (NCC5-568).

References

Brooks, F. 1982, *The Mythical Man-Month*, Addison-Wesley, 42, 44

Scheiffler, R.W. & Gettys, J. 1992, *The X Window System*, Digital Press, 5

ASTRONOMICAL DATA ANALYSIS SOFTWARE AND SYSTEMS XIV
ASP Conference Series, Vol. 347, 2005
P. L. Shopbell, M. C. Britton, and R. Ebert, eds.

Extending and Customising DS9

De Clarke and Steven L. Allen

UCO/Lick Observatory, University of California, Santa Cruz, CA 95064

William A. Joye

Harvard-Smithsonian CFA, 60 Garden Street, Cambridge, MA 02138

Abstract. UCO/Lick Observatory adopted SAOImage DS9 as the standard real-time and quick-look display for DEIMOS (commissioned Summer 2002) and subsequent instruments. Lick software staff made several modifications and extensions to DS9, taking advantage of its open architecture and the ease of adding Tcl/Tk code to the DS9 core at runtime. We worked in close collaboration with DS9 author Bill Joye on both core modifications and external 'plug-ins'. Our poster describes features added to DS9 (in the core or externally), and a strategy for integrating DS9 with real-time image acquisition. The Lick software team invites other DS9 users to form a community and discuss code-sharing, APIs and other DS9 issues.

1. Introduction

SAOImage DS9[1] is a Tcl/Tk GUI with a powerful image display widget (descendant of saotng[2]), integrating a number of popular libraries for world coordinate system computations, remote image access via HTTP, and other features (Joye & Mandel 2005). DS9 enjoys increasing popularity among astronomers. One attraction of DS9 is its amenability to modification and extension: sites and users can customise it for particular tasks. This poster[3] discusses the customisation and extension of DS9 generally, and our experiences at UCO/Lick specifically.

2. Approaches to Customising/Extending DS9

DS9 can be customised or extended in several different ways.

1. The obvious and traditional method for a large complex application is to get the source code and modify it. In some cases modification of the DS9 core was necessary to add features important to UCO/Lick; however, we refrained from modifying the core ourselves and instead resorted to the equally traditional method of begging the developer for new features. We did this even though DS9 is an open-source application. The reasons were that (a) the core of DS9 is

[1] `http://hea-www.harvard.edu/RD/ds9/`

[2] `http://hea-www.harvard.edu/saord/saotng/`

[3] `http://spg.ucolick.org/Docs/ADASS/2004/ClarkeEtAl/`

written in C++, yacc and lex and hence is somewhat opaque to the uninitiated: we didn't feel it was worth the learning curve given Joye's excellent support; and (b) unofficial core hacks would have to be reapplied with every new release; releases were fairly frequent and with several architectures to support, the tedious housekeeping of applying local patches after each release was not attractive.

2. Joye provides a fairly simple "extend the Analysis menu" toolkit for adding image processing "mini pipelines" to DS9. One can add to the Analysis menu (at startup, via a config file) commands which invoke any command-line image processing tools. DS9 passes the image to the external tool; the output from that operation (as text) can be fed back to a popup window for the DS9 user to review. This is a nice flexible feature, but we wanted more interactivity, more visible on-image user feedback and tools rather than background bit-processing.

3. Joye provides a set of XPA "access points" which enable local or remote processes to exchange image regions with DS9 and to send DS9 commands. This is more powerful than option 2 and more interactive; but we found that not every DS9 internal command we wanted to use had an access point, and we didn't like to keep asking the author to create more and more XPA access points just for us – not when a better alternative was readily available.

4. Our "better alternative" was to get at DS9 directly via Tcl/Tk code. The running DS9 application is a Tk interpreter with many specialised libraries and commands added. Extra Tcl code can easily be loaded at startup time to add new features or modify existing features in the interpreter; or a separate Tk process on the same X display can use Tk `send` to send commands to the DS9 interpreter and thus "drive" it from outside. Since we have many years of Tcl/Tk coding experience we found direct Tcl-hacking to be the quickest and easiest way to modify and extend DS9. The price of this quick, dirty and cheap approach was that our modifications were vulnerable to any changes the author might make to internal DS9 Tcl command syntax. We accepted that price and forged ahead, adding many upgrades and features to the base DS9.

3. Our Experience So Far

Our DS9 project started in 2000 or so when we were looking around for a real time display for DEIMOS, a mosaic-detector spectrograph. Our existing, legacy RTD was incapable of handling mosaic images and was not trivially extensible. When we heard about DS9 and saw it demonstrated at a conference, it seemed like a promising candidate; but at that time, DS9's support for mosaic images was incomplete and there was no provision for live readout. Over the next couple of years we worked closely with author Bill Joye, who provided us with many upgrades of type 1 (above): full support for multi-extension FITS files (MEF), multi-WCS images, FITS files in shared memory segments, etc. These core upgrades (of general utility for many DS9 users) provided the hooks or basic features on which we could build our own site-specific features, which we'll enumerate and discuss below.

3.1. Mosaic images

While DS9 already supported MEF for mosaic images, its heuristics were not general enough to recognise the DEIMOS flavour of MEF. Joye fixed this early in our evaluation period so that we could reassure ourselves by loading faked DEIMOS images from disk.

3.2. Multiple WCS support

Many WCS are implicit in a DEIMOS mosaic image: one pixel system per amplifier (8 to 16) and one "whole mosaic" pixel system, plus sky, mask coordinates (mm), detector surface coordinates etc. DS9 originally offered only a handful of coordinate systems (sky, physical, pixel), but Joye added support for the 26 additional WCS prescribed by the FITS standard (Greisen & Calabretta 2002). This enabled us to display both amplifier coordinates and our whole-mosaic or "pane" coordinates in real time. We are still working on the transforms for mask and wavelength WCS. We could not practically have implemented WCS support outside the core.

3.3. Live Readout

Live readout, which appears a thorny problem, was actually not very difficult to implement from outside the core once there was support for reading FITS images from shared memory (shmem) segments. Our client/server, keyword-based instrument control system permits real-time telemetry and control from Tcl/Tk applications by means of a Tcl extension; this in turn meant that our DS9 could be equipped with a keyword API via one `load` command, and thereafter could monitor the image capture system status and respond to events. The image capture system needed only to be modified to write the current image into shmem instead of a disk file, and to publish the details (the `shmid`, size of shmem segment, the last row read out) during readout.

An external Tk app, very small and lightweight, monitors the exposure subsystem at all times during observing. When it detects the start of an exposure it uses Tk `send` to invoke some Tcl code loaded into DS9 at startup time. The invoked code sends DS9 into a fairly busy state, looping and polling keywords, watching for the start of image readout to shmem. Using the shmid read from a keyword, it invokes internal DS9 Tcl commands (provided by Joye) to read the image headers and update selected rectangles of the image buffer based on the keyword which reports 'last row read out.' The loop is not particularly fast, but the observer sees a decent refresh rate with stripes of image painting at about 1Hz.

Both Tk `send` and preloaded Tcl code are used to implement this feature. The preloaded code (at startup) includes the loading of a Tcl extension (written in C) which adds the realtime keyword API. No site-specific modifications to the core are needed, only invocations of quite general "load from shmem" functions.

3.4. Improved File Chooser

We are very impressed with most of DS9, but have to admit that no one really liked the default file chooser. We had a Tcl-based file chooser of our own, from the legacy RTD, which users did seem to like. Since DS9 is "just Tcl" at the top level, it seemed only natural to try retrofitting the familiar file chooser to the new RTD. This was fairly trivial.

The familiar file chooser was a standalone Tk app which talked to the old RTD via the X server. This mechanism was excised in the new version, and replaced with a few Tk `send` commands which instruct DS9 to load whichever flavour of FITS file has been detected. It was then simple, at startup time, to modify the DS9 top menubar and add a button that would `exec` the standalone chooser. Several benefits were realised: the customised chooser stays up even if DS9 is stopped and restarted, and remains pointed at the same image directory; it remains up after an image is selected and loaded; it shows the user more files at a glance; and it is able to scan the FITS headers for selected keywords on a casual click (without loading into DS9) so that the user can easily search for a particular image or browse for images of interest. Recently it was further customised with the live keyword API so that it can automatically point to the current live image readout directory. All these features save precious observing time.

We have not used any other RTD which could so easily and cheaply have been retrofitted with a custom file chooser.

3.5. Subraster Readout Definition

For a recent instrument project (an upgrade to HIRES), we added a long-desired feature: the user, by means of keywords, can define up to 16 subrasters or regions of interest (ROI), and save only those ROI as a MEF (one HDU per region). If only some areas of the image are of interest, significant savings in disk space can be realised. The definition of these ROI in keyword syntax is cumbersome, so we wanted the user to be able to draw them on the RTD display (as rubberband boxes) and automatically set the relevant keywords.

DS9 offers built-in annotation overlay including rectangular regions, so the rubberband-box mechanism was ours for free. All we had to do was find a way to turn the keyword values into regions (show the user the existing ROI definitions) and vice versa (allow the user to create, resize, delete the regions and rewrite the keywords accordingly). This took little time and made multiple subrasters very easy to use.

3.6. Basic Centroiding and Stats

DS9 comes with no image math or image reduction facility, only markup tools such as contouring, coordinate grids, rulers etc. Our users were accustomed to a very simple, venerable "stats box" utility which allowed them to draw a rectangular region around an object and compute the centroid and some second-order statistics. We took the C code for the algorithm and turned it into a Tcl extension, then added Tcl code into DS9 at startup time that would load the "stats" extension and add commands to the Analysis menu. These commands would invoke more code added to DS9 at startup time, for selecting and computing stats for multiple regions. Further refinements were added later, such as saving and reloading stats boxes, logging, TSV export, and semi-automated statistics gathering on multiple images in one directory.

The stats box required the most coding of our DS9 extensions and enhancements, but the strategy was simple and robust: code any math-intensive functionality in C as a Tcl extension; load the Tcl extension at startup time along with any additional Tcl code and menu customisations. No communication with other processes was required.

4. Conclusions

Our conclusion after three years with DS9 is that our original hopes were justified. It is easy to modify, customise and extend without altering the core distribution. Earlier versions were sometimes difficult to build from source, but Bill Joye accepted improvements to the configuration and make procedures that have streamlined the build process. When we compare the addition of fairly superficial hacks and ancillary applications to DS9, with the task of writing our own RTD or re-writing the venerable legacy RTD, the labour and time savings are significant. Having local Tcl/Tk expertise certainly helps, but is not required: we suspect that many things we are doing with Tcl/Tk "code overlays" at startup time could also be done via the XPA API, by those who don't wish to acquire the Tcl/Tk skills.

Our feeling is that DS9 will become the *de facto* standard for RTD and image analysis for the astronomy community, at least for the next few years. We felt that it was perhaps time for DS9 users to form a coding community, with the goal of addressing the potential problems outlined above: first, to provide support, tutorials and advice for new users, and to help distinguish real bug reports from naive user errors. A community willing to answer newbie questions on a mailing list or web site could protect the small core development team from some of the "costs of success." It also seemed like a good idea to start exploring code-sharing and see if any of our local customisations and enhancements would be useful to others – and what code we might adopt from others in return.

Part of this poster, therefore, is an invitation to DS9 users (advanced and beginner) to join a mailing list (`DS9@ucolick.org`) and start talking to one another.

Acknowledgments. We'd like to thank Bob Kibrick (UCO/Lick) for his patient alpha and beta testing of new DS9 customisations; thanks also to the HIRES and DEIMOS support teams at Keck Observatory who helped users become comfortable with the new RTD; and to the many contributors to DS9 – authors of the several libraries which comprise the core application.

References

Greisen, E.W., & Calabretta, M.R. 2002, A&A, 375, 1061

Joye, W.A., & Mandel, E. 2005, in ASP Conf. Ser., Vol. 347, ADASS XIV, ed. P. L. Shopbell, M. C. Britton, & R. Ebert (San Francisco: ASP), 110

ASTRONOMICAL DATA ANALYSIS SOFTWARE AND SYSTEMS XIV
ASP Conference Series, Vol. 347, 2005
P. L. Shopbell, M. C. Britton, and R. Ebert, eds.

Porting the Starlink Software Collection to GNU Autotools

Norman Gray[1]

Starlink Project, Rutherford Appleton Laboratory, Chilton, Didcot, OX11 0QX, UK. <http://www.astro.gla.ac.uk/users/norman/>

Tim Jenness

Joint Astronomy Centre, 660 North A'ohoku Place, University Park, Hilo, HI 96720, USA. <http://www.jach.hawaii.edu/~timj/>

Alasdair Allan,[1] David S. Berry,[1] Malcolm J. Currie, Peter W. Draper,[1] and Mark B. Taylor[1]

Starlink

Brad Cavanagh

Joint Astronomy Centre

Abstract. The Starlink software collection currently runs on three different Unix platforms and contains around 100 separate software items, totaling 2.5 million lines of code, in a mixture of languages. We have changed the build system from a hand-maintained collection of makefiles with hard-wired OS variants to a scheme involving feature-discovery via GNU Autoconf.

As a result of this work, we have already ported the collection to Mac OS X and Cygwin. This had some unexpected benefits and costs, and valuable lessons.

The Starlink Software Collection (Draper 2005) is a large set of data-reduction applications, written over about two decades in a variety of languages. Until recently the source code was managed, and the products distributed, using a system that worked, but which was rather ad hoc and opaque, and which was correspondingly rather expensive to maintain. For a variety of technical, political, and financial reasons we made the decision to overhaul this build system, resulting in an CVS-plus-autoconf setup which is much more familiar to current users and developers. In Section 1. we elaborate the motivations for this and the problems we faced; in Section 2. we briefly describe the resulting system; and in Section 3. we list some of the lessons we learned, in the hope that these will be of use to other projects considering the same important move.

[1]Alternate affiliations: NG, University of Glasgow; AA, University of Exeter; DSB, University of Central Lancashire; PWD, University of Durham; MBT, University of Bristol

1. Background

Starlink[1] was established originally in the late seventies, as a way of supplying UK astronomy with hardware ('astronomers will never need more than six VAXes...'), naturally along with the data-analysis software to go with it, and the system management to make it all work smoothly.

We switched the hardware and software to Unix in the early nineties, ending up via a sequence of lesser migrations (plus Digital/Compaq rebrandings) supporting the collection on Solaris, Tru64 and RedHat Linux. The port to Unix meant that some software was dropped at this time, and the rest preened quite extensively. The build system we are now moving away from was designed in this period.

Starting in the late nineties, Starlink slimmed down (the diet was not voluntary), dropped the hardware and management provision, and focused on maintaining and developing the large legacy codebase – now renamed the 'Classic' collection, in the best marketing traditions – and on developing new software, mostly in Java, in tune with more recent developments in astronomical software. Though it is not now actively developed, the classic software is still depended upon by astronomers world-wide, as well as being a valuable resource for pipelines (Cavanagh et al. 2003; Currie 2004) and application engines in the VO. For this reason, and because we expect the community will have to take a larger role in its curation, we need to make the classic software as accessible as possible.

The biggest problem with the software was that there was an awful lot of it! Even with ongoing pruning, we have ended up with about 100 components, totaling around 1800 kSLOC written by the project or curated by it, in various languages including Fortran, C, C++, Perl and Tcl/Tk. We add to this another 700 kSLOC of thirdparty code, some tweaked, and all built at the same time. Not included in this total is about 300 kSLOC of Java, built separately. Just to put this in context, David A Wheeler's `SLOCCount` would have us believe that that's worth over $100M and, according to his analysis of the RH7.1 distribution (Wheeler 2001), it appears that 1.8 MSLOC is larger than anything in that distribution except the kernel, Mozilla and XFree86. Busy types, aren't we?

As mentioned above, the build system we had was idiosyncratic, with code scattered amongst the distributed developers, and a set of source and compiled build products maintained by cut-and-paste of template makefiles. This system did work, and was a reasonable design for the early nineties, but it required discipline and effort to maintain, and worked largely because most users obtained their software through pre-built tape and CD distributions, in many cases installed by system managers who were also at that time employed by the Starlink Project (whose complaints could therefore be bought off by donations of beer). As Starlink's role changes, it is important that this software set appear more normal, and be maintainable (without medication) by a broader range of developers.

One of the reasons the build system was odd was because the collection had a large number of conventions about installation locations, plus documentation

---

[1] `http://www.starlink.ac.uk`

and code-generation tools, all of which had to be catered for when building software. In order to avoid major rewrites of the code base, it was necessary to duplicate these conventions and tool support in the new build system.

Portability was not expected to be a major problem (and wasn't in fact), since the software was already being supported on three Unix platforms; as well, some of the older applications had already been through a port from VMS to Unix. Also, the project has benefited from generally good software practices over the years, so the code is mostly clean, conservative, and well documented; the project's long-standing obsession with code-standards has certainly paid off, with interest. Known portability issues were isolated, though these were handled by including platform-specific versions of key routines, selected by the user at build-time; this limited the target platforms to those explicitly supported by the project.

2. The Outcome

With this project now largely complete, we have made a number of significant improvements to the collection.

All the source code is now conveniently available in a single CVS repository, with anonymous access (`:pserver:anonymous@cvs.starlink.ac.uk:/cvs` with password 'starlink').

The collection now uses the GNU autotools throughout. Libtool handles the mind-bending details of building static and shared libraries, with the result that the collection now builds and installs the shared libraries that were too much trouble before. Automake generates the makefiles, respecting all of our installation conventions and generating support for our internal build tools. And autoconf generates the `./configure` scripts which test the capabilities of the build system and adapt the makefiles and other files appropriately. It turned out to be impractical to use standard autoconf and automake, and so the former was extended with new macros, and the latter adapted with new logic. Because we are using automake, the amount of text we have to put into a `Makefile.am` is substantially reduced, with very little cut-and-pasted boilerplate.

As a result of using the autotools, the entire 2.5 MSLOC collection now builds successfully on Linux (RHEL is our current test platform), Solaris, Windows (using Cygwin), and Mac OS X, even though the latter two were not explicit goals of this porting project. It should be portable to other Unix-like systems, and in particular POSIX systems which include an X server.

3. Lessons and Warnings

The Fortran support in autoconf is rather slim. A large proportion of the autoconf extensions addressed the rather inadequate support for Fortran in the standard autoconf distribution, and these new macros – for testing for intrinsics, open specifiers, support for `%val` and the like – will be offered back to the autoconf mainline.

The port to OS X was easier, in some ways, than we expected, partly because the OS X system compiler is a modified GCC. However Apple's GCC installation does not include `g77`, so we needed to install the Fink/OpenDarwin port of that.

This causes a number of linking problems because of slight differences in the code generated by the two GCC back-ends. This is the `restFP/saveFP` problem: see (Gray 2004) for a summary. Be warned: the OS X linker has some very fixed ideas about the organisation of code.

Let's say that again. The OS X linker can in some circumstances resolve library symbols lazily at runtime, and as a result is particularly alert to duplications, and will find more duplicated globals than you thought you had symbols. It doesn't like common blocks at all. And it has very certain opinions about how libraries should be put together. GNU `libtool` can hide many of the details from you, but be prepared to spend time tweaking your library code.

We couldn't automatically convert our old build system to the new one, because it was mostly hand-maintained. But that turned out not to be a problem in fact: disciplined coding in the past meant that packages could generally be ported to the new system with little thought, and this turned out to be a small part of the effort.

The original plan was to autoconf everything with only necessary code changes. However, it was impossible to stop ourselves refactoring and tidying, sometimes rather extensively. This is both a warning to other projects that they won't be able to stop developers doing this, and a benefit, in the sense that a lot of code-hygiene tasks that have been too boring, confusing or risky in the past, become a lot less so when you're adjusting everything anyway. In addition, merely gathering all our code into a coherent repository turned out to be a non-trivial but valuable exercise. For historical reasons, including the fact that systems like CVS were less common ten years ago, the master copies of code were held by the (distributed) developers. As a consequence of the audit, all the code is now in one place, we have identified modules of generated code and the sometimes developer-specific tools used to produce them, and we rescued some code from backup oblivion, as developers retired or left the project. This consolidation overlaps with the refactoring work, since the consequent perception that code is owned in common means that the developers are more willing to criticise each others' code, and with appropriate consultation refactor or fix it.

Starlink is a well-run project with disciplined developers, but we were surprised by the number of individually minor hurdles we had accumulated over a decade, despite our care, enough that working with the code would have been daunting for a non-insider. Given that we last that long, we should probably plan to have another buildsystem overhaul in a decade or so.

We should have bowed to the inevitable, and started patching the autotools earlier, since delaying this meant some functionality had to be implemented twice. Our initial expectation was that we could provide all of our extra boilerplate using autoconf macros (using the now partly deprecated `aclocal` extension mechanism), but as soon as this boilerplate consisted of extra makefile rules, the scripting involved in supporting this became extremely complicated and obviously fragile. Patching automake is not a trivial undertaking, but the resulting increase in robustness, and the control it gave us over the build system, made it worthwhile. Though it is undeniably complicated, the automake code is well written, and reasonably easy to patch. Even if we had not patched it, it would probably be wise, in a project of this size, to keep a repository copy of the automake and autoconf distributions and mandate their use when bootstrapping the source tree; apart from the increase in long-term security, there are a number of potentially nasty version-skew problems that this practice evades. This admittedly adds an extra step for a developer interested in working on CVS sources, but since the bootstrapping of the CVS sources is necessarily elaborate anyway, and since it is only the more sophisticated developers who would work on CVS rather than distributed sources, this is probably not too much of a disadvantage in practice.

As a general point, we can recommend the use of automake. Though there is a rather steep learning curve, and one does naturally suspect it of being dangerously clever, this turns out not to be a problem *as long as* you are willing to build your project the way automake thinks it ought to be built. Since the (GNU) coding standards it implements are both conventional and reasonably sensible, this is not the imposition it appears. However it follows that adapting a pre-existing makefile to automake can be frustrating and hard, and it is better in this case to start again, and produce the much shorter `Makefile.am` automake source file from scratch.

In parallel with this porting project, we worked through the internal bureaucracy involved in applying a GPL licence to the code where possible. This turned out to be harder than we expected, largely because of the donations, in the past, of code written in gentler times, with bizarre licences (such as 'public domain, for academic use only'!). We ended up conceding defeat on this, as it would clearly be a huge effort to work through all the contributed code, identifying first authors then consistent conditions for each module. We have adopted the pragmatic policy of stating that everything in the collection with an explicit CCLRC copyright is GPL; users concerned about the remainder should get in touch with us and we'll try to work it out.

The project took a *lot* longer than we expected (surprise!). It took around six person-months to get the initial system up and running, and then another six to adapt the bulk of our code to the new system.

References

Cavanagh, B., Hirst, P., Jenness, T., Economou, F., Currie, M. J., Todd, S., & Ryder, S. D. 2003, in ASP Conf. Ser., Vol. 295, ADASS XII, ed. H. E. Payne, R. I. Jedrzejewski, & R. N. Hook (San Francisco: ASP), 237

Currie, M. J. 2004, in ASP Conf. Ser., Vol. 314, ADASS XIII, ed. F. Ochsenbein, M. Allen, & D. Egret (San Francisco: ASP), 409

Draper, P. W., et al. 2005, in ASP Conf. Ser., Vol. 347, ADASS XIV, ed. P. L. Shopbell, M. C. Britton, & R. Ebert (San Francisco: ASP), 22

Gray, Norman 2004, "OSX and the restFP/saveFP problem", `http://www.astro.gla.ac.uk/users/norman/note/2004/restFP/` (cited November 2004)

Wheeler, D. A. 2001 (updated 2002), "More than a Gigabuck: Estimating GNU/Linux's Size", `http://www.dwheeler.com/sloc/redhat71-v1/redhat71sloc.html` (cited November 2004)

ASTRONOMICAL DATA ANALYSIS SOFTWARE AND SYSTEMS XIV
ASP Conference Series, Vol. 347, 2005
P. L. Shopbell, M. C. Britton, and R. Ebert, eds.

Requirements for a Future Astronomical Data Analysis Environment

Preben Grosbøl and Klaus Banse

European Southern Observatory, Germany

Doug Tody, William Cotton, and Tim J. Cornwell

National Radio Astronomical Observatory, USA

Daniel Ponz

European Space Agency, Spain

Janne Ignatius

CSC, Finnish IT Center for Science, Finland

Peter Linde

Lund Observatory, Sweden

Thijs van der Hulst

Kapteyn Astronomical Institute, Netherlands

Vadim Burwitz

Max-Planck-Institut für extraterrestrische Physik, Germany

David Giaretta

Starlink Project, United Kingdom

Fabio Pasian and Bianca Garilli

Istituto Nazionale di Astrofisica, Italy

William Pence

National Aeronautics and Space Administration, USA

Dick Shaw

National Optical Astronomical Observatory, USA

Abstract. Most of the systems currently used to analyze astronomical data were designed and implemented more than a decade ago. Although they still are very useful for analysis, one often would like a better interface to newer concepts like archives, Virtual Observatories and GRID. Further, incompatibilities between most of the current systems with respect to control language and semantics make it cumbersome to mix applications from different origins.

An OPTICON Network, funded by the Sixth Framework Programme of the European Commission, started this year to discuss high-level needs for an astronomical data analysis environment which could provide a flexible access to both legacy applications and new astronomical resources. The main objective of the Network is to establish widely accepted requirements and basic design recommendations for such an environment. The hope is that this effort will help other projects, which consider to implement such systems, in collaborating and achieving a common environment.

1. Why a New Common Software Environment?

The major issues facing astronomers who want to analyze their data may be summarized as follows:

- there exist many different systems but:
  - designed several decades ago,
  - largely incompatible with respect to scripting language used
  - complicated to share data between them (due to keyword semantics)
- current systems provide only rather limited interfaces to Web services, Virtual Observatories, archives and databases
- it is difficult to fully exploit available computer resources e.g., GRID

Due to old (often monolithic) designs, the current generation of systems will be difficult to upgrade to remedy the problems listed above. A new environment seems to be the better option. It would have to fulfill the following criteria:

- usage of important legacy applications from current systems
- easy and attractive for users to develop new tasks
- increase the ability to collaborate and of sharing of software
- support easy access to resources such as archives, Virtual Observatories, and GRID computing
- open, stable and well controlled interface specifications
- define minimum implementation but allow full featured version
- specify open, free base system but also provide interface to commercial software

2. High Level requirements

The success of a system/environment for data analysis is closely linked to the following attributes or points:

stability: Users have to be sure that their 'investment' in learning and using an environment also pays back in the long run.

well tested system: Although the system must be subject to its own regression tests, only actual users will be able to find subtle issues in application procedures.

availability of new state-of-the-art applications: This is only achievable if it's easy and attractive for users to develop new tasks in it.

support: Even with good documentation, high stability and excellent testing, the surrounding software world is not fixed demanding a constant although limited support.

collaboration: Modern science is often done in international teams. Features to make such collaborations as easy as possible must be provided.

up-to-date features: Error propagation and hypothesis testing are examples of features which a modern environment must support. Also easy access to compare results with models is important.

ease of installation: The environment must be trivial to install on most commonly used systems to ensure it can be used by 'normal' astronomers.

compatibility: Future data analysis will rely on access to Virtual Observatory data and Grid computer resources which must be fully supported.

3. What defines a new Environment?

The concept of what an environment is differs significantly depending on whom you talk to. In this context, an environment must have the following properties:

- execute tasks in a transparent way so that users can select easily the appropriate computing resources e.g., desktop, GRID or supercomputer
- provide a standard for parsing information between tasks
- offer a high level scripting language for flow control
- access data transparently no matter where they are located such as in databases or through Virtual Observatories
- make a set of standard services available such as display of data, define a standard interface to services provided

A UNIX system with a standard shell would satisfy many of these criteria although arguably not in an optimal way. The issue of defining a new environment may well be to find the minimum acceptable one.

4. From Dream to Reality

High level requirements, beautiful concepts and solid architectural designs are always good to have but not worth much if they are not realized in an actual implementation. The OPTICON Network is currently only funded to establish the high level definitions but several organizations and projects have concrete plans for implementations of such an environment for data analysis, for example:

ALMA: There is an explicit need for providing an off-line data analysis environment for ALMA data.

ESO: The entry of Finland into ESO has made it possible to start a pilot project with the aim of providing better tools for the analysis of ESO data to its community.

The OPTICON Network offers a forum where general requirements and design concepts can be discussed as it includes people associate to these and other similar projects. There is a strong feeling within the Network that a global environment with identical interfaces must be established although implementation

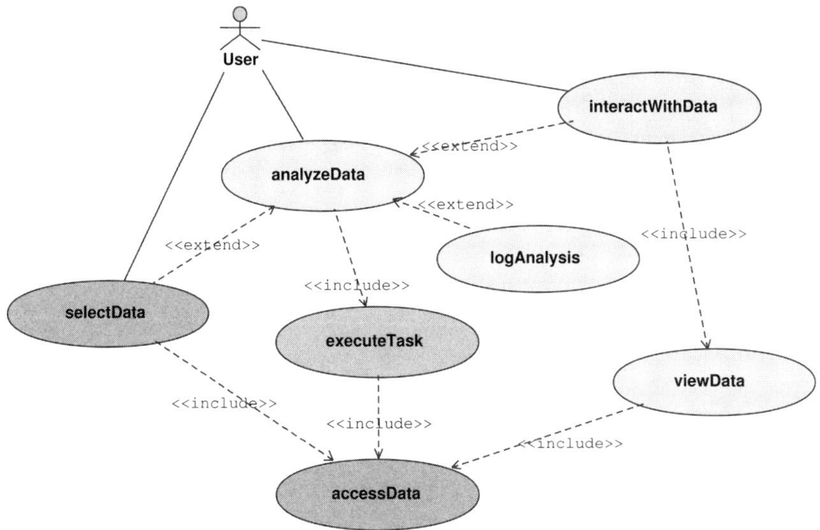

Figure 1. UML diagram of typical Use Case

details may differ. However, in the end only the funding organizations can make this wish come through.

5. OPTICON Network on Future Astronomical Software Environments

OPTICON (Optical Infrared Coordination Network for Astronomy) includes a working group to discuss Future Astronomical Software Environments (FASE) for data analysis. The high level objectives for the FASE working group are:

- identify areas for which a common FASE is desirable and feasible
- establish high level requirements for a FASE
- draft interface and design recommendations for these areas
- learn from past experience

The network has currently 15 members from European institutes and 6 associated members from USA.

6. Top level User Scenario

Several user scenarios were considered including data analysis and development of tasks by both teams and individuals. A typical case for an astronomical user is shown as an UML diagram in Fig. 1. The Use Cases are summarized as:

analyzeData: provides a high level scripting language for controlling the flow of the data analysis that is execution of specific tasks

selectData: lets the user view possible data sets and select appropriate ones

interactWithData: offers the ability to interact with the data by graphics or other means

logAnalysis: logs all actions and results providing the user with a comprehensive view over what has been done

executeTask: executes individual tasks either locally or in a Grid context

viewData: displays data to the user

accessData: provides transparent access to data either locally or at remote sites through the Virtual Observatory interface.

Several of these Use Cases (e.g., accessData, selectData) will be based on the concepts and tools provided by Virtual Observatories while others (e.g., executeTask) will use parts of Grid technology.

7. How can you give your input?

It is important that potential users of the environment participate as much as possible in laying down its requirements, both scientific and technical. Although one would not be able to satisfy all wishes, it is essential to have a detailed view of what the community would like to see and work with. The two main channels for this are discussions on:

- the TWiki: `<http://archive.eso.org/opticon/twiki/bin/view/Main>`
- the Majordomo e-mail list `<fase@eso.org>`

Acknowledgments. This work is supported by OPTICON which is funded by the European Commission under Contract no. RII3-CT-2004-001566.

ASTRONOMICAL DATA ANALYSIS SOFTWARE AND SYSTEMS XIV
ASP Conference Series, Vol. 347, 2005
P. L. Shopbell, M. C. Britton, and R. Ebert, eds.

Multidrizzle: Automated Image Combination and Cosmic-Ray Identification Software

Robert Jedrzejewski, Warren Hack, Chris Hanley, Ivo Busko, and Anton M. Koekemoer

Space Telescope Science Institute, Baltimore, MD 21218

Abstract. MultiDrizzle, a task which automatically combines images taken at different pointings and removes cosmic rays and other image defects, is now in use in the Advanced Camera for Surveys calibration pipeline at STScI. This poster describes the motivation for the development of this capability, details the steps taken in MultiDrizzle processing and describes the design decisions made in implementing MultiDrizzle in the automatic HST calibration pipeline. Interested users can go to `http://stsdas.stsci.edu/multidrizzle` for more information.

1. Motivation

Since its introduction in March 2002, the ACS calibration pipeline has combined images taken at slightly different pointings using the PyDrizzle[1] software. This Python application automates the process of combining ACS images taken at different dither positions. While this has simplified the task of running the Drizzle task (see Fruchter & Hook 2002), it was not designed to do anything more than combine images that were already assumed to have been cleaned of cosmic rays and other image defects. However, many ACS users started specifying observing programs that took only one image at each dither position, assuming that the cosmic ray identification and removal could be done in post-pipeline processing. While this is true (see the Dither Handbook[2] (Koekemoer et al. 2002a) for details), the process is time-consuming and presents a bewildering set of calibration options to the unwary user.

2. History

The development effort for MultiDrizzle is led by Anton Koekemoer, who wrote the prototype version (Koekemoer et al. 2002b) in consultation with A. Fruchter and other members of the Dither Working Group at STScI. It was conceptually based on earlier tasks written individually by Koekemoer, Fruchter, and others, all of which aimed to simplify the interface to the tasks needed for processing dithered data.

[1] `http://www.stsci.edu/resources/software_hardware/pydrizzle`

[2] `http://www.stsci.edu/instruments/wfpc2/Wfpc2_driz/dither_handbook.html`

The fundamental goal behind the design of MultiDrizzle was to provide a seamless "one touch" interface to the variety of dither-related routines, while also providing powerful flexibility through a range of parameters. The prototype was written in Python, acting as an interface to the various IRAF cl scripts and tasks set out in the Dither Handbook (Koekemoer et al. 2002a):

- Remove sky background from the individual images
- Remove static defects from individual images
- Determine output WCS by examining individual image headers
- Drizzle each image onto individual aligned outputs
- Median-filter images to generate cleaned reference image
- Blot median image back to geometry of each input image
- Compare each input image with blotted reference, masking significantly discrepant pixels
- Drizzle each masked input image onto a final output product

The prototype version was released as part of STSDAS 3.2 in March 2004. This version then served as a baseline for a completely redesigned version by the Science Software Branch at STScI, aiming to retain the fundamental functionality, but making the code more robust and streamlined. The goal of this version was to run in the automatic HST calibration pipeline for ACS images. As such, the requirements were:

- Reduce (or remove) dependence on IRAF tasks
- Perform most image manipulations in memory rather than reading to and writing from disk
- Provide detector-specific reduction parameters
- Limit total amount of memory required for image combination
- No significant speed reduction compared to prototype code
- Robust enough to handle all anticipated associations

3. Implementation

MultiDrizzle for the pipeline is implemented in Python, with some C extensions for performance-critical sections. Most of the calls to IRAF tasks in the prototype version of MultiDrizzle involved either array manipulation (for example, determination of the sky background by iteratively clipping the image histogram, and combining the images using either the median task or some concatenation of imcalc calls) or header keyword getting/setting. This was handled in the redesigned version by using numarray objects to represent the data and bad pixel masks, and by using PyFITS to handle file I/O and header keyword manipulation. This also has the benefit of allowing image operations to be done in memory, so that numerous read/write operations are unnecessary and makes the redesigned version significantly faster than the prototype.

Some image operations could not be handled efficiently by using pure Python code operating on numarray objects; for these operations (image statistics, image combination), C extensions were written. The total memory usage was limited by using an image iterator that operates on only part of an image at once, so that, for example, in the image combination step, it is not necessary to have all of the images in memory at once. This means that the total memory usage is determined by the size of the output image, not by the number of images that are to be combined (which can grow to be very large in the case of programs like the Ultra-Deep Field).

The detector-specific section of the code allows different behavior depending on whether the detector is the Wide Field Channel, High Resolution Channel or Solar Blind Channel. Using standard object-oriented methods, the detector-specific code could be isolated, making it relatively straightforward to add the capability of processing other instruments than ACS. Other detector-specific behavior includes choosing different reduction parameters depending on the channel in use; this is handled using a reference file that lists the MultiDrizzle parameters for the observational configuration selected. In this way, for example, the sky subtraction could be handled differently for narrow-band images since most data of this type has a negligible background and so sky subtraction is not necessary.

An example of the product that the new MultiDrizzle generates in the HST ACS calibration pipeline is shown in Figure 1.

Since most of the code is either pure Python or C extensions, it can be ported to platforms other than the standard Unix that IRAF runs on. We have successfully built and run MultiDrizzle on a system running Microsoft Windows.

MultiDrizzle was installed in the ACS calibration on September 21, 2004. The acquisition of a new multi-processor system for running the HST operational pipeline and On-the-Fly Reprocessing means that there is enough CPU capacity to support this compute-intensive application.

MultiDrizzle is included in STSDAS 3.3, which was released in November 2004. As well as providing support for ACS images, it also processes WFPC2, STIS imaging and NICMOS data.

References

Fruchter, A. S. & Hook, R. N. 2002, PASP, 114, 144

Koekemoer, A. M. et al. 2002a, *HST Dither Handbook V2.0*, (STScI:Baltimore)

Koekemoer, A. M., Fruchter, A. S., Hook, R. N., & Hack, W. 2002b, HST Calibration Workshop, eds. S. Arribas, A. Koekemoer, & B. Whitmore (STScI:Baltimore), 339

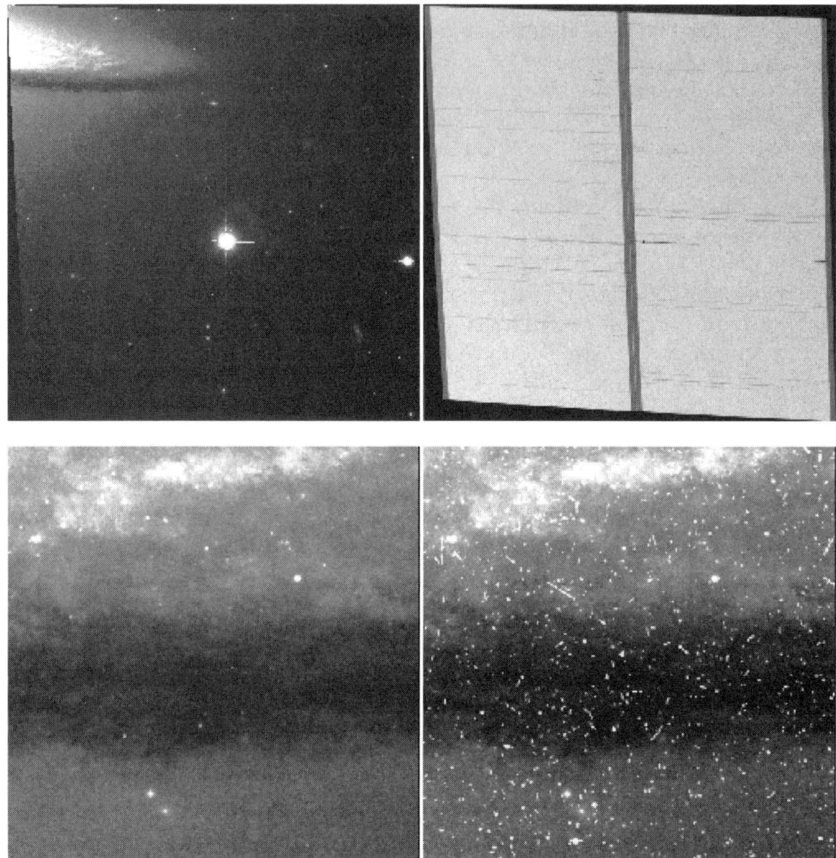

Figure 1. An example of the cleaned MultiDrizzle product from the ACS Archive Pipeline, for a dithered 4-exposure observation of the nearby galaxy NGC 4594. The top panels show the final full-frame drizzled and weight images (left and right, respectively). The bottom left panel shows a close-up of the output image from MultiDrizzle, while the bottom right panel shows the sum of all the accumulated cosmic rays originally present in the exposures

ASTRONOMICAL DATA ANALYSIS SOFTWARE AND SYSTEMS XIV
ASP Conference Series, Vol. 347, 2005
P. L. Shopbell, M. C. Britton, and R. Ebert, eds.

Harmonic Development of an Arbitrary Function of the Moon/Sun/Planets Coordinates to Poisson Series

Sergey M. Kudryavtsev

Sternberg Astronomical Institute, Moscow State University, Universitetsky Pr. 13, Moscow, 119992, Russia

Abstract. A new algorithm for the spectral analysis of an arbitrary function of the Moon/Sun/planets coordinates tabulated over a long period of time is proposed. Expansion of the function to a Poisson series is directly made where the amplitudes and arguments of the series' terms are high-degree time polynomials (as opposed to the classical Fourier analysis where the terms' amplitudes are constants and the arguments are linear functions of time).

1. The algorithm of expansion to Poisson series

Let $f(t)$ be an arbitrary function tabulated by its numerical values over an interval of time $[-T, T]$ with a small sampling step.

Over the same interval we shall build an analytical representation of the function by a finite h-order Poisson series of the following form

$$f(t) \approx \sum_{k=1}^{N} \left[\left(A^c_{k0} + \cdots + A^c_{kh} t^h \right) \cos \omega_k(t) + \left(A^s_{k0} + \cdots + A^s_{kh} t^h \right) \sin \omega_k(t) \right] \quad (1)$$

where $A^c_{k0}, \cdots, A^s_{kh}$ are constants and $\omega_k(t)$ are some pre-defined arguments which are assumed to be q-degree polynomials of an independent variable t

$$\omega_k(t) = \nu_k t + \nu_{k2} t^2 + \cdots + \nu_{kq} t^q. \quad (2)$$

For this we find the projections of $f(t)$ on a basis generated by functions

$$\mathbf{c}_{kl}(t) \equiv t^l \cos \omega_k(t), \quad \mathbf{s}_{kl}(t) \equiv t^l \sin \omega_k(t) \quad (k = 1, 2, \cdots, N; l = 0, 1, \cdots, h) \quad (3)$$

through numerical computation of the following scalar products

$$A^c_{kl} = < f, \mathbf{c}_{kl} > \equiv \frac{1}{2T} \int_{-T}^{T} f(t) t^l \cos \omega_k(t) \chi(t)\, dt \quad (4)$$

$$A^s_{kl} = < f, \mathbf{s}_{kl} > \equiv \frac{1}{2T} \int_{-T}^{T} f(t) t^l \sin \omega_k(t) \chi(t)\, dt \quad (5)$$

by using the definition

$$< f, g > \equiv \frac{1}{2T} \int_{-T}^{T} f(t) \bar{g}(t) \chi(t)\, dt \quad (6)$$

$\bar{g}$ is the complex conjugate to the g function; $\chi(t) = 1 + \cos \frac{\pi}{T} t$ is the Hanning filter chosen as the weight function. The proper choice of arguments $\omega_k(t)$ depends on the specific task (e.g. they can be multipliers of Delaunay arguments and/or planetary mean longitudes, etc.)

However, the basis functions $\mathbf{c}_{k_1 l_1}, \mathbf{s}_{k_1 l_1}, \mathbf{c}_{k_2 l_2}, \mathbf{s}_{k_2 l_2}, \cdots$ are not usually orthogonal. So, we have to perform an orthogonalization process over the expansion coefficients in order to improve the quality of representation of Eq. (1) and avoid superfluous terms. For this procedure we used the algorithm developed by Šidlichovský & Nesvorný (1997). Equations (7)-(13) present the algorithm which we have generalized as indicated below.

Let $f(t)$ be a tabulated complex function and let $\{\mathbf{e}_i\}_{i=1,2,\cdots,M}$ be a set of M basis functions [in our study equal to the complete set of $\mathbf{c}_{kl}(t)$, $\mathbf{s}_{kl}(t)$ so that $M = 2 \times N \times (h+1)$]. The function $f(t)$ is developed in the basis $\{\mathbf{e}\}$ as

$$f(t) = \sum_{i=1}^{M} A_i^{(M)} \mathbf{e}_i + f_M(t) \tag{7}$$

where $A_i^{(M)}$ is a coefficient at $\mathbf{e}_i$ after expanding $f(t)$ over M basis functions, and $f_M(t)$, the difference between the original function and its approximation by M terms, proves to be minimal. Let us define the projections $F_i \equiv < f_{i-1}, \mathbf{e}_i >$ and $Q_{ij} \equiv < \mathbf{e}_i, \mathbf{e}_j >$. The original algorithm by Šidlichovský & Nesvorný (1997) employs a certain normalized basis $\{\mathbf{e}\}$ where the latter scalar product (Q_{ij}) is always a real-valued function. We here expand their result to the case of an arbitrary non-normalized basis $\{\mathbf{e}\}$ where Q_{ij} can take complex values as well.

Thus coefficients $A_i^{(M)}$ are iteratively calculated as follows. At the first step

$$\alpha_{11} = \frac{1}{\sqrt{Q_{11}}}; \quad A_1^{(1)} = \alpha_{11}^2 F_1; \quad f_1(t) = f_0(t) - A_1^{(1)} \mathbf{e}_1 \tag{8}$$

where $f_0(t) \equiv f(t)$ and α_{ij} are hereafter some calculated complex constants.

At the mth step, for every $j = 1, 2, \cdots, m-1$ we compute the following complex coefficients:

$$B_j^{(m)} = -\sum_{s=1}^{j} \bar{\alpha}_{js} Q_{ms} \tag{9}$$

$$\alpha_{mm} = \left(Q_{mm} - \sum_{s=1}^{m-1} \bar{B}_s^{(m)} B_s^{(m)} \right)^{-\frac{1}{2}} \tag{10}$$

(by construction the coefficient α_{mm} takes on a real value for every m)

$$\alpha_{mj} = \alpha_{mm} \sum_{s=j}^{m-1} B_s^{(m)} \alpha_{sj} \tag{11}$$

$$A_m^{(m)} = \alpha_{mm}^2 F_m; \quad A_j^{(m)} = A_j^{(m-1)} + \alpha_{mm} \alpha_{mj} F_m \tag{12}$$

$$f_m(t) = f_{m-1}(t) - \alpha_{mm} F_m \sum_{i=1}^{m} \alpha_{mi} \mathbf{e}_i \tag{13}$$

where $\bar{\alpha}_{js}$ and $\bar{B}_s^{(m)}$ are complex conjugate values of the relevant quantities.

For the selected basis of Eq. (3), the projections $F_i \equiv < f_{i-1}, \mathbf{e}_i >$ are numerically calculated according to Eqs. (4), (5). The values for scalar products of the basis functions $Q_{ij} \equiv < \mathbf{e}_i, \mathbf{e}_j >$ can be found analytically through the following steps.

Step 1. As far as trigonometric functions can be represented in exponential form, we shall further deal with definite integrals of the form

$$I_n(\nu) \equiv \frac{1}{2T}\int_{-T}^{T} t^n e^{i\nu t}\left(1+\cos\frac{\pi}{T}t\right)dt = I_n^a(\nu)+I_n^b(\nu)+I_n^c(\nu) \tag{14}$$

where $i \equiv \sqrt{-1}$ and

$$I_n^a(\nu) \equiv \frac{1}{2T}\int_{-T}^{T} t^n e^{i\nu t}dt \tag{15}$$

$$I_n^b(\nu) \equiv \frac{1}{4T}\int_{-T}^{T} t^n e^{i(\nu+\frac{\pi}{T})t}dt \tag{16}$$

$$I_n^c(\nu) \equiv \frac{1}{4T}\int_{-T}^{T} t^n e^{i(\nu-\frac{\pi}{T})t}dt \tag{17}$$

It is easy to find

$$I_n^a(0) = 2I_n^b\left(-\frac{\pi}{T}\right) = 2I_n^c\left(\frac{\pi}{T}\right) = \begin{cases} 0 & \text{if } n \text{ is odd} \\ \frac{T^n}{n+1} & \text{if } n \text{ is even} \end{cases} \tag{18}$$

Otherwise, if $n = 0$

$$I_0^a(\nu) = \frac{\sin(\nu T)}{\nu T}; \quad I_0^b(\nu) = -\frac{\sin(\nu T)}{2(\nu T+\pi)}; \quad I_0^c(\nu) = -\frac{\sin(\nu T)}{2(\nu T-\pi)} \tag{19}$$

If $n \geq 1$ we calculate the integrals iteratively

$$I_n^a(\nu) = \frac{i}{\nu T}\left(TnI_{n-1}^a - T^n\psi(\nu)\right) \tag{20}$$

$$I_n^b(\nu) = \frac{i}{\nu T+\pi}\left(TnI_{n-1}^b + \frac{1}{2}T^n\psi(\nu)\right) \tag{21}$$

$$I_n^c(\nu) = \frac{i}{\nu T-\pi}\left(TnI_{n-1}^c + \frac{1}{2}T^n\psi(\nu)\right) \tag{22}$$

where

$$\psi(\nu) = \begin{cases} \cos\nu T & \text{if } n \text{ is odd} \\ i\sin\nu T & \text{if } n \text{ is even} \end{cases} \tag{23}$$

Step 2. We partially expand the exponential function of the argument of Eq. (2) to a power series of t by assuming smallness of the second and further items in the right-hand side of Eq. (2) with respect to the first term. (In particular, it is true for all Delaunay arguments and planetary mean longitudes frequencies.) This task can easily be performed by means of a computer algebra system. (We have used the MAPLE V program package by Waterloo Maple Software.) The result is as follows (where the maximal degree of the polynomial argument, q, has been restricted to a value of 4)

$$e^{i(\nu t+\nu_2 t^2+\nu_3 t^3+\nu_4 t^4)} = e^{i\nu t}\left[1 + i\nu_2 t^2 + i\nu_3 t^3 + \left(i\nu_4 - \frac{1}{2}\nu_2^2\right) t^4 + \cdots\right] \tag{24}$$

In our study we obtain an expansion [Eq. (24)] up to terms proportional to t^{24}. Then we define a new integral function of a polynomial argument $\omega(t)$ as

$$J_n(\omega) \equiv \frac{1}{2T}\int_{-T}^{T} t^n e^{i(\nu t+\nu_2 t^2+\nu_3 t^3+\nu_4 t^4)}\left(1+\cos\frac{\pi}{T}t\right) dt \tag{25}$$

By combining Eqs. (14) and (24) we can write

$$J_n(\omega) = I_n(\nu) + i\nu_2 I_{n+2}(\nu) + i\nu_3 I_{n+3}(\nu) + \left(i\nu_4 - \frac{1}{2}\nu_2^2\right) I_{n+4}(\nu) + \cdots \tag{26}$$

Step 3. We expand trigonometric functions to exponential ones, and then the required scalar products can be calculated as

$$\begin{aligned} < \mathbf{c}_{k_1 l_1}, \mathbf{s}_{k_2 l_2} >= & \frac{1}{2T}\int_{-T}^{T} t^{l_1+l_2}\left(\frac{e^{i\omega k_1}+e^{-i\omega k_1}}{2}\right)\left(\frac{e^{i\omega k_2}-e^{-i\omega k_2}}{2i}\right) \\ & \times\left(1+\cos\frac{\pi}{T}t\right) dt \\ = & \frac{1}{4i}\{-J_{l_1+l_2}(\omega_{k_1}-\omega_{k_2}) - J_{l_1+l_2}(-\omega_{k_1}-\omega_{k_2}) \\ & +J_{l_1+l_2}(\omega_{k_1}+\omega_{k_2}) + J_{l_1+l_2}(-\omega_{k_1}+\omega_{k_2})\} \end{aligned} \tag{27}$$

and analogously

$$\begin{aligned} < \mathbf{s}_{k_1 l_1}, \mathbf{c}_{k_2 l_2} >= & \frac{1}{4i}\{J_{l_1+l_2}(\omega_{k_1}-\omega_{k_2}) - J_{l_1+l_2}(-\omega_{k_1}-\omega_{k_2}) \\ & +J_{l_1+l_2}(\omega_{k_1}+\omega_{k_2}) - J_{l_1+l_2}(-\omega_{k_1}+\omega_{k_2})\} \end{aligned} \tag{28}$$

$$\begin{aligned} < \mathbf{s}_{k_1 l_1}, \mathbf{s}_{k_2 l_2} >= & \frac{1}{4}\{J_{l_1+l_2}(\omega_{k_1}-\omega_{k_2}) - J_{l_1+l_2}(-\omega_{k_1}-\omega_{k_2}) \\ & -J_{l_1+l_2}(\omega_{k_1}+\omega_{k_2}) + J_{l_1+l_2}(-\omega_{k_1}+\omega_{k_2})\} \end{aligned} \tag{29}$$

$$\begin{aligned} < \mathbf{c}_{k_1 l_1}, \mathbf{c}_{k_2 l_2} >= & \frac{1}{4}\{J_{l_1+l_2}(\omega_{k_1}-\omega_{k_2}) + J_{l_1+l_2}(-\omega_{k_1}-\omega_{k_2}) \\ & +J_{l_1+l_2}(\omega_{k_1}+\omega_{k_2}) + J_{l_1+l_2}(-\omega_{k_1}+\omega_{k_2})\} \end{aligned} \tag{30}$$

This is the technique of spectral analysis allowing expansion of an arbitrary tabulated function of the Sun/Moon/planets coordinates to Poisson series, with the arguments being high-degree polynomials of time.

2. Testing the algorithm

The proposed method leads to an improvement in accuracy of harmonic development of a tabulated function over a long-term interval and essential reduction of the approximation series' length.

To test the algorithm, we calculated the Earth-Moon distance on every day within [1000BC-5000AD] by using the ELP2000-85 analytical theory of lunar motion (Chapront-Touzé & Chapront 1988), and then made the spectral analysis of the tabulated values by using the new algorithm. As a result, all coefficients of the second-order Poisson series representing the lunar distance in the ELP2000-85 theory (320 terms) have been successfully found by our algorithm. The maximum deviation of the lunar distance calculated by the reconstructed series from that given by the exact ELP2000-85 model does not exceed 1.5 centimeters over the whole interval of six thousand years.

3. Examples of application of the new method

- Harmonic development of the Earth tide-generating potential (TGP) by Kudryavtsev (2004). The resulting accuracy of the new TGP development in the time domain exceeds that of any known harmonic development of the TGP by a factor of at least 3;
- High-accurate analytical expansion of the perturbing function due to the Moon, Sun and planets in satellite motion task (Kudryavtsev 2002);
- New analytical representation of the Moon ephemeris adjusted to LE-405/406 (Standish 1998) numerical ephemeris (under development).

Acknowledgments. The financial aid provided to the author from the ADASS XIV Organizing Committee is sincerely acknowledged. The work is supported in part by grant 02-02-16887 from the Russian Foundation for Basic Research.

References

Chapront-Touzé , M., & Chapront, J. 1988, A&A, 190, 342

Kudryavtsev, S. 2002, IAA Transactions, 8, 112

Kudryavtsev, S. 2004, J. Geodesy, 78, 829

Šidlichovský, M., & Nesvorný, D. 1997, Cel. Mech. Dyn. Astr. , 65, 137

Standish, E.M. 1998, JPL IOM 312.F-98-048

ASTRONOMICAL DATA ANALYSIS SOFTWARE AND SYSTEMS XIV
ASP Conference Series, Vol. 347, 2005
P. L. Shopbell, M. C. Britton, and R. Ebert, eds.

aXedrizzle - Spectral 2D Resampling using Drizzle

Martin W. Kümmel, J. R. Walsh, S. S. Larsen, and R. N. Hook

Space Telescope – European Coordinating Facility,
Karl-Schwarzschild-Str. 2, D-85748 Garching b. München, Germany

Abstract. The aXe spectral extraction software was designed to extract spectra from slitless grism images such as those taken with the Advanced Camera for Surveys on HST. In aXe version 1.4 we implemented aXedrizzle, a new technique to resample 2D spectra which uses the "traditional" drizzle software. aXedrizzle forms deep 2D spectra by coadding the 2D spectra from the individual, dithered images. This method has several advantages compared to the approach of making the 1D extraction on each image and then coadding the 1D spectra. aXedrizzle is the first implementation of the drizzle code to combine spectral data, and the technique can also be applied in other reduction packages for spectral data.

1. Slitless Spectroscopy on the Hubble Space Telescope (HST)

After the demise of STIS (Space Telescope Imaging Spectrograph), the Advanced Camera for Surveys (ACS) is currently the only instrument on the HST which is capable of doing optical and UV spectroscopy. The ACS has three channels (WFC – Wide Field Channel, HRC – High Resolution Channel, and SBC – Solar Blind Channel), and each channel is capable of delivering slitless spectroscopic images by inserting a grism (G800L) or prisms (PR110L, PR130L, and PR200L) into the optical beam. In total, the five possible combinations of ACS channel and dispersing element offer low resolution ($R \sim 100$) spectroscopy from the UV to the far-red wavelength regime. Table 1 gives an overview of the spectroscopic modes of ACS and lists important parameters such as wavelength range, spectral resolution, pixel size, and field of view (FOV).

2. The aXe Spectral Extraction Software

The Space Telescope – European Coordinating Facility (ST-ECF) is responsible for the support of the spectroscopic modes of ACS. A key component of this support is the development of the aXe software package. aXe is a spectroscopic data extraction software package that was designed to handle large format spectroscopic slitless images such as from the ACS. It is distributed as part of the IRAF/STSDAS software package in the subpackage "stsdas.hst_calib.acs.axe". An earlier version (aXe-1.1) was presented at ADASS XII (Pirzkal et al. 2002).

As data input, aXe needs a grism/prism image, a corresponding direct image, and a catalogue which lists the objects detected on the direct image. Driven by the object catalogue, the various aXe tasks extract wavelength and flux calibrated 1D spectra for each object from the grism image.

Datasets, such as those obtained with ACS, often consist of several images with small position shifts (dithers) between them. The direct approach of co-adding the 1D spectra extracted from each slitless image to form a combined, deep spectrum has several disadvantages:

- The data must be rebinned twice, once when extracting the spectrum from the image and again when combining the individual 1D spectra.
- A complex weighting scheme is required to transport the number and contribution of cosmic ray affected pixels through the summation in the cross-dispersion direction and the combination of the individual 1D spectra.
- Low level information on the cross dispersion profile is lost when many 1D extracted spectra are combined to a deep spectrum. Also problem detection and error tracking is more difficult on a 1D than a 2D spectrum.

Channel	Disperser	Wav. Range [Å]	Resolution [Å/pixel]	Pixel Size [mas/pixel]	FOV [arcsecond]
WFC	G800L	5500 – 10500	38.5	50×50	202×202
HRC	G800L	5500 – 10500	23.5	28×24	29×26
HRC	PR200L	1600 – 3900	20[@2500Å]	28×24	29×26
SBC	PR130L	1250 – 1800	7[@1500Å]	34×30	35×31
HRC	PR110L	1150 – 1800	10[@1500Å]	34×30	35×31

Table 1. The spectroscopic modes of ACS.

3. aXedrizzle

Extracting the 1D spectra from MultiDrizzle-resampled images shown in Figure 1 does not deliver optimal results. The description of the spectra, e.g., the trace angle α in Figure 1, as well as the spectral wavelength solution vary with the object position on the chip. The MultiDrizzle procedure corrects the spatial geometrical distortion of the ACS images but does not correctly handle the spatial variation of the dispersion for slitless spectral images. Applying the corrections for ACS geometric distortion leads to distortion of the spectra but without beneficial effects on the dispersion solution.

It is possible, however, to correct the field dependence of the slitless spectra in local solutions, where only a small part of the 2D spectral image is resampled. With our new aXedrizzle technique a deep, co-added 2D spectral image is formed for each object by resampling all spectra in the individual images. For the resampling and co-addition of the individual spectra we use the "traditional" drizzle software (Fruchter & Hook 2002)

The advantages of aXedrizzle as applied to slitless spectra can be summarized:

- Regridding to a uniform wavelength scale and a cross-dispersion direction orthogonal to the dispersion direction is achieved in a single step.
- Weighting of different exposure times per pixel and cosmic-ray affected pixels are correctly handled through the drizzle weights.
- The combined 2D spectra can be viewed to detect problems and the deep spectra can reveal fainter features which were subsumed in the 1D spectra.

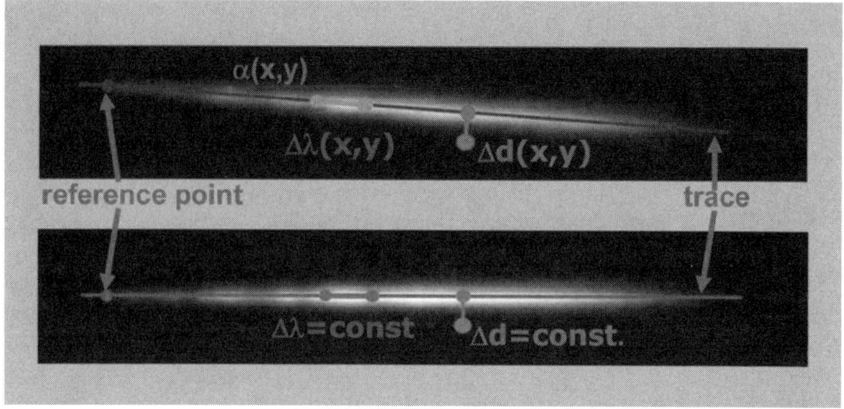

Figure 1. A 2D spectrum before (top) and after (bottom) drizzle resampling.

To illustrate this scheme, Figure. 1 shows the slitless spectrum of a bright, $F850W = 20.6\,\mathrm{mag}_{AB}$ late type star before and after the resampling, in the upper and lower panels, respectively. In the upper panel the trace angle α, the wavelength solution (expressed as wavelength increment $\Delta\lambda$), and the pixel scale in the cross-dispersion direction (Δd) all vary with the object position on the chip. For all the 2D spectral images of this object on the various grism images a set of individual drizzle coefficients is derived to drizzle each individual 2D spectrum onto a single, common drizzled image as seen in the lower panel. The drizzle coefficients are computed such that the combined 2D grism image has a constant wavelength dispersion and a constant pixel scale in the cross-dispersion direction. After the resampling the trace angle is zero, and the combined drizzle image resembles an ideal (distortion-free) long slit spectrum with the dispersion direction parallel to the x-axis and cross-dispersion direction parallel to the y-axis. The final extraction of the 1D spectra from the deep, coadded 2D grism images is done using aXe tools or standard long slit extraction tasks.

aXedrizzle is the first implementation of the drizzle code to combine spectral data. In principle, it is also possible to apply the aXedrizzle resampling scheme and the aXe software to ground based multi-object spectroscopy (MOS). In a pilot study we are currently working on establishing a pipeline to reduce MOS data taken with the FORS2 instrument at the VLT.

4. aXe Version 1.4

The aXedrizzle technique is one of the major extensions of the new aXe version 1.4 (Kümmel et al. 2004a), which was released on October 11, 2004 on the aXe webpage (`http://www.stecf.org/software/aXe`). aXe-1.4 is also part of the new STSDAS 3.3 which was released in November 2004. In a pilot study, the aXedrizzle reduction scheme was just applied to reduce the data from the Hubble Ultra Deep Field (HUDF) HRC parallels (see Kümmel et al. 2004b and the preview pages `http://www.stecf.org/UDF/HRCpreview.html`).

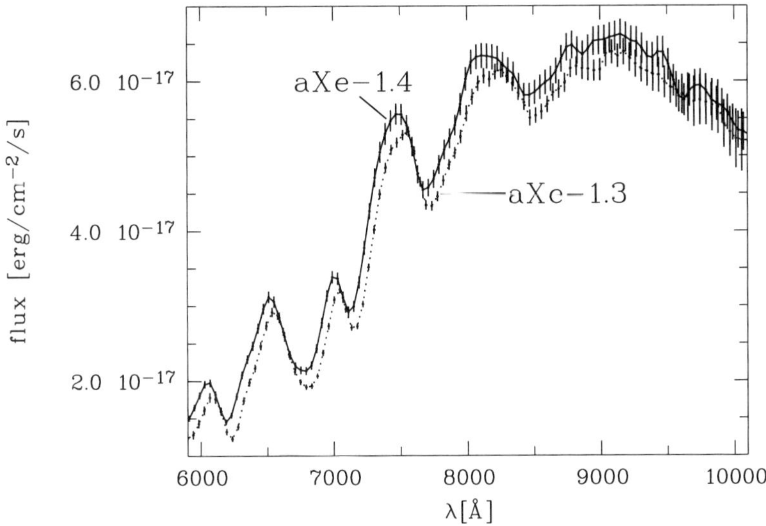

Figure 2. Comparison of aXe-1.3 and aXe-1.4 extractions for the same observed spectrum.

To give a quantitative comparison, Figure 2 shows the spectrum of the star J033242.42-274758.8 ($F850W = 19.1\,\mathrm{mag}_{AB}$), taken in the Grism ACS Program for Extragalactic Science (GRAPES, Pirzkal et al. 2004). While the lower spectrum shows the result of a reduction with aXe-1.3 and subsequent coaddition of the 1D spectra extracted from the individual grism images, the upper line shows the spectrum of the same object as extracted using aXe-1.4. The difference in the flux level and the small wavelength shift are caused by a different width of the extraction box and different wavelength calibration, respectively. The overall agreement of the spectra is very good and the aXe-1.4 reduction provides in addition advantages, such as the combined 2D spectrum for better error detection (see Section 3).

5. With aXe2web to the World Wide Web

Deep ACS WFC grism images can contain spectra of many hundreds of objects, and visual checking of each spectrum is very tedious. For this reason we developed aXe2web (Walsh & Kümmel 2004[1]), a tool which produces browsable web pages for fast, yet thorough, examination of spectra. This additional task to the aXe package produces an html summary using the input catalogue and the aXe output. Each object produces a line in an html table which lists impor-

---

[1]also `http://www.stecf.org/software/aXe/#axe2html`

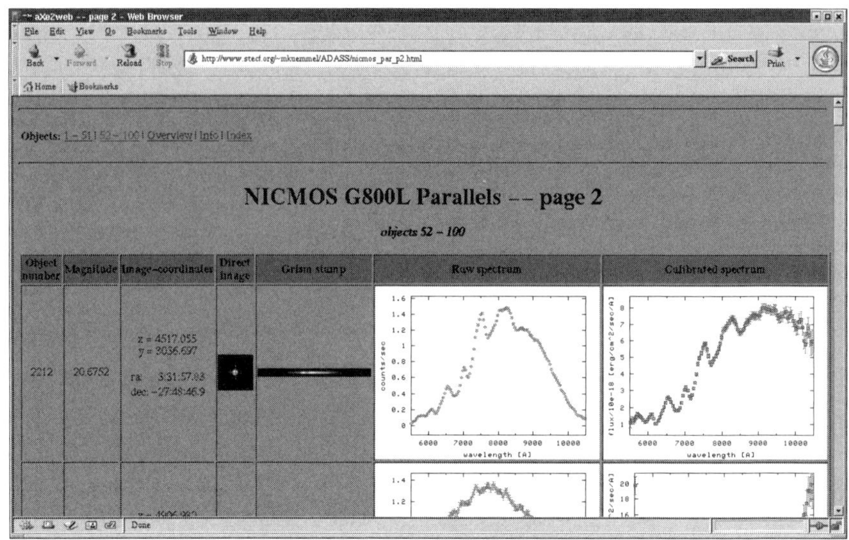

Figure 3. The aXe2web view of J033157.83-274846.9.

tant object parameters (e.g., reference number, Right Ascension, Declination, brightness), a cut-out image showing the direct object, an image showing the 2D spectrum, a 1D extracted spectrum in counts and the same in flux units. Figure 3 shows the line for the object J033157.83-274846.9 (also shown in Figure 2). An overview and an index page accompany the object pages which carry all the information. The overview page contains the basic information for each object. The index page includes a table with the ordered reference number of all objects. Direct links from both the overview page and the index page point to the corresponding locations of the objects in the object pages. With aXe2web, aXe spectra can be easily published on the web without further interactive work.

References

Fruchter, A. S., & Hook, R. N. 2002, PASP, 114, 144

Kümmel, M., Walsh, J., Larsen, S., & Hook, R. 2004a, ST-ECF Newsletter, 37, 14

Kümmel, M., Walsh, J., Larsen, S., & Hook, R. 2004b, ST-ECF Newsletter, 36, 10

Pirzkal, N., Pasquali A., Hook, R.N., Walsh, R., & Albrecht, R. 2002, in ASP Conf. Ser., Vol. 295, ADASS XII, ed. H. E. Payne, R. I. Jedrzejewski, & R. N. Hook (San Francisco: ASP), 485

Pirzkal, N., Xu, C., Malhotra, S., et al. 2004, ApJS, 154, 501

Walsh, J., & Kümmel, M. 2004, ST-ECF Newsletter, 35, 9

ASTRONOMICAL DATA ANALYSIS SOFTWARE AND SYSTEMS XIV
ASP Conference Series, Vol. 347, 2005
P. L. Shopbell, M. C. Britton, and R. Ebert, eds.

A Survey of Software for the Manipulation and Visualization of 3-D Sub-mm Spectral Line Datasets

J. Leech and T. J. Jenness

Joint Astronomy Centre, 660 N. A'ohōkū Place, University Park, Hilo, Hawaii 96720. Email: j.leech@jach.hawaii.edu

Abstract. 3-dimensional spectral line datasets are the end products of observations from single-dish telescopes and interferometers using widely differing instrument technologies over a broad wavelength range. Consequently, a range of software packages exist for the data reduction, analysis and visualization of such 3-D data sets. These packages vary greatly in scope, generality, age, maturity and the availability of support and further development effort. Here we survey a range of relevant software packages emphasising packages suitable for the handling of submillimetre data sets generated by sub-mm focal plane arrays such as HARP/ACSIS on the JCMT. The conclusions are broadly relevant to future software development and re-use for a wide variety of astronomical instruments and wavelength regimes.

1. Introduction

This survey has been motivated by the imminent arrival of a 350 GHz 16-element focal plane array receiver (HARP) and its digital autocorrelation spectrometer (ACSIS) at the JCMT (Dent 2000). In combination, these will generate large ($\sim$ several GByte) data cubes in standard FITS and/or NDF formats. We needed to identify manipulation/visualization software with a short learning curve which we could recommend to users with little or no experience in using traditional astronomical data reduction packages. Thus, this survey excludes the traditional interferometer oriented packages such as AIPS and MIRIAD/GILDAS – the pros and cons of these should be familiar to users with a background in interferometry. Commercial packages with support for astronomical data import/export (e.g., IDL) are not considered here. Traditional spectral-line specific tools with some useful but limited map making capabilities, such as SPECX, CLASS and COMB are also not discussed here.

2. Some Typical User Requirements

- Strong, on-going development and support with open licensing (e.g., GPL).
- Reliable and robust handling of astronomical file formats (e.g., FITS).
- Ability to efficiently handle large data files (e.g., $>$ 4GB).
- Complete handling of astronomical co-ordinates, projections and units.
- Ease of use e.g., for graduate student on first observing run at the JCMT.
- Intuitive user interface, powerful scripting capabilities.
- Radio astronomy friendly - knows about dual sideband receivers, $T_{\rm sys}$ as a measure of expected noise, standard rest frames etc.

- Bad pixel (region) masking.
- Smoothing and decimation of spectral cubes.
- The mosaicking of spectral cubes.
- Spectral baseline fitting and removal across cubes.
- Mathematical operations on entire cubes (and regions).
- Truncation of spectral cubes.
- Display arbitrary 2D slices of 3D cubes (pixel and contour representations).
- Calculation of statistics over cubes (sub-cubes and regions).
- Calculation and display of 2D integrated intensity maps from 3D cubes.
- Calculation and display of higher-order moments e.g., velocity fields.
- Any 3D rendering capability is a plus.

3. Complete DR and Visualization Packages

3.1. AIPS++

AIPS++ (McMullin et al. 2004) is a large, general purpose software package for the manipulation and visualization of astronomical data. It was developed by a consortium of observatories, and saw its first public release in 1999. The consortium was disbanded in 2003 and software effort and support reorganized amongst the member institutions. Support of the central code base remains the responsibility of NRAO and the source code is freely available under the GNU Public License.

On a technical level, AIPS++ consists of a scripting language, glish, which acts as a software "glue" between pre-compiled C++ clients. The GUI is provided via the use of glish bindings to Tk widgets and PGPLOT primitives. The software is organized around the concept of distinct tools which can be created and destroyed via the glish command line or AIPS++ GUI. A wide range of tools exist for handling single dish and interferometric data, as well as general purpose tools for image analysis, astrometry etc. In addition, there are tools to perform telescope specific calibration and visualization tasks.

2D and 3D image manipulation and visualization is handled via the `image` tool. FITS data cubes can be read in and converted to the AIPS++ table-based image format. AIPS++ images can be viewed via the image viewer tool, which enables 2D slice (rasters and contours) and 1-D spectral profile display. Image tool functions include masking, mosaicking, image moment calculation, smoothing and decimation, region selection and statistics finding. Glish scripting provides the flexibility to combine operations to perform powerful, user defined 3D image manipulations.

Overall, the reception of AIPS++ by the astronomical community has been somewhat mixed and widespread adoption slow. It has developed a reputation for being non-user friendly, probably due mostly to the poor integration of the GUI and command line interaction and the lack of conceptual integration between tools. Features such as full support for the draft spectral FITS WCS standards, rotations between pixel and sky frame axes and support for handling spectra from dual-sideband receivers have not yet been implemented. AIPS++ and glish have arguably found more success as a lower level toolkit for the development of facility-specific data reduction software such as the parallel real-time data reduction system for the ACSIS spectrometer, which is built around glish messaging and AIPS++ libraries. Plans for future development of the AIPS++ code base will be targeted more towards application for particular instruments, such as ALMA. The glish scripting language has the disadvantage of a small user community and low manpower availability for core language support and development. Thus, much work is being done to decouple the AIPS++ C++ libraries from glish, so that they can be used with other, powerful, wide user scripting languages, most notably Python. Unfortunately, this focus on core software infrastructure and ALMA-specific goals by the AIPS++ development team means there is less effort for more "user-centric" feature requests and bug fixes. Also, it is hard to recommend AIPS++ to users with the knowledge that the UI (scripting language and GUI) will change rapidly over the next few years.

3.2. Starlink

The Starlink (Draper et al. 2005) Software Collection consists of an array of tools targeted towards astronomical applications at a wide range of wavelengths. Much of the software is mature and has a wide user base, particularly from the UK astronomical community. Recent work includes moving the code base into CVS, simplifying build procedures using autoconf, ensuring Mac OS X compatibility and migration to the GPL license.

One strength of Starlink is its hierarchical data format NDF (N-dimensional Data Format). NDF can support data of arbitrary dimension and can contain variance and data quality arrays in additional to image data. AST data objects describing WCS (World Co-ordinate System) information provide sophisticated astrometry handling. AST also handles spectral co-ordinate axes, such as the one-to-many mappings which arise in dual-sideband heterodyne systems. Starlink includes a `convert` application to handle conversion to and from NDF and other formats such as FITS. `convert` allows one to specify easily the required FITS header convention e.g., classic AIPS, WCS paper III compliant etc. Direct manipulation of 3D NDF data cubes is achieved mostly through the Starlink `kappa` package. It consists of 180 general purpose command-line programs for manipulation, image processing and visualization of NDF data. While `kappa` includes no GUI frontend other than a PGPLOT display window, its commands are reasonably self-explanatory and well suited to use with scripts (e.g., UNIX shell, Perl, etc.).

Starlink offers the `datacube` package to handle higher level, 3D spectral data specific visualizations. `datacube` is a collection of scripts, run from the command line, which use lower-level Starlink applications, such as `kappa`, to achieve visualization tasks. `datacube` allows slice-by-slice steps through 3D cubes, interactive plotting of spectra at given sky positions, as well as the generation of velocity and line-peak maps following interactive Gaussian fitting to spectra within a cube. 3D data cubes from contiguous areas of the sky can be mosaicked together into larger datacubes using `makemos`, part of the Starlink `ccdpack` package. `makemos` can handle 3D NDF data easily and allows the user fine control over what happens to any regions of overlap between component subcubes.

To summarize, applications provided as part of the Starlink collection can meet most of the typical user requirements described above. However, using Starlink software does involve some user-friendliness related disadvantages. Firstly, the required functionality comes from a variety of different software packages, with separate documentation. Secondly, the user interaction with most of the relevant software is almost exclusively via the command line. This second disadvantage is shared by most "traditional" astronomical data reduction packages (AIPS, MIRIAD etc.), and at least maintains flexibility at the expense of a steeper initial learning curve.

4. 3-D Visualization Software

As opposed to general purpose FITS viewers which can display 2D slices of 3D data (e.g., fv, GAIA, and DS9), the following packages offer more sophisticated 3D visualization capabilities.

4.1. Karma

Karma (Gooch 1996) is a package specifically targeted for the visualization of 3D astronomical data sets. It consists of a suite of visualization tools as well as a programmers library and API distributed under the GPL and LGPL licenses. While Karma is mostly the work of single developer, the project is mature and continues to be actively supported. Karma has two main tools useful for visualization, namely `kviz` and `xray`. `kviz` enables 2D slice viewing in raster and contour mode, spectral profile display, image annotation, higher order image moment generation and display, renzogram display and image statistics calculation. `xray` is a tool which enable powerful 3D volume rendering of datacubes. User interaction with Karma is via a fast graphical user interface.

4.2. OpenDX

OpenDX (`http://www.opendx.org`) is the open source version of IBM's Visualization Data Explorer. It is aimed at the 3D visualization of scientific data from a wide range of sources. Users include mechanical and electrical engineers, medical imaging professionals as well as astronomers. OpenDX is available for Unix platforms (inc. Mac OS X and Windows (under Cygwin). Visualizations are specified in OpenDX by preparing a "Visual Program" in a GUI-based Visual Program editor. These visual programs specify the interconnections of data-driven visualization tools. These tools handle primitive visualization operations such as file import, statistics calculation, surface rendering and colour table setup. There is great flexibility in combining these elements to perform a wide variety of visualizations.

For the visualization of 3D astronomical datacubes Starlink provides a variety of Visual Programs for common datacube visualization, such as displaying sequences of 2D slices of 3D cubes and displaying 3D isosurfaces. Starlink also provides software to convert NDF datacubes into the native OpenDX .dx file format. OpenDX is a well featured and flexible program for 3D visualization suitable for astronomers with more specialized visualization needs. The majority of observers will probably continue to look to conceptually simpler, astronomy-specific software packages for the most of their needs for datacube visualization.

5. Useful Scripting Language Extensions

PDL (`http://pdl.perl.org`) is a Perl module, written in C, which gives Perl the ability to manipulate large N-dimensional data sets efficiently. The astronomer-oriented development team provide support for I/O in FITS and NDF formats.

Python has become a popular scripting language for scientific applications. Python modules exist for array manipulation (such as `numeric` and `numarray`) and the PyFITS module enables FITS I/O.

6. Some Concluding Observations

- No single software package meets all of the user requirements in Sec. 2.
- It is clear that a modern GUI which is well integrated with scripting capabilities requires considerable design and implementation effort.
- FITS has the advantage of file format simplicity but it is arguably too easy to develop multiple keywords/conventions. Publication of the FITS Spectral WCS standard is a helpful step here.
- The hierarchical NDF format is more sophisticated than FITS but has a steeper learning curve poor and end-user adoption outside of the UK.
- Much can be learned by comparing user requirements for 3D software across the whole electromagnetic spectrum. Such comparisons will encourage software reuse and prevent a fragmentation of software effort.
- Continuing support and development of a software package is essential.
- The control that open-source licenses such as GPL give the user community over the software is highly derirable.
- The JCMT will recommend Starlink software and Karma to users for routine handling of ACSIS data. Some support for AIPS++ users will exist in the form of cookbooks. Support is expected for users wanting to import ACSIS data into older packages such as AIPS, CLASS and SPECX.
- Future developments include the Euro3D (`http://www.aip.de/Euro3D/`) project which aims "to provide standard software for the visualization and analysis of datacubes" which aims for a first public software release is scheduled for December 2004. The developers stated aim is to produce tools which are "independent of data source", so hopefully the software will be useful at sub-mm/radio as well as optical/IR wavelengths.

References

Dent, W.R.F., et al. 2000, in Imaging at Radio through Submillimeter Wavelengths, ASP Conf. Ser., Vol. 217 (San Francisco:ASP), 33

Draper, P.W., et al. 2005, in ASP Conf. Ser., Vol. 347, ADASS XIV, ed. P. L. Shopbell, M. C. Britton, & R. Ebert (San Francisco: ASP), 22

Gooch, R.E. 1996, in ASP Conf. Ser., Vol. 101, ADASS V, ed. G. H. Jacoby & J. Barnes (San Francisco: ASP), 80

McMullin, J.P., Golap, K., & Myers, S.T. 2004, in ASP Conf. Ser., Vol. 314, ADASS XIII, ed. F. Ochsenbein, M. Allen, & D. Egret (San Francisco: ASP), 468

ASTRONOMICAL DATA ANALYSIS SOFTWARE AND SYSTEMS XIV
ASP Conference Series, Vol. 347, 2005
P. L. Shopbell, M. C. Britton, and R. Ebert, eds.

WAX : A High Performance Spatial Auto-Correlation Application

Serge Monkewitz and Sherry Wheelock

Infrared Processing and Analysis Center, California Institute of Technology, Mail Code 100-22, CA 91125

Abstract. We describe the algorithms employed by WAX, a spatial auto-correlation application written in C and C++ which allows for both rapid grouping of multi-epoch apparitions as well as customizable statistical analysis of generated groups. The grouping algorithm, dubbed the swiss cheese algorithm, is designed to handle diverse input databases ranging from the 2MASS working point source database (an all sky database with relatively little coverage depth) to the 2MASS working calibration source database (a database with sparse but very deep coverage). WAX retrieves apparitions and stores groups directly from and to a DBMS, generating optimized C structures and ESQL/C code based on user defined retrieval and output columns. Furthermore, WAX allows generated groups to be spatially indexed via the HTM scheme and provides fast coverage queries for points and small circular areas on the sky. Finally, WAX operates on a declination based sky subdivision, allowing multiple instances to be run simultaneously and independently, further speeding the process of merging apparitions from very large databases. The Two Micron All Sky Survey will use WAX to create merged apparition catalogs from their working point and calibration source databases, linking generated groups to sources in the already publicly available all-sky catalogs. For a given 2MASS source, this will allow astronomers to examine the properties of many related (and as yet unpublished) 2MASS extractions, and further extends the scientific value of the 2MASS data sets.

1. Overview

The publicly released Two Micron All Sky Survey[1] (2MASS) catalogs, accessible via the NASA/IPAC Infrared Science Archive[2] (IRSA) website, were generated by running source extraction software on image data from many roughly rectangular regions on the sky. Many of these regions (scans) overlap each-other, meaning that a significant number of 2MASS sources were observed more than once (sometimes hundreds or thousands of times). In the 2MASS catalogs, multiple extractions (apparitions) corresponding to a single source were resolved by picking a *best*[3] apparition for catalog membership. Consequently, data for many

[1]`http://www.ipac.caltech.edu/2mass/overview/access.html`

[2]`http://irsa.ipac.caltech.edu/`

[3]`http://www.ipac.caltech.edu/2mass/releases/allsky/doc/sec5_4.html`

2MASS extractions are unavailable and astronomers cannot examine the set of extractions corresponding to each source.

The Working Auto-Correlation[4] (WAX) software addresses these issues by generating groups of apparitions likely to correspond to distinct astronomical sources. Groups are made available as a catalog in conjunction with the database of extracted apparitions. Furthermore, the grouping algorithm employed is conservative: if an apparition is assigned to more than one group, then the apparition and its containing groups are flagged as confused. Attempts at resolving confusion are deferred so as not to impose a particular algorithm on the astronomer.

2. Architecture

WAX is a portable C/C++ application which performs I/O directly to and from a relational database management system (RDBMS). As of this writing, support is limited to the Informix RDBMS. By avoiding intermediate representations of input and output data, disk space requirements and time spent on I/O are both drastically reduced. Another benefit is that output can be queried and served immediately, without requiring the intervention of a database administrator.

The grouping algorithm (dubbed the swiss cheese algorithm) is hardwired, and places minimal constraints on the input RDBMS table: a per-apparition unique identifier and position suffice to run WAX. The output consists of between two and four tables :

Group Catalog: A table containing a unique identifier, apparition count, and confusion flag for each group. Other attributes (such as an average position) may also be computed by a user supplied plug-in.

Link Catalog: A table containing group/apparition identifier pairs. This table allows identifiers for all the apparitions belonging to a particular group to be retrieved. Similarly, identifiers for all groups containing a particular apparition can be retrieved.

Singleton Catalog: An optional table containing groups with just a single member apparition.

Grouped Apparition Catalog: An optional table mapping each multiply-assigned apparition to a *preferred* group (as determined by a user specified plug-in).

WAX itself does not perform computation of group attributes or choose *preferred* groups for apparitions. Instead, this task is left to a plug-in. At compile-time, the plug-in provides a retrieval and output table column specification. This specification is used to generate C structures corresponding to rows in the input and output tables. The specification is also used to generate high performance database I/O code (ESQL/C). At run-time, groups and apparitions are passed

[4] `http://irsa.ipac.caltech.edu/applications/2MASS/WAX/docs/html/`

to the plug-in via the generated data structures, allowing for computation of arbitrary group attributes (e.g. average position, mean magnitudes, etc.). Because the complexity of I/O and data representation is hidden from the plug-in, its software interface is small and simple: the entire interface is specified with just five C function prototypes. To further simplify plug-in implementation, libraries are provided for common tasks such as computing the observational coverage and spatial index of a position.

Taken together, these features allow the WAX application to be tailored to specific apparition databases. Although initially developed to generate 2MASS data products, WAX is general enough to support other missions, and is expected to be employed by the Wide-field Infrared Survey Explorer[5] (WISE) team.

3. The Swiss Cheese Algorithm

The swiss cheese algorithm first computes both a *density* and a *centroid* for each apparition on the sky. The *density* for an apparition a is defined as the number of apparitions within a user specified angular distance θ of a. The *centroid* is defined as the average position of the apparitions contributing to the *density* of a.

Next, the apparitions are sorted into decreasing *density* order. Each apparition in the resulting queue is said to be a *seed*; that is, an unprocessed apparition from which a group may be generated. A *group* is formally defined as the set of apparitions within a user specified distance ϕ of a *seed centroid*.

Groups are constructed as follows :

1. Generate a *group* g from the *seed* s at the head of the queue.
2. Apparitions assigned to g are removed from the queue of *seeds*. Note that s (the head of the queue) will always be removed.
3. While the queue is non-empty, repeat steps 1 and 2.

In some circumstances, the algorithm assigns a given apparition to more than one group. Apparitions belonging to two or more groups (as well as groups containing such apparitions) are said to be *confused*. Figure 1 illustrates the swiss cheese algorithm and show how *confusion* arises.

Since the apparition databases being processed are far too large to fit in memory, they must be partitioned in some way. The WAX software splits the sky into declination bands to reduce the working set of apparitions, and, instead of processing *seeds* strictly in density order, works in spatial order while maintaining *density* order results. This allows for efficient computation of *densities*, *centroids*, and *groups*, each of which involve finding apparitions within some small radius of a position. Furthermore, it allows individual declination bands to be processed in parallel (non-adjacent bands are guaranteed to be data-independent). Figure 2 depicts the spatial subdivision scheme and traversal order used by WAX.

Groups are generated and *seeds* are discarded if and only if doing so does not violate the density ordering constraints imposed by the swiss cheese algorithm :

1. A *group* is generated around a *seed centroid* if and only if the *seed* cannot be a member of any *group* generated from a *denser seed*.

---

[5] `http://www.astro.ucla.edu/~wright/WISE/`

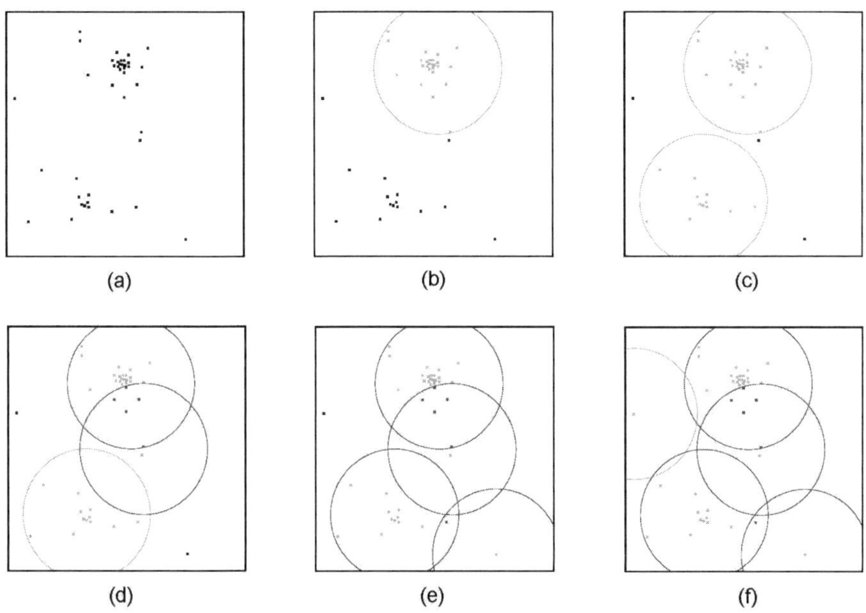

Figure 1. The swiss cheese algorithm operating on a small region of the sky. (a) The unprocessed sky. (b-c) The first and second *groups* are generated. Unprocessed apparitions are drawn in black, unconfused apparitions and *groups* in light grey. (d-f) As more *groups* are generated, confusion - drawn in dark grey - appears.

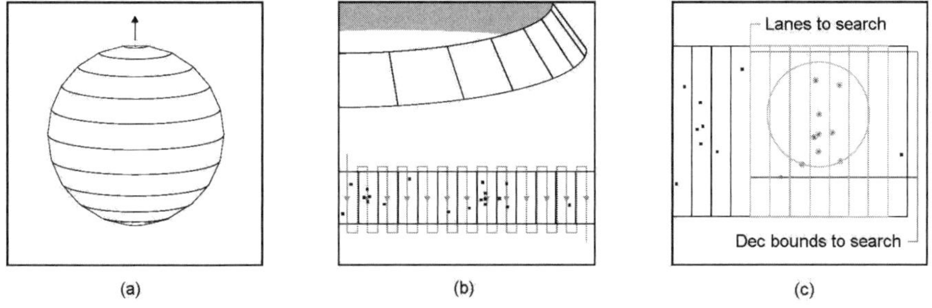

Figure 2. (a) The sky is subdivided into declination bands. (b) Each band is subdivided into lanes which are traversed in spatial order, indicated by the light grey line. (c) Only a small number of apparitions are considered to find a *density*, *centroid*, or *group*.

2. When a *group* is generated around a *seed* s, then an apparition assigned to s is discarded from the set of *seeds* if and only if it is less *dense* than s.

Because the sky cannot be considered in its entirety, there are cases where processing *seeds* in spatial order will produce different results than *density* order. In such cases, WAX attempts to minimize the deviation incurred and also flags each *group* generated out of order. Note that such cases did not occur during 2MASS data processing, and have yet to be encountered in practice.

4. Performance

The WAX software has been used to generate group catalogs for the 2MASS Point, Extended, 6X, and Calibration Working Source Databases. In particular, a pre-filtered copy of the 2MASS Working Point Source Database containing roughly 800 million apparitions was processed by 4 instances of WAX running in parallel on a single 4 CPU machine in approximately 4.5 days.

ASTRONOMICAL DATA ANALYSIS SOFTWARE AND SYSTEMS XIV
ASP Conference Series, Vol. 347, 2005
P. L. Shopbell, M. C. Britton, and R. Ebert, eds.

Image Segmentation and Denoising Based on Shrira-Pesenson Equation

M. Pesenson, M. Moshir, D. Makovoz, D. Frayer, and D. Henderson

Spitzer Science Center (SSC), California Institute of Technology

Abstract. We propose a nonlinear partial differential equation to control the trade-off between smoothing and segmentation of images. Its solutions approximate discontinuities, thus leading to detection of sharp boundaries in images. The performance of the approach is evaluated by applying it to images obtained by the Multiband Imaging Photometer for *Spitzer* (MIPS), 70μm imaging band.

1. Introduction

The principal goals of image enhancement are noise removal, or smoothing, and recognition of objects of interest, which is also called image segmentation. These operations are not independent and their connection is the source of major difficulties encountered in the field of image processing. The trade-off is inevitable, and a balance between these two desirable but incompatible objectives depends on a specific application. The traditional approach to the noise removal problem ignores this connection by simply convolving an image with a Gaussian. Though such approach effectively removes noise, at the same time, it drastically smears point sources and blurs boundaries and edges. The first breakthrough in tackling this problem was achieved by Witkin (1983), who recognized that convolving an image with a Gaussian, is equivalent to solving a Cauchy problem for the linear partial differential equation (PDE) of diffusion with the noisy image as initial condition, thus explaining the blurring of boundaries, as one would expect from diffusion. This insight has led to the construction of multi-scale representations of image data.

2. A Multi-scale Representation of Images by Using PDEs

The multi-scale representations of image data are obtained by embedding the given image into a one-parameter family of derived images. This family should be parameterized by a scale parameter and be generated in such a way that fine-scale structures are successively suppressed when the scale parameter is increased. Such construction allows to obtain a separation of the image structures in the original image, such that fine scale image structures only exist at the finest scales in the multi-scale representation thus simplifying the task of object detection. This objective can be achieved by employing the aforementioned connection between image processing and partial differential equations. Starting with pioneering works of Rudin (1987) and Perona & Malik (1987), nonlinear filtering based on nonlinear PDEs has become very useful in multiscale description of images, image segmentation, edge detection, image enhancement (Sapiro

2001). This approach is a state of art based on the design and analysis of PDEs. Perona and Malik (1987) proposed a nonlinear diffusion equation with the coefficient of diffusion D as a nonnegative function of the magnitude of local image gradient. The desirable D decays when the gradient grows and increases when the gradient decays. A different PDE was introduced by Rudin (1987) as means for nonlinear denoising of images without blurring sharp boundaries. He used a modification of a nonlinear equation describing evolution of shock waves (thus preserving or even sharpening objects). The approach suggested by Rudin had to deal with the following complications: the original equation was not symmetric in space, and it required a two-dimensional generalization. We use here an equation which was derived in Shrira and Pesenson (1983) (see also Pesenson 1991) to describe diffraction and stability of multi-dimensional shock waves and solitons. The equation has later become known as Shrira-Pesenson(SP) equation (Kivshar & Pelinovsky 2000). The SP equation can be written in the following form $I_{tt} - c^2 \Delta I + (I_t^2 + I_y^2 + I_z^2)_t - \nabla(D \nabla I_t) = 0$. It turns out, that this equation naturally overcomes the aforementioned difficulties. Indeed, its two dimensional, rotationally invariant and thus, following Rudin (1987), can be used to develop a nonlinear filter.

3. Mosaics, Completeness and Reliability

Mosaics of two astronomical observation requests - AOR 3865856 and AOR 6070016 (NGC300) - are presented on Figures 1 and 2. Images were obtained by MIPS, 70 microns imaging band (Rieke et al. 2004). Mosaic size is 0.9 deg x 0.5 deg. Effective exposure time for one AOR is 30 sec. Effective exposure time for four AORs is 120 sec. Basic calibrated data (BCD) is from SSC pipeline version 10.0. Mosaic image is made by MOPEX (Makovoz & Khan 2004). There are three images in each figure, showing from top to bottom respectively - the mosaic before filtering, after a linear filtering (point sources are smeared), and the mosaic after the nonlinear filtering (point sources remain sharp). One of the goals of nonlinear filtering is to improve point source extraction. We applied point source extraction software included in MOPEX (Makovoz & Khan 2004), to a single AOR3865856 before and after filtering. The results were compared with a true list of sources from an image obtained by combining four AORs for the same field of view as AOR3865856. Figure 3 shows the completeness and Figure 4 shows the reliability of the point source extraction (the solid line corresponds to the filtered image, and the dashed line to the non-filtered image) vs. logarithm of true flux in μmJy. Here completeness is defined as the ratio of the number of matched sources to the number of true sources; reliability is defined as the ratio of the number of detected sources to the number of matched sources. The improvement is in the highest flux range, reaching 100% reliability.

4. Conclusion

It can be seen from the completeness and reliability plots (Figs. 3,4), that the nonlinear filtering ensures better completeness and reliability for higher fluxes. Stopping filtering at less coarse levels of scale (smaller times, in PDE‘s termi-

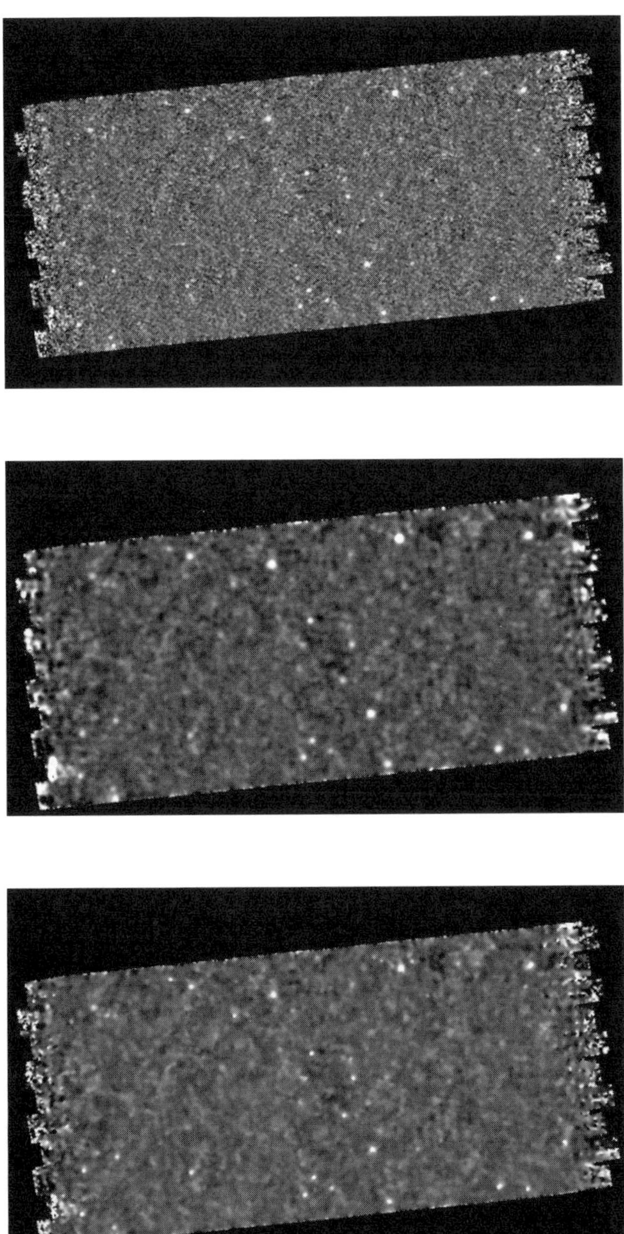

Figure 1. One AOR mosaic. From top to bottom - before filtering, after a linear filtering (point sources are smeared), and after the nonlinear filtering (point sources remain sharp).

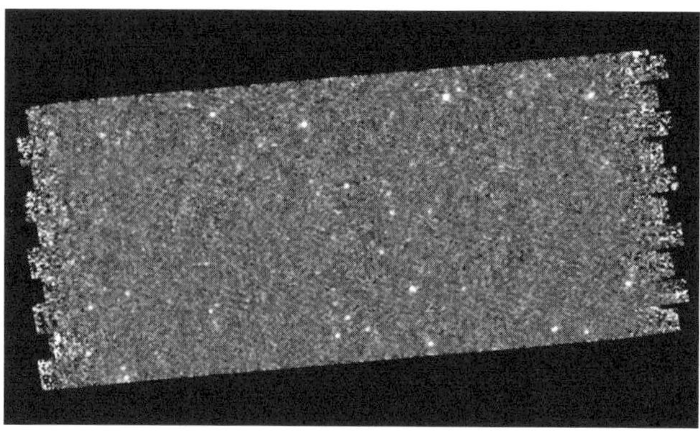

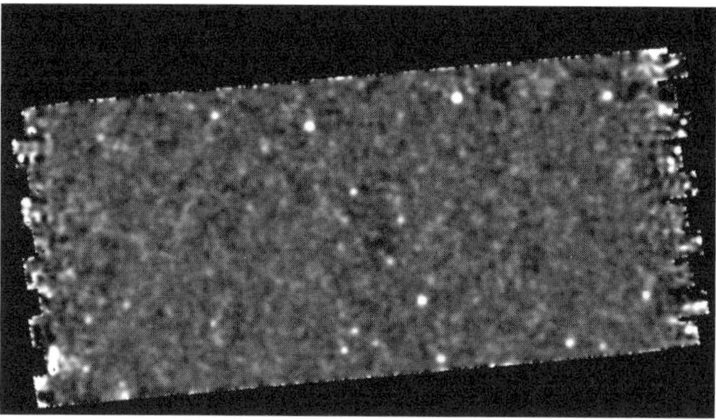

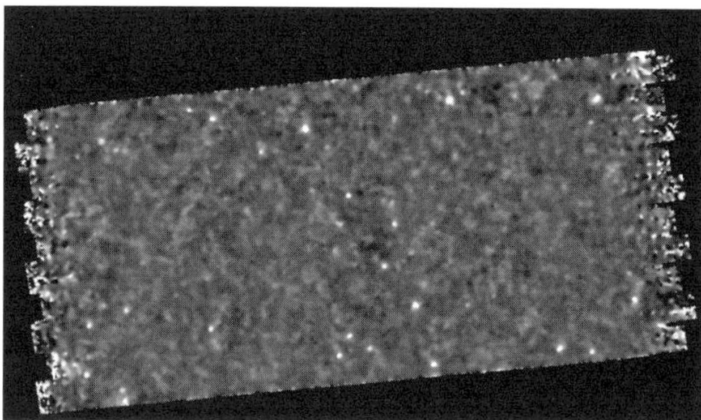

Figure 2. NGC300 mosaic. From top to bottom - before filtering, after a linear filtering (point sources are smeared), and after the nonlinear filtering (point sources remain sharp).

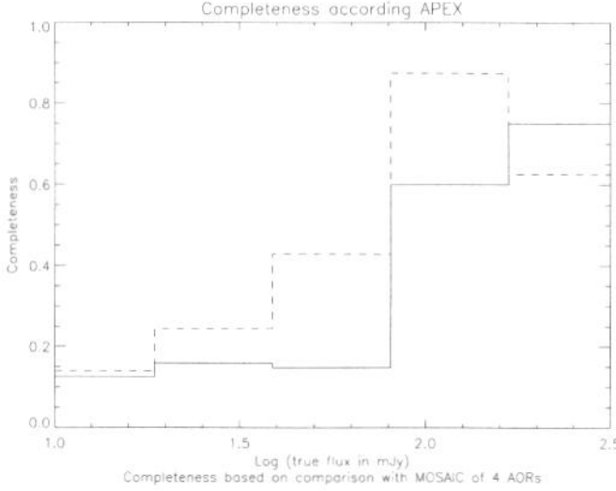

Figure 3. Completeness vs. log of true flux in μmJy.

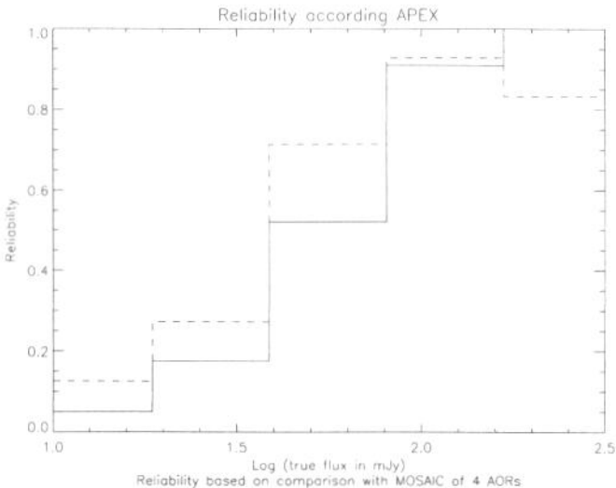

Figure 4. Reliability vs. log of true flux in μmJy.

nology) will increase completeness and reliability of the point source extraction at lower flux levels as well.

Acknowledgments. This work was carried out at the *Spitzer* Science Center, with funding from NASA under contract 1407 to the California Institute of Technology and the Jet Propulsion Laboratory. One of us (M.P.) wants to thank Tetsuyasu Uekuma for his insights and encouragement.

References

Kivshar, Y., & Pelinovsky, D. 2000, Physics Reports, 331, 117

Makovoz, D., & Khan, I. 2004, in ASP Conf. Ser., Vol. 347, ADASS XIV, ed. P. L. Shopbell, M. C. Britton, & R. Ebert (San Francisco: ASP), 81

Perona, P., & Malik, J. 1987, IEEE Trans. Pattern Anal. Mach. Intell., 12, 629

Pesenson, M. 1991, Physics of Fluids, A3, 12, 301

Rieke, G., et al. 2004, ApJS, 154, N1

Rudin, L. 1987, Images, Numerical Analysis and Shock Filters, Technical report 5250, CIT

Sapiro, G. 2001, Geometric Partial Differential Equations and Image Analysis (Cambridge University Press: England)

Shrira, V., & Pesenson, M. 1983, In Nonlinear and Turbulent Processes in Physics, Ed. R. Sagdeev, Vol. 2, (Harwood Academic, NY)

Witkin, A. 1983, In Proc. Intern. Conf. on Artificial Intelligence

ASTRONOMICAL DATA ANALYSIS SOFTWARE AND SYSTEMS XIV
ASP Conference Series, Vol. 347, 2005
P. L. Shopbell, M. C. Britton, and R. Ebert, eds.

An Analysis Package for Bolometer Ground Testing

Bernhard Schulz, Lijun Zhang, and Ken Ganga

Infrared Processing and Analysis Center, California Institute of Technology

Hien Nguyen and Warren Holmes

Jet Propulsion Laboratory, California Institute of Technology

Abstract. ESA's Herschel Space Observatory, to be launched in 2007, will be sensitive to far infrared wavelengths beyond 60 μm. The longer wavelength interval between 200 and 670 μm will be covered by SPIRE, a combination of broad band camera and Fourier transform spectrometer. SPIRE will use exclusively spiderweb bolometers as detectors, which are manufactured and tested at JPL. We describe a data analysis package developed at the NASA Herschel Science Center at IPAC in support of the testing activity, which expects to cover 12 detector arrays with between 24 and 144 channels each. The package consists of a widget based viewer allowing immediate display and limited processing of the 193 recorded data channels in the lab and a suite of subroutines and scripts, allowing fast and flexible pipeline data reduction.

1. Software Structure

BoloLibrary is a software package written in IDL (interactive data language) to support the analysis of test data from bolometers, as used in SPIRE, one of three instruments on board of the planned Herschel Space Observatory. The bolometer arrays undergo a series of performance tests in a laboratory cryostat at JPL (Nguyen et al. 2004). The raw data products emerging from the test facility BoDAC are files that start with a short ASCII header, followed by a binary data array. This array consists of one channel with time information and 192 signal channels. These files and some additional ASCII parameter files are the input to the programs described in this paper. The following major data analysis tasks are presently performed using *BoloLibrary*:

- Signal Viewing
- Noise Analysis
- Optimum Bias Determination
- Time Constant Determination
- Load Curve Analysis

Figure 1 shows the data flow starting from the initial binary files produced by BoDAC to the various analysis results. In an initial conversion the data is written to a file in the more widely known FITS format. This step is also used to change channel names and units if necessary, to remove amplification factors of the BoDAC electronics, and to convert the signals of temperature sensors into units of Kelvin. Two editable ASCII files and a temperature conversion routine

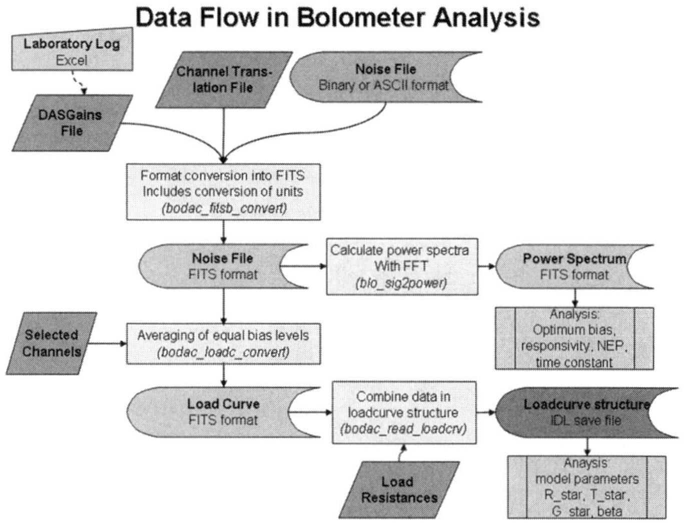

Figure 1. The general data flow from the initial BoDAC files to the analysis results.

contain the necessary information, allowing for enough flexibility in an often changing laboratory environment. In our configuration the signal channels of the resulting FITS "noise" file contain either voltages in [Volts] or temperatures in [K]. The time channel is always in [seconds] starting at the beginning of the measurement. Date and other relevant metadata are found in the FITS header.

Depending on the experiment, the noise file is either transformed into a power spectrum or translated into a load curve file. A noise file is transformed into a load curve file if the bias voltage was changed during the time interval covered by the input file. In this case the respective program averages data over intervals of constant bias voltage. The channel containing the time information is replaced by the bias voltages, and the data is stored again in another FITS file, including uncertainties estimated from the scatter of the averaged data. Since the detector arrays vary in size, only the channels listed in an ASCII file "SelectedChannel.txt" are included into the load curve file. Channel selection and averaging greatly reduce the size of a load curve file and save disk space.

Load curves taken at different temperatures need to be analyzed as a whole. Therefore data from many load curve files are collected in a pointer array of IDL data structures and stored in an IDL save file. There are several procedures provided in *BoloLibrary* for dark load curve data analysis, specifically to work with these structures.

2. Laboratory Data Viewer

An invaluable tool to view and assess the quality of newly obtained data in the lab as well as when performing offline analysis, is the data viewer called

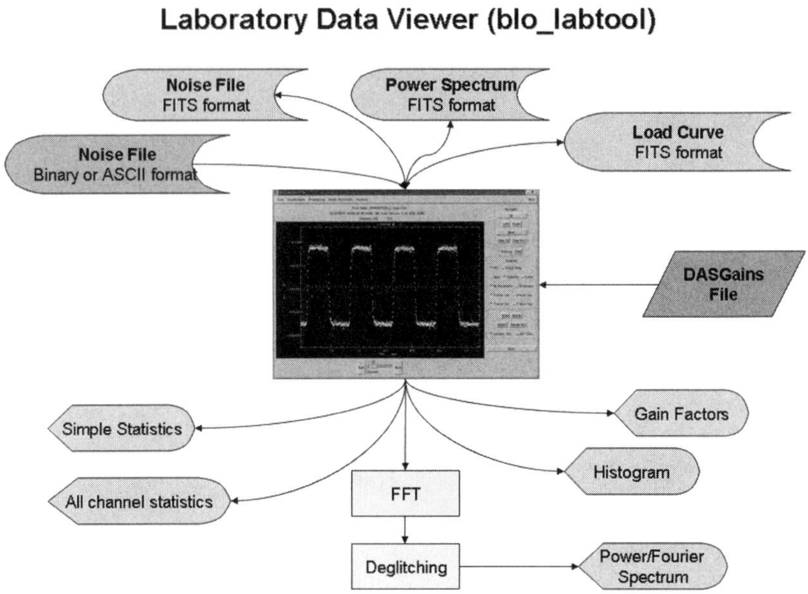

Laboratory Data Viewer (blo_labtool)

Figure 2. Input and output of the data viewer.

Blo_LabTool. Figure 2 shows the data flow chart of *Blo_LabTool*, which has an intuitive GUI and is started from the IDL commandline. It allows to rapidly view and manipulate raw data, show derived power spectra, save these and more. Data files in various formats,(.bin, .fits, .txt) can be opened from the "File"menu. Highlights of the viewer are as follows:

- Show the signal versus time plots channel by channel
- Navigate the plots and zoom in and out
- Change between symbols, dots and lines in the display
- De-select and re-select data points
- Allow multiple windows
- Calculate power spectrum of selected data range and display in new window
- Calculate and display a histogram
- Display uncertainty data if available
- Calculate and display statistics of selected data points
- Determine amplitude ratios of sine signals to measure amplification factors
- Apply conversion factors to data
- Derive load curves interactively
- Display a HTML help file in a web browser

3. Batch Analysis

Although the viewer is very important to check on certain aspects of the data, the large number of channels requires more flexibility than a GUI based tool is usually able to offer. The "production" of analysis results is controlled by IDL scripts, dedicated to individual datasets that call subroutines from a library that forms the basis of this software. Besides allowing to easily repeat processing steps, the scripts serve also as documentation for the analysis of a particular dataset. In the following we briefly touch upon each of the analysis items that are currently implemented. Several items require the data first to be transformed into power spectra.

3.1. Noise Analysis & 1/f Knee Frequency

The power spectrum derived from the output voltage of an undisturbed dark bolometer consists of a plateau in the middle, a rising power spectrum to lower frequencies that goes with 1/frequency and a high frequency roll-off. The important parameters determined by the software are the plateau level and the location of the 1/f knee, i.e. the frequency where the spectrum has reached $\sqrt{2}$ times the plateau level. A typical challenge is to automatically eliminate the microphonic lines that otherwise affect the plateau measurement.

3.2. Optimum Bias & Time Constant

These experiments involve a hot external blackbody that sends an infrared beam through a neutral filter onto the detector array. The beam is chopped with an adjustable frequency so that the detectors are illuminated alternately by the blackbody and the 300 K environmental radiation. The power spectrum of the bolometer voltage shows a line at the position of the chopper frequency. Its strength is a measure for the difference in intensity between the two infrared illumination levels. The optimum bias is determined by finding the maximum line strength while varying the bias voltage of the detector. The time constant is determined by observing the decrease in line strength when the chopper frequency is increased.

The analysis uses modules to first derive the chopper frequency for each data file in a selected clean channel, and then determine the line strength for all channels depending on either bias or chopper frequency. Simple ASCII tables are used for the IO of intermediate products. Since the coverage by different bias voltages is relatively sparse, a 2nd order polynomial fit around the maximum line is used to determine the maximum with somewhat higher accuracy. By fitting $P(f) = \frac{A}{\sqrt{1+2\pi f \tau^2}}$ to the line strength depending on chopper frequency f, the thermal time constant τ of each bolometer is determined. A is the amplitude at low frequencies and a free fit parameter.

3.3. Load Curve Analysis

A dark load curve is obtained by putting the bolometer in a light-tight chamber and measuring the output voltage across the bolometer as a function of input bias voltage. Taking load curves at a variety of bath temperatures allows to determine 4 fundamental parameters R_*, T_*, G_0, and β that are required to model bolometer operation. Our models follow general theory as described by

Richards (1994), Rieke (1994), Sudiwala at al. (2002), Woodcraft et al. (2002), Nguyen et al. (2004), and others. The temperature dependence of the bolometer resistance is described as

$$R = R_* \exp \sqrt{\frac{T_*}{T}}. \tag{1}$$

After correction of voltage offsets based on the symmetry of the load curve w.r.t. the zero point, the resistances of the bolometers at zero power are determined from the slope of the load curves at zero bias. A linear fit determines the first two parameters.

The thermal conductance is modeled as

$$G(T) = G_0(T/T_0)^\beta \tag{2}$$

so that the electric power can be expressed as

$$P = \frac{G_0}{(\beta+1)} T_0^{-\beta} (T^{\beta+1} - T_{bath}^{\beta+1}). \tag{3}$$

Here T is the bolometer temperature, T_{bath} is the bath temperature, and $T_0 = 0.3$ K is a constant. While the load curves used to determine the zero power resistances only need to cover the smallest biases sufficient to fit a line, the parameters G_0 and β require a bias coverage that goes sufficiently beyond the turnover points, where the bolometer voltage drops due to the increasing electrical power dissipated in the bolometer and the resulting drop in resistance. The parameters are obtained by tailored nonlinear fitting routines in the library.

References

Nguyen, H. T. et al. 2004, SPIE proceedings, submitted

Richards, P. L. 1994, J. Appl. Phys. 76, 1-24

Rieke, G. H. 1994, Detection of Light, Cambridge University Press

Sudiwala, R. V. et al. 2002, Int. Jour. of Infrared and Millimeter Waves, 23, 4, 545

Woodcraft A. L. et al. 2002, Int. Jour. of Infrared and Millimeter Waves, 23, 4, 575

ASTRONOMICAL DATA ANALYSIS SOFTWARE AND SYSTEMS XIV
ASP Conference Series, Vol. 347, 2005
P. L. Shopbell, M. C. Britton, and R. Ebert, eds.

Improved Reduction Algorithm for ISOPHOT-S Chopped Observations

Bernhard Schulz

California Inst. of Technology / IPAC, MC 100-22, Pasadena, CA 91125, USA

Abstract. The astronomical data gathered by the extrinsic photoconductors on-board the ISO satellite were greatly affected by the high energy radiation environment in space. In addition so called transients with long time constants make automated pipeline processing of the data a challenge. ISOPHOT-S is the high sensitivity spectrometer of the ISOPHOT instrument, featuring two 64 element Si:Ga arrays for the wavelength range of 2.5 to 11.6 microns. It collected a wealth of almost 300 observations of extragalactic nuclei at a spectral resolution of ~ 100. Most of those observations were performed using the chopped observing mode, justifying a dedicated effort to revise and improve the automatic processing techniques in order to derive a homogeneously reduced dataset with realistic uncertainties. The improvements compared to the original standard pipeline processing include techniques like ramp subdivision, smoothed sigma kappa deglitching, spike filtering and accounting for a non-Gaussian signal distribution due to the glitch residuum.

1. Introduction

ISOPHOT-S is a low resolution, high sensitivity infrared spectrometer with fixed gratings and a $24''$x $24''$aperture. It was part of ISOPHOT on board of ESA's Infrared Space Observatory (ISO), which was launched Nov 1995 and operated until May 1998[1]. The wavelength coverage comprised two ranges, 2.49 - 4.90 μm (SS), and 5.86 - 11.65 μm (SL) in two 64 pixel Si:Ga arrays (Lemke et al. 1996; Laureijs et al. 2003; Acosta-Pulido & Abraham 2003). The small detector currents were integrated in cold amplifiers, and appear in the downlinked data as integration ramps that are sampled in equidistant time intervals (Figure 1). Observations were executed in terms of Astronomical Observing Templates (AOTs), which in the case of PHT-S always started with a 32 sec dark measurement. This procedure helped to mitigate memory effects of the detectors by starting each observation under the same conditions. The following sky measurement was conducted either in staring mode, chopped mode or raster mode. This paper will concentrate on the data reduction techniques for the chopped mode measurements between a source and one or two background positions, being the instrument's most important mode for observations of faint extragalactic sources.

[1] `http://www.iso.vilspa.esa.es`

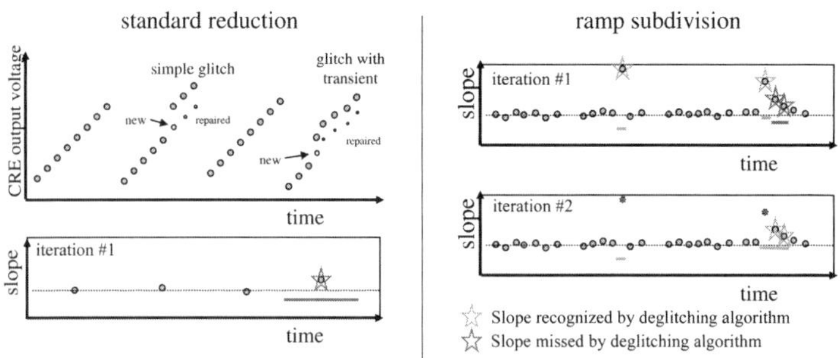

Figure 1. Upper left: Schematic integration ramps affected by a glitch and a spike with transient. Lower left: Effect of old reduction scheme suffering from spike "repair". Right: New scheme using slopes from subdivided ramps. Less integration time is lost by eliminated glitches.

2. The Detector Signal

The ISOPHOT-S data comes in equidistant samples of a slowly rising voltage that is reset in regular intervals. Ideally the photocurrent is determined simply by fitting the slope. Unfortunately high energy particle hits (glitches) create breaks in the integration ramps, and in a number of cases change the responsivity of the detector element for up to minutes (spike). The frequency of time intervals between glitches with a given duration falls exponentially towards longer intervals. Half of the events are less than 8.25 sec apart and most glitches are 1.5 sec apart. Therefore it is almost impossible to find an integration ramp, which typically lasts 16 to 32 sec for this AOT, that is not broken by glitches.

The standard pipeline processing (OLP) applied the so called two threshold deglitching described in Laureijs et al. (2003). To avoid ramp fragments of different sizes, the algorithm corrects the remaining valid datapoints after a glitch using the estimated slope and fits an overall slope to the entire ramp. This effective insertion of artificial datapoints leaves the calculation of uncertainties less reliable and may introduce larger errors when many readouts in a ramp are flagged after a spike event.

3. Ramp Subdivision

To avoid the statistical problems associated with either artificial datapoints or different size ramp fragments, in this approach integration ramps are broken up into small subramps of equal duration (1 sec). This interval is matched to the distribution of time intervals between glitches as mentioned above.

Before slope fitting no effort is made to "repair" ramps that are obviously glitch affected. This has the advantage that, although the resulting slope distribution is wider due to the smaller effective integration time, the glitch affected subramps become much more prominent outliers in the distribution that can be easily removed by a Sigma-Kappa deglitching algorithm as used by OLP or PIA (Gabriel 1997). If a subramp is removed, a much smaller amount of integration time is sacrificed as if an entire integration ramp of 32 sec must be eliminated.

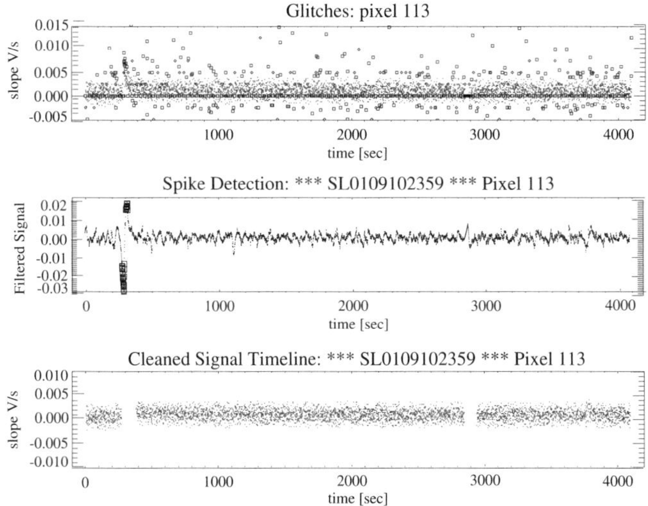

Figure 2. (Upper: Slopes with glitches. Middle: correlation function indicating spike. The second spike is detected by method a). Lower: Cleaned signal.

Without "repairs" the dataset remains "cleaner" and more homogeneous in a statistical sense, since all ramps are of the same size. A drawback of the shorter subramps is the magnification of discrete slope levels due to the digital representation of the data. Therefore the signal determination can not use medians as discussed below.

4. Deglitching

The deglitching algorithm used here differs from the one used in the standard processing, in that it first establishes a signal baseline by smoothing the data with a 60 sec boxcar, calculates a standard deviation σ from this baseline, and discards all points more distant than $\pm 2.5 \times \sigma$. This procedure is repeated until no slopes drop out anymore. The baseline has the advantage of better following longterm drifts that are common in these detectors. This reduction step removes all small energy glitches, however does not catch spike events, which render a longer continuous interval of slopes useless (Figure 2).

The observations described here were made by switching continuously between source and background (chopping) with a minimum of 4 subsequent integration ramps on each position and at least 2 chopper cycles. The background position is located either on one side (rectangular mode) or on opposite sides of the source (triangular mode). Most of the time the signal difference is much smaller than the noise, however it can still broaden the distribution and lead to a systematic lower signal difference due to undetected glitches in the valleys between on-source integrations. In a second phase, a provisionally calculated average difference between on- and off-source pointing is temporarily removed. This enables another deglitching pass to find missed glitches in the valleys between chopper plateaux.

5. Despiking

Strong events (spikes) that change the detector responsivity for up to a few minutes, so far had to be removed manually. Here we describe a new algorithm, that detects and removes spikes automatically. To avoid hiding the spike signature, slopes within 5 σ that were flagged as glitches before, are temporarily included back into the analysis. Two indicators have been found to work most successfully: a) If the ramps were saturated for at least 8 sec, indicated by an uninterrupted sequence of readouts flagged bad and set to zero, the following 90 sec are flagged. b) The slope signal is correlated with a prototype exponential decay signal with a time constant of 90 sec, and the smoothed correlation function is searched for extreme outliers. If at least 5 instances of the correlation function within 90 sec exceed 3 times its standard deviation , all slopes within -9 and +63 sec from this point in time are discarded. See Figure 2.

6. Signal Determination

Once the data is cleaned from glitches and spikes, the on-source and off-source signals must be determined. Sometimes the detector signal shows strong drifts, typically after previous bright illumination. The resulting baseline is fitted with a simple exponential drift model (Huth & Schulz 1998) and removed before determination of the signal. Since the energy distribution of glitches drops continuously to zero, smaller glitches can not be separated from other statistical noise. Therefore and due to residual transients in the data, the slope distribution is not symmetric and shows a small tail towards higher values. The median is closer to the real peak of the distribution, but suffers from the discrete nature of the data (Schulz et al. 2002), since the median can be only an existing value. The best results providing the most reliable signals and uncertainties are obtained by automatically creating separate histograms for on-source and off-source slopes and fitting Gaussians. The difference of their center positions determines the source signal. The uncertainty of each signal is given as the standard deviation of the fitted Gaussian, divided by the square root of the total number of valid slopes.

7. Flux Calibration

The transformation of signals into point source fluxes is still done using the signal dependent relation given by Acosta-Pulido & Abraham (2003). The results are already quite good, however another improvement may be obtained by subjecting the calibration stars to the same processing, to remove any algorithm dependent systematic effects from the flux calibration. This step is still under way at the time of writing.

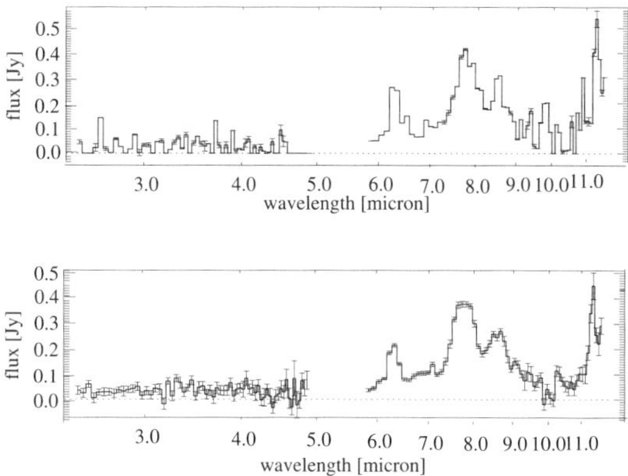

Figure 3. Example of a spectrum taken from the archive (top) and the same data reduced automatically with the described method (bottom).

8. Comparison

In Figure 3 we compare spectra derived from the same raw data set that were obtained by ISO observing the starburst galaxy NGC701. The upper one is a standard pipeline product as found in the ISO archive. The lower is produced automatically using the methods described above. The fluxes of the pipeline product are strongly affected by undetected glitches, and the uncertainties appear unrealistically small. The lower spectrum is less noisy and uncertainties are more consistent with the pixel to pixel noise. The subdivision technique was already used to reduce the Seyfert spectra presented by Clavel et al. (2000); however, the despiking was only automated recently, allowing fully automatic reduction of all chopped ISOPHOT-S observations in this way.

References

Acosta-Pulido, J. & Abraham, P., 2003, ESA SP-481

Clavel, J. et al. 2000, A&A, 357, 839-849

Gabriel, C., Acosta-Pulido, J., Heinrichsen, I., Skaley, D., Morris, H.,& Tai, W.M. 1997, in ASP Conf. Ser., Vol. 125, ADASS VI, ed. G. Hunt & H. E. Payne (San Francisco: ASP), 108

Huth, S. & Schulz, B. 1998, in ASP Conf. Ser., Vol. 145, ADASS VII, ed. R. Albrecht, R. N. Hook, & H. A. Bushouse (San Francisco: ASP), 212

Laureijs, R.J., Klaas, U., Richards, P.J.,Schulz, B.,& Abraham, P. 2003, ESA SP-1262

Lemke, D. et al. 1996, A&A, 315, L64-L70

Schulz, B. et al. 2002, A&A, 381, 1110-1130

ASTRONOMICAL DATA ANALYSIS SOFTWARE AND SYSTEMS XIV
ASP Conference Series, Vol. 347, 2005
P. L. Shopbell, M. C. Britton, and R. Ebert, eds.

Monte Carlo Image Analysis in Radio Interferometry MC-FIT : A Bayesian Approach to Object Detection

U. Rau[1] and T. J. Cornwell

National Radio Astronomy Observatory, Socorro, NM 87801

Abstract. Image analysis, such as component fitting of radio interferometric images has traditionally been based on likelihood techniques applied to deconvolved images. The analysis usually ignores uncertainties arising from fitting flux components to extended emission as well as from the process of deconvolution itself. One would therefore like to estimate the properties of components representing the entire emission present in the raw, dirty image. In practice, this is not feasible given the large dimensionality of the parameter space. We present a Bayesian approach in which we fit elliptical Gaussian components to sub-regions of the dirty image, taking into account the point spread function. Our method samples the posterior distribution to estimate the relative probabilities and uncertainties associated with the number of flux components and their parameters. This information can be used to augment the process of object detection and characterization. We compare the performance of this approach to the standard methods.

1. Introduction

There are two sources of uncertainty which are generally ignored in traditional methods of fitting combinations of Gaussian, disk or point source models to brightness distribution patterns in deconvolved radio interferometric images.

1. Incomplete sampling of the spatial coherence function makes deconvolution a non-linear inverse problem with no unique solution. Algorithms that use different approximations are suited to different emission patterns and can converge to different results. Moreover, the only estimate of deconvolution error is the residual image which holds no information about the possible existence of multiple solutions with the same goodness-of-fit.
2. The objective function involved in fitting a linear combination of overlapping flux components to extended emission is often multi-modal and any likelihood technique (unconstrained and constrained minimization) is subject to an inherent uncertainty.

These uncertainties can be formally quantified and estimated to form confidence limits for the individual fitted components. First, the processes of deconvolution and component fitting can be combined by representing an image as a set of parameterized flux components, taking into account the point spread function, and solving for component parameters that fit the emission in the dirty image. This, coupled with a Monte Carlo approach that samples the posterior

[1]Department of Physics, New Mexico Institute of Mining and Technology,Socorro

distribution derived from the Bayesian formulation of deconvolution, can provide information about relative probabilities and uncertainties associated with the number of flux components and their parameters.

We present the MC-FIT algorithm, in which an image is represented as a collection of elliptical Gaussian flux components, and Monte Carlo sampling is performed on the number of flux components and their parameters required to accurately represent observed emission patterns.

2. MC-FIT Algorithm

Bayes theorem formulates a posterior distribution as the product of a likelihood distribution representing goodness of fit, and a prior distribution representing a-priori information. In MC-FIT, χ^2 is used as the goodness of fit criterion to evaluate the likelihood distribution, and a-priori information is provided in the form of probability distribution functions for each type of parameter. The Bayesian posterior distribution is constructed with the following a-priori information:

1. Position parameters are allowed to vary uniformly within the region of interest. Prior to the fitting of emission inside a given region, the emission outside the region is removed from influence by masking out and subtracting from the observed data, visibilities corresponding to parts of a previously deconvolved image.
2. Scale parameters are allowed to vary as Ase^{-Bs} where s represents the scale and A, B are suitable constants. This approximates the distribution of scales observed in a typical image, where most features correspond to small scales and a comparatively smaller number of large scale features exist.
3. The amplitude is allowed to vary uniformly within a range, and the position angle (defined here as the angle between the horizontal axis and the first specified axis of the ellipse) is allowed to vary uniformly between 0 and 90 degrees.
4. Image based priors such as entropy, emptiness and positivity can be applied by including them during the evaluation of the posterior distribution function.
5. Sampling on the number of components is achieved by defining a distribution for the number of components (uniform or Poussin) within a specified range.

Various techniques exist for efficiently sampling the resulting Bayesian posterior distribution in the multi-dimensional parameter space. MCMC techniques generate samples by following a Markov chain through the parameter space based on a transition matrix whose limiting distribution is the target posterior distribution. For this work, we used BayeSys[2]: an application that implements a MCMC algorithm along with selective annealing to sample a specified posterior distribution.

---

[2] `http://www.maxent.co.uk/`

3. Tests on Simulated Synthesis Data

The MC-FIT algorithm was applied to a simulated synthesis data set corresponding to a VLA C array observation of a source composed of four overlapping elliptical Gaussian features (Figure 1).

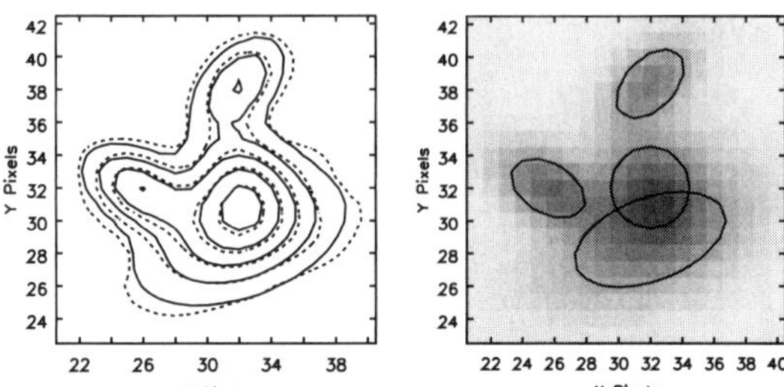

Figure 1. Tests on simulated synthesis data (SNR=100): [Left] original image (solid line) and dirty image (dashed line). [Right] Most probable model image with 3σ contours for the corresponding set of Gaussian components.

The true parameters of the elliptical Gaussian components used to compute the sample data are given in the following table. Inspection of Figure 2 shows that all parameter histograms peak at the correct values. Increased noise levels result in slower annealing convergence and wider distributions. In the absence of any significant flux components, the parameter histograms reflect the shape of the prior distributions.

Component	x_0	y_0	Amplitude	σ_x	σ_y	Position Angle
1	5.0	4.5	80	1.0	0.5	20
2	4.0	5.0	100	0.3	0.5	60
3	5.0	6.0	80	0.5	0.3	45
4	5.0	5.0	100	0.5	0.5	45

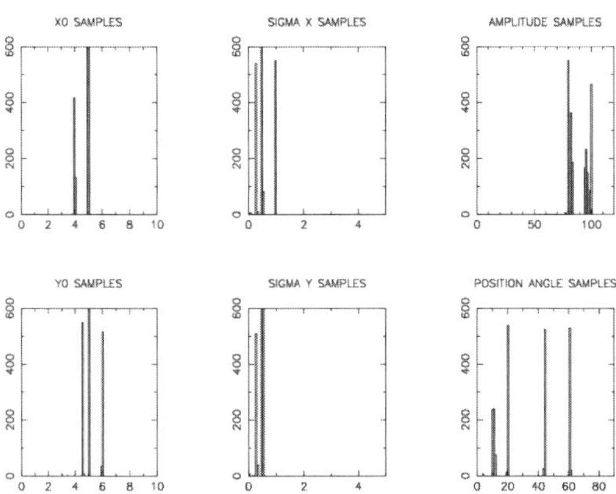

Figure 2. Histograms of the obtained Gaussian parameter samples.

		Normalized χ^2
1	Gaussian fits to CLEAN restored image	2.023
2	Gaussian fits to MEM restored image	1.996
3	MC-FIT mode image	1.122
4	CLEAN model image	1.113
5	MEM model image	1.114

The above Table lists values of normalized χ^2 for various model images. For rows 1 and 2, model images were obtained via the traditional method of fitting Gaussian components to deconvolved, restored images. Row 3 shows the normalized χ^2 for the image formed from the most probable set of component parameters produced by MC-FIT. Values in rows 4 and 5 were computed directly from the CLEAN and MEM un-restored model images. These un-restored minimum residual images represent the best-fit (minimum χ^2) images produced by the deconvolution algorithm before being convolved with the CLEAN beam to form restored images.

This data shows that if an image can be decomposed into elliptical Gaussian components, the MC-FIT algorithm is capable of producing a component list that represents the image almost as well as the (numerically optimal) minimum χ^2 CLEAN and MEM model images.

4. Conclusions

A Bayesian approach to component fitting in image analysis can thus be used to obtain estimates for parameters of flux components in the emission along with associated uncertainty estimates. This technique can be applied to non-synthesis as well as synthesis data, by computing the likelihood function appropriately, and is generic over any parameterized image representation. The running time of this algorithm is proportional to the number of components being fitted ($O(N)$), the number of parallel sample streams ($O(N)$), and the image size ($O(N^2 log_2 N)$ for an $N \times N$ image). The number of iterations required to reach sampling convergence depends on the type of prior information provided to the algorithm, the signal to noise ratio of the data, the area under the CLEAN beam, and the complexity of the brightness distribution.

ASTRONOMICAL DATA ANALYSIS SOFTWARE AND SYSTEMS XIV
ASP Conference Series, Vol. 347, 2005
P. L. Shopbell, M. C. Britton, and R. Ebert, eds.

Making the Most of Missing Values: Object Clustering with Partial Data in Astronomy

Kiri L. Wagstaff

Jet Propulsion Laboratory, California Institute of Technology, Pasadena, CA 91109, `kiri.wagstaff@jpl.nasa.gov`

Victoria G. Laidler

Computer Sciences Corporation, Space Telescope Science Institute, Baltimore, MD 21218, `laidler@stsci.edu`

Abstract. Modern classification and clustering techniques analyze collections of objects that are described by a set of useful features or parameters. Clustering methods group the objects in that feature space to identify distinct, well separated subsets of the data set. However, real observational data may contain missing values for some features. A "shape" feature may not be well defined for objects close to the detection limit, and objects of extreme color may be unobservable at some wavelengths.

The usual methods for handling data with missing values, such as *imputation* (estimating the missing values) or *marginalization* (deleting all objects with missing values), rely on the assumption that missing values occur by random chance. While this is a reasonable assumption in other disciplines, the fact that a value is missing in an astronomical catalog may be physically meaningful. We demonstrate a clustering analysis algorithm, KSC, that a) does not impute values and b) does not discard the partially observed objects. KSC uses soft constraints defined by the fully observed objects to assist in the grouping of objects with missing values. We present an analysis of objects taken from the Sloan Digital Sky Survey to demonstrate how imputing the values can be misleading and why the KSC approach can produce more appropriate results.

1. Introduction

Clustering is a powerful data analysis tool that divides a set of objects into a number of distinct groups based on a problem-independent criterion, such as maximum likelihood (the EM algorithm) or minimum variance (the k-means algorithm). In astronomy, clustering has been used to analyze both images (POSS-II, Yoo et al. 1996) and spectra (IRAS, Goebel et al. 1989). Notably, the Autoclass algorithm identified a new subclass of stars based on clustering results (Goebel et al. 1989).

However, most clustering algorithms require that all objects be fully observed: objects with missing values for one or more features cannot be clustered. This is particularly problematic when analyzing astronomical data sets, which often contain missing values due to incomplete observations or varying survey depths. Missing values are commonly handled via *imputation*, where the gap is "filled in" with an inferred value. While appropriate for some domains, this approach is not well suited to astronomical data sets, because a missing value may well be physically meaningful. For example, the Lyman break technique (Giavalisco

2002) can identify high-redshift galaxies based on the *absence* of detectable emissions in bands corresponding to the FUV rest frame of the objects. In such cases, imputing missing values is misleading and can skew subsequent analyses of the data set.

We propose the use of a clustering approach that avoids imputation and instead fully leverages all existing observations. In this paper, we discuss our formulation of the missing data problem and expand on the KSC algorithm originally presented by Wagstaff (2004). We compare KSC analytically and empirically to other methods for dealing with missing values. As a demonstration, we analyze data from the Sloan Digital Sky Survey, which contains missing values. We find that KSC can significantly outperform data imputation methods, without producing possibly misleading "fill" values in the data.

2. Clustering Astronomical Objects with Missing Values

Missing values occur for a variety of reasons, from recording problems to instrument limitations to unfavorable observing conditions. In particular, when data is combined from multiple archives or instruments, it is virtually certain that some objects will not be present in all of the contributing sources. Little and Rubin (1987) identified three models for missing data. When values are Missing At Random (MAR) or Missing Completely At Random (MCAR), imputation may be a reasonable approach since the values may be inferable from the observed values. The third type of missing values are Not Missing at Random (NMAR); that is, the value itself determines whether it is missing. This is precisely what happens when objects fall below a detector's sensitivity threshold. There is no way to impute these values reliably, because they are *never* observed.

There are three major approaches to handling missing values when clustering. The first, *marginalization*, simply removes either all features or all objects that contain missing values. The second method, *imputation*, attempts to "fill in" any missing values by inferring new values for them. The advantages and drawbacks of two marginalization and three imputation techniques are summarized in Table 1. Finally, some recent methods (e.g. Browse, Skillicorn & McConnell 2003) avoid both of these approaches and instead seek to incorporate all observed values (no marginalization) without inferring the missing ones (no imputation). Our work falls into this category.

In our approach, we divide the data features into two categories. We use the fully observed features for clustering, and we use the partially observed features (with missing values) to generate a set of *constraints* on the clustering algorithm. We have previously shown that constraints can effectively enable clustering methods to conform to supplemental knowledge about a data set (Wagstaff et al. 2001). We use the KSC ("K-means with Soft Constraints") clustering algorithm, proposed by Wagstaff (2004), to incorporate information from the partially observed features as a source of information that supplements the fully observed features. Before discussing our experimental results, we will describe what we mean by constraints and briefly outline the KSC algorithm.

A *soft constraint* between two objects indicates a preference for, or against, their assignment to the same cluster. We represent a soft constraint between objects o_i and o_j as a triple: $\langle o_i, o_j, s \rangle$. The strength, s, defines the nature

Table 1. Comparison of marginalization and imputation methods.

Technique	Advantages	Drawbacks
Feature Marginalization: Omit features with missing values	Simple, reduces data set size	Lose information about every object
Object Marginalization: Omit objects with missing values	Simple, reduces data set size	Lose individual objects
Mean Imputation: Replace each missing value with data set mean	Simple	Likely to be inaccurate; mean value may never truly occur
Probabilistic Imputation: Replace with random value according to data set distribution of values	Inferred values are "real" (actual observations)	Inferred values may have no connection to the objects
Nearest Neighbor Imputation: Replace with value from the nearest neighbor	Inferred values are "best possible guess"	Inferred values may still be inappropriate (unobservable)

and confidence of the constraint. A positive value for s indicates a preference towards clustering o_i and o_j together; a negative value suggests that they should be assigned to different clusters.

The KSC algorithm is based on the basic k-means algorithm first proposed by MacQueen (1967). While the k-means algorithm seeks to minimize the total variance, V, of a proposed partition P, KSC minimizes the combination of the variance and a penalty, CV, for constraint violations:

$$f(P) = (1-w)\frac{V(P)}{V_{max}} + w\frac{CV(P)}{CV_{max}} \quad (1)$$

The CV penalty is calculated as the sum of the squared strengths, s, of all constraints violated by the partition P. The quantities V and CV are normalized by their maximum possible values. The user-specified weight parameter w indicates the relative importance of variance versus constraint violations; a good value can be chosen based on performance on a small labeled subset of the data.

3. Experimental Results

The Sloan Digital Sky Survey (SDSS) contains observations of 141 million galaxies, quasars, and stars (as of data release 3), many with missing values. We selected a small subset of this data for our experiments. The features we used were brightness (psfCounts), texture, size (petroRad), and shape (M_e1 and M_e2). 42 of the 1507 objects we analyzed have missing shape features. We clustered this data set based on the three fully observed features and generated constraints based on all of the features, F. For each pair of objects o_1, o_2 that had observed shape features, we generated a constraint with strength proportional to the observed distance between o_1 and o_2. More precisely, we calculated $s = -\sqrt{\sum_{f \in F}(f(o_1) - f(o_2))^2}$, where $f(o)$ indicates the observed value for object o in feature f. In this case, s is negative because the distance between the objects indicates a preference against grouping them together. The result was 1,072,380 constraints that were "mined" from the data set.

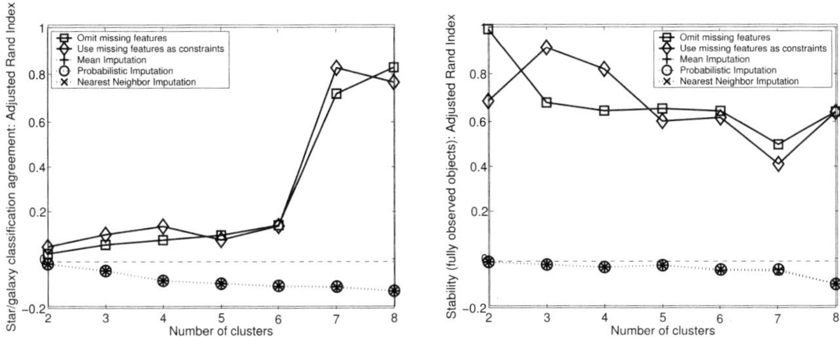

Figure 1. Empirical comparison of imputation, marginalization, and KSC methods. Figure (a) shows agreement with true star/galaxy separation; figure (b) shows stability of clustering assignments for the fully observed objects when partially observed objects are included. In both figures, the dashed line shows the agreement expected by random chance.

3.1. Separating Stars and Galaxies

We calculated the agreement between the partition created by each method and the "true" star/galaxy classification (based on SDSS labels), using the Adjusted Rand Index (ARI), which was proposed by Hubert and Arabie (1985). An ARI of 0 indicates the amount of agreement expected by randomly assigning the same number of items to the specified number of clusters, with the same number of items per cluster. An ARI of 1 indicates perfect agreement.

Figure 1(a) shows performance results for partitions that contain two to eight clusters. All three imputation methods negatively impact performance, producing results that are actually worse than that expected by random chance. We observe that when the shape features are omitted (marginalization), agreement steadily increases as the number of clusters goes up. The clustering method is able to assign unusual objects to their own clusters and more accurately separate stars and galaxies in the rest. With eight clusters, we see an improvement of more than 0.9 over the imputation methods. Including the observed shape information as constraints is competitive with, and sometimes superior to, marginalization. We expect that if shape information were more relevant for separating stars from galaxies, even higher agreement would be observed.

3.2. Impact on Fully Observed Objects

In Figure 1(b), we perform a stability analysis of the fully observed objects in the data set. Since only 42 of the 1507 objects have missing values, their inclusion should result in little change in the cluster assignments for the fully observed objects. We find that this is true for marginalization and for KSC (agreement for the fully observed objects stays high), but again the imputation methods do not perform well. Imputing the missing values seems to significantly impact the placement of other objects in the data set.

4. Conclusions

In this paper, we have demonstrated that the imputation methods commonly used to cluster objects when some feature values are missing can be particularly misleading when applied to astronomical data. Our empirical results with SDSS data show that imputation methods prevent the correct separation of stars and galaxies, while KSC with constraints generated from the fully observed objects performs much better (up to 90% improvement). In addition, KSC minimizes the impact on cluster assignments for the fully observed objects.

When there are a only a few observed values for a given feature, imputation methods are even less reliable because they have too little information from which to infer the missing values. This is exactly the case where KSC is most efficient, and effective, since the runtime require to generate and enforce all of the constraints is proportional to the number of fully observed objects. We plan to explore this further in future experiments.

Acknowledgments. We would like to thank Marie desJardins for her suggestions. This work was supported in part by NSF grant IIS-0325329. Funding for the Sloan Digital Sky Survey (SDSS) has been provided by the Alfred P. Sloan Foundation, the Participating Institutions, the National Aeronautics and Space Administration, the National Science Foundation, the U.S. Department of Energy, the Japanese Monbukagakusho, and the Max Planck Society. The SDSS Web site is http://www.sdss.org/.

References

Browse, R. A., Skillicorn, D. B., & McConnell, S. M. 2003, Using competitive learning to handle missing values in astrophysical datasets, in Workshop on Mining Scientific and Engineering Datasets, SIAM Data Mining Conference, 1

Giavalisco, M. 2002, Lyman-break galaxies, ARA&A, 40, 579

Goebel, J., Volk, K., Walker, H., Gerbault, F., Cheeseman, P., Self, M., Stutz, J., & Taylor, W. 1989, A&A, 222, L5

Hubert, L., & Arabie, P. 1985, Journal of Classification, 2, 193

Little, R. J. A., & Rubin, D. A. 1987, in Statistical analysis with missing data (John Wiley and Sons)

MacQueen, J. B. 1967, Some methods for classification and analysis of multivariate observations, in Proceedings of the Fifth Symposium on Math, Statistics, and Probability (Berkeley, CA: University of California Press), 281

Wagstaff, K. 2004, Clustering with missing values: No imputation required, in Classification, Clustering, and Data Mining Applications (Springer), 649

Wagstaff, K., Cardie, C., Rogers, S., & Schroedl, S. 2001, Constrained k-means clustering with background knowledge, in Proceedings of the Eighteenth International Conference on Machine Learning (Morgan Kaufmann), 577

Yoo, J., Gray, A., Roden, J., Fayyad, U. M., de Carvalho, R. R., & Djorgovski, S. G. 1996, in ASP Conf. Ser., Vol. 101, ADASS V, ed. G. H. Jacoby & J. Barnes (San Francisco: ASP), 41

ASTRONOMICAL DATA ANALYSIS SOFTWARE AND SYSTEMS XIV
ASP Conference Series, Vol. 347, 2005
P. L. Shopbell, M. C. Britton, and R. Ebert, eds.

JDBC Driver for AIPS++ Tables

Jason Ye

Brown University, Box 4087, Providence, RI 02912

Wesley K. Young and Boyd Waters

National Radio Astronomy Observatory, Socorro, NM 87801

Abstract. Integrating Java with AIPS++ can provide many advantages that cannot be realized with AIPS++ alone. Beyond simplifying architecture and code, use of Java in astronomical processing is promising because of its standardized nature, widely available tool packages and its exceptional GUI rendering abilities. We have implemented a Java Database Connectivity (JDBC) driver for the AIPS++ table system. This allows us to use the Table Query Language (TaQL), which is similar to SQL, to query and manipulate the database from Java and provides a standard interface between the AIPS++ table system and future Java applications.

1. Motivation

In a previous project, we implemented a Java program that communicated with the AIPS++ tables via Glish and Perl scripts. These scripts simply relayed the query to an AIPS++ table server that accesses the actual tables. In order to simplify the overall design and increase efficiency, we sought to eliminate all Glish and Perl components and to re-implement this system using only Java and C++. Furthermore, we desired to provide a standard interface so that the AIPS++ databases are easily accessible to any Java programmer.

2. Design

The overall design consists of three components: the AIPS++ database, a JDBC driver, and a Java program which uses the driver. The JDBC driver can be further divided into the C++ server side and the Java client side (see Figure 1). These components communicated with each other over the network via TCP/IP. This was accomplished by extending an existing application developed by Young called, ATABD. Communicating between C++ and Java necessitated some sort of serialization protocol in order to transmit information between the components. For this purpose, we used Virtual Observatory (VO) tables. Developed specifically to store astronomical data, VO tables were the obvious choice for the point of integration because its XML format allowed easy transfer over the network.

In anticipation for many simultaneous users, a multithreaded server was developed, allowing a single machine to accommodate many clients. Furthermore,

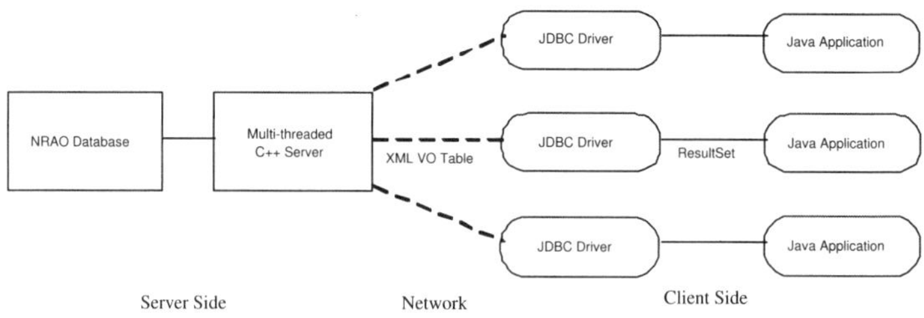

Figure 1. Architecture

the server application can be deployed on either a dedicated machine across the network or locally. The former option is better for security and multiple users. The latter is better for speed because the cost of network transfer is now essentially nonexistent.

For a typical transaction, the parent Java program submits a TaQL query via the JDBC client. The client packages the query as XML and transmits it to the server. Upon receipt, the server gathers the pertinent data using the C++ function calls on the database. The result is packaged in XML and returned to the client, which parses it into an object called the DataSet.

3. The DataSet

The DataSet object is an implementation of the java.sql.ResultSet interface. Unlike the typical commercial JDBC driver, which only allows scrolling in a certain direction, the DataSet allows forward scrolling, backward scrolling and jumping to any row in the DataSet. A smart caching technique allows the DataSet to choose which rows to cache depending on how the DataSet is currently being used. For example, if the DataSet detects heavy use of the previous() function, it will cache the rows such that more rows numbered less than the current row are obtained from the database. Of course, this caching mechanism is invisible to the user. To accomplish this, the DataSet queries the database via the JDBC client and gathers the new rows on demand.

The large size of the average observational data set requires a caching scheme to increase efficiency and prevent OutOfMemoryErrors. Maintaining a large cache size means less database accesses, but also means that each database access requires more time. Conversely, a smaller cache size requires less time per access, but more accesses. This tradeoff is addressed by allowing the user to set the cache size to optimize for the particular use.

4. Simplicity

One of our primary goals was to deliver an easy way to interface a Java program with the AIPS++ database. Whereas previous efforts required a complex system of scripts and intermediary components, we can now write a complete database application in just a few lines. Moreover, the syntax used is simple, standardized and complies with the syntax of the java.sql package. Here is an example:

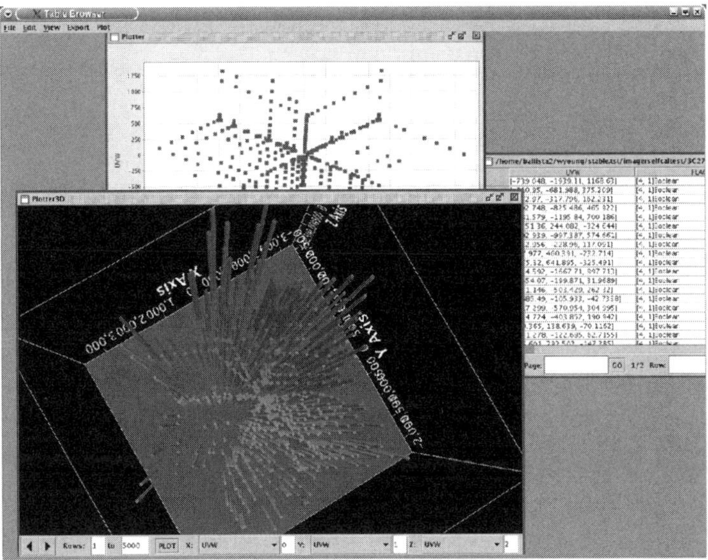

Figure 2. TableBrowser

```
Class.forName("TableDriver");
TableConnection con =
   DriverManager.getConnection("host", "username", "password");
QLStatement stmt = con.createStatement();
DataSet ds = stmt.executeQuery("select * from aTable");
ds.beforeFirst();
while(ds.next()){
   System.out.println(ds.getString(1));
}
```

5. Extensions

Besides the JDBC driver, two projects were implemented to demonstrate the usability of the JDBC driver. In the first project, we implemented a table viewing and editing application, entitled TableBrowser, using the Swing graphics package (Figure 2). It was modeled after an existing Glish version but demonstrated dramatic improvements in performance due to faster data acquisition from the AIPS++ database and Java's superior graphical rendering abilities. New features introduced include exporting data into VO Table format, 2-dimensional graphing of table data, and 3-dimensional graphing of table data. The latter was made possible by using an open source package based on Java3D.

The second project consisted of integrating the driver with a tool called ConeSearch. Using inputs of right ascension, declination and radius, ConeSearch searches the database and outputs astronomical objects within that region in space. It was rebuilt as a Java web servlet that connected to the AIPS++

database via the JDBC driver. The conversion from the Java-Glish-Perl-C++ protocol to JDBC was required minimal effort.

6. Future Work

We have introduced a procedure for connecting Java programs to the AIPS++ tables that is both efficient and easy to implement. Future projects, especially those consisting of extensive graphical user interfaces, can benefit by combining the graphical rendering capabilities of Java and the computational power of Glish.

The JDBC driver standardizes AIPS++ database access from Java. This allows a project to quickly adapt open source code not specifically built for astronomical purposes, saving both time and resources. Collaboration among the astronomy community can result in the development of reusable data analysis applications.

Recently, some AIPS++ data including the Cone Search Archive has been exported to other databases such as Oracle. If this trend continues, existing software developed specifically for the AIPS++ tables may become obsolete. However, Java applications using the JDBC driver are easily portable simply by replacing this JDBC driver with the driver distributed by the commercial database.

Finally, the server side of this driver provides a point of integration for the AIPS++ tables and any language supporting XML and TCP/IP. Perl, Glish and other scripting languages can connect as a client to the multi-threaded C++ server and process table data in programs that do not require heavy use of graphical rendering.

Part V
The Virtual Observatory: Handling the Data

ASTRONOMICAL DATA ANALYSIS SOFTWARE AND SYSTEMS XIV
ASP Conference Series, Vol. 347, 2005
P. L. Shopbell, M. C. Britton, and R. Ebert, eds.

Science with Virtual Observatory Tools

P. Padovani[1]
European Southern Observatory, Karl-Schwarzschild-Str. 2, D-85748 Garching bei München, Germany, Email Paolo.Padovani@eso.org

Abstract. The Virtual Observatory is now mature enough to produce cutting-edge science results. The exploitation of astronomical data beyond classical identification limits with interoperable tools for statistical identification of sources has become a reality. I present the discovery of 68 optically faint, obscured (i.e., type 2) active galactic nuclei (AGN) candidates in the two GOODS fields using the Astrophysical Virtual Observatory (AVO) prototype. Thirty-one of these sources have high estimated X-ray powers ($> 10^{44}$ erg/s) and therefore qualify as optically obscured quasars, the so-called QSO 2. The number of these objects in the GOODS fields is now 40, an improvement of a factor > 4 when compared to the only 9 such sources previously known. By going ~ 3 magnitudes fainter than previously known type 2 AGN in the GOODS fields the AVO is sampling a region of redshift – power space much harder to reach with classical methods. I also discuss the AVO move to our next phase, the EURO-VO, and our short-term plans to continue doing science with the Virtual Observatory.

1. Astronomy in the XXI Century

Astronomy is facing the need for radical changes. When dealing with surveys of up to $\sim 1,000$ sources, one could apply for telescope time and obtain an optical spectrum for each one of them to identify the whole sample. Nowadays, we have to deal with huge surveys (e.g., the Sloan Digital Sky Survey [SDSS[2]], the Two Micron All Sky Survey [2MASS[3]], the Massive Compact Halo Object [MACHO[4]] survey), reaching (and surpassing) the 100 million objects. Even at, say, 3,000 spectra at night, which is only feasible with the most efficient multi-object spectrographs and for relatively bright sources, such surveys would require more than 100 years to be completely identified, a time which is clearly much longer than the life span of the average astronomer! But even taking a spectrum might not be enough to classify an object. We are in fact reaching fainter and fainter sources, routinely beyond the typical identification limits of the largest telescopes available (approximately 25 magnitude for 2 - 4 hour exposures), which makes "classical" identification problematic. These very large surveys are also producing a huge amount of data: it would take more than two

[1] Astrophysical Virtual Observatory Scientist

[2] `http://www.sdss.org/`

[3] `http://www.ipac.caltech.edu/2mass/`

[4] `http://wwwmacho.anu.edu.au/`

months to download at 1 Mbytes/s (a very good rate for most astronomical institutions) the Data Release 3 (DR3[5]) SDSS images, about a month for the catalogues. The images would fill up $\sim$ 1,300 DVDs ($\sim$ 650 if using dual-layer technology). And the final SDSS will be about twice as large as the DR3. These data, once downloaded, need also to be analysed, which requires tools which may not be available locally and, given the complexity of astronomical data, are different for different energy ranges. Moreover, the breathtaking capabilities and ultra-high efficiency of new ground- and space-based observatories have led to a "data explosion", with astronomers world-wide accumulating more than one Terabyte of data per night (judging from some of the talks at this conference, this is very likely to be an underestimate). For example, the European Southern Observatory (ESO)/Space Telescope European Coordinating Facility (ST-ECF) archive is predicted to increase its size by two orders of magnitude in the next eight years or so, reaching $\approx 1,000$ Terabytes. Finally, one would like to be able to use all of these data, including multi-million-object catalogues, by putting this huge amount of information together in a coherent and relatively simple way, something which is impossible at present.

All these hard, unescapable facts call for innovative solutions. For example, the observing efficiency can be increased by a clever pre-selection of the targets, which will require some "data-mining" to characterise the sources' properties before hand, so that less time is "wasted" on sources which are not of the type under investigation. One can expand this concept even further and provide a "statistical" identification of astronomical sources by using all the available, multi-wavelength information without the need for a spectrum. The data-download problem can be solved by doing the analysis where the data reside. And finally, easy and clever access to all astronomical data worldwide would certainly help in dealing with the data explosion and would allow astronomers to take advantage of it in the best of ways.

2. The Virtual Observatory

The name of the solution is the Virtual Observatory (VO). The VO is an innovative, evolving system, which will allow users to interrogate multiple data centres in a seamless and transparent way, to make the best use of astronomical data. Within the VO, data analysis tools and models, appropriate to deal also with large data volumes, will be made more accessible. New science will be enabled, by moving Astronomy beyond "classical" identification with the characterisation of the properties of very faint sources by using all the available information. All this will require good communication, that is the adoption of common standards and protocols between data providers, tool users and developers. This is being defined now using new international standards for data access and mining protocols under the auspices of the recently formed International Virtual Observatory Alliance (IVOA[6]), a global collaboration of the world's astronomical communities.

---

[5] `http://www.sdss.org/dr3/`

[6] `http://ivoa.net`

One could think that the VO will only be useful to astronomers who deal with colossal surveys, huge teams and Terabytes of data! That is not the case, for the following reason. The World Wide Web is equivalent to having all the documents of the world inside one's computer, as they are all reachable with a click of a mouse. Similarly, the VO will be like having all the astronomical data of the world inside one's desktop. That will clearly benefit not only professional astronomers but also anybody interested in having a closer look at astronomical data. Consider the following example: imagine one wants to find *all* the observations of a given source available in *all* astronomical archives in a given wavelength range. One also needs to know which ones are in raw or processed format, one wants to retrieve them and, if raw, one wants also to have access to the tools to reduce them on-the-fly. At present, this is extremely time consuming, if at all possible, and would require, even to simply find out what is available, the use a variety of search interfaces, all different from one another and located at different sites. The VO will make all this possible very easily.

3. The VO in Europe and the Astrophysical Virtual Observatory

The status of the VO in Europe is very good. In addition to seven current national VO projects, the European funded collaborative Astrophysical Virtual Observatory initiative (AVO[7]) is creating the foundations of a regional scale infrastructure by conducting a research and demonstration programme on the VO scientific requirements and necessary technologies. The AVO has been jointly funded by the European Commission (under the Fifth Framework Programme [FP5]) with six European organisations participating in a three year Phase-A work programme. The partner organisations are ESO in Munich, the European Space Agency, AstroGrid (funded by PPARC as part of the United Kingdom's E-Science programme), the CNRS-supported Centre de Donnees Astronomiques de Strasbourg (CDS) and TERAPIX astronomical data centre at the Institut d'Astrophysique in Paris, the University Louis Pasteur in Strasbourg, and the Jodrell Bank Observatory of the Victoria University of Manchester. The AVO is the definition and study phase leading towards the Euro-VO - the development and deployment of a fully fledged operational VO for the European astronomical research community. A Science Working Group was also established to provide scientific advice to the project.

The AVO project is driven by its strategy of regular scientific demonstrations of VO technology, held on an annual basis in coordination with the IVOA. For this purpose progressively more complex AVO demonstrators are being constructed. The current one, a downloadable Java application, is an evolution of Aladin (Ochsenbein et al. 2005), developed at CDS, and has become a set of various software components, provided by AVO and international partners, which allows relatively easy access to remote data sets, manipulation of image and catalogue data, and remote calculations in a fashion similar to remote computing (see Figure 1).

---

[7] `http://www.euro-vo.org`

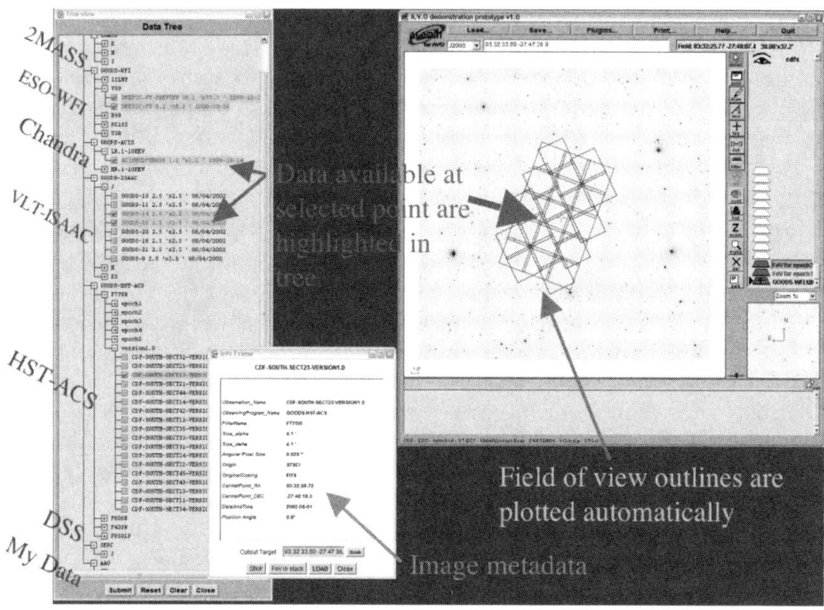

Figure 1. The AVO prototype in action. An ESO/WFI image of the GOODS southern field, overlaid with the HST/ACS data field of view outlines. The "data-tree" on the left shows the images available in the Aladin image server. Data available at selected coordinates get highlighted in the tree. Metadata information is also accessible. The user's own data can also be loaded into the prototype. This is based on the use of IVOA agreed standards, namely the Data Model, descriptive Metadata, and data interchange standards.

4. Doing Science with the AVO

The AVO held its second demonstration, 'AVO 1st Science', on January 27 - 28, 2004 at ESO. The demonstration was truly multi-wavelength, using heterogeneous and complex data covering the whole electromagnetic spectrum. These included: MERLIN, VLA (radio), ISO [spectra and images] and 2MASS (infrared), USNO, ESO 2.2m/WFI and VLT/FORS [spectra], and HST/ACS (optical), XMM and Chandra (X-ray) data and catalogues. Two cases were dealt with: an extragalactic case on obscured quasars, centred around the Great Observatories Origin Deep Survey (GOODS) public data, and a Galactic scenario on the classification of young stellar objects.

The extragalactic case was so successful that it turned into the first published science result fully enabled via end-to-end use of VO tools and systems, the discovery of ~ 30 high-power, supermassive black holes in the centres of apparently normal looking galaxies.

5. Discovering Optically Faint, Obscured Quasars with VO Tools

How did we get a scientific paper out of a science demonstration? The extragalactic science case revolved around the two GOODS fields (Giavalisco et al. 2004a), namely the Hubble Deep Field-North (HDF-N) and the Chandra Deep Field-South (CDF-S), the most data-rich, deep survey areas on the sky. Our idea was to use the AVO prototype to look for high-power, supermassive black holes in the centres of apparently normal looking galaxies.

Black holes lurk at the centres of active galaxies (AGN) surrounded by dust which is thought to be, on theoretical and observational grounds (see, e.g., Urry & Padovani 1995; Jaffe et al. 2004), distributed in a flattened configuration, torus-like. When we look down the axis of the dust torus and have a clear view of the black hole and its surroundings these objects are called "type 1" AGN, and display the broad lines (emitted by clouds moving very fast close to the black hole) and strong UV emission typical of quasars. "Type 2" AGN, on the other hand, lie with the dust torus edge-on as viewed from Earth so our view of the black hole is totally blocked by the dust over a range of wavelengths from the near-infrared to soft X-rays. The optical/UV spectrum of type 2 AGN is characterized by emission lines much narrower than those of quasars, as they are emitted by clouds which are further away and therefore move more slowly.

While many dust-obscured low-power black holes, the Seyfert 2s, have been identified, until recently few of their high-power counterparts were known. This was due to a simple selection effect: when the source is a low-power one and therefore, on average, closer to the observer, one can very often detect some features related to narrow emission lines on top of the emission from the host galaxy, which qualify it as a type 2 AGN. But when the source is a high-power one, a so-called QSO 2, and therefore, on average, further away from us, the source looks like a normal galaxy. Until very recently, QSO 2s were selected against by quasar surveys, most of which were tuned to find objects with very strong UV emission. The situation has changed with the advent of Chandra and XMM-Newton, which are providing a sensitive window into the hard X-ray emission of AGN.

5.1. The Method

The two key physical properties that we use to identify type 2 AGN candidates are that they be obscured, and that they have sufficiently high power to be classed as an AGN and not a starburst. Our approach was to look for sources where nuclear emission was coming out in the hard X-ray band, with evidence of absorption in the soft band, a signature of an obscured AGN, and the optical flux was very faint, a sign of absorption. One key feature was the use of a correlation discovered by Fiore et al. (2003) between the X-ray-to-optical ratio and the X-ray power, which allowed us to select QSO 2s even when the objects were so faint that no spectrum, and therefore no redshift, was available.

We selected absorbed sources by using the Alexander et al. (2003) X-ray catalogues for the two GOODS fields, which provide counts in various X-ray bands. We define the hardness ratio $HR = (H - S)/(H + S)$, where H is the hard X-ray counts ($2.0 - 8.0$ keV) and S is the soft X-ray counts ($0.5 - 2.0$ keV). Szokoly et al. (2004) have shown that absorbed, type 2 AGN are characterized by $HR \geq -0.2$. We adopt this criterion and identify those sources which have $HR \geq -0.2$ as absorbed sources. We find 294 (CDF-S: 104, HDF-N: 190) such absorbed sources which represent $35^{+3}_{-2}\%$ of the X-ray sources in the Alexander catalogues. Note that increasing redshift makes the sources softer (e.g., at $z = 3$ the rest-frame $2 - 8$ keV band shifts to $0.5 - 2$ keV) so our selection criterion will mistakenly discard some high-z type 2 sources, as pointed out by Szokoly et al. (2004). The number of type 2 candidates we find has therefore to be considered a lower limit.

The optical counterparts to the X-ray sources were selected by cross-matching the absorbed X-ray sources with the GOODS ACS catalogues (29,599 sources in the CDF-S, 32,048 in the HDF-N). We used version v1.0 of the reduced, calibrated, stacked, and mosaiced images and catalogues as made available by the GOODS team[8]. The GOODS catalogues contain sources that were detected in the z-band, with BVi photometry in matched apertures (Giavalisco et al. 2004b).

We initially searched for optical sources that lay within a relatively large threshold radius of $3.5''$ (corresponding to the maximal 3σ positional uncertainty of the X-ray positions) around each X-ray source. This was done using the cross match facility in the AVO prototype tool using the "best match" mode. Since the $3.5''$ radius is large relative to the median positional error, and given the optical source density the initial cross match inevitably includes a number of false and multiple matches. To limit our sample to good matches, we use the criterion that the cross match distance be less than the combined optical and X-ray 3σ positional uncertainty for each individual match. Applying this distance/error < 1 criterion we limit the number of matches to 168 (CDF-S: 65, HDF-N: 103). These matches are all within a much smaller radius than our initial $3.5''$ threshold, with most of the distance/error < 1 matches being within $1.25''$ (and two matches at 1.4 and $1.5''$). The estimated number of false matches we expect to have is small, between 8 and 15%.

[8]http://www.stsci.edu/science/goods/

Previously classified sources and their spectroscopic redshifts are available from Szokoly et al. (2004) for the CDF-S and Barger et al. (2003) for the HDF-N. Derivation of X-ray powers for these objects is straightforward[9]. For the unclassified sources we estimated the X-ray power as follows: we first derived the $f(2-10keV)/f(R)$ flux ratio (converting the ACS i magnitudes to the R band), and then estimated the X-ray power from the correlation found by Fiore et al. (2003), namely $\log L_{2-10} = log f(2-10keV)/f(R) + 43.05$ (Fiore, p.c.; see their Figure 5). Note that this correlation has an r.m.s. of ~ 0.5 dex in X-ray power. We stress that our estimated X-ray powers reach $\sim 10^{45}$ erg/s and therefore fall within the range of the Fiore et al. (2003) correlation. On the other hand it should be pointed out that our sources are much fainter than the objects which have been used to calibrate the Fiore et al.'s correlation.

The work of Szokoly et al. (2004) has shown that absorbed, type 2 AGN are characterized by $HR \geq -0.2$. It is also well known that normal galaxies, irrespective of their morphology, have X-ray powers that reach, at most, $L_{\rm x} \lesssim 10^{42}$ erg/s (e.g., Forman, Jones, & Tucker 1994; Cohen 2003). Therefore, any X-ray source with $HR \geq -0.2$ and $L_{\rm x} > 10^{42}$ erg/s should be an obscured AGN. Furthermore, following Szokoly et al. (2004), any such source having $L_{\rm x} > 10^{44}$ erg/s will qualify as a type 2 QSO.

5.2. Results

Out of the 546 X-ray sources in the GOODS fields, 203 are absorbed ($HR \geq -0.2$). Out of these we selected 68 type 2 AGN candidates, 31 of which qualify as QSO 2 (estimated X-ray power $> 10^{44}$ erg/s). We note that the distribution of estimated X-ray power covers the range $5 \times 10^{42} - 2 \times 10^{45}$ erg/s and peaks around 10^{44} erg/s (see Figure 2). The number of QSO 2 candidates, therefore, is very sensitive to the dividing line between low- and high-luminosity AGN, which is clearly arbitrary and cosmology dependent. For example, if one defines as QSO 2 all sources with $L_{2-10} > 5 \times 10^{43}$ erg/s, a value only a factor of 2 below the commonly used one and corresponding to the break in the AGN X-ray luminosity function (Norman et al. 2002), the number of such sources increases by $\sim 50\%$. We also note that, based on the r.m.s. around the Fiore et al. (2003) correlation, the number of QSO 2 candidates fluctuates in the $13-54$ region. The number of type 2 AGN, on the other hand, can only increase, as all our candidates have estimated $\log L_{2-10} > 42.5$.

Our work brings to 40 the number of QSO 2 in the GOODS fields, an improvement of a factor ~ 4 when compared to the only nine such sources previously known. As expected, being still unidentified, our sources are very faint: their median ACS i magnitude is ~ 25.5, which corresponds to $R \sim 26$ (compare this to the $R \sim 22$ typical of the CDF-S sources with redshift determination). The QSO 2 candidates are even fainter, with median i magnitude ~ 26.3 ($R \sim 26.8$). Therefore, spectroscopical identification is not possible, for the large majority of objects, even with the largest telescopes currently available. We have used our estimated X-ray powers together with the observed fluxes to derive redshifts for our type 2 candidates (tests we have performed on the type 2 sources with

[9]Throughout this paper we adopt a cosmological model with $H_0 = 70$ km s^{-1} Mpc^{-1}, $\Omega_{\rm M} = 0.3$, and $\Omega_\Lambda = 0.7$.

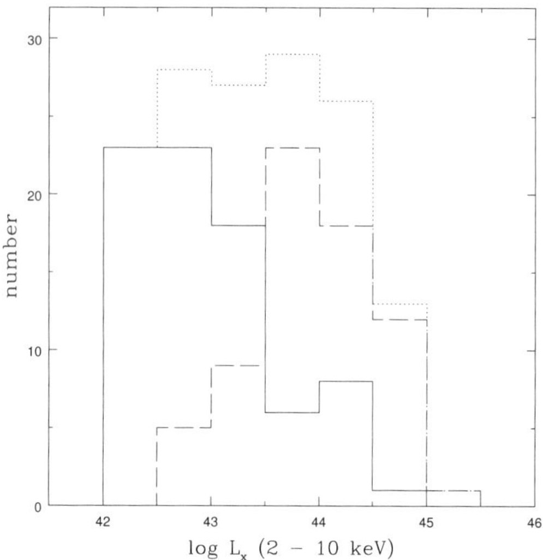

Figure 2. The X-ray power distribution for our new type 2 AGN candidates (dashed line), previously known type 2 AGN (solid line), and the sum of the two populations (dotted line). QSO 2 are defined, somewhat arbitrarily, as having $L_{2-10\mathrm{keV}} > 10^{44}$ erg/s.

spectroscopic redshifts show that this method, although very simple, is relatively robust). Our type 2 AGN are expected to be at $z \approx 3$, while our QSO 2 should be at $z \approx 4$. By using VO methods we are sampling a region of redshift - power space so far much harder to reach with classical methods. For the first time, we can also assess how many QSO 2 there are down to relatively faint X-ray fluxes. We find a surface density > 330 deg^{-2} for $f(0.5 - 8keV) \geq 10^{-15}$ erg cm^{-2} s^{-1}, higher than previously estimated.

Figure 2 shows the X-ray power distribution for our new type 2 AGN candidates (dashed line), previously known type 2 AGN (solid line), and the combined sample (dotted line). It is interesting to note how the distributions are very different, with the already known type 2 AGN peaking around $L_{\mathrm{x}} \sim 10^{43}$ erg/s and declining for luminosities above $\sim 3 \times 10^{43}$ erg/s, while our new candidates are rising in this range and peak around $L_{\mathrm{x}} \sim 10^{44}$ erg/s. To be more quantitative, while only $\sim 1/5$ of already known type 2 AGN have $\log L_{\mathrm{x}} > 43.5$, $\sim 3/4$ of our candidates are above this value. This difference is easily explained by our use of the X-ray-to-optical flux ratios to estimate X-ray powers and by the fact that our candidates are on average ~ 3 magnitudes fainter than previously known sources. Our method is then filling a gap in the luminosity distribution, which becomes almost constant in the range $10^{42} \lesssim L_{\mathrm{x}} \lesssim 3 \times 10^{44}$ erg/s. This also explains the fact that the number of QSO 2 candidates we find is $\gtrsim 3$ times larger than the previously known ones.

The identification of a population of high-power obscured black holes and the active galaxies in which they live has been a key goal for astronomers and will lead to greater understanding and a refinement of the cosmological models describing our Universe. The paper reporting these results has been recently published (Padovani et al. 2004).

The AVO prototype made it much easier to classify the sources we were interested in and to identify the previously known ones, as we could easily integrate all available information from images, spectra, and catalogues at once. This is proof that VO tools have evolved beyond the demonstration level to become respectable research tools, as the VO is already enabling astronomers to reach into new areas of parameter space with relatively little effort.

The AVO prototype can be downloaded from the AVO Web site[10]. We encourage astronomers to download the prototype, test it, and also use it for their own research. For any problems with the installation and any requests, questions, feedback, and comments you might have please contact the AVO team at twiki@euro-vo.org. (Please note that this is still a prototype: although some components are pretty robust some others are not.)

6. Near Future AVO Science Developments

The AVO is promoting science with VO tools through two further developments: a Science Reference Mission and the next science demonstration.

6.1. The AVO Science Reference Mission

The AVO team, with input from the Science Working Group, is putting together a Science Reference Mission. This will define the key scientific results that the full-fledged EURO-VO should achieve when fully implemented and will consist of a number of science cases covering a broad range of astronomical topics, with related requirements, against which the success of the EURO-VO will be measured.

6.2. The Next AVO Science Demonstration

The next and last AVO science demonstration is to be held in January 2005 at the European Space Astronomy Centre (ESAC; formerly known as VILSPA). Preparations are still on-going so the details are not fully worked out yet but it is firmly established that we will be dealing with two scenarios. The first, on star formation histories in galaxies, will revolve around the European Large-Area ISO Survey (ELAIS), which covers five different areas of the sky over ~ 10 deg^2. The second, on the transition from Asymptotic Giant Branch to Planetary Nebulae, will be the strongest one on the science side and should produce a new list of stars in this very interesting transitional phase.

On the technical side, the science demonstration will see the rollout of the first version of the EURO-VO portal, through which European astronomers will gain secure access to a wide range of data access and manipulation capabilities. Also, we will demonstrate the use of distributed workflows, registry harvesting, and the wrapping of sophisticated astronomical applications as Web services.

The AVO demonstration will also mark the transition from the AVO to the EURO-VO. Funding for the technology part of the EURO-VO, VO-TECH, has

---

[10]`http://www.euro-vo.org/twiki/bin/view/Avo/SwgDownload`

been secured from the European Community at a level of 6.6 million Euros, which will translate into 12 Full Time Equivalent (FTEs). Twelve more FTEs will be provided by the partners, which include Edinburgh, Leicester, and Cambridge in the UK, ESO, CDS, and INAF in Italy.

7. Summary

The main results of this paper can be summarized as follows:

- The Virtual Observatory is happening because it has to! If it does not, we will not be able to cope with the huge amount of data astronomers are being flooded with.
- Astronomy can and *is being* done with Virtual Observatory tools, which are now mature enough. Real science results are being produced and papers are being published.
- The Astrophysical Virtual Observatory, soon to be EURO-VO, is committed to the pursuit of science with Virtual Observatory tools through scientific demonstrations, science papers, and a Science Reference Mission.

Acknowledgments. The obscured quasar paper was done in collaboration with Mark Allen, Piero Rosati, and Nic Walton. It is a pleasure to thank the AVO team for their superb work, without which the paper would have not been possible, and the many people who have produced the data on which the paper is based, particularly the GOODS, CDF-S, Penn State, and HELLAS Teams.

References

Alexander, D. M., et al. 2003, AJ, 126, 539

Barger, A. J., et al. 2003, AJ, 126, 632

Cohen, J. G. 2003, ApJ, 598, 288

Fiore, F., et al. 2003, A&A, 409, 79

Forman, W., Jones, C., & Tucker, W. 1994, ApJ, 429, 77

Giavalisco, M., et al. 2004a, ApJ, 600, L93

Giavalisco, M., et al. 2004b, in preparation

Jaffe, W., et al. 2004, Nature, 429, 47

Norman, C., et al. 2002, ApJ, 571, 218

Ochsenbein, F., et al. 2005, in ASP Conf. Ser., Vol. 347, ADASS XIV, ed. P. L. Shopbell, M. C. Britton, & R. Ebert (San Francisco: ASP), 193

Padovani, P., Allen, M. G., Rosati, P., & Walton, N. A. 2004, A&A, 424, 545

Szokoly, G. P., et al. 2004, ApJS, in press [astro-ph/0312324]

Urry, C. M., & Padovani, P. 1995, PASP, 107, 803

ASTRONOMICAL DATA ANALYSIS SOFTWARE AND SYSTEMS XIV
ASP Conference Series, Vol. 347, 2005
P. L. Shopbell, M. C. Britton, and R. Ebert, eds.

Interoperability in Action: the Aladin Experience

François Ochsenbein, Pierre Fernique, François Bonnarel, Mark Allen, Thomas Boch, Françoise Genova, and André Schaaff

Centre de Données astronomiques de Strasbourg, Observatoire de Strasbourg, UMR 7550, 11 rue de l'Université, 67000 Strasbourg, France

Abstract. Since its first version created 8 years ago, Aladin has been continuously developed, and has been recently used as the main component of the Astrophysical Virtual Observatory prototype, demonstrating the power of VO interoperability standards for enabling real science. But interoperability is not just a set of standards used to describe accurately the data – it is also a means by which programs and tools can interact and cooperate. Several ways of communicating with and from Aladin have been developed, including a scripting language, Java interfaces, and plug-ins. The paper focuses on the lessons learned during the Aladin experience in several aspects: the strengths and weaknesses of the programming language (Java); the importance of the standards and protocols, their surprises and their challenges; the difficulties of actual interoperation; and the necessity of maintaining compatibility with the users' material.

1. Introduction

Aladin was born 8 years ago as a small applet named *Aladin-lite*, devoted to a visual correlation of sky images with the astronomical objects described in catalogues and databases. Since then, Aladin has been continuously developed to provide good visualisation of catalog data overlaid on the observed pixels, and to be easily applicable to the holdings of existing image and data servers (see e.g., Bonnarel et al. 2000). Today, various institutes and data centers use Aladin to visualize some of their data (Simbad, VizieR, NED, XMM-Newton, CFH, CADC, IVBS, LEDAS,...). Aladin was moreover used recently in AVO *(Astrophysical Virtual Observatory)* demonstrations as the main component of the AVO prototype, demonstrating the usefulness of VO interoperability standards for actual scientific discoveries (Padovani et al. 2004; Fernique et al. 2004). Aladin usage currently represents about 800 sessions per day.

We will discuss the two main points which, in our mind, explain Aladin's success story: the historical context of Aladin development, and the choice of a technology well adapted to its purpose.

2. The Context of Aladin Development

The emergence of the first digital sky surveys about a decade ago introduced a new possibility of a visual comparison between the data stored in the databases managed by the CDS with the actual sky atlases; and it happened that this need joined a real need of the astronomical community that could by summarized by "get all astronomical data on my desktop with a simple toolkit to compare and correlate them". In other words, it is a wish for a Virtual Observatory before the VO.

It also happened that Aladin was born just after the Web revolution, when astronomical institutes and data centers became all accessible via a unique way which characterizes the Web paradigm: a universal addressing scheme (the URL) plus a universal protocol (HTTP over IP) in an open world, transformed this dream of *data available on everybody's desktop* into a reality. Without this revolution, a tool like Aladin would not have had any chance to reach its goal.

3. The Suited Technology

Aladin basically consists in a graphical tool able to display and manipulate images and catalog data associated to a capability of communicating with data servers. Originally the choice was limited to the programming language and the used libraries, notably the graphical library used in the user interface. With the Web revolution, the extensive development of the browser capabilities introduced such possibilities of manipulating remote data and images: HTML pages/forms, with Frames and/or Javascript, using Java natively or via a plug-in.

Each technology has its own advantages and disadvantages, and the final choice among the available technologies is always a compromise notably on the following factors:

- *interactivity* — how easily it is to manipulate the data
- *performance* — the speed and size of the data which can be manipulated
- *installation easiness* — the tool is easily downloadable and does not require computer science skills or computer-dependent tuning
- *connectivity factor* — how easy it can be integrated to other existing tools, within other environments
- *development easiness* — how easy it is to maintain and improve the tool.

Eight years ago, we chose Java as the language for Aladin, and we offered two possibilities: a native Java Applet that can be launched directly in any Web browser, and a Java standalone application requiring an actual installation.

We will explain why this technology choice seemed to be a good compromise for Aladin in regards of the various factors specified above.

3.1. Interactivity

A high level of interactivity is required as soon as we want to "play" with the images and data, like zooming, cropping, overlaying... The Java native multi-threading and event programmation presents decisive advantages for reaching easily a high level of interactivity. From these Java capabilities, we were able to build an intuitive user interface.

3.2. Performance

Java is not the best choice to get high performances in terms of CPU and memory management: it is 2 to 10 times slower than an equivalent program written in C. The Java Virtual Machine moreover requires a fixed memory size, which may cause some trouble in actual implementations.

We however notice a very positive evolution with the *Just-In-Time* (JIT) Java method (cpu native compilation during a Java application running critical code, such as intensively used loops), and also a real improvement of the memory management, and the addition of "native" libraries (such as the JAI library).

In this context, Aladin features are always close to the limit of what is permitted by the Java limitations in CPU and memory: the color composition and the computation of image contours could reasonably be introduced only in 2002; the visualisation of catalogued parameters *(filters)* was added in 2003, and the access to the full dynamics, the multiview and the blinking modes were introduced in 2004.

3.3. Installation Easiness

The installation easiness is closely related to the compatibility factor. It can be easy to prepare a package that runs well on a well-known platform such as a PC running Windows, but having this package running on the large variety of the existing user configurations becomes a challenge. And seen from the user's side, the installation can turn into a nightmare or simply become impossible if some existing configurations have been neglected.

The platform compatibility is an intrinsic advantage offered by Java, as the Java bytecode is interpreted by a *Java Virtual Machine* (JVM) which exists for all kinds of platforms. But the great idea of Java was the *applet* mode: the applet consists in Java code which runs in a Web browser without any installation. For many years Aladin was essentially used in this applet mode, with a very good user feedback.

Unfortunately, the current evolution tends to minimize the attractivity of the applet concept: there are now various JVMs which can react differently, and the JVMs included in browsers are generally fairly old, and do not implement the new libraries. Moreover Microsoft's XP Operating System is removing totally JVM from its Web Browser, meaning that the users will have again to install their Java environment. In other terms, the Applet concept is now dying. In this new context, we are looking for WebStart technology hoping that it can offer a good alternative for Aladin.

3.4. Connectivity Factor

Whatever clever and smart a tool can be, it has to be connected to other tools and servers. This *connectivity factor* expresses the easiness of exchanging data and information with other tools or servers.

Aladin uses extensively the URL/HTTP concept to access remote servers either for getting the remote data/images or for retrieving the related Web pages — we would call this paradigm the current "royal way" of linking a tool to the rest of the world. In order to avoid any hard-coded URL, Aladin uses a registry mechanism called *GLU* (Fernique et al. 1997), which not only does the URL substitution, but chooses the most accessible site among mirrors, performs the required parameter substitutions, etc... The rapid development of the Virtual Observatory protocols also lead us to implement as fast as possible the standards emerging in this context: VOTable, SIA, ConeSearch, Skynode, ...

The connectivity with other Java applications is generally easy — or briefly said "Java likes Java" — and for instance SpecView (Busko 2000) and VOSpec

(Osuna et al. 2005) can communicate with Aladin. The connectivity to and from the other tools is also possible through an *Application Program Interface* (API), and two have been developed around Aladin: a scripting language which drives Aladin, and a specific *ExtApp* Java interface which implements methods to interact with Aladin, currently used by VOPlot (Kale et al. 2004).

3.5. Development Easiness

Java is a great language, and has strong advantages for its maintainability, robustness, and reusability. But the performance factor and the installation easiness can involve difficult development strategic choices. One concern is to maintain a compatibility with the actual users' configurations — and for instance we noticed that 40% of Aladin users are still using the now old JVM 1.1. It therefore means that we have to adapt our code to stay compatible with these old configurations; it also implies that we should not use native library such as JAI, which are not known in old configurations. Some Java classes are also notoriously slow and/or heavy (e.g., Image class, or the WebServices library), requiring additional code in order to keep acceptable performances.

4. Aladin Version 3

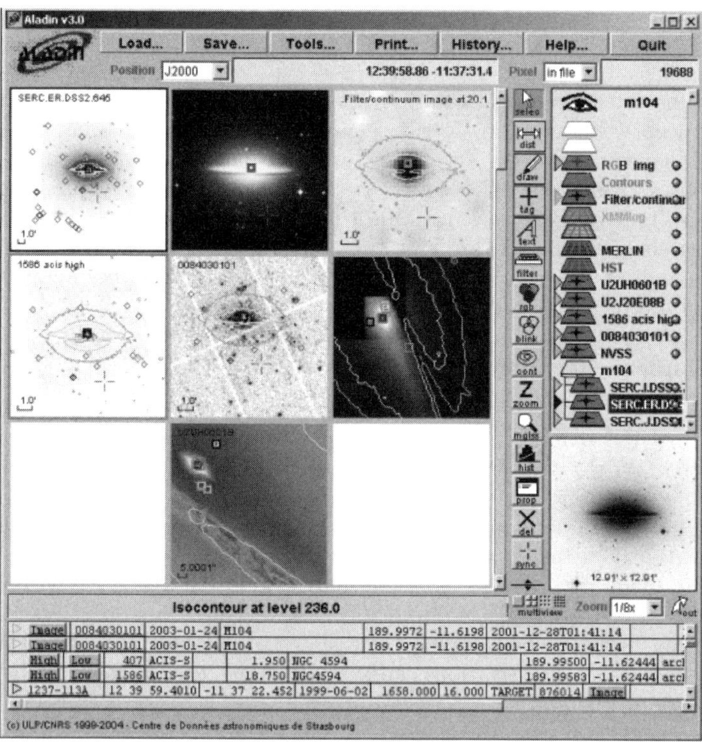

Figure 1. An example of the Multiviews in Aladin V3

The new Aladin V3 release illustrates the compromises presented above. The progress in the development of the VO standards and the improvements of the Java performances made possible the implementation of the following new features:

- Aladin V3 can manage several views simultaneously, each having its own projection for overlays; this new capability is illustrated in Figure 1.
- Aladin V3 keeps the full pixel depth — the previous versions restricted the dynamics to 8 bits for obvious memory limitations. New capabilities of rendering this extended dynamics have been added: scaling functions (log, sqrt, linear, squared), image profiles, etc...
- a sequence of images can now be displayed as a movie.
- Aladin V3 implements the new VO registry standard, giving a unified access to all the resources described in the VO registry — an actual implementation of a *VO discovery* tool.

5. Futures

The context is continuously changing, new standards will continue to emerge, and Aladin will continue to evolve accordingly. The two main challenges that we foresee for the near future are most likely the development of the Web Services (the SOAP standards) such as SkyNode, and the connection with astronomical integrated environments such as IRAF, IDL or MIDAS.

References

Bonnarel, B., et al. 2000, A&AS, 143, 33

Busko, I. 2000, in ASP Conf. Ser., Vol. 216, ADASS IX, ed. N. Manset, C. Veillet, & D. Crabtree (San Francisco: ASP), 79

Fernique, P., Ochsenbein, F., & Wenger, M. 1997, in ASP Conf. Ser., Vol. 145, ADASS VII, ed. R. Albrecht, R. N. Hook, & H. A. Bushouse (San Francisco: ASP), 466

Fernique, P., et al. 2004, "Toward an International Virtual Observatory", Proceedings of the ESO/ESA/NASA/NSF Conference

Kale, S., et al. 2004, in ASP Conf. Ser., Vol. 314, ADASS XIII, ed. F. Ochsenbein, M. Allen, & D. Egret (San Francisco: ASP), 350

Osuna, P., et al. 2005, in ASP Conf. Ser., Vol. 347, ADASS XIV, ed. P. L. Shopbell, M. C. Britton, & R. Ebert (San Francisco: ASP), 198

Padovani, P., Allen, M. G., Rosati, P., & Walton, N. A. 2004, A&A, 424, 545

ASTRONOMICAL DATA ANALYSIS SOFTWARE AND SYSTEMS XIV
ASP Conference Series, Vol. 347, 2005
P. L. Shopbell, M. C. Britton, and R. Ebert, eds.

VOSpec: A Tool for Handling Virtual Observatory Compliant Spectra

Pedro Osuna, Isa Barbarisi, Jesus Salgado, and Christophe Arviset

European Space Astronomy Centre, ESA, Villafranca del Castillo, 28080 Madrid, Spain

Abstract. There are lots of tools for spectra analysis and display in the astronomical community. However, there is a need for a "VO-enabled" tool that can superimpose spectra coming from different projects within the VO and in VO-format.

One of the main reasons for the current absence of such a tool is the fact that astronomical spectra are not as well defined as, e.g., images are. The IVOA has made a significant effort in trying to define a Simple Spectral Access Protocol by which part of this problem is solved, giving the rules to be able to handle spectral data in the VO context.

By defining an algorithm based on dimensional analysis for the superimposition of spectra coming from VO resources, we at ESAC have been able to create a tool that can superimpose VO spectra that declare their metadata in a specific way. To do this, VO SSAP-compatible resources will only have to give three new extra fields in their VOTable SSAP results for the tool to be able to superimpose the spectra. At the time of writing this abstract, the ESAC is planning to propose these three extra metadata fields for acceptance at the IVOA as optional inputs in the definition of the Simple Spectral Access Protocol.

1. Access to Spectra in the Virtual Observatory context: the Simple Spectrum Access Protocol

At the time of this presentation, the International Virtual Observatory Alliance has not published yet the Simple Spectrum Access Protocol specification document. ESAC proposed in the IVOA Workshop held in Strasbourg in October 2003 the inclusion of a Simple Spectrum Access Protocol that would be very much like the Simple Image Access Protocol (SIAP) already published by the IVOA but suited for Spectrum access. ESAC was asking for the inclusion of three extra parameters, to be described later, to allow for the automatic handling of units, a never-ending problem in the astronomical spectrum world. The VOSpec tool was built following the assumptions that the IVOA would publish an SSAP along the lines described above, which we believe is already the case. In view of the absence of a specification, we will describe here how we understand the basic mechanism of the Simple Spectrum Access protocol and how we have implemented it to show the usefulness and functionality of the VOSpec tool.

To access spectra through the SSAP, a data provider has to implement a service that accepts a call with the following basic input parameters: **POS** and

SIZE, where POS describes the position of the object under investigation as an RA,DEC pair in decimal degrees, and SIZE is the size in degrees around the position in the Sky where results are sought.

An example of a valid SSAP request would be something like:

http://pma.standby.vilspa.esa.es:8080/aio/jsp/siap.jsp?POS=304,38&SIZE=0.3

where the example corresponds to the SSAP service for the Infrared Space Observatory data at ESAC.

The response from the SSAP service is a VOTable containing the matches for the query plus the relevant metadata for them. One of the compulsory entries in those metadata is the pointer to the real spectra for the matches.

2. SSAP metadata extension: dimensional parameters

One of the biggest problems in handling spectra is the fact that the units have to be parsed to be able to compare different spectra. This is normally cumbersome and very prone to errors. Even though there have been attempts to homogenise the units names, it is very difficult, if not impossible, to get agreement from everybody to use the same unit names. Even after agreeing (the IAU gives a recommendation on `http://www.iau.org/IAU/Activities/nomenclature/units.html`), people tend to make mistakes or to even ignore the standards and use different names for the same units. The very simple case of the micron is a very good example: names such as "um", "micron", "micra", and "mu" are common for this unit, making the automatic handling of spectra coming from different data providers a very difficult task.

Pedro Osuna and Jesus Salgado proposed a solution to this problem to the IVOA using dimensional analysis. The proposal, as was done to the community, can be found at `http://ivoa.net/forum/dal/0311/0098.htm`. A more detailed explanation can be found at `http://esavo.esa.int/vodocs/DimensionalAnalysisSpectrum.pdf`.

According to this approach, the metadata for each spectrum should be accompanied by the following three extra parameters:

- FIELD ID="AXES" ucd=VOX:Spectrum_Axes
- FIELD ID="DIMEQ" ucd=VOX:Spectrum_Dimeq
- FIELD ID="SCALEQ" ucd=VOX:Spectrum_Scaleq

where the terminology used is that of the VOTable standard (the standard output format for the Simple Spectrum Access Protocol; more details at: `http://cdsweb.u-strasbg.fr/doc/VOTable/`).

The three parameters above represent:

- The wavelength and flux axes declaration. In the case of spectra in fits format, these would correspond to the keyword names in the fits corresponding to the main wavelength and fluxes. In the case of VOTable or ASCII spectra, they would correspond to the FIELD ID declarations for those two columns.
- The spectrum units dimensional equation and the spectrum units scale. This is needed for the software to be able to automatically handle units,

do multi-wavelength analysis and spectrum superimposition (see references above).

3. A working example: the HST case

As an example on how these extra parameters (the DIMEQ and the SCALEQ) are calculated, we give a working example from HST data. Spectra coming from HST are in ergs per squared centimeter, second, and Angstrom. The dimensional units are, therefore, calculated as follows:

$$[HST] = \left[\frac{erg}{cm^2 s \mathring{A}}\right] = \frac{10^{-7} M L^2 T^{-2}}{10^{-4} L^2 T 10^{-10} L} = 10^7 M L^{-1} T^{-3} \tag{1}$$

According to the proposed algorithm to go from these units to the reference ones in VOSpec (arbitrary ones, chosen to be Jy) the following calculation will have to be done (automatically by the software):

$$\left.\begin{array}{l} Jy \quad dimensions : [Jy] = M T^{-2} ScalingFactor 10^{-26} \\ HST \quad dimensions : [HST] = M L^{-1} T^{-3} Scaling factor 10^7 \end{array}\right\} \tag{2}$$

$$\frac{[Jy]}{[HST]} = \frac{M T^{-2}}{M L^{-1} T^{-3}} = L^1 T^1 \tag{3}$$

$$[\lambda]^n [c]^m = L^n L^m T^{-m} == L^1 T^1 \tag{4}$$

$$\left.\begin{array}{r} n - m = 1 \\ -m = 1 \end{array}\right\} \Longrightarrow \left\{\begin{array}{l} m = 1 \\ n = -2 \end{array}\right. \Longrightarrow [Jy] = [HST] \frac{\lambda^2}{c} \frac{10^7}{10^{-26}} \tag{5}$$

Therefore, once the dimensional and scaling factors are given, we are able to change units from one to the other automatically without string parsing (the client only has to be able to solve a linear system of equations).

4. VOSpec fact sheet

The VOSpec is the ESAC's answer to the necessity to have a VO-enabled Spectrum Display tool, that implements the aforementioned algorithm. The VOSpec can:

- Access the Registry to get information on available SSAP servers for the region of interest.
- Use available Web Services when needed (e.g., the SIMBAD name resolver from the target name resolution)
- Display VOTable information (standard for VO data)
- Get Spectra from SSAP servers - Display and superimpose spectra
- Accept spectra in both VOTable and FITs binary table formats
- Handle units automatically through the dimensional analysis proposal above - Able to do SEDs and multi-wavelength analysis
- Polynomial, blackbody, and Gaussian fitting
- And more functionality is on the way...

5. VOSpec: An Interoperable Tool

As explained above, the VOSpec is a tool immersed in the VO context. The following is the general layout of the tool, where its different interoperability aspects are shown:

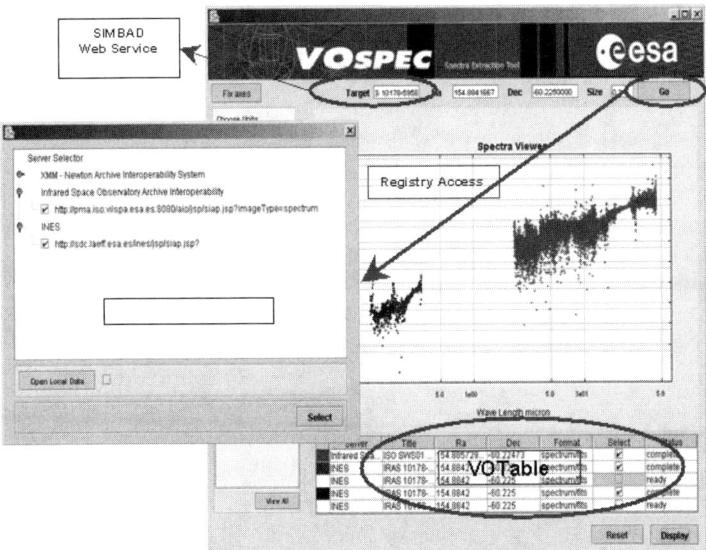

6. VOSpec: Automatic Unit conversion

Thanks to the aforementioned dimensional algorithm, VOSpec can automatically handle changes in units. In the image, the conversion is done from meters in the wavelength axis and Janskys in the flux axis to Angstroms in the wavelength axis and ergs/cm^2/s/Å in the flux axis:

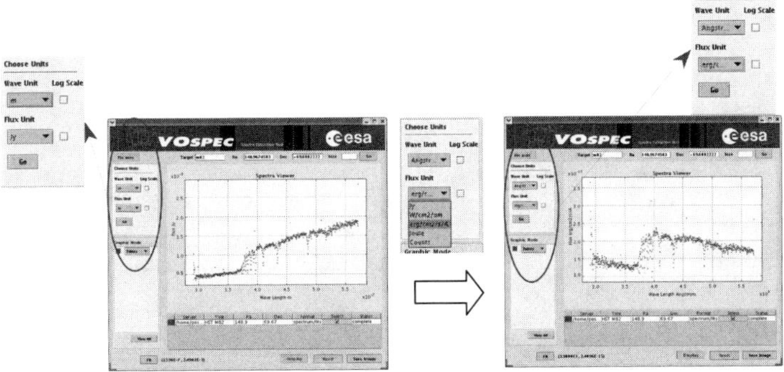

7. VOSpec: Multi-wavelength analysis and superimposition

VOSpec is able to display SEDs as soon as data are well formatted in an SSAP-like wrapper. Data providers are encouraged to prepare their data in an SSAP manner to be able to display them in VOSpec. A brief "How To" document can be found at:

http://pma.standby.vilspa.esa.es:8080/vospec/VOSpecHowTo.html

In the example, the following data can be seen:

- ISO SWS data (Infrared) (available through SSAP format)
- IUE data (Ultraviolet) (available through SSAP format)
- HST data (UV) (Manually VO-formatted) (After this talk, STECF people at Garching made their data available through SSAP and therefore, available automatically to the VOSpec.)
- ISO-PHT data (Infrared) (available through SSAP format)
- XMM-Newton data (X-Ray) (manually VO-formatted)

Data have been conveniently VO-enabled whenever the data were not accessible through SSAP.

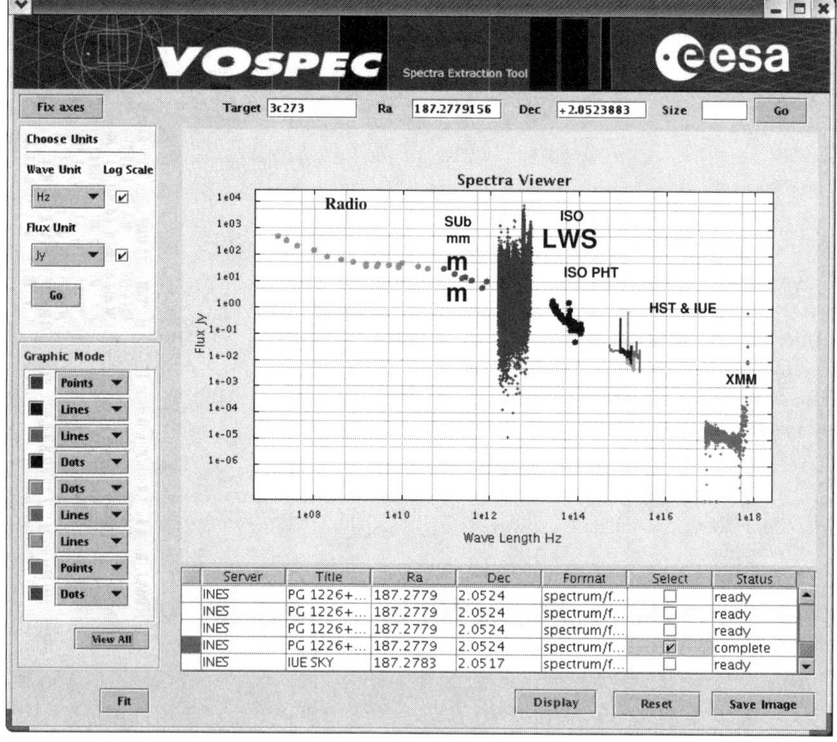

ASTRONOMICAL DATA ANALYSIS SOFTWARE AND SYSTEMS XIV
ASP Conference Series, Vol. 347, 2005
P. L. Shopbell, M. C. Britton, and R. Ebert, eds.

Integrating Legacy Code Into Virtual Observatories – A Test Case

Jörg M. Colberg, Ryan Scranton, and Andrew J. Connolly

University of Pittsburgh, Department of Physics and Astronomy, 3941 O'Hara Street, Pittsburgh PA 15260, USA

Andrew W. Moore

Carnegie Mellon University, 5000 Forbes Avenue, Pittsburgh PA 15213, USA

Abstract. As virtual observatories are being built, the integration of existing astronomical analysis codes is becoming increasingly important. This is particularly true for widely used standard applications. In these cases, creating what amounts to on–site data–mining facilities is clearly beneficial for a wide variety of users, including, but not limited to, users with very little current knowledge of the underlying technologies or users interested in using the machinery for educational purposes. Given the nature of most legacy codes, there are a number of problems that have to be solved. Here, we present an application that makes a very fast and efficient n–point correlation function code interact with the Sloan Digital Sky Survey. We discuss the basic ideas and setup, technical problems encountered, and possible extensions.

1. Introduction

Over the past few years, the size of data has grown along with the increasing sophistication of experiments. For most major astronomical experiments the number of people working on it has grown beyond small groups, which are located at the same or just a handful of places. What is more, funding agencies have required experiments to make their data publicly available after a very short span of time.

At the conception of the Sloan Digital Sky Survey (SDSS), it was realized that much could be gained from embracing industry–strength databases and, later, modern internet technologies to distribute and enable access to the data.

However, using relational databases comes at a price. The vast majority of astronomers have never heard of, let alone worked with SQL. What this means is that efforts have to be made to enable possible users to access the data without requiring them to learn SQL – at least initially. This includes providing the data in a format that a user can work with.

Once one is considering a situation where large amounts of data need to be accessed by remote users one is basically thinking of a Virtual Observatory (VO). There is a lot to be gained from remote access to data: In the world of VO's cross–correlating data from different archives is easy provided data formats are transparent and compatible.

Making data public is only part of the solution. It is desirable to provide a set of standard tools that are commonly used. These tools enable possible users to more efficiently extract new information from archives ("data mining"). Set up correctly, these tools can then also be hooked up to create even more complicated mechanisms to work with large, distributed data sets.

In this work, we will discuss the implementation of one such standard tool for the SDSS. To be more precise, we will describe the implementation of a very fast and efficient n–point correlation function code into the context of the SDSS archive.

2. The SDSS Archive

The Sloan Digital Sky Survey (York et al. 2000) is an automated multi–colour galaxy redshift survey that, at its end, will cover a region of about 8000 square degrees. After the initial processing, the data are stored in a relational database, Microsoft SQL Server.

Data are being released incrementally to the public in data releases (DR's). Data Release 1 (DR1), which we have been using while developing the software discussed here, contains the photometry for more than 84 million objects or, after classification, more than 31 million galaxies and 20 million stars and it uses about 565 GB of disk space. The code is now set to run on the latest data release (DR3 at the time of this writing).

3. n–point

Spatial correlation functions have been widely used in astronomy since at least the 1970's (see e.g. Peebles 1993). Because of the computational complexities of the problem the two–point correlation function has been explored most widely. For a set of points, the two–point correlation function gives the excess probability of finding a neighbour at distance r for each point in the set. So far, higher order correlation functions have not been used as much because they are computationally very expensive. Lately, the three–point correlation function has received some attention.

Computing the correlation function involves finding neighbours at any given distance. Writing an efficient and fast code to compute correlation functions is challenging. The most promising *ansatz* uses trees, spatially hierarchical data structures. Finding nearest neighbours in a tree is straightforward and very fast. In this work, we are working with n–point[1], a code based on a kd–tree (e.g. Moore 1991).

4. Implementing n–point as a Webservice for the SDSS

Webservices allow standardized access to code running on a remote computer. In a nutshell, a webservice is a piece of code that runs on a remote computer

[1] `http://www.autonlab.org/autonweb/showSoftware/127/`

(accessed via the internet) and that looks like any piece of code on a local computer. Webservices are tied in with unified data standards which, in this case, means the use of XML.

Microsoft's solution for web technologies is called .NET. .NET contains a web application server and a framework for the development of static/dynamic web pages and of webservices, including easy interactions with SQL Server. .NET is centered on three programming languages, C#, C/C++, and VisualBasic.NET, of which we have used the former two. C# shares with Java its most important aspects, in that it is very heavily object oriented, and it uses garbage collection. The .NET variant of C/C++, called Managed C/C++, also uses garbage collection and can interact with C#.

Before discussing the technical details of the implementation we want to outline the workflow of the application and introduce the various components/pieces that have to interact.

Angular selection parameters (either ra, dec or λ, η) are obligatory; physical ones such as cuts in magnitude are optional. For a given set of parameters, four database queries have to be run to get the raw data. The first query selects the actual galaxy data, using the spatial and (if provided) physical constraints given by the user. The second and third query are based on the spatial constraints only and find the information that describes which part of the desired region in the sky has to be masked out as it lies outside the survey area or contains a star or other unwanted objects (see below). The fourth query, which is also only based on spatial constraints, retrieves information about the systematics (such as seeing etc.). These queries are run from the main part of the code, which is written in C#. Connecting to the SQL Server database, running the queries, and extracting the data from the results can be done very easily and efficiently in C#. The main part of the code, after having obtained the data from the database, then calls all the other pieces, passing data back and forth.

Once the mask data are available, the effective area, which lies inside the area specified by the user is constructed. The webservice takes as input parameters ranges for ra, dec or λ, η. If this region contains areas that lie outside the covered survey area or if there are stars or other objects, which require masking part of the area, this part of the code determines the actual area where data can be kept and for which random data have to be generated. It uses a fairly complex pixelization scheme coded by one of us (R.S.), which makes use of two further external software packages: "Mangle" (Hamilton & Tegmark 2004) to do an initial coordinate transformation and the GNU Scientific Library[2] for some of the more complex data types.

The resulting mask is used to filter the galaxies returned by the database to exclude those galaxies that lie outside of the area covered by the mask. Using the same area, where galaxies are allowed, a sample of random points is created. The size of the random sample can be specified by the user. Typically, a size of five times the size of the (filtered) galaxy sample is used.

The filtered galaxies and random points are then sent to the n–point code to compute the data–data, data–random, and random–random counts as a function of the angles specified by the user. Despite the use of a very efficient *kd*–tree, the

[2] `http://www.gnu.org/software/gsl/`

n–point code is still slow for very large angles. Because of that, for angles larger than 0.5 degrees, we compute the counts by using the pixelized galaxy data: For these large angles, galaxy counts in cells are computed where the cell sizes are much smaller than 0.5 degrees on the side.[3] This speeds up the computation of the counts on large scales significantly. The resulting correlation function is then computed from the counts.

One of the main challenges for the integration of existing analysis code is that typically different programming languages have to be bridged. In our case, n–point is written in (unmanaged) C.[4] The interaction between n–point and .NET's C/C++ did not turn out to be as easy as we had hoped it would be, the problem being the difference between managed and unmanaged memory. We had to *map* n–point data to data used inside the .NET framework and vice versa. We separated passing data between C# and Managed C/C++ from passing data between Managed C/C++ and regular C. The two parts of the problem were solved as follows.

As mentioned above, C# and Managed C/C++ are supposed to be able to talk each other and, in fact, making it work for simple data types is trivial. Passing arrays back and forth is more complicated. Assume we have a Managed C/C++ method `doTheNPoint` that accepts an array of `double`'s and returns a string. In Managed C/C++, that method would have to be defined as

```
String *doTheNPoint(double (*data) __gc[]).
```

Note how Managed C/C++ has a data type `String`, and also note the extra `__gc[]` behind the `double` array. `__gc[]` indicates that the array `data` is subject to garbage collection. This method is called from C# using

```
String result = npoint.doTheNPoint(ref data),
```

where we have assumed that the object `npoint` and the `double` array `data` have been declared and instantiated properly elsewhere. Note how this passes *pointers* back and forth between C# and Managed C/C++. For each case where we have to call legacy code we first call a method written in Managed C/C++, which receives data from C#. In what follows, we will refer to these Managed C/C++ bits as *wrappers*.

Calling the actual legacy code from the wrapper is more complicated. The legacy code is written in regular C so its memory is not subject to garbage collection. Inside the wrapper, managed memory has to be mapped to non–managed memory. Once that is achieved the data are being passed using pointers. Assume we are inside the aforementioned Managed C/C++ method `doTheNPoint`. The first step is to allocate unmanaged data memory through

```
IntPtr datapointer = Marshal::AllocHGlobal(...),
```

where the size of the memory needed is specified inside the allocation call. Afterwards, the data is copied from the managed to the unmanaged memory using

```
Marshal::Copy(*data, 0, datapointer, (...)),
```

[3]Note how this procedure is very similar to what is being done in N–body simulations. For small scales, the gravitational forces are computed for particle pairs. For large scales, the gravitational field on a mesh, onto which the particles are binned, is computed.

[4]The same is true for Mangle and GSL.

with information about the size of the data again omitted. Lastly, a pointer is constructed that will be used in the call of the legacy C code through

```
double *pdatapointer = (double *) datapointer.ToPointer(),
```

so that, finally, the legacy C code can be called through

```
double *result = npt(pdatapointer),
```

where we assume that we're calling a C method called `npt` that has been properly defined elsewhere.

5. Discussion and Extensions

During the course of the integration of the legacy codes we encountered a set of pitfalls. For example, while it is indeed possible to have different programming languages interact with each other, this can get fairly complicated for non–trivial data types. Unfortunately, it turned out to be quite difficult to find how these interactions can be coded in the documentation. The interaction with other developers on news groups provided invaluable help. However, once those technical problems were solved, .NET turned out to be a stable and robust environment to work in. As could be expected, the bulk of the time was spent on wrapping the legacy code so that it could be used inside the .NET environment. Similar problems can be expected for a Java environment.

We intend to integrate other codes from the Auton Lab[5] code suite. As these use the same interfaces as n–point, the integration will be easy and straight–forward, and the codes – such as an anomaly and a filament finder – will add further data mining power to the SDSS database.

The n–point code can be accessed through http://npoint.phyast.pitt.edu. For the final version of the code, there will be a web form interface and an actual webservice through which the code can be accessed.

Acknowledgments. We thank Andrew Hamilton for allowing us to use his excellent software package "Mangle".

References

Hamilton, A. J. S., Tegmark M. 2004, MNRAS, 349, 115

Moore, A. 1991, A tutorial on kd-trees, University of Cambridge Computer Laboratory Technical Report No. 209

Peebles, P. J. E. 1993, Principles of Physical Cosmology, Princeton University Press

York, D. G. et al. 2000, AJ, 120, 1579

[5]`http://www.autonlab.org`

ASTRONOMICAL DATA ANALYSIS SOFTWARE AND SYSTEMS XIV
ASP Conference Series, Vol. 347, 2005
P. L. Shopbell, M. C. Britton, and R. Ebert, eds.

The Multimission Archive at the Space Telescope Science Institute in the Context of VO Activities

Inga Kamp, Randall Thompson, Alberto Conti, Dorothy Fraquelli, Tim Kimball, Karen Levay, Bernie Shiao, Myron Smith, Rachel Somerville, and Richard L. White

Space Telescope Science Institute, 3700 San Martin Drive, Baltimore, MD 21218

Abstract. In the past year, the Multimission Archive at the Space Telescope Science Institute (MAST) has taken major steps in making MAST's holdings available using VO-defined protocols and standards, and in implementing VO-based tools. For example, MAST has implemented the Simple Cone Search protocol, and all MAST mission searches may be returned in the VOTable format, allowing other archives to use MAST data for their VO applications. We have made many of our popular High Level Science Products available through Simple Image Access Protocol (SIAP), and are implementing the VO Simple Spectral Access Protocol (SSAP). The cross correlation of VizieR catalogs with MAST missions is now possible and illustrates the integration of VO services into MAST. The user can easily display the results from searches within MAST using the plotting tool VOPlot. MAST also participates in the NVO registry service. Thus, the user can harvest MAST holdings simultaneously with data from many other surveys and missions through the VO DataScope Data Inventory Service.

1. Introduction

The Multimission Archive at Space Telescope Science Institute is NASA's UV-optical science archive center. It contains the Hubble Space Telescope archive plus more than a dozen additional active, planned, and legacy mission datasets. The total MAST data volume exceeds 20 Terabytes, making it one of the most significant astronomical collections available on-line.

MAST is an important node in the upcoming Virtual Observatory and we are continually improving our data discovery and retrieval abilities. MAST has taken major steps in making its holdings available using VO-defined protocols and standards and in implementing VO-based tools. All MAST mission searches may be returned in the VOTable format, allowing other archives to use MAST data for their VO applications.

2. MAST Webservices

Currently, MAST provides three different forms of webservices: HTTP GET Requests, an RSS news service, and a dataset verifier SOAP service. Additional information can be found at `http://archive.stsci.edu/vo/mast_services.html`. HTTP GET Requests allow search parameters to be included in the URL. As such, they can be called from within programs to automate data searches. Cur-

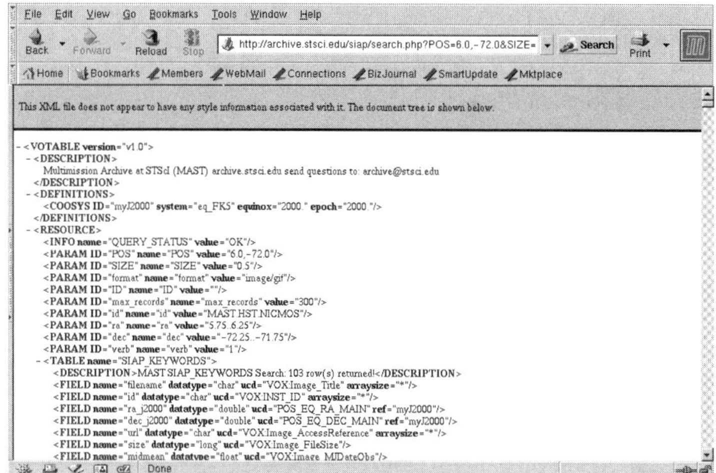

Figure 1. Example of the Simple Image Access Protocol: the command `http://archive.stsci.edu/siap/search.php?POS=6.0,-72.0&SIZE=0.5&format=gif&ID=nicmos` returns the above VOTable containing the information on all matching datasets.

rently, MAST provides this service for Mission Searches, Simple Cone Searches and the Simple Image Access Protocol. The results are returned in VOTable XML format. The result of the Mission Searches can also be returned as an Excel spreadsheet or as comma-separated values, which can simplify ingesting results into user-written programs. Figure 1 shows the VOTable returned for a positional search in the archived NICMOS images: the search radius is 0.5 arcsec and the search is restricted to data in gif-format.

The RSS news service offers to search a list of MASTs most recent news items and returns the result in RSS XML-format. The output can also be displayed in HTML.

MAST has one SOAP-based web service, which is accessed via the ADS. The web service allows users to enter a data set name and obtain a verification and link to the archival website of this dataset. More information about the ADEC naming conventions can be found at `http://archive.stsci.edu/pub_dsn.html`. The service can be accessed from the web form at: `http://ads.harvard.edu/ws/DataVerifier`, or using the latest ADEC naming conventions: `http://vo.ads.harvard.edu/dv/DataVerifier.cgi` (Figure 2).

3. VOPlot at MAST

Users may plot search results from any MAST mission or VizieR catalog search with the new JAVA-based graphical display tool called VOPlot (developed within the Indian VO project in collaboration with CDS). The example in Figure 3 shows the galactic distribution of STIS UV observations with the Echelle gratings.

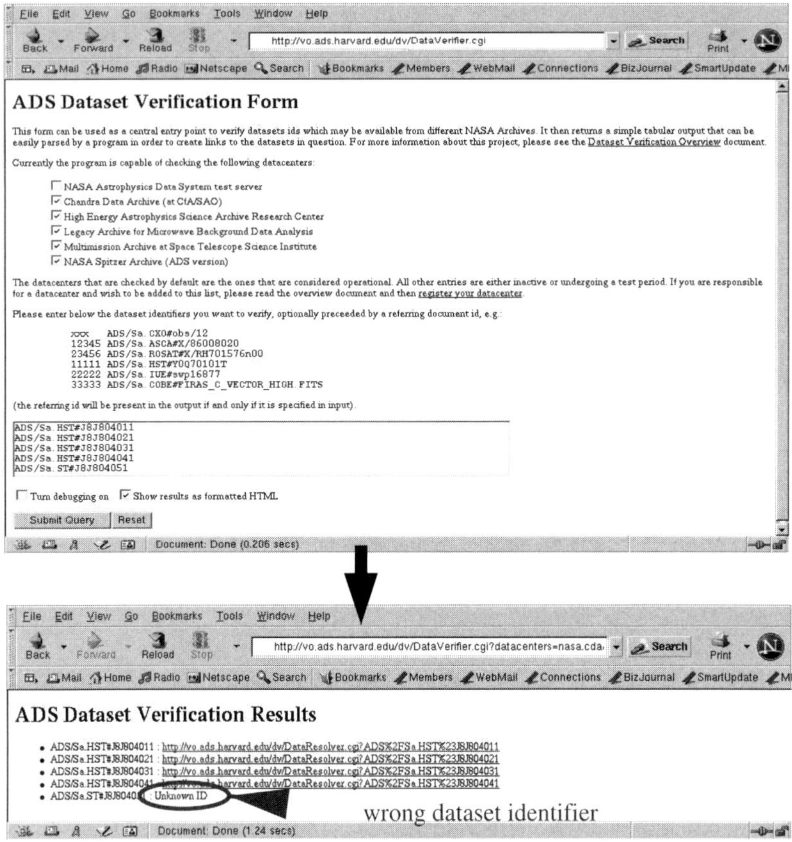

Figure 2. Example of the SOAP-based web service: enter the dataset names in the verification form and submit the request. The result contains the verification of the dataset name as well as links to these datasets.

4. MAST and VizieR

Users may now search the entire set of 4,000 VizieR catalogs and cross correlate the results with any MAST mission. This uses a VizieR web service which communicates using the VOTable standard. The example shows the result of a cross correlation between the Vega-type star catalogue and archived FUSE observations. Enter the catalog information at `http://archive.stsci.edu/vizier.php` and cross correlate it with the FUSE data in MAST. Figure 4 shows how to navigate through the cross correlation. In addition, it provides a cutout of the returned information on the available FUSE datasets for Vega-type stars from the specified catalog.

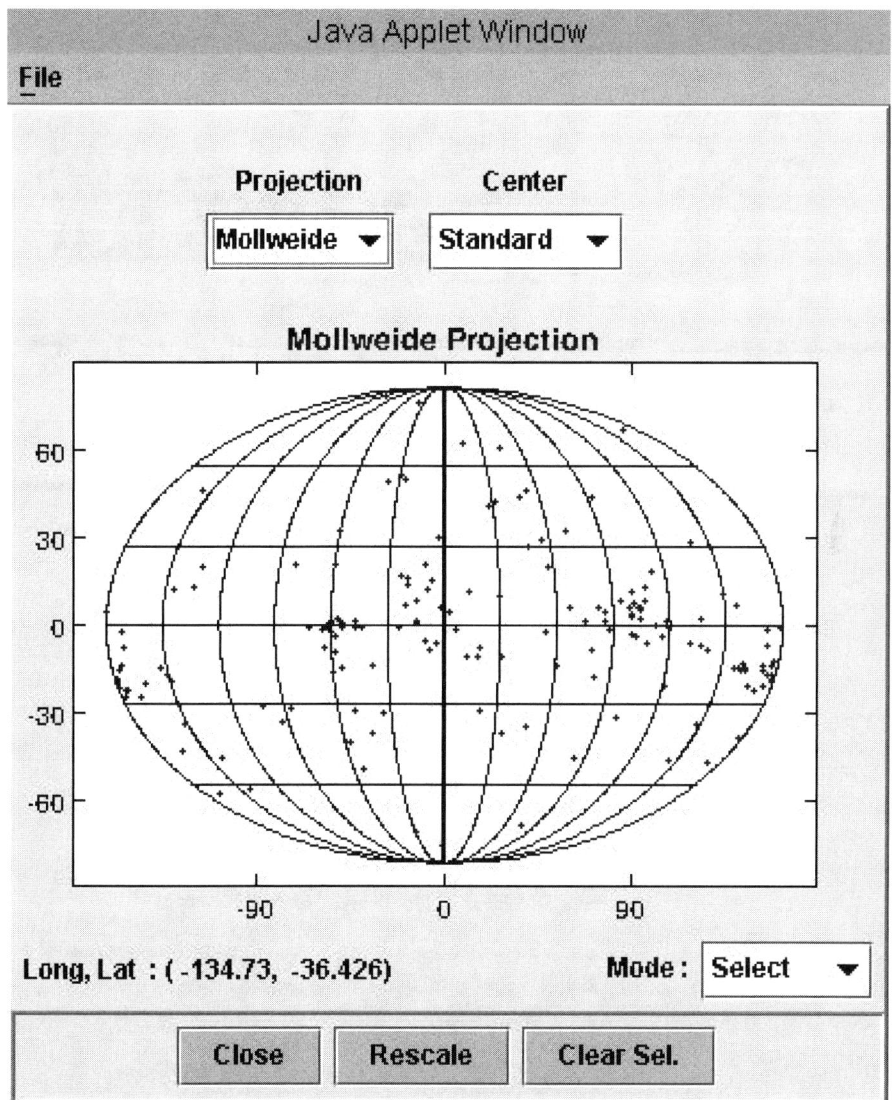

Figure 3. The galactic distribution of STIS high-resolution UV observations: the graphical display tool VOPlot is called from the result page of a MAST archive query.

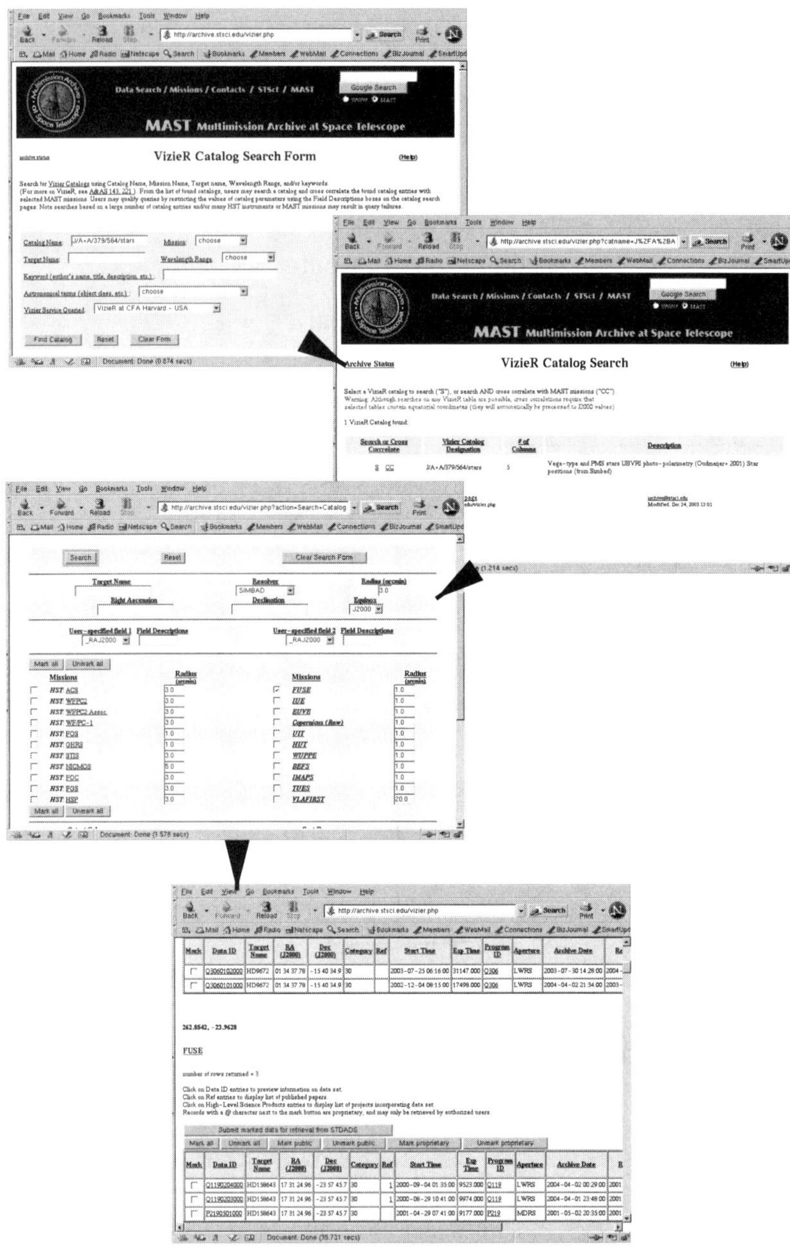

Figure 4. Provide the catalog information for the cross correlation at `http://archive.stsci.edu/vizier.php`. After choosing the cross correlation option for the catalog of Vega-type stars, the FUSE mission was selected and the last page shows a pane of the resulting web table.

ASTRONOMICAL DATA ANALYSIS SOFTWARE AND SYSTEMS XIV
ASP Conference Series, Vol. 347, 2005
P. L. Shopbell, M. C. Britton, and R. Ebert, eds.

ROME (Request Object Management Environment)

Mihseh Kong, John C. Good, and G. Bruce Berriman

Infrared Processing and Analysis Center, California Institute of Technology, Mail Code 100-22, CA 91125

Abstract. Most current astronomical archive services are based on an HTML/ CGI architecture where users submit HTML forms via a browser and CGI programs operating under a web server process the requests. Most services return an HTML result page with URL links to the result files or, for longer jobs, return a message indicating that email will be sent when the job is done.

This paradigm has a few serious shortcomings. First, it is all too common for something to go wrong and for the user to never hear about the job again. Second, for long and complicated jobs there is often important intermediate information that would allow the user to adjust the processing. Finally, unless some sort of custom queueing mechanism is used, background jobs are started immediately upon receiving the CGI request. When there are many such requests the server machine can easily be overloaded and either slow to a crawl or crash.

Request Object Management Environment (ROME) is a collection of middleware components being developed under the National Virtual Observatory Project to provide mechanism for managing long jobs such as computationally intensive statistical analysis requests or the generation of large scale mosaic images. Written as EJB objects within the open-source JBoss applications server, ROME receives processing requests via a servelet interface, stores them in a DBMS using JDBC, distributes the processing (via queuing mechanisms) across multiple machines and environments (including Grid resources), manages real-time messages from the processing modules, and ensures proper user notification.

The request processing modules are identical in structure to standard CGI-programs – though they can optionally implement status messaging – and can be written in any language. ROME will persist these jobs across failures of processing modules, network outages, and even downtime of ROME and the DBMS, restarting them as necessary.

1. Introduction

All of us are familiar with retrieving data via a Web browser when the system we are accessing is busy, the request can time-out or take forever, leaving us with nothing to do but wait. Some data retrieval systems handle such situations by accepting a request as a background batch job, returning an acknowledgement of the request immediately and emailing the user when the job is done. Neither approach provides the user with a means of monitoring the progress or interacting with the processing of his request.

From the data provider's point of view, a choice has to be made as to whether to build the service as a blocking CGI job or a batch job. Furthermore, when hundreds, thousands, or even tens of thousands of requests pour into an application system (some requiring intense computation or complicated DBMS searching), the system will inevitably reach its limit, and grind to a halt.

ROME is a set of software components that sit between client applications and data retrieval/computational services and manage client requests. From a user's point of view, ROME provides them with the ability to submit, monitor and to some extent interact with their processing requests. For the service provider, ROME stages pools of requests in a secure, fault tolerant system so that the server machine that performs the computation and the DBMS search will not become overloaded.

In this paper, we will describe the high level design of ROME and present request submission scenarios from both a user's and a data retrieval processor's point of view.

2. Design

ROME consists of "Request Manager" and "Request Processor" functionalities. The Request Manager (RM) is a collection of servlets and Enterprise Java Bean (EJB)-based services whose function is to manage information relating to user requests. The Request Processor is a lightweight engine for overseeing the operation of a set of processing applications and conveying status information back to the Request Manager (and ultimately the user).

The ROME request submission mechanism operates through standard client interfaces (Web forms and JAVA GUIs) which collect user parameters with a standard HTTP Web interface. ROME takes responsibility for scheduling and overseeing the execution of request although it contains no processing code itself for efficiency reasons. The request processing is done by CGI programs on a collection of servers.

From a user's point of view, all of the jobs submitted are "batch" and all they see in response to their submission is an acknowledgement (one job ID for each request). There is a Web form for polling and a JAVA GUI client interface to view the job status with immediate and continuous feedback. These job monitor mechanisms allow some user interaction with job executions. For example, a user may submit a set of requests simultaneously, then monitor those jobs' status using ROME's request monitor JAVA GUI. Based on the results for a few job threads, the user may decide that he does not need to complete some or all of the other jobs and can send interrupts to abort them. Similarly, after monitoring some long-running requests for a while he may decide to add email notification and disconnect.

In order to accomplish the above senario for possibly thousands of simultaneous requests, we need a system that not only maintains the database of requests and messages efficiently but is also capable of persisting data in case of the system failure. ROME achieves this by employing the Enterprise Java Bean (EJB) technology (*e.g.*, the open source EJB system JBoss), coupled with a robust DBMS infrastructure (in our case INFORMIX). The EJB framework is an industry standard for business component development and provides an abstraction between component transaction monitors (CTMs) and distributed object services. EJBs reside in a specialized JAVA VM (EJB server) which is responsible for providing the infrastructure for managing such things as transactions, persistence, concurrency, and security.

In one respect our problem is quite different from the standard business transaction. Unlike those systems, our concern is mainly for long-lived queries. In their case, the actual processing of information (which often involves rapid database interaction) can perfectly well be handled within the EJB code itself. In ours, we wish to decouple the actual processing, which could take hours or even days, from the EJB server so the EJB container can do what it suppose to do best and not become a processing bottleneck.

We have done this by separating the components that processes the requests (Request Processor JVM) from the system that manages them (Request Manager JVM) so that the latter can remain lightweight and responsive to request and message handling. With this separation, the Request Management (RM) is responsible for accepting requests and messages, synchronizing them to the database, and retrieving them when requested by a Request Processor (RP). A Request Processor is responsible for managing the worker threads that actually process the requests (usually by means of an external "CGI" process).

Our design (see Figure 1: ROME Architecture Diagram) persists the requests by writing their states to the database so they can be recovered after a system failure. It balances load by having several request processors (RP) external to the RM, each of them having multiple worker threads to run the application programs that actually process the query. Figure 1 shows at least one instance of all the components that comprise or interact with ROME. The steps involved in a typical processing scenario are:

1. A user submits a query,
2. A worker thread in the RP picks up the request,
3. The worker thread starts a copy of a processing application,
4. The worker thread sends information about itself to the RM,
5. The worker thread sends input parameters to the application,
6. The application sends messages to the worker thread,
7. The worker thread sends message to the RM,
8. The RM sends message to the user,
9. The user queries for information pertaining to a certain request.

3. Current Status

ROME has been fully implemented using the open source EJB container JBoss and Informix DBMS and tested within the IRSA testbed. We are currently conducting intense load testing with 20000 requests to ensure its robustness. We are planing to implement a version that interfaces with MYSQL DBMS so that ROME does not rely on any vendor specific software.

The current implementation schedules the requests with first in/ first serve policy plus a priority flag for slow jobs, but it is up to the user and ROME operator to set job priorities. A more sophisticated scheduling algorithm is desirable but not developed.

ROME is designed for long-lasting jobs; if a user has submitted hundreds of such jobs, he can monitor them all at once using ROME's job status monitor

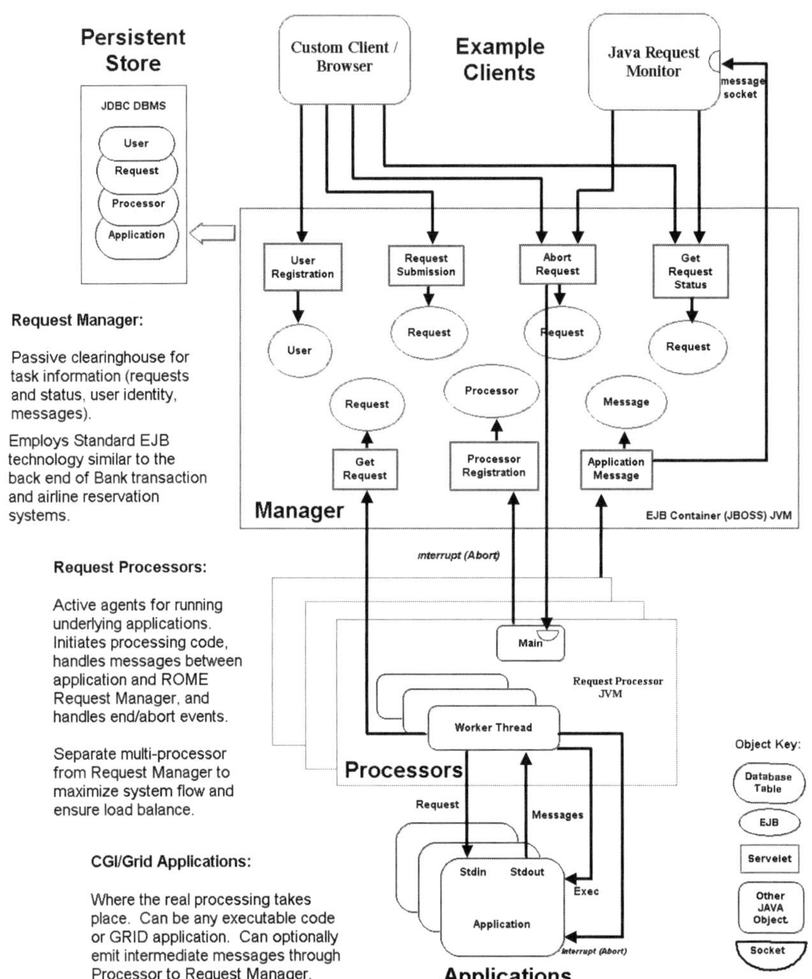

Figure 1. Architecture Diagram

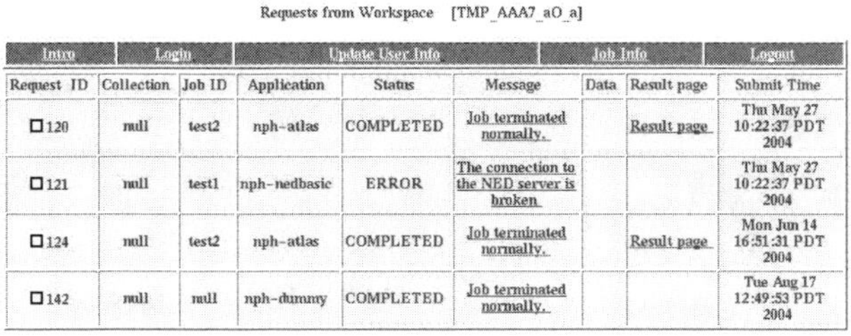

Requests from Workspace [TMP_AAA7_aO_a]

Intro | Login | Update User Info | Job Info | Logout

Request ID	Collection	Job ID	Application	Status	Message	Data	Result page	Submit Time
☐ 120	null	test2	nph-atlas	COMPLETED	Job terminated normally.		Result page	Thu May 27 10:22:37 PDT 2004
☐ 121	null	test1	nph-nedbasic	ERROR	The connection to the NED server is broken			Thu May 27 10:22:37 PDT 2004
☐ 124	null	test2	nph-atlas	COMPLETED	Job terminated normally.		Result page	Mon Jun 14 16:51:31 PDT 2004
☐ 142	null	null	nph-dummy	COMPLETED	Job terminated normally.			Tue Aug 17 12:49:53 PDT 2004

Figure 2. Job status

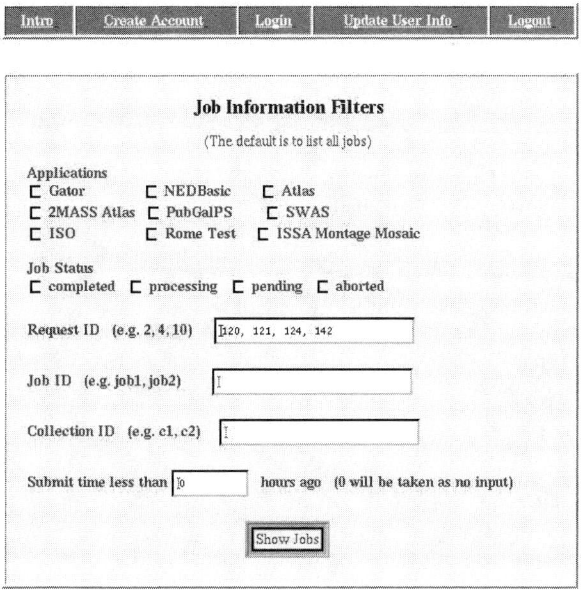

Intro | Create Account | Login | Update User Info | Logout

Job Information Filters

(The default is to list all jobs)

Applications
☐ Gator ☐ NEDBasic ☐ Atlas
☐ 2MASS Atlas ☐ PubGalPS ☐ SWAS
☐ ISO ☐ Rome Test ☐ ISSA Montage Mosaic

Job Status
☐ completed ☐ processing ☐ pending ☐ aborted

Request ID (e.g. 2, 4, 10) [120, 121, 124, 142]

Job ID (e.g. job1, job2) [ ]

Collection ID (e.g. c1, c2) [ ]

Submit time less than [0] hours ago (0 will be taken as no input)

[Show Jobs]

Figure 3. Job filter

page. Figure 2 shows a job status page and Figure 3 a job filter pages for locating the jobs currently of interest to the user.

Acknowledgments. ROME is funded by the National Virtual Observatory, under NSF Cooperative Agreement AST01122449.

ASTRONOMICAL DATA ANALYSIS SOFTWARE AND SYSTEMS XIV
ASP Conference Series, Vol. 347, 2005
P. L. Shopbell, M. C. Britton, and R. Ebert, eds.

VO Software: Overview of the Past Twelve Months of Prototypes & Tools

M. C. Leoni

European Southern Observatory (ESO), Karl-Schwarzschild-Straße 2, Garching bei München, D-85748, Germany

T. A. McGlynn

Goddard Space Flight Center, NASA

R. D. Williams

California Institute of Technology

Abstract. Thanks to the increased effort in the development and spread of the "VO novel", the past year saw a spur in the creation of software for the Virtual Observatory: including both prototypes and released tools, produced by the national VO project members of the International Virtual Observatory Alliance (IVOA).

This work describes completely new developments as well as existing software adapted to the VO. It highlights their main features and the way they interact and interoperate using common industry standards as well as astronomy-specific ones developed within the IVOA.

1. Introduction

Thanks to the increased effort in the development and spread of the "VO novel", the past year saw a spur in the creation of software for the Virtual Observatory: including both prototypes and released tools, produced by the national VO projects, members of the *International Virtual Observatory Alliance* (IVOA).

Following is a comprehensive list of software for the VO - as of this writing. Highlights are on main features and the way each one interacts and interoperates with the others using common industry standards as well as astronomy-specific ones developed within the IVOA.

2. Formats & Protocols

All the VO software uses these protocols to interoperate with each other. Up to now, the main point in the structure is the creation of a registry containing all the available resources (described using the VOResource[1] metadata standard [Hanisch, 2004]), thus allowing software to automatically discover and use what is needed.

---

[1] General term referring to a VO element with a name and a unique identifier.

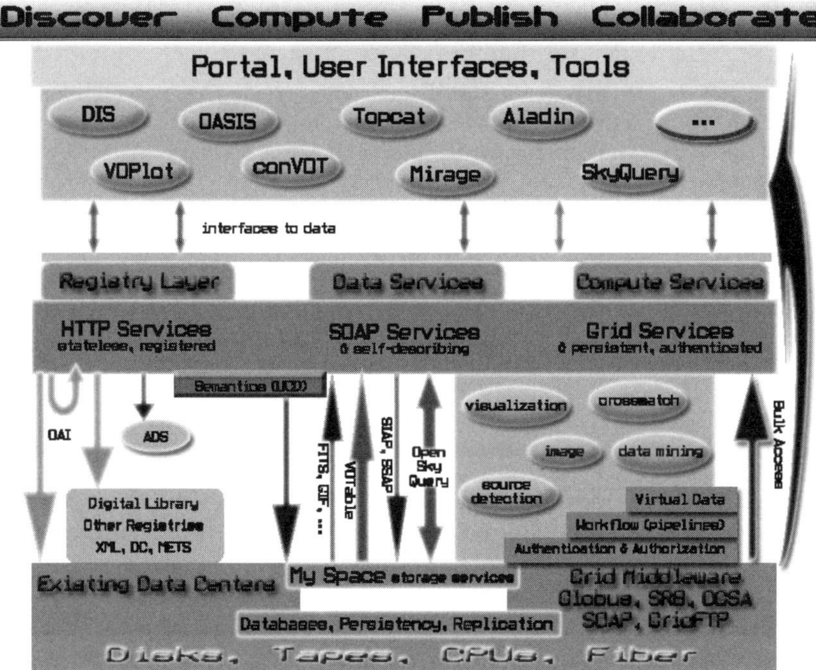

Figure 1. Overview of the Virtual Observatory architecture.

Besides using well known industry-level protocols, the VO is developing other standards to help the interoperability and to improve performance.

VOTable – (Virtual Observatory Table)
VOTable is a XML-dialect developed to ease the transfer of astronomical data, in particular tabular data: "The VOTable format is an XML standard for the interchange of data represented as a set of tables" (Ochsenbein et al. 2004; Williams et al. 2004). A VOTable document contains metadata, a description of the data themselves, and allowances to keep metadata and data stored separately. A reference to remote data (typically a URL link) is given in order to retrieve them. This is a key feature in a distributed-computing environment, like the Grid, where the idea is to minimize the transfer of big datasets, such those used in astronomy.
SIAP – (Simple Image Access Protocol)
From the IVOA document (Tody & Plante 2004): SIAP "defines a prototype standard for retrieving image data from a variety of astronomical image repositories through a uniform interface".
SSAP – (Simple Spectrum Access Protocol)
Similar to the previous one, this protocol allows the retrieval of spectrum data through a standard interface.
ADQL – (Astronomical Data Query Language)

XML translation of SQL[2] queries to be exchanged among VO servers.
SOAP & **XML** – (Simple Object Access Protocol, eXtensible Markup Language)
XML is a pared-down version of SGML[3], designed especially for Web documents, that allows designers to create their own customized tags. SOAP represents a minimal set of conventions for invoking code using XML over HTTP[4].

The separation of protocols is not as definite as depicted in the overall architecture picture (Figure 1). Due to the existing strong relationships amongst applications, some of them may actually fall under different parts of this representation (or contain software with functionalities belonging to more than one area).

3. Tools

Since *VOTable* (see Section 2.) is the main format to exchange data, most of the tools first developed regard the manipulation of such VO standard-based files. The list includes visualization tools, filters, as well as software libraries.

JAVOT – Java parser for VOTable: http://us-vo.org/VOTable/JAVOT.
Mirage [VO-enabled] – Java-based wrapper for Mirage, a tool for exploratory analysis and visualization of classification and proximity structures of multi-dimensional numerical data from an arbitrary domain of study:
http://skyservice.pha.jhu.edu/develop/vo/mirage/.
Specview – java application for 1-D spectra visualization and analysis of astronomical spectrograms:
http://www.stsci.edu/resources/software_hardware/specview.
TOPCAT – interactive graphical viewer and editor for tabular data:
http://www.starlink.ac.uk/topcat/
VOTFilter – enables the import of VOTable documents into OpenOffice Calc spreadsheet; written in java, this XML filter can be downloaded from:
http://www.china-vo.org/VOTable Filter for OpenOffice Calc.htm.
VOPlot – a java visualization tool, to manipulate VOTable data. Available as standalone application or a web-based one, the latter being an applet. More info at http://vo.iucaa.ernet.in/ voi/voplot.htm
VOSpec – Java applet tool to handle VO spectra:
http://pma.standby.vilspa.esa.es:8080/vospec/index.html.
VOTable Java Streaming Writer – java package to convert files in different formats (e.g. ASCII) to VOTable:
http://vo.iucaa.ernet.in/ voi/votableStreamWriter.htm.
VOTool – package for allowing the IRAF system to handle VOTable:
http://iraf.noao.edu/projects/votool/

---

[2]Structured Query Language

[3]Standard Generalized Markup Language

[4]HyperText Transfer Protocol

Xmatch (CDS) – Aladin plug-in:
http://www.euro-vo.org/twiki/bin/view/Avo/CDSXMatchPlugin.

4. Grid and Web Services

Several Web Services have been created, both wrapping already existing services as well as exposing new functionalities. Grid services are in the making to allow workload distribution and resource sharing.

ACE – (Astronomical Catalogue Extraction) Based on the well-known SExtractor tool, this web service exposes the same capabilities to the web world: http://wiki.astrogrid.org/bin/view/Astrogrid/AVODemoDesign
Aladin Image – gives image URL locations or images based on DIME:
http://cdsweb.u-strasbg.fr/cdsws/aladin_image.gml
Astronomical Coordinates – provides astronomical coordinate translation:
http://cdsweb.u-strasbg.fr/cdsws/astroCoo.gml
Cosmic Distances – XML web sservice to calculate different distance measures in a given cosmology: http://voservices.net/Cosmology/ws_v1_0/Distance.asmx
Filter Profiles – access profiles of well-known astronomical instruments:
http://voservices.net/filter/
GLU Resolver – resolves catalogue tags through GLU:
http://cdsweb.u-strasbg.fr/cdsws/glu_resolver.gml
Grist – Grid services for data mining in astronomy: http://grist.caltech.edu/
Name Resolver – resolves names for VizieR, Simbad, NED:
http://cdsweb.u-strasbg.fr/cdsws/name_resolver.gml
NED – XML Web Service interface to NED:
http://voservices.net/NED/ws_v1_0/NED.asmx
SDSS Cas – XML Web Service interface to the SDSS Catalog Archive Server:
http://voservices.net/CasService/ws_v1_0/CasService.asmx
SDSS Image Cutout – SDSS DR1 finding chart tool:
http://skyserver.sdss.org/dr1/en/tools/chart/chart.asp
Spectrum – portal for VO Web Services to access data from SDSS DR1 and 2dFGRS: http://voservices.net/spectrum/
UCD – resolves UCD tag, gives the complete list of UCDs, gives the list of UCD contained in a catalogue: http://cdsweb.u-strasbg.fr/cdsws/ucdClient.gml
VizieR WS – Web Service to access VizieR catalogues:
http://cdsws.u-strasbg.fr/axis/services/VizieR

5. Prototypes & more

Aladin for AVO – Interactive sky atlas, to visualize images, superimpose catalogues, interactively access data & info from on-line databases:
http://www.euro-vo.org/twiki/bin/view/Avo/SwgDownload
Astrogrid – The suite developed covers all the needs a data centre has to publish its data archive:
http://www.astrogrid.org/maven/docs/SNAPSHOT/astrogrid-software-site/
Atlasmaker – Grid-based software for building atlases of astronomical images:

http://www.cacr.caltech.edu/projects/nvo/atlasmaker/
Carnivore – VO registry with publishing, querying and harvesting capabilities (uses XML database): http://mercury.cacr.caltech.edu:8080/carnivore
DataScope – (Data Inventory Service) Quick look at all image, observation and catalogue data available for a given position in the sky:
http://heasarc.gsfc.nasa.gov/vo/
Montage – Grid software for mosaics computation:
http://montage.ipac.caltech.edu/
NCSA Registration Portal – portal that allows data providers to make their resources VO visible: http://nvo.ncsa.uiuc.edu/nvoregistration.html
NVO Registry – searchable registry, exposes keyword/SQL interface:
http://nvo.stsci.edu/VORegistry
Open SkyQuery – based on Microsoft.net technology, offers SOAP services to perform distributed astronomical queries using VOTable and ADQL:
http://openskyquery.net/Sky/skysite/
Pegasus – (Planning for Execution in Grids) system to map and execute workflows on the Grid: http://pegasus.isi.edu/
SDSS Sky Server – provides access to SDSS DR3 data:
http://cas.sdss.org/astro/en/
SkyQuery.net – prototype of a federated database application:
http://www.skyquery.net/main.htm
SkyView – panchromatic image atlas, generates images of any part of the sky at wavelengths in all regimes from radio to gamma-ray: http://skys.gsfc.nasa.gov/
VirtualSky – portal to a panchromatic sky atlas: http://www.virtualsky.org/

References

Hanisch, R. 2004, `http://www.ivoa.net/Documents/latest/RM.html`

Ochsenbein, F., Williams, R., Davenhall, C., Durand, D., Fernique, P., Giaretta, D., Hanisch, R., McGlynn, T., Szalay, A., Taylor, M., & Wicenec, A. 2004, `http://ivoa.net/Documents/latest/VOT.html`

Tody, D.,& Plante, R. 2004, `http://www.ivoa.net/Documents/latest/SIA.html`

Williams, R., Hanisch, B., Linde, T., McDowell, J., Moore, R., Ochsenbein, F., Ohishi, M., Rixon, G., Szalay, A., & Tody, D. 2004, `http://www.ivoa.net/Documents/latest/IVOArch.html`

ASTRONOMICAL DATA ANALYSIS SOFTWARE AND SYSTEMS XIV
ASP Conference Series, Vol. 347, 2005
P. L. Shopbell, M. C. Britton, and R. Ebert, eds.

XML Data in the Virtual Observatory

R. G. Mann

Institute for Astronomy, University of Edinburgh, Royal Observatory, Blackford Hill, Edinburgh, EH9 3HJ, United Kingdom

R. M. Baxter, R. Carroll, and Q. Wen

National e-Science Centre, University of Edinburgh, 15 South College Street, Edinburgh, EH8 9AA, United Kingdom

O. P. Buneman, B. Choi, W. Fan, R. W. O. Hutchison, and S. D. Viglas

School of Informatics, University of Edinburgh, Appleton Tower, Crichton Street, Edinburgh, EH8 9LE, United Kingdom

Abstract. XML is the lingua franca of the Web services world and so will play a major role in the construction of the Virtual Observatory. Its great advantages are its flexibility, platform-independence, ease of transformation, and the wide variety of existing software that can process it. An obvious disadvantage in its use as an astronomical data format is its verbosity; the number of bytes taken up writing the XML tags can easily outnumber those constituting the actual astronomical data. The verbosity of XML in this regard is a problem in many other disciplines, and computer scientists are developing more generic solutions to that found in the VOTable specification. In this paper we describe two of these projects currently underway in Edinburgh, which focus on the compression and querying of XML, and a technology for representing the structure of a binary file in XML, enabling it to be read as if it were XML.

1. Introduction: what's wrong with VOTable?

One of the first fruits of the nascent Virtual Observatory (VO) movement was the draft specification for VOTable[1], an XML standard for the exchange of astronomical data in tabular form. The goal in designing a new data format was to aid interoperability between the distributed data centres coming together in the VO. The choice of XML as the basis for that format was determined by its central place in the Internet world, together with the existing XSLT[2] technologies for transforming XML documents.

The flexibility of XML as a way of storing data comes at a cost, namely the overhead in data volume of the tags surrounding every data value in an XML document. The definers of VOTable understood that this overhead would mean that VOTable would be a prohibitively expensive format for storing large quan-

[1] See Ochsenbein et al. (2004) for V1.1 of the VOTable specification

[2] `http://www.w3.org/TR/xslt`

tities of astronomical data. So, in addition to the pure XML version of VOTable, in which every data value is fully tagged inside a `<TABLEDATA>` element, they provided two additional ways of storing the tabular data: (i) a FITS Serialization, in which the metadata describing the table are presented in fully tagged XML in the VOTable document, but the data values themselves are stored in an external FITS binary file, which is referenced within the VOTable document; and (ii) a Binary Serialization, in which the table of data values was stored in a binary format within the VOTable document, using some suitable encoding.

These two additional options were pragmatic choices on the part of a group addressing the pressing needs of the VO community, but they lack some elegance, as well as each having specific design problems. For example, the VOTable specification leaves unclear what a parser should do if, when manipulating a data set using the FITS serialization, it finds that the metadata in the `<FIELD>` elements of the VOTable document do not match those in the header of the linked FITS file. Also, application of the Binary serialization is made at the level of the full table, thereby rendering the whole table unreadable by humans. At a more practical level, the additional complexity of the FITS and Binary variants of VOTable meant that few data centres published data using them, nor, at least initially, did many applications support the ingestion of them; although some (e.g., TOPCAT[3]) now do.

While the VO community was debating the pros and cons of binary and XML data formats, the same issues were being addressed in a number of other, data–rich scientific disciplines, and computer scientists began to devise generic solutions to these problems. In this paper we introduce to the VO community two of these, developed at the University of Edinburgh: the first, VX (Buneman et al. 2004), circumvents the verbosity of the pure XML variant of VOTable by decomposing the XML document into a vectorized form which admits efficient querying; while the second, BinX (Carroll et al. 2004), is based on a language for describing binary data and their layout in XML.

2. VX: Vectorizing XML

Vertical partitioning is a well-known technique for optimizing query performance in relational databases. An extreme form of this technique, which we call *vectorization*, is to store each column separately. Liefke and Suciu (2000) applied the idea of vectorization to XML documents, decomposing them into *vectors* (the sequences of data values appearing under all paths bearing the same sequence of tag names) and the *skeleton* which stores the tree–like structure of an arbitrary XML document. This work yielded XMILL, a tool for compressing XML documents, but the aim of VX is somewhat broader than simple compression – as it aspires to developing techniques whereby large XML repositories can be efficiently queried – and it also differs in some technical aspects from that of Liefke & Suciu (2000); it does not compress the vectors and uses a different method for compressing the skeleton.

---

[3]`http://www.starlink.rl.ac.uk/TOPCAT`

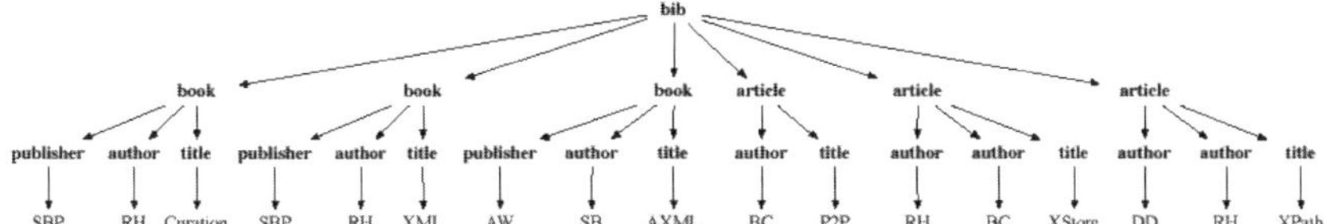

Figure 1. The tree representation of an XML document.

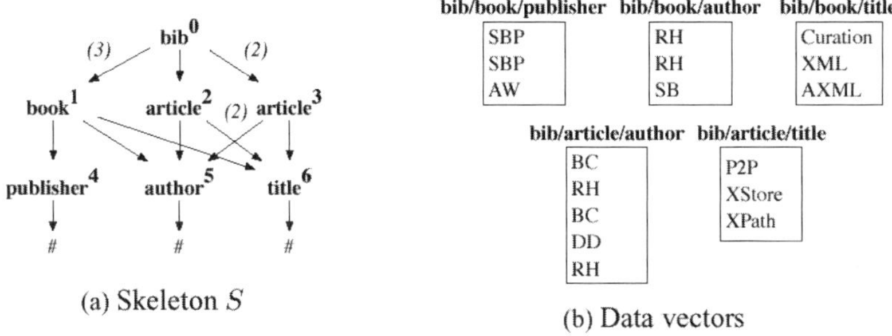

Figure 2. Its VX decomposition: (a) skeleton; and (b) vectors

Those interested in the details of VX should consult the paper by Buneman et al. (2004), but some of its essential features can be understood from comparing Figures 1 and 2, which represent, respectively, the original, tree form of an XML document, and its vectorized decomposition. The numbers next to the edges in 2(a) denote structures which appear more than once in a depth-first traversal of the tree. Queries on the XML document can then be run by a processor which holds the skeleton in main memory and reads data values from the vectors on disk only when needed. The current processor works with a fragment of XQuery that can express all relational conjunctive queries. The regular structure of a tabular dataset yields a trivial skeleton under VX decomposition, allowing very quick querying. Tests (Buneman et al. 2004) on the PhotoObj table of the SDSS EDR SkyServer database (Stoughton et al. 2002) show that the VX prototype is comparable with the EDR SkyServer for those queries where SQL Server does not make heavy used of indexes, since these have not been implemented in VX yet.

3. BinX

BinX (Carroll et al. 2004) has been developed by the *edikt*[4] project at the National e-Science Centre, as a solution to a dilemma expressed within the VO community, and many other data-rich disciplines: *We want the flexibility of storing our data as XML, but we also want the compactness of storing our data in a binary format.* BinX comprises an XML annotation language which de-

[4]`http://www.edikt.org`

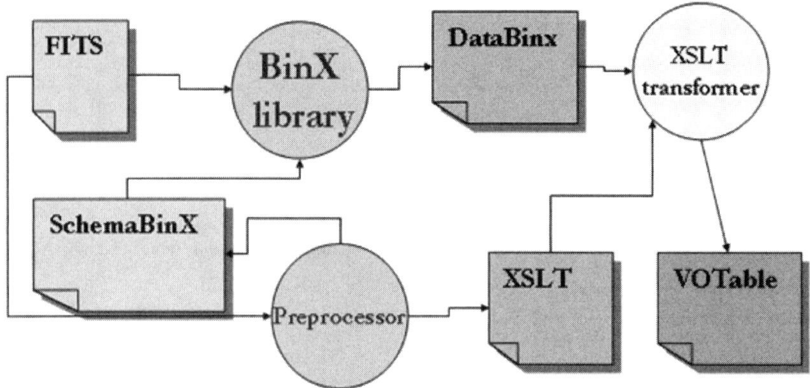

Figure 3. FITS–to–VOTable conversion using BinX.

scribes the data types and structures in binary data files and an associated library for making use of that language in the manipulation of XML and binary data files. The BinX language can describe primitive data types – such as characters, bytes, integers (16, 32, and 64 bits), floating point numbers (32 and 64 bits) and strings – as well as arrays, sequences (structs), and unions, and allows the user to define a type based on an array, sequence or union.

The description of the binary data is stored in an XML file known as a SchemaBinX document. This contains metadata describing each data element within the binary data file, as well as a reference to the binary data file itself: like the FITS variant of VOTable, BinX separates metadata (stored as XML) from bulk data (which remains in a compact binary format). The availability of a SchemaBinX descriptor enables one to manipulate a binary file as if it were an XML document – e.g., run an XPath query over it – via the BinX library. A second representation of the binary data is the DataBinX document, in which the schema information describing a binary data file is supplemented by all the actual data values.

Figures 3 and 4 show schematically one use to which BinX has been put: converting between FITS binary tables and VOTable files (or any other format which can be specified using an XSLT transformation of the DataBinX file). Note that the VOTable-to-FITS conversion is more complicated, due to the ASCII header. This conversion works well, but the intermediate DataBinX files can be large, so one possible development would be to combine VX and BinX. A working group[5] of the Global Grid Forum is extending the ideas of BinX in the development of a Data Format Description Language (DFDL), and BinX may morph into a reference implementation of that standard, as it develops.

[5]`http://http://forge.gridforum.org/projects/dfdl-wg`

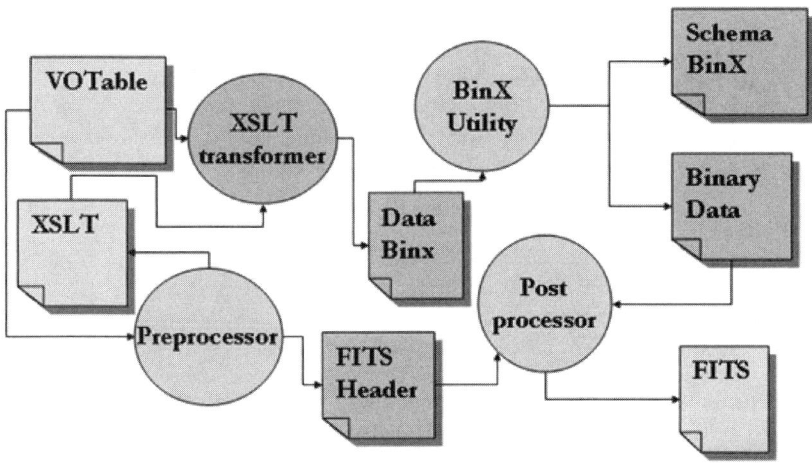

Figure 4. VOTable–to–FITS conversoin using BinX.

4. Summary and Conclusions

As the VO moves from development to production, and people begin wanting to move large volumes of tabular data around, the pure XML version of VOTable will not suffice, and experience suggests that the FITS and Binary serialized versions will not be used widely within the community. Therefore the VO community must think about how it will handle large volumes of tabular data, and the purpose of this paper is to introduce to that community two research projects which may point toward promising solutions to that problem. Those interested in learning more about VX or BinX should contact Bob Mann (*rgm@roe.ac.uk*).

References

Buneman, P., Choi, B., Fan, W., Hutchison, R.W.O., Mann, R.G., & Viglas, S.D. 2004, accepted for publication in Proc. of The 21st International Conference on Data Engineering (ICDE 2005). Preprint available from `http://homepages.inf.ed.ac.uk/v1bchoi/paper.pdf`

Carroll, R., Virdee, D., & Wen, Q. 2004, in Proc. of The UK e-Science All Hands Meeting 2004, `http://www.allhands.org.uk/proceedings/papers/124.pdf`

Liefke, H., & Suciu, D. 2000, in Proc. ACM SIGMOD International Conference on Management of Data

Ochsenbein, F., et al. 2004, IVOA Recommendation, `http://www.ivoa.net/Documents/latest/VOT.html`

Stoughton, C., et al. 2002, AJ, 123, 485

ASTRONOMICAL DATA ANALYSIS SOFTWARE AND SYSTEMS XIV
ASP Conference Series, Vol. 347, 2005
P. L. Shopbell, M. C. Britton, and R. Ebert, eds.

Creating Data that Never Die: Building a Spectrograph Data Pipeline in the Virtual Observatory Era

Douglas J. Mink, William F. Wyatt, John B. Roll, Susan P. Tokarz, Maureen A. Conroy, Nelson Caldwell, Michael J. Kurtz, and Margaret J. Geller

Smithsonian Astrophysical Observatory

Abstract. Data pipelines for modern complex astronomical instruments do not begin when the data is taken and end when it is delivered to the user. Information must flow between the observatory and the observer from the time a project is conceived and between the observatory and the world well past the time when the original observers have extracted all the information they want from the data. For the 300-fiber Hectospec low dispersion spectrograph on the MMT, the SAO Telescope Data Center is constructing a data pipeline which provides assistance from preparing and submitting observing proposals through observation, reduction, and analysis to publication and an afterlife in the Virtual Observatory. We will describe our semi-automatic pipeline and how it has evolved over the first nine months of operation.

1. Introduction

Hectospec (Fabricant et al. 1998) is a 300-fiber spectrograph which takes advantage of the 1-degree field available on the 6.7-meter MMT telescope on Mt. Hopkins in Arizona. The fibers are positioned by robots, so the entire observing process takes place under computer control. In this paper, we look at the entire life cycle of an observation made by the Hectospec, enlarging the concept of a "pipeline" to include all of the processes from the time an astronomer gets an idea for a scientific observation to the permanent archiving of the data in a way which is accessible to the world. This process is not a closed system, and in the era of the world-wide Virtual Observatory, connects to database and archives outside of our observatory at many points. Figure 1 shows the flow of data from the mind of the astronomer to its resting place in the Virtual Observatory.

2. Preparing to Create Data

Complete information about the Hectospec is available online, including the full "Hectospec Observers Reference Manual" (Fabricant & Caldwell 2004). This instrument needs object positions accurate to 0.25 arcsecond, and most recent large catalogs, such as the USNO-B1.0, 2MASS PSC, GSC II, and UCAC2, are adequate (Mink et al. 2003). Good positions for uncatalogued objects can be taken from images which are aligned with any of these catalogs using the WCSTools software (Mink 1996), which can access the catalogs locally or over the Internet.

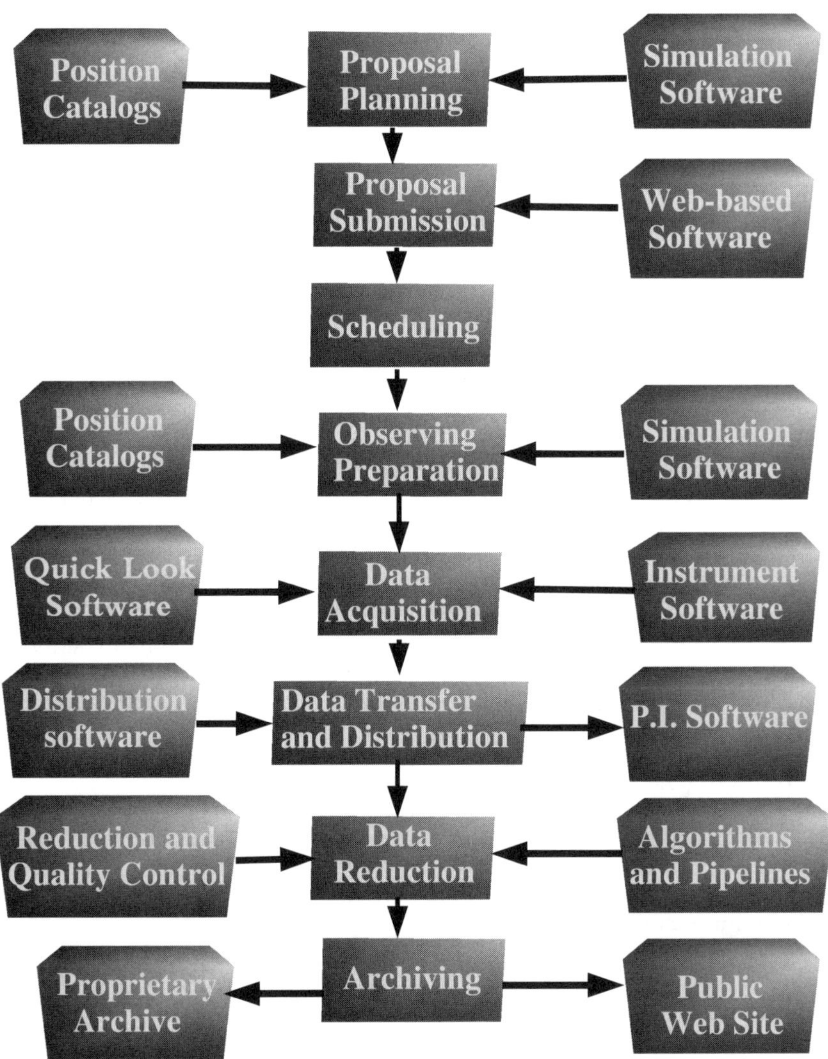

Figure 1. Hectospec Pipeline

Application for time on the Hectospec is entirely Web-based, including voting by the Time Allocation Committee, using Perl software based on that used by NOAO (Gasson and Bell 2002). Abstracts and metadata submitted in this process remain attached to data taken via the program numbers assigned by the TAC when the telescope schedule is put online. The schedules are archived, as are the abstracts and metadata.

An astronomer with time allocated on the Hectospec must set up fiber positions using an interactive program (Roll et al. 1998). Sky positions and guide stars are found automatically using one of the large all-sky catalogs mentioned above. Multiple exposures are often needed to get spectra of all of the program objects in a field. Since the exact time of the observation is not known by the astronomer, final fiber assignments are made the night of the observation by the telescope staff so that exact circumstances, such as differential refraction across the focal plane, can be taken into account.

3. Taking Data

In addition to the bias and dark frames, flat field images are taken of a white field on the telescope dome (to find out where the spectra are on the image) and the sky (to calibrate fiber throughput). The spectrograph, which is in a separate temperature-controlled room from the telescope, is stable enough that a single set of calibration lamp spectra works for an entire night. Each field is exposed at least three times so that cosmic rays can be eliminated without damaging similar-looking emission lines in the spectra. Each of the two CCDs of the detector is read out through two amplifiers, and the four images which are created are stored in multi-extension FITS files along with the fiber position maps as FITS table extensions. The data is immediately transferred from Arizona to Cambridge, Massachusetts, where processing and archiving takes place. Figure 2 shows examples of the three major kinds of images.

4. Reducing Data

Hectospec data is reduced using the image processing and fiber extraction features of IRAF, controlled by locally-written IRAF CL scripts. We have found that a modular system has made it easy to update processing techniques as we have found ways to improve them.

After combining the multiple images of a single pointing to eliminate cosmic rays and merging the amplifiers, spectra are extracted into individual files using a variation on the IRAF DOFIBERS task. Wavelength calibration is more complex than normal because the fibers are alternately staggered which allows them to be more closely packed. Throughput calibration is very important, so that sky background spectra from one fiber can be subtracted from image plus sky spectra in another fiber. This requires both careful calibration and fine-tuning by scaling night sky emission lines individually for each fiber before subtracting the background. The object, sky, and object + sky spectra are saved in IRAF Multispec FITS files.

As part of our reduction pipeline, we compute redshifts for each object spectrum using our own XCSAO program (Kurtz & Mink 1998). The results are

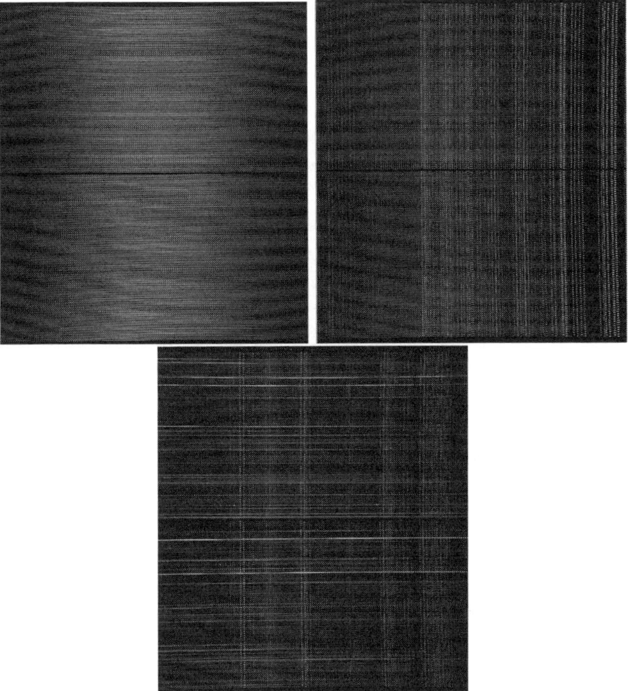

Figure 2. Hectospec dome flat spectra, used as aperture mask, calibration lamp spectra, showing staggered fibers, and object spectra, dominated by night sky emission lines

written into the image header of each object spectrum file. All of the reduction software is available as IRAF, ksh, and C software.

5. Archiving and Distributing Data

The raw data, reduced data in a single IRAF Multispec FITS file per pointing, and reduced data as one background-subtracted IRAF Multispec FITS file per fiber are archived. Investigators are given access to all versions of their data. As scientific papers which use the data are published, that data will be moved into a public archive, where it will be accessible both through our instrument web site, searchable by sky position, and through the Astrophysical Data System, accessible by observing program.

6. Conclusions

The Virtual Observatory is a tool as well as an archive. Data published on the VO can not only be used in its own right, but can point toward data needed to fill gaps. Catalogs available through the VO can be used to calibrate images and suggest blank areas of sky for background removal as well as suggest positions to observe. Data which is destined to be shared via the Virtual Observatory must be well-documented. Not only information about the hardware used to obtain the data, but also the processing which it undergoes must be linked in some way to the data. Knowledge of the process from proposal to archive should be maintained. Every process which changes the data must leave some description of what it has done. Not just keywords, but their meanings must be accessible to future users of the data.

References

Fabricant, D. G., Hertz, E. N., Szentgyorgyi, A. H., Fata, R. G., Roll, J. B., Zajac, J. M. 1998, in Optical Astronomical Instrumentation, ed. Sandro D'Odorico Proc. SPIE Vol. 3355, 285

Fabricant, D. G., & Caldwell, N. 2004, Hectospec Observers Reference Manual

Gasson, D., & Bell, D. 2002, in ASP Conf. Ser., Vol. 281, ADASS XI, ed. D. A. Bohlender, D. Durand, & T. H. Handley (San Francisco: ASP), 457

Kurtz, Michael J., & Mink, Douglas J. 1998, PASP, 110, 934

Mink, D. J. 1996, in ASP Conf. Ser., Vol. 125, ADASS VI, ed. G. Hunt & H. E. Payne (San Francisco: ASP), 249

Mink, D. J., Brown, W. R., Kurtz, M. J. 2003 in ASP Conf. Ser., Vol. 314, ADASS XIII, ed. F. Ochsenbein, M. Allen, & D. Egret (San Francisco: ASP), 141

Roll, J. B., Fabricant, D. G., McLeod, B. A. 1998, in Optical Astronomical Instrumentation, ed. Sandro D'Odorico Proc. SPIE Vol. 3355, 324

ASTRONOMICAL DATA ANALYSIS SOFTWARE AND SYSTEMS XIV
ASP Conference Series, Vol. 347, 2005
P. L. Shopbell, M. C. Britton, and R. Ebert, eds.

MaxBCG: The Importance of Database Systems for the Virtual Observatory and the Grid

María A. Nieto-Santisteban, Alexander S. Szalay, Aniruddha R. Thakar, and William O'Mullane

Department of Physics & Astronomy, The Johns Hopkins University, Baltimore, MD 21218, USA. Email: nieto, szalay, thakar, womullan@pha.jhu.edu

Jim Gray

Microsoft Research, San Francisco, CA 94105, USA. Email: gray@microsoft.com

James Annis

Experimental Astrophysics, Fermilab, Batavia, IL 60510, USA. Email: annis@fnal.gov

Abstract. The Maximum-likelihood Brightest Cluster Galaxy algorithm, MaxBCG, finds galaxy clusters within the Sloan Digital Sky Survey (SDSS). We have coded a SQL version of the algorithm and compared the performance with an earlier Tcl-C-file-based implementation. The SQL database implementation runs 20 to 40 times faster. Our prototype and tests illustrate the benefits of combining computing clusters and database systems for data-intensive astronomical applications. Taking a step further, we extend our clustering conception to grid environments.

1. Introduction

The MaxBCG algorithm solves the specific astronomical problem of locating clusters of galaxies in the SDSS catalog. It searches for galaxy clusters over a wide range of redshifts and masses and provides precise redshifts and good mass estimators. The search relies on the fact that the brightest cluster galaxies (BCG) in most clusters have remarkably similar luminosities and colors (Gladders & Yee 2000). Conceptually, the MaxBCG algorithm moves a cylinder around in a 5-dimensional space, calculating the cluster likelihood at each point. The 5-space is defined by two spatial dimensions, Right Ascension (RA) and Declination (Dec); two color dimensions, g-r and r-i; and one brightness dimension, i.

2. The File Approach

MaxBCG (Annis et al. 2002) was originally developed in Tcl using the SDSS Astrotools package. MaxBCG was designed to take advantage of the intrinsic parallel nature of the problem by using a divide-and-conquer strategy which breaks the sky in 0.5×0.5 deg^2 target fields and surrounds them by 0.25 deg buffer areas in order to search the neighborhood. The application was run in parallel on a 10-CPU Condor cluster called TAM (Terabyte Analysis Machine) at Fermilab. Each 0.25 deg^2 field would take 1000 s to be processed.

The same code was integrated with the Chimera Virtual Data System[1] created by the Grid Physics Network (GriPhyN) project to test grid technologies. Both implementations used hundreds of thousands of data files served from the SDSS Data Archive Server (DAS) to the computing nodes.

3. The Database System Approach

All the data required by MaxBCG are available in the SDSS Catalog Archive Server (CAS), the relational database system hosting the SDSS catalog. At JHU, we have implemented a SQL version of MaxBCG and added two main improvements: bigger buffer areas and finer redshift steps (z-steps). This version, fully implemented on MS SQL Server 2000 in SQL, takes advantage of the indexing and searching capabilities of the database management system, and finds galaxy clusters much faster.

The SQL implementation can run either on a single SQL Server or on a cluster of SQL servers. Using the Zones strategy (Gray et al. 2004), the data are distributed among servers and arranged so each server is completely independent from the others. We present results for a target area of 11 deg $\times$ 6 deg inside a buffer area of 13 deg $\times$ 8 deg but in principle the target area can be much larger. Bigger target regions give better performance because the relative buffer area (overhead) decreases (Nieto-Santisteban et al. 2004).

4. Time Performance

We have measured raw execution times for three implementations of the MaxBCG algorithm. A single TAM node processed a 0.25 deg^2 field in 1000 s. For larger fields, TAM execution time scales linearly with the number of nodes that are used. The SQL Server implementation processed a much larger 66 deg^2 field in 18,635 s with a single server and 8,988 s with a three-node cluster. The SQL Server implementation used finer z-steps and a larger buffer around each field, which increased the computational difficulty by a factor of approximately 25.

These raw execution times must be scaled to a common hardware and software configuration before comparing performance. Table 1 lists the differences between the two hardware and software configurations. The last column gives scale factors that convert TAM raw execution times to predicted times for the hardware environment and software configuration used in the SQL Server test case. We divide the TAM raw execution time by 8 to account for the fact that the dual-CPU nodes running SQL Server were each 4 times faster than the single CPU in each TAM node. In fact, TAM spends much of its time waiting for data from disk, so this is being generous to the TAM system, which had a comparable disk subsystem. We also multiply TAM raw execution times by 264

[1] `http://www.griphyn.org/chimera/`

Table 1. Cluster Configurations and Scale Factors: time scale factors for converting the TAM test case to the SQL Server test case. A single TAM CPU takes 1000 seconds to process each 0.25 deg^2 field.

	TAM	SQL Server	Scale Factor
CPUs used	1	2	0.5
CPU	600 MHz	2.6 GHz	~0.25
Target field	0.25 deg^2	66 deg^2	264
z-steps	0.01	0.001	25
Buffer	0.25 deg	0.5 deg	
		Total Scale Factor:	**825**

for to account for the much larger field that was covered in the SQL Server test case. Finally, we multiply raw TAM execution times by 25 to account roughly for the significant increase in processing time that is required to use finer z-steps and a larger buffer zone around each field. Combining all of these individual factors yields a global scale factor of 825, which must be applied to the TAM raw execution time of 1000 s.

Table 2. Performance Comparison: Scaled TAM versus measured SQL server performance for a target field of 65 deg^2.

Cluster	Nodes	Time (s)	Ratio
Scaled TAM	1	825,000	44
SQL Server	1	18,635	
Scaled TAM	5	165,000	18
SQL Server	3	8,988	

Table 2 gives the predicted TAM and measured SQL Server execution times for the SQL Server test case with large field, fine z-step, and large buffer region. The predicted TAM execution times for this difficult test case are simply the measured execution time of 1000 s for the simple test case, scaled by the factor of 825 from Table 1. Results are given for single-node systems and clusters, assuming TAM execution time scales inversely with the number of nodes. The last column contains the ratio of predicted TAM to measured SQL Server times for tests with one node or a cluster of nodes.

Table 2 implies that the relational database solution is about 40 times faster per node than the file-based approach. For the specific cluster configurations considered here the SQL Server approach is about 20 times faster than TAM. Even if one were willing to wait 20 times longer, TAM nodes do not have enough memory to handle z-steps of 0.001 and a buffer of 0.5 deg.

5. The Future

We are already working on a grid-enabled version that uses different instances of the SDSS CasJobs[2] system (Fermilab and JHU) and avoids transferring large volumes of data through the network. In the future, when a user submits an application, upon authentication and authorization, the SQL code (500 lines) will be deployed on the available grid nodes hosting the CAS databases and CasJobs system. The system will automatically perform the data space partitioning and task scheduling. As nodes become free, they will process the piece of the sky assigned. We foresee some of these data grid nodes being portals to HPC Beowulf clusters hosting database management systems. The performance of such a grid system will be outstanding!

6. Conclusions

- **Lesson 1:** Using database management systems and SQL we can improve the computational performance of data-intensive astronomical applications.
- **Lesson 2:** In grid environments, database systems are typically used only to access and integrate the data but not to perform analytic or computational tasks. Data is accessed and moved to the computational nodes.

It would be useful for the Virtual Observatory if grid nodes were able to support database management systems, so that SQL applications could be deployed as easily as traditional grid applications coded in C, Fortran, etc.

An extended and detailed version of this work, including the MaxBCG SQL code, can be downloaded.[3]

Acknowledgments. This work was funded by the NASA Applied Information Systems Research Program – Grant NRA0101AISR.

References

Annis, J., Zhao, Y., Voeckler, J., Wilde, M., Kent, S., & Foster, I. 2002, in Proceedings of the 2002 ACM/IEEE conference on Supercomputing (IEEE: Baltimore), 16

Gladders, M. D. & Yee, H. K. C. 2000, AJ, 120, 2148

Gray, J., Szalay, A. S., Thakar, A. R., Fekete, G., , Nieto-Santisteban, M., O'Mullane, W., Heber, G., & Rots, A. H. 2004, Microsoft Technical Report MSR-TR-2004-32 (ftp://ftp.research.microsoft.com/pub/tr/TR-2004-32.pdf)

Nieto-Santisteban, M., Gray, J., Szalay, A., Annis, J., Thakar, A., & O'Mullane, W. 2004, Microsoft Technical Report MSR-TR-2004-81 (ftp://ftp.research.microsoft.com/pub/tr/TR-2004-81.pdf)

[2] http://casjobs.sdss.org/casjobs

[3] ftp://ftp.research.microsoft.com/pub/tr/TR-2004-81.pdf

ASTRONOMICAL DATA ANALYSIS SOFTWARE AND SYSTEMS XIV
ASP Conference Series, Vol. 347, 2005
P. L. Shopbell, M. C. Britton, and R. Ebert, eds.

VWhere: A Visual, Extensible 'where' Command

Michael S. Noble

Center for Space Research, Massachusetts Institute of Technology

Abstract. In this paper we describe *VWhere*, a S-Lang guilet which augments the computational power of *where* with the point-click ease of a Gtk-based visual interface. The result is a new mechanism for interacting with datasets, which unifies the constraint evaluation cycle implicit within observational analysis and provides a number of compelling advantages over traditional tool-based methods.

1. The Problem: Exploratory Analysis

The act of working raw data into a form from which its innate properties may be more readily discerned, referred to here as *exploratory analysis*, is of fundamental importance to scientific inquiry. For example, from an astrophysical dataset $\boldsymbol{D}$ one may wish to examine events within some region $\mathcal{R}$, after time $\mathcal{T}$, but only if they originated from detectors $\mathcal{I}$ and $\mathcal{J}$, and were above energy $\mathcal{E}_0$ but below energy $\mathcal{E}_1$. Such constraint sets $\boldsymbol{C} = \{C_0, ..., C_i\}$ are derived by iterating through "*What If?*" cycles, modeled in the large as

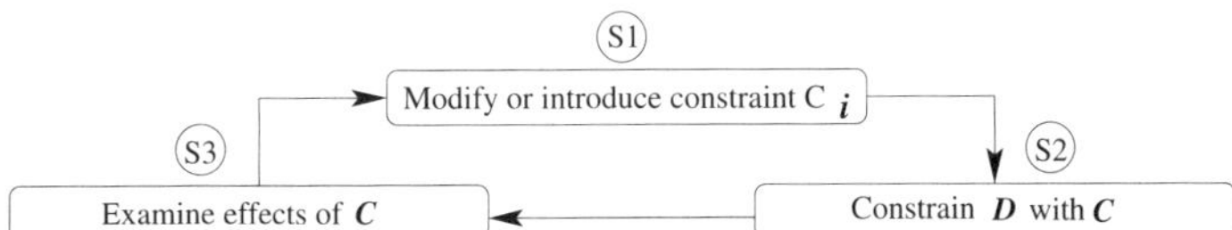

Stages (S1) and (S3) may involve numerical and visualization tasks, such as fitting models and plotting residuals. Stage (S2) may be manifested by applying filters to columns and images, or grouping, binning, et cetera. This process plays out whether $\boldsymbol{D}$ is a single observation or multiple observations spanning a number of wavelengths and telescopes.

1.1. A Traditional Solution: Command Line Tools

Many existing astronomical analysis systems support data exploration via some combination of command line tools and interactive applications. The tools in this model operate as file transformers (*file in* —> *file out*), and characterize the exploration process as one in which (S1), (S2), and (S3) are performed by

- multiple invocations of distinct programs (e.g., `fcopy/dmcopy/XSpec`)
- requiring many passes over files in $\boldsymbol{D}$
- during which a number of intermediate files are produced

While this approach certainly works, and provides bookkeeping advantages if the tools record their operation (e.g., as comments in a FITS header), it is clumsy for ad-hoc inquiry and generates file litter. It is also needlessly consumptive of time and computational resources (at times enormously so, as corroborated by Davis et al. (2005). Moreover, it is easy to mistakenly use the filtering syntax of one system when executing tools from another. Finally, traditional tools are static in function, incapable of being selectively extended by the end-user via the importation of a module or script.

1.2. A New Solution: Interactive 'where' Command

To anyone who has analyzed data within array-oriented systems such as IDL™, CIAO/ISIS, or PyRAF, the *where* command should need little introduction. Couched in intuitive syntax, it provides a powerful mechanism for filtering arrays via arbitrary expressions. For example, consider plotting from an event list all photons whose position (specified by X, Y arrays) fell within a circle of radius 10 centered at the origin

```
isis> i = where( X^2 + Y^2 < 10);
isis> plot(X[i], Y[i]);
```

The *VWhere* "guilet" detailed herein augments this computational power with a graphical interface that permits constraints to be specified, manipulated, and evaluated visually. In this model constraint sets are represented as a combination of plots and the region filters cumulatively applied to them. Plots are generated from the axes of a dataset ***D*** resident in memory, or from new axes derived by arbitrary analytic combinations of them, all without incurring any additional I/O overhead or tool execution costs per iteration.

2. Exploring Cygnus X-1

Let us now walk through portions of a published research scenario, in which *VWhere* was used to correlate two observations (from Chandra and the RXTE All-Sky Monitor) of black hole candidate Cygnus X-1. We begin by loading *VWhere* into ISIS

```
isis> require("vwhere");
```

done explicitly here to underscore the fact that it is *not* linked directly into ISIS but is rather a standalone module that can be loaded at runtime into *any* application which embeds the S-Lang interpreter. To maximize benefit to the community, while still promoting freedom of choice, we've also begun to investigate means by which S-Lang modules may be utilized within other interpreters, such as Tcl or Python. After loading the data into a structure (not shown), we launch *VWhere*

```
isis> indices = vwhere(data);
```

and generate a lightcurve by plotting 3 fields: `time` versus `lchan + mchan`. Our objective is to look for times where Cyg X-1 is emitting harder X-rays during low count states. Each such plot is generated in its own tabbed subwindow, and may be panned, zoomed, deleted, printed, or cosmetically altered in many of the usual manners one would expect within a plotting GUI. Arbitrarily complex filters may then be constructed by laying regions upon plots. Pressing *Done*

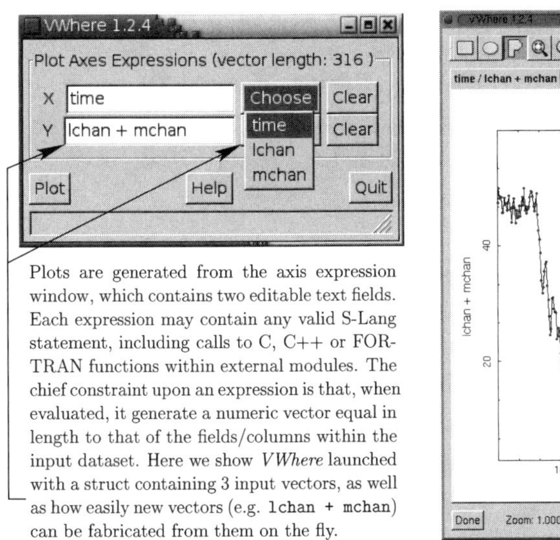

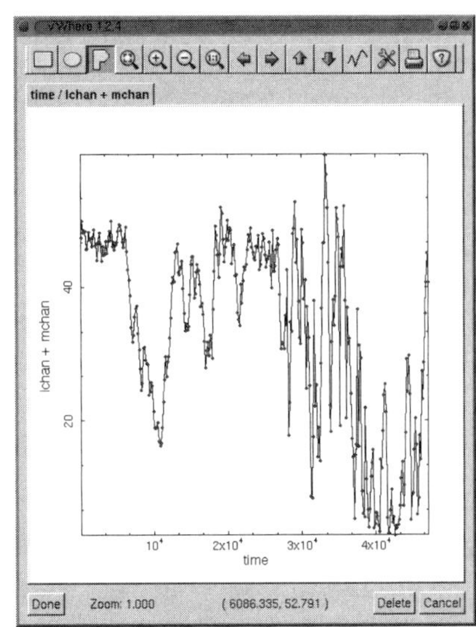

Figure 1. A lightcurve for Cygnus X-1, and its axis expressions.

returns to the caller a list of indices representing all points selected by the applied filters (the same semantics as *where*).

Notice that the lightcurve above exhibits a good deal of variability. The color intensity diagram in the first panel of Figure 2 shows that as count rate drops (lower X value) CygX-1 emission is getting harder (smaller Y value), except for a spike at the very lowest rates. So we select points within that spike, by applying a polygonal region, and look at the lightcurve again in the second panel. The points filtered away are drawn grayed out, revealing that the hard, low count-rate selection comes from a deep dip near the end of the lightcurve.

3. Benefit Analysis

3.1. Unifying the Constraint Cycle

This example demonstrates how *incremental filtering* in *VWhere* unifies stages (S1), (S2), and (S3) of the constraint evaluation cycle, dramatically shrinking the time and effort required to discern subtle patterns within data, and without any package-specific filtering syntax. With *VWhere* it is instantly clear how cuts applied to portions of a dataset, or new axes derived from it, affect other portions. As a result, the process of constructing constraint sets is considerably more fluid. It is also far more powerful, since by virtue of the extensibility endowed by S-Lang, *VWhere* may import a wide range of C, C++, or FORTRAN modules, and rather easily using the SLIRP code generator[1]. This is simply not feasible with traditional tools. Finally, since *VWhere* operates directly upon

[1] `http://space.mit.edu/cxc/software/slang/modules/slirp/`

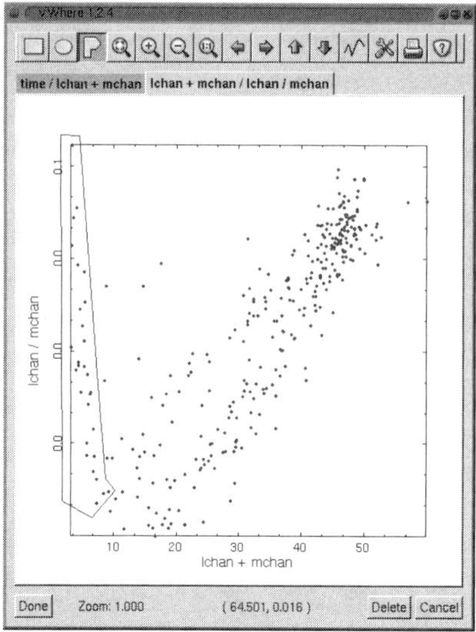

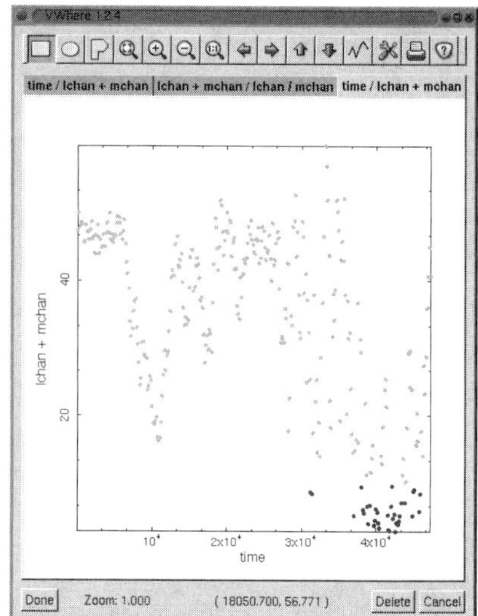

Figure 2. A polygon filter applied to the color intensity diagram, and its incremental effect upon the original lightcurve.

arrays already loaded within its host application, memory is conserved while I/O overhead and runtime performance are improved by many factors.

3.2. Real Modularity

Drawing from our experience as both user and developer of major astronomical packages, we note a tendency in extant systems whereby algorithms or libraries are internally coupled in ways which sharply curtail their use elsewhere. Such systems, while open in the superficial sense that source code is publicly available, are in a deeper sense closed by their complex web of internal dependencies, which accrete to form a monolithic structure impenetrable to the practicing scientist and difficult to navigate even for professional programmers.

Potential users must confront the prospect of downloading, sifting through, and picking apart tens or hundreds of thousands of lines of software simply to obtain a small library or algorithm. This effectively prevents the use of novel *portions* of a system apart from its *whole*. Conversely, tightly coupled architectures betray a distinct evolutionary disadvantage, in that they also tend to inhibit the incorporation of software developed externally to the system.

VWhere is one of the earliest examples of our response to this problem. Our methodology shuns monolithic constructions in favor of modular, extensible components that are orthogonal in function, and which may either be woven together to form larger systems or used completely standalone. Other fruits of this effort are described on our modules page[2] and in Davis et al. (2005).

---

[2] `http://space.mit.edu/cxc/software/slang/modules/`

3.3. SLgtk

As argued elsewhere (e.g., Primini & Noble 2005), S-Lang is an excellent open-source language for scientific scripting, particularly for those with FORTRAN, C, or IDL experience. SLgtk augments its core strengths (e.g., built-in support for array-based numerics) with an importable module that makes it possible to construct sophisticated graphical interfaces directly from the S-Lang interpreter. Most of SLgtk is generated automatically by SLIRP.

3.4. Guilets

A driving force behind the development of SLgtk has been the notion of the *guilet*, by which we mean visual software of a small, scriptable nature which may be easily embedded within other applications, even those with a primarily textual interface. This fosters the use of graphical interfaces where they are appropriate and beneficial, without committing the entire application to such.

In code size and development cycle the typical guilet is considerably smaller than traditional GUIs (applications designed with a primarily visual interface, coded in a compiled language, and explicitly utilizing low-level toolkits such as Xt, Motif, or the Windows Foundation Classes). Moreover, since guilets are scripts, they are more amenable to changing requirements, rapid feature evolution, and even customization by the end-user. This can be vital to scientists who:

- are interested in advanced analysis not yet supported by existing systems
- use systems with proprietary periods, but are unable to wait for upgrades
- find it advantageous to concoct working prototypes for developers, to either supplement or supplant paper specifications

In point of fact, *VWhere* provides modern analysis and plotting capability in only 1650 lines of scripts, and has been enhanced on numerous occasions with turnaround times as short as 2 hours.

4. Summary

This paper describes *VWhere*, which extends the S-Lang *where* command with a Gtk-based visual interface. We have argued that by integrating the specification, accumulation, and visualization of analytical constraints *VWhere* presents a facile mechanism for exploring data, one which eclipses tool-based methods for such, particularly in terms of performance, extensibility, and ease of use.

Acknowledgments. This work was supported in part by the Chandra X-Ray Center contract SV1-61010 from the Smithsonian Institution. The author would like to thank colleagues at MIT and the Harvard-Smithsonian CFA (especially Michael Nowak), as well as those within the open source community, for various contributions and commentary.

ASTRONOMICAL DATA ANALYSIS SOFTWARE AND SYSTEMS XIV
ASP Conference Series, Vol. 347, 2005
P. L. Shopbell, M. C. Britton, and R. Ebert, eds.

New NED XML/VOtable Services and Client Interface Applications

Olga Pevunova, John Good, Joseph Mazzarella, G. Bruce Berriman, and Barry Madore

Infrared Processing and Analysis Center, California Institute of Technology, Mail Code 100-22, CA 91125

Abstract.
The NASA/IPAC Extragalactic Database (NED) provides data and cross-identifications for over 7 million extragalactic objects fused from thousands of survey catalogs and journal articles. The data cover all frequencies from radio through gamma rays and include positions, redshifts, photometry and spectral energy distributions (SEDs), sizes, and images. NED services have traditionally supplied data in HTML format for connections from Web browsers, and a custom ASCII data structure for connections by remote computer programs written in the C programming language. We describe new services that provide responses from NED queries in XML documents compliant with the international virtual observatory VOtable protocol. The XML/VOtable services support cone searches, all-sky searches based on object attributes (survey names, cross-IDs, redshifts, flux densities), and requests for detailed object data. Initial services have been inserted into the NVO registry, and others will follow soon. The first client application is a Style Sheet specification for rendering NED VOtable query results in Web browsers that support XML. The second prototype application is a Java applet that allows users to compare multiple SEDs. The new XML/VOtable output mode will also simplify the integration of data from NED into visualization and analysis packages, software agents, and other virtual observatory applications. We show an example SED from NED plotted using VOPlot.

The NED website is: `http://nedwww.ipac.caltech.edu`.

1. Goals

- To simplify access to NED data from client computer programs.
- Follow NVO protocols by providing the output in XML VOTable format
- Simplify the integration of data from NED into visualization and analysis packages, software agents, and other virtual observatory applications.

2. Example Query

NED queries can be performed from the NED Object Search form, which can be found off the NED front page: click on “Objects” and then “By Name”. A sample query is as follows for position 142.2167 10.80306 EQ and radius of 0.1 degrees: `http://nedwww.ipac.caltech.edu/cgi-bin/nph-NEDobjsearch`. A sample output VOTable document tree is shown in Figure 1. This NED resource

```
- <VOTABLE version="v1.0">
  + <DEFINITIONS></DEFINITIONS>
  - <RESOURCE type="results">
    - <DESCRIPTION>
        Results from query to NASA/IPAC Extragalactic Database (NED).
      </DESCRIPTION>
      <INFO name="QUERY_STATUS" value="OK"/>
    + <LINK content-role="query" content-type="char"></LINK>
    - <TABLE ID="NED_MainTable" name="Main Information Table for Searching NED within 6.0 arcmin of 142.216700, 10.803060">
      - <DESCRIPTION>
          Main information about object (Cone Search results)
        </DESCRIPTION>
      + <FIELD ID="main_col1" name="count" ucd="ID_NUMBER" datatype="int"></FIELD>
      + <FIELD ID="main_col2" name="name" ucd="ID_MAIN" datatype="char" arraysize="30"></FIELD>
      - <FIELD ID="main_col3" name="ra" ucd="POS_EQ_RA_MAIN" datatype="double" unit="degrees">
          <DESCRIPTION> Right Ascension in degrees (J2000) </DESCRIPTION>
        </FIELD>
      + <FIELD ID="main_col4" name="dec" ucd="POS_EQ_DEC_MAIN" datatype="double" unit="degrees"></FIELD>
      + <FIELD ID="main_col5" name="objtype" ucd="OBJ_TYPE" datatype="char" arraysize="*"></FIELD>
      + <FIELD ID="main_col6" name="velocity" ucd="VELOC_HC" datatype="double" unit="km/sec"></FIELD>
      - <FIELD ID="main_col7" name="redshift" ucd="REDSHIFT_HC" datatype="double">
        - <DESCRIPTION>
            Heliocentric redshift Z-value for the object, if it exists in NED
          </DESCRIPTION>
        </FIELD>
      + <FIELD ID="main_col8" name="distance" ucd="DISTANCE" datatype="double" unit="arcmin"></FIELD>
      + <FIELD ID="main_col9" name="refnumber" ucd="REF_NUMBER" datatype="int"></FIELD>
      + <FIELD ID="main_col10" name="notesnumber" ucd="NOTES_NUMBER" datatype="int"></FIELD>
      + <FIELD ID="main_col11" name="photnumber" ucd="PHOT_NUMBER" datatype="int"></FIELD>
      + <FIELD ID="main_col12" name="posnumber" ucd="POS_NUMBER" datatype="int"></FIELD>
      + <FIELD ID="main_col13" name="znumber" ucd="Z_NUMBER" datatype="int"></FIELD>
      + <FIELD ID="main_col14" name="assocnumber" ucd="ASSOC_NUMBER" datatype="int"></FIELD>
      - <DATA>
        - <TABLEDATA>
          - <TR>
              <TD>1</TD>
              <TD>NVSSJ092852+105232</TD>
              <TD>142.220375</TD>
              <TD>10.875806</TD>
              <TD>RadioS</TD>
              <TD/>
              <TD/>
              <TD>4.370000</TD>
              <TD>0</TD>
              <TD>0</TD>
              <TD>1</TD>
              <TD>0</TD>
              <TD>0</TD>
              <TD>0</TD>
            </TR>
          </TABLEDATA>
        </DATA>
      </TABLE>
    </RESOURCE>
</VOTABLE>
```


Figure 1. Document Tree

has been validated using the cone search validator at `http:/nvo.ncsa.uiuc.edu/VO/services/csvalidate.html`.

3. Query Results

Presentation of NED SED results for radio galaxy 3C 279 is displayed by the VOTable Style Sheet available through DataScope[1]. Figure 2 shows 3C 279 within DataScope. The VOTable output from Figure 2 from NED is plotted using VOPlot in Figure 3.

NED data in VOTable format can also be converted to a simple ASCII file as comma separated values (CSV).

---

[1]`http:/skyview.gsfc.nasa.gov/cgi-bin/tam/vot/datascope/init.pl`

The NASA/IPAC Extragalactic Database

VOTable for resource NED_search_photo_data near 187.706,12.391

See also:

Metadata -- Registry information describing service.
Raw VOTable -- Cached information at DataScope server.
-- http://skyview.gsfc.nasa.gov/tempspace/vot/187.706000_12.391000_.25/NED_search_photo_data_1.xml
VOPlot -- Plot data with VOPlot
VOStat -- Analyze data with VOStat
Directly invoke service again -- XML result will not be cached.

Resource: Results from query to NASA/IPAC Extragalactic Database (NED).

Info Elements (for all tables in this resource)	
Info. Key	**Info Value**
QUERY_STATUS	OK

PhotometricData for 3C 279 Published and Homogenized [Frequency, Flux Density] Units

count	bandpass	PHOT_measurement	PHOT_uncertainty	units	frequency	NED_PHOT_measurement	NED_PHOT_uncertainty	NED_PHOT_units	refcode
ID_NUMBER	VOX:BandPass_ID	PHOT	PHOT_UNCERTAINTY	PHOT_UNITS	FREQUENCY	NED_PHOT_MEASUREMENT	NED_PHOT_UNCERTAINTY	NED_PHOT_UNITS	REFER_BIBCODE
1	EGRET 4-10 GeV	1.8E-11	+/-1.2E-11	Jy	1.48E+24	1.80E-11	+/-1.20E-11	Jy	1996ApJ..461..698H
2	EGRET 2-4 GeV	9.6E-11	+/-2.5E-11	Jy	6.70E+23	9.60E-11	+/-2.50E-11	Jy	1996ApJ..461..698H
3	EGRET (0.1-5 GeV)	2.3103E-10	+/-8.3300E-12	Jy	6.17E+23	2.31E-10	+/-8.33E-12	Jy	1995ApJS..101..259T
4	EGRET 1-2 GeV	1.08E-10	+/-0.25E-10	Jy	3.35E+23	1.08E-10	+/-2.50E-11	Jy	1996ApJ..461..698H
5	EGRET 0.5-1 GeV	3.1E-10	+/-0.4E-10	Jy	1.67E+23	3.10E-10	+/-4.00E-11	Jy	1996ApJ..461..698H
6	EGRET 300-500 MeV	5.6E-10	+/-0.6E-10	Jy	9.25E+22	5.60E-10	+/-6.00E-11	Jy	1996ApJ..461..698H
7	EGRET 150-300 MeV	7.9E-10	+/-0.7E-10	Jy	5.02E+22	7.90E-10	+/-7.00E-11	Jy	1996ApJ..461..698H
8	EGRET 100-150 MeV	1.19E-9	+/-0.13E-9	Jy	2.94E+22	1.19E-09	+/-1.30E-10	Jy	1996ApJ..461..698H

Figure 2. Radio Galaxy 3C 279 DataScope

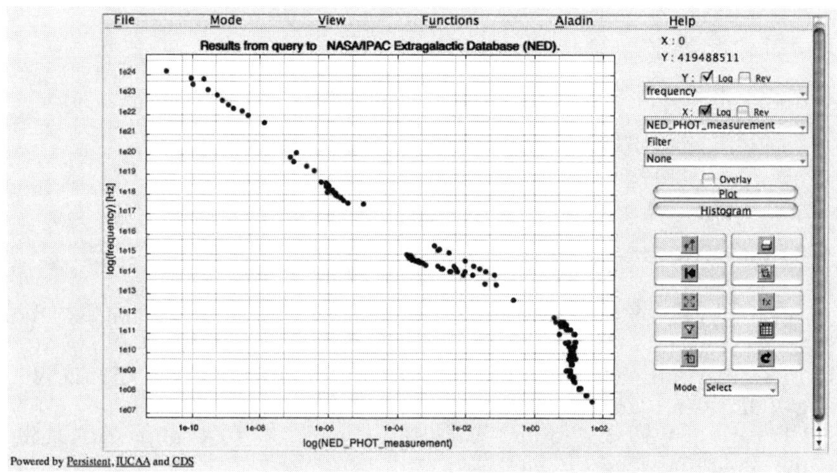

Figure 3. VOPlot of Radio Galaxy 3C 279 using VOTable results

4. NED Images

50,000 FITS compliant NED images can now be queried by position and areal coverage using the NVO Simple Image Access Protocol (SIAP). Access to these images can be found through the web-based http portal and through the NVO SIAP services. The front page is displayed in Figure 4. FITS image search results can be downloaded to disk or sent to On-line Science Information System (OASIS - The Infrared Science Archive's data visualization tool) for viewing.

Acknowledgments. This work is the result of a collaboration between NED, IRSA and the NVO consortium. O. Pevunova, J. Good, and G. B. Berriman were supported by NSF/ITR NSF Cooperative agreement AST 0122449. J. Mazzarella and B. Madore were supported by the Jet Propulsion Laboratory, California Institute of Technology, under contract with NASA.

NASA/IPAC EXTRAGALACTIC DATABASE

Help | NED Home
| NED Help | IRSA Home

NED Image Data Atlas

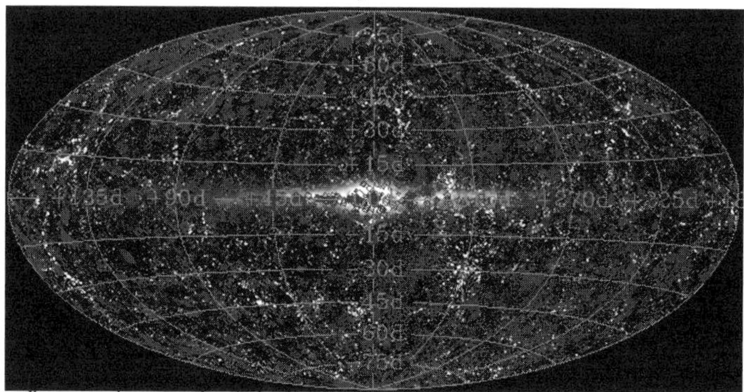

You can get a close-up map of a region by selecting a point in red on the above image, or by typing a coordinate below.

The NASA/IPAC Extragalactic Database (NED) is built around a master list of extragalactic objects for which cross identifications of object names and positions are established using data published in journal articles and survey catalogs covering all wavelengths. Detailed photometry, positions, and redshift data are assimilated, entered with uncertainties when available, and linked to bibliographic references, abstracts and online literature. NED also maintains an image archive that currently contains over 1.9 million images, maps and links to external images. NED's image archive focuses on unique FITS images acquired from authors of journal articles, but also includes cut-outs from very large sky surveys such as 2MASS and the Digitized Sky Survey. NED currently contains data for over 7.5 million objects, and the content is continually being updated. Revised versions of the public database are released every few months.

Images in NED are represented above as overlays in red on the 2MASS all-sky image above. Either type in a coordinate or click on any red region to get a close-up of the area; the size of the search region is adjustable. Please note that the above all-sky grid is in Galactic coordinates, however, the default coordinate search is done using Equatorial J2000, unless specified otherwise within the coordinate input string. This service is a collaborative project between the NED group and the Infrared Science Archive (IRSA) group at the Infrared Processsing and Analysis Center (IPAC).

Searching for NED images is done using **Atlas**. Click here for more information and instructions on using Atlas

Coordinate/Object:

Size (deg): [maximum 12.5] 2.0 Images must cover coordinate

Submit Reset

Coordinate Examples: 10.68469 +41.26904 eq | 00h 42m 44.33s +41d 16m 08.5s Equ J2000 | 121.17432

Figure 4. Front page to search NED images using position and areal coverage. The same service can also be used through the NVO, using SIAP.

ASTRONOMICAL DATA ANALYSIS SOFTWARE AND SYSTEMS XIV
ASP Conference Series, Vol. 347, 2005
P. L. Shopbell, M. C. Britton, and R. Ebert, eds.

K-12 Education with the National Virtual Observatory

M. Jordan Raddick, William O'Mullane, and Alexander S. Szalay

Department of Physics and Astronomy, The Johns Hopkins University, Baltimore, MD 21218

Carol Christian

Space Telescope Science Institute, Baltimore, MD 21218

Abstract. We have created the first K-12 education activity using data provided by the National Virtual Observatory (NVO). The activity, "Adopt an Object," was suggested by Heidi Kaiter, a middle school science teacher from Concord, MA. It is designed for middle school students, but could be adapted for high school and Astro 101 students as well.

Each group of students selects a well-known sky object to "adopt" for detailed study. Students use the NVO's "DataScope" to look up multiwavelength images, observations, and catalog data for their object. The activity ends with a brief oral presentation and written report summarizing what each group has learned about their object.

The activity includes a complete teachers guide, with a lesson plan, additional resources, and correlations to national education standards. We are currently inviting K-12 teachers to field test the activity in their classes. We are also developing several other activities for K-12 and college teachers using NVO data.

The activity discussed here is available online at `http://www.voservices.net/nvoedu`.

1. History

For the past three years, we have developed the SkyServer web site[1], which makes the entire dataset of the Sloan Digital Sky Survey available free of charge to teachers and the general public. SkyServer has been a successful site, getting more than 2 million hits per month. With the development of applications for the National Virtual Observatory (NVO)[2], we are ready to apply our experience with SkyServer to create an outreach product for NVO.

The project described in this paper began with an E-mail to a mailing list of about 200 teachers and amateur astronomers signed up to receive updates on SkyServer. The E-mail explained the promise of NVO – easy online access to all astronomy data ever taken – and asked what teachers would do with such a tool. Six of the teachers' responses were developed into six "education cases" – real-life examples of how teachers envision turning NVO tools into viable classroom activities.

---

[1] `http://skyserver.sdss.org/`

[2] `http://www.us-vo.org/`

2. Education Cases

1. Adopt an Object - a group of middle school students chooses an object in the sky (star, star cluster, galaxy, or nebula) and uses NVO tools and web resources to learn about the object. Each group prepares a short presentation and a written report about its object.
2. A New Nebula - advanced high school students study McNeil's Nebula, a variable reflection nebula discovered by an amateur astronomer in 2004. Can students find the nebula in older images?
3. Image Catalog (with Captions) - a retired high school physics teacher gives public lectures to schools and astronomy clubs. He is looking for a catalog of astronomy images with descriptive captions to use in his presentations.
4. Olympic Coordinates - a high school physics teacher and a middle school geography teacher design a competitive event for the National Science Olympiad. The event deals with using various celestial coordinates to describe positions of objects in the sky.
5. Color-Magnitude Diagrams - a high school astronomy teacher teaches stellar properties by making two color-magnitude diagrams – one for a random sample of stars and one for stars in a cluster.
6. The Moon and Planets - an elementary school special education teacher has students look at the moon and visible planets with the naked eye and with a small refracting telescope, then look at Hubble Space Telescope images of the same objects.

All of the education cases can be done with current NVO tools except number 6 (because planetary data can not be easily searched in the NVO's current framework, which requires celestial coordinates). We are developing the education cases into full projects; the first available is Case 1 – Adopt an Object. It is available online at `http://www.voservices.net/nvoedu`.

3. DataScope

The NVO has developed a simple tool called the DataScope[3] (formerly the Data Inventory Service), a web-based application for retrieving data on a specific object or sky position (McGlynn et al. 2003). DataScope builds on lower-level NVO services like the Cone Search and SIAP protocols, as well as the registry (Greene et al. 2003). DataScope also links transparently to Aladin (Boch, Fernique, & Bonnarel 2003) and OASIS (Good, Kong, & Berriman 2003), Java applets for displaying images and overlaying catalog data.

To use the DataScope, enter a name or sky position in its initial search window and click "Submit Request." As results return from DataScope's query of NVO services, the screen changes into a list of all available data for the object or position, divided into Images (GIF/FITS), Observations (pointings with instruments), and Objects (entries in surveys and catalogs). The page refreshes every 15 seconds with more results from the registry. The DataScope is easy to use, and it offers powerful access to many different types of data. It is directly

---

[3] `http://heasarc.gsfc.nasa.gov/vo/`

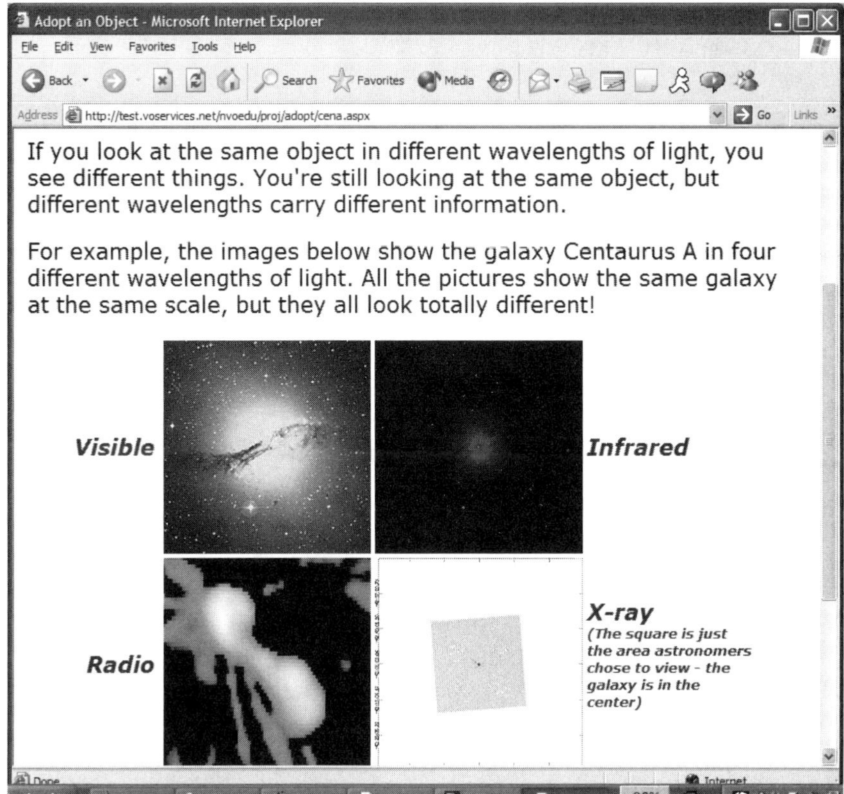

Figure 1. A screenshot from the Adopt an Object student project

linked from the activity we have developed, and the activity includes a user's guide to the DataScope.

4. Adopt an Object - Student Project

We have designed a student activity for the Adopt an Object project. It begins with a description of how telescopes allow us to see many more sky objects than can be seen with the naked eye. The activity continues with an introduction to various objects that can be seen in the sky.

The activity teaches students that visible light is not the only way of studying the sky: other wavelengths of light can offer different information. As an example, we show the active galaxy Centaurus A in four wavelengths of light side-by-side: visible, infrared, radio, and x-ray. Figure 1 shows this comparison.

Each group of students then chooses an object from the following list, or selects another object they are interested in.

Stars: Alpha Centauri, Eta Carinae
Star Clusters: Pleiades, Hyades, M5

Nebulae: M20, M42, M78
Galaxies: M31, M33, M51, NGC4753
Quasars: 3C273

Students learn how to use the DataScope, then freely explore the DataScope to find as much information about their object as they can. Students also explore other sites, such as `Astronomycenter.org`, the SEDS Messier Database, and the Chandra X-Ray Observatory photo album, looking for more images, data, and other information about their object.

5. Adopt an Object - Teacher Reference Pages

Because the Adopt an Object student activity is so open-ended, it requires teachers to be well-trained and prepared to guide students through their research and exploration. We have developed teacher's pages for the activity, including a complete lesson plan, books and web sites for further information, a grading rubric, and correlations to the National Science Education Standards (NSES). Figure 2 shows the overview section of the teacher reference pages.

6. Future Work

We will field-test the Adopt an Object activity in early 2005, beginning in the Concord, MA classroom of the teacher who initiated the project. The field testing will advise further developments of future NVO educational activities and tools. We will develop student and teacher activities based around the other education cases described here, using tools that the NVO has already developed.

References

Boch, T., Fernique, P., & Bonnarel, F. 2004, in ASP Conf. Ser., Vol. 314, ADASS XIII, ed. F. Ochsenbein, M. Allen, & D. Egret (San Francisco: ASP), 221

Good, J. C., Kong, M., & Berriman, G. B. 2003, in ASP Conf. Ser., Vol. 295, ADASS XII, ed. H. E. Payne, R. I. Jedrzejewski, & R. N. Hook (San Francisco: ASP), 89

Greene, G., Hanisch, B., Gaffney, N., & O'Mullane, W. 2004, in ASP Conf. Ser., Vol. 314, ADASS XIII, ed. F. Ochsenbein, M. Allen, & D. Egret (San Francisco: ASP), 285

McGlynn, T., et al. 2004, in ASP Conf. Ser., Vol. 314, ADASS XIII, ed. F. Ochsenbein, M. Allen, & D. Egret (San Francisco: ASP), 319

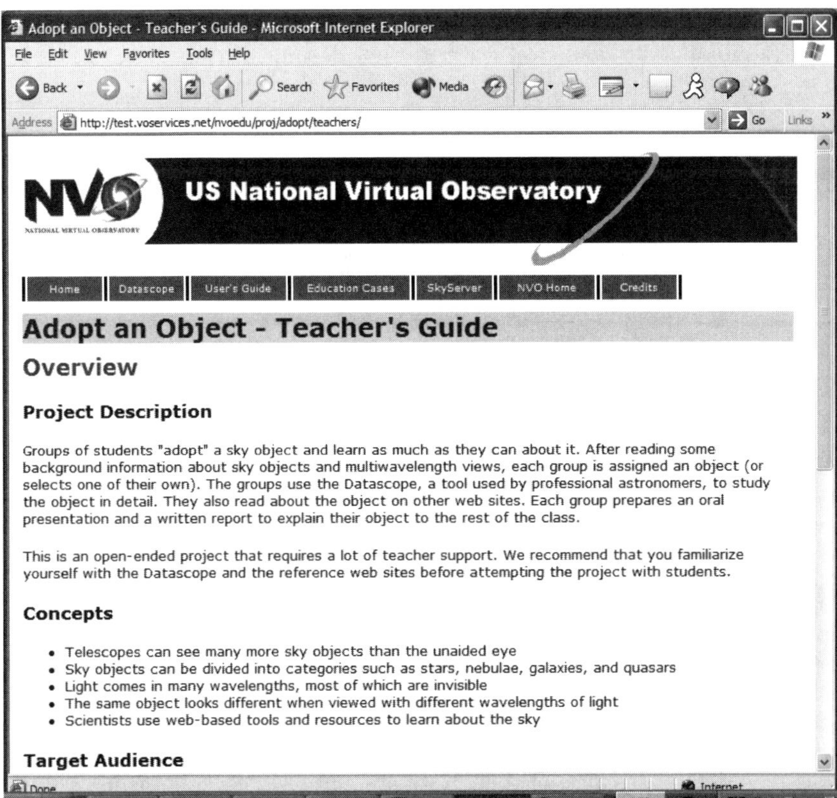

Figure 2. A screenshot from the Adopt an Object teacher reference pages

ASTRONOMICAL DATA ANALYSIS SOFTWARE AND SYSTEMS XIV
ASP Conference Series, Vol. 347, 2005
P. L. Shopbell, M. C. Britton, and R. Ebert, eds.

Development of a Virtual Observatory Prototype in ALMA-Japan

T. Sawada

Nobeyama Radio Observatory, Minamimaki, Minamisaku, Nagano 384-1305, Japan

K. Tatematsu and R. Kandori[1]

National Astronomical Observatory of Japan, Mitaka, Tokyo 181-8588, Japan

K. Nakanishi, K.-I. Morita, and K. Sunada

Nobeyama Radio Observatory, Minamimaki, Minamisaku, Nagano 384-1305, Japan

Abstract. We report the present status of a prototype VO development in the Japanese side of the Atacama Large Millimeter/submillimeter Array. The prototype has been constructed based on our science requirements and science cases. Since we consider that pipeline data reduction and calibration from raw data will be crucial in future VOs, pipeline reduction for existing radio and near-infrared telescopes is implemented into the prototype. Interactive data reduction is also available.

1. Introduction

The Japanese side of Atacama Large Millimeter/submillimeter Array (ALMA-Japan) is going to develop a Virtual Observatory, ALMAJ-VO. It will be one of the key functionalities of the ALMA Regional Center in Japan, and ALMA data will be quite important even for other VOs. Though present astronomical databases generally store well-reduced and calibrated 2-dimensional images, the data produced with ALMA will be delivered to the observer after a pipeline process by default because of their huge amount. Moreover, pipeline reduction will be claimed even for non-ALMA data in order to satisfy a variety of research interests and serendipity. Accordingly we consider that pipeline reduction from *raw* data will become crucial to future VOs. Treating 3-dimensional data is also an essential feature in ALMAJ-VO since radio line observations produce fully 3-dimensional data.

We are performing prototype development for ALMAJ-VO. We have summarized our science requirements and science cases. Data processing pipelines for existing radio and near-infrared (NIR) telescopes are being developed in order to evaluate system design and interfaces. We have also included an interac-

[1]Graduate University for Advanced Studies (Sokendai), Mitaka, Tokyo 181-8588, Japan

tive operation in our prototype, because we regard it as an essential feature for astronomical data reduction. We are also interested in data abstraction (i.e., making catalogues) of 3-dimensional data, because it will be a key technology for efficient VO astronomy with ALMA, which produces radio data at a high rate. Here we report the present status of our activities. The program is being carried out in collaboration with Astronomical Data Analysis Center, National Astronomical Observatory of Japan (NAOJ).

2. Science Requirements

Science requirements have been listed with priorities, such as:

- **Well-defined coordinates, velocity etc.** Axes of 3-dimensional data, such as $(d\alpha, d\delta)$ in global sinusoidal projection, WCS, and radio definition of velocity, should be strictly defined and treated.
- **Data quality.** Data quality, such as map noise on primary-beam corrected interferometric data, should be well defined.
- **Pipeline process.** As mentioned in Section 1, automated (or interactive, if necessary) pipeline process of reduction and calibration from raw data is crucial. Though interactive reduction is essential even for VO, the problem is how to implement it.
- **Versatile data display.** Various styles of display of data up to 3-dimension are needed: such as moment maps, velocity channel maps, profile maps, and overplotting.
- **Analysis tool.** Efficient data analysis tools (e.g., cross-correlation between maps, complicated region selection) should be available.
- **Catalog tool.**

3. Science Cases

We have studied science cases on the basis of our expertise in radio astronomy at Nobeyama Radio Observatory (NRO). In total, sixteen cases have been listed, and we have selected the following three cases as prototyping references:

- **Galaxies.** Relations between characteristics of galaxies (e.g., Hubble type, bar) and molecular gas in the inner kiloparsecs using Nobeyama Millimeter Array CO data together with optical, infrared, radio continuum, and HI data.
- **Molecular clouds in the inner Galaxy.** Relations between the Galactic structure (spiral arms) and cloud properties using NRO 45-m CO survey and other data.
- **Globules.** Depletion, cloud structure, and star formation using radio (NRO 45-m) and NIR (Nagoya–South Africa Infrared Survey Facility [IRSF] 1.4-m) data.

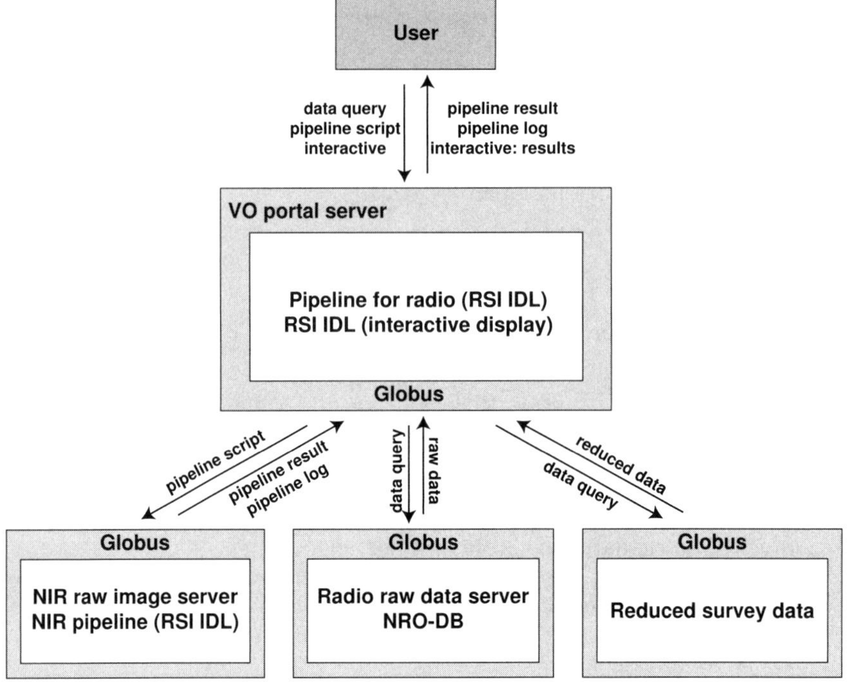

Figure 1. A schematic illustration of the prototype configuration.

4. An Overview of the Prototype

4.1. Query and User Interaction

Configuration of the prototype is illustrated in Figure 1. The user accesses the VO portal server via a Web browser and initiates a request. VNC is used when interactive operation is needed. The portal server and data servers (radio-raw[2], NIR-raw[3], and reduced) are connected with Globus.

Pipelines for radio and NIR data are being developed. The radio pipeline deals with data taken with 25-beam receiver array BEARS (Sunada et al. 2000, Yamaguchi et al. 2000) on NRO 45-m telescope, while the NIR pipeline processes data with JHK_s camera SIRIUS (Nagashima et al. 1999, Nagayama et al. 2003) on IRSF 1.4-m telescope. When the user requests reduced data, the data are transferred from the reduced data server to the portal server. On the other hand, if raw data are requested, the data are processed through a "default" pipeline, then the results and a log are returned to the user. The log, which describes the pipeline parameters, is reusable for the next request: if the user is not satisfied

[2]`http://nrodb.nro.nao.ac.jp/`

[3]IRSF/SIRIUS data are not open to the public. Our own data and server are used for the prototype development.

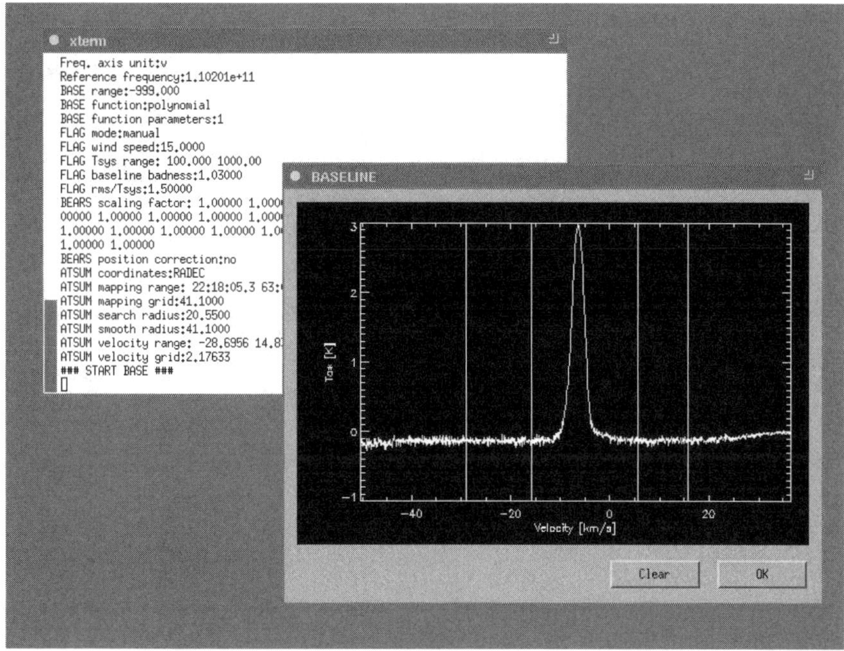

Figure 2. A screenshot of the radio pipeline. Interactive baseline subtraction is running.

with the results, they may edit the reusable log and re-submit a request. When necessary, interactive operations (e.g., data flagging, selection of data area) can be done.

4.2. Radio Pipeline

Radio pipeline consists of the following procedures: (1) baseline subtraction, (2) flagging bad data, (3) coadding, and (4) making data cube. In the default process, parameters for these procedures are automatically determined: e.g., baselines are subtracted using preset channel ranges of spectrometers, and bad data are flagged out using some header parameters which relate to the data quality. By editing the reusable log, the user can change these parameters or request interactive operation (Figure 2). Three-dimensional FITS cube, JPEG images, and reusable log are output.

4.3. NIR Pipeline

NIR pipeline consists of the following procedures: (1) dark frame subtraction, (2) flat frame calibration, (3) dithering & combine, (4) sky frame subtraction, (5) stellar detection, and (6) photometry. Calibrated J, H, K_s FITS images, JPEG images, list of magnitudes, colors, coordinates, and reusable log are output.

Examples of pipeline-processed radio and NIR images are shown in Figure 3.

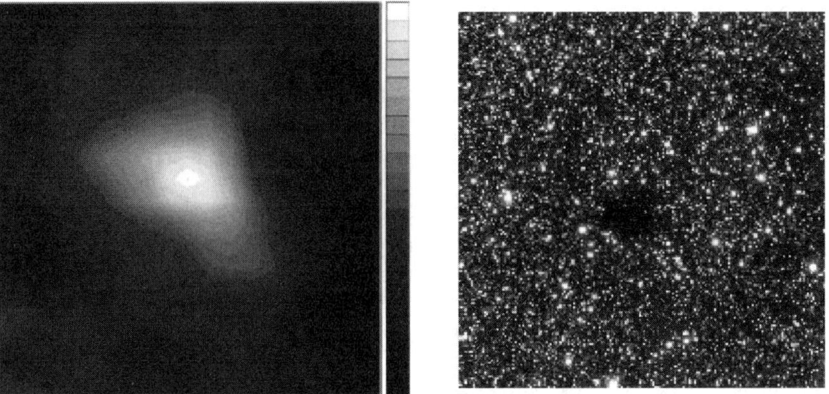

Figure 3. Examples of pipeline-processed images. (Left) radio integrated intensity map of a globule; (right) JHK_s composite image.

5. Summary

We are developing prototypical pipelines for some Japanese original contents to prepare for VO in ALMA Regional Center in Japan. Radio and NIR pipelines are being developed. User-specified process and interactive operation have been implemented. We are collaborating with JVO group in NAOJ in part.

Acknowledgments. This work was supported by the JSPS Core-to-Core Program. The IRSF/SIRIUS project was initiated and supported by Nagoya University, National Astronomical Observatory of Japan, and University of Tokyo in collaboration with South African Astronomical Observatory under a financial support of Grant-in-Aid for Scientific Research on Priority Area (A) No. 10147207 and No. 10147214, and Grant-in-Aid No. 13573001 of the Ministry of Education, Culture, Sports, Science, and Technology of Japan.

References

Nagashima C. et al. 1999, in Star Formation 1999, ed. T. Nakamoto (Nobeyama Radio Observatory), 397

Nagayama, T., et al. 2003, in Proc. SPIE, 4841, 459

Sunada, K., et al. 2000, in Proc. SPIE, 4015, 237

Yamaguchi, C., et al. 2000, in Proc. SPIE, 4015, 614

ASTRONOMICAL DATA ANALYSIS SOFTWARE AND SYSTEMS XIV
ASP Conference Series, Vol. 347, 2005
P. L. Shopbell, M. C. Britton, and R. Ebert, eds.

A Metadata Layer to Enable VO-compliant Access to the ESO/ST-ECF Archive

D. Sforna[1] and A. Micol[1]

ESA/ESO Space Telescope European Coordinating Facility, Karl-Schwarzschild-Str. 2, D-85748 Garching, Germany

Abstract. The Metadata Layer is built around a centralized repository from where metadata can be retrieved. The existence of this repository is a key point for offering archive services in the VO frame, allowing us to provide a better data characterization in a comprehensive and homogeneous way. At the same time we are also building for our services a web service interface, together with the traditional HTML pages designed for human consumption. The ST-ECF Instrumental Characteristics Service (Micol 2005) is one example where both the metadata layer and the new VO interface are already in place. The service provides a match between characteristics of standard bandpasses and the ones of real filters mounted on ESO/HST instruments. The output for matching filters is offered both via a web service and in an HTML page. The web service is used to retrieve at once in VOTable format all the data for the matching filters together with a set of standardized metadata. We will show usage examples for the Instrumental Characteristics and other ST-ECF services, focusing on the new access possibilities and on the advantages of a hybrid interface.

1. The Metadata Layer

The Metadata Layer provides the ability to store standardized metadata (ucd [Ortiz et al. 1999], utype, unit, relevant links, textual description) associated with data, services, database objects, and so on. Once the metadata are stored in our database, services can make use of such a layer for a number of purposes, the main one being interoperability. As an example, the Instrumental Characteristics Service (Micol 2005) makes use of the Metadata Layer to return results in VOTable format. The Metadata Layer we developed is made up of two main components:

- A relational database (SYBASE).
- An in-house developed API implementing the "CRUD"[2] logic on the storage.

Our main intent is to associate to each of the existing HST and ESO database tables (for example the observing logs of the telescopes) all those pieces of meta-information for which SYBASE is not providing direct support. Since in a VOTable each piece of information (the VOTable FIELDs, PARAMs, and

[1]Affiliated with the Space Telescope Division of the European Space Agency, ESTEC, Noordwijk, Netherlands.

[2]Create, Read, Update, Delete.

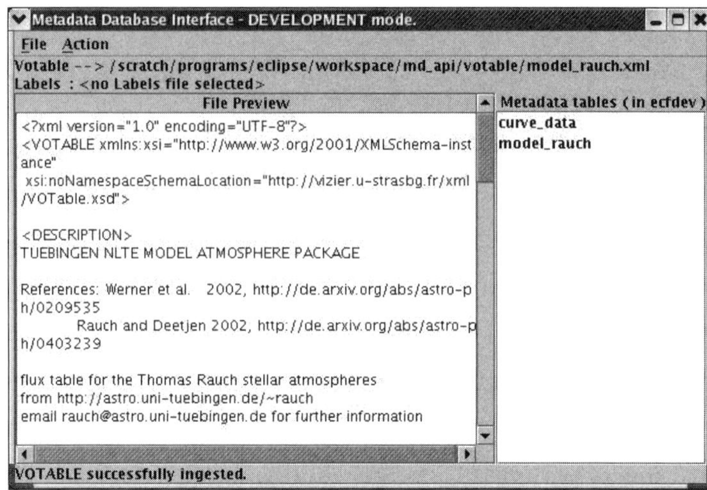

Figure 1. A new set of metadata is successfully stored using the software GUI.

GROUPs) comes with its own metadata (ucd, units, utypes, datatypes, description, links, etc), a VOTable is a natural way to associate to each of the columns in our existing HST and ESO databases the necessary metadata. Our Metadata Layer API is hence based on the VOTable paradigm.

The API is equipped with a basic GUI interface, shown in Figure 1, which accepts as input a VOTable[3] in order to populate the metadata layer and to store the association between such metadata and the corresponding piece of data or service. To offer metadata to the external user, we have developed applications that are able to provide the output in VOTable format. With the same format for input and output, it is straightforward to extract metadata, correct or enhance them, and store back the modified version in the database.

Having a centralized system is a key point to avoid duplication of efforts and a maintenance nightmare, but it is not mandatory to manage it through dedicated software. On the other hand it is easier to ensure the data integrity using a software interface. This is particularly important in the current, still experimental, situation: being able to reconfigure the metadata of a data table with just a click of the mouse is worth the effort allocated for software development.

2. Storing Metadata with the Metadata API

Associating data with metadata is conceptually easy[4], but it is an error-prone activity. Using a software interface eases the problem, filtering out most of the procedural errors that can arise. The required steps to store a new set of metadata with the software interface are:

---

[3]Using a standard format in input helps also in easily retrieving metadata from external sources.

[4]The difficult issue here is coming up with the correct set of metadata.

1. The operator starts the interface and loads the metadata to ingest (in VOTable format).
2. The VOTable is shown in the GUI before the storing operation, so the operator can check if it is the correct one.
3. The Software API ingests the metadata in the selected Database.

During this process, a number of checks are enforced by the software to help the operator and to avoid storing erroneous metadata. Any inconsistency stops the process before the changes to the database are committed.

3. Retrieving Metadata with the Metadata API

The Metadata Database needs to be accessible to the archive services that want to present the users with enhanced products. For this purpose we integrated into the Metadata Layer "helper" applications: each helper on one side shares a common interface with the Metadata Database, and on the other side is tailored to handle the specific input format of a given archive service.

Note that the input coming from the archive service is usually closely related to a data request and needs not to be extended in order to address the Metadata Database: the answer to the question "which metadata are relevant for the required data" is transparently provided by the helper application.

The helpers' workflow can be summarized as follow:

1. Figure out which metadata are useful for the required data or service.
2. Retrieve the data.
3. Retrieve the metadata.
4. Combine data and metadata in a VOTable and return it as output.

4. Web Services for the Metadata Database

The above mentioned helpers are coded in Java, therefore it is straightforward to expose them as Web Services using the Axis SOAP Engine[5]. For Java packages, Axis provides facilities to generate automatically the SOAP handler code (for both the server and the client) and the Web Service WSDL[6].

Web services have many advantages that make them ideal for being used in the VO frame:

- They allow building a bigger application as a federation of smaller ones.
- They allow interoperability across different software programming languages, operating systems, and so on.
- They allow data to be processed where they reside.
- They can be called programmatically.

On the other hand, the amount of technologies and standards that are needed for using web services cannot be ignored: it is a burden that the users might not be willing to take, unless they have good reasons to do so. Therefore we decided to start offering web service interfaces to our archive services together with the

---

[5] `http://ws.apache.org/axis/`

[6] An introduction to web services and related technologies can be found at `http://ws.apache.org/axis/java/reading.html`

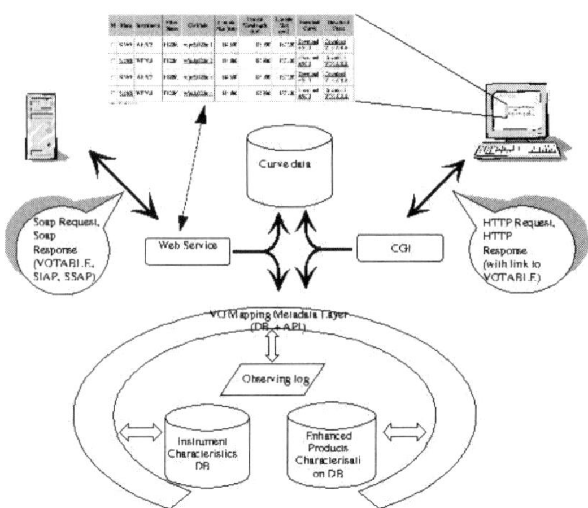

Figure 2. Both CGI and Web Service interfaces of the Instrumental Characteristics Service take advantage of the Metadata Database to offer enhanced products. The observation modes in the HTML table are the same used as input to the Web Service, as highlighted by the dotted arrow.

usual human-oriented HTML interface. In this way the users can familiarize themselves with the archive service, and turn to the web service interface at a later stage, after evaluating if it is worth doing so for their needs.

In the following, we present an example that shows how we implemented this concept.

5. A Practical Example

The ST-ECF Instrumental Characteristics Service has a HTML web page interface and a web service interface (Figure 2).

Both interfaces offer the possibility to get an output VOTable: from the HTML result page there is a link to download VOTables as XML files, while the web service returns the VOTable as a string. In the final release the interface will also offer directly the VOTable[7].

We chose to expose as a web service the VOTable writer that uses as input the "observation mode" to select the required transmission curve. The observation mode is basically a string used to uniquely identify an instrument configuration that has a one-to-one relationship with the transmission curve and is therefore a straightforward input parameter for a service offering access to transmission curves.

---

[7] An implication of using web service technology is that the VOTable encoding depends upon the client used: for instance, in the case of a Java client the VOTable will be a javabean object, while in the case of a Perl client it will be a hash table.

The "human-oriented" interface starts a step above, giving the user the opportunity to define query conditions in order to reach the point where one or more observation modes are selected. The same query can be given as input to a web service, and we actually have in the Metadata Layer an SQL handler as well. The plan is to extend it and offer an ADQL-based web service to interrogate HST and ESO Observation Logs. The Web Service is currently running behind the ESO firewall, and it will soon move to an operational server.

6. Conclusions

During the last months we put in operation a Metadata Layer, backed by a software API. The Metadata Layer allows us to build interoperable services and offer enhanced products to the community, keeping at the same time centralized the development and maintenance effort.

An example of its use is in our Instrumental Characteristics Service, where we also decided to offer two parallel interfaces: a web service and a standard HTML web page. In this way the users have always available the "safe haven" of a known interface, but should it be to their advantage, they can also call programmatically the web service.

The next steps in our roadmap are extending the metadata coverage on our data, increasing the number of services that make use of these metadata and are compliant with protocols like SIAP/SSAP[8], and making our services available through the VO Registry.

References

Micol, A. 2005, in ASP Conf. Ser., Vol. 347, ADASS XIV, ed. P. L. Shopbell, M. C. Britton, & R. Ebert (San Francisco: ASP), 543

Ortiz, P. F., Ochsenbein, F., Wicenec, A., & Albrecht, M. 1999 in ASP Conf. Ser., Vol. 172, ADASS VIII, ed. D. M. Mehringer, R. L. Plante, & D. A. Roberts (San Francisco: ASP), 379

[8]See the International Virtual Observatory Alliance (IVOA) technical documents: `http://www.ivoa.net/Documents/latest/`

ASTRONOMICAL DATA ANALYSIS SOFTWARE AND SYSTEMS XIV
ASP Conference Series, Vol. 347, 2005
P. L. Shopbell, M. C. Britton, and R. Ebert, eds.

Structured Query Language for Virtual Observatory

Yuji Shirasaki, Masatoshi Ohishi, Yoshihiko Mizumoto, Masahiro Tanaka, Satoshi Honda, and Masafumi Oe

National Astronomical Observatory of Japan, 2-21-1 Osawa, Mitaka City, Tokyo 181-8588, Japan

Naoki Yasuda

Institute for Cosmic Ray Research, University of Tokyo, 5-1-5 Kashiwa-no-Ha, Kashiwa City, Chiba 277-8582, Japan

Yoshifumi Masunaga

Ochanomizu University, 2-1-1 0tsuka,Bunkyo-Ku, Tokyo 112-8610, Japan

Abstract. Currently two query languages are defined as standards for the Virtual Observatory (VO). Astronomical Data Query Language (ADQL) is used for catalog data query and Simple Image Access Protocol (SIAP) is for image data query. As a result, when we query each data service, we need to know in advance which language is supported and then construct a query language accordingly. The construction of SIAP is simple, but it has a limited capability. For example, there is no way to specify multiple regions in one query, and it is difficult to specify complex query conditions. In this paper, we propose a unified query language for any kind of astronomical database on the basis of SQL99. SQL is a query language optimized for table data, so to apply SQL to image and spectral data, the data structures need to be mapped to table-like structures. We present a specification of this query language and an example of the architecture for the database system.

1. Introduction

At present, two kinds of data query language are defined as standards in the Virtual Observatory. One is the parameter query, which is used for image data search and is called "Simple Image Access Protocol" or SIAP. Search criteria are specified by a set of "key" and "value" pairs. Another one is a structured query language, which is used for catalog data query and called "Astronomical Data Query Language" or ADQL (Yasuda et al. 2004). ADQL can specify more complex search criteria than SIAP does. It also has the ability to join multiple tables, and it can select values derived from the DB columns. So it is worthwhile to adapt the ADQL to perform an image query. The high flexibility of the ADQL syntax, however, raises difficulties to develop an ADQL-compliant data service. It is known that incompatibility among various DBMS products is due to the complexity of the SQL specification. The success of the Virtual Observatory project depends on uniformity of interfaces of all the astronomical data services, so we need to avoid the complexity and make it simple and easy to implement.

2. Syntax Specification

In order to realize interoperability among the distributed data archives, we need to define a standard query language. As the standard query language must be supported by all of the data services, its specification should be simple and clear for easy implementation. On the other hand, some data services require more sophisticated query syntax to allow users to specify more efficient search criteria. Thus we define a basic syntax, which must be supported by all the data services, and allow several enhancements on the basic syntax.

2.1. Basic syntax

The following restrictions are applied to the "select" statement of the SQL99 specification.

1. Only a column name or "*" can be specified in the selection list, so algebraic expressions and functions are not allowed. So the role of the "select" clause is just to specify the desired columns. The calculation of the derivative from those columns should be done on a portal server or by users themselves.
2. Only one table is specified in the "from" clause, so any type of table join syntax is not allowed. The join of the tables should be done on a portal server or by users themselves.
3. Allowed comparison operators and predicates are `=`, `<`, `>`, `>=`, `<=`, `<>`, `within`, `contains`, `overlaps`, `LIKE`, and `BETWEEN`. Other operators are not allowed. The operators `within`, `contains`, and `overlaps` are not specified in SQL99, but are introduced to specify a region on the sky, since such search criteria are fundamental for an astronomical data service. These three operators compare two values of geometry data type, which is described in the next subsection.
4. Functions `point()`, `circle()`, and `box()` are introduced to express a region on the sky, and `distance()` is also introduced to describe the proximity of two coordinates.
5. Allowed logical operators are `AND` and `NOT`. `OR` is not allowed as the mixture of `AND` and `OR` makes the search condition too complex.
6. Tables and columns may have alias names.

2.2. Geometry Data Type

The geometry data type is introduced to represent a point or a region in the sky. A point is always expressed by two coordinate values. In most of the cases a single coordinate value is meaningless and only a pair have a physical meaning. So it is natural to have a column which has a pair of values to express the coordinate of celestial object, then it becomes simple to describe a search criterion. For example, if the coordinate values are prepared in two separate columns, the search condition can be expressed as follows: `point(t1.ra, t1.dec) within box((t2.ra, t2.dec), 1.0, 1.0)`. On the other hand, if the coordinate values are prepared in one single column, the same statement can be expressed more compactly as: `t1.point within t2.region`. We define "Point", "Circle" and "Box" data types. Expressions of these data types in SQL are summarized in Table 1.

Table 1. Expressions of the geometry data type value

Data Type	Examples of Expression
Point	`Point(23., +10., 'FK5')`, `Point(23., +10.)`, `(23., +10.)`
Box	`Box((23.0, +10.0), 1, 1)`, `((23.0, +10.0), 1, 1)`
Circle	`Circle((23.0, +10.0), 1)`, `((23.0, +10.0), 1)`

2.3. Example SQLs for Basic Syntax

The following SQL shows an example of a catalog data query:

```
Select g.ra, g.dec, g.mag_r
From   galaxy as g
Where  Point(g.ra, g.dec) within Circle((24.3, +5.0), 1.0)
       and g.mag_r < 24
```

This query describes "from a table named galaxy select right ascension, declination, and R-band magnitude of celestial objects located in the circle whose center coordinate is (ra, dec) = (24.3 deg, +5.0 deg) and radius is 1.0 deg." In the "galaxy" table, coordinates of the objects are prepared in two separate columns, "ra" and "dec", so the Point() function is required to describe the search region. In the "galaxy2" table of the next SQL example, the coordinates are prepared in a single column, "point".

```
Select g.point, g.mag_r
From   galaxy2 as g
Where  g.point within Circle((24.3, +5.0), 1.0) and g.mag_r < 24
```

The following SQL shows an example of an image data query:

```
Select img.filter, img.imageURL
From   imageData as img
Where  img.region = BOX((24.3, +5.0), 0.1, 0.1)
```

This is equivalent to the following URL-based SIAP:

```
http://jvo.nao.ac.jp/image?POS=24.3,5.0&SIZE=0.1&FORMAT=VOTABLE
```

In the above SQL, the column `region` is used as a parameter to specify the region of interest, and the access URL of the cut-out image is returned as data of `imageURL` column. Those columns are actually not present in the relational table on the DBMS, but the user can specify it, as if it virtually exists. So these columns are called "virtual columns", and its table is called a "virtual table", or a *view* in RDB terminology.

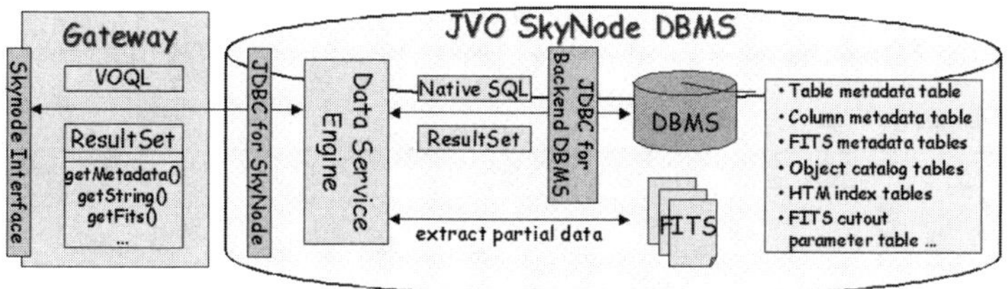

Figure 1. Architecture of JVO Skynode designed to be compatible with the proposed VO Query Language (VOQL).

2.4. Syntax Enhancement

The following enhancements may be applied as optional features.

1. Multiple tables enhancement: Multiple tables can be queried with a single SQL. External tables provided in a form of VOTable must be treated in the same manner as the original tables of the data service. The external table is specified as EXT::<fileNumber>.<resourceName>.<tableName>. A join of the predicate and sub-query are not mandatory.
2. Unit support enhancement: Any numeric value may be followed by a unit. Unit conversion must be carried out by the data service.
3. Algebraic expression enhancement: An algebraic expression can be specified in the selection list and search criteria in the "where" clause.
4. Logical operator `OR` enhancement: Support for a mixture of `AND`, `NOT`, and `OR` in the "where" clause.
5. Object data type enhancement: A column may have structured data and access methods.
6. Use of an identifier for specifying a table name (portal): A table name is expressed by an identifier of dot notation to specify the table uniquely in the VO. A dot character in the identifier must be escaped by a backslash. For example, `naoj:sxds.v0\.1.galaxy` where `naoj` is an authority name, `/sxds/v0.1` is a catalog resource path name and `galaxy` is a table resource name. This is a feature dedicated to the portal data service.
7. UCD (portal): A UCD (Derriere et al. 2004) can be used in place of a column name. This is a feature dedicated to the portal data service. The portal service searches data resources which have a UCD specified in the SQL and translate to the column name, using the column metadata collected from the data services.
8. Omission of "From" part (portal): This is a feature dedicated to the portal data service. The portal searches the data resources from the registry according to the query condition. In this case, a UCD must be used for describing columns and a query condition.

3. JVO Skynode Architecture

We have designed an architecture to implement a data service compatible with the proposed query language (Figure 1). The "gateway" in the figure provides interfaces to access the skynode (Yasuda et al. 2004) services. The received

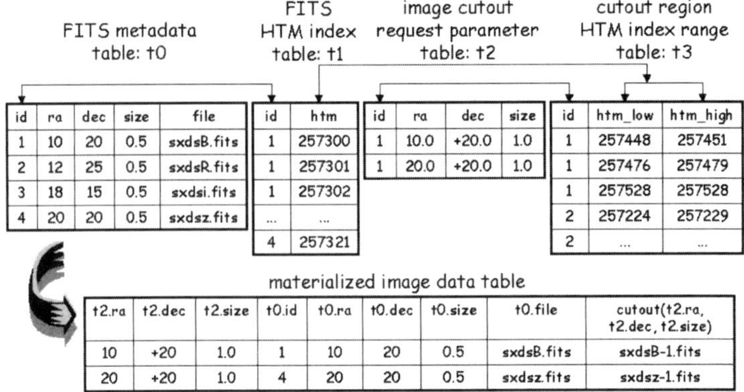

Figure 2. An example of creating a virtual image table.

query is transfered to the JVO SkyNode DBMS, and the result is returned in the form of a `ResultSet` Java object. The access interface is provided by JDBC, which is a standard interface to access a DBMS in the Java environment. The celestial object catalog is stored in a relational data base (RDB) system, such as PostgreSQL, MySQL, Oracle, etc., and FITS files are managed in the unix file system, with their metadata stored in the RDB. HTM indexing (Kunszt et al. 2000) is used to perform a fast data search. The query condition related to the search region on the sky is converted to a condition on the HTM index ranges at the "Data Service Engine", and then the modified SQL is submitted to the DBMS. In the case of image data query, it is necessary to create the virtual image table as follows: First, parameters specifying the image cut-out regions are stored in a table t2 (Figure 2), and the corresponding HTM index range table t3 is also created. Then the FITS metadata table t0 is joined with table t2 with the intervention of the HTM index tables t1 and t3, creating the virtual image table shown at the bottom of Figure 2.

Acknowledgments. This work was supported by the JSPS Core-to-Core Program and Grant-in-aid Information Science (15017289 and 16016292) carried out by the Ministry of Education, Culture, Sports, Science, and Technology of Japan.

References

Derriere, S. et al. 2004, in ASP Conf. Ser., Vol. 314, ADASS XIII, ed. F. Ochsenbein, M. Allen, & D. Egret (San Francisco: ASP), 315

Kunszt, P.Z., Szalay, A.S., Csabai, I., & Thakar, A. R. 2000, in ASP Conf. Ser., Vol. 216, ADASS IX, ed. N. Manset, C. Veillet, & D. Crabtree (San Francisco: ASP), 141

Yasuda, N. et al. 2004, in ASP Conf. Ser., Vol. 314, ADASS XIII, ed. F. Ochsenbein, M. Allen, & D. Egret (San Francisco: ASP), 293

ASTRONOMICAL DATA ANALYSIS SOFTWARE AND SYSTEMS XIV
ASP Conference Series, Vol. 347, 2005
P. L. Shopbell, M. C. Britton, and R. Ebert, eds.

Japanese Virtual Observatory (JVO) Prototype 2

Masahiro Tanaka, Yuji Shirasaki, Satoshi Honda, Yoshihiko Mizumoto, and Masatoshi Ohishi

National Astronomical Observatory of Japan, 2-21-1 Osawa, Mitaka, Tokyo 181-8588, Japan

Naoki Yasuda

Institute for Cosmic Ray Research, University of Tokyo, 5-1-5 Kashiwanoha, Kashiwa, Chiba, 277-8582, Japan

Yoshifumi Masunaga

Ochanomizu University, 2-1-1 Otsuka, Bunkyo, Tokyo, 112-8610, Japan

Yasuhide Ishihara, Katsumi Abe, and Jumpei Tsutsumi

Fujitsu Ltd., 1-9-3 Nakase, Mihama, Chiba, 261-8588, Japan

Hiroyuki Nakamoto, Yuusuke Kobayashi, Tokuo Yoshida, and Yasuhiro Morita

Systems Engineering Consultants Co. Ltd., 9-8 Sakuragaoka, Shibuya, Tokyo, 150-0031, Japan

Abstract. We describe the architecture of the Japanese Virtual Observatory (JVO) prototype system version 2. JVO aims at seamless access to astronomical data archives stored in distributed data servers as well as data analysis environment. For this purpose, it is important to establish a framework for access to remote servers, including remote procedure calls (RPCs) and data transfer. A data request for distributed database is described in the JVO Query Language. The JVO system parses the query language and decomposes it into individual remote procedures, such as the retrieval of catalog images and spectra, cross matching, and workflow generation. Based on this workflow, remote procedures are called. For RPCs of JVO prototype system 1, we employed Globus toolkit 2 (GT2). However, latency time of GT2 RPCs was too long for successive short-time jobs. Therefore, we employed Globus toolkit 3 (GT3) for JVO prototype system 2. As a result, we find that Grid Service in GT3 improves performance of RPC. In addition to Grid Service, Reliable File Transfer (RFT) is used for efficient data transfer. Astronomical data stored in distributed servers are discovered through a registry server which provides metadata discussed in the IVOA registry working group and is built using a XML database.

1. Introduction

JVO[1] is a project of National Astronomical Observatory of Japan. The objective of JVO is to provide astronomers with seamless access to huge data archives produced with astronomical telescopes of Japan such as Subaru telescope. JVO is also participating in IVOA[2] (International Virtual Observatory Alliance) to discuss standard protocols for the purpose of accessibility to astronomical data archives from/to the world.

We have developed JVO prototypes iteratively; the JVO prototype version 1 (Proto 1) was developed in 2002, the prototype version 2 (Proto 2) in 2003, and the prototype 3 is under development. In the course of Proto 1 development, we defined a JVO Query Language (JVOQL; Mizumoto et al. 2003) based on SQL by extending cross-matching and image retrieval functionalities. And we constructed the Proto 1 system, which consists of distributed database and analysis servers federated through the Grid (Ohishi et al. 2004, Shirasaki et al. 2004). As a result, we confirmed that Proto 1 accepts properly query commands described in JVOQL and actually operates as a distributed database. However, we found several problems, including the performance of remote execution and data discovery. These issues are improved in JVO Proto 2 as described in this paper.

2. System Design

The system configuration of JVO Proto 2 is shown in Figure 1, which is basically same as Proto 1. Technical components adopted for JVO prototypes are compared in Table 1, and described in the following sections.

Table 1. Technical components adopted for JVO prototypes

	Registry	Grid middleware	remote execution	data transfer
Proto 1	UDDI	Globus v.2	globus-job-run	GridFTP
Proto 2	XML DB	Globus v.3	Grid Service	RFT & SFS

3. Execution Controller

User's query commands described by JVOQL are processed by the **Controller**. The Controller cooperates with the following subcomponentsL **JVOQL Parser**, **Scheduler**, and **Executer**. These components are written in Java language and installed in the portal server. The JVOQL parser analyzes the syntax of the JVOQL command and decomposes it into query elements. The Java code to parse the syntax of the JVOQL is generated with JavaCC (Java Compiler Compiler). The Scheduler receives an output of the parser, and generates a workflow which consists of job elements for individual servers. Based on this

[1] `http://jvo.nao.ac.jp/`

[2] `http://www.ivoa.net/`

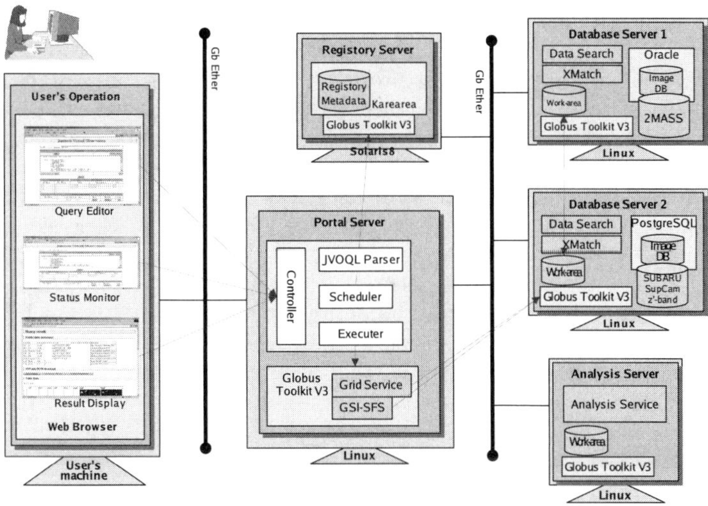

Figure 1. JVO system configuration

workflow, the Controller calls the Executer which calls procedures on remote machines.

4. Data Discovery

When the Controller performs a query for distributed data servers, it is necessary to find the location of the requested data. Such information is retrieved from a Registry server. For Proto 1, we constructed a registry server using UDDI (Universal Description, Discovery, and Integration). UDDI is designed to find services in the Web Services framework, however, UDDI is not necessarily suitable for search services of complicated astronomical metadata. Therefore, we redesigned the registry system for Proto 2. The contents of metadata are defined so as to keep compatibility with the metadata standard proposed in IVOA. We stored these metadata into an XML database product KAREAREA, and enabled metadata search by XPath. We plan to use the OAI-PMH (Open Archives Initiative Protocol for Metadata Harvesting) for exchanging metadata between data servers and a registry server.

5. Remote Execution

Standard protocols for remote procedure calls (RPCs) are necessary for federation of astronomical databases distributed all over the world. For this purpose, we considered to apply standard Grid middleware, **Globus Toolkit**, developed by Globus Alliance[3], and investigated its applicability to JVO. We constructed

[3]`http://www.globus.org/`

Proto 1 using Globus Toolkit version 2 (GT2). However, the result showed that elapsed time to execute a basic JVOQL example was more than 10 minutes (Ohishi et al. 2004). This is probably because the `globus-job-run` command used for Proto 1 is not designed for a series of light-weight, i.e., pseudo-realtime, procedure calls. To improve performance, we reimplemented the RPCs of JVO Proto 2 using a **Grid Service** introduced in the Globus Toolkit 3 (GT3). The Grid Service is based on the Web Services. After the implementation into Proto 2, we examined two queries which take 2.3 and 13 seconds as actual processing times in remote servers, respectively. The result is that elapsed times including RPCs are 2.8 and 16 seconds, respectively. The measured overhead time is only around 30 ms. Thus the performance of Proto 2 is much improved compared with Proto 1. This result shows that Proto 2 system using Grid Service can be a basis of practical virtual observatory systems.

6. Data Transfer

We employed two data transfer protocols for JVO Proto 2; one is the Reliable File Transfer (RFT), and the other is the Self-certifying File System (SFS)[4]. The RFT is one of the GT3 services, and it provides interfaces for recoverable file transfer using the GridFTP protocol. The RFT enables a portal server to issue file copy commands between two remote hosts. This function is useful for multi-server operation like cross-matching (XMatch). The SFS is a secure network file system over the Internet. We find the SFS does not provide data transfer service between remote hosts, which is a benefit of the RFT, since SFS server and client processes cannot coexist on a single machine.

As mentioned above, we adopted Grid Service for remote execution, and RFT and SFS for data transfer. However, the use of different protocols brings complicated implementation. Furthermore, IVOA standard protocols like SIAP (Simple Image Access Protocol) are implemented with single protocols like HTTP or Web Services. Therefore, we are considering using HTTP or Web Services for RPCs and data transfer in the next prototype and operational JVO systems.

Acknowledgments. This work was supported by the JSPS Core-to-Core Program and Grant-in-aid "Information Science" (15017289 and 16016292) carried out by the Ministry of Education, Culture, Sports, Science and Technology of Japan.

References

Mizumoto, Y., et al. 2003, in ASP Conf. Ser., Vol. 295, ADASS XII, ed. H. E. Payne, R. I. Jedrzejewski, & R. N. Hook (San Francisco: ASP), 96

Ohishi, M., et al. 2004, in ASP Conf. Ser., Vol. 314, ADASS XIII, ed. F. Ochsenbein, M. Allen, & D. Egret (San Francisco: ASP), 296

Shirasaki, Y., et al. 2004, in ASP Conf. Ser., Vol. 314, ADASS XIII, ed. F. Ochsenbein, M. Allen, & D. Egret (San Francisco: ASP), 46

[4]`http://www.fs.net/`

Part VI
The Virtual Observatory: Grid

ASTRONOMICAL DATA ANALYSIS SOFTWARE AND SYSTEMS XIV
ASP Conference Series, Vol. 347, 2005
P. L. Shopbell, M. C. Britton, and R. Ebert, eds.

Deploying the AstroGrid: Science Use Ready

Nicholas A. Walton
Institute of Astronomy, University of Cambridge, Madingley Road, Cambridge, CB3 0HA, UK

Paul A. Harrison and Anita M. S. Richards
Jodrell Bank Observatory, University of Manchester, Macclesfield, Cheshire SK11 9DL, UK

Martin C. Hill
Institute for Astronomy, University of Edinburgh, Royal Observatory, Blackford Hill, Edinburgh, EH9 3HJ, UK

Abstract.
AstroGrid (see `http://www.astrogrid.org`), a UK eScience project with collaborating groups drawn from the major UK data archive centres, is creating the UK's Virtual Observatory.

In order to ensure rapid user feedback AstroGrid is fully deploying each iteration release, connected to relevant data and application products, in such a fashion as to allow scientific use of that release. The early users are primarily the science advisory group and now the beta tester communities of the project.

The scientific functionality of the current (Sep. 2004) deployment is highlighted. This includes access to a sophisticated workflow capability. Its use in allowing image extraction from multiple image datasets, input photometry file creation, redshift determination, and visualisation for the outputs to allow discovery of high redshift objects is described. Further examples show the use of the latest Astrophysical Data Query language standard and how it is being used to enable large data queries of remote databases in searching for low mass objects in the Pleiades with results returned to a virtual "MySpace" user storage area, where further visualisation and processing can be performed.

We note how the deployed system is being tested by the science community, and how that comment and feedback is vital in informing the project as to future releases including the January 2005 "AstroGrid Release 1".

1. Introduction

AstroGrid[1], a UK eScience project with collaborating groups drawn from the major UK data archive centres, is creating the UK's first virtual observatory (VO). Together with the other major world-wide virtual observatory projects (see the International Virtual Observatory Alliance[2]), AstroGrid is creating a

[1] `http://www.astrogrid.org/`

[2] `http://www.ivoa.net`

set of co-operating and interoperable software systems that will: allow users to interrogate multiple data centres in a seamless and transparent way; provide powerful new analysis and visualisation tools; and give data centres and providers a standard framework for publishing and delivering services using their data. This paper describes how the project is currently deploying its intermediate product releases, in advance of its "Release 1" in January 2005.

2. Science Scope and System Design for AstroGrid

A key goal for AstroGrid and indeed all the VOs, is enabling science: both new science which would have been difficult or impractical to undertake without VO systems, or faster and more effective science.

In designing the system, science drivers have been important in setting priorities for development of the various required capabilities. In the case of AstroGrid, it produced a detailed Phase-A report[3], setting out the science requirements for the project.

AstroGrid has been constructed in a modular fashion, allowing for the creation of a system to meet the challenges imposed by the science drivers. The main elements of the system are described elsewhere in these proceedings: Portal and Workflow, Registry, Dataset Access, Common Execution Architecture (Harrison et al, 2004), and MySpace. An overview of the project's architecture can be found on the AstroGrid Wiki site.

The **AstroGrid Science Advisory Group**[4] ensures a continuing conduit of information to and from the scientific user community. They have advised on alterations to the development priorities of the system through the project lifetime, as new opportunities or issues have arisen. They have been able to comment on and provide feedback as to the functionality delivered by the iterative releases of the AstroGrid product.

As AstroGrid enters it second three year project phase, where operationally deploying the AstroGrid VO framework becomes the main focus of activity, the Science Advisory Group will play an increasingly important role in highlighting, through example usage, the value of the infrastructure for research.

The **beta test programme** attempts to ensure that early usage feedback is received from end users. Currently the beta test community is based around the SAG, but with the increasingly functional system now being released, the programme is being expanded. Further information is available at:
`http://wiki.astrogrid.org/bin/view/Astrogrid/BetaTesting`

3. AstroGrid Deployment Strategy

The main goal of the deployment strategy was to provide rapid operational feedback to the development process. AstroGrid has an iterative release cycle (Linde et al. 2004) with each release being deployed during the following "iteration".

---

[3] `http://wiki.astrogrid.org/bin/view/Astrogrid/PhaseAReport`

[4] `http://wiki.astrogrid.org/bin/view/Astrogrid/AGSAG`

Each deployment release involves the software components being installed on the "AstroGrid Testbed". Associated release documentation and example science walk throughs provide help in the use of the system to the end user. A dedicated feedback reporting system has been installed (implemented via the AstroGrid Bugzilla ticketing system) to facilitate feedback from the "beta-testers".

AstroGrid, as a member of the Astrophysical Virtual Observatory[5], is aligning elements of its deployment with the annual demonstrator release of the AVO. For instance, its contributions to the AVO First Light demonstrator are described in Quinn et al (2004).

The **"AstroGrid Testbed"** is composed of hardware and software resources made available across the Atsrogrid participating institutions. In early releases the system was deployed on a limited range of machines. Since the iteration 5 deployment, the testbed has expanded in scope to link the AstroGrid consortium. For instance in the current deployment, applications are provided on machines in Manchester and Cambridge, data resources are located through Cambridge and Edinburgh, the workflow and portal are deployed on machines in Leicester, and so forth. All the participating resources are hosted in institutions attached to the high speed SuperJanet4[6] academic research network.

4. AstroGrid and the Iteration 5 & 6 Deployments

The AstroGrid iteration 5 deployment was released August 2004. This release, conforming to internationally[7] agreed interoperability standards, presented a single sign-on entrance, with user MySpace, basic workflow capabilities and query access to a limited number of data sets - primarily the full INT Wide Field Survey (McMahon et al. 2001) catalogues.

The recently released iteration 6 deployment represents the baseline AstroGrid, with all elements of functionality in evidence. In particular a more significant range of data resources and applications are accessible - with the user being able to generate sophisticated workflows, including flows and sequences, logic and conditionals.

Many **helper applications** are now available to handle data and results stored in the users MySpace. These are externally provided, some being existing software, some newly provided specifically to take advantage of standardised VO interfaces. AstroGrid is maintaining a list of these helper Applications[8].

In the context of the **AVO 1st Light demonstrator** AstroGrid developed the web service wrapper to the SExtractor source extraction application (Bertin & Arnouts, 1996). More recently, the workflow together with access to applications (catalogue extraction using SExtractor (Bertin & Arnouts, 1996), a table cross federator, and the hyper-z statistical redshift determinator (Bolzonella, Miralles, & Pelló 2000)) were integrated into the AVO 1st Science prototype. This

---

[5]`http://www.euro-vo.org`

[6]`http://www.superjanet4.net/`

[7]`http://www.ivoa.net`

[8]`http://wiki.astrogrid.org/bin/view/Astrogrid/VoResources`

was successfully used to enable the discovery of a large population of obscured QSO (Padovani et al., 2004).

AstroGrid is playing a central role in the development of the forthcoming Jan 2005 AVO demonstrator, where a number of new applications (including spectral synthesis codes) will be made accessible, embedded in a sophisticated workflow, to allow the dynamic comparison of modeled and observed spectral energy distributions of cluster and field galaxies.

In **conclusion**, AstroGrid is nearing the completion of its first phase, and its Release 1 will be available end Jan 2005 from `http://www.astrogrid.org/release`. Rapid release of the product and involvement of the community in "doing" science is proving critical as AstroGrid moves from "development" to "deployment" in its second phase 2005–2007.

References

Bertin, E., Arnouts, S., 1996, A&AS, 117, 393

Bolzonella, Miralles, & Pelló 2000, A&A, 363, 476

Harrison, P. A., Winstanley, N., Taylor, J. D., 2004, in ASP Conf. Ser., Vol. 347, ADASS XIV, ed. P. L. Shopbell, M. C. Britton, & R. Ebert (San Francisco: ASP), 291

McMahon, R. G., Walton, N. A., Irwin, M. J., Lewis, J. R., Bunclark, P. S., Jones, D. H. 2001, NewAR, 45, 97

Padovani, P., Allen, M. G., Rosati, P., Walton, N. A., 2004, A&A, 424, 545

Quinn, P. J., et al., 2004, in ASP Conf. Ser., Vol. 314, ADASS XIII, ed. F. Ochsenbein, M. Allen, & D. Egret (San Francisco: ASP), 304

ASTRONOMICAL DATA ANALYSIS SOFTWARE AND SYSTEMS XIV
ASP Conference Series, Vol. 347, 2005
P. L. Shopbell, M. C. Britton, and R. Ebert, eds.

Reliable, Automatic Transfer and Processing of Large Scale Astronomy Datasets

Tevfik Kosar, George Kola, and Miron Livny

Computer Sciences Department, University of Wisconsin-Madison, 1210 W. Dayton Street, Madison, WI 53706

Robert J. Brunner[1] and Michael Remijan

NCSA, University of Illinois at Urbana-Champaign, Champaign, IL 61820

Abstract. Astronomers are increasingly obtaining larger datasets, particularly in the optical and near-infrared. Unfortunately, the technologies to process large amounts of image data and share the data and the results with collaborators spread around the globe, have not kept pace with the data flow. In the past, this type of software has required significant human involvement to deal with failures. We have designed and implemented a fault-tolerant system that can process large amounts of astronomy images using idle CPUs on desktops, commodity clusters and grid resources. It reliably replicates data and results to collaborating sites and performs on-the-fly optimization to improve throughput. It is highly resilient to failures and can recover automatically from network, storage server, software and hardware failures. To demonstrate the capabilities of this framework, we have successfully processed three terabytes of DPOSS images using idle grid resources spread across three organizations.

1. Introduction

In different fields of astronomy, ranging from astrometry to star formation, from cosmology to planetary sciences, especially where optical and infrared techniques are used, there is a need to process large amounts of images and to make them available at different locations for collaborative analysis. While the compute intensive nature of the tasks and the inherent parallelism make them good candidates for distributed processing, the data intensive nature creates a set of new challenges. These include hardware and software failures, network overheads, and human hours needed to baby-sit the whole process. These applications do not perform well under existing distributed scheduling systems and may end up not making forward progress depending on the failure characteristics.

In this work, we have designed and implemented a system that can process astronomy images fully automatically and can handle off-site replication in a fault-tolerant manner. Our system takes care of transferring the original images to the compute nodes, processing them and transferring the generated results to a set of destinations and handles all the failures that may occur during the

[1]Also affiliated with the Astronomy Department

process. Our system enables researchers to manage all steps including the actual processing of the images and movement of the data in an automated manner without any human interaction.

A unique feature of our system is the separation of the data movement from the computation. We treat the data movement as a full-fledged job and schedule it. This separates the data-movement failures from processing failures and allows us to optimize the data movement by intelligent scheduling.

As a case study, we show how our system was used to process 3 Terabytes of DPOSS (Djorgovski et al. 2001) image data in less than 6 days making use of Grid resources, and discuss the contributions of this system to the whole astronomy community.

2. Methodology

Existing distributed scheduling systems do not work well for data intensive applications that require large amounts of data to be moved across wide area networks pre- and post-processing. Distributed scheduling systems in the absence of a distributed file system allocate the processor, stage-in the input data, perform the computation and stage-out the result of the computation. If the data transfer in the stage-out fails, the computation is re-done because the stage-out is part of the computational job and the whole job is re-tried. This decreases end-to-end performance in case of intermittent network outages.

In our model we separate computation and data placement into separate full-fledged jobs and schedule them appropriately. To handle the dependencies we use a Directed Acyclic Graph (DAG) model where we represent jobs as nodes and dependencies between jobs as directed arcs. For instance, if a job A should be executed before job B, we draw an arc from A to B. A node in the DAG is executed only after all its parent nodes are executed. If enough processors are available, computation is performed immediately after data becomes available. In this sense, the model is similar to data flow computing.

We use a computation scheduler such as Condor (Litzkow et al. 1988) to schedule the jobs on the compute cluster. It is not necessary to replace the existing computation scheduler on a cluster as our computational scheduler has the ability to use existing scheduling systems while giving users a uniform interface. This is very useful if we want to use heterogeneous grid resources. We schedule the data transfer using a specialized data placement scheduler, Stork (Kosar & Livny 2004). We have added features to make the scheduler take into account network and host characteristics to optimize the transfer.

We determine the characteristics of the storage server to tune the optimal concurrency level. This information can be generated by running our profilers on the storage server, and we do it automatically if we are allowed to execute on the storage server. The profilers look at CPU load and I/O characteristics of the server at different concurrency levels and determine reasonably close to optimal values for our configuration.

We use DAGMan (Thain et al. 2002) to handle the dependencies. We have enhanced it to submit the data placement jobs to Stork and computation jobs to Condor. This DAG model is flexible and any arbitrary computation can be performed to determine the set of destinations.

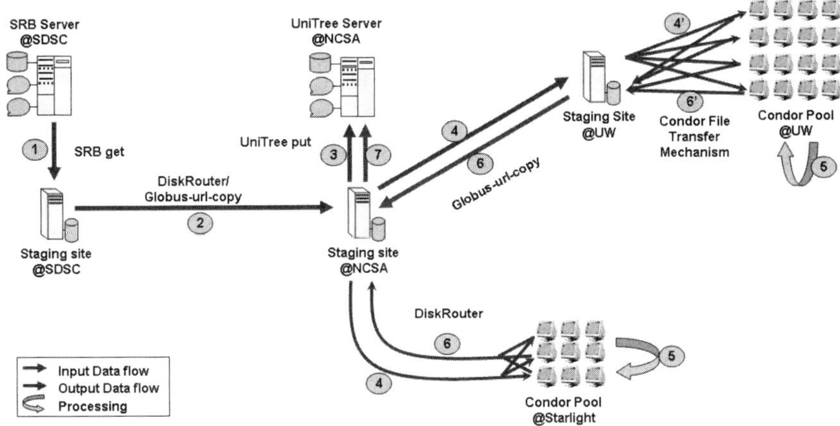

Figure 1. NCSA DPOSS image replication and processing pipeline.

3. Case Study

First, we wanted to transfer the Digital Palomar Sky Survey (DPOSS) image data residing on SRB mass storage system at SDSC in California to the UniTree mass storage system at NCSA in Illinois. Then we wanted to process these data using the Grid resources at different sites, and transfer the output data back to the UniTree server at NCSA.

The total data size was around three terabytes (2611 files of 1.1 GB each). Since there was no direct interface between SRB and UniTree at the time of the experiment, the only way to perform the data transfer between these two storage systems was to build a data transfer pipeline. In this pipeline, we set up two staging nodes between the source and destination storage systems. The first staging node was at the SDSC site very close to the SRB server, and the second staging node was at the NCSA site near the UniTree server.

We had some disk space limitations on the staging nodes. This implied that whenever we were done with a file at a staging node, we had to remove it from there to create space for the transfer of another file.

We had access to over a thousand CPUs at the UW-Madison Computer Science Department condor pool and eight CPUs at Starlight. We had to use different strategies for the two domains. At UW-Madison, we used desktop CPUs, so an interactive user could evict our job when he started using the machine. In this case, directly transferring data to the compute node was not such a good idea, as an eviction would result in re-transfer over the wide-area network. We handled this by using a staging node at UW-Madison. At Starlight, the eviction was more static. There would be slots where users would use all the CPUs for experiments. Starlight lets us use the CPUs when there is no scheduled experiment. This meant that if an experiment started, we would be unable to get the Starlight CPUs for a long time.

Processing time for the initial analysis of a single image varies from half-an-hour to two-and-a-half hours. After initial analysis, a more detailed processing

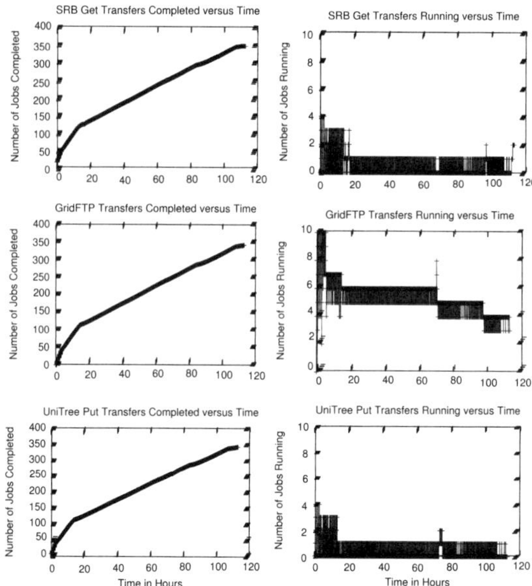

Figure 2. Shows automated failure and flow control in the pipeline.

is required, taking up to an order of magnitude longer for some of the images. The wide area transfer took less than three minutes.

Figure 2 shows how the pipeline adapts the flow control on the fly. Around 4 hours, GridFTP encounters some wide-area failures and the pipeline lowers the number of concurrent transfers to seven. Close to 20 hours, SRB refuses new connection and the pipeline responds by trying to maintain a single connection. This affects the next hop and the number of concurrent GridFTP transfers drops to six. After that, UniTree accepts more connections and then slows down and this causes GridFTP to drop the number of concurrent transfers to five because of space limitations at the NCSA cache node.

We have processed three terabytes of data using our system in less than six days. During the transfer and processing of three terabytes of data, we had a wide variety of failures. The source and destination mass-storage systems stopped accepting new transfers, third-party transfers hang, wide-area network outages observed, and software upgrades scheduled. All of these failures were recovered automatically, and the transfers were completed successfully without any human interaction.

4. Related Work

GFarm (Morita et al. 2003) provides a global parallel file system with online petascale storage. Their model specifically targets applications where data primarily consists of a set of records or objects that are analyzed independently. Gfarm takes advantage of this access locality to achieve a scalable I/O bandwidth using a parallel file system integrated with process scheduling and file distribution.

OceanStore (Kubiatowicz et al. 2000) aims to build a global persistent data store that can scale to billions of users. The basic idea is that any server may create a local replica of any data object. These local replicas provide faster access and robustness to network partitions. Both Gfarm and OceanStore require creating several replicas of the same data, but still they do not address the problem of scheduling the data movement when there is no replica close to the computation site.

BAD-FS (Bent et al. 2004) builds a batch aware distributed file system for data intensive workloads. However, we are not sure if it would be able to handle computation that for performance reasons prefer to access source data from local disk rather than over a network file system. Further, BAD-FS at present does not schedule wide-area data movement, which we feel is necessary given the size of the astronomy datasets. We feel that incorporating our scheduling system into BAD-FS would help it handle larger datasets.

5. Conclusion

We have designed and implemented a fault-tolerant system that can process large amounts of astronomy images using idle CPUs on desktops, commodity clusters and grid resources. It reliably replicates data and results to collaborating sites and performs on-the-fly optimization to improve throughput. It is highly resilient to failures and can recover automatically from network, storage server, software and hardware failures. Using this system, we have successfully processed three terabytes of DPOSS images using idle grid resources spread across three organizations. This system is freely available to others groups who wish to significantly decrease the time and effort required to perform large-scale data transfer and processing.

References

Bent, J., et al. 2004, Proceedings of NSDI 2004

Djorgovski, S. G., et al. 2001, Wide Field Surveys in Cosmology

Kosar, T., & Livny M. 2004, Proceedings of ICDCS 2004

Kubiatowicz, J., et al. 2000, Proceedings of ASPLOS 2000

Litzkow, M. J., et al. 1988, Proceedings of ICDCS 1988

Morita, Y., et al. 2003, Proceedings of CHEP 2003

Thain, D., et al. 2002, Grid Computing: Making the Global Infrastructure a Reality

ASTRONOMICAL DATA ANALYSIS SOFTWARE AND SYSTEMS XIV
ASP Conference Series, Vol. 347, 2005
P. L. Shopbell, M. C. Britton, and R. Ebert, eds.

A Parallel Code Analysis of Light Curve Sequences of Magnetically Active Close Binaries through the Astrocomp-grid Portal

U. Becciani, A.F. Lanza, and A. Costa

INAF - Osservatorio Astrofisico di Catania, Via S. Sofia, 78 – 95123 Catania, Italy, Email: ube@ct.astro.it

M. Rodonò[1]

INAF - Viale del Parco Mellini, 84 – 00136 Roma, Italy

Abstract. We present some results of the analysis of long-term light curve sequences of magnetically active close binaries by means of a numerical parallel code made available through the Astrocomp web portal. The code looks for the best values of the photometric parameters and evaluates their confidence intervals for eclipsing binaries with cool spots on their surface. Cool spots produce distortions of the light curves that may induce systematic errors in the luminosity ratio, the fractionary radii of the component stars and the inclination of the orbital plane. Our method reduces the impact of spots on the determination of such parameters by simultaneously fitting a long-term sequence of data along which the spots' coverage and distribution show remarkable changes (cf. Rodonò et al. 2001). The large amount of computational work required, is managed by means of a parallel code based on MPI. The code is going to be made available through a web-based user-friendly interface called Astrocomp.

1. Introduction

MARA (Multi light curve Analysis of Rotator Activity) is a code for modeling sequences of optical light curves of single or close binary stars with surface brightness inhomogeneities. The code takes into account ellipsoidicity, gravity-darkening and reflection effects as well as a distribution of cool spots in order to fit the light curve of a close binary system. The determination of the spot surface distribution is an ill-posed problem solved by introducing the Maximum Entropy and Tikhonov regularizations (Lanza et al. 1998). A sequence of optical light curves of a close binary system can be analysed as a single data set to derive the best values of the photometric parameters, i.e., the luminosity ratio, the fractionary radii of the components and the inclination of the orbital plane of the system. If the light curve sequence spans an entire activity cycle, the systematic errors introduced by the cool spots in the determination of the photometric parameters can be significantly reduced. The code was developed using the MPI library. Each processor analyses a light curve with a given set of parameters.

[1]Universitá di Catania - Dipartimento di Fisica e Astronomia, Via S. Sofia, 78 – 95123 Catania, Italy

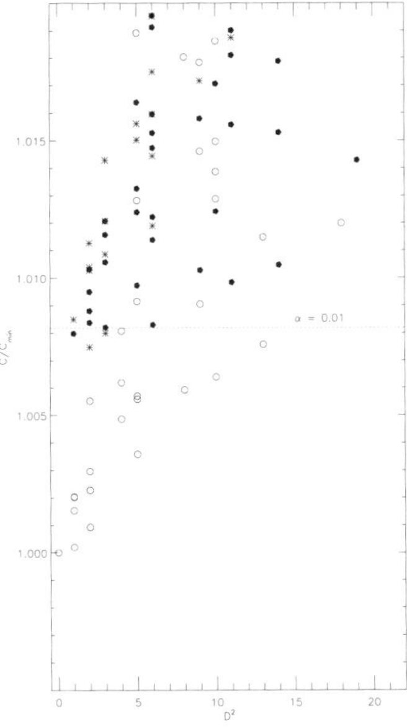

Figure 1. The statistics of the ratio of variance C/C_{min} versus the square of the Euclidean distance D^2 from the minimum C_{min} in the neighbourhood of C_{min}. The different symbols indicate different values of the photometric parameters. The confidence level corresponding to a probability of $\alpha = 0.01$ is indicated (see Rodonò et al. 2001 for details).

The Dynamic Load Balance of the run is achieved by sharing information on the overall computation status among processors.

2. Analysis of a Sequence of Light Curves of RS CVn

A sequence of 16 V-band light curves of the proto-type active binary RS CVn (K2IV + F5V) was analysed to test our computational approach. It spans an entire activity cycle sampling the full range of spot coverage as observed in the system from 1963 to 1993 (Rodonò et al. 1995). Each light curve of the sequence was modeled by solving for the Maximum Entropy and Tikhonov spot distributions with given sets of photometric parameters (Rodonò et al. 2001). A total number of 2205 parameter combinations were tried to effectively sample the parameter space. The total CPU time was of 3475 hours on a 32-processor Cray-T3E. The best values of the photometric parameters were determined by minimizing the total χ^2 of the light curve sequence while their confidence intervals were established by means of the ratio-of-variance method (Lampton et al. 1976). In Figure 1, the distribution of the ratio of variances is shown versus a measure of the Euclidean distance of the trial set of parameters from the best-fit set in the 4-D photometric parameter space (see Rodonò et al. 2001 for details).

2.1. The Parallel Procedure

Our parallel procedure distributes light curves with a fixed set of photometric parameters to each processor to minimize the objective functionals Q_{ME} and Q_T of the Maximum Entropy and Tikhonov regularized best fits, respectively. There is no parallelism to analyze a single light curve, but the parallelism consists of the distribution of single curve analysis among the processors. At the beginning of the calculation, the number of curves to be analyzed is compared with the number of processors available for the parallel execution and a first static assignment of the work load is done. The code assigns a number of light curves *nlc_pe* to each processor (hereafter PE), according to:

$$nlc\_pe_k = INT(\frac{n_l}{npes}) + 1 \cdot (f_c(pe_k)), \tag{1}$$

where $f_c(pe_k) = 1$ if $pe_k < \text{mod}(n_l, npes)$ and $f_c(pe_k) = 0$ if $pe_k \leq \text{mod}(n_l, npes)$, $pe_k = [0, npes - 1]$ is the PE number, n_l the number of light curves and $npes$ the total number of processors. If $npes > n_l$ then $nlc\_pe_k = 0$ for $pe_k \geq n_l$ in the initial static assignment phase. However, in the subsequent stage, any free PE will immediately begin to help the other PEs in the framework of the load balance procedure in order to optimize the overall performance of the light curve assignment (see Becciani et al. 2004 for details).

The parallel procedure makes use of two status matrices *sum_chime* and *sum_chitik* to keep information on the light curve analyses already completed and those that are still awaiting to be done as the computation goes on. At the end of the run, the matrices will contain the ratio-of-variance statistics of the sequence of light curves for each set of the photometric parameters.

3. Mara for Astrocomp Users

Astrocomp[2], developed by the INAF-Catania Astrophysical Observatory, University of Roma "La Sapienza" and ENEA, is a portal based on WEB technologies, to manage and use codes for astrophysical applications (Costa et al. 2003). The portal is a user-friendly application which allows to submit a job in a grid of computing systems. The computational platforms presently available through Astrocomp are the CINECA MPP systems, i.e., the IBM SP with 512 PEs, the SGI ORIGIN 3800 with 64 PEs, the IBM Linux cluster with 128 PEs, and the IBM SP system of INAF-Catania Astrophysical Observatory with 32 PEs. Registered Astrocomp users can freely use the above systems within their allocated time quota. In order to use MARA through Astrocomp, a user should select the MARA code section and access the pages of the code that allow him to create all the input parameter files, upload his own light curve sequence, and create a job to submit the run in the available MPP systems (Figure 2). The user can monitor the job status and download the output data files at the end of the run. Moreover, the light curve best fits and the corresponding spot maps can be visualized as postscript or jpeg files by means of a dedicated utility based on IDL (Figure 3).

---

[2]`http://www.astrocomp.it`

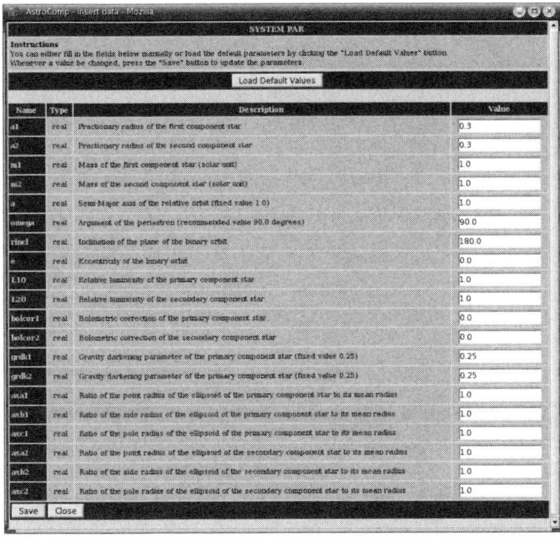

Figure 2. The Astrocomp mask to upload the set of parameters for each light curve analysis.

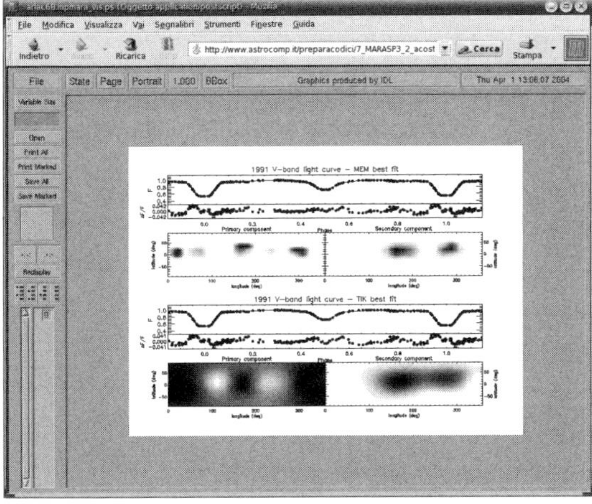

Figure 3. The visualization of a postscript file showing the best fit and the Mercator maps of the components of a close binary system for the Maximum Entropy solution (upper panels) and the best fit and the Mercator star maps for the Tikhonov model (lower panels), respectively.

Astrocomp makes it possible to use MARA without downloading and installing the code and provides some free computing time on MPP systems to run the code.

Acknowledgments. We are grateful to V. Antonuccio, R. Capuzzo-Dolcetta, P. Miocchi, V. Rosato and P. Di Matteo for the development of the Astrocomp Portal and the inclusion of MARA in the portal. Research on stellar physics at Catania Astrophysical Observatory of the *INAF (Istituto Nazionale di Astrofisica)*, and at the Dept. of Physics and Astronomy of Catania University is funded by *MIUR (Ministero dell'Istruzione, Universit e Ricerca)* and *Regione Sicilia*, whose financial support is gratefully acknowledged. The extensive use of computer facilities at the Catania node of the Italian Astronet Network is also gratefully acknowledged.

References

Becciani, U., Lanza, A. F., Costa, A., Rodonò, M. 2004, Comp. Phys. Comm., submitted

Costa, A., Becciani, U., Antonuccio, V., et al. 2003, MEM Sait Supp. 1, 89

Lampton M., Margon B., & Bowyer S. 1976, ApJ, 208, 177

Lanza, A. F., Catalano, S., Cutispoto, G., Pagano, I., & Rodonò, M. 1998, A&A, 332, 541

Rodonò, M., Lanza, A. F., & Becciani, U. 2001, A&A, 371, 174

Rodonò M., Lanza A. F., & Catalano, S. 1995, A&A, 301, 75

ASTRONOMICAL DATA ANALYSIS SOFTWARE AND SYSTEMS XIV
ASP Conference Series, Vol. 347, 2005
P. L. Shopbell, M. C. Britton, and R. Ebert, eds.

Integration of an MPP System and Astrophysical Applications in the INFN-GRID

A. Costa, A. S. Calanducci, and U. Becciani

INAF - Osservatorio Astrofisico di Catania, Cittá Universitaria, Via S.Sofia, 78 I-95123 Catania Italy

Abstract. We are going to present the middleware changes we have made to integrate an IBM-SP parallel computer into the INFN-GRID and the results of the application runs made on the IBM-SP to test its operation within the grid. The IBM-SP is an 8-processor 1.1 GHz machine using the AIX 5.2 operating system. Its hardware architecture represents a major challenge for integration into the grid infrastructure because it does not support the LCFGng (Local ConFiGuration system Next Generation) facilities. In order to obtain the goal without the advantages of the LCFGng server (RPM based), we properly tuned and compiled the middleware on the IBM-SP: in particular, we installed the Grid Services toolkit and a scheduler for job execution and monitoring. The testing phase was successfully passed by submitting a set of MPI jobs through the grid onto the IBM-SP. Specifically the tests were made by using MARA, a public code for the analysis of light curve sequences, that was made accessible through the Astrocomp portal, a web based interface for astrophysical parallel codes. The IBM-SP integration into the INFN-GRID did not require us to stop production on the system. It can be considered as a demonstration case for the integration of machines using different operating systems.

1. Our Starting Scenario

The following machines were initially installed in our grid site:

N.1 Computing Element: *gridce.ct.astro.it*
N.1 Storage Element: *gridse.ct.astro.it*
N.1 User Interface: *gridui.ct.astro.it*
N.1 LCFGng server: *grid1.ct.astro.it*
N.2 Worker Nodes: *gridwn1.ct.astro.it, gridwn2.ct.astro.it*

All these nodes are i386 architecture machines running Linux Redhat 7.3 as their Operating System. They represent the first quantum grid at the Astrophysical Observatory of Catania. The middleware used is INFN-GRID 2.2.0.

2. How To Build a Grid: The Classic Way

The configuration and installation of an INFN-GRID site usually relies on the LCFGng[1] system (Local ConFiGuration system Next Generation), developed by

[1]`http://www.lcfg.org/`

the Department of Computer Science at Edinburgh University. LCFG provides a configuration scripting language and a central repository of software (rpms) and configuration specifications, from which individual Linux machines can be automatically installed and configured. Changes to the central specifications automatically trigger corresponding changes in the actual configuration of individual nodes. LCFGng uses XML/HTTP for transporting configuration profiles. The latest release and its documentation is designed to be suitable for public use (under the GPL license).

3. Our Goal

Our goal was to determine procedures to integrate an IBM SP parallel computer into the INFN-GRID as a worker node. The following is the actual configuration of this system:

N.3 IBM SP Power 3 nodes (8 CPUs per node)
N.1 IBM SP Power 4 node (8 CPUs)

It was not possible to use an LCFG server to install an IBM SP system as a worker node for at least two reasons: First, the LCFG version of INFN-GRID initializes machines, reinstalling the OS and all needed applications; second, the repository managed by the LCFG contains rpm packages compiled for the i386 architecture and Linux Redhat 7.3: so the architecture and OS of the target machines are definitely different.

4. IBM SP4 Configuration

We compiled and installed OpenPBS 2.3.16 and pbs_mom. Some slight changes were required to the OpenPBS installation scripts. We used "-enable-sp2" to compile special code to deal with the SP high speed switch. We installed IBM Grid Toolbox V3 for AIX (the IBM version of Globus Toolkit 3), and finally we created VOs users:

Ex: *inaf001* is the username used by *gridce.ct.astro.it* to map the following certificate:
/C=IT/O=INFN/OU=Personal Certificate/L=INAF Catania/CN=Alessandro Costa/Email=acosta@ct.astro.it.

The population of */etc/grid-security/certificates/* with CA certificates and Certification Revocation Lists required the creation and execution of a cron copy from the Computing Element. Moreover, we needed to make some minor changes: mktemp command compilation and installation, and the creation of a list containing ssh keys of trusted hosts were required on AIX 5.2.

5. Tweaking Computing Element

In the Computing Element we modified the PBS server configuration adding our IBM SP4 machine as worker node. In particular we added the following string at the end of */var/spool/pbs/server_priv/nodes*:

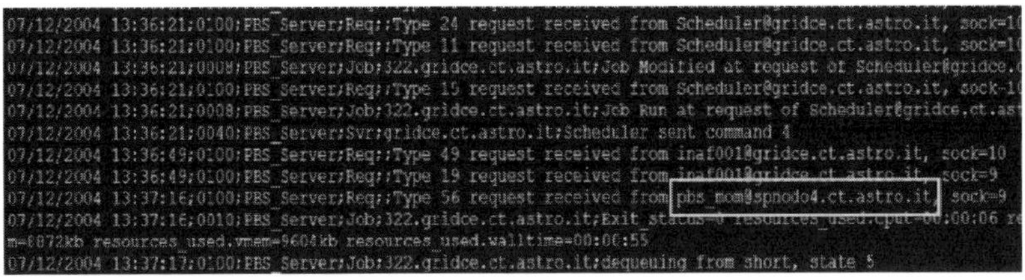

Figure 1. PBS_Server log file in the Computing Element

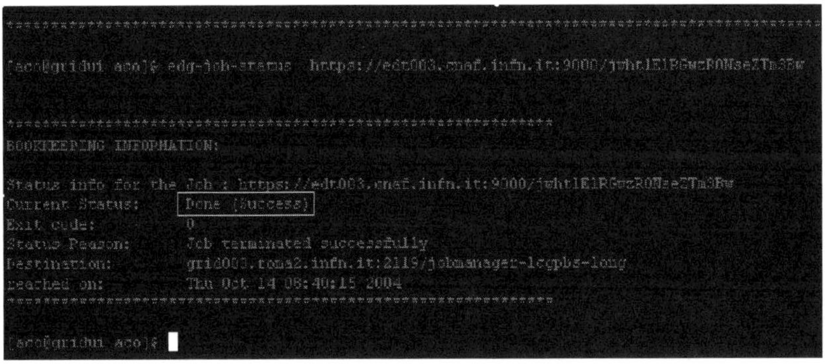

Figure 2. Status of the submitted job

sp4.ct.astro.it np=8 ibmsp

In addition, we made another little change in */opt/globus/lib/perl/Globus/-GRAM/submit-helper.pl* (belonging to the package: *lcg-extra-jobmanagers-1.1.6-1*). It is worth emphasizing that these changes have no side effects on the other linux worker nodes.

6. Testing Phase

For testing purpose, we made use of MARA, a public code for the analysis of light curve sequences. The input sandbox of submitted jobs usually contains the executable, the light curve and the parameter file. Figure 1 shows the log file on the Computing Element after the job submission. Figure 2 shows the status of the submitted job. Finally, Figure 3 shows a diagram of the Mara job output.

7. Conclusions

We have been able to add a Worker Node to an already working grid site without the use of an LCFGng server and without any data loss or interruptions of the MPP system. These procedures can be easily extended to different architectures/OSs other than Power4/AIX. Moreover we have identified the strictly necessary software components to transform a computer into an INFN-GRID

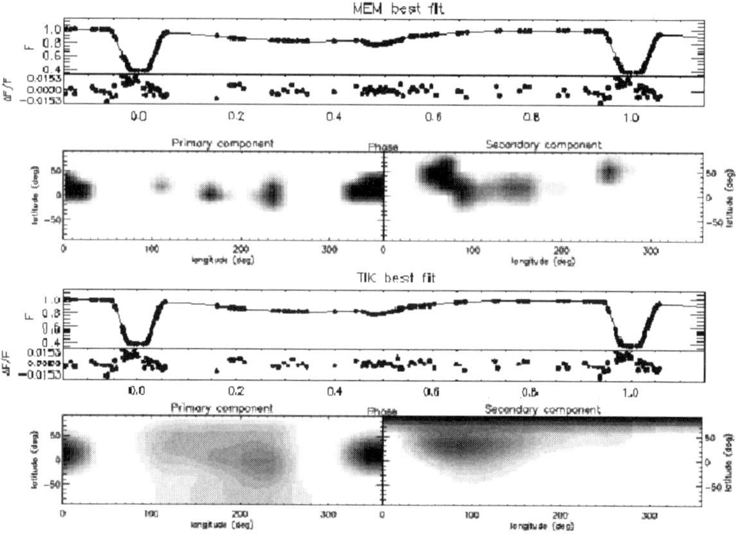

Figure 3. Output Results of a MARA job

Worker Node: a Job Executor (pbs_mom if you use OpenPBS as a Local Resource Manager), and Globus components dealing with GSIFTP file transfers, certificates verification and GASS cache management.

ASTRONOMICAL DATA ANALYSIS SOFTWARE AND SYSTEMS XIV
ASP Conference Series, Vol. 347, 2005
P. L. Shopbell, M. C. Britton, and R. Ebert, eds.

The AstroGrid Common Execution Architecture (CEA)

Paul Harrison and Noel Winstanley

AstroGrid, Jodrell Bank Observatory, Manchester University, UK

John D. Taylor

AstroGrid, Royal Observatory Edinburgh, UK

Abstract. The UK Virtual Observatory (VO) project AstroGrid (see `http://www.astrogrid.org` and related talks at this meeting) began in 2001 and is nearing the successful completion of its first release in December 2004.

This paper describes AstroGrid's Common Execution Architecture (CEA). This is an attempt to create a reasonably small set of interfaces and schema to model how to execute a typical astronomical application within the VO.

The CEA has been designed primarily to work within a web services framework, with the parameter passing mechanism layered on top of this so that the web interface for all applications is described by a single constant piece of WSDL - the differences between applications are expressed by the registry entries for each application. Within AstroGrid we have created pluggable components that can wrap legacy command-line applications, HTTP GET/POST applications and databases as CEA compliant web services, which when combined with the Astrogrid Workflow component make distributed processing within the VO a reality. See `http://www.astrogrid.org/maven/docs/snapshot/applications/` for current information.

1. Introduction

The Common Execution Architecture (CEA) is an attempt to create a reasonably small set of interfaces and schema to model how to execute a typical Astronomical application within the Virtual Observatory (VO). In this context an application can be any process that consumes or produces data, so in existing terminology could include:

- A Unix command line application
- A database query
- A web service

The CEA has been primarily designed to work within a web services calling mechanism, although it is possible to have specific language bindings using the same interfaces. For example Astrogrid has a Java implementation of the interfaces that can be called directly from a Java executable.

1.1. Motivation

The primary requirements motivating the creation of this architecture are:

- To create a uniform interface and model for an application and its parameters. This has twin benefits:

1. It allows VO infrastructure writers a single model of an application for which to develop code.
2. Application writers know what they have to implement to be compatible with a VO Infrastructure.

- To provide a higher level description than WSDL 1.1 can offer.
- Restrict the almost limitless possibilities allowed by WSDL into a manageable subset.
- Provide specific semantics for some astronomical quantities.
- Provide extra information not allowed in WSDL - e.g. default values, descriptions for use in a GUI etc.
- To provide extensions with the VO Resource schema (See the IVOA WG) that can describe a general application
- To provide asynchronous operation of an application - This is essential as the call tree that invokes the application cannot be expected to be active for extremely long lasting operations - e.g. a user from a web browser invokes a data-mining operation that takes days
- Provide callback for notification of finishing.
- Provide polling mechanisms for status.
- To allow for the data flow to not necessarily have to follow the call tree. In a typical application execution the results are returned to the invoking process - In a VO scenario, it can be useful if the application can be instructed to pass the results on to a different location

1.2. Origins

Amongst the VO specifications there was no existing model for applications defined at the level at which this design attempts to address. In the VOResource schema an application is defined as a Service with the interface definition. The interface definition either relies on referring to a WSDL definition of the service, or on other schema extending the service definition to provide some specific detail as in the case of a Simple Image Access service. There is no general definition of an application in the resource specification.

It is clear that the WSDL model of an interface has had a large influence on the design of the CEA, but recall that the CEA is intentionally layered on top of WSDL, so that CEA controls the scope and semantics of operations. There is only one WSDL definition for all applications, so as far as web services are concerned the interface is constant. CEA works by transporting meta information about the application interface within this constant WSDL interface.

2. Interfaces

The 3 WSDL ports that are used to interact within CEA are briefly described below:

CommonExecutionConnector This is the main port that is used to communicate with the application. The main operations in this port are:

- **init** - this will initialize the application environment - returns and executionId by which any particular execution run can be referenced later.

- **registerResultsListener** - any number of services can register themselves as wanting to receive the results from the run when they are available as long as they implement the ResultsListener port below
- **registerProgressListener** - any number of services can register themselves as wanting to receive status messages during the run as long as they implement the JobMonitor port below
- **execute** - will actually start the asynchronous execution of the application specified in the init call.
- **queryExecutionStatus** - this call can be used to actively obtain the execution status of a running application, rather than passively waiting for it as a JobMonitor
- **abort** - will attempt to abort the execution of an application
- **getExecutionSummary** - request summary information about the application execution
- **getResults** - actively request the results of the application execution, rather than passively waiting for them as a ResultsListener.
- **returnRegistryEntry** - this returns the registry entry for the particular CommonExecutionConnector instance - this will probably be removed from this interface to be replaced by the equivalent operation in the standard VO service definitions.

JobMonitor The only operation in the JobmMonitor port is the monitorJob operation, which expects to receive a message with the job-identifier-type (as specified in the original init operation of the CommonExectutionConnector port) and a status message

ResultsListener The only operation is the putResults on the ResultsListener port. This accepts a message that contains a job-identifier-type and a result-list-type, which is just a list of parameterValues.

2.1. Objects

The objects that participate in CEA can be split into two groups

1. Those used to describe the application in the registry
   - *Application* the overall application, which has a series of
   - *Parameter* which are the detailed descriptions of the parameters and their types.
2. Those used to describe the application in the WSDL interface
   - *Tool* - An instance of an application with real parameter values.
   - *ParameterValue* which is used to pass a values to a Tool.

Application As this model depicts, an application in CEA is really quite a simple entity, consisting of one or more interfaces that consist of zero or more input parameters and zero or more output parameters.

Parameter Definition The description of the parameters and the parameter values are probably the heart of the CEA. It is the model for the parameters that allow us to add semantic meaning, and to give the flexibility in how the parameters are transported. The implementation is still in its infancy, but it is hoped that the parameter definition will be extended to encompass any data models that the VO produces.

Tool The tool represents the full collection of parameters that are passed to a particular interface of an application and the results that are returned.

ParameterValue The parameterValue model is simple but powerful representation of the parameters that are passed to an application. The parameterValue element has a Value subelement with a string representation of the value as well as having 2 attributes

- **name** - The name of the parameter.
- **indirect** - This describes whether the value element of the parameter should be used as is (indirect="false"), or if the value of the parameter represents a URI from which the actual value should be fetched (indirect="true"). It has not been defined what is the minimum set of transport mechanisms a service should understand to be CEA compliant, but the different sorts of transport mechanism are expected to include
  - SOAP messages
  - http get/put
  - SOAP attachments
  - ftp/gridftp
  - MySpace
  - local filestore

3. Astrogrid Implementation

3.1. CEA

AstroGrid has created a set of software that implements the CEA. The main server components that implement the CommonExecutionConnector web service are called ExecutionControllers and there are currently two specialized adapter controllers for

- Wrapping legacy command line applications in the framework.
- Wrapping legacy GET/POST style HTTP applications in the framework.

There are plans for a further adapter for other web services themselves as well as the possibility of creating new applications that directly implement the CEA.

3.2. Job Execution System (JES)/Workflow

The CEA only specifies how to call applications, there is another component, JES[1], with the responsibility for actually organizing the order of execution of a sequence of applications. . The job system in AstroGrid is driven from the workflow which has an xml schema representation. The unit of execution within the workflow is a Step, which can itself be part of a Sequence (a set of actions performed in series) or a Flow (a set of actions performed in parallel). The Step contains a Tool which is the full CEA description of a particular execution instance of an application. This paper does not discuss the full complexity of the

[1] `http://www.astrogrid.org/maven/docs/SNAPSHOT/jes/`

workflow which also includes a scripting component known as JEScript (based on the Java scripting language Groovy) which allows workflows to be arbitrarily complex.

ASTRONOMICAL DATA ANALYSIS SOFTWARE AND SYSTEMS XIV
ASP Conference Series, Vol. 347, 2005
P. L. Shopbell, M. C. Britton, and R. Ebert, eds.

A Java Thick Client User Interface for Grid Processing

Ted Hesselroth

Spitzer Science Center

Abstract. A user interface (CAPRI) which is configurable at runtime has been developed which allows application features to be maintained and upgraded on a central server, available to users without the need for reinstalling software. The user interface is specified by an XML file accessed through a URL and parsed by the open-source SWIX library, which returns a completely laid-out container with the applications controls. A set of generic model-view-controller-actions classes are also instantiated by the CAPRI package based on parsing of the input XML file. Hierarchical relationships present in the XML file are reflected in membership relationships among the classes. An event-driven architecture with a central event handler allows for convenient extensibility. Client/server software is based on the Java Web Services package with SOAP message passing. The server has access to data and computing resources and brokers the requested computation. Sun Grid Engine software is used to manage the cluster of processing nodes. This application has been deployed at the Spitzer Science Center to allow rapid interactive processing of science data.

1. Introduction

Grid architectures that are to be accessible to novice users require graphical user interfaces. But software developed for efficient data reduction differs substantially from that used for GUIs. The requirements for each are in fact widely divergent.

Data reduction systems need to run as fast possible, and usually are run on dedicated machines. Since they are non-graphical, the binaries that comprise them may be compiled from disparate programming languages. Processing may be distributed and may access resources such as database servers and file servers. The systems are tightly, stably managed behind a firewall and require a long lead time to get up and running. Processing is automated.

GUIs need to run quickly only compared to the user's perception. The machine will be used simultaneously for other tasks. The environment in which they will run is only known to within a few parameters such as operating system and minimum clock speed or RAM. The graphical presentation requires selecting one language which is strong in graphics. All resource access is expected to be through the GUI and preferably from one server. The GUI needs to be easy to install and immediately usable with no training. Processing is customized by the user.

Any software architecture benefits from homogeneous interfaces, reusable components, and a unified internal structure, in so far as these are possible. But every program has some "entropy", that is, lines of source code that are unique to it. These differences represent functionality supported by, but not predictable from, the architecture. A GUI manifests homogeneous conditions in that it consists of components such as buttons and panels that are supplied from a stock library. A GUI is also highly entropic in that the appearance and arrangement of these components can be infinitely varied.

Furthermore, in the analysis of scientific data, the data reduction itself reflects the state of knowledge of the instruments involved and the phenomena being observed. This changes over time, and a well-designed data reduction system will handle modifications with minimal impact. Likewise, an associated GUI for custom processing needs to be expandable to encompass new routines as they are realized. The entropy, or arbitrariness, is thus not only high in one particular implementation of the interface, but over time as well.

In order to accommodate this high degree of disorder while preserving the homogeneity and unity that can also exist, it is advantageous to collect the more arbitrary portion of the program into a separate entity which can be handled in a specialized way. The XML format is suitable for encapsulating data which has some degree of structure but is otherwise variable. This paper describes a software architecture that has been built using XML files as "entropy sinks" to contain the highly variable parts of the software.

2. The Architecture

CAPRI[1] is an acronym for Configurable Application for Pipeline Reprocessing Interfaces. The package was written in order to support custom reprocessing of Spitzer Science data on a local grid cluster. Processing on the user's system is also supported.

The CAPRI software package utilizes XML files which contain descriptions of the user interfaces for various data reduction packages available through the Spitzer Science Center. The packages are written in C or C++, and CAPRI provides a java graphical user interface for controlling the data reduction software. One user interface has been released: SPICE[2] (Spitzer IRS Custom Extraction). An interface for the Spitzer Mosaicker, MOPEX, is planned. Each interface is based on an XML file which describes its buttons, checkboxes, textfields, labels, etc, and their layout.

Here is the sequence of events that occurs when CAPRI is used to submit a job to a local cluster:

1. The user starts CAPRI, either as an application or as an applet in a browser.
2. CAPRI fetches description of interface from server.
3. CAPRI creates user interface based on XML description.
4. User selects data and processing parameters using graphical controls.
5. User presses Run.
6. Settings are converted to SOAP message and sent to server.
7. Server converts SOAP to shell script using XSL transform.
8. Server forwards the script to the Grid Engine Manager.

---

[1] `http://ssc.spitzer.caltech.edu/postbcd/capri`

[2] `http://ssc.spitzer.caltech.edu/postbcd/spice.html`

9. Grid Engine Manager dispatches job to grid node.
10. Upon completion of job, server sends results back to client.
11. CAPRI displays the results and saves them to disk

The steps which make the most use of the CAPRI architecture are the conversion of XML into the GUI (step 3) and subsequent conversion of user input to a SOAP message (step 6), so these are discussed in more detail below.

2.1. XML Parsing and Object Creation

In CAPRI, there is a tight coupling of the xml description and the program itself. As far as there is order in the schema, functions within CAPRI can reflect it. The unity presented by the schema is translated into a unified framework within the programming model.

When an xml file containing a description of the gui is parsed by CAPRI, top-level nodes with the tagname "function" cause the instantiation of objects of class "Function" in the application. If a "function" node contains child nodes of type "function", then these cause child Function objects to be instantiated in an array belonging to the parent Function. Thus the hierarchy contained in the the xml file is represented in the hierarchy of objects in the application. This allows the reuse of code in configurations. Figure 1 shows an example partial XML file and the Function objects created in CAPRI based on it.

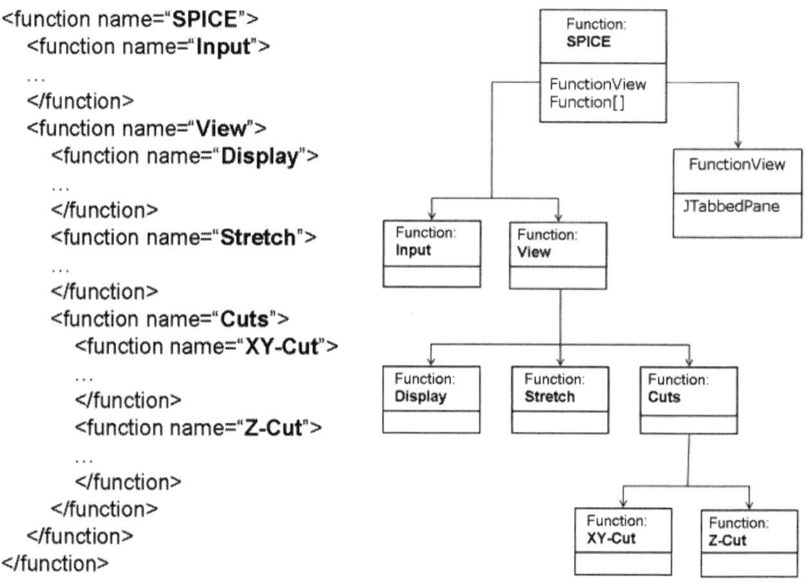

Figure 1. The internal structure of Functions generated by XML snippet shown above

The reuse of code has the advantage that each function specified in the xml file will automatically provide a basic level of support which can be utilized by components. Chief among these is the event-passing mechanism between functions. The implementation of a core event mediator architecture allows Functions to be developed independently; any calls to and from other Functions are handled by a simple and transparent set of calls to event registration, fire, and receive methods.

The structure imparts organization to the program itself, but its main purpose is to organize the GUI's visible appearance. The graphical layout of child components within parent components which is seen by the user corresponds to child nodes and parent nodes within the xml document. For the high-level "function" tag, the visual component invoked is a tabbed pane. The hierarchy of Functions is seen by the user as tabbed panes within tabbed panes, allowing for quick and intuitive selection from one Function to another by the well-known gesture of clicking on the tab of the Function that is to be used.

The open-source package SWIX[3] is used to generate the java swing objects for the lower-level components. These follow a similar hierarchical pattern: buttons and text fields within panels, etc. SWIX is given as input an xml node specifically describing the swing components and their layout. Each specific component type has a tagname, such as "button", "textfield", "panel", and so forth. Custom components may also be registered and given tagnames. They are placed in the GUI layout in the same way as standard java swing components. Node attributes are used to specify properties such as size, font, initial value, etc. An important additional attribute is one with name "key". This provides a string which will become the key for that component in a hash table which stores the user inputs. A simple example of a button, textfield, and menu is shown below:

```
<function name="SPICE">
  <function name="Input">
    <gui_input actionclass="SettingsController">
      <panel>
        <button id="b1" text="Open" action="directoryChooser"/>
        <textfield id="t1" key="Input Directory" action="tfAction"/>
        <menu>
          <menuitem id="m1" key="Ch1" text="Ch 1" action="orderAction"/>
          <menuitem id="m2" key="Ch2" text="Ch 2" action="orderAction"/>
        </menu>
      </panel>
    </gui_input>
  </function>
</function>
```

Each Function also contains a base FunctionAction object which provides a standard set of actions. The action classes form the intermediate layer between the GUI and the rest of the program. Each swing component specifies by name exactly one action listener object and invokes its "actionPerformed" method when the user interacts with that component. The "actionPerformed" method can be customized for each component, but it is sufficient for most processing to have a generic action class for each type of component. For example, for a component which can be either selected or unselected, such as a checkbox, put or remove the component's key into the hash table according to the selection

[3] http://www.swixml.org

state, or for a component which has a text value, put the component's key into the hash table with the text as its value.

2.2. Conversion of User Inputs to SOAP

When the "Run" button is pressed, the hash table is consulted to obtain the settings that the user specified through the GUI's components. A SOAP message is constructed which carries the hash table information verbatim, using attributes "key" and "value", e.g.,

```
<setting key="Input Directory" value="/scr/data"/>
```

Furthermore, SOAP allows for attachments of any file format to be included with the SOAP message. These are specified as input files by the user in the same way as for any other setting. Output files from the server come as attachments to the server's return SOAP message, and are saved using names specified by the user.

When the SOAP message is received by the server, it can be parsed and used in any way that is convenient for the process which is to be invoked. In the implementation described here, it is desired to run on job on a grid cluster. Therefore on the server end, the message is handled using XSL to convert the message to a shell script, which is submitted to the grid job manager. This allows for increased security, as the user does not need direct access to the grid cluster, but only to the server receiving the SOAP message.

3. Conclusion

Using existing technologies, it is possible to bridge the gap between the requirements of processing systems and those of interactive users. There is a requirement of some familiarization on the part of users and developers in the use of XML. However, this is compensated for by substantially reduced development times for expanding capability and by the stability of the core CAPRI code, which does not have to be reworked for the addition of a new user interface. The use of XML to contain the highly variable parts of the software allows the use of the organizing power of XML schema and custom editors. A side benefit is the ability to construct the user interface at runtime based on a description loaded over the network, thus combining the power of Java with the convenience of web-based browsing.

Acknowledgments. This work was carried out at the *Spitzer* Science Center, with funding from NASA under contract 1407 to the California Institute of Technology and the Jet Propulsion Laboratory.

ASTRONOMICAL DATA ANALYSIS SOFTWARE AND SYSTEMS XIV
ASP Conference Series, Vol. 347, 2005
P. L. Shopbell, M. C. Britton, and R. Ebert, eds.

EdSkyQuery-G: A Data Federation Grid Architecture for Astronomy

Brian Hills, Aileen Campbell, Alan Gray, Rob M. Baxter,[1] and Arthur S. Trew[1]
eDIKT, University of Edinburgh

Bob G. Mann and Martin C. Hill
Royal Observatory of Edinburgh

Tom P. Sugden
EPCC, University of Edinburgh

Abstract. EdSkyQuery-G extends ideas developed by SkyQuery and the Global Grid Forum's Data Access and Integration Services working group to build an open-standards data federation tool for astronomy. EdSkyQuery-G's aim is to use Eldas (from eDIKT at the UK's National e-Science Centre) and OGSA-DAI (from EPCC at the University of Edinburgh and IBM) interchangeably, building a true loosely-coupled distributed data federation architecture based on emerging web and Grid service standards.

1. Introduction

Data access and integration within a Grid environment has been the subject of much attention in recent years. The Global Grid Forum[2] is currently seeking to define standards in this area through the Data Access and Integration Services (DAIS)[3] working group with input from both the academic community and key industry players. The current focus of middleware authors is the provision of a federated view of data in order to access *and* integrate data that may exist in large, distributed and heterogeneous data resources. This will encourage a wider adoption of the Grid by both scientists and the business community.

In astronomy terms, the identification of observations of particular celestial objects in multiple, large (often multi-terabyte) heterogeneous databases distributed around the world lies at the heart of the Virtual Observatory concept. The EdSkyQuery-G project aims to provide a scalable architecture for doing this through extending the ideas prototyped in the SkyQuery .NET web service developed at Johns Hopkins University (see Malik et al. 2003). EdSkyQuery-G is being developed by the eDIKT group at the National e-Science Centre in Ed-

---

[1]EPCC, University of Edinburgh

[2]`http://www.ggf.org/`

[3]`https://forge.gridforum.org/projects/dais-wg`

inburgh in collaboration with the UK's AstroGrid project (see Lawrence 2003) and the OGSA-DAI development team. It is hoped that the development of EdSkyQuery-G will help inform the specification of the International Virtual Observatory Alliance's OpenSkyQuery data integration standard (IVOA 2004).

1.1. EDIKT, Eldas and OGSA-DAI

EDIKT is a team of software engineers at the UK's National e-Science Centre in Edinburgh set up to support a wide range of e-Science projects through good software and data engineering. One of eDIKT's goals is to look at the scalability and robustness of emerging data management technologies and their applicability to the serious challenges of e-Science; eDIKT has thus adopted Java 2 Enterprise Edition (J2EE) technologies as foundations for Grid middleware rather than the basic web server Java containers (e.g., Apache tomcat) adopted by most other middleware projects. EDIKT's Eldas[4] package (Campbell et al. 2003) is a Grid-enabled data access software stack designed and built using Enterprise Java Beans (EJBs), with both OGSA-based Grid service[5] and industry-standard web service interfaces.

OGSA-DAI[6], developed by EPCC, IBM, Oracle and the Universities of Newcastle and Manchester, is an open source database access and integration middleware package designed as a reference implementation of the DAIS specifications (see Anjomshoaa et al. 2004). OGSA-DAI was designed to dovetail with the emerging Globus web services versions (Globus Toolkit v3.x and v4.x)[7] and thus adopted the tomcat web server container architecture.

Both Eldas and OGSA-DAI provide Grid service interfaces to a variety of databases according to the DAIS Web Services Data Access and Integration specifications (WS-DAI and related specs — Antonioletti et al. 2004). One of the design goals of EdSkyQuery-G is to explore the use of both data access packages within a single distributed network of databases and to compare and contrast their performance in support of a major astronomy application.

2. The EdSkyQuery-G Federation Architecture

As with the original SkyQuery architecture (Malik et al. 2003) we employ a wrapper-mediator model which includes three core components:

- the Query Builder, a graphical client which enables astronomers to enter complex queries (analogous to SkyQuery's *Clients*);
- the Service Manager, a federation service that receives queries from the query builder, processes queries and interacts with data access services (analogous to SkyQuery's *Portal*);

---

[4]`http://www.edikt.org/eldas/`

[5]`https://forge.gridforum.org/projects/ogsa-wg`

[6]`http://www.ogsa-dai.org.uk/`

[7]`http://www.globus.org/`

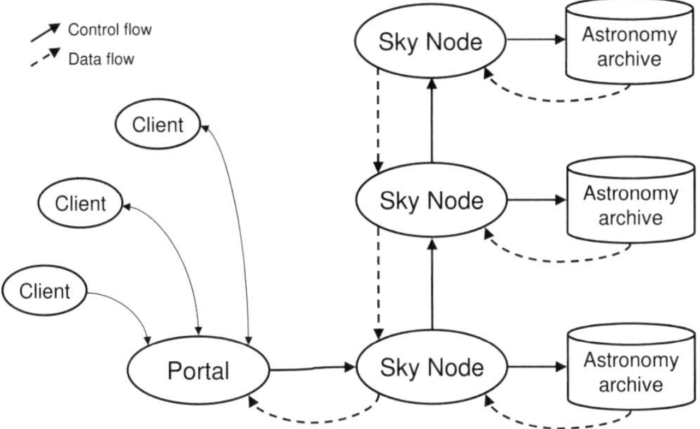

Figure 1. The original SkyQuery architecture (from Malik et al. 2003).

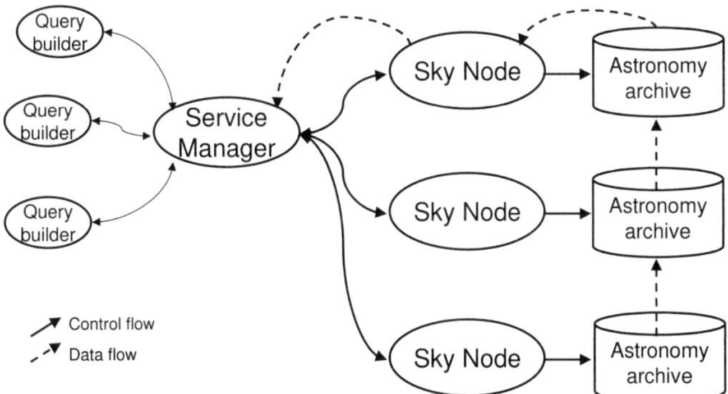

Figure 2. The EdSkyQuery architecture. Contrast the control and data flow patterns between the Portal/Service Manager and the Sky Nodes in this diagram with those of Figure 1.

- the Data Access Services or Sky Nodes, used to perform queries and cross-database joins. The underlying data access and integration services in EdSkyQuery-G are provided by Eldas and OGSA-DAI.

Riccardi (2004) describes two architectures which may be used to facilitate data access for SkyQuery using DAIS services. The original SkyQuery employs a recursive approach whereby a node executes its part of the query plan and passes the results and the plan onto the next node, hence the control and data paths are identical (Figure 1). In contrast we have opted to use an iterative model for control flow as shown in Figure 2. In this model, the Service Manager governs the control path, interacting with each node, whilst data is transferred directly between the servers hosting the databases. The Service Manager may either return the full set of results to the Query Builder or, in the case of a large set of results, a handle to the results file. The handle will enable the

Query Builder to retrieve the results when required using an appropriate data transport mechanism. We believe the benefits of this iterative approach include:

- scalability: there is no need to return sets of results directly back to the Service Manager and Query Builder. This both minimizes the volume of data being transferred between components and avoids potential Java Virtual Machine (JVM) memory problems;
- robustness: the Service Manager can intelligently handle error conditions; e.g., if a particular database is unavailable the Service Manager may report a sensible error to the user or even try and use a replica database. Handling error conditions using the recursive model is more complex;
- usability: astronomers wish to perform complex queries which may take many minutes, hours or even days to perform. It is therefore essential that users receive regular status updates on the progress of a particular task. As the Service Manager controls the processing it can send updates back to the Query Builder, for example "Time t: now joining partial results with database *Ed-SSA*".

The other key distinguisher between EdSkyQuery-G and SkyQuery is our use of Java and open Grid standards rather than proprietary Microsoft products. Through Eldas and OGSA-DAI EdSkyQuery-G can work automatically with a wide range of database platforms and operating systems. By this route we hope to be able to support the OpenSkyQuery initiative (IVOA 2004) in defining open standards in this area.

3. Current Status

EdSkyQuery-G currently exists as an end-to-end working prototype based on two instances of IBM's DB2 database running on Microsoft Windows XP and 2003 Server. At the moment both are managed by Eldas with the Service Manager deployed separately in a JBoss EJB container. One of the databases contains a 10GB trial sample from the SDSS-EDR archive[8] while the second is a similar-sized sample from the SuperCOSMOS Science Archive[9] at the Royal Observatory of Edinburgh.

We plan to introduce an OGSA-DAI managed data source at a later date, although variations in implementation permitted by the WS-DAI specifications have raised interesting interoperability issues which will need to be addressed. For application builders such as ourselves the difficulties are compounded by the fact that neither data access middleware platform has been able to keep pace with the changes in WS-DAI. Thus one of the main results of EdSkyQuery-G may well be to feed back experience to the middleware and specification authors.

The focus of our design prototyping so far has been on the performance of data transfer between SkyNodes. In EdSkyQuery-G the potential need to move large volumes of data between databases during query execution is challenging in terms of overall performance. We have investigated and prototyped a number of different intermediate formats and transport protocols, settling on using the

---

[8] `http://www.sdss.org/`

[9] `http://surveys.roe.ac.uk/ssa/index.html`

Sun XML WebRowSet (Sun 2004) metadata description with a comma separated value (CSV) file for the intermediate formats, and direct database-to-database exports and imports using stored procedures for transport. Use of the Globus GridFTP engine for data transport was ruled out because of the current absence of a GridFTP server for the Windows operating system.

4. Conclusions and Further Work

So far in EdSkyQuery-G we have shown the viability of federating potentially arbitrary astronomy databases using a true Grid service-oriented architecture. We anticipate the use of both Eldas and OGSA-DAI will demonstrate their utility as middleware platforms for large scale astronomy projects, although interoperability challenges remain even between two packages implementing the same DAIS specifications. In due course we hope to extend our prototype to include enhanced query capability and more generic components, making the system of potential interest for e-Science more widely.

Acknowledgments. Our thanks to Charaka Palansuriya at EPCC and Patrick Dantressangle of IBM for help and assistance in configuring DB2, and to the Systems teams at NeSC and EPCC for their continued support.

References

Anjomshoaa, A., et al. 2004, *The Design and Implementation of Grid Database Services in OGSA-DAI*, in Proc. UK e-Science All Hands Conference, Nottingham, UK, September 2004

Antonioletti, A., et al. eds. 2004, *Web Services Data Access and Integration (WS-DAI)* and related GGF DAIS working drafts, September 2004[10]

Campbell, A., et al. 2003, *Eldas (Enterprise Level Data Access Services)*, in Proc. UK e-Science All Hands Conference, Nottingham, UK, September 2003

IVOA VOQL Working Group 2004, IVOA SkyNode Interface, Version 0.7.4[11]

Lawrence, A. 2003, *AstroGrid: The UK's Virtual Observatory*, in Proc. UK e-Science All Hands Conference, Nottingham, UK, September 2003

Malik, T., et al. 2003, *SkyQuery: A Web Service Approach to Federate Databases*, in CIDR 2003, Asilomar, CA, USA, January 5th–8th, 2003

Riccardi, G. 2004, *DAIS Data Service Interactions and the SkyQuery Portal*[12]

Sun Microsystems Inc. 2004, Sun WebRowset XML Schema Definition[13]

---

[10] `https://forge.gridforum.org/projects/dais-wg`

[11] `http://www.ivoa.net/internal/IVOA/IvoaVOQL/SkyNodeInterface-0.7.4.pdf`

[12] `http://www.nesc.ac.uk/~greg/skyquery.pdf`

[13] `http://java.sun.com/xml/ns/jdbc/webrowset.xsd`

ASTRONOMICAL DATA ANALYSIS SOFTWARE AND SYSTEMS XIV
ASP Conference Series, Vol. 347, 2005
P. L. Shopbell, M. C. Britton, and R. Ebert, eds.

Grist: Grid-based Data Mining for Astronomy

Joseph C. Jacob, Daniel S. Katz, Craig D. Miller, and Harshpreet Walia
Jet Propulsion Laboratory, California Institute of Technology, Pasadena, CA 91109-8099

Roy D. Williams, S. George Djorgovski, Matthew J. Graham, and Ashish A. Mahabal
California Institute of Technology, Pasadena, CA 91125

G. Jogesh Babu and Daniel E. Vanden Berk
The Pennsylvania State University, University Park, PA, 16802

Robert Nichol
ICG, University of Portsmouth, PO1 2EG, UK

Abstract. The Grist project is developing a grid-technology based system as a research environment for astronomy with massive and complex datasets. This knowledge extraction system will consist of a library of distributed grid services controlled by a workflow system, compliant with standards emerging from the grid computing, web services, and virtual observatory communities. This new technology is being used to find high redshift quasars, study peculiar variable objects, search for transients in real time, and fit SDSS QSO spectra to measure black hole masses. Grist services are also a component of the "hyperatlas" project to serve high-resolution multi-wavelength imagery over the Internet. In support of these science and outreach objectives, the Grist framework will provide the enabling fabric to tie together distributed grid services in the areas of data access, federation, mining, subsetting, source extraction, image mosaicking, statistics, and visualization.

1. Overview

The Grist[1] project (`http://grist.caltech.edu/`) is enabling astronomers and the public to interact with the grid projects that are being constructed worldwide, and bring to flower the promise of easy, powerful, distributed computing. Our objectives are to understand the role of service-oriented architectures in astronomical research, to bring the astronomical community to the grid – particularly TeraGrid, – and to work with the National Virtual Observatory (NVO) to build a library of compute-based web services.

[1]Part of this research was carried out at the Jet Propulsion Laboratory, California Institute of Technology, and was sponsored by the National Science Foundation through an agreement with the National Aeronautics and Space Administration.

The scientific motivation for Grist derives from creation and mining of wide-area federated images, catalogs, and spectra. An astronomical image collection may include multiple pixel layers covering the same region on the sky, with each layer representing a different waveband, time, instrument, observing condition, etc. The data analysis should combine these multiple observations into a unified understanding of the physical processes in the Universe. The familiar way to do this is to cross-match source lists extracted from different images. However, there is growing interest in another method of federating images that reprojects each image to a common set of pixel planes, then stacks images and detects sources therein. While this has been done for years for small pointing fields, we are using the TeraGrid to perform this processing over wide areas of the sky in a systematic way, using Palomar-Quest[2] (PQ) survey data. We expect this "hyperatlas" approach will enable us to identify much fainter sources than can be detected in any individual image; to detect unusual objects such as transients; and to deeply compare (e.g., using principal component analysis) the large surveys such as SDSS, 2MASS, DPOSS, etc. (Williams et al. 2004).

Grist is helping to build an image-federation pipeline for the Palomar-Quest synoptic sky survey (Djorgovski et al. 2004), with the objectives of mining PQ data to find high redshift quasars, to study peculiar variable objects, and to search for transients in real-time (Mahabal et al. 2004). Our PQ processing pipeline will use the TeraGrid for processing and will comply with widely-accepted data formats and protocols supported by the VO community.

2. Service-Oriented Architectures for Astronomy

The Grist project is building web and grid services as well as the enabling workflow fabric to tie together these distributed services in the areas of data federation, mining, source extraction, image mosaicking, coordinate transformations, data subsetting, statistics – histograms, kernel density estimation, and R language utilities exposed by VOStatistics[3] services (Graham et al. 2004), – and visualization. Composing multiple services into a distributed workflow architecture, as illustrated in Figure 1, with domain experts in different areas deploying and exposing their own services, has a number of distinct advantages, including:

- Proprietary algorithms can be made available to end users without the need to distribute the underlying software.
- Software updates done on the server are immediately available to all users.
- A particular service can be used in different ways as a component of multiple workflows.
- A service may be deployed close to the data source, for efficiency.

Interactive deployment and control of these distributed services will be provided from a workflow manager. We expect to use NVO services for data access – images, catalogs, and spectra – as well as the NVO registry for service discovery.

---

[2] `http://www.astro.caltech.edu/pq/`

[3] `http://vostat.org/`

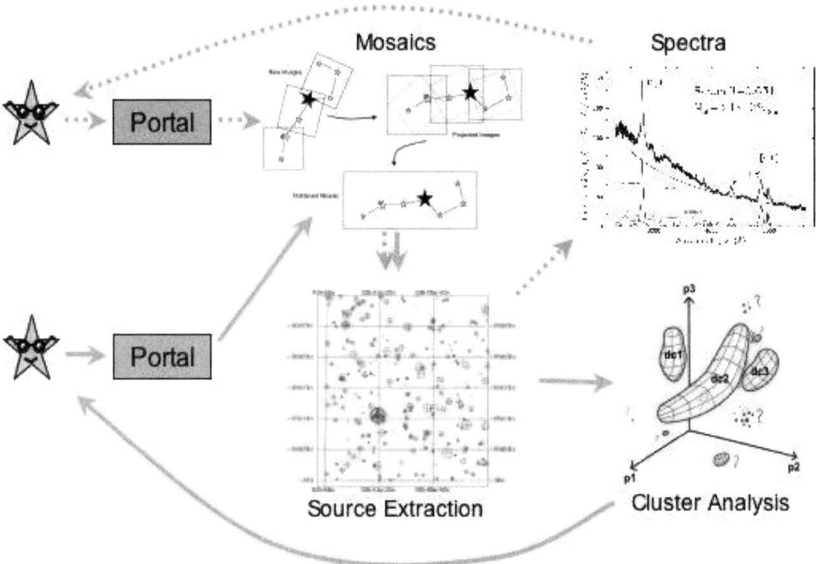

Figure 1. Grist will deploy a library of interoperable services, which may be composed in different ways for astronomical data mining (e.g., two distinct workflows are indicated by the solid and dashed arrows).

3. Graduated Security

As described in Section 2., much of the pipeline and mining software for Grist will be built in the form of web services. One of the reasons for building services is to be able to use them from a thin client, such as a web browser. However, for such services to be able to process private data or use high-end computing, there must be strong authentication of the user. The VO and Grid communities are converging around the idea of X.509 certificates as a suitable credential for such authentication. However, most astronomers do not have such a certificate, and we don't want to make them go through the trouble of getting one unless it is truly necessary. Therefore, we are building services with "graduated security", meaning not only that small requests on public data are available anonymously and simply, but also that large requests on private data can be serviced through the same interface. However in the latter case, a certificate is necessary. Thus the service "proves its usefulness" with a simple learning curve, but requires a credential to be used at full-strength (see illustration in Figure 2).

4. Palomar-Quest Data Mining

A key science-driven workflow we are constructing is illustrated in the schematic in Figure 3. The primary objectives are to search for high redshift quasars and optical transients in data from the Palomar-Quest sky survey. The pipeline

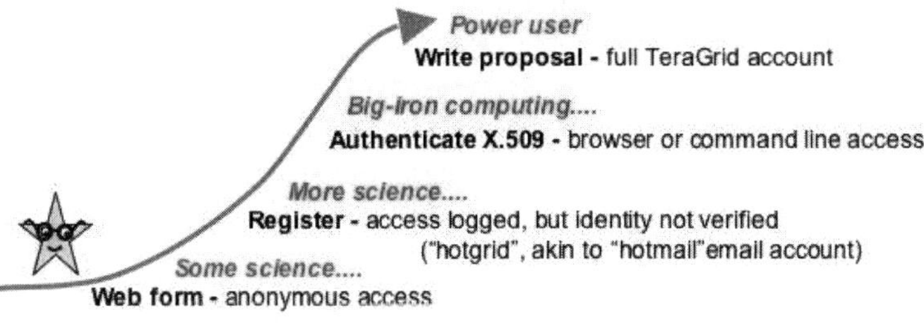

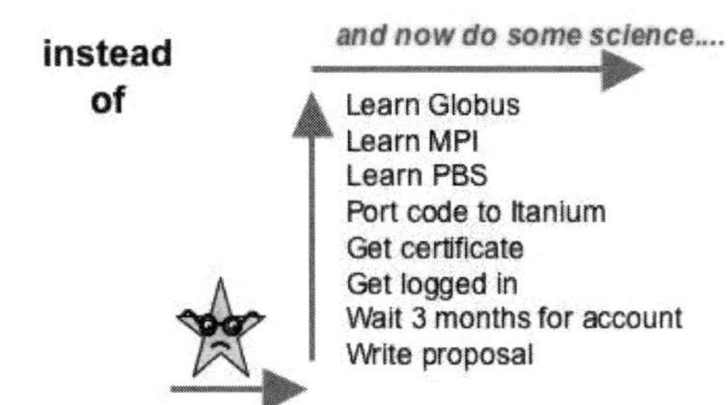

Figure 2. "Graduated security" will shorten the hurdles that stand in the way of scientists who would like to take advantage of the power of computational grids for their research.

begins by federating multiwavelength datasets, and matching objects detected with the z filter with catalogs at other frequencies. Cluster analysis performed on the resulting color-color plots (e.g., i-z vs. z-J) yield new quasar candidates, and outliers may indicate the presence of other objects of interest.

Single epoch transients are indicated by objects that are detected in one filter but not others. An object that is detected in the reddest filter is of special interest since it could be a highly obscured object or a high redshift quasar. For multi-epoch transient search, illustrated in the lower part of Figure 3, we compare new data with a database of past epochs to detect new transients or other variable objects.

As described above, a primary objective of the PQ survey is the fast discovery of new types of transient sources by comparing data taken at different times. Such transients should be immediately re-observed to get maximum scientific impact, so we are experimenting with "dawn processing" on the TeraGrid, meaning that data is streamed from the telescope to the compute facility as it is taken (rather than days later). The pipeline itself is being built with streaming protocols so that unknown transients (e.g., newly identified variables or asteroids) can be examined within hours of observation with a view to broadcasting an email alert to interested parties.

5. Summary

Grist is developing a library of interoperable grid services for astronomical data mining on the TeraGrid, compliant with Grid and VO data formats, standards,

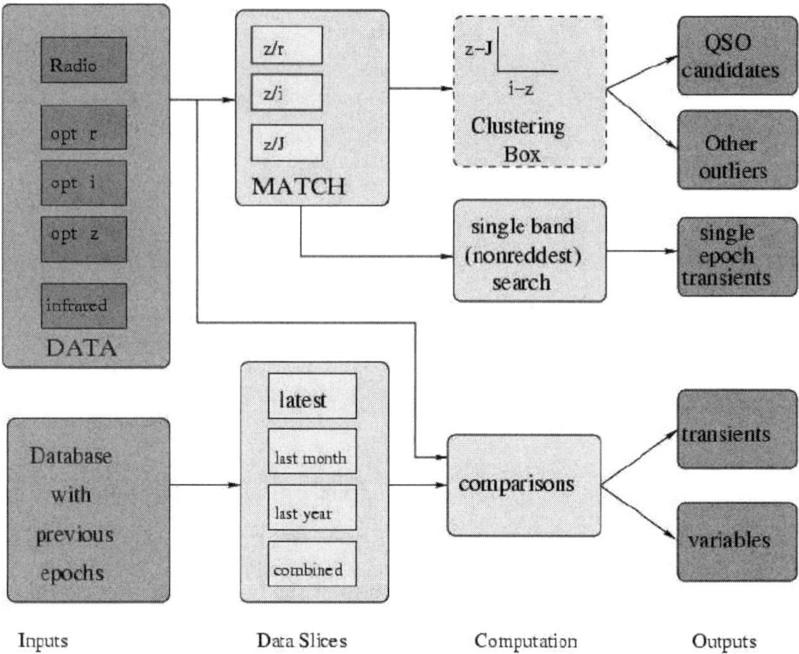

Figure 3. A schematic pipeline to look for quasars, transients and other variables. Combining multi-filter information with multi-epoch datasets through a set of well established techniques will yield a rich set of astronomically interesting objects.

and protocols. For ease of use, Grist services are built with graduated security, requiring no more formal authentication than is appropriate for a given level of usage. Grist technology is part of a Palomar-Quest data processing pipeline, under construction, to search for high red-shift quasars and optical transients. More information on Grist can be found on our project web site at `http://grist.caltech.edu/`.

References

Djorgovski, S. G., et al. 2004, BAAS, 36, 805

Graham, M. J., et al. 2005, in ASP Conf. Ser., Vol. 347, ADASS XIV, ed. P. L. Shopbell, M. C. Britton, & R. Ebert (San Francisco: ASP), 394

Mahabal, A., et al. 2005, in ASP Conf. Ser., Vol. 347, ADASS XIV, ed. P. L. Shopbell, M. C. Britton, & R. Ebert (San Francisco: ASP), 604

Williams, R. D., et al. 2004, in ASP Conf. Ser., Vol. 314, ADASS XIII, ed. F. Ochsenbein, M. Allen, & D. Egret (San Francisco: ASP), 364

ASTRONOMICAL DATA ANALYSIS SOFTWARE AND SYSTEMS XIV
ASP Conference Series, Vol. 347, 2005
P. L. Shopbell, M. C. Britton, and R. Ebert, eds.

The VST Data Reduction: An Application for the GRID Infrastructure

M. Pavlov, J.M. Alcalá, A. Grado, E. Cascone, G. Capasso, and V. Manna

INAF-Osservatorio Astronomico di Capodimonte

F. Pasian

INAF-Osservatorio Astronomico di Trieste

L. Benacchio

INAF-Osservatorio Astronomico di Padova

Abstract. It is foreseen that the VLT Survey Telescope (VST) will start operations by the end of 2005. VST will provide large data volumes that will require parallel processing for their processing. In this contribution we present strategies for the possible gridization of the data reduction and calibration pipeline of the VST data. Preliminary results on the performance of the data processing within GRID are also presented.

1. Introduction

The VLT Survey Telescope (VST) is a cooperation between ESO, INAF- OAC-Capodimonte and the European OmegaCam consortium for the realization, construction and operation of a wide-field imaging facility. The telescope will have an aperture of 2.6m and will be equipped with the OmegaCam camera. This is a mosaic camera with 32 CCDs (2k$\times$4k chips, 15μm pixels). The corrected field of view of the system will be 1 square degree with a plate scale of 0.21 arcsec/pix. The Sloan photometric system (u', g', r', i', z) will be adopted, but the Johnson B and V filters will be also available. It is foreseen that VST will start operations by the end 2005 at Paranal, in Chile. The VST+OmegaCam system implies an enormous data flow that calls for adequate planning, archiving, scientific analysis and support facilities: the expected VST data flow will be about 150 Gbyte of raw data per night.

The Italian project Enabling platforms for high performance computational GRIDS oriented to scalable virtual organisations started in 2002. The work package 10 (WP10) of this project, leaded by Prof. L. Benacchio, is devoted to Astrophysical applications (see also Vuerli et al. 2005). In the framework of this WP, the INAF- OA-Capodimonte is studying the possibility to use the GRID infrastructure for the VST data reduction. Given the embarrassingly parallel nature of VST data processing, the porting of "Distributed Storage" subsystem and "parallel executions" subsystem to the GRID is suitable and feasible.

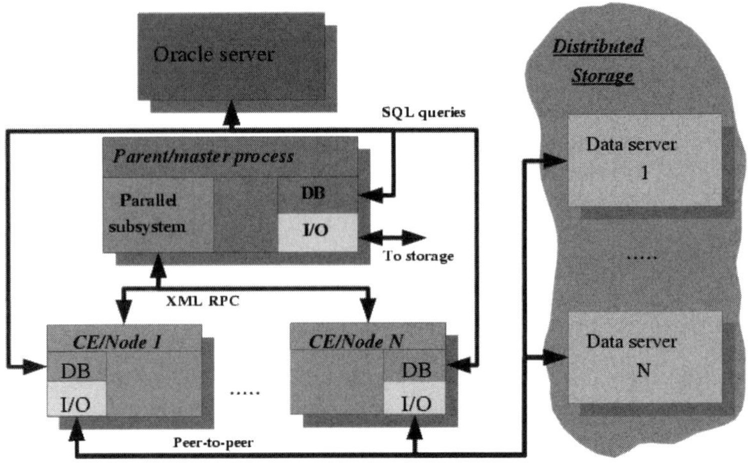

Figure 1. Astro-Wise processing model

2. Strategies for VST Data Processing

Parallel processing and distributed storage are very important components of the future data processing system for the VST data reduction . The strategies for the processing of the VST images will be based on the guidelines of the Astronomical Wide-Field Imaging System for Europe (ASTRO-WISE: `http://www.astro-wise.org`), an European consortium aimed at providing the European community with an astronomical survey system, facilitating astronomical research, data reduction, and data mining based on the new generation of wide-field sky survey cameras. The data reduction will be performed on Beowulf-type architecture (c.f. Cascone et al. 2002; `http://www.na.astro.it/beowulf`). For details on the VST data model, we refer the interested reader to the paper by Grado et al. (2004).

3. An Application for the GRID Infrastructure

The goal of this task is to study the use of GRID technology for the processing of Wide-Field astronomical Images that will be obtained with VST. The Astro-Wise pipeline design includes parallel processing, distributed storage and federated database concepts. In order to reach our goals we have to deal with two important issues, which refer to Figure 1:
i) Distributed storage: currently, the Astro-Wise pipeline uses an original "peer-to-peer" mechanism to have access to the distributed data storage. We have studied the possibility to integrate, in the Astro-WISE pipeline, the standard elements (storage element and replica manager) required by the GRID infrastructure. It is indeed possible to use such elements after minor modifications of the pipeline. This is of crucial importance for users of both the GRID and the Astro-Wise pipeline, because the GRID infrastructure shall allow standard

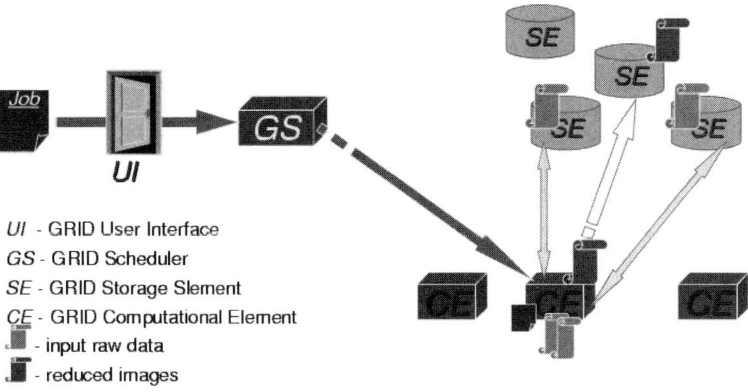

Figure 2. Typical GRID image reduction: data flow.

access to data storage, not only within the pipeline, but also for other astronomical GRID applications (e.g. data mining, other type of image processing, etc.);
ii) Parallel or distributed computation: the current version of the Astro-Wise pipeline uses an original XML based scheduler to distribute the jobs in the different computation nodes. An intensive study of the feasibility to incorporate in the Astro-Wise pipeline the standard elements of GRID infrastructure (resource broker and scheduler) has been performed. Contrary to the previous point, it was found that a considerable effort is necessary in order to achieve the successful integration of the pipeline into the GRID. This is due to the fact that the Astro-Wise pipeline uses an object oriented approach, which keeps track of the different classes of objects during the image processing. In the case of remote executions, the pipeline maintains the relationship between the different objects using a special mechanism, which is implemented in the Astro-Wise scheduler. Therefore, one important task that must be afforded in the near future is the implementation of a similar mechanism when the GRID scheduler will be used.

4. Preliminary Results

We have perform a series of tests in order to verify the performance of the data processing within the GRID context. As it was mentioned above, and as can be seen from Figure 1, there are two main components of the VST data reduction system: distributed storage and parallel processing. To perform our exercises, we developed a simplified pipeline prototype (Figure 2). This prototype uses the GRID Storage Elements (SE) to store the input raw images, as well as the pre-reduced calibration output frames. The standard GRID job scheduler is used in order to process the data in parallel (32 CCD images are processed independently and in parallel). At the same time, this prototype allows to measure some critical overheads of the GRID execution.

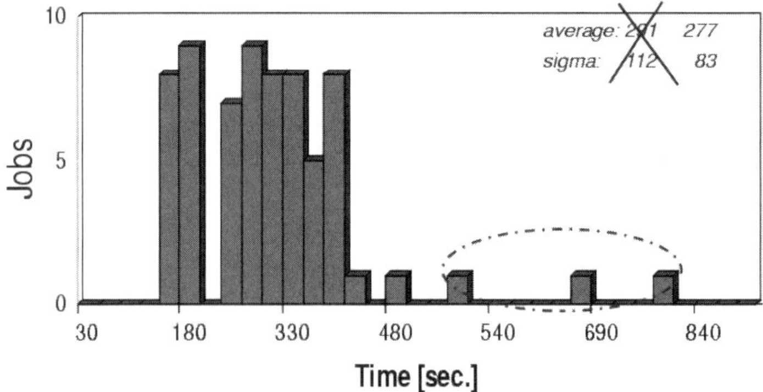

Figure 3. GRID scheduler overhead ("zero execution" jobs).

The results, reported below, were obtained by running our test application on INFN-GRID (http://grid-it.cnaf.infn.it/).

4.1. Parallel processing

As it is shown in Figure 3, GRID job executions have an additional overhead. Such overhead was measured by using a "zero executional time" task. The results are illustrated in Figure 3. An average overhead of about 5 minutes exists. This is significant for small/fast computational tasks or in applications that require a fast response (interactive application). However, for "batch" applications such delay is not an important issue.

The second critical parameter is the "scalability for parallel execution". To measure this parameter, our prototype calibration pipeline was used (Master Bias/Flat creation). As one can see from Figure 4 left panel, the execution time for 32 jobs launched in parallel is 23 times faster (~700s) than when performing the same number of jobs sequentially (32*500s).

4.2. Distributed Storage

An additional overhead, that might be very significant for applications processing large data volumes, is the one due to Input/Output (I/O) operations through Wide Area Network (WAN). The combination of a relatively fast network connection between GRID Computational Element(CE) and GRID Storage Element (SE) with the optimisation of GRID scheduler CE-SE "distances", can significantly minimise such kind of overheads. We find that for 5 raw-bias frames, the GRID execution adds about 70s (see Figure 4 right panel) to the total execution time. In comparison with the job scheduler overhead this is not significant. Note, however, that in this particular case the job scheduler CE-SE optimisation was not used.

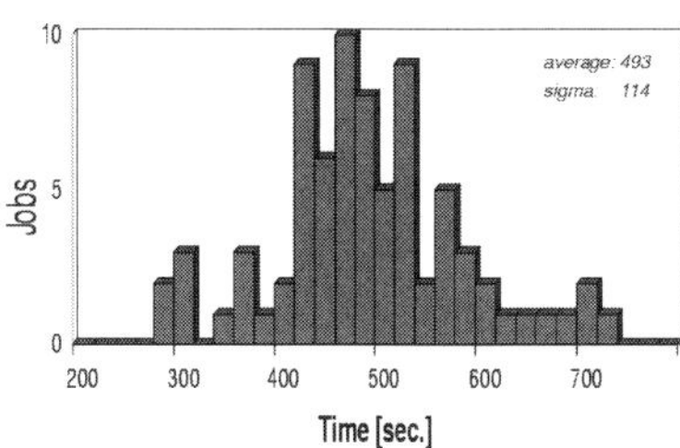

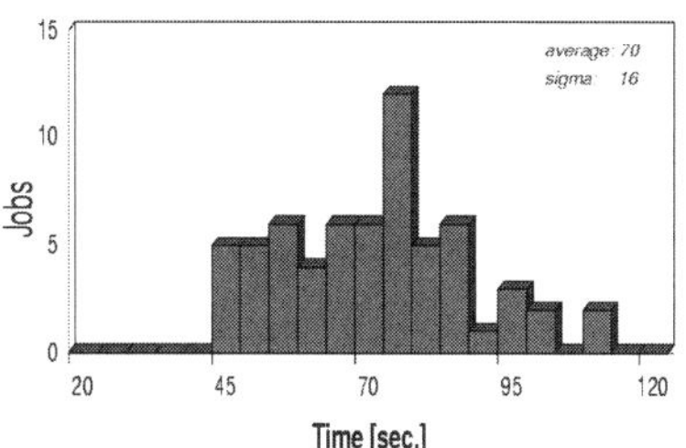

Figure 4. Job execution times (left) and I/O overhead (right).

5. Conclusions

A study of the different possibilities for the integration of the GRID job scheduler in the Astro-Wise pipeline was done. Our tests demonstrate a good scalability of GRID parallel execution. On the other hand, we also find, that regardless of the application execution time, there is a significant overhead of about 5 minutes for whatever job. Other tests regarding the SE were performed. It is found that the usage of GRID SE, as distributed storage, is reliable, flexible and robust. Since the VST images will require a huge storage, the use of distributed GRID storage is a very suitable solution. We also find that the I/O overhead, typically about 1 min., is not critical for the data reduction application.

Acknowledgments. M.Pavlov acknowledges a grant from the FIRB Grid.it (WP10) project.

References

Cascone, E., Grado, A., Pavlov, M., & Capasso, G. 2002, SPIE 4848, 328

Grado, A., Capaccioli, M., Silvotti, R., Pavlov, M., et al. 2004, AN 325, 601

Vuerli, C., Pasian F., Baruffolo, A., Benacchio, L., et al. 2005, in ASP Conf. Ser., Vol. 347, ADASS XIV, ed. P. L. Shopbell, M. C. Britton, & R. Ebert (San Francisco: ASP), 334

ASTRONOMICAL DATA ANALYSIS SOFTWARE AND SYSTEMS XIV
ASP Conference Series, Vol. 347, 2005
P. L. Shopbell, M. C. Britton, and R. Ebert, eds.

Access Control in AstroGrid Software

Guy Rixon

Institute of Astronomy, Madingley Road, Cambridge CB3 0HA, UK

Kevin Benson

Mullard Space Science Laboratory, Holmbury St. Mary, Dorking RH5 6NT, UK

David Morris

Institute of Astronomy, Madingley Road, Cambridge CB3 0HA, UK

Abstract. Access control is important for many areas of the Virtual Observatory and particularly for the AstroGrid system which exploits user-writeable, on-line storage. We describe AstroGrid's current arrangements for access control and outline the likely evolution of AstroGrid's system as IVOA develops standards for security arrangements.

1. Introduction

Some resources in the virtual observatory (VO) have restricted access:

- some archived data not publicly available;
- some services have usage quotas or are available only to researchers from certain organizations;
- some data-sets are created and owned by individual users; these must be kept on-line for the use of their owners but must not be made public.

AstroGrid's[1] implementation of the VO depends on writeable, on-line storage via the "MySpace" (Davenhall et al. 2004) facility. Files in MySpace are created and owned by end-users of AstroGrid. Hence, AstroGrid has an urgent need for access control and has taken the lead in defining a security architecture.

2. Single Sign-on via Community Services

AstroGrid provides single-sign-on (SSO) features.

- Sign on to the whole of AstroGrid once per session with a single password.
- No need to remember a password for each service.
- Users register with an on-line community.
- No need to for end-users to register with each target service.
- No need for each data-centre to manage registrations and membership.

[1] `http://www.astrogrid.org/`

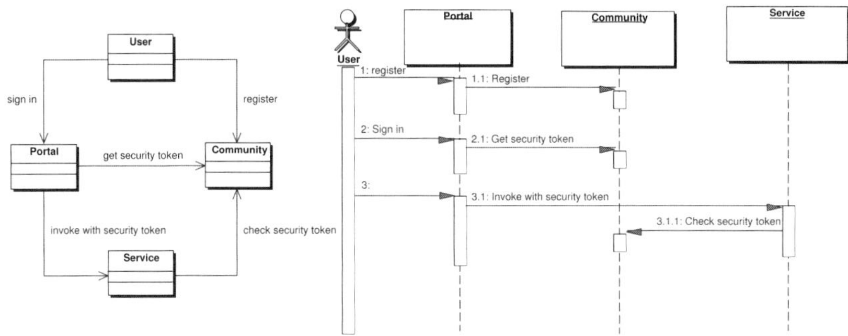

Figure 1. Class and sequence diagram showing the SSO process.

AstroGrid implements SSO via a community package that provides web services. Users interact with the community via one of several web portals or via interactive, desktop applications (see Figure 1). Signing in at the community loads into the portal or application the security tokens needed to authenticate to AstroGrid services. Because the community services are separate from the portals, users are not tied to a single user interface; community membership can be used from any AstroGrid-compliant programme.

3. Programme-to-programme Authentication

AstroGrid provides Java class-libraries allowing a SOAP client to authenticate a user's identity to a SOAP service. Message-level security is used: this avoids the need for transport-level security; services can use unencrypted HTTP. JAX-RPC/Axis message handlers add security credentials to outgoing messages and check credentials on incoming messages. This handler mechanism is hidden behind the SecurityGuard facade (Figure 2) to simplify application code.

Credentials are passed in the headers of the SOAP message according to the OASIS WS-Security standard[2]. . They combine the account name with a single-use password issued by the community service. Services check the password by contacting the community service; this avoids the need for services to maintain lists of passwords and any need for identity certificates or external certificate authorities.

4. Authorization

Authorization (i.e. deciding what a named user can do at a service) in Astro-Grid's 2004 system is very basic. There is no library support; service authors must provide their own authorization code. The MySpace service understands

[2]`http://docs.oasis-open.org/wss/2004/01/oasis-200401-wss-soap-message-security-1.0.pdf`

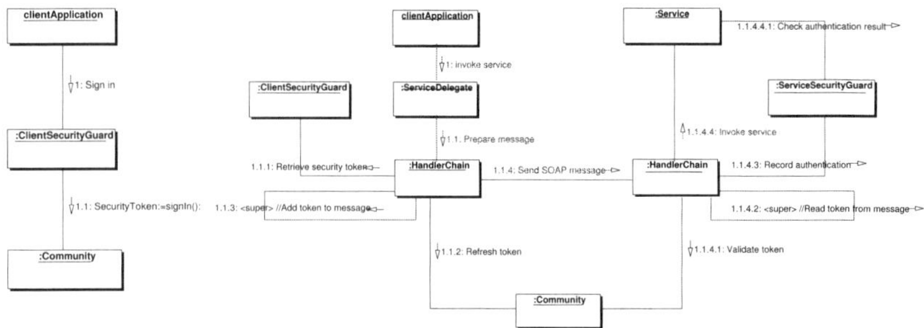

Figure 2. Use of the "Security Guard" facade. Left diagram: signing in; right diagram: authenticating to a service.

ownership of files: the owning account has full access to files while other accounts have no access.

AstroGrid's 2005 system will have much better support including these features.

- File-sharing support in MySpace: users can allow controlled access to their files by other users.
- Library support for local management of authorization databases.
- Authorization in the community service: service providers assign access to the community as a whole, not to individual users.
- Group support: service providers can define a group of users within a community to receive access rights.
- Delegation support: an application can authorize a service to delegate to a subsidiary service in the user's name.

A service provider has the choice of managing authorization policy locally or of delegating that function to the community service (Figure 3). With local authorization, the service provider must do more administration and must supply more customized software but gains flexibility in applying authorization. With authorization delegated to the community, the service provider gives up flexibility to reduce the costs of settings up and of running the service

5. IVOA and Interoperability

The current SSO system is unique to AstroGrid. It must evolve into a standard that works through the international VO. IVOA[3] is developing this standard.

- SOAP Message-level security is still used.
- The WS-I profile for security[4] is followed; this is based on OASIS's WS-Security.
- The community concept is retained.

[3] http://www.ivoa.net/

[4] http://www.ws-i.org/deliverables/workinggroup.aspx?wg=basicsecurity

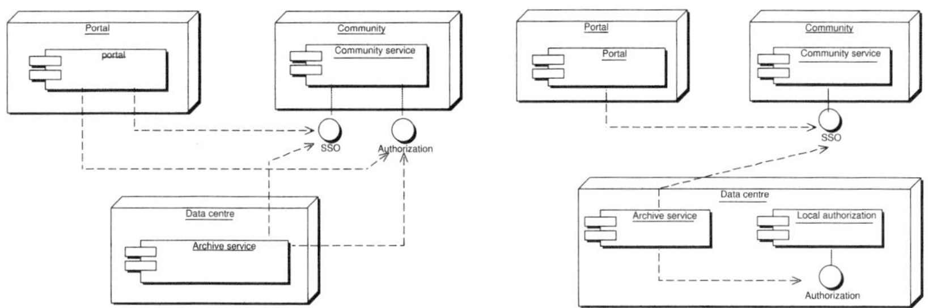

Figure 3. Alternatives for authorization. Left diagram: authorization is handled in the community, data centre does not maintain authorization records. Right diagram: data centre handles its own authorization.

- Cryptographic tokens (public-private key pairs and X.509 certificates) replace the single-user passwords.
- Communities become certificate authorities and issue identity certificates.
- The new credentials are to be interoperable with grid services and GridFTP.
- Gateway services provide access to non-SOAP services and alternate SSO arrangements such as Shibboleth[5].

Acknowledgments. Many in AstroGrid have advised on design of the access-control software for which we are grateful. We thank our colleagues in IVOA for their input to the security debate. AstroGrid is funded by the Particle Physics and Astronomy Research Council of the UK.

References

Davenhall, A. C., Qin, C.L., Noddle, K. T., & Walton, N.A. 2004, in ASP Conf. Ser., Vol. 314, ADASS XIII, ed. F. Ochsenbein, M. Allen, & D. Egret (San Francisco: ASP), 330

[5]http://shibboleth.internet2.edu/

ASTRONOMICAL DATA ANALYSIS SOFTWARE AND SYSTEMS XIV
ASP Conference Series, Vol. 347, 2005
P. L. Shopbell, M. C. Britton, and R. Ebert, eds.

Prototypes of a Computational Grid for the Planck Satellite

Giuliano Taffoni, Giuliano Castelli, Riccardo Smareglia, Claudio Vuerli, Andrea Zacchei, and Fabio Pasian

National Institute for Astrophysics, OATs, Trieste, Italy

Davide Maino

University of Milano, Milan, Italy

Giancarlo Degasperis

University of Rome "Tor Vergata", Rome, Italy

Salim G. Ansari and Jan Tauber

European Space Agency, ESRIN, Holland

Thomas Ensslin

Max Planck Institute for Astrophysics, Garching, Germany

Roberto Barbera

National Institute for Nuclear Physics, Catania, Italy

Abstract. A prototype of a Computation Grid is designed to assessing the possibility of developing pipeline setup for processing Planck Satellite data. The amount of data collected by the satellite during its sky surveys requires an extremely high computational power both for reduction and analysis. For this reason a Grid environment represents an interesting layout to be considered when processing those data.

1. Introduction

The ESA Planck satellite mission will fly in 2007. This experiment is aimed to map the microwave sky performing at least two complete sky surveys with an unprecedented combination of sky and frequency coverage, accuracy, stability and sensitivity (Tauber 2000). Planck is composed of a number of microwave and sub-millimeter detectors which are grouped into a high frequency instrument (HFI) and a low frequency instrument (LFI) (Pasian 2002) and covers a frequency range from 30 up to 850 GHz.

All levels data processing is assign to the two Data Processing Centers (DPCs), one for the LFI centralized at OAT in Trieste, Italy and one for the HFI distributed between Paris, France, and Cambridge, UK. Both DPCs share a site producing an Early Release Compact Source Catalog (IPAC, Pasadena, USA)

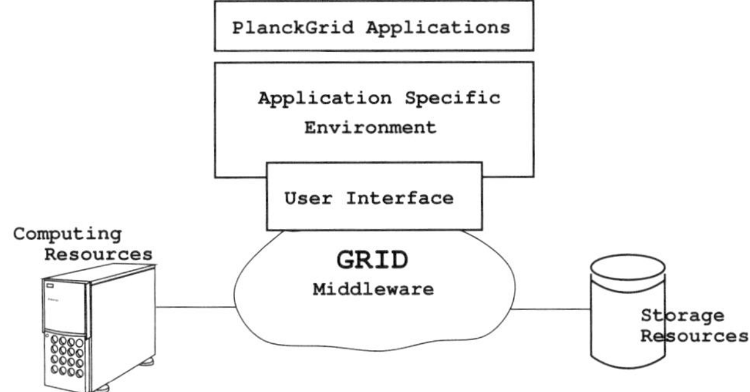

Figure 1. Structure of the Planck@Grid application deployment.

and a site gathering and documenting the final results of the mission, located at MPA, in Garching, Germany. The amount of data produced by the whole mission and by the necessary post-processing is a challenging task both in terms of storage and computational needs (Bond et al. 1999). For example only the LFI DPC is in charge to process 100 GB of data. PlanckGrid is a project whose main goal is to verify the possibility of using a Grid Technology to process Planck Satellite data (Smareglia et al. 2004). The project is exploring scientific and technical problems that must be solved to develop GRID data reduction applications and make them available to the Planck community.

In this paper we describe the prototype of a specialized environment based on the GRID middleware, to support Planck Applications (see Figure 1). This environment must guarantee: retrieval of data from a storage located outside the GRID with http or ftp protocol, distribution of the Planck software (LevelS, Level1 and Level2) and libraries and storage and replica of raw and reduced data with a secure access policy.

This project coordinates two main initiatives:

- ESA and INAF-OATs joint collaboration;
- INAF/GILDA (a test-bed Grid infrastructure setup to host test-bed applications that at a later stage will be proposed as test-bed for EGEE).

2. The GRID Environment

GRID computing enables the virtualization of distributed computing and data resources such as processing and storage capacity to create a single system image, granting users and applications seamless access to vast IT capabilities. The whole software that underlines the fundamental grid services, such as information services, resource discovery and monitoring, job submission and management, brokering and data management and resource management represents the GRID middleware. The middleware builds upon an number of open source solutions like Globus Toolkit 2 (GT2, Foster & Kesselman 1997) and EDG libraries (Ghiselli 2002). In the case of the ESA-INAF collaboration the grid middleware

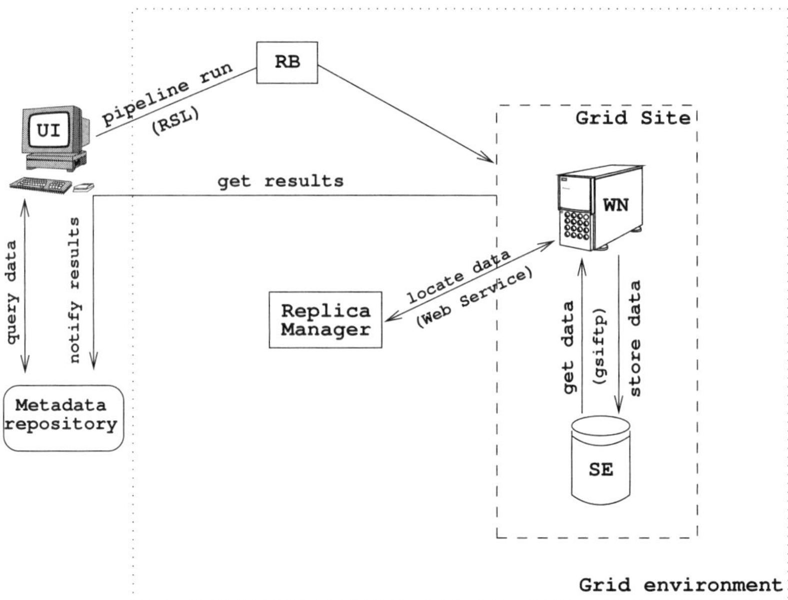

Figure 2. Structure of the Planck simulation workflow. The application environment check on a metadata repository if a simulation was already done using an XML description of the Cosmological and instrumental parameters. If the data exists the application download/reduce them otherwise a new simulation started. The simulated data are stored on the Grid. The application get the metadata that describe the simulated data and store them on the metacatalog repository for future download or post-processing.

is based on GT2 and ESA supplied a proprietary workflow tool: GridAssist[1]. This workflow tool acts as a resource broker of the computational and storage resources. It is already well tested on GAIA Grid (Ansari 2004).

In the case of INAF-GILDA collaboration, the production GRID is supplied by the Istituto Nazionale di Fisica Nucleare (INFN) of Catania. Grid INFN Laboratory for Dissemination Activities (GILDA[2]) was developed as part of the Italian INFN Grid project and the Enabling Grids for E-science in Europe (EGEE[3]) project as testbed for the EGEE applications. The EGEE middleware based on LCG (Robertson 2001) provides some basic services like a User Interface (UI), a full equipped job submission environment and a data Replica Manager (RM, Kunszt et al. 2003). The UI is a computer of the GRID system that allows users to submit jobs, access DB and store/replicate files. Storage elements and Worker nodes are the GRID resources in charge of data management and computing. Parallel and scalar computing are supplied by the system.

---

[1] `http://tphon.dutchspace.nl/grease/public`

[2] `https://gilda.ct.infn.it`

[3] `http://egee-intranet.web.cern.ch/`

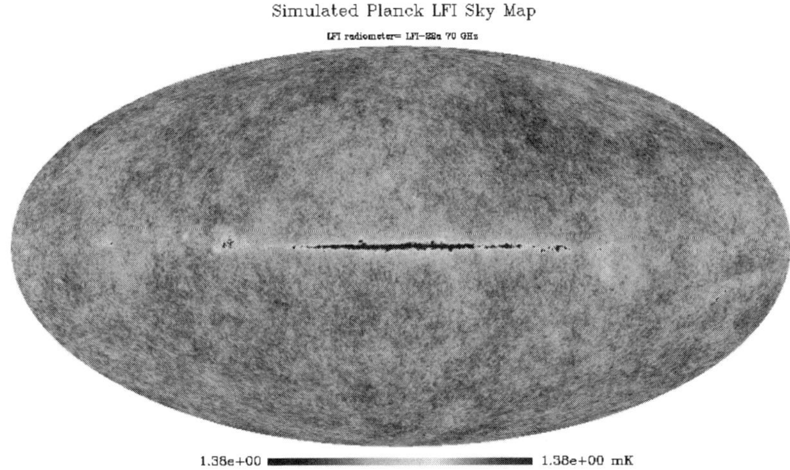

Figure 3. A map of the sky simulated via Grid for a 70 GHz LFI channel.

3. Planck@Grid

One of the primary issue for the DPCs is to define, design and run a complete simulation of the Planck mission to test the data analysis pipelines. The simulation software must mimic the Planck observing procedure, and any source of systematic effect related to the detectors in a realistic way. As a first test for the PlanckGrid project we concentrate on the simulation software (SW).

The ESA/INAF project setup a GRID of 3 sites (ESTEC, OATs and ESRIN) managed by the GT2 and by the GridAssist that acts also as the application environment to run the Planck Applications. In the INAF/GILDA GRID a workflow is built upon the EGEE middleware. The interaction with the GRID middleware is requested for: authentication, data movement, Planck SW distribution, resource selection, and computing (see Figure 2).

3.1. Working Testbed

We successfully port Planck mission simulation SW (LevelS) on LCG/EGEE Grid. SW is supported by a set of Linux shell scripts that allow to interface the simulation pipeline with Grid services. We use the GRID Job Description Language (JDL, Pacini 2003) to submit the numerical calculations and access to the RM data service. Our prototype distributes the LevelS SW on the Grid (RM), selects the available resources, and runs the pipeline, stores the simulated data and assures data access to the Planck users (all the users joining Planck Virtual Organization - VO). A meta-data schema is used to describe the output (parameters, date,size, etc.) and make easy data recovery and post-processing. An example of the simulation results is shown in Figure 3.

We also test the reduction SW on the simulated data. As an example we use the destriping procedure as described by Maino et al. (2002). Our input files are the simulated Time Ordered Data distributed on Grid. We design a pipeline that interacts with RM to locate raw data, identify the computing and storage resource suitable for its needs (using the GRID Resource Broker) and finally process raw data (using the GRID job submission tools). The output is stored on Grid and signed on the metadata repository. It can be accessed by Planck Users (VO).

4. Conclusions and Future Work

We successfully run a simulation of the Planck mission for LFI and we store the simulated mission data on GRID. This data is described by a set of XML files. The metadata description is still on a prototype stage and more work is required to identify the final semantic. The metadata description for the simulation files is stored on GRID. Data files are available for all the Planck users together with their XML description. We plan to port the whole simulation architecture on EGEE GRID to simulate the whole mission also for HFI. This requires to deploy a stable "application specific layer" and define the metadata description. We also plan to run simulations for different values of the parameters (Cosmological and Instrumental) and to test on GRID the simulated raw data reduction. This implies to port on GRID also the Level2 reduction software and to extend the metadata semantic also to reduced data.

Acknowledgments. This work is done with the economical support of the Italian Government and in particular of MIUR.

References

Ansari, S. G., 2004, in ASP Conf. Ser., Vol. 347, ADASS XIV, ed. P. L. Shopbell, M. C. Britton, & R. Ebert (San Francisco: ASP), 429

Bond, J.R., Crittenden, R.G.,Jaffe, A.H., & Knox, L., 1999, in Computers in Science & Engineering, 1, 21

Foster, I. & Kesselman, C., 1997, Intl J. Supercomputer Applications, 11, 115

Ghiselli, A., 2002, in TERENA Networking Conference, 503

Kunszt, P., Laure, E., Stockinger, H., & Stockinger, K. 2003, in Lecture Notes in Computer Science, Springer-Verlag Heidelberg Vol. 3019, ed. R.Wyrzykowski et al., 848

Maino, D., Burigana, C., Grski, K. M., Mandolesi, N., & Bersanelli, M. 2002, A&A, 387, 356

Pacini, F. 2003, DataGrid-01-TEN-0142-0_2

Pasian, F. 2002, MmSAI, 74, 502

Robertson, L. 2001, CERN/2379/rev

Smareglia, R., Pasian, F., Vuerli, C., & Zacchei, A., 2004, in ASP Conf. Ser., Vol. 314, ADASS XIII, ed. F. Ochsenbein, M. Allen, & D. Egret (San Francisco: ASP), 674

Tauber J.,A. 2000, in IAU Symposium, 204, 40

ASTRONOMICAL DATA ANALYSIS SOFTWARE AND SYSTEMS XIV
ASP Conference Series, Vol. 347, 2005
P. L. Shopbell, M. C. Britton, and R. Ebert, eds.

WebCom-G: Implementing an Astronomical Data Analysis Pipeline on a Grid-type Infrastructure

Seathrún Ó Tuairisg, Michael Browne, John Cuniffe, and Andrew Shearer
Computational Astrophysics Laboratory, Information Technology Department, National University of Ireland, Galway, Ireland

John Morrison and Keith Power
Centre for Unified Computing, Department of Computer Science, University College Cork, Ireland

Abstract. The recent upsurge in astrophysical research applications of grid technologies, coupled with the increase in temporal and spatial sky-coverage by dedicated all-sky surveys and on-line data archives, have afforded us the opportunity to develop an automated image reduction and analysis pipeline. Written using Python and Pyraf, the Python implementation of the IRAF package, this has been tailored to act on data from a number of different astronomical instruments. By exploiting inherent parallelisms within the pipeline, we have augmented this project with the ability to be run over a network of computers. Of particular interest to us is an investigation into the latency penalties in running the pipeline within a cluster and between two clusters. We have used a condensed graph programming model, the Grid middle-ware solution WebCom-G, to realize Grid-implementation. We describe how a re-organisation of such an astronomical image analysis structure can improve operational efficiency and show how such a paradigm can be extended to other applications of image processing. It is intended to use this project as a test bed for eventually running our image processing applications over a grid network of computers, with a view toward possible implementation as part of a virtual observatory infrastructure.

1. Introduction

Recent advances in astronomy have revolutionised the manner and depth at which we image the night sky, increasing the information potential which can be gleaned from scientific exposures. Systematic sky-survey archives are now stored at multiple sites and accessed across the globe via high-speed networks. Although modern astronomical research is increasingly geared at the efficient mining and processing of these massive, high-density, distributed data-banks, most astronomical applications have not evolved to deal effectively with this new computational paradigm. Rather than re-code existing structures to enable them to more efficiently deal with the current rigours of astronomical data processing, we propose to take advantage of inherent parallelisms in our data process flows to allow execution across a network of computing resources, using the recently developed middle-ware application WebCom-G (Morrison et al.

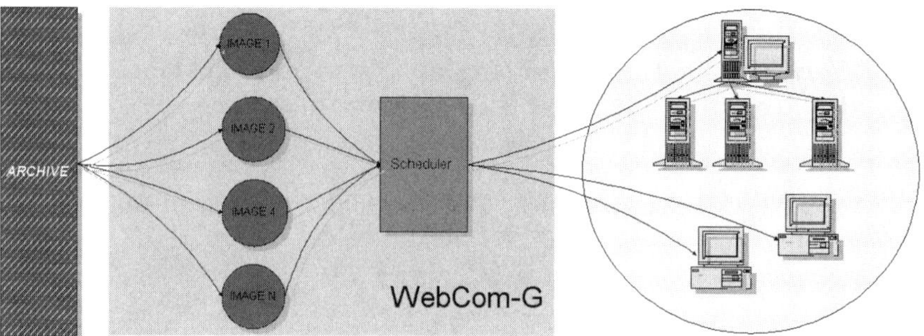

Figure 1. A schematic of the data processing flow, showing how WebCom acts as the link between the data archive and the processing nodes, which are executed across distributed computing resources.

2001). Our scientific example is a comprehensive survey of supernova remnants in M31.

2. The Scientific Application

The search for supernova remnants involves mining the data for Hα-bright and morphologically distinct objects (Magnier et al. 1995). A brief schematic of the main parts of the data processing flow, including object detection and subtraction, is outlined in Figure 1. The pipeline was written in Python and utilised Pyraf (the Python implementation of IRAF) tasks. Increased complexity was added by optimizing the point-spread function(PSF)-modeling stage. This involves the subtraction of fainter stars surrounding the stars used for creating the PSF to create a better PSF model. The execution time per image for a single processor is typically several hours, although this can vary significantly depending on stellar density and the sensitivity limits applied to the object detection.

The M31 images derive from the Wide Field Survey Isaac Newton Group archive(Lewis, Bunclark & Walton 1999) and are 4-chip (2048×4100 pixels) frames, covering 0.29 degrees squared. Each 4-chip image is 67MB (integer type) giving a total of ~25GB of raw data for the ~400 images being analysed. Processing incurs a two-fold data volume increase which include derivative images and photometry files. These are pre-reduced images, simplifying the processing load.

3. WebCom-G

The WebCom-G meta-computer is being developed with the purpose of providing users with a light interface to the Grid for their applications. It uses condensed graphs (Morrison 1996) to provide an integrated solution from applications to hardware. A graph node fires when all dependencies are accounted for and a free resource becomes available. Our application graph (Figure 1) is a series of nodes representing each image in our archive. More complicated structures in our data pipeline can also be graphed to increase parallelism. WebCom-

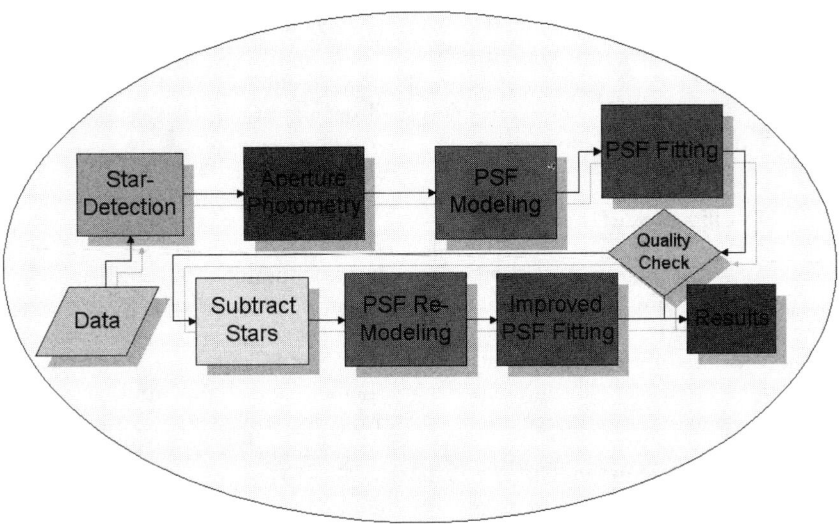

Figure 2. A schematic of the application data flow as executed by WebCom-G. This node is executed on a single machine and can act on one or more images.

G is compatible with existing grid middle-ware tools such as Globus, Condor and Cactus and security features are currently being implemented.

4. Hardware and Software

We allowed a 32-node PC cluster (dual 2.4GHz Pentium 4 processors, 2GB memory per node, interconnected via Gigabit Ethernet and running Red Hat Linux) to act as a WebCom client. A shared file system minimised data transfer times between nodes. WebCom-G assigns nodes to machines on an application-available basis by interrogating a database. The astronomical pipeline was written using Python/Pyraf to encapsulate the procedures from the original IRAF pipeline (Ó Tuairisg et al. 2004). Thus software availability is the only limitation on accessible computer resources. WebCom embeds each node in the graphed application with its own scripts, which handle scheduling and resource allocation.

5. The Method In Practice

We found this approach to be a simple, flexible and robust way of obtaining significant performance increases in our data reduction and analysis processes. It is a flexible approach because it allows us execute each node (Figure 2) on one or more images. Increased parallelism can be obtained with minimal recoding by further resolving our graph into separate nodes to be executed simultaneously across different machines. We ran our application across a shared file system, thus by-passing the network bandwidth bottleneck (although this is likely to be a factor when using more widely distributed computing resources). It should be noted that because each image samples a different region of the galaxy, and so execution time varies per node, increased data-splitting (within each image) would incur better performance.

6. Conclusions and Future Work

We have described a novel method of running astronomical data processing applications, with minimal re-coding, across a network of distributed computing resources using WebCom-G. We benefit by obtaining significant performance increases, yet do so by re-using existing applications. In our example we act on a split data archive, but through expressing operations in WebCom-G's graphical format, we can exploit increasingly fine-grained parallelisms within our processes. This approach is also perfectly tailored to utilise current Grid6 technologies, which exploit distributed, low-cost computing resources. The Virtual Observatory concept, recently gaining currency (Pierfederici et al. 2001; Quinn et al. 2002; Szalay 2001) could benefit from this approach of increasing performance through re-using existing astronomical applications. We also intend to extend this technique to other scientific applications, including medical imaging.

Acknowledgments. We acknowledge the support of Science Foundation Ireland under the WebCom-G program.

References

Lewis, J.R., Bunclark, P.S, & Walton, N.A. 1999, in ASP Conf. Ser., Vol. 172, ADASS VIII, ed. D. M. Mehringer, R. L. Plante, & D. A. Roberts (San Francisco: ASP), 179

Magnier, E.A., et al. 1995, Astronomy & Astrophysics Supplement, 114, 215

Morrison, J. 1996, PhD Thesis, Technische Universiteit Eindhoven

Morrison, J., Power, D., & Kennedy, J. 2001, Journal of Super Computing, 18, 47-63

Ó Tuairisg, S., Butler, R., Golden, A., Shearer, A., Voisin, B., & Micol, A. 2004, in ASP Conf. Ser., Vol. 314, ADASS XIII, ed. F. Ochsenbein, M. Allen, & D. Egret (San Francisco: ASP), 444

Pierfederici, F., Benvenuti, P., Micol, A., Pirenne, B., & Wicenec, A. 2001, in ASP Conf. Ser., Vol. 238, ADASS X, ed. F. R. Harnden, Jr., F. A. Primini, & H. E. Payne (San Francisco: ASP), 141

Quinn, P.J., Benvenuti, P., Diamond, P.J., Genova, F., Lawrence, A. & Mellier, Y. 2002, in Virtual Observatories, Proc. SPIE, 4846, 1-5

Szalay, A.S. 2001, in ASP Conf. Ser., Vol. 238, ADASS X, ed. F. R. Harnden, Jr., F. A. Primini, & H. E. Payne (San Francisco: ASP), 3

ASTRONOMICAL DATA ANALYSIS SOFTWARE AND SYSTEMS XIV
ASP Conference Series, Vol. 347, 2005
P. L. Shopbell, M. C. Britton, and R. Ebert, eds.

Astronomical Database Related Applications in the Grid.it Project

Alessandra Volpato,[1] Giuliano Taffoni,[2] Serena Pastore,[1] Claudio Vuerli,[2] Andrea Baruffolo,[1] Riccardo Smareglia,[2] Giuliano Castelli,[2] Fabio Pasian,[2] and Leopoldo Benacchio[1]

National Institute for Astrophysics, 00136 Rome, Italy

Edgardo Ambrosi and Antonia Ghiselli

National Institute for Nuclear Physics - CNAF - Bologna, Italy

Abstract. We describe the activity done in the contest of the Grid.it project to access Astronomical Catalogues and Archives through grid environment. Different approaches are tested: job oriented, web service oriented. The crucial aspect of this work is the development and deployment of an IVOA compliant Data Source Engine model in the grid middleware.

1. Introduction

The Astronomical Observatories of Trieste and Padova are involved in the Grid.it project aiming at 'Enabling platforms for high-performance computational grids orientated to scalable virtual organizations'. In this framework we explore the use of grid technologies for developing astrophysical applications. The testbed for grid applications is the Italian INFN Production Grid for Scientific Applications based on the LCG-2 (LHC Computing Grid, Robertson 2001) distribution, which is the evolution of the software produced by the European DataGrid project[3]. Within Grid.it[4], the OAPd group focused on the portability to the Grid of an existing system for the consultation of large astronomical catalogues, currently serving on the net the Second Guide Star Catalog (GSC-II) (Benfante et al. 2000), while the Astronomical Observatory of Trieste (OATs) is dealing with the similar problem of integrating in the Grid the archive of observational data from the Italian Galileo National Telescope (TNG) (Smareglia et al. 2003) and, as a further step, to provide processing for the data retrieved through grid-enabled pipelines. Since in the grid infrastructure based on LCG an adequate model for DBMSes is not available, INFN and INAF institutes are collaborating to design and implement an architecture to integrate DBMSes in the existing

[1]Osservatorio Astronomico di Padova, Padova, Italy

[2]Osservatorio Astronomico di Trieste, Trieste, Italy

[3]`http://www.eu-datagrid.org/`

[4]`http://grid-it.cnaf.infn.it/`

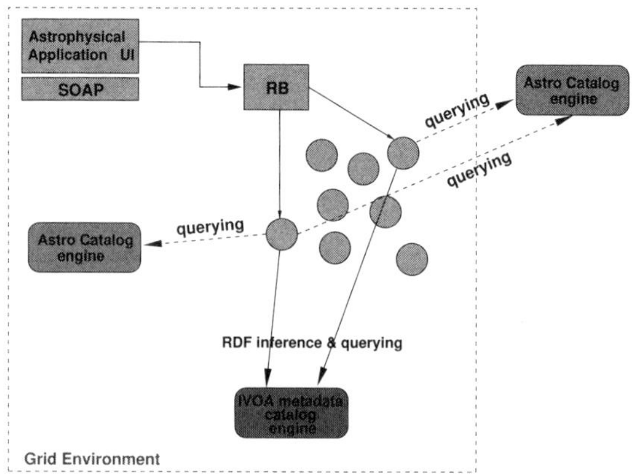

Figure 1. Grid DSEs schema, illustrating the Grid procedure to locate and query an Astronomical Database with IVOA standards. The Grid environment is equipped with a metadata engine that is first located by the Resource Broker. The metadata source engine is accessed by a Grid node and reply to the node the location of the Astrophysical source engine. The grid node then query the Astrophysical DB that can be internal or external to the grid environment.

grid infrastructure. In this paper we describe our effort in modeling the DBMS on the grid environment and the work done by OAPd and OATs to guarantee the access to archive and catalogues using the pre-DBMSes grid infrastructure.

2. A Data Source Engine model for LCG

In the current grid architecture, the Information Services (IS) encompass semantic and interaction models for Grid resources. They are able to retrieve run-time data and information from the defined Grid node elements: Computing Elements (CE) and Storage Elements (SE). However, to publish and discover Astrophysical data sources and services it is necessary to add to the IS the Data Source Engine (DSE) as new grid resource entity. The basic assumption driving this work is that a DSE can be modeled in terms of a batch system architecture and can be evolved to a new Grid resource analog to a Computing Element. The deployment of a DSE implies the definition of the information schema to advertise on the Grid Monitoring and Discovering System the DSE resources, recording not only physical characteristics but also relevant semantic informations, with the additional constraint of compliance with the Resource and Data-Service schemas proposed by IVOA (Hanisch & Quinn 2003). Moreover, the Resource Specification Language instruction set must be enhanced to manage query activities and the Local Job Manager (LJM) capabilities must be extended to deal with the newly defined 'query jobs'. As we want to be compatible with the IVOA standards, we must supply the integration with the LJM of a library to allow query jobs to be specified in terms of IVOA compliant ADQL documents. Finally Grid Resource Information Index Backend should be interfaced with the Grid-DSE IP and the WN MOM should be evolved in order to

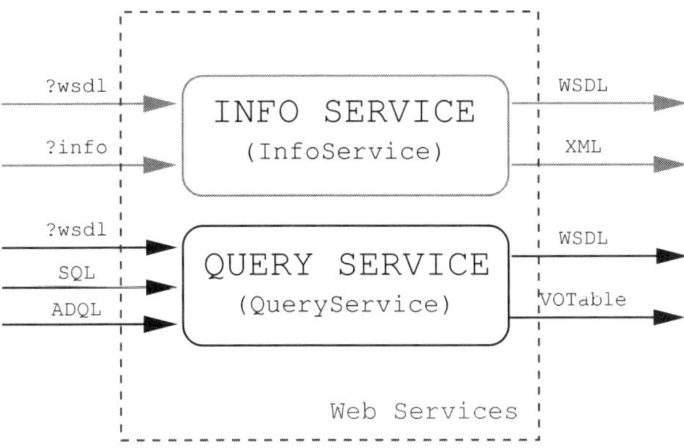

Figure 2. The Web Services implemented by the OaPd to access the GSC-II astronomical catalogue.

reflect DSE backend capabilities. Currently, a prototype solution is undergoing the test phase: it involves the integration of the new G-DSE schema in the IS, thus allowing the discovery of existing Astrophysical Data Sources within the Grid.

3. Astronomical Archives and Catalogues: the grid access

The goal is to develop a grid-integrated system for accessing large astronomical catalogues and archives. This is the first step to use the grid to access to visualize Astronomical data and process them.

3.1. Web Service Approach

The OaPd built a prototype system to access the GSC-II astronomical catalogue (Benfante et al. 2001) following the web services (WS) paradigm (Newcomer 2002). The system developed is made up of two web services: a Meta-data Access service and a Query service (see Fig. 2). Meta-data Access Service retrieves metadata information about catalogues by connecting with a "lightweight" DB running on the same machine as the service container. This service answers the info request with an XML file containing the catalogue metadata in an IVOA compliant format. Query Service answers queries expressed in SQL or ADQL with a VOtable containing the result set. The two services are installed on a WN connected to the Production Grid; they are deployed within an Apache web server equipped with Tomcat and Axis, and secured by EDG security packages (an extension of the standard Grid Security Infrastructure specific for web applications). The interaction between a client and the services is carried out by SOAP messages over https. Currently, the services can be accessed either interactively, through JSP, or in batch mode, requesting the execution of a command line client application by submitting a Job Description Language (JDL) file from a machine with the User Interface Grid software installed.

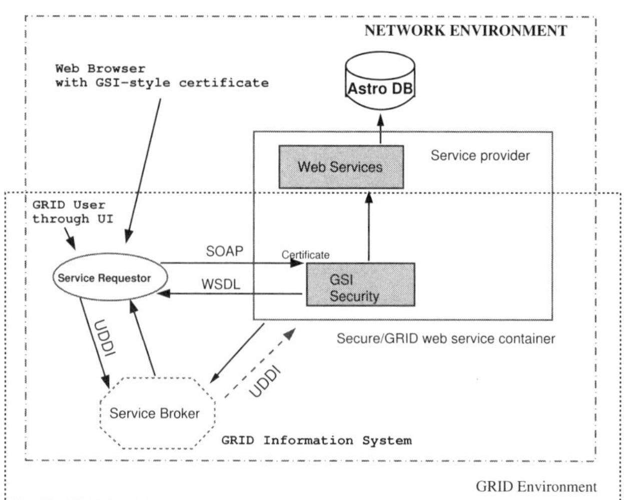

Figure 3. An illustration of a client node accessing the GSC-II astronomical catalogue through a Web Service grid infrastructure.

3.2. A Job-oriented approach

The OATs developed an application specific layer on the grid middleware to access Astronomical archive and produces on-the-fly calibration for the requested data (Taffoni et al. 2004). The testbed for this application is the Long Term Archive of Telescopio Nazionale Galileo (LTA-TNG) located at OaTS (Smareglia 2003). We designed the connection service as a client/server grid application based on a query service (QS), a client service (CS) and a reduction driver. The QS is a java application installed on a CE which uses the OJDBC driver to connect with the DB machine. Security is guaranteed by username/password and direct cable connection. The QS is queried by the CS, which is installed on each UI and allow users of INAF VO to access data. It is constituted by a set of shell script wrapping the globus job submission tools. Copy and registering of the query output and/or images is implemented as a Java application that uses the Replica-Manager client java classes to access the Replica-Manager grid service to copy and register the files. A logical file name (LFN) for each file must be supplied by the user. A reduction driver was implemented. At this stage our interest concerns the possibility of calibrating a set of Astronomical images removing the instrumental signature. The calibration driver is based on Eclipse. To deploy the "data calibration on demand" service we must first install in the grid WN the Eclipse software. Then a collection of shell scripts are used to run a reduction pipeline. The pipeline is submitted to the grid using JDL . We customized the UI, designing a graphical application to manage the access and reduction services. It is based on Java Swing. It allows users to make the query, display the results, save the query on a grid file, save data on grid files. It compiles the JDL file to run the reduction and submits it.

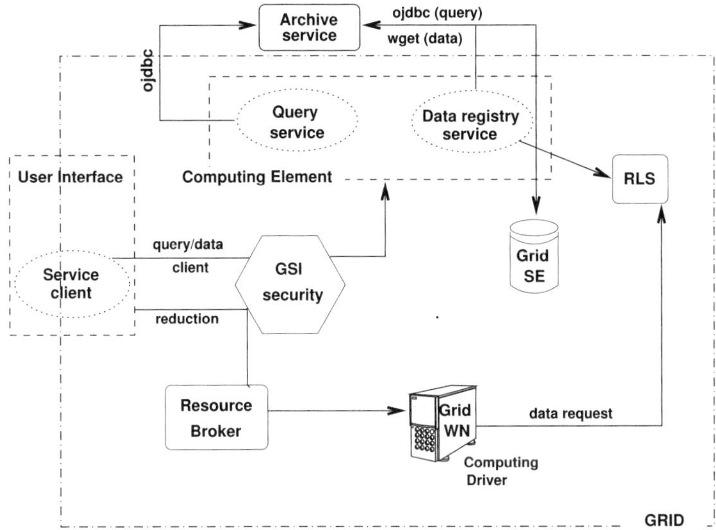

Figure 4. The job-oriented data access. We show the work flow for a job-oriented database access with a calibration driver to call on demand. The security is based on GSI and OJDBC. A graphical user interface allow users to make query and reduction

4. Conclusions and future work

We verified the possibility of accessing Astronomical data through a grid environment and to perform simple data processing. Our tests show that a crucial need is the integration in the Grid of a DSE to collect and query metadata information on the resource available in a IVOA format. Our future work will be centered on the complete modification of the Grid IS to finally integrate the DSE and on the refinement of the application developed to take full advantages of the presence of a DSE metadata collector.

Acknowledgments. This work is done with the support of the Italian Government and particularly of MIUR.

References

Benfante, L., Volpato, A., Baruffolo, A., & Benacchio, L., 2001, in ASP Conf. Ser., Vol. 238, ADASS X, ed. F. R. Harnden, Jr., F. A. Primini, & H. E. Payne (San Francisco: ASP), 160

Hanisch R., J. & Quinn P., L., 2003, `http://www.ivoa.net/pub/info/`

Newcomer E., 2002, Understanding Web Services, Addison-Wesley

Robertson L., 2001, CERN/2379/rev

Smareglia, R., Becciani, U., Caproni, A., Gheller, C., Guerra, J. C., Lama, N., Longo, G., Pasian, F., & Zacchei, A, 2003, MmSAI, 74, 514

ASTRONOMICAL DATA ANALYSIS SOFTWARE AND SYSTEMS XIV
ASP Conference Series, Vol. 347, 2005
P. L. Shopbell, M. C. Britton, and R. Ebert, eds.

Grid-related Activity in Progress at INAF

C. Vuerli and F. Pasian

INAF-Osservatorio Astronomico di Trieste

A. Baruffolo and L. Benacchio

INAF-Osservatorio Astronomico di Padova

J. M. Alcalá and E. Cascone

INAF-Osservatorio Astronomico di Capodimonte, Naples

U. Becciani

INAF-Osservatorio Astrofisico di Catania

A. Fontana

INAF-Osservatorio Astronomico di Roma

G. Longo

Università di Napoli

R. Tagliaferri

Università di Salerno

Abstract. INAF, the Italian National Institute of Astrophysics, is currently involved in several Grid-related projects, at both the national and international level. This paper gives an overview on these projects; for each of them a short description is given together with its current status and other relevant information like allocated resources, collaborations, work in progress, results obtained so far, and future plans.

1. GRID.IT

GRID.IT is a multi-disciplinary national project with the purpose of developing and testing a national Grid infrastructure for the Italian research community. Astrophysics has been grouped within the Research Unit 4/WP10 (`http://web.pd.astro.it/wp10/`) of the project and contributes with three applications listed below:

1. Access astrophysical archives and catalogues (namely the pilot archive of the Telescopio Nazionale Galileo and the GSC-II).
2. Gridify the tools for the processing of wide-field astronomical images that will be obtained with the VLT survey telescope (VST).

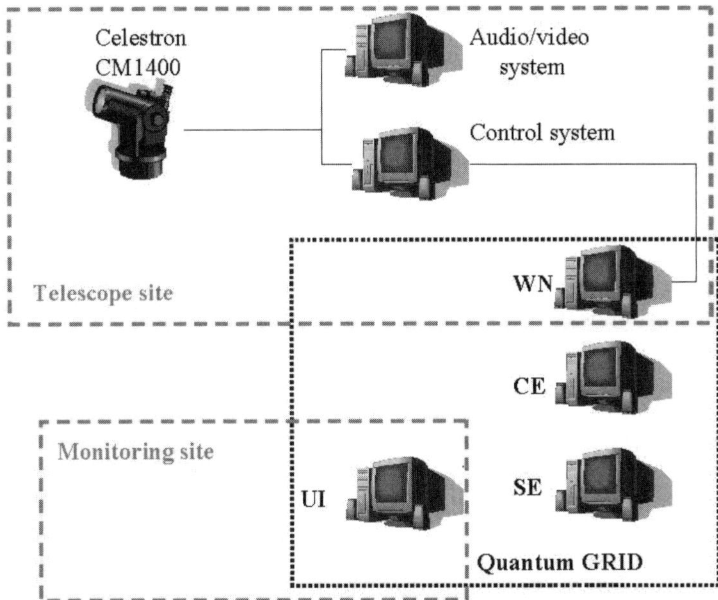

Figure 1. Hardware layout for Application 3

3. Gridify the remote monitoring of observational structures. Users will be able to check the status of the telescope and related instruments and to collect scientific data from a Grid-UI.

Application 1 (Volpato et al. 2005) has been developed through a strict collaboration between Astronomical Observatories of Padova and Trieste. Additional work was necessary in this case to make the GRID.IT middleware able to access databases. Some work is in progress at INFN-CNAF to modify the middleware so that databases become well-known entities on the Grid, although external to it. A pool of metadata characterizing each database is handled by Grid services so that they are published and made available to applications. Applications will be able to build queries by specifying values for a subset of metadata; queries are then submitted to the Grid; the list of databases able to fulfill the user query are returned to the Grid-UI; finally, the application will choose and contact one of the proposed databases. In the meantime Padova and Trieste built two ad-hoc solutions to circumvent the access problem and to go ahead with the work on application 1. Trieste, in addition, will work on a gridified pipeline to process data extracted from the TNG archive; to do that, contact will continue with people working on application 2.

Application 2 (Pavlov et al. 2005) is under development at the Capodimonte Observatory. The goal of this task is to study the use of GRID technology for the processing of Wide-Field astronomical Images that will be obtained with the VLT survey telescope (VST). The image processing of the VST data will be based on the Astronomical Wide-Field Imaging System for Europe (Astro-

Wise[1]) data reduction pipeline. The Astro-Wise pipeline design includes parallel processing, distributed storage and federated database concepts. A preliminary study of the different possibilities for the integration of the GRID job scheduler in the Astro-Wise pipeline has shown a good scalability of GRID parallel execution. For example, parallel execution of 32 jobs (one job per one CCD in the case of the OmegaCam mosaic camera) is 23 times faster than sequential execution of the same number of jobs (see also Pavlov et al. 2005).

The goal of application 3, under development at INAF-OATs, is the gridification of "distributed monitoring of remote observations" (see Figure 1). The test-bed for this application is a CELESTRON CM1400 telescope with its own control/acquisition system. The application will be developed through intermediate steps; in the basic release the GRID system will interact with the control system through an ad-hoc built web-service whose role is delivering commands from the Grid-UI to the control system and pick up telemetry and scientific data to be passed back to the Grid-UI. Further refinements of the application will include a RTA/QLA pipeline, an archive system for telemetry and scientific data, a user interface and finally an "advanced" version of the application with a direct interface to the control system (if feasible).

2. DRACO

DRACO[2], the Datagrid for Italian Research in Astrophysics and Coordination with the Virtual Observatory, is a concept aiming at providing the scientific community with a distributed multi-functional environment allowing the use of specialized (observational, computing, storage) Grid nodes. DRACO is based on the national cooperation named IG-BIGEST (Italian Grid for Business Industry Government E-Science and Technology); the participation in this structure of the astrophysical community is granted by INAF. DRACO provides the framework through which the Italian astrophysical community can participate in the IVOA, the International Virtual Observatory Alliance[3] effort. A reference person, (F. Pasian of INAF-OATs) has been nominated by INAF as its contact point for IVOA activities. DRACO can be considered as an extension of the FIRB project. Two additional INAF institutes (the Astronomical Observatory of Rome and the Astrophysical Observatory of Catania) plus two Universities (Naples and Salerno), in fact, are part of the Draco project, in addition to those already participating in the FIRB project. Draco, moreover, aims at using Grid technology for processing and mining huge amounts of data. Some participating institutes contribute test-bed applications while some others are validating dedicated technological solutions. DRACO is also the project allowing the Italian community to participate in the Virtual Observatory. Some considerations about the Draco project are: a) applications to be ported on Draco may require a great amount of resources (computing resources as well as storage resources); b) because these applications are rather complex, they need to be tested before

---

[1]`http://www.astro-wise.org/`

[2]`http://wwwas.oat.ts.astro.it/draco/`

[3]`http://www.ivoa.net/`

being ported on a production Grid; c) special resources may be required to be accessed by the Grid (e.g., databases); d) some applications ported in Draco will migrate to EGEE as a further step. For these reasons Draco has chosen to adopt GILDA as Grid infrastructure. GILDA is a test Grid for EGEE applications. The Draco project has been organized in three thematic areas: the "Grid Middleware", "Simple Applications" and "Advanced Applications". The pool of astrophysical applications to be ported on Draco is currently under discussion. Good candidates, however, have already been identified (e.g. a complete Planck pipeline using the simulations software of Planck).

3. Planck Gridification

Two main activities are related with the gridification of pipelines built to process Telemetry and Science data at the two Planck DPCs, both during the pre-launch phase of the project as well as during the operative phase of the mission.

3.1. Planck-Grid

Some initial considerations about the Planck data processing are:

- INAF-OATs is setting up and will host the Planck/LFI Data Processing Centre.
- Institutes being part of the Planck/LFI consortium are contributing the prototype version of the software that will be integrated at LFI DPC to create the data reduction pipelines of the Planck mission.
- The simulations software (level "S") has been jointly developed by groups of the two Planck consortia (LFI and HFI). This software can be used to simulate the entire Planck mission.
- The gridification of Planck pipelines are an excellent test-bed for any GRID infrastructure as they require considerable resources in terms of computing power and data storage space.
- Planck consortia are strongly interested in gridifying DPC software; it could gain great advantage in running on Grid infrastructures in terms of parallelization and then of time required to get a complete pipeline run.
- A test to bring a partial pipeline on the Grid was successfully carried out in June 2004. G. Taffoni and G. De Gasperis (Taffoni et al. 2005) brought on the GRID.IT middleware the map-making software, one of the basic components of the Planck pipeline. Test results were encouraging so it is worth to proceed in this direction.

In May 2004 a joint meeting ESA-OATs at ESRIN (Frascati) was dedicated to the Planck processing gridification. It was decided to set up a mini-grid ESA-OATs for this purpose. A dedicated Grid infrastructure eas selected because: a) in this way it is possible to play with different combinations of Grid components with no limitations, so that the best configuration to run Planck pipelines can be identified; b) it is possible to have full control on the Grid user certification. Experiments within the Planck-Grid project are currently in progress. Because the Planck pipeline gridification is an application that is moving towards EGEE, the Planck-Grid project could come together in EGEE in the next months.

3.2. Planck Pipelines on EGEE

The project *Enabling Grids for E-science in Europe* (EGEE, `http://public.eu-egee.org/`), is funded by the European Commission and aims to build on recent advances in grid technology and develop a service grid infrastructure in Europe which is available to scientists 24 hours-a-day. The two Planck consortia decided to propose a test-bed application for EGEE, using the Planck pipelines built with the already available simulation software. In Dec 2003 the Kick-off meeting of the EGAAP (EGEE Generic Applications Advisory Panel) took place in Paris. The astrophysical community was strongly encouraged to propose applications for EGEE. Among those, the gridification of Planck pipelines is certainly one of the most attractive for EGEE. In June 2004 the first EGAAP meeting took place at CERN. At that time the Planck community was not yet ready to port the simulation pipelines to EGEE. In the meantime GILDA (`http://gilda.ct.astro.it/`), the EGEE test-bed Grid infrastructure was set up, and people were encouraged to bring their applications there prior to porting to the EGEE. Meanwhile, the first experiments of porting partial Planck pipelines to the Grid were carried out successfully at OATs. As a result of these two events: a complete Planck pipeline built by assembling the simulation software will be brought onto GILDA soon,and a complete Planck pipeline will be officially proposed as an EGEE test-bed application at the second EGAAP meeting (Den Haag, Netherlands, 25 Nov 2004). The entire Planck community is watching this experiment with great interest. A Planck-WG8, in fact, was set up, and its goal is to investigate the gridification of Planck data processing. WG8 met in Edinburgh in Jan 2004, and the LFI DPC is keeping in touch with their efforts. While gridifying Plack pipelines, we discovered that the network infrastructure is currently unable to support efficiently moving the required amount of data onto the Grid (from one SE to another). For now, we will move computation rather than data, taking into account the "nearest CE" parameter.

4. Conclusions

Several Grid-related projects are in progress at INAF involving a small number of specialists. The Grid seems to be a promising solution for a considerable number of astrophysical problems, so the plan is to start initiatives within INAF to find new applications to bring onto the Grid, and to demonstrate to the astrophysical community that Grid technology is a valid tool for their work.

References

Pavlov, M., Alcalá J. M., Grado, A., Cascone, E., Capasso, G., Manna, V., Pasian, F., & Benacchio, L. 2005, in ASP Conf. Ser., Vol. 347, ADASS XIV, ed. P. L. Shopbell, M. C. Britton, & R. Ebert (San Francisco: ASP), 311

Taffoni, G., De Gasperis G., Maino, D., Zacchei, A., Castelli, G., Vuerli, C., Pasian, F., Ansari, S., & Tauber, J. 2005, in ASP Conf. Ser., Vol. 347, ADASS XIV, ed. P. L. Shopbell, M. C. Britton, & R. Ebert (San Francisco: ASP), 320

Volpato, A., Taffoni, G., Pastore, S., Vuerli, C., Baruffolo, A., Smareglia, R., Castelli, G., Pasian, F., & Benacchio, L. 2005, in ASP Conf. Ser., Vol. 347, ADASS XIV, ed. P. L. Shopbell, M. C. Britton, & R. Ebert (San Francisco: ASP), 329

Part VII
The Virtual Observatory: Mining and Analysis

ASTRONOMICAL DATA ANALYSIS SOFTWARE AND SYSTEMS XIV
ASP Conference Series, Vol. 347, 2005
P. L. Shopbell, M. C. Britton, and R. Ebert, eds.

OpenSkyQuery and OpenSkyNode - the VO Framework to Federate Astronomy Archives

William O'Mullane, Tamás Budavári, Nolan Li, Tanu Malik, María A. Nieto-Santisteban, Alexander S. Szalay, and Aniruddha R. Thakar

Department of Physics & Astronomy, The Johns Hopkins University, Baltimore, MD 21218, USA

Masatoshi Ohishi

National Astronomical Observatory of Japan, 2-21-1 Osawa, Mitaka, Tokyo 181-0015, Japan

Ramon Williamson

NCSA/UIUC, Champaign, IL 61820

Abstract. OpenSkyNode and ADQL are the major new steps in the Data Access layer of the Virtual Observatory. OpenSkyQuery (OSQ) allows cross matches between catalogs on registered nodes and supports the upload of lists of sources to be cross matched. This system utilizes the IVOA's nascent standard Astronomical Data Query Language (ADQL).

1. Introduction

OpenSkyQuery was mooted last year (Budavári et al. 2004). This year we present a working system utilizing the new Astronomical Data Query Language (ADQL) (Yasuda et al. 2004) to send queries to distributed nodes. Figure 1 below depicts the architecture of the system. Individual SkyNodes implement the SkyNode interface to either basic or full level. These nodes are registered in the Virtual Observatory registry from where the OpenSkyQuery portal may find them (automated discovery). Users may interact with the nodes directly or use the portal to make higher level queries involving more than one node.

2. OpenSkyNode and ADQL

The OpenSkyNode IVOA working draft specification[1] (currently 0.7.4) defines an interface to astronomical data that accepts ADQL(0.7.4) and exposes meta-data. Additionally for full compliance the node must support XMatch(cross match) by implementing the ExecutePlan interface. ADQL[2], an IVOA pro-

[1] `http://www.ivoa.net/internal/IVOA/IvoaVOQL/SkyNodeInterface-0.7.4.pdf`

[2] `http://www.ivoa.net/internal/IVOA/IvoaVOQL/WD_ADQL-0.9.pdf`

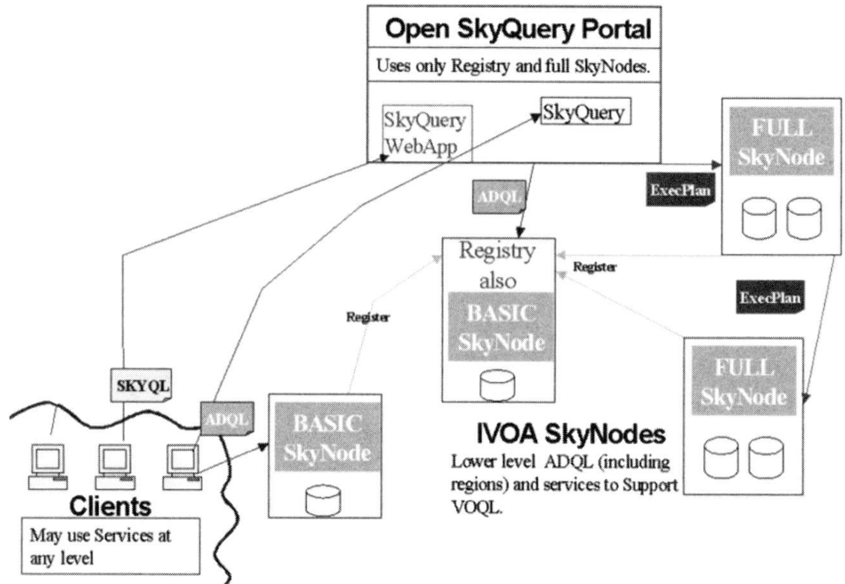

Figure 1. OpenSkyQuery Architecture with Full and Basic nodes

posed recommendation (currently 0.9), is an SQL[3] based syntax for expressing queries. ADQL includes regions, XMatch, and math functions. The main additions from 0.7.4 are to allow for XPATH type expressions that enable easier querying of data model based interfaces as well as minor bug fixes in the grammar. For more information about SkyNodes and ADQL and to get the latest specifications visit the IVOA VOQL twiki page[4]. It is worth mentioning that two formats exist for ADQL:

1. ADQL/x - an XML format, a parse tree meant for machines; and
2. ADQL/s - a string based format which is meant for humans.

A simple WebService and web page parser for translation may be found at JHU[5]. ADQL may be used by multiple services in the VO Data Access Layer (DAL) e.g., upgraded CONE/SIAP services. Shirasaki et al. (2005) present ideas on this and other modifications for ADQL for 2005.

3. OpenSkyQuery Portal

OpenSkyQuery is a demonstration portal that uses multiple nodes to perform distributed queries including XMatch of small areas. Figure 2 shows the query

[3] http://www.contrib.andrew.cmu.edu/~shadow/sql/sql1992.txt

[4] http://www.ivoa.net/twiki/bin/view/IVOA/IvoaVOQL

[5] http://openskyquery.net/adqltranslator/

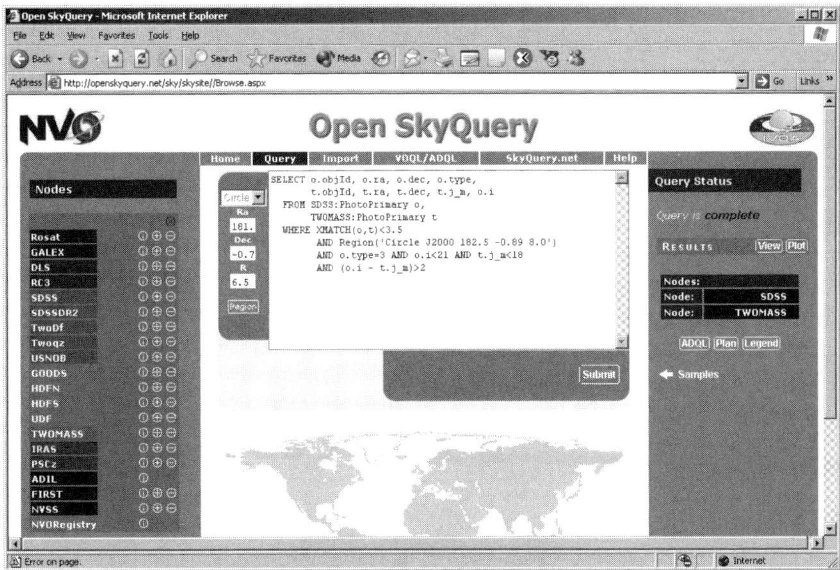

Figure 2. OpenSkyQuery portal, the query screen

screen where the main action occurs. Clicking on the (+) near the node name builds a query with an XMatch statement in the query panel. After submission, on the right, a query status is displayed and regularly updated.

All nodes displayed in the left panel are looked up from the registry. An implementor may create a SkyNode and add it to the registry, and it will appear in this list. Clicking on the (i) gives a summary of the registry information for the node, clicking on the node name lists the tables for the node and clicking on a table lists it's columns. Such schema browsing is supported by the SkyNode interface.

3.1. What the portal can do

One feature of the portal which is very useful is the Upload functionality. Clicking on Import in the top menu bar will take one to the upload screen. Here tables, in a variety of formats, may be uploaded. These appear in a special node called MyData on the query screen. The NVO Summer School walks through an example of uploading and cross matching the Abell catalogue[6].

The other main functionality is to enable expression of queries concerning values in more than one catalogue. The Brown Dwarf Search[7] uses the difference between magnitudes in SDSS and 2MASS to find candidate brown dwarfs. With VOPlot now integrated results from a query may be instantly analyzed.

---

[6] `http://chart.stsci.edu/twiki/bin/view/Main/ColorSegregation`

[7] `http://chart.stsci.edu/twiki/bin/view/Main/BrownDwarf`

The portal and nodes demonstrate that ADQL is an adequate language for these tasks. The integration of VOPlot for analysis of results further demonstrates the world wide collaboration in the VO community.

4. SOAP Services

We have found that SOAP services provide a very clean API for any system. In this system the portal site sits upon a set of SOAP services and the SkyNode interface is defined in terms of a WSDL (WebServices Description Language) definition. Users may access these services directly using a SOAP toolkit in their preferred programming language. At JHU we have tried Python and Java (AXIS) clients for WebServices successfully. Others have written Perl clients. More information on this is available at the IVOA Web site[8].

4.1. SkyNode WebService

The SkyNode specification defines interfaces including :

1. Interface for listing tables, columns etc.;
2. Interface for PerformQuery, takes ADQL returns data as VOTable ; and
3. Interface for ExecutePlan, to enable execution of distributed spatial join.

See the full interface at `http://openskyquery.net/nodes/sdss/nodeb.asmx`.

4.2. The Portal WebService

The portal contains many SOAP interfaces including:

1. Interface for distributed query, takes ExecPlan Or String (ADQL/s);
2. Interface to convert ADQL/s to plan; and
3. Interface for reporting query status from the nodes.

See the full set of interfaces at `http://openskyquery.net/Sky/SkyPortal/SkyPortal.asmx`.

Examples of using the portal programmatically are contained in the NVO Summer School software package. To use the package the software must be installed as described on the software page[9], additionally text for walking through the examples are in the course notes[10].

5. SkyNode Implementations

Nominally to implement a SkyNode one would use a SOAP library to generate the server side stubs and then fill in the provided shell with one's implementation. In practice this is quite an undertaking and NVO is attempting to provide reference implementations. Currently we have two implementations.

---

[8] `http://www.ivoa.net/twiki/bin/view/IVOA/WebgridTutorial`

[9] `http://chart.stsci.edu/twiki/bin/view/Main/Software`

[10] `http://chart.stsci.edu/twiki/bin/view/Main/SummerSchoolClientSide`

5.1. .NET FullSkyNode implementation

The first reference implementation of the SkyNode was produced using .NET and this is the version powering most of the current nodes. This is a new interface for any database set up for the old SkyQuery system, and may be downloaded directly from the home page of OpenSkyQuery.net. Detailed information on setting up the database for such a node is provided on SkyServer.org/skynode.

5.2. Java BasicSkyNode

The Apache AXIS library has been used to generate the server stubs for SkyNode and gone on to make a very flexible package which operates with most database systems. This implementation uses XSL to transform the ADQL to the native format, and XSL for many popular databases are provided in the package. This is included in the summer school software package and an example of setting it up to run with MySQL and the Abell catalogue is described in the notes[11].

In either case after setting up a node it must be registered to show up in the portal and to be available to users searching the registry[12].

6. The Future

Currently the queries in the portal and on the nodes all run synchronously. For larger queries or heavier load this will need to be made asynchronous. This will require identification of users on nodes. With identification we would not require the user to login to each node. Rather we will have a single sign on protocol supported by some form of federated trust.

As the plan executes, the result from each node is passed back to the calling node. It may be more efficient to write the answer somewhere and pass instead a handle to the data. This would be facilitated by the VOStore protocol which is a general mechanism for reading and writing data on distributed nodes. This also requires some form of single sign on function. Both of these topics are under discussion in the IVOA Web and Grid Services working group.

We are also investigating scheduling and caching algorithms for the SkyNode queries that would make the preliminary performance queries unnecessary.

References

Budavári, T., et al. 2004, in ASP Conf. Ser., Vol. 314, ADASS XIII, ed. F. Ochsenbein, M. Allen, & D. Egret (San Francisco: ASP), 177

Shirasaki, Y., et al. 2005, in ASP Conf. Ser., Vol. 347, ADASS XIV, ed. P. L. Shopbell, M. C. Britton, & R. Ebert (San Francisco: ASP), 262

Yasuda, N., et al. 2004, in ASP Conf. Ser., Vol. 314, ADASS XIII, ed. F. Ochsenbein, M. Allen, & D. Egret (San Francisco: ASP), 293

---

[11] `http://chart.stsci.edu/twiki/bin/view/Main/SummerSchoolServerSide`

[12] `http://nvo.stsci.edu/voregistry`

ASTRONOMICAL DATA ANALYSIS SOFTWARE AND SYSTEMS XIV
ASP Conference Series, Vol. 347, 2005
P. L. Shopbell, M. C. Britton, and R. Ebert, eds.

An O($N \log M$) Algorithm for Catalogue Crossmatching

Drew Devereux, David J. Abel, Robert A. Power, and Peter R. Lamb

CSIRO ICT Centre, GPO Box 664 Canberra ACT 2601, Australia

Abstract. Catalogue cross-matching is an inherently costly operation. Our algorithm applies filter-refine and plane sweep techniques. Pre-processing consists of a sort by declination, and the active list is a queue indexed by a binary tree. The algorithm is O($N \log M$) in both I/O and processor costs, with only moderate memory requirements. Empirical assessment on catalogues of up to a billion records suggests that the algorithm performs at least an order of magnitude better than the techniques in current use.

1. Introduction

Catalogue cross matching is a fundamental operation in the virtual observatory. However, it is inherently costly, because the catalogues of interest are very large, and the matching operation is intrinsically a join. Current algorithms have costs that are quadratic with problem size and would take days to cross-match the very large catalogues being produced by current surveys.

The obvious approach to cross-matching is to match source records on the basis of positional coincidence. In this paper, we report a fast algorithm for determining the pairs of records, drawn from two catalogues, that are positionally coincident. The algorithm has preprocessing costs of O($N \log N + M \log M$), where N and M are the number of objects in the input catalogues. The matching operation proper has I/O costs of O($N+M$) and processor costs of O($(N+M) \log M$). An empirical assessment of the algorithm shows that, with modest computing equipment, matching of catalogues of a billion records can be performed in under six hours.

2. The Catalogue Cross-Matching Problem

We are given two source catalogues A and T, each with a large number of records in no particular sequence. We assume that the number of records is too large to be held in main memory. Each record specifies the apparent position of a source, taken to be a normally distributed random variable expressed as a mean location in right ascension and declination coordinates, and a standard deviation value expressed as a single angular distance. We assume that the catalogues have been standardised in their coordinate systems and epoch, so that the spatial locations are commensurable. Errors may vary between records in a catalogue.

A pair (a, t) of source records will be declared a match only if we are sufficiently confident that the true positions of a and t are the same. We can be $100\alpha\%$

confident that a and t are positionally coincident only if the angular distance between their means is no more than $z_{\alpha/2}\sqrt{\sigma_a^2 + \sigma_t^2}$.

Note that this problem definition permits the same record to occur in more than one candidate pair. Further processing may be required to further refine the set of matches, and possibly to select the best match for each record.

3. Outline of the Algorithm

Since computation of angular distance is an expensive operation, our problem definition leads to a test that is costly to perform. The key to efficient cross-matching is to perform this test as few times as possible. To achieve this, our algorithm applies two concepts. The first is to use the well-known filter-refine strategy from geospatial databases; the second is an algorithm based on the plane sweep metaphor from computational geometry.

3.1. Filter-Refine

In filter-refine strategies, we defer executing an expensive test by first performing a cheap "filter" test. The filter is designed to rapidly reject "obvious" non-matches at very low cost, but is guaranteed not to reject any true matches. Only those candidates that are accepted by the filter are subject to the full test. Our algorithm uses the filter-refine strategy by decomposing the full test into separate, and somewhat weakened, tests for right ascension and declination.

3.2. Plane Sweep

In plane sweep algorithms, a line is swept through the space, and an event is processed each time the line encounters an object of interest. Often the event processing serves to maintain an *active list*: a list of objects that are of current interest. Tests may then be restricted to members of the active list, which is typically much smaller than the full set of objects. Our algorithm adapts the plane sweep metaphor to the celestial sphere by sweeping a circle of constant declination through the space of records from catalogue T. An active list is maintained to contain those records from catalogue A that fall within the filter's declination bounds for the current record from catalogue T.

3.3. The Algorithm

As a prerequisite to sweeping by declination, both catalogues are sorted by declination. During sorting, we note the largest errors for each catalogue. We then sweep the space of records from catalogue T. As each record t is encountered, we compute upper and lower bounds on declination and right ascension for records that could match t, taking into account the catalogues' largest errors. We then update the active list to contain all and only those members of catalogue A whose mean positions fall within the declination bounds. Finally, we query the active list for all members whose mean positions fall within the right ascension bounds. For each returned record a, we perform the full test on the pair (a, t). For more detail, see Abel et al. (2004) and Devereux, Abel & Power (2004).

3.4. The Active List

The active list is a queue on declination, and a threaded binary tree on right ascension. This ensures that we can update and query the active list in $O(\log N)$ time.

The active list remains small enough to reside in main memory. For a catalogue of a billion objects and even if we assume errors of one arc minute, the active list will still only have a maximum of about 600,000 elements; the memory requirements for this are fairly modest.

4. Results

The initial sort is trivially $O(N \log N + M \log M)$. The sweep is $O(M + N)$ in I/O, and $O((N + M) \log M)$ for processing, where M is the size of the catalogue A, and N is the size of catalogue T.

Performance was empirically assessed by pairwise matching of six catalogues. Preprocessing costs are given in Table 1 and the costs of matching are given in Table 2. Benchmarking was performed on a Dual Pentium Xeon 2Ghz processor with 2Gb of main memory. The disk was five RAID5 storage arrays with 10K RPM UltraSCSI disks with a maximum transfer rate between the RAIDs and the server of 80Mb/s. For full details, see Power & Devereux (2004).

Catalogue	Times	
	elapsed	cpu
1XMM	0:02	0:01
SUMSS	0:02	0:03
Tycho2	0:34	0:33
2MASS	106:00	95:00
USNO A2	45:00	43:00
USNO B1	95:00	80:00

Table 1. Preprocessing Times (mm:ss)

	1XMM	SUMSS	Tycho2	2MASS	USNO A2	USNO B1
1XMM	0:00	0:01	0:04	12:00	13:00	27:00
SUMSS	0:00	0:01	0:03	9:00	11:00	20:00
Tycho2	0:03	0:03	0:08	14:00	15:00	33:00
2MASS	57:00	32:00	34:00	57:00	60:00	112:00
USNO A2	50:00	34:00	19:00	56:00	44:00	116:00
USNO B1	116:00	63:00	103:00	166:00	138:00	226:00

Table 2. Matching Times (mm:ss)

5. Related Work

Reported algorithms for catalogue crossmatching have been based either on exhaustive enumeration or spatial indexing. Malik et al. (2003) report an exhaustive enumeration approach as the basis of the XMATCH operation in SkyQuery. The approach is clearly $O(NM)$ in processor costs, and O(N+M) in I/O only if the catalogues can be held in main memory. Gray et al. (2004) report an exhaustive enumeration approach with zoning for the closely related problem of neighbour finding. Index-based joins have been reported by Kunzst, Szalay & Thakar (2001), Kalpakis et al. (2001), and Page (2003). Preprocessing costs depend on the index to be built, but are in general quite high. Matching costs are reasonably low.

6. Conclusions

We have reported an algorithm for catalogue crossmatching based on the application of a filter-refine strategy and a plane sweep algorithm. Our algorithm is logarithic in both I/O and processor costs, and has low main memory requirements. Empirical performance assessment shows solution times of around six hours elapsed for matching of a billion record catalogue with a half billion record catalogue. This appears to be a significant improvement on current algorithms. Tentative comparisons with the approaches of Gray et al. (2004) and of Page (2003) indicate that for catalogues of moderate size our algorithm is faster by at least an order of magnitude. Consideration of the orders of complexity suggest that the advantage will be higher still for large catalogues.

References

Abel, D. J., Devereux, D., Power, R. A. & Lamb, P. R. 2004, An $O(N \log M)$ Algorithm for Catalogue Matching, CSIRO ICT Centre Technical Report TR-04/1846

Devereux, D., Abel, D. J. & Power, R. A. 2004, Notes on the Implementation of Catalogue Cross Matching, CSIRO ICT Centre Technical Report TR-04/1847

Gray, J., Szalay, A. S., Thakar, A. R., Fekete, G. F., O'Mullane, W., Heber, G., Rots, A. H. 2004, There Goes the Neighborhood: Relational Algebra for Spatial Data Search, Microsoft Research Technical Report 2004-32

Kunszt, P. Z., Szalay, A. S., Thakar, A. R. 2001, in Mining the Sky: Proc. of the MPA/ESO/MPE workshop, 631

Kalpakis, K., Riggs, M., Pasad, M., Puttagunta, V., & Behnke, J. 2001, in ASP Conf. Ser., Vol. 238, ADASS X, ed. F. R. Harnden, Jr., F. A. Primini, & H. E. Payne (San Francisco: ASP), 133

Malik, T., Szalay, A. S., Budavari, T. & Thakar, A. R. 2003, SkyQuery: A Web Service Approach to Federate Databases. In Proc. CIDR'03

Page, C. G. 2003, in ASP Conf. Ser., Vol. 295, ADASS XII, ed. H. E. Payne, R. I. Jedrzejewski, & R. N. Hook (San Francisco: ASP), 39

Power, R. A. & Devereux, D. 2004, Benchmarking Catalogue Cross Matching, CSIRO ICT Centre Technical Report TR-04/1848

ASTRONOMICAL DATA ANALYSIS SOFTWARE AND SYSTEMS XIV
ASP Conference Series, Vol. 347, 2005
P. L. Shopbell, M. C. Britton, and R. Ebert, eds.

Data Mining in Distributed Databases for Interacting Galaxies

Kirk Borne

School of Computational Sciences, George Mason University

Abstract. We present results from an exploratory data mining project to identify classification features of special classes of interacting galaxies (for example, infrared-luminous galaxies) within distributed astronomical databases. Using a variety of data mining techniques, interaction-specific features are learned, to distinguish this class of galaxies from a control sample of normal galaxies. Eventually, the corresponding rule-based feature model of that class of galaxies will be applied to the large multi-wavelength astronomical survey databases that are becoming available. This distributed data mining activity is a prototype science use case for the VO (Virtual Observatory). We specifically apply multi-archive multi-wavelength data to the problem. In a preliminary validation experiment, we recovered exactly the type of object that we hope to find automatically with our data mining tools: a distant hyper-luminous infrared galaxy (HyLIRG), the most luminous class of known galaxies. This particular galaxy was previously known, but we re-discovered it serendipitously.

1. Introduction

The astronomy community is rapidly becoming the beneficiary of huge multi-terabyte databases from a host of sky surveys. These new digital surveys have the potential to produce a wide range of scientific discoveries, but the rich and diverse information content within this "virtual sky" will far exceed the current capacity of data search and research tools. The globally emerging Virtual Observatory (VO) should meet this challenge. Multi-wavelength studies have led to the discovery of several remarkable classes of objects, including: QSOs, GRBs, ULIRGs, XRBs, radio galaxies, submm galaxies, and Type II quasars. The VO should further this legacy of discovery. To facilitate the new type of science enabled by the VO, new techniques in data mining and knowledge discovery in large distributed databases must be developed and deployed. The value of scientific data mining and some early test case results will be discussed in the context of interacting galaxies as the building blocks of the Universe.

The common feature of the new large sky surveys is that they will produce massive databases, containing terabytes and eventually petabytes of data products (e.g., LSST = the Large Synoptic Survey Telescope project). Enormous scientific databases such as these contain information nuggets that could go undiscovered simply because the databases are too large to be thoroughly investigated even with the support of today's most sophisticated database tools. Significant efforts are being made now to extend traditional data access techniques with state-of-the-art research in data mining and visualization. Without this, the scientific research community will be overrun by volumes of data that cannot be assimilated. Scientific discoveries lie hidden among the myriad of entries in not just single databases but most especially among the cross-correlated contents of heterogeneous distributed databases. Techniques are being developed that will be applicable to large, distributed, federated databases within this new distributed astronomical data environment (the VO).

Halem et al. (1999) argued in their "Data Crisis" white paper that the next evolutionary step in the current data explosion is the transformation of data into knowledge-based information. In other words, the end-goal of data mining and archival research in large databases is not the collection and curation of the data bytes alone, but the real goal is the extraction of the rich information and knowledge content represented therein. Thus, the application of data mining (knowledge discovery) functions in general, and distributed data mining services in particular, must become a core feature of the VO for it to succeed.

2. VO Science Use Cases

The 2001 Aspen workshop "New Astrophysics and Cosmology with a Virtual Observatory" produced a "wish list" of science drivers, use cases, and requirements for the emerging VO research environment, including the following:

- Solar System: NEOs, Long-Period Comets, TNOs, Killer Asteroids!
- The Digital Galaxy: find star streams and populations – relics of past and present assembly phases. Identify components of disk, thick disk, bulge, halo, arms, and whatever else.
- The Low-Surface Brightness Universe: spatial filtering, multi-wavelength searches, intersection of the image and catalog domains.
- Panchromatic Census of AGN (Active Galactic Nuclei): complete sample of the AGN zoo, their emission mechanisms, and their environments.
- Precision Cosmology and Large-Scale Structure: hierarchical assembly history of galaxies and structure, cosmological parameters, dark matter, lensing, and galaxy biasing as $f(z)$.
- Precision science of any kind that depends on very large sample sizes.
- "Survey Science Deluxe".
- Search for rare and exotic objects (e.g., high-z QSOs, high-z Sne, L/T dwarfs).
- Serendipity: Explore new domains of parameter space (e.g., time domain, or "color-color space" of all kinds).

We describe here a project that addresses the "hierarchical assembly history of galaxies" through the exploration of multi-wavelength "color-color space" (Borne 2003a; Borne 2003b).

3. A Distributed Data Mining Prototype Study

Galaxy interactions (i.e., collisions and mergers) are the means by which mass assembly occurs in the Universe. Understanding the dynamical history and the "merger family tree" of galaxies will enable us to understand the rate and fate of hierarchical mass assembly processes throughout cosmic time. As a demonstration of some of the key aspects of distributed data mining that can be applied to such astrophysics research problems, we present a case study of one particularly interesting class of galaxies that has not been well studied to-date: the Very-Luminous Infra-Red Galaxies (VLIRGs; Arribas et al. 2004). Data mining in a variety of multi-wavelength multi-project multi-modal (imaging, spectroscopic, catalog) databases will eventually yield interesting new scientific properties, knowledge, and understanding for the VLIRGs as well as for many other objects of astrophysical significance. Very-Luminous IR Galaxies (VLIRGs) are key to the study of galaxy formation and evolution. On one hand, they offer the opportunity to study how the fundamental physical and structural properties of galaxies vary with IR (infrared) luminosity, providing the link between the "boring" set of normal galaxies and the extremely chaotic and dynamic Ultra-Luminous IR Galaxies (ULIRGs; Borne et al. 2000; Colina et al. 2001; Bushouse et al. 2002). On the other hand, VLIRGs are believed to be closely related with recently identified cosmological populations in the sense that VLIRGs should be either the low-redshift analogs or the direct result of the evolution of those cosmologically interesting populations of galaxies that appear only at the earliest stages of cosmic time in our Universe.

Several cosmologically interesting populations have been recently identified by astronomers. These include: (1) the galaxies that comprise the Cosmic Infrared Background (CIB); (2) the high-redshift submillimeter sources at high redshift; (3) the infrared-selected quasars; (4) the "Extremely Red Objects" (EROs) found in Hubble Space Telescope images and other deep-sky surveys; and (5) host galaxies for the extremely luminous Gamma-Ray Burst sources. The primary questions pertaining to these cosmological sources are: What are they? Are they dusty quasars? or dusty starbursting galaxies? or massive old stellar systems? Some of these classes of objects are undoubtedly ULIRGs, while the majority are probably related to the significantly more numerous class of galaxies: the VLIRGs. Our VO data mining project will enable us to identify signatures of VLIRGs that are indicative of this special class, and thus to distinguish between starbursting galaxies, massive ellipticals, and quasars. The results will therefore be applicable to understanding and interpreting the properties of the five cosmological samples listed above. If, as believed, both ULIRGs and VLIRGs represent low-redshift analogs to the high-redshift galaxies, a comprehensive analysis of the VLIRGs is needed. This is clear, for instance, when interpreting the diffuse CIB. Deep submm surveys have confirmed that IR-luminous galaxies are a significant, if not the dominant, contributor to the CIB. However, submm observations have been able to identify only ULIRGs and the most luminous VLIRGs. Taking into account the significantly larger density of VLIRGs in the Universe, it appears likely that VLIRGs could contribute a significant fraction of the infrared radiation at high redshifts. A thorough data mining exercise within the VO across multiple databases will enable a significant research advantage toward resolving these scientific questions.

We initiated a proof-of-concept VO science search scenario by attempting to identify potential candidate contributors to the CIB. This is significant since it has been shown that fully one half of all of the radiated energy in the entire Universe comes to us through the CIB! Our approach (Borne 2000) involved applying the full power of several distributed on-line databases and the linkages between these databases, archives, and published literature. Our search scenario involved finding object cross-identifications among the various distributed source lists and archival data logs. In a very limited sample of targets that we investigated to test our VO data mining approach to the problem, we did find one object in common among three distributed databases (HST, IRAS, and FIRST): a known hyperluminous infrared galaxy (HyLIRG) at moderate redshift harboring a quasar, which was specifically imaged by the HST because of its known HyLIRG characteristics. In this extremely limited test scenario, we did in fact find what we were searching for: a distant IR-luminous galaxy that is either a likely contributor to the CIB, or else it is an object similar in characteristics to the more distant objects that likely comprise the CIB (Borne 2000).

This project will increase our understanding of the VLIRG class (a particular type of interacting/merging galaxy system), as a step toward the ultimate goal of understanding the structure and evolution of the Universe. The success of this limited proof-of-concept study encourages us to pursue even greater database searches and data mining scenarios.

4. The VO: Distributed Data Mining in Action

Distributed data mining (DDM) will be a core function of the VO (e.g., see the DDM resources at `http://www.cs.umbc.edu/~hillol/Kargupta/ddm.html`). In particular, we hope to develop useful functional test results and working algorithms for other scientists to explore the rich distributed astronomical data archives that will comprise the VO. We anticipate that XML-based Web Services will play a key role in our developments.

The prototype project described above focuses explicitly on the challenges imposed by a distributed scientific data mining system. For example, ULIRGs represent one in a million galaxies in the sky (the proverbial "needle in a haystack" – a perfect challenge problem for data mining). Despite their local rarity, the ULIRGs are likely to be extremely significant components of the early Universe, and thus greater knowledge of their properties will enable a greater understanding of our Universe. ULIRGs and VLIRGs are characterized by a unique clustering of their optical, radio, Near-Infrared, and Far-Infrared emission properties in the corresponding multi-dimensional parameter space. Learning algorithms enhance the data discovery rate through classification rule induction and association mining of ULIRG parameters. The main challenge for this and similar VO research programs is to conduct data mining in a distributed environment. Our goal is to address this challenge as a VO pathfinder project.

We are also investigating the application of genetic algorithms and other data mining methods, such as cluster-finding, artificial neural networks, association mining, Bayesian networks, and decision trees.

The steps that we are following are also generally applicable to other distributed data mining projects in the VO. In particular: we compile a training set using ULIRG and VLIRG properties tabulated in existing astronomical data catalogues. We initially use a subset of known ULIRGs to train our data mining algorithm (i.e., Supervised Learning), and then we attempt to "re-discover" the remaining subset in order to validate the classifier. Cluster-finding and association mining algorithms can be used to identify additional unique characteristics of ULIRGs that can be mined in larger data sets (present and future). These intelligent data mining techniques can then be deployed as a virtual "science instrument" within the VO in order to find similar "outliers" within future VO databases. With the full growth of scientific data mining technologies and the full implementation of distributed data system interoperability, along with the integration of these two streams of information technology development, wonderful new scientific discoveries are waiting to be made.

5. Summary

The VO will change the way that astronomical research is conducted in the future. For the first time since astronomers began making observations of the night sky, we will soon have a complete multi-spectral survey database of the entire sky. Even though astronomers will archive the sky, this gigantic "virtual telescope" currently lacks good instrumentation (e.g., tools for distributed data mining). The proven technologies that will result from VO distributed data mining projects may eventually be applied to future deep space missions in the exploration, analysis, and interpretation of their voluminous science data products. As future science mission data streams grow exponentially, it may be desired that an on-board autonomous "intelligent data understanding" capability be used in real-time to mine data streams for autonomous on-board science planning and scheduling. In general, we expect these data mining techniques to enhance various scientific data-collecting/mining/analysis modes in the future, including distributed "sciencecraft" (Satellite Constellations), remote sensors (or sensor webs), and geographically distributed science data archives (e.g., VO).

Acknowledgments. Support for this work was provided in part by NSF through Cooperative Agreement AST0122449 to JHU, and in part by research award 749-10-02 from the NASA CICT Intelligent Systems Program.

References

Arribas, S., Bushouse, H., Lucas, R., Colina, L., & Borne, K. 2004, AJ, 127, 2522

Borne, K. 2000, in Mining the Sky, 671

Borne, K. 2003a, in AIP Conf. Proc., Vol. 666, The Emergence of Cosmic Structure, ed. Holt & Reynolds (AIP), 307

Borne, K. 2003b, in SPIE Data Mining & Knowledge Discovery V, 5098, 211

Borne, K. D., Bushouse, H., Lucas, R.A., & Colina, L. 2000, ApJ, 529, L77

Bushouse, H., Borne, K., Colina, L., Lucas, R., et al. 2002, ApJS, 138, 1

Colina, L., Borne, K., Bushouse, H., Lucas, R., et al. 2001, ApJ, 563, 546

Halem, M., et al. 1999, "Data Survivability Crisis" white paper

ASTRONOMICAL DATA ANALYSIS SOFTWARE AND SYSTEMS XIV
ASP Conference Series, Vol. 347, 2005
P. L. Shopbell, M. C. Britton, and R. Ebert, eds.

A Source Extraction Web Service with Cross Matching Capability

K. S. Krughoff, A. J. Connolly, and J. M. Colberg

Department of Physics and Astronomy, University of Pittsburgh, Pittsburgh, PA

W. O'Mullane

Department of Physics and Astronomy, The Johns Hopkins University, Baltimore, MD

R. D. Williams

Center for Advanced Computing Research, California Institute of Technology, Pasadena, CA

Abstract. Web services provide an excellent means for users to take advantage of algorithms and computing time on machines that are physically displaced from their location. We can construct tools that are essentially platform and language independent by using XML, specifically SOAP, as the communication language between client and server. This allows for a large set of utilities to be built for use by the entire astronomical community via simple client programs.

We present a case study in which we have deployed the popular source extraction code "SExtractor" in a web service. "SExtractor" takes an input FITS image and output a catalog of sources with associated parameters. Our web service provides the additional option of cross matching the resultant catalog with other previously published data sets (i.e. SDSS, 2MASS, FIRST, HDF, etc.) through the SkyPortal web service provided by OpenSkyQuery. This is an important step in the development of web services as we begin to use pre-existing web services to build increasing utility into the set of tools available.

1. Introduction

Recently web services have been getting more attention as a potentially viable way of deploying data and algorithms to the astronomical community as a whole. Although the term web service has a relatively broad definition, we will be referring to programs resident on a server machine that can be interacted with via Simple Object Access Protocol (SOAP) messages over TCP/IP.

Web services allow users to make use of algorithms and datasets that are too large, too complicated, or simply too common to justify reconstructing them nearby physically. Recently, the National Virtual Observatory (NVO), Virtual Observatory India (VO-India), and the Astrophysical Virtual Observatory (AVO) have been making some strides in the direction of making web service a formidable tool for astronomical data access as well as data analysis.

In particular we make use of the OpenSkyQuery and SkyPortal services developed and hosted at JHU for the NVO for cross matching and dynamic discovery of database content. We also utilize VOPlot which is a data plotting tool created by VO-India expressly to handle the VO standard data format, VOTable. An embedded version of Aladin, created by CDS/AVO and hosted at Harvard, provides the data visualization and allows overlay of data and image.

2. Scientific Motivation

The facilities for any astronomer to create a source catalog and cross match it with other survey catalogs exist. However, by making the machinery available for doing this tedious and complex task through the web, we can distill the process so that it can be accomplished with very little time expenditure. The fact that the web service can be accessed both through a simple web interface and called programmatically via automatically generated code means that at all stages of a scientific project this web service can be of use. In early stages, it can be used in feasibility studies. As the project progresses, it can be used to process thousands of images with very little human intervention.

Some of the immediate uses for this service are:

- Photometric calibration by comparison to catalogs with known photometry
- Identification of optical counterparts to sources in other bands
- Search for high redshift objects by looking for photometric dropouts
- Star-galaxy separation
- Variability studies

Currently WSExtractor is being used to identify radio galaxies with no Sloan Digital Sky Survey optical counterpart for infrared observational followup. It is also being used in a feasibility study for use of SExtractor as the source detection software for an upcoming large scale survey.

3. The Machinery

The components that make up this web service can be divided into three distinct modules which need only be connected by an internet connection. These three portions are represented by the three areas separated by dashed lines in Figure 1.

3.1. Client/Web Front End

Automatically generated code to aid in communication with the server can be created by processing the WSDL (Web Service Definition Language) associated with the server application. Currently, only Java classes generated within the Apache Axis framework have been tested.

3.2. Server

The server was written in Java using the Apache Axis framework. The server expects to get an image in the Flexible Image Transport System (FITS). The image must also contain a world coordinate system (WCS) that is readable by the SExtractor software. Ideally, all transfer of data is done in the VOTable format, although the server will return the data in a string if so desired.

3.3. XMatcher

Cross matching is done by uploading the extracted source catalog to the OpenSkyQuery via the SkyPortal web service. The uploaded catalog is then cross matched with any catalog currently available on OpenSkyQuery.net by submitting a query, again through the SkyPortal web service, in valid Astronomical Data Query Language (ADQL). ADQL is based on standard SQL with additional syntax to allow cross matching, and distributed database queries.

4. Communication

All communication between client, server and xmatcher are with SOAP. In general, parameters and return data are contained in the body of the messages with the upload FITS image contained in an attachment to the SOAP message. See Table 1 for a description of the communication between client, server, and xmatcher.

Acknowledgments. The following sites have provided software and useful information in this project.

SExtractor: `http://terapix.iap.fr/soft/sextractor/`
OpenSkyQuery: `http://www.openskyquery.net`
ADQL: `http://skyservice.pha.jhu.edu/develop/vo/adql/`
Java: `http://java.sun.com/`
VOTable Standards: `http://www.us-vo.org/VOTable/`
U.S. Virtual Observatory Services: `http://voservices.org/`
Apache Web Services: `http://ws.apache.org/`

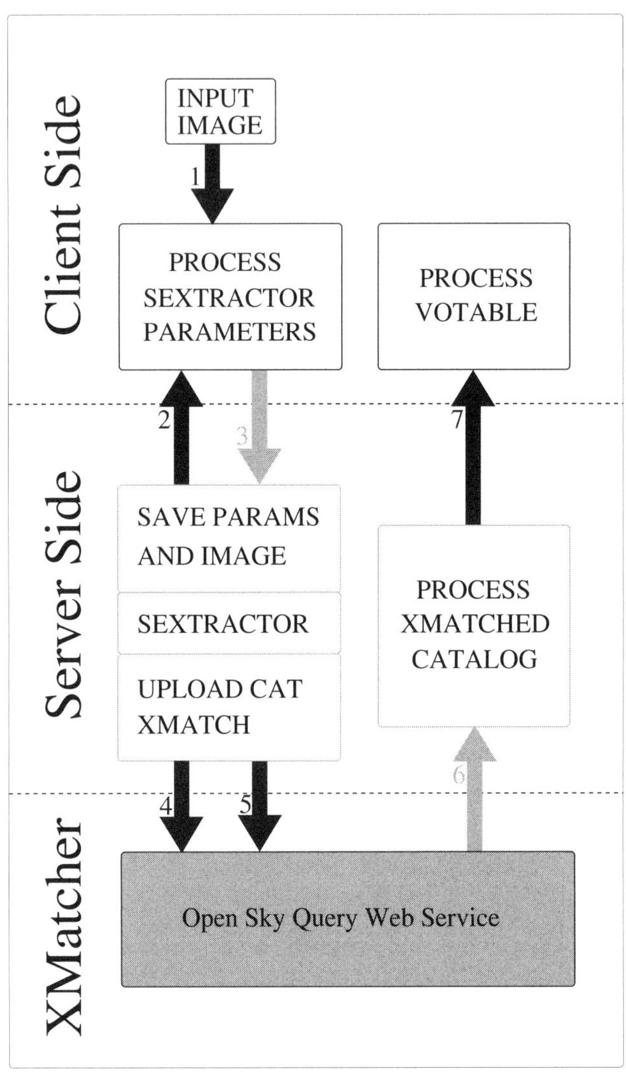

Figure 1. Data flow through web service. Arrows that cross dashed lines represent SOAP messages.

Table 1. This is a description of the communication in the web service data flow. Numbers correspond to those in Figure 1.

Num	Description of Communication
1	The client must read a FITS image. The FITS image must have a WCS that is recognized by SExtractor.
2	The client may request default values for the input parameters and output fields from the server. The client stubs parse the messages into Java classes containing modifiable variables.
3	The client stubs attach the FITS image to an outgoing SOAP message. Optionally, modified input parameters and output fields are sent in the message body. Upon receipt, the server stubs[a]extract the attached image and save it to the local file system. The parameter files necessary for running SExtractor are also read to the local filesystem.
4	After running SExtractor, the server parses the source catalog into an instance of the VOTable[b]class. The VOTable is passed to the SkyPortal web service which loads it into a SQL database and returns the name of the table.
5	Once the VOTable is uploaded and the table name has been returned, the server constructs a valid ADQL query string for cross matching with other catalogs and sends it to the SkyPortal web service.
6	SkyPortal returns a VOTable containing the cross matched catalog. Upon receipt, the server parses the VOTable into a Java VOTable class. At this point any desired post-processing may be done.
7	The server ships the final cross matched catalog back to the client in a VOTable. The client stubs then parse the VOTable into a Java VOTable class. At this point the catalog can be read out to a file, or sent on to other VO enabled web services for further manipulation.

[a] Stubs are java classes which act as proxies for instances of other components. Most of the time, they can be automatically generated from a description of the web service (WSDL).

[b] VOTable is the adopted data transfer protocol for the Virtual Observatory. It is defined in XML making it directly mappable to language specific datatypes.

ASTRONOMICAL DATA ANALYSIS SOFTWARE AND SYSTEMS XIV
ASP Conference Series, Vol. 347, 2005
P. L. Shopbell, M. C. Britton, and R. Ebert, eds.

Distributed Data Mining for Astrophysical Datasets

Sabine M. McConnell and David B. Skillicorn

School of Computing, Queen's University, Kingston, Ontario, Canada

Abstract. Over the past decade, data mining has gained an important role in astronomical data analysis. Traditionally, such analysis is performed on data at a single location. However, one of the main motivational forces behind a virtual observatory is the distributed nature of both data and computational resources. Existing data-mining methods for distributed data are either communication-intensive or result in a loss of accuracy. In this paper, we introduce a general approach to supervised data mining that allows data to remain distributed, but still produces satisfactory results. We demonstrate by applying the approach to a number of astronomical datasets.

1. Introduction

Data mining, as the automated or semi-automated discovery of novel and potentially useful information, has become an important tool for astrophysical datasets over the past decade. This becomes evident very quickly when one considers the widespread range of astronomical applications, utilizing a variety of techniques. Examples for this include star/galaxy separation (Odewahn and Nielsen 1994), classification of stars (Weaver 2000), galaxies (Odewahn 1997), and planetary nebulae (Faúndez-Abans et al. 1996), antimatter search in cosmic rays (Bellotti 1997), detection of expanding HI shells (Daigle et *al.* 2003), and selection of Quasar candidates (Padovani et *al.* 2004).

In a traditional setting, data-mining algorithms are applied to data at a single location. When data is collected in a distributed way, this means that, at some point, the data must be transferred to a single computer. This enables conventional data-mining algorithms to be applied, but is costly in terms of communication, and storage facilities at the central site. Processing power and memory size limits at such a central site will limit the sophistication of algorithms, and the size of the datasets that can reasonably be processed. It also requires each data source to reveal its raw data completely, which may not always be appropriate.

One of the major design goals of the virtual observatory is to achieve universal access to geographically distributed datasets, so new data-mining approaches that can handle distributed data are required. At present, few approaches that can achieve these goals are known.

In this paper, we introduce a method for distributed data mining that both reduces the time needed to produce results while maintaining or even increasing the accuracy compared to the same algorithm applied to centralized data at a single site. We demonstrate the effectiveness of this method using a variety of astronomical datasets.

2. Partitions of astronomical datasets

An astronomical dataset typically contains information about a set of objects, such as galaxies or stars, and a set of attributes or measurements about these objects. Because of the way data is collected, these datasets are partitioned by objects (horizontal partitioning) and by attributes (vertical partitioning).

Horizontal partitioning stems from a variety of factors such as the physical location of the collection instruments (Canadian Galactic Plane Survey vs. Southern Galactic Plane Survey, for example), differing target objects (Condon 1987 and Condon 1990), as well as detection limits (fainter objects cannot be detected by less sensitive instruments). Vertical partitioning stems from the fact that different surveys measure different parameters, for example spectra at different wavelengths. A more subtle form of vertical partitioning occurs because of the variation in resolution caused by technological advances as instruments become more sensitive.

Astrophysical data mining is therefore more complex than in many other domains where data may be partitioned vertically or horizontally, but typically not both. Mining astronomical datasets obtained from ground- or space-based surveys is therefore at one extreme of a spectrum of complexity.

3. Related Work

For horizontally distributed data, where information about different objects appears in different partitions, a number of approaches are known, typically producing similar accuracy to the equivalent centralized algorithm. These may require only minimal communication, but they often require some assumption that the data in different partitions is similar, so that local models are good approximations to the final, global model (Skillicorn 1999). This assumption may not be valid for astrophysical data.

Another successful approach for data partitioned by objects is based on *ensembles*, that is collections of classifiers. In ensemble techniques, predictors are built from local datasets and then combined, most often through a simple voting scheme, to produce a final classification for previously unseen data.

For vertically partitioned data, two approaches, *Meta-Learning* (Chan & Stolfo 1993) and *Collective Data Mining* (CDM) (Kargupta et al. 1999) have been developed. Both techniques are expensive, and the accuracy achieved through CDM, which represents the function to be learned with an approximation as the sum of a set of basis functions together with their corresponding coefficients, is always lower than that achieved on a centralized dataset.

4. A Lightweight Distributed Data Mining Approach

Our approach to building a predictor for data that is both horizontally and vertically partitioned is to build a predictor locally on each partition, send these predictors to a central site and combine their predictions by voting. Hence this is an ensemble technique extended to a more-general form of data partitioning. This is a simple idea, but still outperforms much more sophisticated and expensive alternatives, as we show below.

For the evaluation presented here, we use a decision tree, which classifies the dataset by a hierarchical series of single-attribute relations (Quinlan 1986). This is a supervised data-mining technique in which the goal is to learn how to predict the target class of new data, given training data labeled with the target class. An example of such a classification task is galaxy classification: images of unclassified galaxies are categorized based upon a model derived from known galaxy classifications (Odewahn 1997). For our implementation we used J48, the decision tree implemented in the data mining package WEKA, which is available for download at `http://www.cs.waikato.ac.nz/ml/weka/`. It is important to note that our approach is not restricted to decision trees, but can use any weak learner.

An outline of the algorithm is given below.

```
create training set
create out-of-bag test set
create vertical partitions of the training set
for all vertical partitions do
    create horizontal partitions for the training set
end for
for all partitioned datasets do
    build a model using local data only
    classify the out-of-bag test set
    communicate the results to a centralized site
end for
Combine the local results using voting to obtain overall results
```

Training sets are created by resampling the data with replacement until a set of the original size of the dataset is reached. Typically about a third of the data will never have been sampled, and plays the role of a test set, called an out-of-bag test set. Both simple voting, and weighted voting using test accuracy are used.

This approach was evaluated on 5 astronomical datasets, two of which were taken from the SDSS Data Release 3. Table 1 shows the number of objects, the number of classes (distinct groups of objects) as well as the number of attributes (information about the objects) for each of the datasets.

Name	**Objects**	**Classes**	**Attributes**
Dwarfs	8419	10	3
Galaxy/QSO	11427	10	2
Galaxies	1151	4	3
Galaxy/Star	4192	14	2
Variable Stars	10970	3	12

Table 1. Overview of dataset properties

5. Results and Discussion

Table 2 shows the prediction accuracies for a decision tree given the entire dataset, and the best and worst prediction accuracies for distributed predictors. Predictors were built for all combinations of partitions where the number of horizontal partitions ranges from 1 to 8; and the number of vertical partitions ranges from 1 to the number of attributes. Hence, in the hardest case, predictors were learned from single attributes, and only 1/8th of the objects.

Name	centralized case	best distributed case	worst distributed case
Dwarfs	99.69	98.65	96.61
Galaxy/QSO	80.98	80.60	75.11
Galaxies	77.14	**78.96**	71.60
Galaxy/Star	93.55	**94.96**	87.03
Variable Stars	99.97	99.95	90.03

Table 2. Prediction accuracies

From these results, it can be seen that, for each of the datasets under investigation, the achieved accuracy in the best distributed case is at least as good as that obtained from a centralized approach. For the two datasets which exhibited an increase in accuracy, the result is statistically significant. Even for the worst case, the Variable Stars dataset, the accuracy drop is only 10%, and occurs in the worst configuration when each vertical partition contains only a single attribute. When the partitions contain 2 attributes, the prediction accuracy immediately increases to over 95%.

In general, other advantages of our approach are that it requires less time to build the prediction models, since each is learning from a smaller dataset; and the models are typically smaller than that learned from centralized data, even in aggregate.

6. Summary

Our lightweight distributed data-mining approach, in which local models are built from data partitioned both by attributes and objects and then combined through a simple voting scheme, can achieve performance comparable to, and occasionally better than, that of a centralized predictor for all of the datasets investigated here.

The advantages of our approach are: it is simple, it requires very little communication, it can use any weak learner as the underlying predictor, and it is typically faster than the equivalent centralized version. In addition, no transmission of raw data is necessary, so our approach is also suitable for situations where disclosure of data is not desired, such as, for example, an initial proprietary period after survey completion.

Given the success of data mining in astrophysical datasets, data mining will be an important component of the Virtual Observatory. Modelling from distributed data without collecting it in a single location requires data-mining approaches of the kind described in this paper. The existence of a simple and effective way to build predictors from such distributed data demonstrates additional advantages of the Virtual Observatory over existing centralized data repositories.

References

Bellotti, R. 1997, Applications and Science of Artificial Neural Networks III: Orlando: SPIE3077, 169

Chan, P., & Stolfo, S. 1993, Proceedings of the second international conference on information and knowledge management, Washington, DC, 314

Condon, J.J. 1987, ApJS, 65, 485

Condon, J.J., Helou, G., Sanders, D.B., & Soifer, B.T. 1990, ApJS, 73, 359

Daigle, A., Joncas, G., Parizeau, M., & Miville-Deschênes, M. 2003, PASP, 115, 662

Faúndez-Abans, M., Ormeño, M. I., & de Olivieira-Abans, M. 1996, A&AS, 116, 395

Kargupta, H., Park, B., Hershberger, D., & Johnson, E. 1999, Collective data mining: A new perspective toward distributed data mining, Advances in Distributed Data Mining, Eds: Hillol Kargupta and Philip Chan (AAAI/MIT Press)

Odewahn, S.C., & Nielsen, M. L. 1994, Vistas in Astronomy, 38, 281

Odewahn, S.C. 1997, Applications of Digital Image Processing XX: SPIE 31647, 471

Padovani, P., et al. 2004, A&A, 424, 545

Quinlan, J. R. 1986, Machine Learning 1(1), 81

Skillicorn, D. 1999, IEEE Concurrency, 7(4), 26

Weaver, W. B. 2000, ApJ, 541, 298

ASTRONOMICAL DATA ANALYSIS SOFTWARE AND SYSTEMS XIV
ASP Conference Series, Vol. 347, 2005
P. L. Shopbell, M. C. Britton, and R. Ebert, eds.

Assembly and Classification of Spectral Energy Distributions – A New VO Web Service

Hans-Martin Adorf,[1] Florian Kerber,[2] Gerard Lemson,[1] Alberto Micol,[3] Roberto Mignani,[4] Thomas Rauch,[5] and Wolfgang Voges[1]

[1]*GAVO, Max-Planck-Institut für extraterrestrische Physik, Garching*

[2]*ST-ECF, European Southern Observatory, Garching*

[3]*ST-ECF, European Space Agency, Garching; affiliated with the Space Telescope Division of ESA, ESTEC, Noordwijk, Netherlands*

[4]*European Southern Observatory, Garching*

[5]*Institut für Astronomie und Astrophysik, Universität Tübingen*

Abstract. We report progress on a new Web service for automated supervised object classification, which is being implemented in a collaboration between the GAVO[6] and AVO[7] projects. The new service will eventually carry out the following steps: An input list of sky-positions is used for querying multiple distributed catalogues covering different wavelength ranges. The sources returned are *spatially* matched using a statistical method, and a list of observational spectral energy distributions (SEDs) is assembled. Model SEDs are prepared using a library of theoretical spectra. The observational SEDs are submitted to a statistical classifier that uses the model SEDs for *photometric* template matching. For each observational SED the best-matching model SEDs are identified. – A science case has been selected for testing the capabilities of this Web service.

1. Scientific Motivation

The first scientific result from the European AVO project is the discovery of obscured active galactic nuclei (AGN) by combining multi-wavelength data (Padovani et al. 2004). Similarly, the "pan-chromatic mining for quasars" is a keystone science application of the US-NVO. A data mining tool producing spectral energy distributions (SEDs) will not only facilitate AGN-research, but can also support research in such diverse areas as isolated neutron stars, brown dwarfs, cataclysmic variables, or planetary nebulae (PNe).

In the PN field, for instance, a complete optical-to-IR color coverage is essential in order to better characterize the SEDs of known central stars (CSs), and to derive their astrophysical parameters, i.e. T_{eff} and $\log g$. Such a characterization can also be used to search for CS-candidates of those PNe, for which no CS is presently known – about 30 % of the extended PNe. In addition, SED

[6]`http://www.g-vo.org`

[7]`http://www.euro-vo.org`

characterization can assess the frequency of binarity (indicated by an IR-excess) among PNCSs, which is an important unresolved issue.

In the following we report on work in progress to create a VO-tool for the assembly and subsequent classification of SEDs, which is divised to directly enable scientific investigations. The tool comprises several discernible modules (Figure 1), namely *catalogue query and (positional) matching* with *SED assembly*, *SED transformation*, and *SED classification*, and these are described in some detail below.

2. Catalogue Query and Matching

The catalogue query and matching task, which is solely based on sky-positions and their uncertainties, is itself a multi-stage process: The user provides an input list of RA and Dec coordinates, selects a list of catalogues, and chooses a search radius. For each catalogue a deterministic matching service, provided in our case by CDS/Vizier, is invoked that carries out a simple cone search for each input position. The result is a set of primary match lists, one per catalogue queried. These lists are fed into a statistical matcher that, using GAVOs fuzzy matching algorithm, joins them into a single "fused" master list. The latter contains all plausible match candidates, where each such candidate comprises at most one source from each catalogue.

Sometimes, for a given input position, no counterpart is found in a particular catalogue. In other cases, the primary matching process will be ambiguous, particularly when the search radius is relatively large, which is necessary when the input sky-positions are imprecise.

3. Assembly of the Observational SEDs

Catalogue matching is followed by an assembly process producing spectral energy distribution (SEDs) from the observational data. The assembly in turn takes several steps: For each match-candidate the photometric measurements are collected from the contributing catalogues. When a particular catalogue has no counterpart for a given input position, the corresponding photometric measurements are set to null. Even when the catalogue has a matching source, there may still be no detected photometric flux in a given passband. Clearly, in many if not most situations, SEDs with incomplete ("missing") data will result. Next, for each SED-candidate, unit conversions are applied to the photometric measurements, in order to produce a "well-formed" SED. The resulting SED comprises the "features" which the classifier subsequently operates on.

Note that sometimes such an assembled SED makes no sense, since the photometric measurements are being collected into a "hypothetical SED" assuming that the matching sources from the different catalogues belong to the same underlying physical object. Of course, this hypothesis is sometimes simply wrong, but the error cannot be detected on the basis of uncertain astrometric information alone.

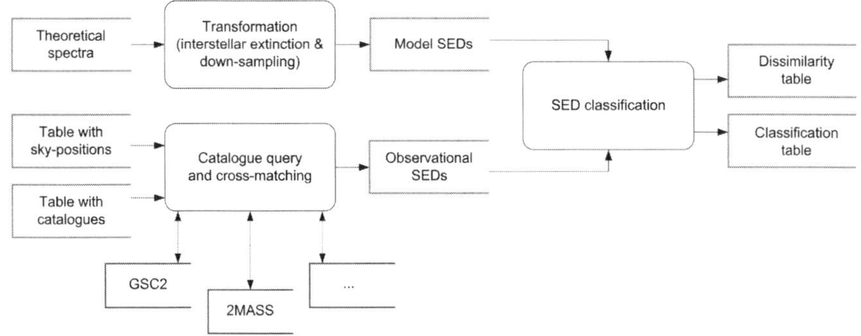

Figure 1. The Gane-Sarson data flow model diagram of the SED classification tool: the *transformer* module produces a set of model SEDs; the *catalogue query, matching & SED assembler* module produces a table of observational SEDs; these are fed into the *SED classification* module which classifies the observational SEDs.

4. Generation of the Model SEDs

Our scheme for classifying the observational SEDs is based on a comparison with model SEDs. We are using a grid of theoretical stellar atmosphere spectra[8], which needs to be transformed into the observational space in order to serve the subsequent classification stage. Since the spectra have a much higher spectral resolution than the observational broad band SEDs, the former have to be degraded (down-sampled) to match the resolution of the latter.

Currently, the flux of each model spectrum is simply extracted at the central wavelengths for each of the 7 filters used (Johnson B, V, R, I, H, J, K). – In the future we plan to transform the theoretical spectra to model SEDs by applying the corresponding filter transmission curves. In addition, interstellar extinction needs to be taken into account.

5. Supervised Classification

The list of observational SEDs (more precisely, *candidate* SEDs) is submitted to a supervised classifier, which uses the library of model SEDs for photometric template matching, similar to the approach employed by Adorf (1986).

We extend this work in several respects: (i) The user may choose the set of fluxes to be used as features during classification. This is important e.g. for PN research. (ii) Due to physical distance and absorption/extinction effects, the fluxes of the model and observational SEDs are not on the same scale. Therefore a scaling factor is estimated, which brings a given model SED onto the same scale as a given observational SED. The factor depends linearly on the observational SED, and non-linearly on the model SED. (iii) The SED classifier offers two algorithms using different dissimilarity metrics: (a) a simple deterministic

[8] `http://astro.uni-tuebingen.de/~rauch`

algorithm based on the Euclidean distance in feature space, which may be used in the absence of uncertainties; (b) a more advanced statistical algorithm based on the Mahalanobis-distance, which is closely related to the χ^2-metric. The variance-covariance matrix required for the Mahalanobis-distance is computed from the photometric uncertainties of the observational SEDs.

For each observed SED a nearest neighbour (NN) algorithm identifies the three best-matching model SEDs, which are reported to the user. In case of the Mahalanobis-distance metric, the best-matching class is identical to the maximum-likelihood classification.

In principle any model SED library, whether pre-stored or user-supplied, may be used for photometric template matching, provided the model SEDs comprise the same features as the observational SEDs. Thus the model SEDs could also be *empirical prototypes* derived from observational data. Often such prototypes will be generated by unsupervised classification. – In principle our classifier could also be used for purposes quite different from SED classification.

6. Reporting and Visualization

The classification results are reported in a classification table, comprising among other items, the number of (non-null) features participating in the classification, the identifiers of the three best-matching models, and the corresponding dissimilarities between the observational and the model SEDs. – Additionally, the full complement of pair-wise dissimilarities can be output upon request. This table may become very large, since it scales with the number of observational SEDs times the number of model SEDs.

An easy assessment of the results is facilitated by a quick-look on-line graphics capability. For each observational SED a chart is offered, displaying the empirical SED and an overplot of the three best matching model SEDs. Additionally, each theoretical spectrum can be retrieved via a URL in graphical form, as well as in ASCII or VOTable format. Also, a Web form[9] allows browsing through the various models according to their physical characteristics.

7. Status and Outlook

The SED classifier module has been implemented by GAVO in Java. The module can be invoked either locally from the command line, or remotely as a Web application[10]. Internally, GAVO's table utility package is used, which is backed by HSQLDB, an embedded 100% pure-Java relational database management system. The user interface is constructed using the JSP-based JavaServer Faces (JSF) framework, which was augmented by GAVO's custom tag library for table input and output. For visualization we employ the JSP-based Cewolf library, which wraps the JFreeChart Java graphics package. – During classifier development we have used a manually compiled table with observational SEDs of

[9] `http://archive.eso.org/wdb/wdb/vo/model_rauch/form`

[10] `http://www.g-vo.org/sedclassifier`

galactic planetary nebulae (Kerber et al. 2003). This table will also be used for the validation of the query, matching & assembly module.

AVO has implemented an initial SED transformation module in Perl. The module can be invoked from the command line, or as a SOAP Web service. – In the near future, we plan to apply interstellar extinction to the model SEDs, with the option of iteratively fitting the extinction parameter. This requirement will probably place higher demands on the SED transformer module.

While GAVO has implemented some Java code for catalogue query, matching, and assembly of empirical SEDs, it is currently part of another Web application[11] that directly feeds the ClassX classifier (McGlynn et al. 2004). Work is under way to refactor this application, and to isolate modules that can feed the SED classifier instead.

For seamless operation we wish to eventually chain our individual modules into a *workflow* (Fig. 1) to be operated from a single user interface. We are also looking forward to the forthcoming Astronomical Data Query Language (ADQL), which should facilitate the formulation of the matching and assembly tasks.

Ultimately we wish to create a VO-compliant tool that assembles observational SEDs in order to optimally estimate the astrophysical parameters of the underlying objects – a goal similar to the one pursued by Bailer-Jones (2002) in the context of the GAIA astrometric satellite project.

Acknowledgments. GAVO is sponsored by the *Bundesministerium für Bildung und Forschung.*

References

Adorf, H.-M. 1986, in Data Analysis in Astronomy II, ed. V. Di Gesú, L. Saasi, P. Craine, J. H. Friedman, & S. Levialdi (New York:Plenum), 61

Bailer-Jones, C. A. L. 2002, Ap&SS, 280, 21

Kerber, F., Mignani, R.P., Guglielmetti, F., & Wicenec, A. 2003, A&A, 408, 1029

McGlynn, T.A., Suchkov, A.A., Winter, E.L., Hanisch, R.J., White, R.L., Ochsenbein, F., Derriere, S., Voges, W., Corcoran, M.F., Drake, S.A., & Donahue, M. 2004, AJ, submitted

Padovani, P., Allen, M. G., Rosati, P., & Walton, N. A. 2004, A&A, 424, 545

[11]`http://www.g-vo.org/classx`

ASTRONOMICAL DATA ANALYSIS SOFTWARE AND SYSTEMS XIV
ASP Conference Series, Vol. 347, 2005
P. L. Shopbell, M. C. Britton, and R. Ebert, eds.

FROG: Time Series Analysis for the Web Service Era

Alasdair Allan

School of Physics, University of Exeter, Stocker Road, Exeter, U.K.

Abstract. The FROG application is part of the next generation Starlink[1] software work (Draper et al. 2005) and released under the GNU Public License[2] (GPL). Written in Java, it has been designed for the Web and Grid Service era as an extensible, pluggable, tool for time series analysis and display. With an integrated SOAP server the packages functionality is exposed to the user for use in their own code, and to be used remotely over the Grid, as part of the Virtual Observatory (VO).

1. What is FROG?

FROG[3] (see Figure 1) is a package for time series analysis and display. It has been designed to provide a simple user interface for astronomers wanting to do time-domain astrophysics but still provide the powerful features found in packages like PERIOD (Dhillon & Privett 1997).

The program is written in Java and runs on a wide range of different platforms including Linux, MS Windows, Solaris and Mac OSX.

As well as a user-friendly graphical interface, FROG also offers a powerful web service API which allows the user to integrate it into their own software environment.

2. Series Manipulation

The package includes a number of tools for manipulation of time series. Amongst other things these allow the user to,

- Combine individual time series
- Detrend series (multiple methods)
- Perform basic arithmetic functions

The data can also be exported directly into the Starlink TOPCAT (Taylor 2005) application for further manipulation if needed.

---

[1] `http://www.starlink.ac.uk`

[2] `http://www.gnu.org/copyleft/gpl.html`

[3] Additional support for FROG users can be found at `http://www.starlink.ac.uk/frog/`

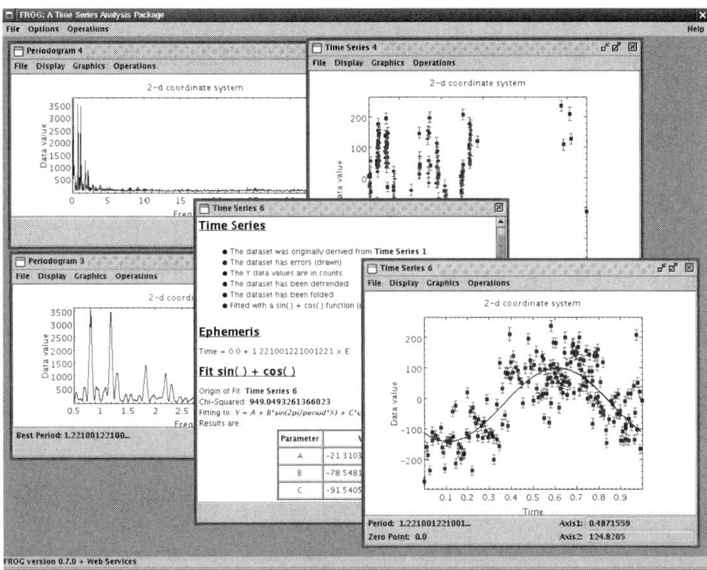

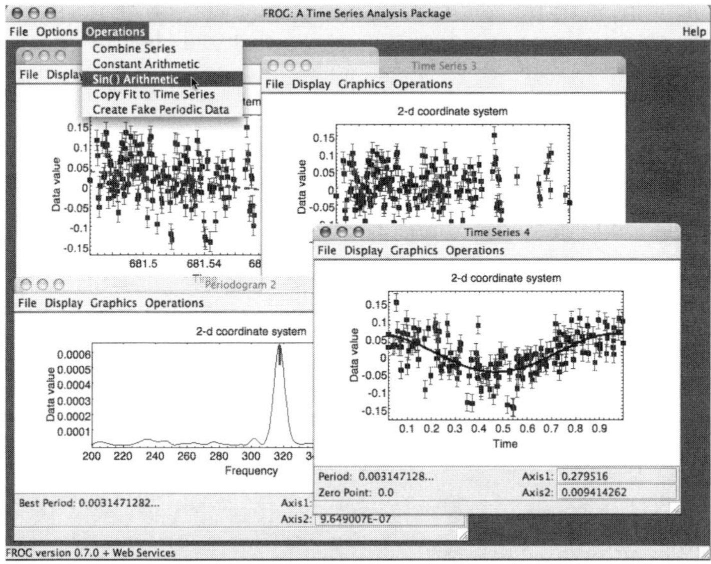

Figure 1. **Top:** The main FROG interface running on a Linux desktop. Time series photometry of IC348 has been analysed and folded around the best-fit period. The fold is shown at the bottom right, along with the associated meta-data. **Bottom:** The FROG interface running under Mac OS X. Here data from V405 Aur has been folded around the best fit period which was automatically identified for the user by the program.

3. Periodograms

While the number of period finding algorithms is currently limited this will be the first focus of new development now work on the user interface is completed.

The application will use the Nyquist frequency of your series to try to predict sensible initial values to start exploration of frequency space for each dataset.

The best (highest power) frequency in the periodogram can be identified automatically, and the originating series can be folded around this period with only two mouse clicks.

4. Folding and Binning Data

The package also supports folding of the time series data, both binned and unbinned. Error bars, if any are available for the unfolded data, are propagated correctly to the folded plot.

5. Fake Data Creation

FROG has extensive facilities for fake data creation. The basic data creation tool allows the user to add Box-Muller noise and data point scatter to a basic sine wave.

The sin() arithmetic tool allows fake data to be created with the same window function as an actual time series. Additionally this tool allows you to add, subtract, multiply and divide existing time series by a user specified sin() function.

6. Fitting Data

The package provides automated curve fitting for time series. In a similar fashion to the period-finding algorithms, the types of fitting functions available are currently limited, but a much more extensive fitting package is planned.

7. Other features?

The package provides publication-quality plots, the user can export to both PostScript and JPEG formats for all the plots. A variety of line styles and color schemes are also available should the user not like the default values chosen by the application.

8. FROG and Web Services

While FROG can be used locally as a normal application, it also exposes functionality via a built in web server as a SOAP web service (see Figure 2). The web service is language neutral and can be accessed using a variety of SOAP toolkits in languages that include Perl, Python, PHP, Java, C, C++ and even Fortran.

This means that FROG can be used from your own code or scripts, either for display, or for data processing, without having to learn Java or write a lot

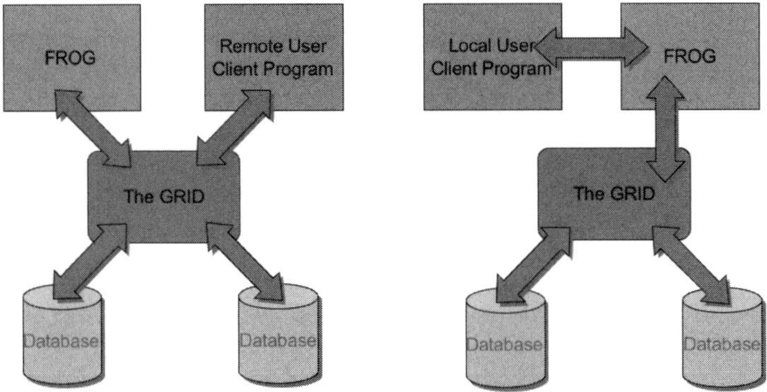

Figure 2. **Left**: FROG can be used by an astronomer located at a remote machine as a processing engine. The user can pass a time series via a SOAP message to the FROG web service interface for processing, e.g. creating a periodogram from the data, with the results being returned to the user in another SOAP message. **Right:** The web-service interface also allows FROG be used locally by the astronomer as a display tool for their own programs. The FROG web service interface will allow the user to load, display and manipulate time series very easily by passing SOAP messages to the FROG web server.

of code; client-side programs can be a short as a few lines depending on the language and toolkit chosen.

9. Plug and Play

There is a long list of features that could be added to FROG, and it is under active development by the Starlink Project. In addition, due to the pluggable nature of the program architecture, it is very easy for users to add support for their own favorite algorithms, or to use the application as a display tool from their own code by creating a FROG instance.

10. Where next?

Support for additional period finding and line-fitting algorithms and functions is next on the development schedule. After this, user-interface improvements and additional file format support are high on the list. FROG development is user driven, having already attracted a small community of users. If it doesn't do exactly what you need, ask! Patches and contributions are also very welcome

11. Obtaining FROG

A binary distribution of the latest version of FROG v0.7.0 is available[4] for RedHat Linux 9.0 and Enterprise Linux 3.0, Debian Linux 3.0R2, Solaris 8/9, Mac OS X 10.3.x and MS Windows.

For unsupported systems a checkout of the Starlink CVS Archive[5] is available, updated nightly, so that the application can be built directly from source for your platform.

References

Dhillon, V.S., & Privett, G.J. 1997, Starlink User Note 167, Starlink Project, CCLRC

Draper, et al. 2005, in ASP Conf. Ser., Vol. 347, ADASS XIV, ed. P. L. Shopbell, M. C. Britton, & R. Ebert (San Francisco: ASP), 22

Taylor, M. 2005 in ASP Conf. Ser., Vol. 347, ADASS XIV, ed. P. L. Shopbell, M. C. Britton, & R. Ebert (San Francisco: ASP), 29

[4]`http://dev.starlink.ac.uk/`

[5]`http://cvs.starlink.ac.uk`

ASTRONOMICAL DATA ANALYSIS SOFTWARE AND SYSTEMS XIV
ASP Conference Series, Vol. 347, 2005
P. L. Shopbell, M. C. Britton, and R. Ebert, eds.

Linking and Tagging Initiatives at the Astrophysical Journal

Greg J. Schwarz[1]

AAS Journals Editorial Staff Scientist

Abstract. At the start of 2005 the Astrophysical Journal will be offering three new linking and tagging options for the electronic edition. The linking projects are in conjunction with partnerships with ADS, CDS, and NED. Authors may use new AASTeX mark up to tag data sets of participating data centers. A common name resolver hosted at ADS verifies the data set tags during copy editing and sends the reader who follows the links to the appropriate data center. A demonstration of this capability can be seen in the September Spitzer ApJ Supplement. In a similar vein we will also support object linking to SIMBAD and NED. For this project we will also query authors to supply the most important objects of their paper during peer review. These object lists will be supplied to the data centers to help them integrate the information into their databases and potentially check for errors. The SIMBAD and NED object web pages that are now created weeks and months after publication at CDS will be also be dynamically added to the navigation bar once they are made available. The final endeavor is a controlled set of major astronomical facility keywords to help organizations track the effectiveness of their telescopes. The ultimate goal of each project is to allow authors to make their papers more useful and to help researchers seamlessly navigate between the journal and the data archives.

1. Introduction

Over the last few years the AAS has begun working with ADS, NED, and CDS to add verified and stable links to outside resources. These projects allow authors of the ApJ and AJ to link to data sets available at participating data centers through ADS and to object names at NED and CDS with new AASTeX tagging macros. ADS and CDS have created name resolvers that are used during copy editing to check author's data set and object names. A third project is a list of acronyms of important astronomical facilities for authors to identify instruments used in their papers. The available acronyms come from a controlled vocabulary list administered by the AAS, checked during copy editing, and given their own SGML tags in the AAS Journal archive. The Journal's search engine isolate the SGML marked keywords which will greatly improve the way organizations identify papers using their facilities.

All three undertakings are completely voluntary. However they are designed to benefit many different portions of the astronomical community. With a keyword list, encoded in SGML tags, funding agencies and facilities have a better way to derive their metrics. Readers get seamless navigation between the AAS Journals

[1]University of Arizona, Tucson, AZ 85721-0065

and the data centers. The data centers get important information into their archives quickly and more efficiently. With this information the data centers can rapidly construct links from their archives back to the e-journals which helps researchers find the pertinent knowledge regarding their favorite objects and data sets. The increase in the interconnectivity between Journals and data centers means that for a modest amount of effort authors are rewarded with much greater visibility for their papers.

Table 1. Participating Data Centers

Center	Available Missions	**PrivateId**	Availability of Identifiers
LAMBDA	WMAP	Sa.WMAP	Full IDs on-line
	COBE	Sa.COBE	
HEASARC	ASCA	Sa.ASCA	Full IDS on-line
	ROSAT	Sa.ROSAT	
	RXTE	Sa.RXTE	
MAST	HST	Sa.HST	Only **PrivateId** on-line
	IUE	Sa.IUE	
	EUVE	Sa.EUVE	
	FUSE	Sa.FUSE	
	HUT	Sh.HUT	
	UIT	Sh.UIT	
	WUPPE	Sh.WUPPE	
Chandra	Chandra	Sa.CXO	Only **PrivateId** on-line
Spitzer	Spitzer	Sa.Spitzer	Archive not yet available

2. How Data set linking works

Authors tag data set identifiers with the new `‘‘\dataset{}’’` command available in AASTeX 5.2 since January 2004. Valid data set identifiers have the form: ADS/*FacilityId*#**PrivateId**. Each participating data center assigns a unique identification, **PrivateId**, for each data set in their archive. A list of participating data centers, their available mission archives, *FacilityIds*, and the current availability of full valid identifiers in the archive are provided in Table 1. LAMBDA and HEASARC provide full identifiers in a direct archive search while MAST and Chandra only give the **PrivateIds**. The Spitzer archive will be fully integrated in the near future. Authors should contacted the data centers directly if they have questions about individual data set identifiers. *FacilityId* is a facility acronym assigned by ADS. ADS provides a verification service of all identifiers from participating data centers. During copy editing the ADS resolver is used to check identifiers in the `‘‘\dataset{}’’` commands. Authors are prompted to supply new identifiers when the ADS resolver can not find a match. Valid

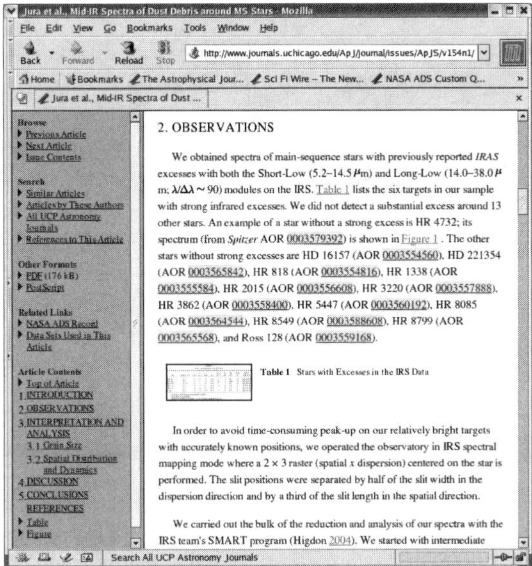

Figure 1. An example of data set links in a paper from the ApJS *Spitzer* issue (Jura et al. 2004).

identifiers appear as links to the ADS verification service in the electronic edition and readers who follow the links are redirected to the appropriate data center by ADS. Twenty-six papers in the Spitzer ApJ Supplement issue (September 2004) already have these links in them (see Figure 1). This was the first issue to have data set links and served as a demonstration of their capabilities. The data set links are also available via the "Data Sets Used in This Article" link in the left navigation frame which means the information can be access from anywhere in the paper. The data sets from each paper are distributed to the data centers via ADS. The data centers will use this information to build links from their archives back to the article.

3. How Object linking will work

Object tagging is a more ambitious project than data set tagging but builds on the same underlying techniques. Objects tagged with the `''\object{}''` macro in AASTeX 5.2 will also be verified during copy editing. CDS has supplied the object resolver, SESAME, which queries both the Simbad and NED databases. Authors will be asked to provide known object names when SESAME does not recognize the tagged name. One important way this project differs with data set linking is in the early author involvement. During peer review authors will be asked to provide their paper's important objects in a web form. Figure 2a shows the initial page of this form where the data is entered. These objects are sent to SESAME for verification and the results, the data center's object name, coordinates, and type, are displayed for the author to confirm (see Figure

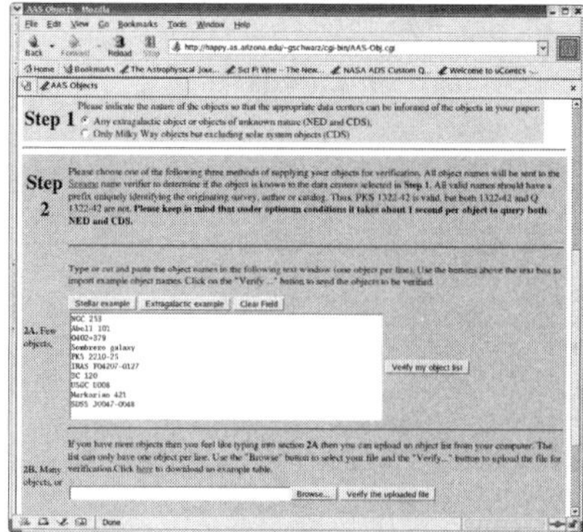

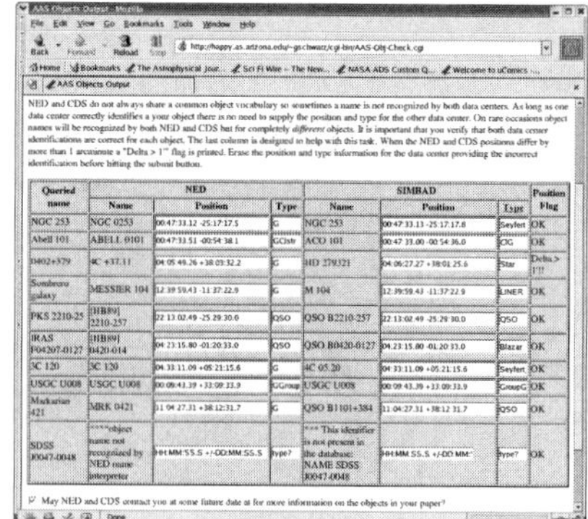

Figure 2. The AAS Journal's object query web page and query results.

2b). For new or unknown objects authors can enter their coordinates and type to help the data centers identify and integrate the objects into their databases. When authors are finished the information will be made available to CDS and NED. Authors not using the ``\object{}'' macro at this point will also be given their object names inside ``\object{}'' tags so they can place them in their final manuscript. In the electronic edition the ``\object{}''s will appear as links similar to how A&A already does object linking but with a few important differences. First, the links will be live as soon as the paper is posted. Second, readers will be able to access both NED and CDS when the object is in both databases. Finally, outside resources at ADS, NED, and SIMBAD will be queried dynamically in the navigation frame while the main text (which is static) loads. Links to these resources, the "ADS record", "NED objects", and "Simbad Objects", will only appear when they are available.

4. Facility keywords

The current controlled acronym list has 279 of the most important astronomical telescopes/platforms. Proposals will be accepted to expand the list for missing or future facilities. The full list is available to authors at `http://happy.as.arizona.edu/~gschwarz/TOOLS/facilitytool.html`. The web page, shown in Figure 3, allows facilities searches based on observational wavelength, location, or solar facilities. Most keywords are just the facility's common name or acronym but some include the telescope's physical location to avoid the possibility of redundancy in the keywords. Authors may include additional text after a keyword surrounded by parenthesis which can be used to cite facility instruments or telescope configurations. For example, the Wide Field Planetary Camera II on Hubble would be designated as HST(WFPC2). Facility keywords used in the research of the paper should appear after the acknowledgments and preceded by the word "Facilit(y—ies):" in italics. Multiple keywords are separated by commas. Unlike subject heading there are no limits to the number of keywords used. The keywords, but not the optional information, will be checked against

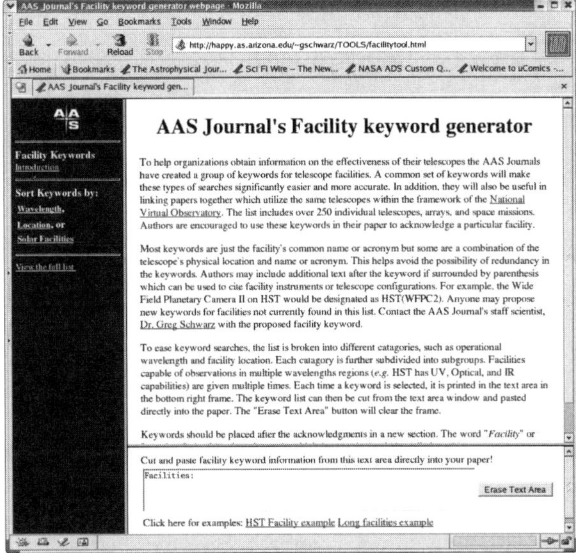

Figure 3. The AAS Journal's Facility keyword generator.

the master list during copy editing. Keywords will receive special SGML tags to make searches with the Journal's search engine easier and more effective.

5. Summary

The AAS Journals will introduce three new projects in January 2005 for authors to increase the content and visibility of their papers. The first, data set linking, was demonstrated in the Spitzer Supplement issue. The five major data centers participating in this project gives authors the ability to tag a large fraction of the data available from NASA missions. Authors will also have the ability to link to all of the objects in the Simbad and NED databases. While the object and data set links will be immediately available to readers as soon as the papers are posted, links to outside resources will be inserted in the navigation frame when they become available. Finally, we have created a set of keywords for common facilities so that funding agencies and organizations can track papers that use their products. None of these projects is mandatory for new submissions but each has tremendous benefits for the entire astronomical community and authors which easily out-weighs the extra work by the author to supply the information.

References

Jura et al. 2004, ApJS, 154, 453

ASTRONOMICAL DATA ANALYSIS SOFTWARE AND SYSTEMS XIV
ASP Conference Series, Vol. 347, 2005
P. L. Shopbell, M. C. Britton, and R. Ebert, eds.

The Chandra Bibliography Database: Bibliographic Statistics

Sarah Blecksmith, John Bright, Arnold H. Rots, Sherry Winkelman, Paul Green, and Mihoko Yukita

Chandra X-ray Center, Smithsonian Astrophysical Observatory, Harvard-Smithsonian Center for Astrophysics, 60 Garden St.,Cambridge, MA 02138 (USA)

Abstract. We have started to generate and catalog bibliographic statistics for the Chandra mission. As a result, the databases associated with the Chandra bibliography continue to expand in scope and purpose. Publication data, including number of citations, are collected for each Chandra proposal and stored in the bibliography database. Scientific metrics based on these data are computed monthly and stored in the metrics database for trending analysis of publication statistics. The design of the system and some results for illustration are presented. In addition, we have discovered some unexpected operational uses for the bibliographic database which will also be discussed.

1. Introduction

The Chandra bibliography database continues to grow as a valuable tool for the Chandra community. We have now added three new descriptive flags to our previous bibcode categorization scheme and have extended the bibliography database to include automated generation of bibliography metrics and metrics cataloging for future trending analysis.

2. The Chandra Bibliography Database

The database is an extensive catalog of all Chandra related publications. For each article, the database keeps the "bibcode" (for direct ADS access), category, date of publication, whether it is refereed or not, the kind of article, the type of publication, the number of citations, the keywords attached to the article, a Chandra proposal connection and Chandra ObsId connections.

The publications in the database have been reviewed and categorized by hand according to the following definitions:

- Presents specific Chandra observations
- Refers to published Chandra results
- Predicts Chandra results
- Describes instrumentation, software or operations
- Cannot be classified

In addition, Table 1 shows the flags for Chandra instruments and types of data usage that are set for each bibcode in the database.

Table 1. Bibliography Flags

ACIS	LETG	EPHIN	*Multi-wavelength
HRC	PCAD	Software	*Follow-up
HETG	HRMA	Operations	*Theory

The flags marked with an * have been added in the last year in order to better classify how papers in the database use Chandra data:

A. Complementary observations

B. Simulations or follow-up analysis

C. Astrophysical theory that explains Chandra results

The database can be searched by any combination of these flags and categories as well as the all of the publication attributes listed above through our web interface at the Chandra Bibliography Search[1].

3. The Bibliography Statistics

To aid in the generation of statistics, a table called `bibstats` was added to the bibliography database. This table contains data for many different kinds of plots of bibliographic statistics. A script runs monthly to calculate the data, populate the table and generate new plots. This database makes trending analysis of the statistical data possible. We now provide bibliographic metrics for internal use and are beginning to make related information available to the public.

For the purposes of these statistics, all refereed Category 1 entries as well as refereed Category 2 publications that have at least one additional flag (A,B or C) are considered Chandra papers. Programs are defined as groups of fully observed ObsIds from the same approved proposal in a single proposal cycle. Citation counts for each bibcode in the database are updated weekly from ADS.

We look at papers, citations, number of programs and the number of programs without papers and citations. The data can be organized by program type, science category, observing cycle and exposure time and normalized by the sum of the program exposure times associated with the Chandra papers.

4. Plots of Statistical Data

We currently generate 71 different statistical plots on a monthly basis and almost all of them can be misleading if not interpreted with care. The plots shown

[1]`http://cxc.harvard.edu/cgi-gen/cda/bibliography`

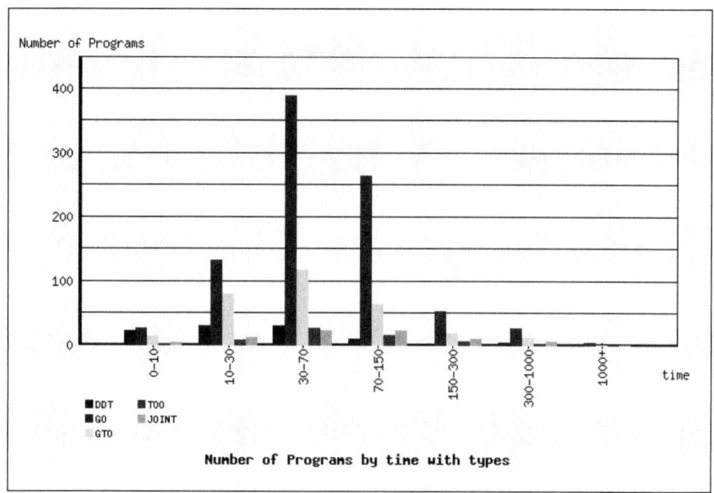

Figure 1. The number of programs by observation time range for each type of program. Note that there are more GO programs than GTO programs for every time bin.

here illustrate some of the difficulties one faces when attempting to interpret bibliographic statistics.

Figure 1 shows the number of programs by observation time for each type of program. We can see that in every time bin, there are more General Observer (GO) than Guaranteed Time Observer (GTO) programs. In Figure 2, we see that there are more papers per ks for the GTO observations for almost every exposure time bin. This was puzzling until we looked at some of the GTO papers and realized that these observations were being done by teams. GTO's outperform the more numerous GO's because GTO observations tend to be done by large groups, whereas GO's are typically one or two people. We connect a paper to a proposal (and thus count it as a program type in the statistics) when either the Principal Investigator or Co-Investigator is an author; even if they are the tenth author out of ten.

Figures 3 and 4 are a good example of a possible interpretation pitfall. In Figure 3, it appears that shorter (less than 30ks) observations have a greater "impact" than long observations. But we can see in Figure 4 that the longer observations appear to have a greater lag time from observation to citation count. Large observations take longer to analyze and publish. Without that information, it would be easy to draw a mistaken conclusion from Figure 3. It's very important to look at the data several ways before you decide what conclusions can be drawn.

5. Lessons Learned

An extensive bibliography database is a very useful tool for observers and the observatory alike. There are considerable payoffs to collecting as much metadata

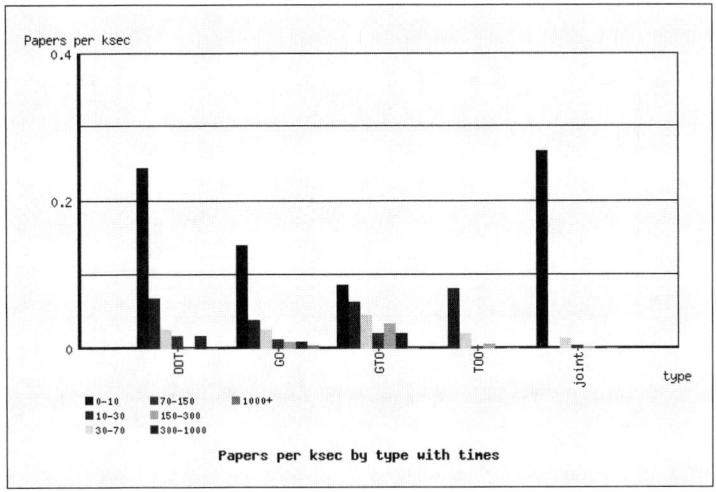

Figure 2. The papers per ks by program type for each exposure time range. There are more papers per ks for the GTO programs in almost every time bin.

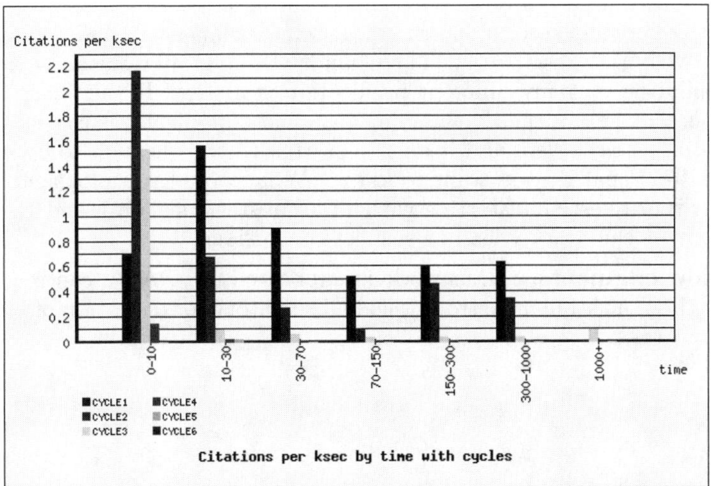

Figure 3. The citations per ks by exposure time for different cycles. In this graph, it appears that shorter observations have a greater "impact". However, there appears to be a longer latency period with large programs.

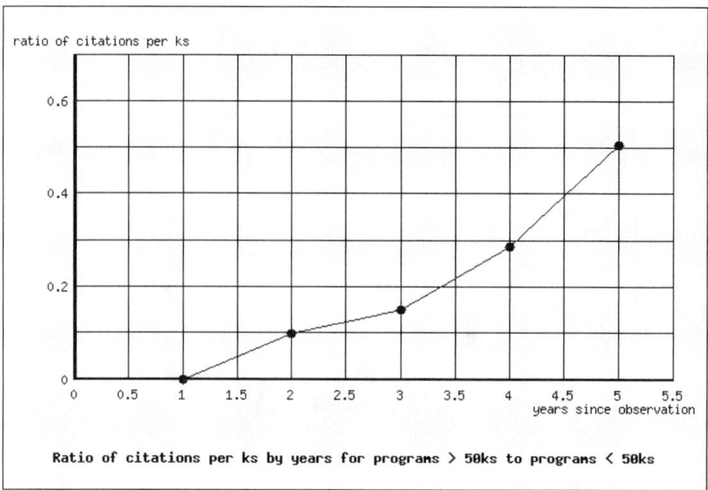

Figure 4. The ratio of citations per ks of large programs to small programs.

as possible, from the beginning. A great deal of advance thought into what the end uses of the data will be is an extremely helpful and time-saving step when planning such a database. Even so, other uses may not initially be apparent. In our case, we have found that we can use the bibliography database to close out grant contracts by finding all of the publications related to a proposal as well as to measure a proposer's effectiveness at getting papers published.

Data concerning the productivity and scientific impact of observations is becoming an important piece of information used for managerial and policy making decisions within an observatory. The Chandra Bibliography Database allows for the compilation of many different publication statistics. In our study of these plots, it has become clear that assumptions and complications can lead the interpretation astray. We found it best to start with the question to be answered by the data. What do you want to know? What would you plot against what to answer that question? Starting with a plot of statistical data and attempting to interpret it can quickly lead to suspect conclusions.

Acknowledgments. This work is supported by NASA contract NAS8-03060 to the Smithsonian Astrophysical Observatory for operation of the Chandra X-ray Center.

ASTRONOMICAL DATA ANALYSIS SOFTWARE AND SYSTEMS XIV
ASP Conference Series, Vol. 347, 2005
P. L. Shopbell, M. C. Britton, and R. Ebert, eds.

MIGALE: Milestones and Roadmap

Igor Chilingarian[1]

Sternberg Astronomical Institute of the Moscow State University, RUSSIA

Philippe Prugniel

CRAL Observatoire de Lyon; GEPI Observatoire de Paris-Meudon, FRANCE

Hector Flores, Jean Guibert, Regis Haigron, Isabelle Jegouzo, Frédéric Royer, Françoise Tajahmady, Gilles Theureau, and Jacques Vétois

GEPI Observatoire de Paris-Meudon, FRANCE

Abstract. The MIGALE collaboration has existed since 2003. We are maintaining several astronomical databases and telescope archives. We are developing and maintaining the PLEINPOT software package that powers all of our database related projects. We developed prototypes of the tools that would become available in the VO framework in the near future and constitute the "virtual instrumentation" – a set of online tools and facilities for data processing and analysis, devoted to the specific scientific area (evolution of galaxies).

1. Introduction

MIGALE (The Multiparametric Virtual Instrument to Study Galaxy Evolution) is a collaboration between French (GEPI Observatoire de Paris, CRAL Observatoire de Lyon), Russian (Sternberg Astronomical Institute), and Bulgarian (Sofia Observatory) laboratories. The scientific goal feeding the project is to study the evolution of galaxies during the last 9 Gigayears.

The goals and main activities of MIGALE have been described elsewhere (Prugniel et al. 2003, 2004). Presently, MIGALE is in charge of the following database projects:

- HyperLEDA (`http://leda.univ-lyon1.fr`), a database containing homogenized data for 3 million galaxies
- GIRAFFE archive (`http://giraffe-archive.obspm.fr`), distributing reduced data obtained with the Giraffe integral field spectrograph on the VLT
- HiGi (`http://klun.obs-nancay.fr/`), a database distributing radio spectral data for spiral galaxies obtained with the Nancay radiotelescope
- ASPID (`http://alcor.sao.ru/db/aspid/`), distributing the raw data obtained using selected instruments on the Russian 6-m telescope since 1992 (including all of the 3D spectroscopic data)

[1]CRAL Observatoire de Lyon, FRANCE – visiting investigator

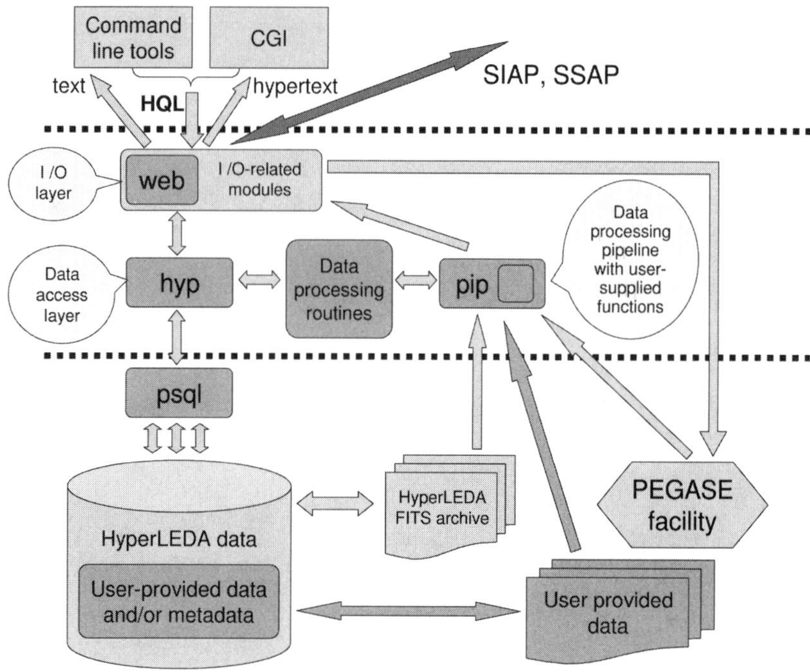

Figure 1. General structure of a PLEINPOT-powered project.

These projects are powered by the PLEINPOT package, distributed by the team.

2. PLEINPOT

PLEINPOT is an open source software package, developed and maintained by MIGALE. It provides a ready-to-use DBMS system based on PostgreSQL, filled with the HyperLEDA data, and name resolving mechanism fed by it. The package is portable: it works on Linux, Sun Solaris, Compaq Tru64 UNIX and Mac OS X platforms.

PLEINPOT has a modular structure containing the following major parts:

- astronomical and statistical functions and procedures
- database access layer
- data processing pipeline
- output formatting routines

PLEINPOT is a platform to build astronomy related database projects. The general structure of a project powered by PLEINPOT is presented in Figure 1. The user provided data and methods to deal with them are indicated by a pink colour. The diagram represents three layers: client applications on top, data storage and data generation tools on bottom and all the data processing mechanisms provided by PLEINPOT in between. The example of a data generation tool is PEGASE, a facility to compute synthetic spectra of galaxies on the fly, that is included in the PLEINPOT distribution.

During the installation the user can choose whether he/she wants to have HyperLEDA data and/or an HFA (HyperLEDA FITS archive) locally or have a remote transparent access to them.

The data processing pipeline ("pip" module) provides a set of tools for online processing of images and spectra: normalization, convolution, unsharp masking, continuum fitting, etc. It is easily extensible with user-written dynamically loadable procedures.

The following APIs are provided: C/C++, Fortran, Java (optional), PHP (optional), Perl (optional), Python (optional). All routines in the library are natively written in C or Fortran; the wrappers for other languages are provided via SWIG (Simplified Wrapper and Interface Generator).

All of the queries from client applications (presently command-line tools or CGIs) are passed to the "web" module, where the HQL (HyperLEDA Query Language) query is parsed and forwarded to the data access layer or data processing routines. We also plan to implement VO protocols (SIAP and SSAP). The output of the system is processed by different formatting routines depending on the requested output type. PLEINPOT supports or will support the following output formats:

- Hypertext document
- Plain text
- VOTable
- SIAP and SSAP (not implemented yet)

The individual output formatting routines shall be replaced by a unified solution. The general idea is to generate multi-part mime documents and pass them to an output producer which will extract various parts of the document and present the VOTable sections according to XSLs.

Building an astronomical database using PLEINPOT allows one to save time and manpower. For instance, the Giraffe archive was built in three weeks by one engineer working full time.

A very nice example of a stand-alone project powered by PLEINPOT and not connected directly to MIGALE is the ELODIE archive (Moultaka et al. 2004) in the Observatoire de Haute Provence (`http://atlas.obs-hp.fr/elodie/`). It is a telescope archive containing homogeneous data: more than 4000 reduced flux-calibrated high-resolution spectra collected with the ELODIE spectrograph at the 1.93m telescope at the Observatoire de Haute Provence.

3. Virtual Instrumentation

MIGALE built the prototypes of the tools that would constitute the virtual instrumentation: a set of online-tools for data reduction, processing, and analysis using VO protocols for data access and exchange.

- DisGal3D - a tool to deconvolve 3D spectral data using high-resolution direct images (for instance HST) as references (Flores et al. 2004). The technique allows to determine the velocity profiles of distant galaxies using very small velocity fields obtained by the means of integral field spec-

troscopy. The example in shows a reconstruction of the velocity field of a distant galaxy obtained with the FLAMES-Giraffe spectrograph at VLT.
- PEGASE.HR - a new generation facility to build high-resolution (R=10000) synthetic spectra of galaxies using the ELODIE.3 library (Le Borgne et al. 2004).
- SPIKeR (Stellar Population and Internal Kinematics Reconstruction tool) - Using PEGASE.HR models with the penalized pixel fitting algorithm (Cappellari & Emsellem 2004), it is possible to extract internal kinematics and put constraints on the properties of stellar populations (Chilingarian et al. 2004).

The online versions of the tools will have significant advantages compared to the standalone versions. Although the software is available, it is quite complex and its installation takes time. The online version allows one to avoid this procedure.

The data may be either extracted from the archive or provided by the user. In the former case the data reduction will be done on the fly (if necessary); this will save time over downloading and processing raw data.

The proposed techniques require rather complex computation, so a large amount of computer time is necessary. The enline versions will propose applying the power of a PC cluster.

Acknowledgments. The authors appreciate the organizing committee for the financial support provided for attending the conference, CNRS for financial support of visits between France and Russia (bilateral collaboration), and RFBR for funding the travel of the presenting author to Pasadena (project #04-02-27117).

References

Cappellari, M. & Emsellem, E. 2004, PASP, 116, 138

Chilingarian, I., Prugniel, P., & Sil'chenko, O., 2004, JENAM meeting (proceedings in press)

Flores, H. et al. 2004, A&A, 420, 31

Le Borgne, D. et al. 2004, A&A, 425, 881

Moultaka, J. et al. 2004, PASP, 116, 693

Prugniel, P. et al. 2003, SF2A Meeting, EdP-Sciences, Conf. Series, 283

Prugniel, P. et al. 2004, SF2A Meeting, EdP-Sciences, Conf. Series, meeting abstract

ASTRONOMICAL DATA ANALYSIS SOFTWARE AND SYSTEMS XIV
ASP Conference Series, Vol. 347, 2005
P. L. Shopbell, M. C. Britton, and R. Ebert, eds.

TOML - Telescope Observation Markup Language

Shaun de Witt, Tim Jenness, and Frossie Economou

Joint Astronomy Centre, 660 N. A'ohōkū Place, University Park, Hilo, HI 96720, USA

Martin Folger

UK Astronomy Technology Centre, Royal Observatory, Edinburgh, Blackford Hill, Edinburgh EH9 3HJ, U.K.

Abstract. Telescope Observation Markup Language (TOML) has been developed by the Joint Astronomy Centre as a means of expressing an astronomical observing program in XML. It provides a means of encoding a program developed using a graphical tool in a language neutral format which can be sent over the web to a database or a colleague who is also working on the same project. This can then be loaded into any tool capable of accepting TOML.

1. Origin of TOML

An early alpha release of TOML came from the JAC Observing Tool work (Folger et al. 2002) which formed a part of the OMP project (Economou et al. 2002) where both telescopes operated by the JAC moved to using flexible scheduling. Early on in this project, it was realised that there was a need to convey the project expressed graphically to a project database. In designing TOML, we have tried as far as possible to have a one-to-one correspondence between XML and graphical element.

2. Why TOML?

The use of a well defined structure has many advantages. In the original concept of the OMP system, users would be able to use the Observing Tool anywhere. Observers could work on their program at the office, at home, or while traveling, without the need to be connected to the internet except for submission to the project database. This requires a well defined format for storage of information.

Some format based on XML was chosen as an ideal choice for this since it is language neutral and the JAC use a number of computer languages in different systems which all require to read some part of the file.

The same well defined format is also used for retrieving projects from a database, the results being returned in TOML format which can be loaded into a suitable tool.

This mode of operation has proved very successful as UKIRT and JCMT have moved their mode of operation to automated scheduling. Observers awarded telescope time now have to prepare their observations as early as possible in order to give them the maximum opportunity to be observed. Since TOML is based on XML, and the JAC observing tool is written in Java and supported for unix, MacOS and Windows, the user has a wide choice of machine on which they can prepare their program, or can use several different machines.

3. Design of TOML

TOML essentially consists of three types of data:

elements - corresponding to basic XML elements and associated attributes. These are the basic building blocks of TOML. On a Graphical Tool, each element would correspond to a specific widget selection;

components - representing unique items which can be wholly represented by a number of elements. Examples are targets, site quality information, instrument information and observations;

folders - these allow groupings (*containers*) and inheritance (*iterators*). Folders can contain elements, components and other folders. For example, an observer may wish to perform the sequence *object, sky, sky, object* several times for the same object. This can be achieved by including the sequence in a repeat folder.

This design allows the definition of TOML to be as telescope agnostic as possible.

Inheritance is seen as essential to the design, and has been built into TOML from the start. For example, the user can specify a configuration for an instrument once, and then use the inheritance to use this configuration in several observations of different targets. Not only does this simplify the definition of TOML, but it also makes life simpler for an observer developing a program, minimising the need to cut and paste identical components.

Grouping is also an essential feature of TOML. For instance, an MSB (Minimum Schedulable Block) is used to group tightly related observation, such as observations of a flat/arc, standard star and science target which should be done at the same time.

In the design and development of TOML, we have a number of design goals in mind.

Flexibility - the same basic building blocks can be used with any telescope.

Extensibility - it is simple to add new instruments, types of folder or component, etc.

Maximise Re-use - break out common elements to minimise the size of the TOML definition.

Accesibility - Any services available through the use of TOML should be accessible both over the web and locally by a user.

4. Definition of TOML

XML allows two possibilities for defining document structures; Document Type Definitions (DTDs) and schemas. Both have advantages and disadvantages, as shown in Table 1.

Table 1. Comparison of DTDs and XML Schemas

DTD	Schema
Simple format, easily human readable	More complex to develop and understand
No way to specify data types or format	Can express data types, patterns and ranges, and can be extended to support new types
Limited cardinality support (1, zero or more, one or more)	Support for any cardinality
No namespace support	Ability to select the use of namespaces or not
Mature, with many develpers experienced in their definition, and many tools available	Less mature. Most parsers now support schemas, but there is still some differences in the interpretation of the standard
Does not allow for context sensitive element definition	Elements can be redefined based on their context

We have chosen to define TOML in terms of a schema. The full definition is available at the JAC schema[1] website.

To aid maintainability and extensibility, we have split the schema into a number of files at different levels in a directory hierarchy. The lowest level specification in the current JAC system is either the file *UKIRT.xsd* or *JCMT.xsd*. These simply include additional schema files in a simply navigable way. If another institution wanted to extend TOML to meet their own needs, they could simply use the schema include mechanism.

5. Using TOML

When implementing TOML, we made a conscious decision to use namespaces, rather than a no-namespace schema. The advantage of using a namespace is that it allows you to distinguish between elements that would have the same name in two different namespaces, for instance a frequency element in a statistic namespace, and a frequency in a physics namespace. We envisage schemas gradually replacing DTDs in many application areas, and see the use of namespaces as a way of future-proofing TOML.

In order to use the TOML schema, it is necessary to include the following in the root element of an XML document.

[1] `http://omp.jach.hawaii.edu/schema`

```
<docRoot xmlns:xsi="http://www.w3.org/2001/XMLSchema-instance"
   xmlns="http://omp.jach.hawaii.edu/schema/TOML">
```

Using this format means that all XML elements which are not qualified by a namespace are assumed to come from the TOML namespace. This was important during development since not only did our observing tool write out XML without a namespace before the final TOML definition, but many other institutions use some form of XML and in both cases this makes the migration to using a schema as simple as possible.

To create a custom extension to TOML is equally simple. Since a schema is an XML document with a root element of `schema`, the top of each schema document should look like the following:

```
<xsd:schema targetNamespace="http://omp.jach.hawaii.edu/schema/TOML"
    xmlns:xsd="http://www.w3.org/2001/XMLSchema"
    xmlns:toml="http://omp.jach.hawaii.edu/schema/TOML"
    elementFormDefault="qualified">
```

Once this is included, then you can reference existing TOML base elements and create your own new elements. So, for example, in the basic specification for a type *positiveDecimal*, which is used to indicate the value associated with an element of this type is any decimal value greater than 0 (we have also defined a *nonNegativeDecimal* type which is any numeric value greater than or equal to zero). To create an element with this type, all you need to specify is the following:

```
<xsd:element name="myElement" type="toml:positiveDecimal/>
```

Finally, in our current structure, you can use the include mechanism to include all previously defined schema.

```
<xsd:schema targetNamespace="http://omp.jach.hawaii.edu/schema/TOML"
    xmlns:xsd="http://www.w3.org/2001/XMLSchema"
    xmlns:toml="http://omp.jach.hawaii.edu/schema/TOML"
    elementFormDefault="qualified">

<include schemaLocation=
    "http://omp.jach.hawaii.edu/schema/TOML/TOMLtypes.xsd/>
<include schemaLocation=
    http://omp.jach.hawaii.edu/schema/TOML/TOMLelements.xsd/>
```

At the JAC we have standardised on using the Xerces-2 parser. We have chosen this since it supports many platforms and is available for Java, C++, and Perl. In addition several sites indicate that Xerces-2 is one of the most fully featured parsers available, offers a small memory footprint for DOM combined with good speed for both SAX and DOM parsing (see, for example,

- http://xmlbench.sourceforge.net and
- http://www.sosnoski.com/opensrc/xmlbench/results.html).

However, during development of the schema, it was found to behave differently to some other parsers for validation. Specifically, XSI allows users to use the include mechanism in files which were themselves included by other files, but this was unacceptable in the version of Xerces we have adopted (2.4.0).

6. Benefits of TOML

There are several benefits associated with the use of a well defined data structure such as TOML. These include:

- On-line and off-line validation, including type and range checking;
- Simple exchange format suitable for media and electronic transfer;
- Easy correspondence between TOML element and GUI widget
- Reduces coding effort for adding new features

Probably the largest benefit from a development and maintainability viewpoint is validation. Since the schema allows type and range checking, this does not need to be built into an Observation Preparation Tool. Also, if something changes, such as a filter becomes unavailable, it can simply be edited out of the definition and anyone attempting to use it will then get a validation error, which can also be sent to a support scientist at the observatory who can communicate the reason for the error to the person developing the program. Since this service can be run over the web, it is possible merely to edit the definition in one place, without the need for an additional software release.

7. Future of TOML

TOML is still under development, and the files *TOMLtypes.xsd* and *TOMLelements.xsd*, which are really just libraries of basic elements and types, will be extended as new instruments are used and if other institutions adapt TOML and make requests for extensions. The JAC are also looking at improving the validation reporting, which currently gives rather obscure error messages. We hope to improve these and make the error clearer to both the project author and support scientist.

As more observatories move to flexible scheduling and need some form of observation preparation tool, we hope to keep the JAC at the forefront of the development of a standard library of XML definitions, and possibly extend TOML to support the work of other observatories.

References

Economou, F., et al. 2002 in ASP Conf. Ser., Vol. 281, ADASS XI, ed. D. A. Bohlender, D. Durand, & T. H. Handley (San Francisco: ASP), 488

Folger, M., et al. 2002 in ASP Conf. Ser., Vol. 281, ADASS XI, ed. D. A. Bohlender, D. Durand, & T. H. Handley (San Francisco: ASP), 453

ASTRONOMICAL DATA ANALYSIS SOFTWARE AND SYSTEMS XIV
ASP Conference Series, Vol. 347, 2005
P. L. Shopbell, M. C. Britton, and R. Ebert, eds.

VOStat: A Distributed Statistical Toolkit for the Virtual Observatory

Matthew J. Graham, S. G. Djorgovski, A. A. Mahabal, and Roy D. Williams

California Institute of Technology, Pasadena, CA 91125, USA

G. Jogesh Babu, Eric D. Feigelson, and Daniel E. Vanden Berk

The Pennsylvania State University, University Park, PA 16802, USA

Robert Nichol

ICG, University of Portsmouth, PO1 2EG, UK

Larry Wasserman

Carnegie Mellon University, Pittsburgh, PA 15213, USA

Abstract. The nature of astronomical data is changing: data volumes are following Moore's law with a doubling every 18 months and data sets consisting of a billion data vectors in a 100-dimensional parameter space are becoming commonplace. Sophisticated statistical techniques are crucial to fully and efficiently exploit these and maximize the scientific return. A long-standing limitation, however, on the range and capability of such analyses has been the paucity of non-proprietary software.

VOStat is the result of a cross-disciplinary collaboration between astronomers and statisticians to meet these challenges; it is a prototype knowledge-based statistical toolkit implemented within the VO paradigm for the entire astronomical community. VOStat consists of an easily extensible distributed web-based framework transparently accessed via a single science endpoint.

An exploratory science application is presented to demonstrate some of the functionality currently offered by VOStat.

1. Introduction

The use of statistics in astronomy is commonplace: of the 15000 astronomical studies carried out each year, 5% explicitly mention "statistics" in their abstract whilst 20% consider variable objects or multivariate datasets. However, the statistical methodologies that are predominantly employed in these studies predate the Second World War:

- Fourier transform (Fourier 1807)
- Least squares (Legendre 1805), Chi-squared (Pearson 1901)
- Kolmogorov-Smirnov (Kolmogorov 1933)
- Principal Component Analysis (Hotelling 1936)

As a whole, astronomy currently produces $\sim$ 1 TB of data per night and within a decade, the LSST[1] alone will produce $\sim$ 13 TB per night. If the prospect of Petabyte-sized data archives were not daunting enough, each data point will occupy a position in a parameter space consisting of several hundred dimensions. Successfully data mining these data sets mandates new sophisticated statistical techniques that are easily accessible to the general astronomical community and implemented in a distributed fashion to take full advantage of the power of the Virtual Observatory and the Grid.

2. VOStat

VOStat[2] is a cross-disciplinary collaboration between astronomers, statisticians and VO scientists specifically to address these issues. It consists of both a pedagogical component to teach astronomers how to apply statistical methods properly, e.g. when is it appropriate to use the Kolmogorov-Smirnov test to determine goodness-of-fit and when is it more appropriate to use the Anderson-Darling test, and a software component to offer them easy access to such methods.

VOStat is implemented upon an extensible distributed web-based framework so that it is simple to expose new functionality, which could be legacy applications, whilst allowing software to run in its optimal environment (hardware and software) without unnecessary porting. By using the VOTable[3] standard as the default data format, data metadata can be easily passed around to permit efficient process initialisation prior to computation whilst the data itself only needs to be transferred when required.

3. Accessibility

The complexity of the distributed network is hidden via access through a single science gateway. Three interfaces to the gateway are being implemented:

- an interactive web form to allow users to play with the software and test data sets
- web services for developers who want to incorporate the software within their own applications in exactly the same way as one does with an external library function
- module/plug-in for popular VO data exploration tools, such as Mirage[4], VOPlot[5] and DataScope[6].

---

[1] `http://www.lssto.org`

[2] `http://www.vostat.org`

[3] `http://www.ivoa.net/Documents/latest/VOT.html`

[4] `http://skyservice.pha.jhu.edu/develop/vo/mirage/default.aspx`

[5] `http://vo.iucaa.ernet.in/~voi/voplot.htm`

[6] `http://heasarc.gsfc.nasa.gov/vo/`

4. Functionality

VOStat currently provides access to selected functionality from the open source statistics package R[7] and multi-resolutional k-dimensional trees. The types of activity that can be carried out are:

- descriptive statistics (e.g. boxplot)
- two- and k-sample tests (e.g. Wilcoxon rank-sum)
- density estimation (e.g. kernel smoothing)
- correlation and regression (e.g. P.C.A.)
- censored data (e.g. survival)
- multivariate classification (e.g. H clustering)
- outlier detection (e.g. k-d trees)

5. Example: Outliers in Color-Color Space

The detection of outliers in a high-dimension parameter space will certainly be a common data mining activity with very large data sets. To illustrate how this might be achieved with VOStat, we took a sample of 1000 randomly-selected objects from the Palomar-Quest synoptic sky survey (Graham et al. 2004), each with 9 colors.

Boxplot A boxplot (Figure 1) shows the relationships between the colors in terms of the mean, median, overlap and outliers.

K-means clustering Figure 2 shows the relationships between the colors having identified 5 clusters in the data using K-means clustering.

Probability density association Figure 3 shows the $B-R$ vs. $R-I$ distribution for the objects identified as having the lowest probability of being associated with a cluster in the parameter space, i.e. highest probability of being an outlier, using k-d trees.

Visual inspection Figure 4 shows BRI plots of the highlighted object in Figure 3. It can clearly be seen to be a B-band dropout.

Acknowledgments. This work is supported in part by the NSF grant DMS-0101360.

References

Graham, M. J., et al. 2004, in ASP Conf. Ser., Vol. 314, ADASS XIII, ed. F. Ochsenbein, M. Allen, & D. Egret (San Francisco: ASP), 14

---

[7] `http://www.r-project.org/`

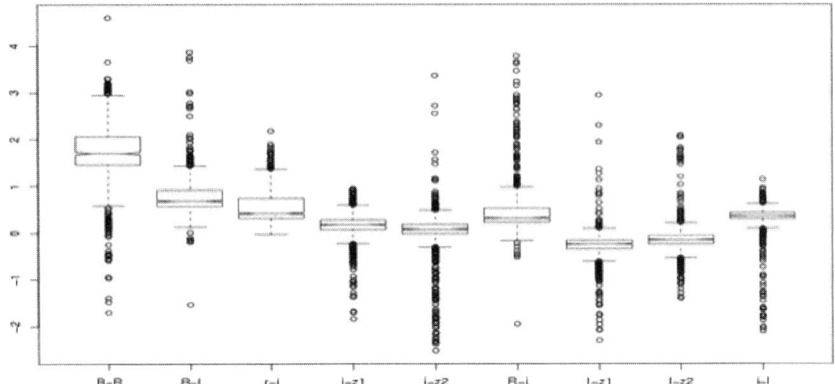

Figure 1. Boxplot of the data

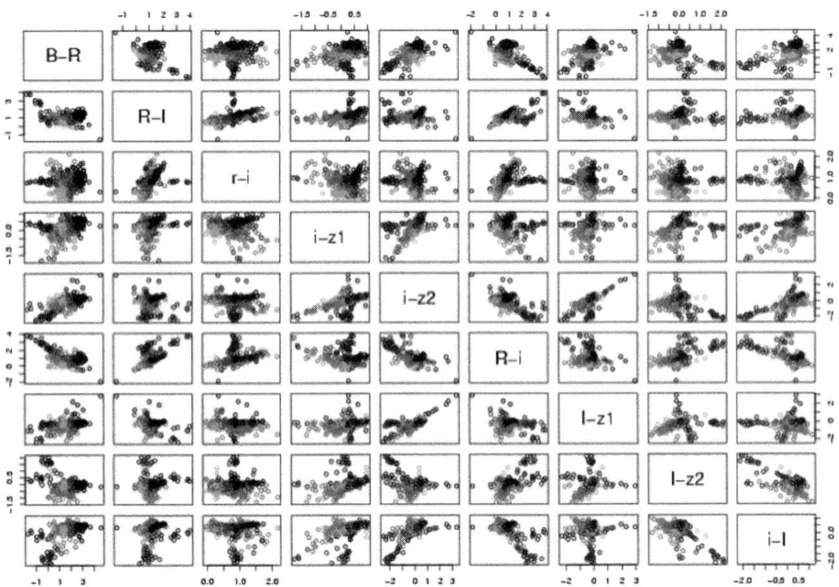

Figure 2. Plot of colors vs. colors having identified 5 clusters with K-means algorithm

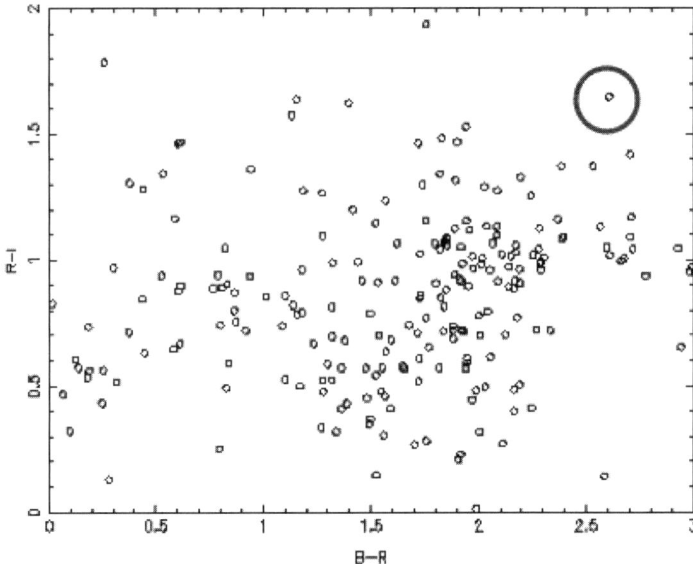

Figure 3. Color-color plot for lowest density association objects

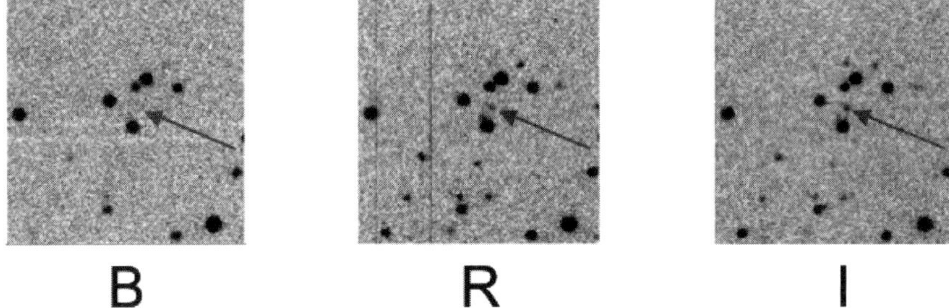

Figure 4. Visual plot of highlighted object

ASTRONOMICAL DATA ANALYSIS SOFTWARE AND SYSTEMS XIV
ASP Conference Series, Vol. 347, 2005
P. L. Shopbell, M. C. Britton, and R. Ebert, eds.

WBUCS: A Web Simulator for Deep Galaxy Fields

Daniel K. Magee, Rychard J. Bouwens, and Garth D. Illingworth

Astronomy Department, University of California, Santa Cruz, CA 95064

Abstract. Today's deep high resolution multiwavelength surveys contain a wealth of information about galaxies at different epochs. Fully exploiting this information to trace out galaxy assembly requires the ability to archive galaxy samples and resimulate these samples at different redshifts. Here we describe some of the tools developed for these ends as well as a simulator now available on the web for demonstration use. Based upon a pixel-by-pixel modeling of object SEDs and their selection volumes, this simulator will provide users with the ability to make realistic multicolor simulations of galaxy fields from galaxy samples at all redshifts: from $z \sim 0$ samples selected from the Sloan Digital Sky Survey to $z \sim 1-6$ samples selected from the Great Observatories Origins Deep Survey. Users only need specify the passbands, noise, and PSFs, or equivalently the exposure times on well-known instruments like HST or ground-based telescopes. As such, this simulator will provide the community with a real world virtual observatory, useful both at the proposal stage and for making comparisons with observations in hand. The engine for the simulator is the well-known BUCS (Bouwens' Universe Construction Set) library developed for galaxy evolution studies, while the web interface is analogous to that used for the popular web archives. The purpose of this poster is to introduce this effort to the computational community as a whole.

1. Introduction

Over the last few years, a wide variety of multi-wavelength high-angular resolution data has become available on different galaxy samples. From the more local Sloan data (Abazajian et al. 2003) to 2MASS (Skrutskie et al. 1997) to deeper, high resolution GOODS data (Giavalisco et al. 2004), observations now span an increasingly wide redshift range. Moreover, as a result of the full complement of telescopes available from Chandra to HST to Spitzer, these same observations are blessed with a wavelength baseline that spans over 8 orders of magnitude in wavelength.

Yet, despite the availability of an increasing rich set of data, the current generation of public simulation software is still very rudimentary in form. Packages like Terapix's Skymaker (Erben et al. 2001), IRAF's artdata (Tody 1993), or Space Telescope's Synphot allow for the simulation of deep galaxy fields, but only using relatively simple exponential or $r^{1/4}$ profiles and only assuming one has an input catalog available to use with the above simulators. Of course, there is no guarantee with these simulations that the inputs will be an accurate representation of the observations. In fact, the use of model profiles essentially assures us that they cannot be.

It is in this context that we have been developing a set of software to archive galaxy samples at a variety of redshifts and resimulate them on demand (Bouwens, Illingworth, & Magee 2005). By taking our inputs directly from large galaxy samples with high-angular resolution data, we are essentially assured that our inputs will be an extremely accurate and model-independent representation of the observations. What is novel in this effort as apart from our previous work on this subject (e.g., Bouwens et al. 1998a,b; Bouwens et al. 2003; Bouwens et al. 2004a) is our porting these capabilities to the web so that all astronomers can use our samples and web simulator to resimulate their own observations. We see this as a model for how things will work in the future.

2. What is WBUCS?

WBUCS (Web BUCS) is a simple web interface to the BUCS (Bouwens Universe Construction Set; Bouwens et al. 2004b, Bouwens, Illingworth, & Magee 2005). One of the primary capabilities of BUCS is the ability to simulate deep galactic fields for a variety of different ground or space-based imagers. In contrast to most simulation tools currently available, BUCS features the use of real galaxy templates extracted in representative numbers from high-quality multicolor observations of deep galaxy fields.

BUCS places these templates on the simulated frames with the same surface (or space) densities as that found in the original data. For our default field galaxy option (which draws objects from the Hubble Ultra Deep Field (HUDF)), individual templates are allotted surface densities equal to exactly 1/area of the survey, which for the HUDF ($\sim$ 13 arcmin^2) means a surface density of 0.08 arcmin^{-2}.

In creating the actual output images for each input, BUCS recalculates the appearance of each galaxy using the best-fit pixel-by-pixel spectral energy distributions and then resamples it onto the output frame. The results are then smoothed to the PSF of the output image, and noise added to reproduce the general noise level calculated to be present in the final output image. Both the noise and PSF present in the input templates are considered in making these final two corrections.

By preserving both the surface densities and pixel-by-pixel morphological details of each input object, BUCS is able to produce remarkably model-independent representations of the field. This provides a powerful complement to the similarly quantitative, but visually less salient results of the standard exposure time calculator available throughout the astronomical world.

3. Web Interface

WBUCS provides a simplified interface to the BUCS simulation package. Through simple forms, the user has the ability to select up to several different combinations of the basic observational parameters for each simulation. These include

- Telescope (HST)
- Instrument (ACS-WFC and ACS-HRC)
- Filter (SDSS griz, Johnson BVI, Narrow V, Broad V)
- Total Exposure Time

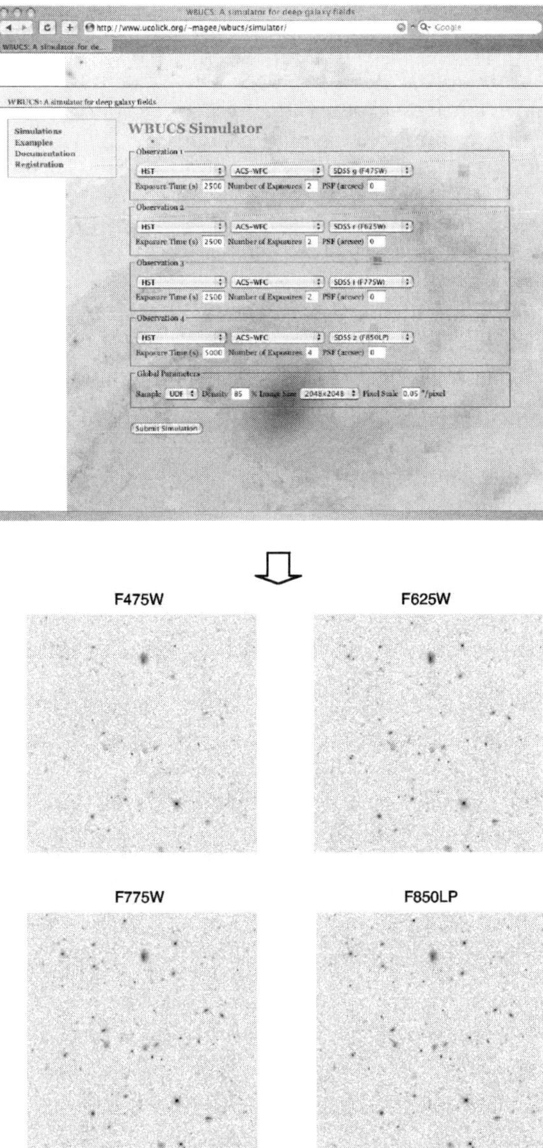

Figure 1. Users request simulations of deep galaxy fields through simple forms like the one shown at the top of this figure. The output is composed of simple fits images of the field, as viewed through different passbands (shown at the bottom of this figure). Once the request has been received by our servers, it is placed in the queue and an e-mail is sent out summarizing the request. After the job is completed, the simulated images are placed at an ftp site and the user is notified as to the location where they can download them.

- Number of Exposures (Read Outs)

This allows a user to simulate the same galaxy field with a variety of different instruments and filters (as shown in Figure 1) or to simulate a field with the same filter at several depths. As of this year's meeting, only the HST ACS Wide Field Camera and High Resolution Camera were supported. The user also has the option of setting a number of global parameters for the simulation. These include

- Model (e.g., HUDF)
- Density (surface density of objects with respect to the HUDF)
- Image Size (512×512, 1024×1024, 2048×2048)
- Pixel Scale (arcsec/pix)

Once a simulation is submitted the user will be notified by e-mail when the simulation is finished and ready for retrieval via FTP. The user also receives an e-mail upon initial submission of the request summarizing its details and verifying that it is in the queue.

4. Future Developments

There are plans to add a number of other features to WBUCS to enhance its functionality. These include:

- Support for a number of other popular instruments and telescopes in space and around the world.
- A wide range of different complete galaxy samples as inputs (from $z \sim 0$ samples to high-z dropout samples)
- Ensemble of well-known fields (e.g., WFPC2 HDFN, HUDF) as inputs.
- Use of real galaxy clusters as inputs.
- Ability to simulate data with known properties.

Acknowledgments. We are grateful to Narciso Benítez, Emmanuel Bertin, Tom Broadhurst, Fred Courbin, Michele de la Pena, Harry Ferguson, Marijn Franx, Andy Fruchter, Jon McCann, and Gerhardt Meurer for input important for development of the BUCS libraries that are essential to the running of this interface. We acknowledge fruitful collaborations with other members of the ACS GTO team and support from NASA grant NAG5-7697.

References

Abazajian, K., et al. 2003, AJ, 126, 2081

Bouwens, R.J., Broadhurst, T.J., & Silk, J. 1998a, ApJ, 506, 557

Bouwens, R.J., Broadhurst, T.J., & Silk, J. 1998b, ApJ, 506, 579

Bouwens, R.J., et al. 2003, ApJ, 595, 589

Bouwens, R.J., et al. 2004a, ApJ, submitted

Bouwens, R.J., et al. 2004b, ApJ, in preparation

Bouwens, R.J., Illingworth, G.D., & Magee, D.K. 2005, in ASP Conf. Ser., Vol. 347, ADASS XIV, ed. P. L. Shopbell, M. C. Britton, & R. Ebert (San Francisco: ASP), 100

Erben, T., Van Waerbeke, L., Bertin, E., Mellier, Y., & Schneider, P. 2001, A&A, 366, 717

Giavalisco, M., et al. 2004, ApJ, 600, L93

Skrutskie, K., et al. 1997, in ASSL Vol. 210: The Impact of Large Scale Near-IR Sky Surveys, ed. F. Garzn, N. Epchtein, A. Omont, B. Burton, & P. Persi (Dordrecht: Kluwer), 25

Tody, D. 1993, in ASP Conf. Ser., Vol. 52, ADASS II, ed. R. J. Hanisch, R. J. V. Brissenden, & J. Barnes (San Francisco: ASP), 173

ASTRONOMICAL DATA ANALYSIS SOFTWARE AND SYSTEMS XIV
ASP Conference Series, Vol. 347, 2005
P. L. Shopbell, M. C. Britton, and R. Ebert, eds.

Configurable Format Conversion Tool from XML to FITS with Binary Extensions

K. Matsuzaki, H. Baba,[1] K. Shuto, and A. Miura

Institute of Space and Astronautical Science, Japan Aerospace Exploration Agency, 3-1-1 Yoshinodai, Sagamihara, Kanagawa, Japan

H. Nakamoto, R. Kanamori, and Y. Kobayashi

Systems Engineering Consultants Co., 22-14, Sakuragaokacho, Shibuya,Tokyo, Japan

Abstract. We have developed a prototype of a configurable format conversion tool from XML document to FITS document with binary extensions. Structure of input data of this tool is specified by an XML schema language – RELAX NG – and that of output is also specified by a schema described in XML. This tool is configured by a definition of simple mapping created from schema of input and output. Since RELAX NG is a general purpose XML schema language and this tool accepts mapping of arbitral data item in XML data structure, we expect this tool will be widely used in data processing for astronomy.

1. Introduction

In development of complex data systems, designers often encounter several document formats determined by external requirements. For example, applications of the Virtual Observatory (VO) must handle XML documents[2] received from web services and may create FITS file with Binary Extensions (Cotton, Tody, & Pence 1995) used by observers downstream. Although conversion of the document format itself does not add new information, the development and testing of software takes some amount of time. To date, there are few solutions to connect modern XML technologies with traditional FITS file processing easily (in the case of development of data systems). Thus, we have decided to develop a configurable format conversion tool from XML document to FITS document with binary extensions, looking toward the upcoming era of VO. We have named this tool XML2FITS.

In order to minimize incoming effort for format conversion in general, we have designed conversions with XML2FITS to be a subset of specifications of conversions that cover much wider document formats (e.g., text and binary document). Although we deal with various types of document formats in our framework of conversions, there were no suitable schema languages which covered the wide range of document formats. Thus, we started with the definition of a schema

[1]Waseda University

[2]`http://www.w3.org/TR/2000/REC_xml-20001006`

```
<element name="data"
datatypeLibrary="http://www.w3.org/2001/XMLSchema-datatypes">
    <zeroOrMore>
        <element name="rec">
            <attribute name="t">
                <data type="dateTime"/>
            </attribute>
            <element name="Voltage">
                <data type="double"/>
            </element>
        </element>
    </zeroOrMore>
</element>
```

Figure 1. Sample of schema of XML document

language. We call the schema language, Mapping Schema. In the next section, we describe the parts of Mapping Schema required for understanding and usage of XML2FITS.

2. Schema of Input and Output

In development of software, the description of interface is often the most important. We can strictly describe interfaces of input and output with schema language, but only to a point. Parameter specification of format conversion on our framework also starts from a description of schema of input and output.

For XML documents, there is a standard schema language RELAX NG (ref. 2), which is endorsed by regular tree grammar. We have designed Mapping Schema so that the schema of the XML document is compatible with RELAX NG. Specifically, we intend to extend the philosophy of RELAX NG carefully into other document formats. In the case of RELAX NG, several syntaxes for presentation of a schema exist. In this paper, we concentrate our description on schema presented by XML (the XML syntax). In the rest of this section, we describe how we divide the language of RELAX NG into common portions of Mapping Schema and XML document specific portions. We also describe how we introduce the FITS document portion.

Mapping Schema consists of a nest of schema elements. In the XML syntax, one schema element is presented by one element of XML. There are several species of schema element. In the XML syntax, each species of schema element is specified by the name of the XML element. A schema element can have various attributes. In the XML syntax, the name and value of an attribute is presented by the name and value of the attribute of the XML element, respectively. All of the document formats share species of schema elements in the core definition of Mapping Schema. Each individual document format has additional species of schema elements. Allowed combinations of nests of schema elements are defined by the core and format specific portions.

```
<BTE
datatypeLibrary="http://www.w3.org/2001/XMLSchema-datatypes">
    <zeroOrMore>
        <BTEData ttype="t" tunit="sec">
            <data type="dateTime"/>
        </BTEData>
        <BTEData ttype="doubleItem" tunit="Volt"
           ttypeComment="Double Precision Value">
            <data type="double"/>
        </BTEData>
    </zeroOrMore>
</BTE>
```

Figure 2. Sample of schema of FITS document

The structure element is the type of schema element that specifies conditional appearance of information items; structure element belongs to the core definition. There are several species – `zeroOrMore`, `oneOrMore`, `optional`, `choice`, and so on. Some document formats have their own specific ways to present some of these structures. For example, a FITS document has several rows in a binary extension. Repeat (of row) is described by `zeroOrMore` in a schema.

A container is a kind of schema element that specifies how data is held by an instance of schema (i.e., input of XML2FITS). Each type of document format has its specific containers. There are two species in XML portion of Mapping Schema – `element` and `attribute`. There are several species in the FITS portion of Mapping Schema – `primaryHeaderValue`, `primaryHeaderComment`, `BTEHeaderValue`, `BTEHeaderComment` and `BTEData`.

A subcontainer is a kind of schema element which specifies data itself. Structure element belongs to the core definition. There are species `data` and `value`. Schema element `data` has an attribute `type`. It is possible to restrict the allowed range of data with schema element `param`. Allowed value for `type` and `param` can be extended by an external datatype library.

We show a sample of schema for XML documents presented by Mapping Schema in Figure 1 and a sample for FITS documents in Figure 2. These examples of schema use an external datatype library which refers to the simple type defined by W3C XML Schema[3]. These examples hold equivalent tables with rows consisting of a time and voltage. The two documents differ in the explanation column. The XML document only gives some hint by name of element; the FITS document with binary extension can have unit and comments in the header area.

[3]`http://www.w3.org/TR/2001/PR-xmlschema-2-20010330`

3. Definition of Mapping

We want to preserve the data itself between the input and output of format conversion. In a schema of Mapping Schema, intrinsic structure and order of data can only be defined by schema elements in the core definition. In such a core definition, only schema can be easily obtained from a specific document format by removing the container preserving schema elements inside.

When we keep schema elements in the core the same between input and output, we can easily combine two schema files into a file of simple mapping definition. Conversion of XML2FITS is specified by such a single mapping definition. When you use this tool for simple mapping, intrinsic manipulation of data must be decoupled with the format conversion and must be performed either before input or after output in any other technology (e.g., using XSLT for XML).

Each schema element of a simple mapping definition belongs to one of three spaces – 1) space for input data which does not appear on output, 2) space for data transferred from input to output and 3) space for output data which does not appear on input. From a simple mapping definition, we obtain a schema of input data by selecting schema elements in the two former spaces. We obtain a schema of output data by selecting schema elements in latter two spaces. In the XML syntax, we allocate one XML namespace[4] for one space. A URI which can be processed by a given tool may differ from one tool to another. For example, the prototype of XML2FITS uses the following three spaces respectively:

- http://isas.jaxa.jp/ums/xml2fits/20040825/xml
- http://isas.jaxa.jp/ums/xml2fits/20040825
- http://isas.jaxa.jp/ums/xml2fits/20040825/fits

We show a sample of a simple mapping definition in Figure 3. This definition specifies a mapping between a valid XML document of schema of Figure 1 and valid FITS document of schema of Figure 2. Compared with general purpose computer language, the roles of simple mapping definition is very limited. When we use these definitions, we can shorten the period of development while keeping the reliability of the system. This definition of conversion does not specify a direction of conversion. It will be possible to convert from FITS file to XML with this specification when tools like FITS2XML become available.

4. Implementation and Application

A prototype of this tool was built in September 2004. We are planning to use this tool for the pipeline processing of telemetry status of satellites Astro-F and Solar-B which ISAS will launch in 2005-2006. Generated FITS file will be served for astrophysical community via web interface DARTS hosted by ISAS. We will create simple definition mapping by transformation of existing ISAS database of telemetry definition which also minimizes effort for development. After our internal testing of XML2FITS, we will start distribution of the tool itself toward the astronomical community.

---

[4] `http://www.w3.org/TR/2000/REX-xml-20001006`

```
<fits:BTE
    xmlns:fits="http://isas.jaxa.jp/ums/xmlfits/20040825/fits" >
    <element name="data"
      xmlns="http://isas.jaxa.jp/ums/xmlfits/20040825/xml">
        <ums:zeroOrMore
          xmlns:ums="http://isas.jaxa.jp/ums/xmlfits/20040825"
datatypeLibrary="http://www.w3.org/2001/XMLSchema-datatypes">
            <element name="rec">
                <fits:BTEData ttype="t" tunit="sec">
                    <attribute name="t">
                        <ums:data type="dateTime"/>
                    </attribute>
                </fits:BTEData>
                <element name="Voltage">
                    <fits:BTEData ttype="doubleItem" tunit="Volt"
                      ttypeComment="Double Precision Value">
                        <ums:data type="double"/>
                    </fits:BTEData>
                </element>
            </element>
        </ums:zeroOrMore>
    </element>
</fits:BTE>
```

Figure 3. Sample of simple mapping definition

5. Conclusions

We have developed a prototype of a configurable format conversion tool from XML document to FITS document with binary extensions. This tool can be fully configured by definition of simple mapping of Mapping Schema and allows the design and development of individual applications away from manual programming of format conversion.

References

Cotton, W. D., Tody, D. B., & Pence, W. D. 1995, Binary Table Extension to FITS, A&AS, 113, 159

Document Schema Definition Languages (DSDL) – PARTS2 Regular-grammar-based validation – RELAX NG, ISO/IEC FDIS 19757-2 (RELAX NG)

ASTRONOMICAL DATA ANALYSIS SOFTWARE AND SYSTEMS XIV
ASP Conference Series, Vol. 347, 2005
P. L. Shopbell, M. C. Britton, and R. Ebert, eds.

Spreading DIRT with Web Services

Marc W. Pound, Mark G. Wolfire, and N. S. Amarnath

Astronomy Department, University of Maryland, College Park, MD 20742

Raymond L. Plante

National Center for Supercomputing Applications, University of Illinois Urbana-Champaign, Urbana, IL 61801

Abstract. Over the past 5 years, we have created and developed the Web Infrared Tool Shed (WITS), a collection of web pages and programs ("tool boxes") for the analysis of astronomical objects that emit primarily in the wavelength range 2 m to 1 mm. WITS provides the astronomical community with a uniform set of models over a large parameter space of source properties. One of the toolboxes is the Dust InfraRed Toolbox (DIRT), a comprehensive Java applet for modeling astrophysical process in circumstellar dust shells around young and evolved stars. The DIRT database currently holds about 600,000 model spectral energy distributions. To facilitate greater flexibility in retrieval and visualization of these models, we aim to provide access to the model database through a variety of Web Services. We describe here a simple service that retrieves a requested model and returns it as a VOTable. This is the first step towards interoperability with a Theory VO. We also describe the classes of Web Service that a theory-based VO should provide.

1. Theory in Virtual Observatories

The primary emphasis of the current VO projects is storage and retrieval of, and discovery with, *observational* data collections. Little effort has been put into representing *theoretical* data collections in a VO. This is primarily because the observational data collections produced by sky surveys are large and fairly uniform even across energy bands, whereas theoretical data collections are smaller and less uniform as they are typically produced by individuals rather than consortia. Thus, there is an enormous diversity in the type, format, and information content of the data. To investigate how theoretical data could not only be accessed via a VO, but usefully compared to observational data therein, requires a large, uniform theoretical data collection. Over the past 5 years, we have developed just such a collection.

It is essential that if the VO is to achieve its full potential, a theory VO and interface must be created to analyze the enormous data flow in the same easily accessible, seamless, and user-centered manner.

2. The Real DIRT

Our theoretical data collection consists of the model outputs from the radiation transfer code of Wolfire & Cassinelli (1986). Each of our model data sets has 7 input parameters and the outputs consist of a least 20 related physical properties. The wavelength grid has 83 separate points while the spatial grid has 200 separate points. We currently hold approximately 600,000 models.

The Dust InfraRed ToolBox (DIRT) - a part of WITS[1] is a Java applet for modeling astrophysical processes in circumstellar shells around young and evolved stars. DIRT has been used by the astrophysics community for about 5 years.

Using DIRT's novel Java interface, users can access the database of radiation transfer models. A flexible, multi-dimensional plotter allows users to view the models, rotate them, tag specific parameters with color and symbol size, filter models by windowing specific parameters, overlay observed data, and probe individual model and data points. Thus, an observer can rapidly, visually, determine the appropriate range of model parameters that match the data. To aid users in finding the best fit model to their data we provide a flexible interface for data input, and a χ^2 fitting routine that searches the selected models for the best fit. An early version of DIRT is discussed in Pound et al. (2000) with recent updates presented in Amarnath, Pound, & Wolfire (2003). The initial stages of our computer cluster used to generate models is presented in Teuben et al. (2000). DIRT has been singled out by the VO community as one of the leading candidates for the theory VO (Teuben et al. 2002).

3. Web Services and the VO Environment

To make theory an integral part of VO-based research, there are two parallel perspectives we need to consider: one of end-user functionality and one of infrastructure. From the first perspective, there are four classes of services that theory-based VO application should provide in order to be useful to the astronomical community:

1. *Discover models* – Astronomers need to be able to locate models based on physical parameters, much like they locate data based on observational parameters.
2. *Simulate observations* – The convolution of models with real instrument response functions is necessary for direct comparison of models with observations.
3. *Benchmark models* – A set of models over a representative sampling of parameter space can serve as a standard for verification against independent modeling efforts.
4. *Compare models with models* – Intercomparison of models within a model database as well as comparison with similar models calculated by others provides important insight to the modeling process and the input physics.

The infrastructure perspective considers how such services can be made readily available and practical to use in a VO environment. The use of common

[1] `http://dustem.astro.umd.edu`

data formats is essential to enabling the comparison of data that comes from different places. Unfortunately, there are no common formats for astrophysical theory data. Another important bit of infrastructure needed to support the VO is a common service framework, and the framework adopted by the VO community is based on Web and Grid Services. By deploying our services within this framework, we make it easier to allow distributed services to work together. Much of the task of integrating theory into the VO is one of visibility. This goes beyond just registering theory data and services in the VO Registry framework (Plante et al. 2004) along side their observational counterparts. Through strategic implementation of standard services, theoretical resources will be easier to incorporate into larger integrating services.

4. How Does DIRT Fit in?

We want to provide non-interactive access to models, allowing DIRT to be plugged into other VO-enabled services and applications. Thus, an important step toward interoperability would be to export our DIRT models in VOTable format, making it easier for users to compare our models with other data in other software packages. While adoption of VOTable for theory data would be a major advance, it would not be a panacea for two reasons: not all the standard semantic metadata needed for theory data has been defined, and the general VOTable structure permits multiple ways of storing the same type of data.

To that end, we have written a very simple Dirt Model Server Web Service to retrieve a single model from the database and return it as a VO Table. We used NVO Summer School[2] software and course notes to write a simple Java server and client. The infrastructure provided by the NVOSS software generates a WSDL based on the server code.

The server has methods to set the 7 parameters needed to specify a model and a source distance by which to scale the model fluxes. The *getReply()* method contacts the model database server via a `java.net.URLConnection`, calling a CGI perl script to query the database for the requested model. The method then returns the selected model as a VO Table.

```
DirtModelServer (wsdl)

   void setGrainType(String)
   void setDensity(String)
   void setLuminosity(String)
   void setEffectiveTemperature(String)
   void setOuterRadius(String)
   void setInnerRadius(String)
   void setExtinction(String)
   void setDistance(float)

   String getReply()
```

By providing DIRT models as VOTables through a simple Web Service we enable other applications to import the models and use them.

---

[2]`http://www.us-vo.org/summer-school/`

5. The Future of DIRT+Web Services

We plan to retool DIRT to become the first application to take advantage of the Web Service interface to theoretical models (assuming we get funding!) In doing so, DIRT will satisfy the essential requirements for functionality and infrastructure support when we provide the following functionality in a coherent set of service operations:

- returning groups of related models based on a user query of physical parameters
- enabling users to compare our models with those computed by other groups
- enabling others to develop analysis tools using our model database as the backend
- returning an individual model as a VOTable
- accepting observational data and finding the best fitting models with user-selected fit constraints
- computing and verifying user-initiated models
- providing on-the-fly convolution of models with user-specified instrument profiles
- providing user-originated χ^2 fitting of models to observations

Our improvements to DIRT will lay the groundwork for new ways in which theoretical models can be used to analyze data. Individual astronomers often confine their work to small portions of the electromagnetic spectrum because it is simply too difficult to learn all the techniques necessary to observe at vastly different wavelengths. Similarly, it is a rare astronomer who is an expert in both theory and observation. The promise of the observational VO is to let astronomers cross such traditional wavelength boundaries, allowing those who are not experts with data in particular wavelengths to nonetheless use those data effectively to make new discoveries. The promise for a theory VO is to break down the wall between theory and observation. With DIRT+Web Services, we remove the first bricks.

References

Amarnath, N. S., Pound, M. W., & Wolfire, M. G. 2003, in ASP Conf. Ser., Vol. 295, ADASS XII, ed. H. E. Payne, R. I. Jedrzejewski, & R. N. Hook (San Francisco: ASP), 381

Pound, M. W., Wolfire, M. G., Mundy, L. G., Teuben, P. J., & Lord, S. 2000, in ASP Conf. Ser., Vol. 216, ADASS IX, ed. N. Manset, C. Veillet, & D. Crabtree (San Francisco: ASP), 628

Teuben, P. J., DeYoung, D., Hut, P., Levy, S., Makino, J., McMillan, S., Portegies Zwart, S., & Slavin, S. 2002, in ASP Conf. Ser., Vol. 281, ADASS XI, ed. D. A. Bohlender, D. Durand, & T. H. Handley (San Francisco: ASP), 7

Teuben, P. J., Wolfire, M. G., Pound, M. W., & Mundy, L. G. 2000, in ASP Conf. Ser., Vol. 216, ADASS IX, ed. N. Manset, C. Veillet, & D. Crabtree (San Francisco: ASP), 644

Wolfire, M. G., & Cassinelli, J. P. 1986, ApJ, 310, 207

ASTRONOMICAL DATA ANALYSIS SOFTWARE AND SYSTEMS XIV
ASP Conference Series, Vol. 347, 2005
P. L. Shopbell, M. C. Britton, and R. Ebert, eds.

A New SIMBAD Software and System : SIMBAD 4

Marc Wenger and Anaïs Oberto

Centre de Données astronomiques de Strasbourg, France

Jean-Philippe Lejal

Université de Nancy, France

Sylvaine Jaehn and Benoît Baranne

Université Technologique de Belfort/Montbelliard, France

Olivier Dellicour and Julien Deprez

Haute Ecole Rennequin Sualem, Liège, Belgium

Abstract. The current version of Simbad has been running since 1990. It was designed for command line queries and evolved to client/server mode. At the same time, its contents increased from 700,000 objects to more than 3,000,000, and the number of queries exploded from a few hundred to an average of 15,000 daily. Fifteen years later, it was time to design a new version taking into account the evolution of technologies in languages, DBMS, networking, and software architecture. This paper presents the architectural design of the new Simbad system. The choices made are analyzed and justified. The integration of SIMBAD 4 in the Virtual Observatory functionalities is also presented.

1. Introduction

SIMBAD[1] (Wenger et al. 2000) is an astronomical Database built around cross-identifications, bibliography and other basic data and measurements for more than 3,000,000 astronomical objects of any type. It was developed and is maintained by the Centre de Données astronomiques de Strasbourg.

The database can be queried by identifiers, coordinates and several criteria applied on *basic data*. SIMBAD is mainly accessed by the Web and a package of C routines. The query by mail facility allows very large queries in a non interactive way.

[1] `http://simbad.u-strasbg.fr/Simbad`

2. History

SIMBAD was preceded by the CSI (Catalogue of Stellar Identifications) created in 1972. Up to now, three versions of the database software have been developed and the fourth release is under development. Their characteristics are:

Period Release	Platform/OS Languages	Access mode	Networks
1972–1981 CSI	IBM/OS360 PL1/Assembler	Keypunched cards printouts	leased lines
1981–1990 SIMBAD 2	Univac/Exec 8 Assembler/PL1	Command line mode	packet switching networks
1990–2005 SIMBAD 3	DEC,SUN/Unix C/PERL	Command line mode XSimbad interface (Motif) Web interface C language package	SPAN Internet
2005–... SIMBAD 4	PC/Linux JAVA	Web Interface Java interfaces Web services Java classes (RMI)	Internet

3. SIMBAD 4 choices

The main choices for the SIMBAD 4 software are summarized here:

- Multi-tier architecture implemented on several PCs running Linux.
- JAVA language for all development - client side, as well as server side.
- PostgreSQL database management system.

These choices will be described in more details in the following sections.

4. Java language

Using one language for all components of a software has the advantage of avoiding duplication of modules required in several components (i.e., application server and graphical user interface client).

Java is currently the only language adapted for development of both server functionalities (servlets, JSP pages, application classes) as well as for graphical user interfaces. Many integrated features allow easy developments, even for sophisticated applications.

Finally, fast and overall robust developments make JAVA a language of choice for long term developments.

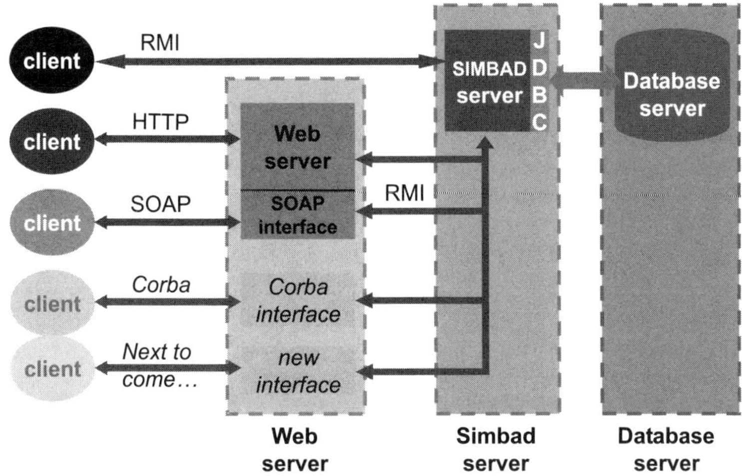

Figure 1. The SIMBAD4 architecture

5. PostgreSQL Database Management System

SIMBAD 3 is based on a home-made database management system including many concepts of object oriented database management features. Now that several vendors are offering true object DBMS, it was tempting to use one of them. Unfortunately, in spite of interesting capabilities for dealing with heterogeneous data, it remains difficult to be confident in the longevity of these vendors, because of the small interest in such systems.

Therefore, we decided to come back to a relational system. In spite of some difficulties in dealing efficiently with heterogeneous, linked together data, it has the advantage of being very stable and normalized. PostgreSQL is the most compliant to SQL standards open source DBMS and was therefore chosen.

6. Multi-tier architecture

The SIMBAD 4 software is organized around four layers (Figure 1):

- A client layer, with several kind of clients for different purposes.
- A communication layer containing a web server, a servlet server, a Web Service interface, and when needed, specialized interfaces dealing with new or exotic protocols.
- An application layer, containing all the application *intelligence*. This layer is accessed only through the RMI protocol. It will therefore remain easy to maintain, and will not interfere with new client technologies.
- A database layer, managing the basic SQL queries to the database. This database server communicates only with the application server through the standard JDBC interface.

These different layers will be implemented on PCs running Linux.

7. Modularity

The fast evolution of communication protocols (sockets, CORBA, RMI, SOAP, ...) and technologies like database management systems (Relational, Object Oriented DBMS) or user interfaces (X/Motif, Web browsers, Java language) enforces the need for building a highly modular application in which every component can easily be replaced by a new one, independently of the others.

Object oriented development, provided by the JAVA language, also enforces modularity by the strong encapsulation of classes.

8. Clients

SIMBAD 4 will have access to several kind of clients :

- Web browsers, served by JSP and servlet technologies. This remains the simplest access way for end users, requiring no preliminary installation.
- Java user interfaces, allowing more sophisticated interactions.
- Web services, well suited for interoperability between VO components.
- Simbad client package, a legacy from SIMBAD 3 offering access to the SIMBAD data in C language applications.
- Client for batch queries, useful for large or repetitive queries.

9. Vendor Independence

A SIMBAD release lives 10-15 years. The first two releases were strongly vendor dependent (IBM, Univac). Switching between them required a significant amount of work each time.

Unix, as a hardware independent OS, facilitates the move from one platform to another. Today, JAVA language enforces this independence, by bringing its own environment (Virtual Machine) to every kind of platform.

JDBC, the JAVA Database Interface, allows one to access any relational DBMS, as long as only standard relational features are used. Moreover, the open source DBMS PostgreSQL grants portability against new OS releases and against commercial vendors unwanted evolution (or disappearance !).

10. New Features

SIMBAD 4 includes the following new features:

- Implementation of two kinds of links between objects : *hierarchies* between a main object and components, and *associations*, linking two object for any kind of reason.
- Reference filtering by authors and title content to limit the output to the most significant references.
- More flexible syntax for identifiers of astronomical objects. Graphical interfaces will also provide on the fly help for building an identifier.

- Enhanced possibilities for sampling, including constraints on measurement fields, which are not currently accessible.
- Deep integration with the other CDS services : Vizier[2] (Ochsenbein et al. 2000) offering access to the data in catalogues having been cross-identified; Aladin[3] (Bonnarel et al. 2000) which will allow object plotting and image overlay facilities.
- Scripting language for batch queries
- VO[4] enabled. SIMBAD 4 will integrate all the required technologies:
  - UCDs[5] (Derrière et al. 2004) for each piece of data. They will be available for queries, especially when measurement fields are involved.
  - VOTABLE[6] (Ochsenbein et al. 2004) output. Output from SIMBAD can hardly fit into one table. Therefore, an astronomical object is considered as a RESOURCE, containing several TABLES with homogeneous data - basic data, identifiers, bibliographical references, and measurement tables.
  - Web Service accesses (name resolvers, other kind of queries using the future Astronomical Query Language defined for VO applications).
  - Description for registries

  The modular design enables an easy integration of any new standard or further evolution of current standards.

11. Conclusion

We hope that with this design, the SIMBAD software will be able to keep up-to-date with the fast evolutions driven by the Virtual Observatories developments. The new Simbad Software should be operational mid-2005.

References

Bonnarel, F., Fernique, P., Bienaym, O., et al. 2000, A&AS, 143, 33

Derrière, S., Gray, N., McDowell, J.C., et al. 2004, in ASP Conf. Ser., Vol. 314, ADASS XIII, ed. F. Ochsenbein, M. Allen, & D. Egret (San Francisco: ASP), 315

Ochsenbein, F., Bauer, P., & Marcout, J. 2000, A&AS, 143, 23

Ochsenbein, F., Williams, R., & Davenhall, C. 2004, Toward an International Virtual Observatory, Proceedings of the Conference in Garching, Edited by P.J. Quinn, and K.M. Gorski (Springer), 118

Wenger, M., et al. 2000, A&AS, 143, 9

---

[2]`http://vizier.u-strasbg.fr/viz-bin/VizieR`

[3]`http://aladin.u-strasbg.fr/aladin.gml`

[4]`http://www.ivoa.net/`

[5]`http://cdsweb.u-strasbg.fr/doc/UCD.htx`

[6]`http://cdsweb.u-strasbg.fr/doc/VOTable/`

ASTRONOMICAL DATA ANALYSIS SOFTWARE AND SYSTEMS XIV
ASP Conference Series, Vol. 347, 2005
P. L. Shopbell, M. C. Britton, and R. Ebert, eds.

RADAR: A Fast, Scalable, and Distributable Archive Inventory Service

Anzhen Zhang, Thomas H. Jarrett, Anastasia Alexov, G. Bruce Berriman, John C. Good, Mihseh Kong, Naveed D. Tahir-Kheli, and Serge Monkewitz

Infrared Processing and Analysis Center, California Institute of Technology, Mail Code 100-22, CA 91125

Abstract. The NASA/IPAC Infrared Science Archive (IRSA) has recently deployed the Recursive Archive Digest and Reference (RADAR) service, which returns an inventory of IRSA's holdings in response to a spatial query, and offers one-click download of data and links to IRSA's data access services. RADAR also supports inventories and data access from remote archives; the current implementation supports access to the Multi-mission Archive at STScI (MAST) Spectral and Image Scrapbook and NEDBasic Data. When complete, RADAR will maintain the results of multiple queries in "data collections" and will provide tools that will allow users to augment collections, remove data from them, modify search criteria, resubmit jobs, and check job status. RADAR is supported by an evolution of IRSA's component based architecture. It utilizes a fast estimation service and runs under the Request Management Environment (ROME) funded by NVO.

1. Introduction

RADAR (Recursive Archive Digest and References; `http://irsa.ipac.caltech.edu/applications/RadarSvc`) was developed by the NASA/IPAC Infrared Science Archive (IRSA) to return inventories of the archive's growing holdings in response to a spatial query. It supports fast inventories of the archive holdings, generation of multiple inventories in the same browser session, one-click packaging and download of data, and links to custom services for individual datasets to support further exploration of the data.

This paper describes the design challenges faced by RADAR, and how they were met to deploy an easily maintainable service that retains its performance as the archive holdings grow.

2. Design and Architecture

RADAR serves as a proxy application, through which the user can make complex data retrieval requests to IRSA's public datasets. IRSA is a multi-mission archive. It hosts huge source catalogs which are stored in the database, and smaller source catalogs that reside on spinning disk. It curates large image datasets and smaller spectral datasets, which are stored on spinning disks while their metadata are stored inside the database; some of these data are in FITS format, while others are JPEG or PostScript files. The archive thus curates a broad range of data formats, data storage, and data access mechanisms. Providing generic software which manages these different data format and access mechanisms in a manner that appeared uniform to the user was a major challenge in designing RADAR.

In addition to the diversity of the holdings, IRSA frequently updates its data holdings and releases new datasets according to data providers' delivery schedules. Some of them may be updated several times a year. IRSA deploys custom interfaces for all new datasets, and updates existing interfaces when data are updated. RADAR thus needs to respond dynamically to updates to the data holdings, while at the same time incorporating inventories of new datasets.

Based on those considerations, RADAR is driven by a master table of datasets. Any updates to an existing dataset only require a modification to its data pointers in the table. Every new dataset release can be incorporated by adding one entry in the master table. This design provides flexibility in supporting updates to the archive holdings and minimizes the additional development and test burden needed to support them.

RADAR is accessible as a web form. It receives requests through a web server (Apache) which are handed off to C-based CGI programs called "applications". Separation of the processing into applications is overkill for RADAR, where coordinated processing across a set of applications / datasets was desired, either to collect data from various sources in response to a higher level request (e.g., for an area on the sky) or as part of a distributed processing paradigm. RADAR is meant to run data access applications simultaneously. Therefore, RADAR has adapted middleware called the Request Management Environment (Kong, Good & Berriman 2004) to take advantage of multi-threaded processing of applications and provide scalable performance as the archive holdings grow.

Most IRSA applications are CGI programs that can accomplish a request in real time. They usually respond with HTML (generated by the CGI program) encapsulating or referencing the data which was generated in the workspace. RADAR requires that every application provide a mode which performs this same processing without standard output; instead, it generates a resultant output file in HTML which will be used by the RADAR Collection Viewer (see Section 6). In this mode, each application provides status information in XML format, which facilitates communication between RADAR, ROME, and itself.

Some applications may take some time to complete (such as long database queries, 2MASS images and, in future, custom image mosaics). In those cases, the applications provide their own status pages. Those pages provide detailed information about every process step and its status in real time. A polling mechanism is in place to check status via a browser. Similar status messages are also passed to the RADAR Collection Viewer through ROME during data processing.

Figure 1 shows a functional flow diagram for RADAR. It shows the relationship between RADAR components and IRSA data access applications. The RADAR components are described in the following sections.

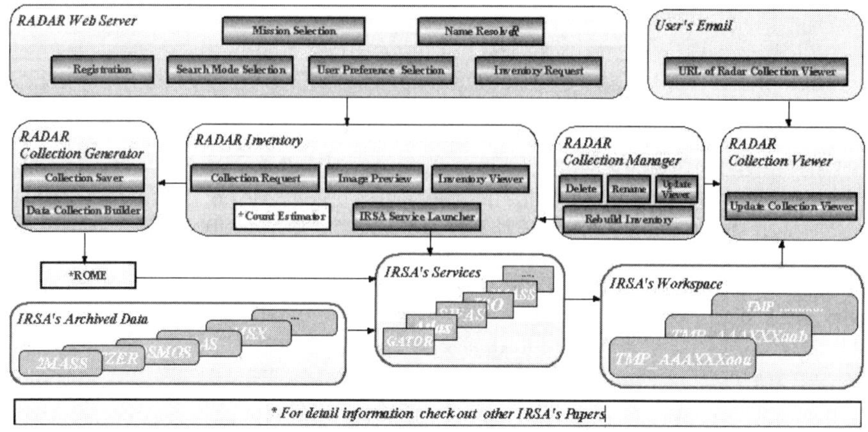

Figure 1. Functional Flow Chart for RADAR

3. Data Collection Generator

A data "collection" in RADAR is defined as any set of data, metadata, references to data, ancillary information, and the organizational structure associated with storing the data for a specific spatial area or target. The collection names must be unique, but otherwise have no naming restrictions. An example of a collection can be all the images, archives, spectra, and catalog subsets associated with a region on the sky plus the metadata describing these data (e.g., image coverage information), arranged into a coherent directory tree.

A collection actually consists of a variety of files. They can be data files or control files for processes run by RADAR, or information and status files which detect the health of software in IRSA. The Collection Generator is responsible for generating a set of control and information files. They record various properties and parameters of a collection. As one example, RADAR derives information from those files and automatically fills in the object name, query criteria, selected datasets, email, and work space information from the last job for a user. Another example is JOBLIST, one of the job control files in XML, which defines a list of datasets for a collection. At RADAR's request, ROME utilizes this file and starts each IRSA CGI program on the list and sends the process status back to RADAR.

The Collection Generator is what makes simultaneous access to multiple datasets possible.

4. Inventory

The Inventory service performs several functions. It checks for data availability at IRSA and quickly returns an inventory of the holdings. The latter utilizes a fast estimation algorithm by taking advantage of pre-calculated statistics on the distribution of sources and image and spectral data across the sky. For catalogs and spectra, the source counts are exact sources, but for some images, such as 2MASS Atlas, the counts are approximate because we assume that the images are point sources found in real database queries. For some datasets where IRSA does not have access to pre-computed statistics, the inventory only indicates that sources are found in the search area (e.g., ISO).

The inventory provides two ways to access data from IRSA. The first is one-click access to datasets selected from the results page, which runs the Collection Generator to perform multiple data access calls simultaneously. When processing is complete, the data are bundled and made available for pick up. The second way is run individual data access services through a browser, specific to a collection or individual application. For every service, the inventory automatically fills index pages with the search criteria input to RADAR. The first method is best for getting a large volume of data simultaneously; the second method is best for data exploration.

Although the inventory return page only lists holdings where data are found, a complete list of all datasets, with or without data meeting the search criteria, is also available.

5. Collection Manager

The Collection Manager is designed to generate and manipulate data collections created earlier. When the Collection Manager is displayed, the user's profile is on the top and all collections belonging to that user are listed. Each collection can be examined simply by clicking on the collection name, and the collection can be updated by entering new search criteria.

6. Data Collection Viewer

The Collection Viewer is responsible for generating a presentation of a data collection in a HTML page suitable for access by a scientist. It summarizes and organizes all the sources which have been found by RADAR together with metadata describing the datasets from which the collection is derived.

The Viewer supports the same data access methods as the RADAR return page: one-click data retrieval and access to IRSA web services, with search parameters filled in.

The collection name is identified at the top of the Collection Viewer, as are the query criteria. During the data retrieval process, the Collection Viewer allows users check the status of processes as they run. This update is accomplished by a set of XML files; some of them generated by RADAR, some of them generated by ROME, and others generated by each application.

7. Summary

RADAR provides a mechanism for generating inventories of input regions quickly, downloading data, and exploring the IRSA archive holdings. The application is highly scalable. It is also distributable and can be extended to provide inventories of remote data sets. Indeed, it already supports a fast inventory of the MAST Spectral and Image Scrapbook. It also supports multiple inventories in a single browser session. When complete, it will maintain these multiple inventories in a permanent workspace assigned to individual users, and will support visualization and update of these "collections."

References

Kong, M., Good, J.C., & Berriman, G.B. 2004, in ASP Conf. Ser., Vol. 347, ADASS XIV, ed. P. L. Shopbell, M. C. Britton, & R. Ebert (San Francisco: ASP), 213

Part VIII
Great Space Observatories

ASTRONOMICAL DATA ANALYSIS SOFTWARE AND SYSTEMS XIV
ASP Conference Series, Vol. 347, 2005
P. L. Shopbell, M. C. Britton, and R. Ebert, eds.

XMM-Newton: Passing Five Years of Successful Science Operations

Carlos Gabriel, Matteo Guainazzi, and Leo Metcalfe

Research and Scientific Support Department of the European Space Agency - European Space Astronomy Centre - P.O. Box 50727 - E-28080 Madrid (Spain)

Abstract. The major ESA X-ray observatory, XMM-Newton, has passed five years of operations in flight. The design concepts driving the activities of the Science Operations Centre at the European Space Astronomy Centre in Spain have shown their validity for this astrophysics long-term mission. Scientific exploitation of data by the astronomical community is facilitated especially through a system resting on 4 pillars:

- scientific data dissemination: from raw telemetry up to processed and calibrated high-level science products, such as images, spectra, source lists, etc;
- dedicated science analysis software: development and distribution of mission specific software, as well as of continuously updated instrument calibration;
- scientific archive: access to data but also to high level information on data contents through state-of-the-art, in-house developed archival facilities;
- documentation: continuously updated documentation on all aspects of spacecraft and instrument operations, data reduction and analysis, with particular focus on the most important scientific results obtained by XMM-Newton, reachable through a comprehensive set of project web pages.

We intend to review all these and related aspects forming the basis of a modern astronomical observatory. Our aim is to show how the innovative but solid elements in the approach to each of the mentioned points contribute to forefront science in the light of some of the outstanding scientific results achieved by XMM-Newton.

1. Introduction

ESA's XMM-Newton observatory exhibits all the characteristics of a modern astrophysical mission. The combination of vast amounts of data gathered by the mission with the objective of maximising product utilisation results in a very user friendly approach. The "individual astronomer on the street" should be able to extract scientific results with a minimum technical effort from his side, concentrating mainly on how to make optimal scientific use of the data observed, free as far as possible from problems posed by data acquisition, reduction, instruments and calibration knowledge. At the same time, the number of astronomers dealing with XMM-Newton data should be maximised. Every observation should be exploitable for both its planned and serendipitous potential.

2. The XMM-Newton Observatory

XMM-Newton (Jansen et al. 2000) was launched on December 10, 1999, by an Ariane 5, in the first successful launch of this rocket type. The three barrel-shaped telescope mirror modules, each of them formed by 58 "Wolter-II" type concentric shells, give rise to an unprecedentedly large effective area in the X-ray domain. This, combined with the simultaneous operations of 5 X-ray instruments localized in the corresponding focal planes, is the main characteristic of XMM-Newton. This is complemented by high quality X-ray instruments: 3 EPIC cameras (Strueder et al. 2001; Turner et al. 2001) and two RGS high resolution spectroscopes (den Herder et al. 2001) as well as by a co-aligned 30 cm optical telescope (OM, Mason et al. 2001) operating simultaneously in the optical and UV domain, including moderate grism spectral resolution. The Ground Segment is distributed among different sites across Europe. The operations are centrally conducted by the XMM-Newton Science Operations Centre (SOC) from ESA's European Space Astronomy Centre (ESAC) in Villafranca del Castillo near Madrid, Spain (the place formerly known as VILSPA). Operations involve mainly the European Space Operations Centre (ESOC) in Darmstadt, Germany (performing spacecraft control and overall commanding), and the Survey Science Centre (SSC), a consortium led by the University of Leicester, responsible for the pipeline processing of all XMM-Newton data, but also deeply collaborating with the SOC in the development of scientific analysis S/W.

3. Maximising Scientific Data Exploitation

The four main aspects contributing in our aim of ensuring a maximum scientific exploitation of the data obtained with the XMM-Newton instruments are:

3.1. Scientific Data Dissemination

XMM-Newton scientific data is distributed at different levels of reduction, from raw telemetry data up to images from EPIC cameras, high resolution spectra obtained with the RGSs, optical / UV images and time series obtained with the OM, source lists in the X-ray and optical domains, and response matrices. Many ancillary products for diagnostics are also part of the derived and distributed products. In addition products recording cross-correlation with X-ray and optical catalogues are produced in the form of tables and finding charts. All of these are obtained through the so-called SSC pipeline and facilitate the analysis and refinement of products by the observers. The SSC pipeline is a subset of the Scientific Analysis System (SAS), running with pre-determined standard parameters. The automatically derived products are carefully screened by a large team of scientists across the world before becoming public, and have to fulfill specific quality criteria. At the time of writing this paper, a new version of the pipeline is in preparation which will be used in early 2005 for full reprocessing of the whole XMM-Newton data archive. This version will incorporate as new products individual low-resolution spectra and light curves, including variability parameterisation, derived from all the sources detected in the FOV of the X-ray cameras showing a sufficient high number of counts. In addition it will be used for automatic derivation of the 2XMM X-ray sources catalogue, expected to contain around 150000 individual sources. The 1XMM catalogue produced by the SSC in 2003 is so far the largest catalogue of X-ray sources derived from observations with CCD energy resolution. 80% of the sources contained in it have not been reported previously as X-ray sources.

3.2. The XMM-Newton Scientific Analysis System (SAS)

SAS (Gabriel et al. 2003) is the software suite for XMM-Newton data analysis. It can run fully interactively (including a complete GUI) but includes also metatasks for automatic or semi-automatic data reduction. Based 100% on free software, it is distributed as binaries produced on many different platforms. The latest version (SAS 6.1) released in November 2004 includes 5 Linux flavours, 3 Solaris and one Mac OS supported platforms.

With SAS, the astronomer can tailor completely the data reduction to the scientific goal. Re-calibration of the data is possible whenever new calibration files have been released. This is especially true in the case of XMM-Newton data due to the dynamic calibration approach, which includes the possibility of single calibration file updates, instead of the traditional calibration versioning system.

Download statistics from the SAS 6.0 version released in March 2004 show that the software serves more than an estimated 1600 users in the worldwide scientific community. To provide potential users of XMM-Newton data with a general introduction to X-ray data analysis and the SAS, the SOC organizes about once a year a 3.5 day workshop, from which more than 50% of the time is devoted to hands-on sessions using SAS, exercising procedures and learning techniques to successfully analyse all types of observation.

3.3. The XMM-Newton Scientific Archive (XSA)

The XSA (Osuna et al. 2004) is among the most advanced scientific archives in the world and is playing a key role in the development of the Virtual Observatory strategies. It is based on ESA's previously developed ISO Data Archive (IDA), making use of the latest technologies (open architecture, Java, XML). It allows powerful and complex queries against the observation catalogue. It has a configurable results display and includes product visualisation tools. It provides customisable product retrieval via a shopping basket as well as a selection of product level for retrieval. Retrieval can be performed either via FTP or CD.

In addition to the individual observations with all their derived products the 1XMM catalogue has been integrated in the XSA. A querying facility on several observation and source-related parameters is part of the XSA, together with source catalogue parameter display, visualisation of finding charts and source images.

The XSA is one of the first scientific archives incorporating Archive Inter-Operability (AIO), a key concept in the Virtual Observatory development. It allows users to access directly the contents of the XSA database by-passing the user interface. So products for a given observation can be retrieved via a server socket (either through a user defined client or an XML file) or through a URL address.

A user registration, which is required solely for administrative purposes, shows that there are so far around 1700 XSA users.

3.4. Documentation

The XMM-Newton SOC puts special emphasis on ensuring that all aspects of spacecraft and instrument operations, data reduction and analysis, as well as calibration and interaction with users are described in up-to-date documentation. A good level of documentation is important to make it easy for observers and data users to understand the data so that the scientific exploitation is maximised. It also ensures transparent operations and serves as a legacy for future scientific missions. A long-term mission like XMM-Newton, which may well operate for 10 to 12 years, also demands comprehensive documentation because of the possible staff turnover during its lifetime. All the documentation is centralised in the XMM-Newton web system, facilitating access and navigation and therefore optimising information dissemination and communication. The key role of the web in a scientific project is consciously re-enforced, making it the main vehicle of communication between the project, the experts and the observers.

4. The Outcome

The quantity and quality of the scientific papers written using XMM-Newton data in a large number of fields, ranging from the study on neutron starts to cosmology, GRB afterglows to supernova remnants, etc, show not only the quality of the data provided by the instruments but also the quality of the ground segment. The number of refereed papers from quarter to quarter has been growing steadily with a 30% increase rate since the first publications.

A cumulative total of 650 refereed publications (as of October 2004), including several Nature and Science Journal articles, make XMM-Newton a very successful mission. Not only the feedback from XMM-Newton observers, but also several facts indicate that everyone of the four aspects referred to in section 3. contribute directly to this success. The large increase of XMM-Newton publications based on public data, for instance, shows the strong influence of the archive. Every single scientific paper dealing with XMM-Newton data is based so far on data reduction performed using SAS. The growth of the XMM-Newton "community" to a number currently above 1500, as the numbers of XSA registered users and SAS downloads indicate, must be attributed partly to the good level of documentation and information of the mission.

References

den Herder, J.W., et al. 2001, A&A, 365, L7

Gabriel, C., et al. 2003, in ASP Conf. Ser., Vol. 314, ADASS XIII, ed. F. Ochsenbein, M. Allen, & D. Egret (San Francisco: ASP), 759

Jansen, F., et al. 2001, A&A, 365, L1

Mason, K.O., et al. 2001, A&A, 365, L36

Osuna, P., et al. 2004, in ASP Conf. Ser., Vol. 347, ADASS XIV, ed. P. L. Shopbell, M. C. Britton, & R. Ebert (San Francisco: ASP), 198

Strueder, L., et al. 2001, A&A, 365, L27

Turner, M.J.L., et al. 2001, A&A, 365, L18

ASTRONOMICAL DATA ANALYSIS SOFTWARE AND SYSTEMS XIV
ASP Conference Series, Vol. 347, 2005
P. L. Shopbell, M. C. Britton, and R. Ebert, eds.

GaiaGrid : Its Implications and Implementation

S. G. Ansari and U. Lammers
Directorate of Science, European Space Agency, ESTEC, 2200 AG Noordwijk, The Netherlands

M. ter Linden
Dept. AS&E/SWE Dutch Space BV, 2303 DB Leiden, the Netherlands

Abstract. Gaia is an ESA space mission to determine positions of 1 billion objects in the Galaxy at micro-arcsecond precision. The data analysis and processing requirements of the mission involves about 20 institutes across Europe, each providing specific algorithms for specific tasks, which range from relativistic effects on positional determination, classification, astrometric binary star detection, photometric analysis, spectroscopic analysis etc. In an initial phase, a study has been ongoing over the past three years to determine the complexity of Gaia's data processing. Two processing categories have materialised: core and shell. While core deals with routine data processing, shell tasks are algorithms to carry out data analysis, which involves the Gaia Community at large. For this latter category, we are currently experimenting with use of Grid paradigms to allow access to the core data and to augment processing power to simulate and analyse the data in preparation for the actual mission. We present preliminary results and discuss the sociological impact of distributing the tasks amongst the community.

1. Introduction

Gaia is an astrometric space mission that will determine the positions and proper motions of 1 billion stars down to the 20th magnitude in our Galaxy. In addition, it will determine photometric properties and radial velocities of a large portion of these stars (see Table 1).

The data processing requirements span a number of algorithms provided by 20 institutes across Europe. The Gaia *Virtual Organisation* is composed of several working groups focused on data analysis aspects ranging from photometric calibrations, radial velocity determinations, astrometric data processing algorithms to classification and fundamental algorithm development.

There are two distinct categories of algorithms: *core* and *shell* tasks. While core tasks operate on the entirety of the data to iteratively determine positions and proper motions of each star as the data is received and processed, shell tasks are basically analysis tools that determine other characteristics, such as photometric properties and radial velocities. The Grid paradigm is used on this latter category.

The objective of the ongoing activity is threefold:

- To provide access to the core data in order to analyse and derive other stellar properties. This maps to the concept of the *Virtual Database*.

Table 1. Summary of the Gaia Mission objectives (Perryman 2003)

Property	Value
Catalogue	$\approx 1 \times 10^9$ stars
	$\approx 0.34 \times 10^6$ at $V = 10$ mag
	$\approx 26 \times 10^6$ at $V = 15$ mag
	$\approx 250 \times 10^6$ at $V = 18$ mag
	$\approx 1000 \times 10^6$ at $V = 20$ mag
Sky Density	
	mean density : 25000 stars deg^{-2}
	maximum density : 3×10^6 stars deg^{-2}
Accuracies	median parallax errors:
	4μas at 10 mag
	11μas at 15 mag
	160μas at 20 mag
Radial Velocity Accuracies	$1 - 10\,\text{kms}^{-1}$ at 17 mag
Photometry	4 broad and 11 medium band filters

- To provide the necessary high-performance computing to simulate the data. This maps to the concept of a *Virtual Supercomputer.*
- To provide a collaborative environment in which the data analysis for the mission can developed and maintained by the Gaia community (see Figure 1). This maps to the concept of a *Virtual Organisation.*

2. Gaia Number Crunching

The Gaia data processing effort will be a numerically intensive task demanding supercomputer-type processing capabilities. Current best estimates for the total number of floating point operations required to generate the final object catalogue are in the range of 130×10^{18} FLOPs (Perryman 2004). This figure is the sum of an estimated 40×10^{18} FLOPs and 90×10^{18} FLOPs for the two processing domains core and shell (see above) respectively, that the whole effort can be broken down into. We illustrate the magnitude of the problem as follows: Assuming a 40 GFLOP/s machine of today as baseline, a start of the processing in 2012 and the validity of Moore's law (doubling of the processing capabilities every 18 months) from now until then, the entire task would require some 3 years of CPU time on a centralised 2012 target hardware.

The main objective of the Gaia core processing is the precise determination of four sets of initially unknown (or uncertain) quantities from repeated elemental CCD-based observations of ≈ 1 billion celestial objects during the mission's lifetime. These quantities are: A: accurate spacecraft attitude at different times, S: parameters (including position, distance and space motion) of observed sources, C: instrument calibration parameters, G: fundamental global parameters (e.g. value of Parameterised Post-Newtonian gravitational parameter γ). Owing to the sheer size of the problem a direct mathematical solution for the unknowns is computationally intractable. The problem is, however, amenable to a least-square minimisation scheme in which the above unknowns are determined in a step-wise, iterative manner. The basic idea is that for the sought solution $(\hat{A}, \hat{S}, \hat{C}, \hat{G})$ the sum of squares of the differences between the observations and

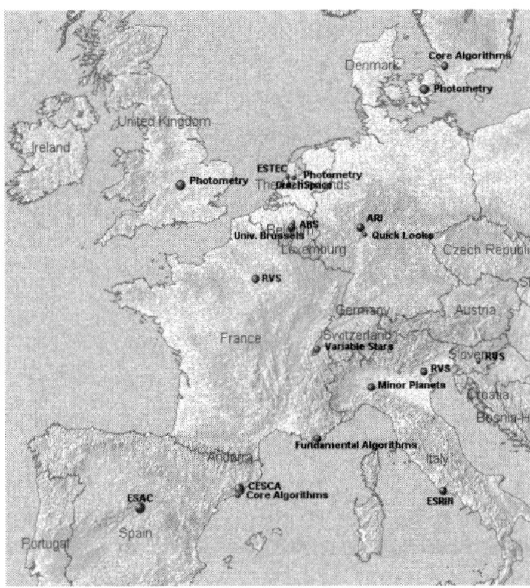

Figure 1. The Gaia Virtual Organisation is composed of about 20 institutes across Europe. The red dots denote data processing centers, while the blue dots denote algorithm providers.

observation data calculated from the basic unknowns becomes minimal:

$$\sum_{\text{observations}} |\text{observed} - \text{calculated}(\hat{A}, \hat{S}, \hat{C}, \hat{G})|^2 = \min \qquad (1)$$

This scheme is called the Global Iterative Solution (GIS) and represents the very heart of the Gaia data analysis problem. The core processing will have to be performed in a centralised manner, i.e., will not be amenable to a Grid computing approach.

The shell processing represents the larger part of the total data reduction problem. In contrast to the core part, the shell algorithms are roughly independent processing tasks and can hence be executed in a distributed manner after on the data that have been processed with the core system. Examples are: Object classification, astrometric binary star analysis, etc. Our current Grid initiative is aimed at demonstrating the practical feasibility of this approach with a particular Grid-enabling application called GridAssist as detailed in the following section.

3. GridAssist

GridAssist is a Grid-based workflow management tool that allows the user to execute workflows in a Grid environment.

GridAssist is based on a client-server architecture. The server runs on a central computer while the client applications run locally on the computer of the end-user. The server, which is called the GridAssist Controller, handles all the calls from the clients and executes the workflows. The client application, called the GridAssist Workflow Tool, is used to create and submit workflows, to monitor progress of running workflows and to maintain the registry.

The GridAssist Controller holds a database and the workflow engine. The database is used to store the registry and the job queue. The registry contains the information about the resources and the services. The workflow engine retrieves jobs from the job queue and executes the jobs on the available resources. The workflow that is submitted with the job is parsed dynamically. First the engine finds the services that can be executed (that have no input, or for which all input is available), then it tries to find a suitable resource that has the service installed to execute the service on. Each of the services can be installed on many resources. For the execution of the services on the Grid nodes and the data transfers between the Grid nodes, the Globus CoG-kit (1) is used. This kit contains an API (Application Programmers Interface) for several programming languages. GridAssist is completely Java based and therefore the Java version of the CoG-kit is used.

The GridAssist Workflow Tool provides the user with an intuitive graphical user interface that makes it easy to construct workflows. One of the first things that is done after starting up the application is to connect to the GridAssist Controller and to retrieve the registry. The services in the registry are presented to the user and can be selected and dragged to a canvas without the necessity of having any prior knowledge about their location or capacity. Data flows can be created by connecting inputs and outputs of the services. Checks are performed whether output of one service can be used as input to another service. The workflow tool is a Java application running on Windows, Linux, Mac and most Unix versions.

The GridAssist Controller is implemented as a standard Web Service. This means that the service is described using WSDL (Web Service Description Language), and that it can be accessed using SOAP (Simple Object Access Protocol). This allows people to write their own clients for the controller, like command line tools or web based interfaces. Also firewalls are usually no problem when using SOAP. The interface between the controller and the Grid environment is based on Globus protocols. For these protocols some specific ports need to be opened in the firewall. However, since this is only necessary for communication between the controller and the (trusted) Grid nodes, and not to the clients, rules can be applied that restrict the access through the firewall.

The Grid environment that can be used by GridAssist is Globus based. The Globus toolkit provides basic Grid services for job management, file transfer, information management and security. The Grid services that GridAssist uses are found in Globus Toolkit 2 and in Globus Toolkit 3 (pre Web Service). These services are currently quite stable and easy to install. The job management service provides a global interface to locally installed job management systems, such as PBS, LSF, SGE and Condor, which are used on many clusters.

4. Conclusions

Preliminary tests show some promising results. The Grid environment used for GaiaGrid consists now of 24 distributed nodes in 8 institutes of 5 countries. Some of the Grid nodes are shared with other Grids. In the Grid environment, a mix of Globus 2 and Globus 3 is used. ESA has set up a Certification Authority (CA) which is used to provide certificates (the Globus security model is based on X.509 certificates) for users and hosts. In case the host already has a host certificate from another trusted CA, this host certificate can be used as well. This is currently the case for two institutes (ESRIN in Italy and Dutch Space). The GaiaGrid environment can easily be extended by other Grid nodes, in most cases the only things to do is to accept ESAGrid certificates, and to authenticate users to use the nodes.

We have successfully run high-performance calculations to simulate binary stars over the five-year observation period of Gaia. The current testbed is also being used to implement an end-to-end astrometric binary star analysis and a variable star analysis. In the coming year we plan to augment the computing power and establish an environment that will allow the Gaia Community to collaborate and to implement all the necessary data processing and analysis elements prior to launch of the spacecraft.

References

Ansari, S.G., Balague-Jordan, Y., Luri, X., & ter Linden, M. 2004, "Applying Grid Technology to Gaia Data Processing", Gaia Symposium, ESA Publications

Laszewski, G. von, Foster, I., Gawor, J., & Lane, P. 2001, "A Java Commodity Grid Kit", Concurrency and Computation: Practice and Experience, 13, 8, 643, `http:/www.cogkit.org/`

Luri, X., Ansari, S.G., Torra, J., Figueras, F., Jordi, C., Llimona, P., & Masana, E. 2003, in ASP Conf. Ser., Vol. 314, ADASS XIII, ed. F. Ochsenbein, M. Allen, & D. Egret (San Francisco: ASP), 653

Perryman, M. 2003, "Gaia Science Capabilities", `http://www.rssd.esa.int/gaia/`

Perryman, M. 2004, "Estimation of Gaia Data Processing FLOPs", Technical Note GAIA-MP-009, June 2004

ASTRONOMICAL DATA ANALYSIS SOFTWARE AND SYSTEMS XIV
ASP Conference Series, Vol. 347, 2005
P. L. Shopbell, M. C. Britton, and R. Ebert, eds.

Spitzer Space Telescope Data Processing and Algorithmic Complexity

Mehrdad Moshir

Spitzer Science Center, California Institute of Technology, Pasadena, CA 91125, Email: mmm@ipac.caltech.edu

Abstract. Automated reduction of a very large body of data from the *Spitzer* Space Telescope requires an intricate and very flexible software system to support more than 50 different pipelines executing on a large distributed computing cluster. Additionally, in order to correct many instrumental signatures and calibration features, a variety of complex algorithms need to be utilized to process and distribute calibrated images and spectra from nearly 5 million instrument frames that are collected by the observatory every year. A sample of the complexities that needed to be accommodated both in system architecture and design as well as signatures encountered and associated algorithms will be described.

1. Overview

The *Spitzer* Space Telescope, launched in August of 2003, has been operating very successfully for well over a year and has returned a significant body of new data in the infrared domain (Werner et al. 2004). The observatory is capable of generating close to 5 million distinct instrument frames every year, and of necessity such a large volume of data needs to be processed in a lights-out fashion. The automated processing of such a voluminous and varied number of datasets in a timely fashion imposes certain complexities on the design and implementation of the system infrastructure as well as its individual modular components. Data collection approach on-board the spacecraft injects further complexity into the ground software and pipelines. In Section 2 we will briefly discuss the data collection environment and its impact on pipeline design. In Section 3 we will describe the operational environment related to downlink processing. In Section 4 we will address the requirements on the system and the approaches adopted. Finally in Section 5 we will touch on some of the signatures and effects that need to be corrected by the ground software.

2. Data Collection

The instruments on board *Spitzer* can collect data in the range of $3.5\mu m$ to $160\mu m$ with the spacecraft either pointed inertially or with the spacecraft performing a sky scan at a constant rate while an internal mirror compensates the motion and keeps the sky fixed on the focal plane. The latter mode of data collection permits large surveys with little pointing overhead, a cartoon of a typical scan observation by the MIPS instrument is seen in Figure 1. The three instrument suites consist of many different array technologies, InSb, Si:As, Si:Sb,

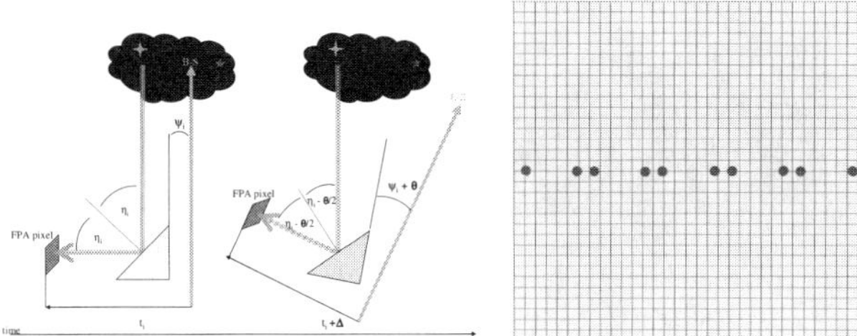

Figure 1. Spacecraft motion compensated by a scan mirror. Image of star is fixed during each integration; 11 overlapping images are made as the scan mirror starting position is ratcheted during the scan leg.

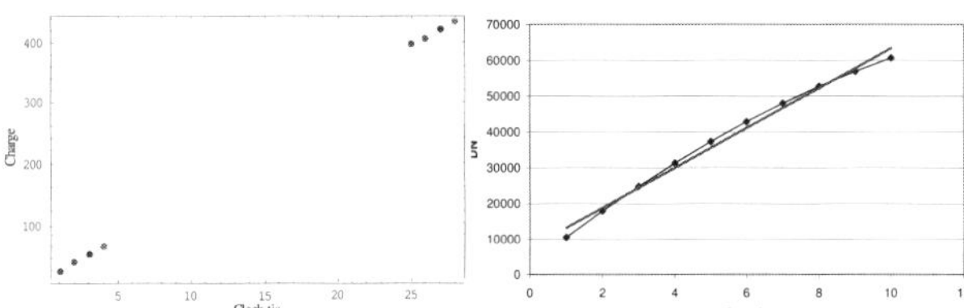

Figure 2. In the left panel FPGAs average the upper set of reads and subtract the average of the lower set of reads and transfer to the C&DH. In the right panel a RAD6000 uses linear regression formula to calculate the slope of the ramp and then transfers it to the C&DH for storage and later transmission to the ground.

Ge:Ga, Stressed Ge:Ga. The array formats range from 256x256 to 2x20 and are capable of integrating from 0.02 to 512 seconds. All of these complexities are controllable by only 8 observing templates (AOTs).

All instruments perform non-destructive reads of the respective arrays. However, due to the limited on-board storage and ground communication bandwidth it is not possible to send down all samples. Depending on the instruments (Ge instruments excluded for now), either FPGAs or a RAD6000 computer perform some "data compression" before transferring the data to the Command and Data Handling computer (another RAD6000). The limited computational power of the RAD6000 does not allow sophisticated or robust estimation, and some undesired side effects may result that will need correction in the pipelines, as will be seen later. An example of data compression by two of the instruments is seen in Figure 2. One of the drawbacks of on-board data compression is that the dispersion in the data is not retained. The issue of how uncertainties are dealt with in *Spitzer* has been discussed previously (Moshir et al. 2003). The data collected by the instruments are compressed and stored by the C&DH and sent to the ground on a pre-determined basis; typically 1GB of compressed data (better than 2X compression) are transferred to the ground every day.

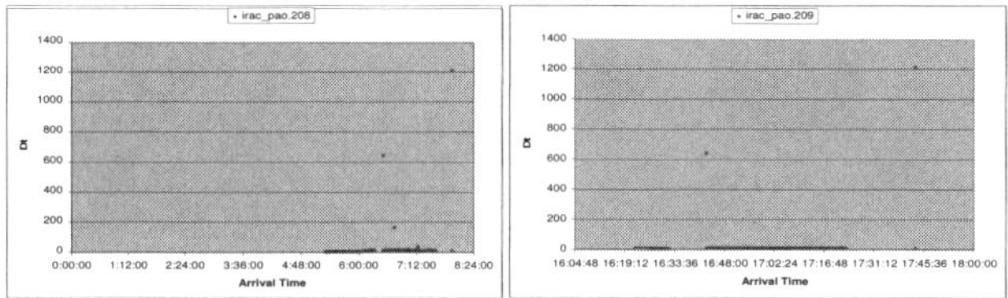

Figure 3. Each point refers to the time separation between one DCE arrival and the next. The left panel shows a morning DSN contact and the right panel shows a typical afternoon contact (both for IRAC).

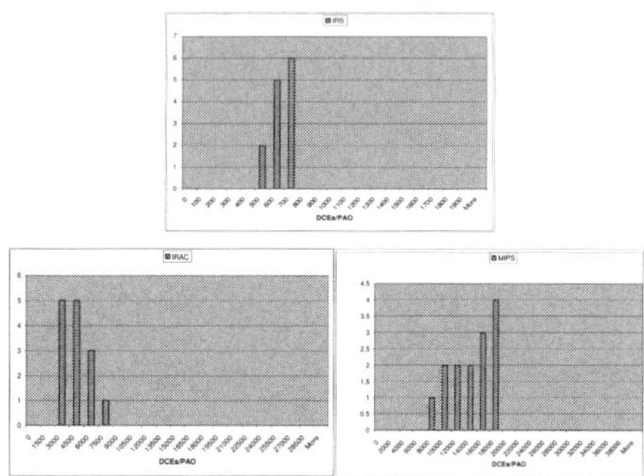

Figure 4. Average DCE rates per day vary by a factor of 20+. To date the peak DCE rates have been 1,160/day, 12,500/day and 35,700/day for IRS, IRAC and MIPS, respectively (1 day $\sim$ 2 PAOs).

3. Operating Environment

The Observatory is radiated a set of observations to be performed autonomously over a given time period; such a period is typically 12 hours long and is called a Period of Autonomous Operation (PAO). Only one instrument operates at a time during a PAO, several such consecutive PAOs form an instrument Campaign. The raw instrument data are referred to as Data Collection Events (DCEs). The DCEs are assembled from decompressed telemetry packets at JPL and sent to the SSC for processing by pipelines. Details of these concepts have been given previously (Moshir 2002).

While a station contact is about 30 minutes long, the data arrival at the SSC is stretched out due to resource sharing at JPL among many other space missions. Typical data arrival duration may range from 1 1/2 to 2 1/2 hours as in Figure 3; the average inter-DCE arrival time is close to 2 seconds (for the cases shown).

As the data arrive at the SSC the ingestion process takes place automatically (event-driven via data arrival). During a day anywhere from 1,100 DCEs to 35,000+ DCE are received. The distribution of data rates for the first 10 months of routine operations is seen in Figure 4. As of campaign 13 of IRS, IRAC and MIPS a total of 102,699, 1,059,065 & 3,005,579 DCEs had been received ($\sim$ 10 months of routine operations).

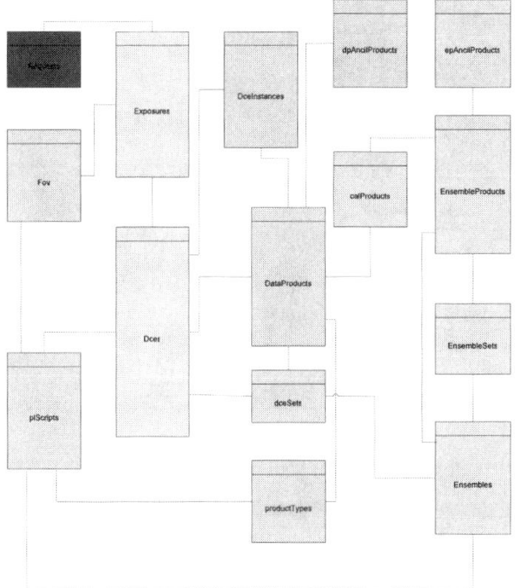

Figure 5. A subset of DB schema just for tracking and running of pipelines. For the whole set of 14 tables, containing 180+ columns, the average growth rate is in excess of 234,000 rows per day

4. System Requirements

The downlink system needs to enable several fundamental capabilities, such as: 1) Book-keeping in a distributed asynchronous system. 2) Automated data reduction of large data sets. 3) Automated calibration retrieval. 4) Support for many pipelines (50+) for many data modes. 5) Rapid reduction of arriving data (5X real time). 6) Timely quality assessment of reduced data. 7) Rapid re-wiring of pipelines when needed (specially during the IOC and SV phases). 8) Re-use of automated pipeline modules in interactive tools. 9) Enhancement of spacecraft pointing reconstruction using science data.

For book-keeping of records and pipeline transactions, among other needs, the system relies on a complex relational DB that contains 1) Uplink/scheduling information and instrument settings. 2) Status of received data and pipeline information as well as pointers to data in the file system. 3) Calibration metadata and file pointers. 4) QA information for the pipelines.

As an illustration a subset of the database schema that deals *only* with tracking and running multiple pipelines is seen in Figure 5. The tables shown have 180+ columns. The main driver, the DCE instances table grows at an average rate of 12,000+ rows per day. For automation of the system it is required to route different data to different pipelines, this is accomplished by creation of "job manifests" that allow the pipeline executive to route the data. Pipelines that require an *ensemble* of data to perfom tasks such as mosaicing, dark estimation, etc. require a way to associate related data for processing. This is accomplished by an automated ensemble maker (Laher & Rector 2004).

Given the job manifests and ensemble sets, the pipelines get started after the ingestion process has been invoked upon data arrival. Pipelines run on

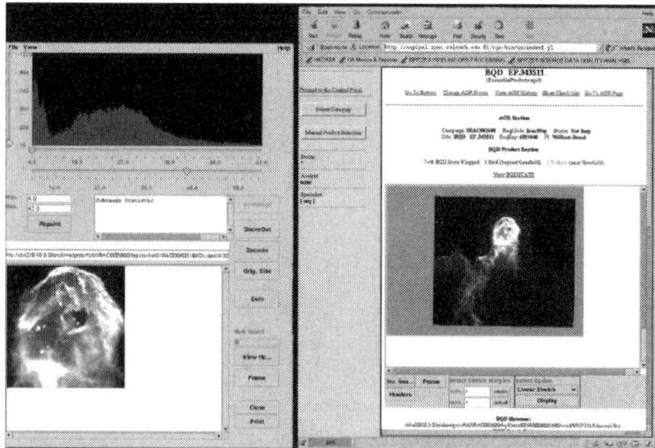

Figure 6. A web-based tool allows the DQA team to inspect any processed dataset to assess validity for release to the observers.

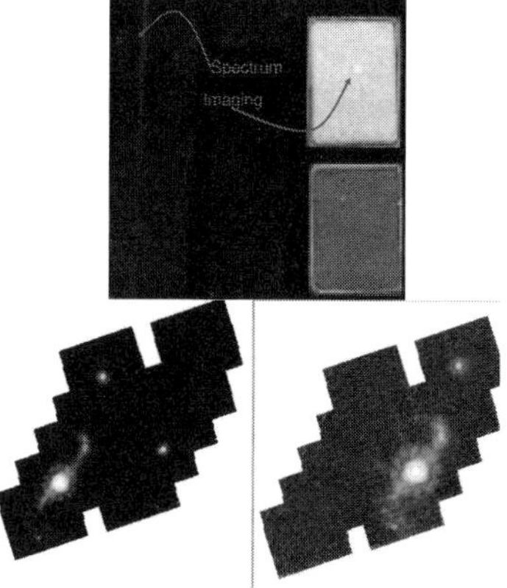

Figure 7. Re-use of IRS instrument for imaging; re-use of software to perform mosaicing of resulting data at $16\mu m$ and $22\mu m$.

many "drones" for each data type and deposit the results in the "sandbox". The pipelines are designed with flexibility in mind and are easy to re-configure (Brandenburg et al. 2004). One set of pipelines performs calibration reduction (darks, flats, etc.), these pipelines populate the Calibration Transfer tables in the database; the "CalTrans" system relies on this database for serving the proper calibration terms to the science pipelines, it ensures that science pipelines use calibration of the epoch in each case (Lee et al. 2004). To support QA functions all pipelines generate significant statistics for each processed DCE and deposit them into QA tables for each instrument. These tables allow the DQA team to easily access and certify data products for release to the observers using a web-based tool (Narron et al. 2003) as seen in Figure 6.

The flexibility of pipeline components allows rapid development of *new* pipelines. For example the IRS instrument, while designed for spectroscopy, also has two apertures in one array that can image the sky. Starting in Cycle 2 of *Spitzer*

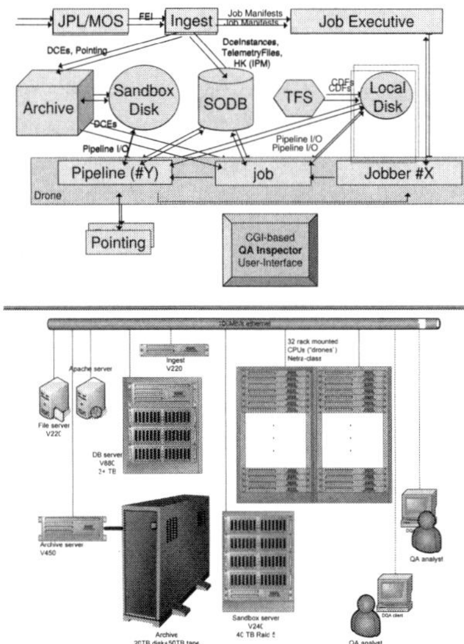

Figure 8. The pipeline data flow diagram (upper panel), and the hardware architecture to support it (lower panel); behind a firewall.

Call for Proposals, users can request imaging at $16\mu m$ and $22\mu m$. To create the pipeline for reducing such data was very simple. Several modules that performed similar tasks for other instruments were re-wired to create a mosaicing capability for these observations, an example seen in Figure 7.

Early in the design of the pipelines it was anticipated that while automation of data reduction was applicable to almost all cases, there would be instances where the *intent* of the observer could not be determined via artificial intelligence and the same pipeline modules would need to support interactive usage. The domain of spectroscopy and spectrum extraction fall into such cases. The individual pipeline modules are easily plugable into interactive tools to allow supervised spectral extractions by an observer (Hesselroth 2004).

To enable processing the data at least five times faster than data acquisition rate, using limited funds for computing machinery ("Netra class" blades), necessitated a large cluster of CPUs. These "drones" are configured to process any two pipelines concurrently, the pipelines get their data from the archive and process them locally then deposit the results into the sandbox for later archiving. The software and hardware architecture to allow this is seen in Figure 8.

5. Instrument Signatures, Data Collection Environment and Algorithmics

The instruments on-board *Spitzer* are all state-of-the-art and not flown before. Unlike the CCDs that have been very well characterized, there are many signatures that need to be understood, characterized and removed. To name a few of complications, we observe that there are on-board "processing" features; we

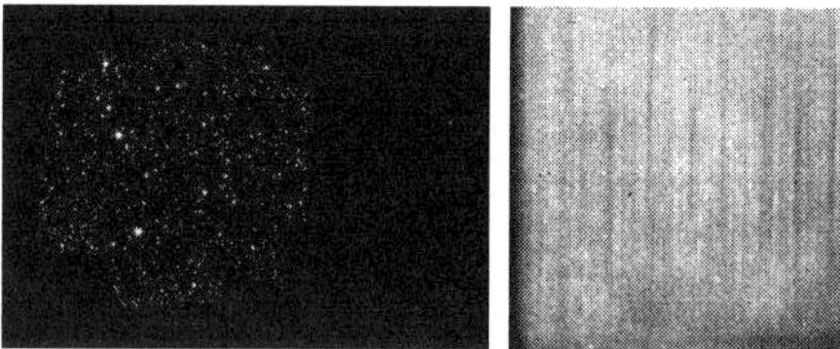

Figure 9. A low background region of sky as seen by IRAC (left). Carefully dithered observations of the same region and rejecting point sources then optimally combining the results yields a very good "sky dark" image (right).

have a shutter-less operation; there is optical cross-talk; instruments are non-linear; there are latent images; there are time-dependent readout signatures; there is readout droop; there is debris on one pick-off mirror that combined with scan mirror motion (Figure 1) results in non-stationary flat-fields; the Ge detectors have time-dependent, background-dependent responsivity changes needing a "self-calibrating" observation strategy; there is long-term memory & "action at a distance"; there are optical distortions; the spacecraft pointing, while good, needs to be further refined using science contents, and so forth.

An example of on-board processing features is that the RAD6000 computer used for partial data reduction (aka "compression") does not have the capacity to account for ramp non-linearities, sample correlation (Fowler 2004) or ramp saturation or a particle hit causing a ramp discontinuity. Nevertheless the pipelines have been designed to correct for these features (Masci et al. 2004).

To operate imaging instruments without deploying the shutter and still obtaining "dark"-corrected calibrated products is non-trivial. To estimate the dark offset one may consider pointing the telescope at a low background part of the sky, say towards the North Ecliptic Pole; unfortunately there are background stars and galaxies everywhere easily visible to the sensitive eyes of the IRAC instruments! However, clever dither patterns and rejecting point sources and outliers and combining the results yields excellent "sky dark" images that, combined with "lab darks", effectively act as though a shutter had been deployed (see Figure 9).

One other example of non-trivial data reduction is pointing refinement. There are some significant drivers for achieving maximal pointing knowledge. Good pointing allows source identification and future follow-on observations, it results in higher S/N in mosaics and point source detections, and it also facilitates potential super-resolution methods (Backus et al. 2004). In the process of pointing refinement it is necessary to remove distortions, since uncorrected distortions will limit how well pointing could be corrected. To characterize distortions a large body of data needs to be analyzed and deviations fitted to a multinomial in array coordinates; these coefficients appear in all of the *Spitzer* products (Shupe et al. 2004). The *Spitzer* imaging pipelines account for distortions (Makovoz & Khan 2004), and the resulting mosaics are distortion-free. An example of the type of distortions encountered and corrected is seen in Figure 10.

Once the distortions in each individual image have been corrected, the point sources within overlapping frames are matched to a good IR astrometric catalog, such as 2MASS, and then the pointing correction for each individual image

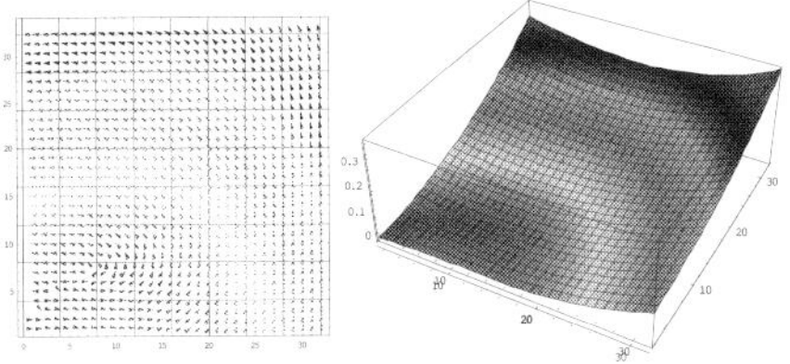

Figure 10. Optical distortion illustration for MIPS Ge, for *one* mirror position.

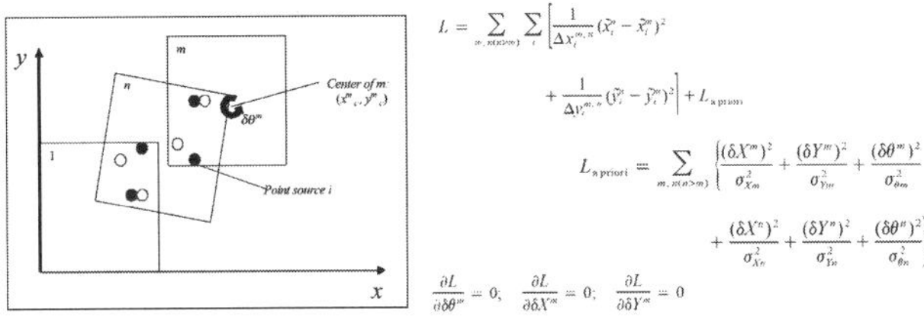

Figure 11. Distances of same source apparitions in overlapping images are minimized while using the prior knowledge about the expected pointing errors, by minimizing the cost function L .

is obtained by minimizing a cost function L, defined by the residual distances between apparitions of same source in overlapping frames as well as against astrometric counterparts while including prior knowledge of the expected pointing uncertainties, as shown in Figure 11.

When such minimization of the cost function takes place, typically it becomes necessary to invert matrices that are several thousand by several thousand; these matrices are usually sparse, and standard libraries are invoked for their solutions (Masci, Makovoz, & Moshir 2004). Pointing corrections for IRAC are significant and reduce the errors from just under $1''$ to $0.1'' - 0.2''$; pointing refinement manifests itself by increasing the S/N in a mosaic as well as sharpening it, Figure 12.

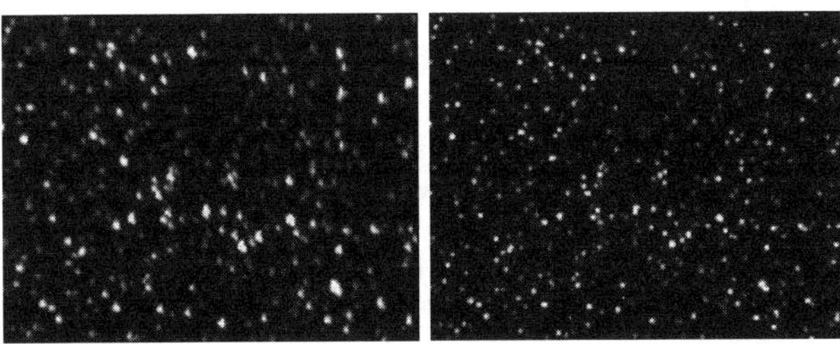

Figure 12. The mosaic on the left was made without pointing refinement; the one on the right was made from same data after performing pointing refinement.

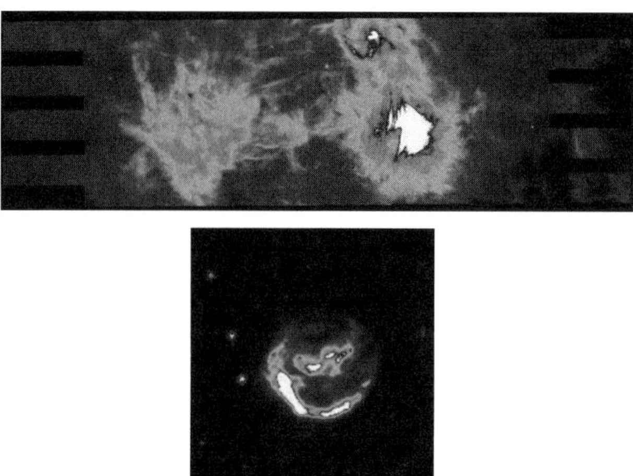

Figure 13. The top mosaic was made from a scan map observation (Perseus molecular cloud from the c2d Legacy Program). Lower mosaic was obtained in dithered mode (Kepler supernova remnant); both at $24\mu m$.

5.1. Scanning, Data Redundancy, and Self-Calibration

In Section 2 pointing modes of observing were discussed. The MIPS suite of instruments is able to observe both in dithered pointed or scan map modes. The complexities of accounting for mirror-dependent distortions and mirror-dependent flat fields are taken care of by the pipelines, and either large area or small area mosaics are obtained (Figure 13). One area that poses new challenges is the reduction of MIPS Ge data because of the well-known tendency of Ge detectors to have non-stationary responsivity. For this reason the Ge instruments provide a self-calibrating observation strategy. At regular time intervals a stimulator flash is turned on and the arrays are exposed to a highly repeatable photon flux. In Figure 1 the redundant coverage of the same point on the sky by different parts of the array was shown for a medium speed scan map. By using such redundancy and tracking the response to stimulator flashes, it is possible to mitigate the Ge responsivity variations, and the data are "self-calibrated" within each observation set (Henderson et al. 2004; Pesenson et al. 2004).

6. Summary

The approach to designing a distributed computing environment with very heavy reliance on a complex database is seen to be capable of meeting the challenges of reducing close to 5 million images per year from *Spitzer* as well as reprocessing them at the same time as the real time arriving data. Although the computing engine for the database exceeds the power and speed of all computing drones used for pipeline processing, the requirement of having a centralized database of all mission data has been met. The approach to modularized and generic design with an eye toward re-use of code in interactive tools has proved to be a significant resource saving measure. The capability to rapidly refine the pipelines during the IOC and SV phases allowed the pipelines to react to on-orbit realities very quickly to produce the best calibrated products. As a result the pipeline products are currently being used for immediate scientific research soon after the start of routine operations (e.g., the issue of ApJ (Supp) 2004, 154, 1).

Acknowledgments. The program described here owes its success to the contributions of many, particularly the efforts of the Downlink development team and the understanding and support of Bill Green during his tenure as the SSC manager. This work was carried out at the *Spitzer* Science Center, with funding from NASA under contract 1407 to Caltech and the Jet Propulsion Lab.

References

Backus, C., et al. 2004, in ASP Conf. Ser., Vol. 347, ADASS XIV, ed. P. L. Shopbell, M. C. Britton, & R. Ebert (San Francisco: ASP), 61

Brandenburg, H., et al. 2004, in ASP Conf. Ser., Vol. 347, ADASS XIV, ed. P. L. Shopbell, M. C. Britton, & R. Ebert (San Francisco: ASP), 575

Fowler, J. 2004, in ASP Conf. Ser., Vol. 347, ADASS XIV, ed. P. L. Shopbell, M. C. Britton, & R. Ebert (San Francisco: ASP), 449

Henderson, D., et al. 2004, in ASP Conf. Ser., Vol. 347, ADASS XIV, ed. P. L. Shopbell, M. C. Britton, & R. Ebert (San Francisco: ASP), 590

Hesselroth, T. 2004, in ASP Conf. Ser., Vol. 347, ADASS XIV, ed. P. L. Shopbell, M. C. Britton, & R. Ebert (San Francisco: ASP), 296

Laher, R., & Rector, J. 2004, in ASP Conf. Ser., Vol. 347, ADASS XIV, ed. P. L. Shopbell, M. C. Britton, & R. Ebert (San Francisco: ASP), 39

Lee, W., et al. 2004, in ASP Conf. Ser., Vol. 347, ADASS XIV, ed. P. L. Shopbell, M. C. Britton, & R. Ebert (San Francisco: ASP), 594

Makovoz, D., & Khan, I. 2004, in ASP Conf. Ser., Vol. 347, ADASS XIV, ed. P. L. Shopbell, M. C. Britton, & R. Ebert (San Francisco: ASP), 81

Masci, F., et al. 2004, in ASP Conf. Ser., Vol. 347, ADASS XIV, ed. P. L. Shopbell, M. C. Britton, & R. Ebert (San Francisco: ASP), 468

Masci, F., Makovoz, D., & Moshir, M. 2004, PASP, 116, 842

Moshir, M., Fowler, J., & Henderson, D. 2003, in ASP Conf. Ser., Vol. 295, ADASS XII, ed. H. E. Payne, R. I. Jedrzejewski, & R. N. Hook (San Francisco: ASP), 181

Moshir, M. 2002, in ASP Conf. Ser., Vol. 281, ADASS XI, ed. D. A. Bohlender, D. Durand, & T. H. Handley (San Francisco: ASP), 336

Narron, B., et al. 2003, in ASP Conf. Ser., Vol. 295, ADASS XII, ed. H. E. Payne, R. I. Jedrzejewski, & R. N. Hook (San Francisco: ASP), 160

Pesenson, M., et al. 2004, in ASP Conf. Ser., Vol. 347, ADASS XIV, ed. P. L. Shopbell, M. C. Britton, & R. Ebert (San Francisco: ASP), 153

Shupe, D., et al. 2004, in ASP Conf. Ser., Vol. 347, ADASS XIV, ed. P. L. Shopbell, M. C. Britton, & R. Ebert (San Francisco: ASP), 491

Werner, M., et al. 2004, ApJS, 154, 1

ASTRONOMICAL DATA ANALYSIS SOFTWARE AND SYSTEMS XIV
ASP Conference Series, Vol. 347, 2005
P. L. Shopbell, M. C. Britton, and R. Ebert, eds.

Analyzing the Cas A Megasecond in Less than a Megasecond

J. E. Davis, J. C. Houck, G. E. Allen, and M. D. Stage

Center for Space Research/CXC, Massachusetts Institute of Technology, Cambridge MA, 02139

Abstract. This spring the Chandra X-ray Observatory concluded its megasecond observation of the Cassiopeia A supernova remnant using its Advanced CCD Imaging Spectrometer (ACIS). From the more than 300 million X-ray events on the ACIS-7 CCD, we created high resolution spatial maps of line emission, Doppler velocities, and temperatures. Although the creation of the maps required extracting more than nine million spectra and performing several hundred thousand spectral fits, we were able to complete the entire mapping process in a matter of days. This task was facilitated by the rapid scripting afforded by the **S-Lang** interpreter as embedded in the Interactive Spectral Interpretation System (**ISIS**).

1. Introduction

The young supernova remnant Cassiopeia A (Cas A) has been observed by the Chandra X-ray Observatory a number of times, including the Chandra First Light Observation, which made the spectacular discovery of the long sought compact remnant from the supernova explosion. Early in the spring of 2004, Chandra performed its deepest imaging observations of the remnant to date (Hwang et al., 2004) through a series of nine long exposures using the observatory's Advanced CCD Imaging Spectrometer (ACIS). With a total exposure time of nearly one million seconds, this set of observations consists of more than 300 million events distributed across the ACIS-7 CCD— almost literally a useful PHA spectrum in every pixel occupied by the remnant.

The availability of such high resolution data on both spatial and spectral scales enabled us to create high resolution spatial maps of Cas A detailing plasma temperatures (Figure 1), line emission (O, Fe, Ne, Mg, Si, S, Ar, Ca, and Ti), and Doppler velocities. This process involved extracting over nine million PHA spectra, and performing hundreds of thousands of spectral fits to individual spectra from adaptively sized regions of the remnant. Although this may sound like a daunting task, we were able to produce such maps at arc-second resolution in a matter of days.

The software involved in this endeavor relied heavily upon the **S-Lang**[1] interpreter as embedded in **ISIS** (Houck 2000), which is distributed as part of CIAO. Coding in an interpreted language also cut the development time to a fraction of what it would have taken using a more traditional compiled language such

[1] `http://www.jedsoft.org/slang/`

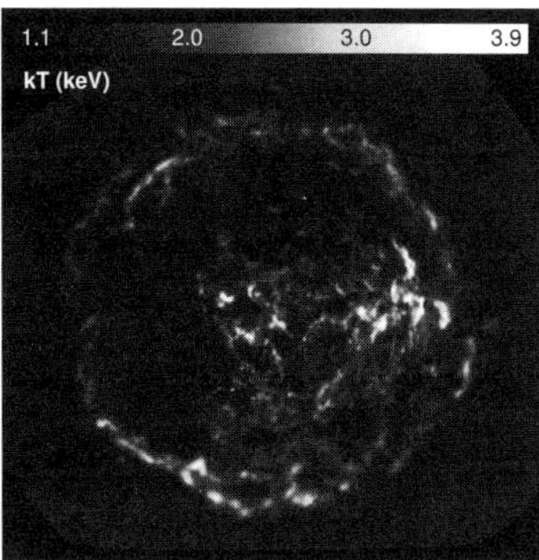

Figure 1. Cas A temperature map based on the bremsstrahlung fits. The kT image is strongly correlated with regions that are dominated by faint, hard continuum emission, likely actually synchrotron radiation instead of bremsstrahlung.

as C or FORTRAN. Because of S-Lang's native support for array-based operations, we did not have to sacrifice execution speed for the convenience of using an interpreter. As described below, we also made extensive use of specialized off the shelf modules to extend the interpreter to allow it to perform such tasks as reading and writing FITS files, creating histograms, and performing distributed computations.

2. Steps involved in creating the Maps

The megasecond observation consisted of nine separate smaller observations at various roll angles and pointings. As such, it was not permissible to simply "add" the data together, rather the pieces had to be kept separate and analyzed jointly. We also included in our analysis the archival data from two earlier Chandra observations of Cas A (obsids 114 and 1952), making 11 data sets in all. The level-2 files were taken as the starting point for the subsequent analysis.

2.1. Exposure Maps

The CIAO tools `mkinstmap` and `mkexpmap` were used to create the exposure maps for each of the 11 datasets. The primary purpose of the exposure maps was for the computation of the correct exposure values used in the fitting process. Hence, our exposure maps included just the effects of exposure time and bad pixels. The energy-dependent effective area and other instrumental effects were not included as these were to be taken into account in the construction of the ARFs.

2.2. Extracting the Spectra

The first step in the spectral extraction stage involved the creation of PHA histograms on an arc-second resolution scale, requiring slightly more than 825,000 histograms for each of the 11 datasets. In order to economically package the more than 9 million PHA histograms, we designed a special binary file format that permitted efficient access to any single spectrum. The resulting file, dubbed a "phacube", consisted of several sections: one for the exposure map, one for the extracted PHA spectra, and a section that tabulates the min, max, and mean chip coordinates of the events contributing to each of the PHA spectra.

One phacube was created for each level-2 event file. To create the cube, a **S-Lang** script (`mkphacube`) was created to read the data from the event file. The script made use of several **S-Lang** modules, including:

- the `cfitsio` module, which was used to read the data from the event files.
- the `histogram` module. This module was used to bin the spectra, and from each histogram's reverse indices, the mean, min, and max chip coordinates of the events contributing to the histogram were computed and stored.
- the `maplib` module, which was used to reproject from one tangent plane to another. The event sky X,Y coordinates and the exposure map were reprojected to a common tangent plane.

On a machine with a 1 Ghz Pentium processor, this process required about 12 minutes to create a phacube for a typical level 2 event file containing over 22 million events, or about one PHA spectrum per millisecond. In stark contrast, a tool-based approach to create the 825,000 PHA spectra would have taken 40,000 times longer, or slightly less than a year to complete, and almost 10 years to process all 11 datasets. This figure was obtained by multiplying the time it took `dmextract` to create a single histogram and then scaling that number by the required number of histograms.

Each phacube had a file size of just over 1.3 GB and contained more than 825,000 PHA spectra binned at a spectral resolution of 0.02 keV, representing events from 0.3 to 8.0 keV. The combined size of the phacubes exceeded 14 GB. With files so large it made no sense to deal with them in a traditional manner employing dynamic memory allocation. Instead, we chose to make use of memory mapping techniques. This was made easy by the `varray` **S-Lang** module, which uses the `mmap` system call to memory map a file and make it available to the interpreter as an ordinary array. The importance of this technique for our analysis cannot be overstated. It allowed our scripts to hop from phacube to phacube reading each one in an almost instantaneous manner without actually allocating any physical memory for the data.

2.3. Adaptive Spatial Extraction

Once the phacubes were created, the next step was the adaptive spatial extraction phase. A spatial grid was specified to define the resolution of the maps. Each grid point was used as an extraction region center. The highest resolution fits used a 466×466 point grid with adjacent grid points separated by 2 sky pixels for a $1''$ resolution. Each of the 217,156 region centers was processed as follows:

- For the specified region center, a square extraction region was adaptively sized until the region contained at least a combined total of 10000 counts from each of the 11 phacubes, with the restriction that no region was allowed to grow beyond 40 pixels in width.

- An exposure value was computed for each of the 11 PHA spectra (one per cube per extraction region) by using the exposure map stored in the corresponding phacube.
- An RMF per spectrum (11 spectra per region) was specified using the average chip coordinates of the events in the extraction region.
- An ARF per spectrum (11 per region) was specified by using the `maplib` module to map from the common tangent plane back to the observation specific tangent plane in order to properly characterize the off-axis position of the extraction region.
- The region specification (center, size) and the 11 RMF, ARF, and exposure triplets were saved to a file.

The adaptive spatial extraction phase took about one day to complete, resulting in a "boxes.dat" file that contained a line per extraction region (217,156 in total). Each line consisted of the specification for the spatial extraction region, and information necessary to compute 11 RMFs and ARFs for each of the 11 spectra, as well as their exposure values. The RMFs and ARFs themselves were not created by this step, only the information required to create them was saved.

2.4. Distributed Computation

The final steps made use of the `PVM` module, which allowed the computation to be distributed across our local network by making use of the Parallel Virtual Machine (PVM) library. We found PVM to be very robust and fault tolerant, and as such it was ideally suited for this sort of master-slave distributed computation. Our "virtual machine" consisted of a single 1.8 Ghz Athlon CPU as the "master", and and 14 dual CPU slave machines, each running Debian ("Woody") Linux. Three of the slave machines had 1 Ghz Pentium processors while the rest utilized 1.8 Ghz Althon CPUs. The virtual machine was configured such that no more than one task was running on a CPU at a time.

2.5. Creating the ARFs and RMFs

The computation of the ARFs and RMFs took slightly more than a day. They were created using the standard CIAO tools `mkarf` and `mkrmf`, and spreading the computational load across the local PVM network.

Because the RMF is calibrated on spatial scales coarser than the typical extraction region size, RMFs were computed on a coarser grid, matching the spatial scale used in the instrument calibration. Similarly, ARFs were computed on a grid chosen to match the size scale of the dither pattern. Bad pixels were accounted for in the exposure map calculation and so were ignored when computing the ARFs.

2.6. Spectral Fitting

As one would expect, the fitting phase of the analysis required the most time. We used the `PVM` module to distribute multiple **ISIS** jobs across the virtual machine as slave tasks. The role of the master was to read the "boxes.dat" file, and for each line in the file, extract PHA spectra for the corresponding region from the phacubes, and dole out those spectra to the first available slave for processing. When a slave was finished with the task, it would pass the results back to the master for saving, and the master would pass it more data for processing. This process would continue until each region in the "boxes.dat" file was handled.

Each subordinate **ISIS** process was passed 11 PHA histograms for joint fitting, along with the names of their respective ARFs and RMFs. The PHA histograms were combined using the **ISIS** `combine_datasets` function, which had the effect of summing the PHA histograms while maintaining the identities of the underlying response functions. Such a procedure was necessary because the X-ray events that contributed to the specified extraction region originate from different regions of the CCD with correspondingly different responses and effective areas.

Our "first generation" maps used a model that consisted of a bremsstrahlung continuum representing plasma, and 14 gaussians representing emission lines from various ions of O, Fe, Ne, Mg, Si, S, Ar, Ca, and Ti. Interstellar absorption was taken into account using a standard photoelectric absorption model (`phabs`).

3. Summary

The primary purpose of this paper is to highlight some of the technical details that went into the production of high resolution spatial maps of the Cas A supernova remnant. The major scientific results will be discussed elsewhere (Stage et al., in preparation).

We have also tried to convey the importance of a scriptable programming environment such as that provided in the CIAO analysis software, and **ISIS** in particular (see also Noble 2005). It is also important to point out that this project would have not been possible using a purely tool-based approach, scripted or otherwise. As noted above, a standard spectral extraction tool would have required nearly ten years to extract the vast number of PHA histograms needed for this project. In contrast, the **S-Lang** script that was used for the extraction completed the task in slightly more than two hours.

Acknowledgments. We would like to thank Una Hwang and the Cas A VLP team for preparing the successful proposal that made the megasecond observation possible. Special thanks go to Mike Nowak and David Huenemoerder for suggesting the catchy title. This work is made possible in part by NASA LTSA grant NAG5-9237 and was also supported under Chandra X-Ray Center contract SV1-61010 from the Smithsonian Institution.

References

Houck, J. C. 2000, in ASP Conf. Ser., Vol. 216, ADASS IX, ed. N. Manset, C. Veillet, & D. Crabtree (San Francisco: ASP), 591

Hwang, et al. 2004, ApJ, in press

Noble, M. 2005, in ASP Conf. Ser., Vol. 347, ADASS XIV, ed. P. L. Shopbell, M. C. Britton, & R. Ebert (San Francisco: ASP), 237

ASTRONOMICAL DATA ANALYSIS SOFTWARE AND SYSTEMS XIV
ASP Conference Series, Vol. 347, 2005
P. L. Shopbell, M. C. Britton, and R. Ebert, eds.

Linearization of Spitzer IRS Data Via Minimization of χ^2 With Correlated Errors

John W. Fowler
Spitzer Science Center, MS 314-6
California Institute of Technology, Pasadena, CA 91125

Abstract. The Spitzer Infrared Spectrograph (IRS) data are taken via read-without-reset measurements to obtain multiple samples forming a photometric "ramp" for each pixel in an echellogram. Each ramp is linearized via a quadratic model. After linearization, a quality-assurance test is performed to determine how linear each pixel's ramp has become. This is accomplished by fitting a straight line to the ramp via χ^2 minimization. The goodness of fit is of primary importance, since this determines whether the inevitable deviations from linearity are statistically significant given the estimated photometric noise. Because the latter is dominated by photon noise which is summed up the ramp, the χ^2 parameter used to measure goodness of fit must include the effects of correlated errors. This paper describes the construction of the full error covariance matrix and its use in the χ^2 minimization.

1. Introduction

One of the challenges in the scientific application of new hardware technology is quantifying the extent to which the hardware behavior is understood. The depth of this understanding is invariably revealed most accurately by the state of the corresponding data-analysis algorithms devoted to the "error analysis", i.e., the mathematical modeling of the uncertainty in the values of any parameters derived from measurements obtained via the hardware. This uncertainty may originate in both the object of measurement (e.g., fluctuations in the arrival rate of photons) and the hardware itself (e.g., dark current drifts). The characterization of uncertainty determines the scientific usefulness of the measurements, so it is desirable to model the hardware as accurately as possible in order to minimize the uncertainty, then to evaluate the irreducible uncertainty accurately enough to make reliable statistical interpretations (e.g., to be confident in judging the difference between an expectable fluctuation and an externally generated "glitch" such as a cosmic ray hit).

It may happen that the errors in two measurements are correlated through mutual dependence on common random events. This can significantly affect statistical interpretations and thus must be taken into account in the error model. A typical example arises in the evaluation of "goodness-of-fit" parameters, the most common of which is the χ^2 statistic. In practice, the effect of ignoring correlations when computing χ^2 is usually underestimation, which tends to suggest misleadingly that the errors are slightly overestimated and thus comfortably conservative. Including the correlations could reveal that the statistical significance of the discrepancies is actually larger than expected, indicating possible

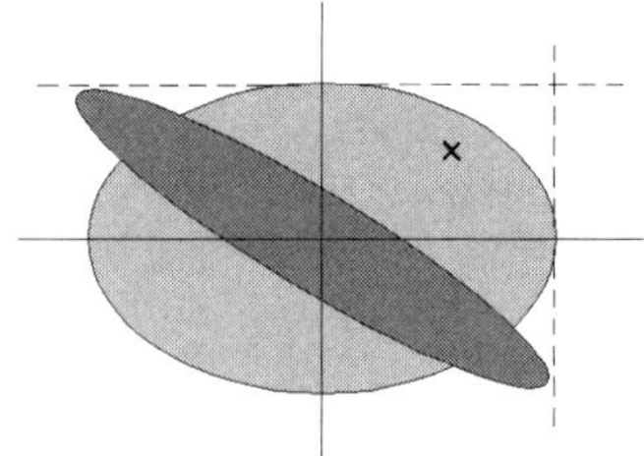

Figure 1. The error ellipse for correlated errors is rotated and has a smaller area than the error ellipse whose covariance matrix has the same diagonal elements but zero off-diagonal elements. Ignoring correlation would underestimate the significance of the deviation of the point marked by X.

problems with the modeling. In any case, an accurate χ^2 value is highly desirable for its ability to signal most problems with the hardware model or the error estimation.

An easily visualized example of correlated errors is that of celestial coordinates whose uncertainty involves an error ellipse that is rotated with respect to the coordinate system. Figure 1 shows a case in which a positive error on the horizontal axis is more likely to accompany a negative error on the vertical axis, i.e., the errors are negatively correlated. The smaller error ellipse includes the correlation information, while the larger unrotated ellipse corresponds to an error covariance matrix with the same diagonal elements but zero off-diagonal elements. The area inside a one-sigma contour is clearly greater when the off-diagonal elements are taken to be zero.

2. Correlated Errors in Spitzer Infrared Spectrograph Ramps

The Spitzer Infrared Spectrograph employs several echelle spectrographs whose measurements are formatted as FITS data cubes, i.e., a stack of N 128×128 image planes, where $4 \leq \mathrm{N} \leq 32$. Each pixel has N nondestructive readouts in the data cube. Since photo-electrons accumulate between each readout, the pixel values generally increase with plane number, and so the set of values is called a "ramp". Ideally each ramp would take the form of a straight line, but noise and nonlinear response prevent perfect linearity. The nonlinear response can be well approximated as quadratic, and each pixel is calibrated accordingly. This calibration is applied as part of the data reduction process, i.e., the quadratic model is inverted to "linearize" the ramp. For quality assessment purposes, the linearized ramp is fit to a straight line, and three goodness-of-fit parameters are evaluated: the linear correlation coefficient, RMS dispersion, and χ^2 statistic.

χ^2 depends on the deviations of the linearized data from the best-fit straight line and on the *a priori* uncertainties in these data. The uncertainties are provided by an error model that incorporates such instrumental effects as calibrated read noise, dark subtraction uncertainty, linearization uncertainty, etc., and "photon noise" caused by "counting-statistics" fluctuations in the incoming photon stream. The photon noise is well approximated as a Poisson process, and the sum of the other uncertainties is well approximated as a Gaussian process.

The Poisson process is generally well into its Gaussian limit also, but the Poisson character is retained in that the variance is equal to the mean, where the mean is the measured number of photo-electrons in the detector well at readout time. Errors in this mean due to Poisson fluctuations at one readout are inherited at the next readout, and so each readout above the first contains the sum of all fluctuation errors below, correlating these errors. The other sources of error occur downstream from the well and are effectively independent from one readout to the next. The error in readout no. i of a ramp is

$$\varepsilon_i = \varepsilon_{pi} + \varepsilon_{ui} = \sum_{j=1}^{i} \Delta\varepsilon_{pj} + \varepsilon_{ui}$$

where ε_{pi} is the total Poisson photon noise at plane i, i.e., the sum of the incremental fluctuations $\Delta\varepsilon_{pj}$ over all lower planes up to and including plane i, and ε_{ui} is the total uncorrelated noise at plane i. We are interested in the expectation value of the product of the errors at any two planes m and n:

$$\varepsilon_m \varepsilon_n = \left(\sum_{j=1}^{m} \Delta\varepsilon_{pj} + \varepsilon_{um} \right) \left(\sum_{k=1}^{n} \Delta\varepsilon_{pk} + \varepsilon_{un} \right)$$

$$= \sum_{j=1}^{m} \left(\Delta\varepsilon_{pj} \sum_{k=1}^{n} \Delta\varepsilon_{pk} \right) + \varepsilon_{um}\varepsilon_{un} + \varepsilon_{um} \sum_{k=1}^{n} \Delta\varepsilon_{pk} + \varepsilon_{un} \sum_{j=1}^{m} \Delta\varepsilon_{pj}$$

Since ε_{um} is uncorrelated with all errors in the ramp other than itself (i.e., ε_{un} when $n = m$), the last two terms on the right will become zero when we take expectation values. Similarly, the second term's expectation value will be zero for $m \neq n$. Furthermore, each incremental photon-noise error $\Delta\varepsilon_{pj}$ is uncorrelated with each other $\Delta\varepsilon_{pk}$ except for $j = k$. For $m = n$, the expectation values are therefore

$$\left\langle \varepsilon_n^2 \right\rangle \equiv v_n = \left\langle \sum_{k=1}^{n} \Delta\varepsilon_{pk}^2 \right\rangle + \left\langle \varepsilon_{un}^2 \right\rangle = \sigma_{pn}^2 + \sigma_{un}^2$$

where v_n is the total error variance at plane n (i.e., v_{nn} with the second index suppressed), the sum of the photon noise variance σ_{pn}^2 and the total uncorrelated noise variance σ_{un}^2 at that plane. This total error variance is computed from the error model and is available at each processing stage. What is not provided and must be reconstructed are the off-diagonal elements of the error covariance matrix, v_{mn}, which we obtain by considering $m > n$:

$$\langle \varepsilon_m \varepsilon_n \rangle \equiv v_{mn} = \left\langle \sum_{k=1}^{n} \Delta\varepsilon_{pk}^2 \right\rangle = \sigma_{pn}^2$$

So the covariance of the error at plane m with that at the lower plane n is the photon noise at plane n. Because of the Poisson character of the photon noise, this is just the number of electrons at plane n, which we will denote y_n to be consistent with the linear equation used in the fitting,

$$y = ax + b$$

where x is the plane number.

3. Linear Fit via χ^2 Minimization With the Full Error Covariance Matrix

We can now construct the full error covariance matrix. The diagonal elements v_n are supplied by the error model, and the off-diagonal elements v_{mn} are just y_n, where $m > n$. For example, a five-plane ramp would have the error covariance matrix:

$$\Omega = \begin{pmatrix} v_1 & y_1 & y_1 & y_1 & y_1 \\ y_1 & v_2 & y_2 & y_2 & y_2 \\ y_1 & y_2 & v_3 & y_3 & y_3 \\ y_1 & y_2 & y_3 & v_4 & y_4 \\ y_1 & y_2 & y_3 & y_4 & v_5 \end{pmatrix}$$

We define the vector u whose i^{th} component is the fitting deviation at plane i:

$$u_i \equiv y_i - ax_i - b$$

With a full $N \times N$ error covariance matrix, χ^2 is

$$\chi^2 = uWu^T$$

where u^T is the transpose of u, and W is the inverse of Ω and has elements w_{ij}. Expanding the vector-matrix-vector multiplication yields

$$\chi^2 = \sum_{i=1}^{N}\sum_{j=1}^{N} w_{ij} u_i u_j$$

Inserting the definition of the components of u, differentiating with respect to the fitting coefficients a and b, and setting the results to zero gives a 2×2 system of linear equations

$$a \sum_{i=1}^{N} w_i x_i + b \sum_{i=1}^{N} w_i = \sum_{i=1}^{N} w_i y_i$$

$$a \sum_{i=1}^{N} x_i z_i + b \sum_{i=1}^{N} w_i x_i = \sum_{i=1}^{N} y_i z_i$$

where

$$w_i \equiv \sum_{j=1}^{N} w_{ij},\ z_i \equiv \sum_{j=1}^{N} w_{ij} x_j$$

This system of equations is easily solved for a and b, and then the value of χ^2 is computed from the equation defining it above.

When applied to well behaved simulation data, the calculations described above reveal the typical features of curve fitting via χ^2 minimization with the inclusion of correlated errors: the coefficients obtained are rather insensitive to whether the error correlation is taken into account, but the value of χ^2 tends to be highly dependent on it. As Figure 1 shows, while the significance of a deviation depends noticeably on which error ellipse is used in the interpretation, the weighted average of many points spread randomly according to the smaller ellipse would not tend to depend strongly on which ellipse supplied the inverse-variance weighting.

Simulation data were generated for five 16-plane ramps with illumination designed to sample the dynamic range of real IRS pixels. The nominal increments in electrons per plane varied linearly from 2,250 to 20,250 from pixel 1 to pixel 5. With a gain of 5.0, this produces a DN of 64,800 at plane 16 of the brightest pixel, close to the limit of an unsigned 16-bit integer. A nominal read noise of 10 electrons/read supplied the uncorrelated non-photon noise. Pseudorandom noises were added to the nominal signal to generate the ramps, and then the linear fits were computed. Numerous trials were performed, and the results shown in the table below are typical.

Table 1. Linear Fits With and Without Error Correlation Included

Pixel	$\Delta e^-/Plane$	χ^2_{uncorr}/N_{df}	$P(\chi^2_{uncorr})$	χ^2_{corr}/N_{df}	$P(\chi^2_{corr})$
1	2250	0.31	0.00719	1.41	0.86001
2	6750	0.24	0.00173	0.86	0.39928
3	11250	0.26	0.00294	1.32	0.81334
4	15750	0.42	0.03064	1.09	0.63516
5	20250	0.29	0.00474	0.98	0.52799

Notes: N_{df} is the number of degrees of freedom, here 14, since there are 16 planes and 2 fitting coefficients; the expectation value of χ^2/N_{df} is unity, and the significance of a given deviation from unity increases with N_{df}; $P(\chi^2)$ is the fraction of all χ^2 that have N_{df} degrees of freedom and are less than or equal to the argument.

For all pixels, χ^2 is too small when error correlation is ignored. $P(\chi^2)$ should be approximately uniformly distributed over multiple trials, and this is clearly not the case unless the error correlation is taken into account. Similar trends are seen in real flight data, but so far these are somewhat diluted by the presence of many uncorrelated errors, some of which are known to be overestimated. Work is underway to refine the error models.

Acknowledgments. This work was performed at the Spitzer Science Center as part of a mission/project managed by Jet Propulsion Laboratory, California Institute of Technology, under a contract with the National Aeronautics and Space Administration.

ASTRONOMICAL DATA ANALYSIS SOFTWARE AND SYSTEMS XIV
ASP Conference Series, Vol. 347, 2005
P. L. Shopbell, M. C. Britton, and R. Ebert, eds.

Post-pipeline Processing of Deep Spitzer Data for the GOODS Legacy Project

David Grumm, Stefano Casertano, Mark Dickinson, and Sherie Holfeltz

Space Telescope Science Institute, Baltimore, MD 21218

Abstract. The GOODS Legacy Project has acquired the deepest images planned to be taken by Spitzer from 3.6 to 24 micron, with effective exposure times of up to 100 hours for IRAC. Optimal analysis of these data requires an effective calibration noise better than a part in 10,000. Starting from the high-quality calibrated frames delivered by the Spitzer Science Center pipeline, we employ a variety of techniques to validate and improve the calibration quality and obtain combined deep images that fulfill the requirements imposed by the GOODS science goals. Post-pipeline processing includes correction for instrumental artifacts, improved alignment, and image combination using drizzle-related techniques. We indicate the improvements obtained with each of the techniques employed.

1. Introduction

The primary data product of the Spitzer Science Center Data Analysis Pipeline is called a Basic Calibrated Data (BCD). Each BCD is an individual calibrated frame with corrected bandwidth, dark current subtracted, nonlinearity-corrected, and flattened. To fulfill the requirements imposed by the Great Observatories Origins Deep Survey (GOODS) science goals, we apply additional post-pipeline processing at STScI to improve the quality of the calibration by making additional corrections for instrumental artifacts, improving the astrometry, and improving the creation of the final coadded images. We describe these aspects of our pipeline.

2. Post-pipeline Refinements

2.1. Instrumental Artifact Corrections

Mux bleed is observed in IRAC channels 1 and 2 wherein a bright source (where signal levels near saturation) leaves a trail in the fast read direction for consecutive readouts. Pulldown causes a decrease in intensity about 5 columns wide on both sides of bright pixels in IRAC channels 1 and 2. A similar effect with the opposite sign is present in channels 3 and 4, but is smaller in magnitude and is not corrected in our pipeline. Images before and after pulldown correction are shown in Figure 1. The vertical artifacts are the pulldown, and the horizontal artifacts are muxbleed. Although muxbleed correction has not been applied to this image, it is applied in our post-pipeline processing.

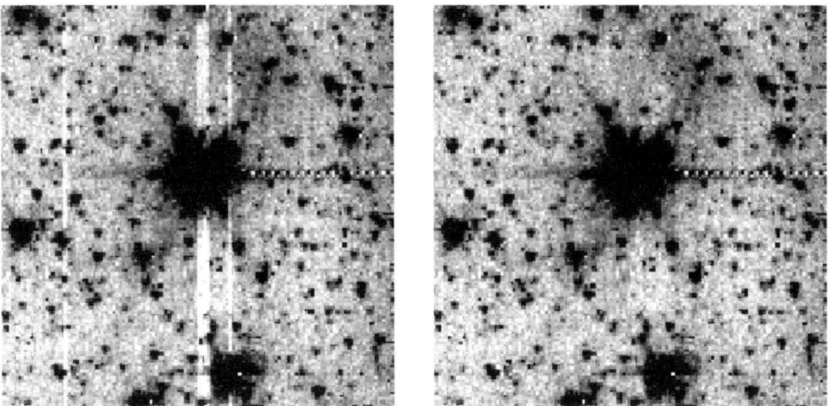

Figure 1. Before and after pulldown correction.

2.2. Astrometry Refinement

An astrometric solution was derived for each Astronomical Observation Request (AOR); solutions were very consistent between the several dozen AORs per field. The effectiveness of the astrometric refinement is shown in Figure 2 for IRAC channel 1. The top plot pertains to RA, and shows the separation of source and initial positions, and the separation of source and final positions; the bottom plot pertains to declination.

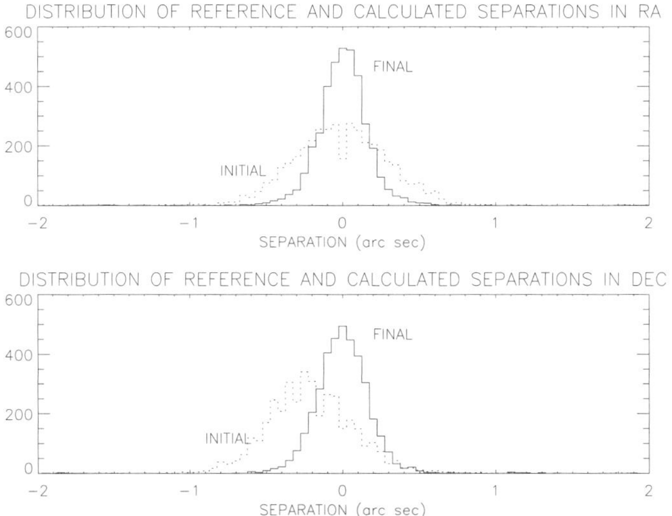

Figure 2. Astrometric refinement for IRAC channel 1

For IRAC channel 1, Figure 3 shows the distortion field with a scale of 5x (i.e., an arrow 1 pixel long corresponds to a 0.2 pixel distortion).

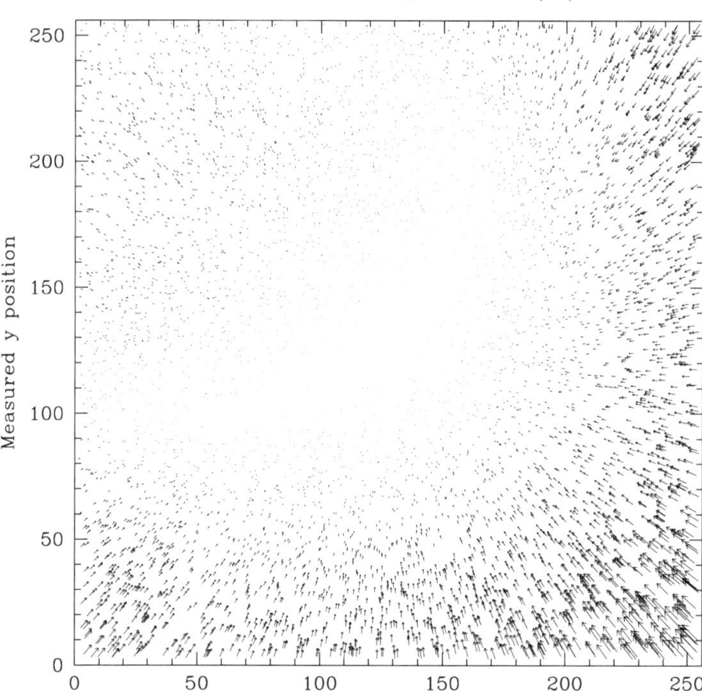

Figure 3. IRAC channel 1 distortion field.

2.3. Image Combination

Images within each AOR were combined into a mosaic using procedures akin to the "multidrizzle" method used for HST/ACS GOODS data processing. The basic routines for projecting pixels from the detector plane to an output image plane uses an implementation of the linear image reconstruction method known as "drizzle" and its inverse operation "blot" (Hook & Fruchter 1996). The following individual steps are performed for each AOR:

1. *Align data images*: use internally consistent astrometric solution derived earlier, and update WCS parameters accordingly.
2. *Drizzle images*: images in each AOR are initially projected to a common tangent plane on the sky using a drizzle scale=0.7 and pixfrac=1.0
3. *Create registered median image*: for each sky location the median for all images is calculated. Pixels observed in fewer than 3 images are excluded because the outlier rejection becomes ineffective in such cases.
4. *Blot median back to each frame*: to create model images, the median image is projected back to the detector frame.
5. *Flag outliers* - for each sky pixel, the noise model accounts for the photon statistics from the sky, source signals, detector readout noise, dark current, and a local gradient term. Data and model pixels that differ by more than

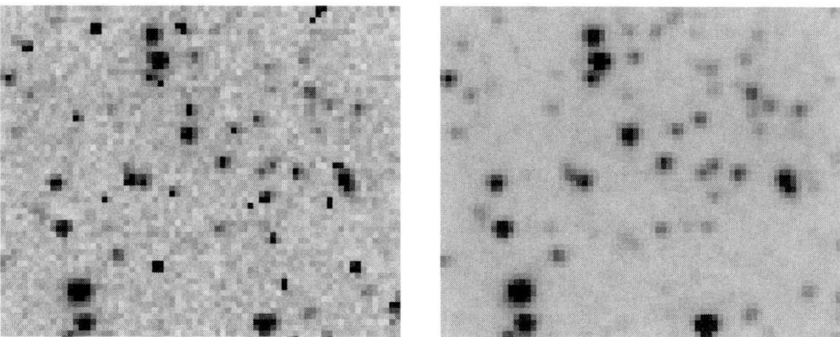

Figure 4. Single data image before CR rejection and its median image

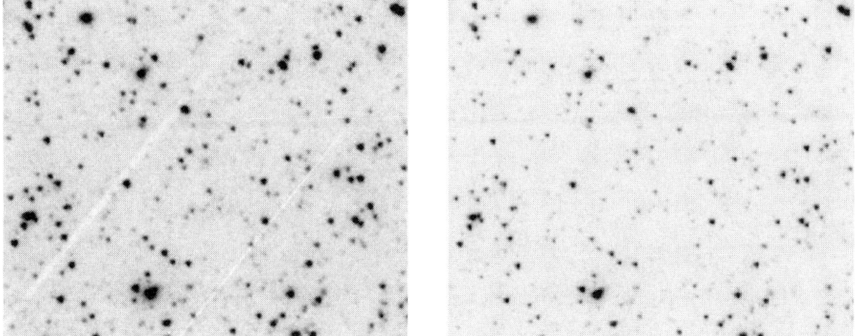

Figure 5. Data from single AOR without instrumental correction, and combination of multiple AORs with correction

the estimated noise are rejected as outliers, allowing masked images to be created.

6. *Create final mosaic for each AOR*: the masked images are drizzled onto a common grid using the "point kernel", which ensures that each input detector pixel per image contributes only to a single output pixel in the drizzled mosaic, so that noise values in adjacent pixels are uncorrelated. Final GOODS mosaics are then created by combining all AORs with exposure time weighting.

A typical single data image and the final median image (after CR rejection) are shown in Figure 4. The images in Figure 5 illustrate the effectiveness of the post-processing after cosmic ray rejection. Figure 6 shows a flow chart of our post-pipeline processing, with the astrometry refinement procedure shown in greater detail. Self-calibration was originally expected to be needed to achieve the required calibration quality (Grumm & Casertano 2002; Grumm & Casertano 2003), but has not yet proved necessary.

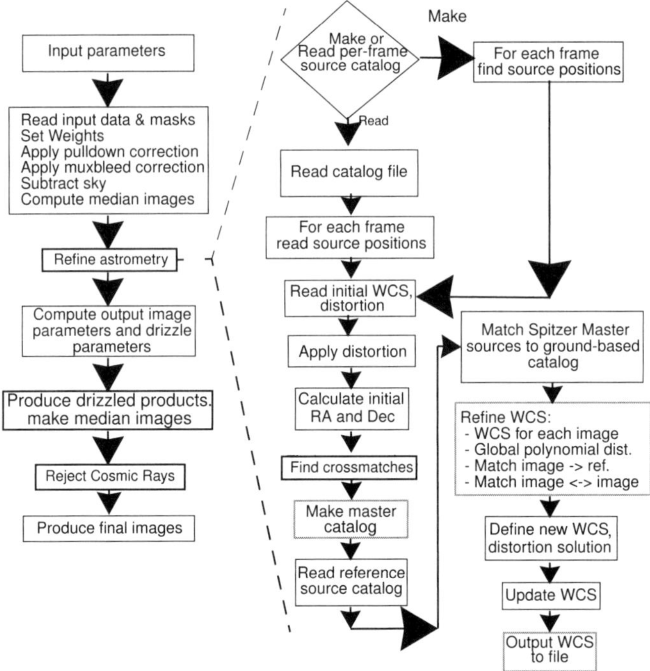

Figure 6. Post-pipeline processing flow chart

3. Status

The first set of combined images produced using the pipeline described here has been made publicly available as part of the first Spitzer Legacy Data Product release. The IRAC CDFS epoch 1 data are available at:

`http://data.spitzer.caltech.edu/popular/goods/20041027_enhanced_v1/`

Acknowledgments. Support for this work, part of the Spitzer Space Telescope Legacy Science Program, was provided by NASA through Contract Number 1224666 issued by the Jet Propulsion Laboratory, California Institute of Technology under NASA contract 1407.

References

Grumm, D., & Casertano, S. 2003, in ASP Conf. Ser., Vol. 295, ADASS XII, ed. H. E. Payne, R. I. Jedrzejewski, & R. N. Hook (San Francisco: ASP), 199

Grumm, D., & Casertano, S. 2004, in ASP Conf. Ser., Vol. 314, ADASS XIII, ed. F. Ochsenbein, M. Allen, & D. Egret (San Francisco: ASP), 562

Hook, R., & Fruchter, A. 1997,in ASP Conf. Ser., Vol. 125, ADASS VI, ed. G. Hunt & H. E. Payne (San Francisco: ASP), 147

ASTRONOMICAL DATA ANALYSIS SOFTWARE AND SYSTEMS XIV
ASP Conference Series, Vol. 347, 2005
P. L. Shopbell, M. C. Britton, and R. Ebert, eds.

Towards the Optimal Association of Hubble ACS Images

Jonas Haase, Richard N. Hook, and Alberto Micol[1]

Space Telescope European Coordinating Facility, Karl-Schwarzschild-Str.2, D-85748 Garching, Germany

Daniel Durand

Canadian Astronomy Data Centre, Herzberg Institute of Astrophysics, National Research Council Canada, 5071 West Saanich Rd., Victoria, BC, V9E 2E7, Canada

Anton M. Koekemoer

Space Telescope Science Institute, 3700 San Martin Drive, Baltimore, MD 21218, USA

Abstract. The ACS Associations Project is aiming to provide combinations of exposures from the Hubble Space Telescope Advanced Camera for Surveys (ACS) similar to those already produced for the older WFPC2 camera. The goal is to associate exposures of a region in the sky with as few restrictions as possible, e.g., regardless of time, science goal or principal investigator of the original observational programmes. Tests on an initial prototype for the ACS Associations are being carried out in order identify methods that will lead to the most interesting science product. Several ways of matching or cross-correlating the science frames, to find their relative shifts and rotations and then correct the astrometric information in the image header world coordinate system (WCS), are being tested against each other and the most stable and effective way of stacking the frames is being sought. At this stage there is also a good opportunity to revisit the implicit assumptions carried over from the WFPC2 Associations and find ways of breaking up the paradigms to allow for more flexible and potentially more useful associations. The association discovering algorithm has been made more complex and the possibility of offering user-defined on-the-fly associations with association strategies matching a user science goal is also being investigated.

1. Overview

The general outline of the ACS Associations effort was first presented at the ADASS XIII conference in 2003 and the project has since moved on to the planning and testing phase. Here we present the current status and highlight the future association definition process and the image registration tests that have been being carried out.

[1]Affiliated with the Space Telescope Division of the European Space Agency, ESTEC, Noordwijk, Netherland

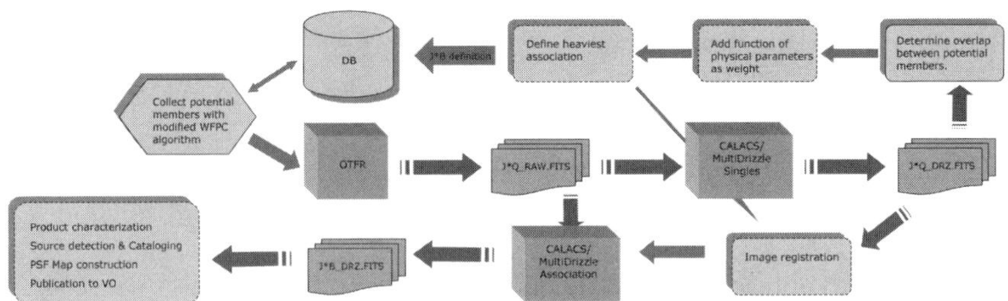

Figure 1. Pipeline schematics, dashed objects represent incomplete or missing stages.

2. The ACS Association Pipeline

The pipeline can be divided into an association discovering stage (the top loop on Figure 1) and an association construction part (bottom loop).

The association discovering algorithm has to find the currently available single HST/ACS frames in the archive and determine which ones in combination will make interesting and scientifically worthwhile associations. Potential members are found with an adapted version of the WFPC2 Association procedure, single members are calibrated and drizzled to remove distortion, geometric overlap is determined and the final association definition is made using several physical parameters as weights.

The association constructor does the stacking of the single members. Extra care has to be taken to ensure the highest possible quality with respect to image registration. It has to be determined whether the offset between frames is given correctly by the WCS information in the header, if not this information has to be corrected (preferably to the right absolute values) before images can be combined. The resulting association product has to be characterized and additional advanced products such as source catalogues, PSF maps and source morphology information should produced to be available at the time of publication to the VO.

2.1. Discovering the ACS Association

The association discovering system is built on top of the time-proven algorithm for WFPC2 associations (Micol et al. 1998). The WFPC2 model mostly relies on the distance between frames and exposure times in the construction of associations and and furthermore constrains frames to have been observed with the same filter and roll angle and be part of the same observing programme. The ACS method has fewer restrictions and introduces a few more steps that hopefully should increase the overall appeal of the resulting scientific product. The basic association definition is:

- Observations from same filter.
- Same position on the sky, original archive associations are deconstructed into their single members.
- No roll angle restriction due to the usage of the multidrizzle package for image stacking.
- Not restricted to single observation program.

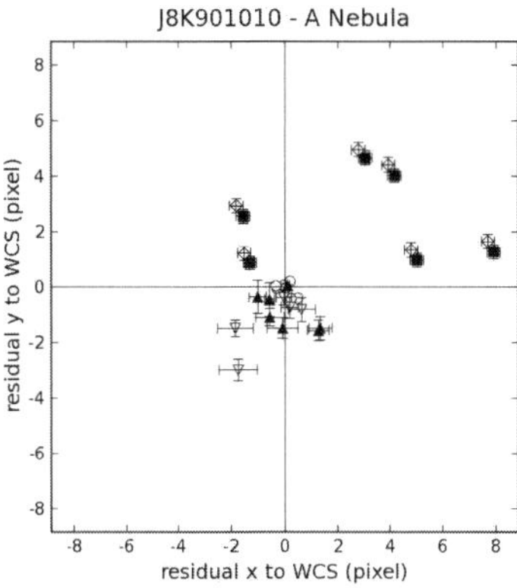

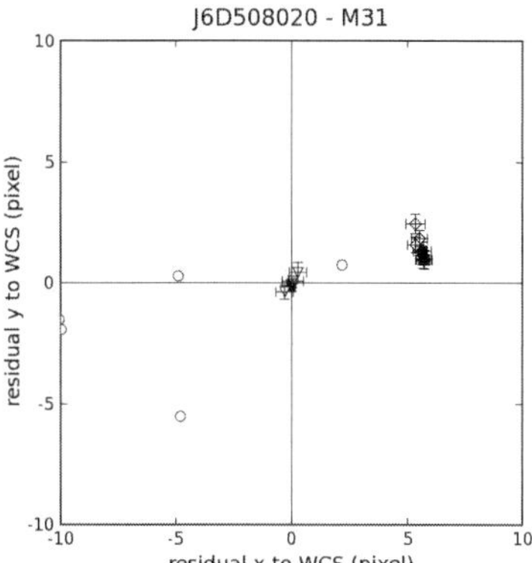

Figure 2. Left: 7 members, Large catalogues (> 1100 objects), many cosmics not cleaned out. Right: 6 members, Match in very good agreement with WCS. Less than 200 rows in catalogues, low S/N. The legend is given in Table 1.

- Created associations must be deeper than the ones in the current HST Science catalogue.

Additional steps to narrow the candidates down to the actual association members:

- The potential members of an association are checked for actual geometric overlap between them.
- Exposure times, amount of overlap, quality of pointing and other physical properties can be introduced as weighting factors in the final association building.
- Conflicting possibilities of associations on the same area on the sky are solved by applying tree-balancing algorithms with a function of weighting factors as node values.
- The possibility is left open to tailor the weighting function to a specific scientific purpose and create custom built associations.

Test Method	Input	Legend
Imcentroid 1	Cleaned catalogues	Black square
Imcentroid 2	Cleaned and recentered catalogues	White diamond
Cross correlation	Images, cleaned and recentered catalogues	White circle
Match 1	Cleaned catalogues	Black triangle
Match 2	Cleaned and recentered catalogues	White triangle

Table 1. Image registration test overview

3. Registration Tests

A number of associations, both the usual archive variety and constructed ones have been run through a series of tests to gauge the robustness of different image registration methods.

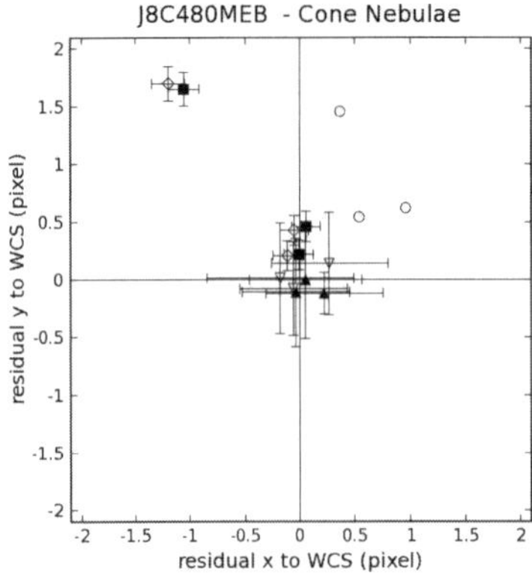

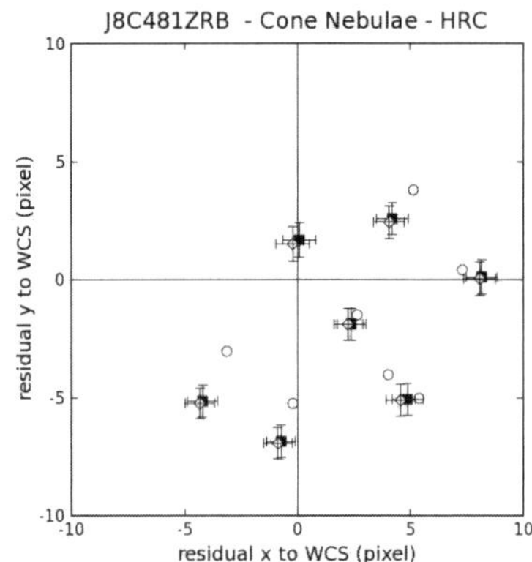

Figure 3. Left: 4 members, many spurious objects confuse the cross correlation and create a high standard deviation with Match (at matchrad=1.0). Right: 8 members, HRC, fairly small input catalogues, lot of noise, even manual matching difficult. The legend is given in Table 1

1. The single members of the associations were calibrated and PyDrizzled in order to get a undistorted image.
2. SExtractor was run on each single member to get a basic catalogue of sources.
3. The catalogues were cleaned by removing detections which did not meet certain criteria. ($class_{star} > 0.8$, $\epsilon < 0.55$, fwhm and $flux_{auto} > 0.0$).
4. An additional catalog was created with the magnitudes from SExtractor and new pixel-space coordinates of objects acquired through iraf.imcentroid.
5. Offsets were determined through cross correlation, Imcentroid shifts and Match. The latter two had either of the two catalogs mentioned above as input (see Table 1).

Match was the only method tried here that can determine scale and rotation between images, which would be needed given the association definition.

The residual shifts of the individual computed offsets compared to the WCS was tested on a number of extreme cases (almost empty fields, tight and relaxed globular clusters and a variety of nebulae). None of the tests yielded a particular stable result or even a discernible systematic pattern (for examples see Figures 2 and 3). Modes of failure:

- Detection of spurious objects - catalogues are easily cluttered with cosmic rays, artifacts from image processing or de-blending.
- Size of catalogues vary from less than hundred to more than thousand rows.
- Insufficient overlap between frames.
- Too few suitable sources - it hardly ever works well with HRC.

It has become abundantly clear that the current rather naive use of these procedures will not work in a reliable fashion and that a more robust pre-processing of the data is required. The chosen method must be able to improve upon the WCS from the observation logs. Some of these additional steps will likely have to be introduced to reduce the modes of failure:

- Added calibration steps to clean the single raw frames for cosmics.
- Improved SExtractor settings.
- A more intelligent and dynamic cleaning of the catalogues.
- Masking of cosmic rays and other artifacts before. cross-correlating.
- Perhaps even building the input catalogue from a combined product of all members and iterate.

4. Conclusions

The definition of the ACS associations will be done with an adapted version of the time-proven WFPC2 Association Algorithm, the use of additional physical parameters in the definition should yield even more interesting associations. Overly simple image registration methods are not helpful in this context and more intelligent and dynamic ways will be investigated. The ACS Associations Project is moving ahead and uniform, deep ACS products should soon be made available through archive and virtual observatory access points.

References

Fruchter, A. S. & Hook, R. N. 2002, "Drizzle: A Method for the Linear Reconstruction of Undersampled Images", PASP, 114, 144

Hook, R., Durand, D., Simard, L., Schade, D., Koekemoer, A., Corbin, M., & Micol, A. 2004, "HST/ACS Associations: The next step after WFPC2", in ASP Conf. Ser., Vol. 314, ADASS XIII, ed. F. Ochsenbein, M. Allen, & D. Egret (San Francisco: ASP), 62

Koekemoer,A. M., Fruchter, A. S., Hook, R. N., & Hack, W. 2002, "MultiDrizzle: An Integrated Pyraf Script for Registering, Cleaning and Combining Images", Proceedings of the 2002 HST Calibration Workshop held at the Space Telescope Science Institute, Baltimore, Maryland, October 17 and 18, 2002. ed. S. Arribas, A. M. Koekemoer, & B. Whitmore, 339

Micol, A., Pirenne, B., & Bristow, P. 1998, "Constructing and Reducing Sets of HST Observations Using Accurate Spacecraft Pointing Information", in ASP Conf. Ser., Vol. 145, ADASS VII, ed. R. Albrecht, R. N. Hook, & H. A. Bushouse (San Francisco: ASP), 45

Valdes, F. G., Campusano, L. E., Velasquez, J. D., & Stetson, P. B. 1995, "FOCAS Automatic Catalog Matching Algorithms", PASP, 107, 1119

ASTRONOMICAL DATA ANALYSIS SOFTWARE AND SYSTEMS XIV
ASP Conference Series, Vol. 347, 2005
P. L. Shopbell, M. C. Britton, and R. Ebert, eds.

Concurrent Science and Wavecal COS Data

Philip E. Hodge

Space Telescope Science Institute, Baltimore, MD 21218

Abstract. Wavelength calibration for the Cosmic Origins Spectrograph (COS) is achieved by taking an exposure (a wavecal) of an internal emission-line lamp and comparing the observed location with the expected location. In order to correct for possible mechanism drifts during exposures which can be more than an hour long, a new observing mode has been proposed whereby the line lamp would be turned on briefly several times during a science exposure. This is practical because the science and wavecal spectra fall on different parts of the detector, and because COS data will normally be taken in time-tag mode, where the time and pixel coordinates are recorded for each detected photon. This paper describes the calibration of such data.

1. Introduction

The Cosmic Origins Spectrograph (COS) is a high throughput, medium dispersion UV spectrograph for the Hubble Space Telescope (HST), built by the Center for Astrophysics and Space Astronomy at University of Colorado, and Ball Aerospace, with the far UV detector system built by UC Berkeley. COS is one of two new instruments for HST (the other is WFC3) that were to have been installed (and hopefully still will be!) during the next HST servicing mission. COS has two detector systems, one for the far ultraviolet (FUV) and one for the near ultraviolet (NUV). The desired configuration is selected by moving the two grating wheels, called Optics Select Mechanisms, or OSM. The default observing mode will be time-tag, for which the raw data file is an events table giving the position and time for each photon that was detected. Accum mode (an image) can also be used but is not relevant for the current discussion. CALCOS, the pipeline calibration program for COS, was written in Python using PyFITS and numarray, with some C extensions. Further information about COS and CALCOS can be found in Hodge (2002) and Hodge (2003).

2. Wavecal

A wavecal is an emission-line lamp exposure for wavelength calibration. This is needed because the location of the spectrum on the detector depends on the position of the OSM, and the OSM is not perfectly repeatable. The wavecal spectrum is also used to determine the location of the science spectrum in the cross-dispersion direction, based on the fixed offset between the wavecal and science apertures. The science aperture is circular, 2.5 arcseconds in diameter, while the wavecal aperture is rectangular, 0.07 by 0.36 arcseconds. The current procedure for taking wavecals ("ordinary" wavecals) is to take one or more exposures that are distinct from the science exposure, for example one wavecal before and one after the science exposure. The problem is that the OSM position can drift. This can be seen very soon after an OSM move or during a long exposure. A solution is described in the next section.

3. Concurrent ("tagflash") wavecal

This new observing mode was proposed by the COS instrument team. The science and wavecal spectra are offset in the cross-dispersion direction (see Figure 1), and control of scattered light is excellent, so it is possible to have the wavecal lamp on during a science exposure. The idea is to use time-tag mode, and to take several (perhaps five) short wavecal exposures during the science exposure. This makes it possible to correct for variations on time scales shorter than an exposure. It also allows more efficient use of telescope time because separate wavecal exposures are not needed. Multiple short exposures are used rather than keeping the lamp on for the entire science exposure to reduce the impact on lamp lifetime. Note that this will only work for time-tag mode; for an accum exposure, the science and wavecal would both just be slightly blurry if the OSM moves during the exposure.

4. Processing tagflash wavecals

The first step is to get the time and duration of each wavecal (each lamp-on interval) from header keywords. During normal operations, these keywords would be set from proposal information. For testing, or as a check, a utility has been written to find when the lamp was on from the count rate, and optionally to set the keywords. The count rate as a function of time is obtained by counting the number of rows in each one-second interval, using only rows with X and Y coordinates in the portion of the detector that is illuminated through the wavecal aperture (FUV burst-detection code was reused for this purpose). A histogram of the count rate shows a large peak near zero (because the background is very low) and a second large peak corresponding to the lamp-on periods. The location of the maximum of the second peak is taken as the lamp count rate, and the lamp is considered to be on when the count rate is greater than half the lamp count rate.

For each time interval during which the wavecal lamp was on, we find and use the subset of rows in the events table corresponding to the current time interval (i.e., we can use Python slice syntax because the events table is sorted by time). We also use only data near the expected location of the wavecal spectrum, temporarily ignoring the science spectrum. Rows flagged as bad in the data quality column are excluded. The positions of the events are obtained from the columns in the events table giving the X and Y pixel coordinates. To find the offset of the wavecal spectrum from nominal in the dispersion direction, we collapse the data along the cross-dispersion direction and cross correlate with a template spectrum (this is the same procedure used for an ordinary wavecal). Collapsing the data is done simply by making a histogram of the pixel coordinates in the dispersion direction. The offset is saved as an attribute, and it is used later for actually correcting the pixel positions. The wavecal spectrum

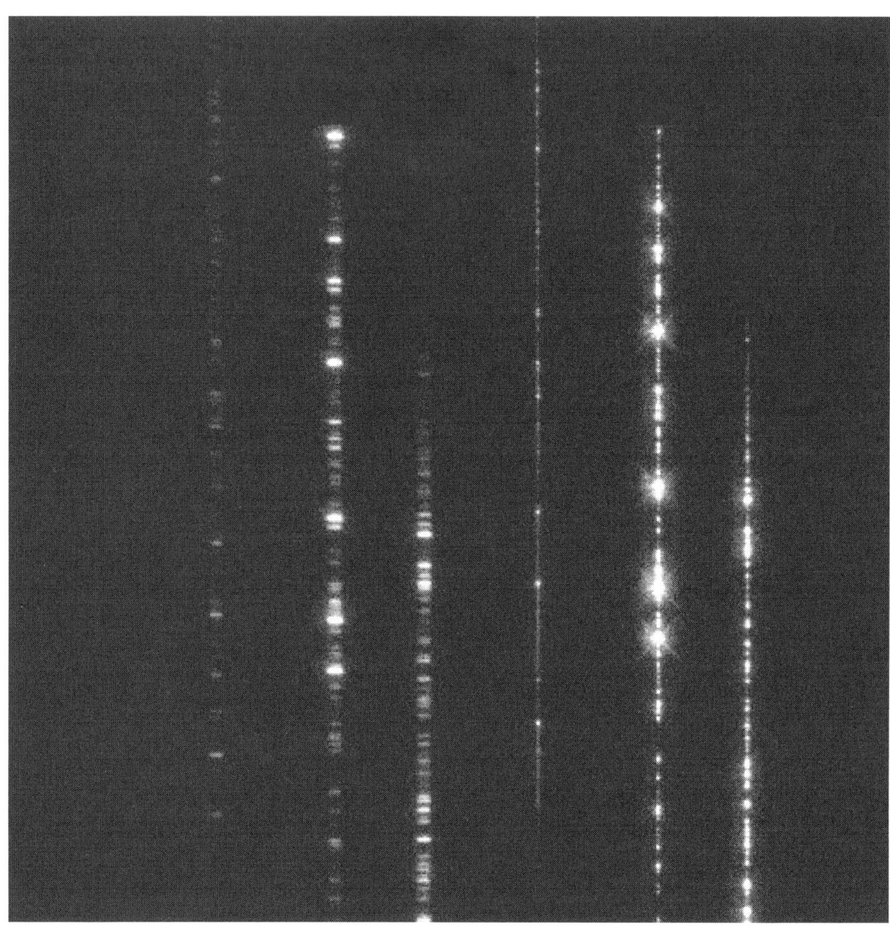

Figure 1. G230L grating, NUV detector, PtNe lamp illumination for both wavecal and science apertures. The three spectra on the left are through the wavecal aperture, and the three on the right are through the science aperture. This is l61h23nvr_03294135032, courtesy the COS instrument team, from thermal-vacuum test on Oct 21, 2003.

for each lamp-on interval is saved in an output table. The location of the wavecal spectrum in the cross-dispersion direction is also determined, by collapsing the data along the dispersion direction, taking into account a possible linear slope of the spectrum, and taking the pixel number of the maximum (nearest pixel is sufficient in this case).

Now that the offset of each of the wavecal spectra is known, the pixel coordinates of the events (both science and wavecal) can be corrected. For each time interval between the wavecals (i.e., between the times at the middle of the lamp-on periods), a Python slice is used to select the corresponding range of rows. The coefficients are computed for a linear change in offset vs. time within the current time interval, and the pixel coordinates are then corrected by subtracting this linear function of time from them. After this, we continue with normal processing of science data, e.g., binning the corrected X and Y coordinates into a 2-D image.

Most of the code is the same for both FUV and NUV. The FUV and NUV channels in COS differ in significant ways, however, and these differences are handled by different methods in subclasses. The dispersion direction for FUV is in the X direction, while it is in the Y direction for NUV; this is handled by simply swapping the X and Y column names. FUV uses two separate detectors for two nearly adjacent regions of the spectrum, and data for each of the two detectors is stored in its own file. NUV has one detector, but three regions of the spectrum are focused onto that detector (see Figure 1, showing three spectra through the wavecal aperture and three through the science aperture). These differences affect the regions over which the corrections will be applied. For FUV the corrections will be applied only to the illuminated region of the detector; there are artificial hot spots for position reference (referred to as electronic stims), and these should not receive any position correction. For NUV, each of the three spectral regions can be offset by a different amount, so different corrections are applied to the various spectral regions.

Acknowledgments. The COS instrument development team proposed this observing mode. They made data from thermal-vacuum tests available to us, and these data were used for testing the changes to the pipeline calibration code that are described here.

References

Hodge, P. E. 2002, in ASP Conf. Ser., Vol. 281, ADASS XI, ed. D. A. Bohlender, D. Durand, & T. H. Handley (San Francisco: ASP), 273

Hodge. P. E. 2003, in ASP Conf. Ser., Vol. 295, ADASS XII, ed. H. E. Payne, R. I. Jedrzejewski, & R. N. Hook (San Francisco: ASP), 203

ASTRONOMICAL DATA ANALYSIS SOFTWARE AND SYSTEMS XIV
ASP Conference Series, Vol. 347, 2005
P. L. Shopbell, M. C. Britton, and R. Ebert, eds.

Processing of 24 Micron Image Data at the *Spitzer* Science Center

Frank J. Masci, Russ Laher, Fan Fang, John W. Fowler, Wen Lee, Susan Stolovy, Deborah Padgett, and Mehrdad Moshir

Spitzer Science Center, California Institute of Technology, Pasadena, CA 91125, Email: fmasci@ipac.caltech.edu

Abstract. The 24μm array on board the *Spitzer* Space Telescope is one of three arrays in the Multi-band Imaging Photometer for *Spitzer* (MIPS) instrument. It provides $5'.3 \times 5'.3$ images at a scale of $\simeq 2''.5$ per pixel corresponding to sampling of the point spread function which is slightly better than critical ($\simeq 0.4\lambda/D$). A scan-mirror allows dithering of images on the array without the overhead of moving and stabilizing the spacecraft. It also enables efficient mapping of large areas of sky without significant compromise in sensitivity. We present an overview of the pipeline flow and reduction steps involved in the processing of image data acquired with the 24μm array. Residual instrumental signatures not yet removed in automated processing and strategies for hands-on mitigation thereof are also given.

1. Introduction

Since the launch of *Spitzer* in August 2003, observations with the MIPS-24μm array have enormously extended our understanding of the infrared Universe. The array has attained sensitivities $\simeq 1.5$ times better than pre-launch estimates (see Rieke et al. 2004 for a review), allowing imaging of star forming regions and high redshift galaxies to sensitivities and spatial resolutions approaching factors of $\simeq 10^3$ and $\simeq 10^2$ respectively better than *IRAS*.

The MIPS-24μm array is a 128×128 pixel Si:As Blocked Impurity Band (BIB) detector and operates in a broad spectral band extending from 21 to about 27μm. Pixels are continuously and non-destructively read out every $\simeq 0.52$ sec over possible integrations ranging from $\simeq 3$ to 30 sec. Due to bandwidth restrictions, the individual samples are not downlinked. Instead, a line is fitted to the ramp samples for each pixel using an on-board linear regression algorithm. The data frames are downlinked in units called Data Collection Events (DCEs) and are packaged by the Multi-mission Image Processing Laboratory (MIPL) into two-plane FITS cubes. The first plane contains the fitted slopes for each pixel and the second, the difference between the first two reads in the ramp, referred to as the first-difference. The first-difference frame effectively increases the dynamic range for slopes derived from ramps which saturate. To further limit data volume, only first-difference values which exceed a nominal threshold (set by the saturation level) are retained and downlinked.

The 24μm array has two data-taking modes. That just described is called "SUR" (for Sample-Up-the-Ramp) and is the primary science mode for this array. The second mode is called "RAW", where all ramp samples are downlinked and received as multi-plane FITS cubes. RAW-mode is only used for engineering purposes. All raw pixel data are represented in signed 16-bit integer format, with SUR-mode data in units of Data Number/read-time (DN/read) and RAW-mode in DN. The popular observing mode is scan-map[1] mode, and results in the highest data volume with typically 11,000 DCEs downlinked per day for the 24μm array alone.

To process such large data volumes, an infrastructure of automated pipelines, running on a cluster of 34 CPUs has been set up at Caltech's *Spitzer* Science Center (SSC) (e.g., Moshir 2001). This paper reviews the reduction steps used to process specifically MIPS-24μm SUR-mode science data at the SSC. A summary of instrumental residuals not yet corrected in automated processing, and which may appear in products distributed to observers, is also given.

2. Pipeline Infrastructure Summary

Raw images undergo several stages of automated processing at the SSC to produce the best calibrated products. The SSC is responsible for archiving, distributing data to observers, and maintaining current information on instrument calibration. For the MIPS-24μm array, eight separate pipelines have been developed for the removal of instrumental artifacts at the DCE level, each specific to the particular data collection mode or flavor of calibration product needed. These are as follows:

1. SUR-mode _science_ (input: two-plane FITS DCE of slope and difference).
2. RAW-mode _science_ (input: multi-plane FITS DCE of sample reads).
3. SUR-mode dark-current _calibration_ (input: ensemble of DCEs).
4. RAW-mode dark-current _calibration_ (input: ensemble of DCEs).
5. Electronic non-linearity _calibration_ (input: ensemble of RAW-mode DCEs).
6. Flatfield (non-uniformity) _calibration_ (input: ensemble of SUR-mode DCEs).
7. Latent-image flagging (input: preprocessed ensemble of BCDs).
8. Pointing reconstruction and Final Product Generation (FPG) on BCD.

On ingestion of data, a pipeline executive ensures that calibration data are processed first, before being employed in the reduction of regular science DCEs. Calibration products are created from ensembles of input DCEs, and a procedural database query tool using a set of predefined rules has been developed (Laher & Rector, 2005). Calibration products are transferred to the science pipelines using database queries handled by a software module called CALTRANS (Lee et al. 2005). The main product resulting from a DCE is a Basic Calibrated Data product, or BCD, with raw pointing and distortion information attached to its FITS header. The final (Post-BCD) processing steps include pointing refinement using astrometric matching (Masci et al. 2004), and mosaicking of ensembles of BCDs to provide seamless final products (Makovoz et al. 2005). Post-BCD processing is not discussed in this paper.

For each processed 24μm science DCE, ten associated BCD image products are archived at SSC, a subset of which are distributed to users. The archived set includes uncertainty images, bit-mask images which summarize the processing status for each pixel (both for the slope and first-difference planes), and processing log files. For a full description, see the *MIPS Data Handbook*[2]. The above

---

[1]For a description of all modes, see http://ssc.spitzer.caltech.edu/documents/SOM/

[2]http://ssc.spitzer.caltech.edu/mips/dh/

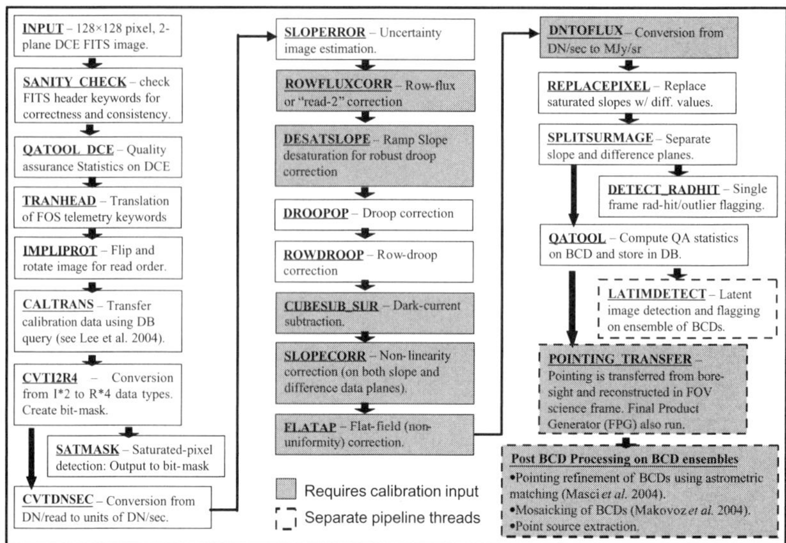

Figure 1. Processing flow in MIPS-24μm primary science pipeline.

suite of pipelines has also been implemented into offline stand-alone versions to facilitate testing and validation of algorithms before deployment in operations.

3. BCD Processing Summary

Figure 1 gives the ordering of reduction steps in the primary 24μm science pipeline. Processing algorithms were designed in collaboration with the MIPS Instrument Team. For a review, see Gordon et al. (2004). Here we give an overview of some of the more important reduction steps and instrumental signatures unique to the 24μm array.

The first step which modifies pixel values is CVTI2R4. This converts the signed 16-bit native raw image data into 32-bit floating point. This step also applies a truncation correction to the on-board computed slopes by adding a constant of 0.5 DN/read to every pixel in the slope plane.

The next step is detection and flagging of "soft" and "hard" saturated pixels (SATMASK). Soft saturation is where the samples saturate somewhere along the ramp, therefore biasing slope estimates. By saturation, we mean that the samples become pegged to the maximum value (+32768) as allowed by the on-board Analog-to-Digital Converter (ADC), which corresponds to $\simeq$ 30% full well. Soft saturated pixels are detected by simply thresholding pixels in the first-difference plane which are above a nominal value. For a 10 sec exposure, this value is $\simeq$ 3000 DN/read. Hard saturation is when the first sample, and all samples thereafter become pegged to the maximum ADC value. This is detected if both the derived on-board slope and first-difference are zero. Both soft and hard saturated pixels are flagged in a processing status mask and propagated downstream.

The SLOPERROR step initializes an uncertainty image using a robust noise model as applicable to slope data. This includes read noise, Poisson noise, and an estimate of the correlation between samples along a ramp from which the slope is derived. For a general overview of uncertainty propagation in SSC pipelines, see Moshir et al. (2003).

The ROWFLUXCORR step corrects a signature known as the "read-2" effect. This effect describes a bias introduced into the on-board slope measurement from a small additive offset in the second sample of every ramp. This offset is seen to vary across the array at the 0.2% level, primarily in the cross-readout (row) direction.

The DESATSLOPE is used to de-saturate slope pixels that were flagged for saturation from above (SATMASK). The de-saturated slope pixels are not propagated downstream. They are only used for robust computation of an effect known as "droop". Droop is an extraneous signal that is added to each pixel (at the $\sim$ 10% level) by the readouts. It is computed and removed by the DROOPOP module. Droop is directly proportional to the total number of counts on the array, including counts which are present above the ADC saturation level. For saturated pixels, slopes are underestimated, and thus for the purpose of computing droop, it is necessary to estimate slopes that would result from ramps which would continue beyond the saturation point. Slopes are de-saturated using the first-difference value and the non-linearity model. The ROWDROOP module corrects a second-order effect which similar to droop but whose signal in a pixel depends on the total signal from all pixels in its row.

The three standard calibrations are next. The dark current on each pixel is removed by subtracting a dark calibration image from both the slope and first-difference planes (CUBESUB_SUR). The dark calibration is computed by performing a symmetric outlier-trimmed average of a few hundred DCEs taken with the scan-mirror in the dark position. The dark current is small, and values range within 0-3 DN/sec. Correction for electronic non-linearity is performed by the SLOPECORR module. The non-linearity is accurately described by a quadratic up to ADC saturation, and the deviation from linearity is typically 10-15%. Flat-fielding is next (FLATAP). Flatfield calibrations exhibit maximum deviations of $\simeq$ 20% from flatness, and are primarily due to dark spots and low-level "blotchiness" in DCEs from absorption by debris on a pick-off mirror. The position and shape of these debris artifacts depend on the angle and scan-rate of the scan-mirror respectively. To correct for these, scan-mirror-dependent flatfields are created offline following each initial campaign processing run and deployed on the operations system before reprocessing of science data.

The final steps include flux-calibration (DNTOFLUX) and pixel replacement (REPLACEPIXEL). This last "pixel-modifying" step looks for all pixels that were flagged for saturation in the mask (from SATMASK), and replaces slope values with first-difference values in the *primary BCD product.* The original processed slope image (with no pixel replacement) is retained as an ancillary product.

4. Possible Instrumental Residuals

Broadly speaking, there are three instrumental signatures that may remain in 24μm BCD products after automated processing with the S11.0 version of SSC pipelines. These can be ameliorated with further hands-on processing, although we expect to automate the corrections once they are sufficiently characterized.

The first residual may arise from inaccurate, scan-mirror dependent flat-fielding. As described above, the process is not yet fully automated. Mismatches between dark spot (debris-artifact) positions and corresponding actual mirror angles can occur. These mismatches lead to bright and dark residual patterns in BCDs, and given sufficient data, can be removed by re-creating flatfields from the BCDs and performing a self-calibration. For the most part, the scan-mirror-dependent flat-fielding is giving excellent results, although observers are encouraged to report anomalous cases to the SSC.

The last two residuals occur when bright, saturating sources are observed. The first is called "readout saturation" and occurs when a saturating cosmic ray or source depresses the output of a single readout channel. Since there are four readout channels, this gives the appearance of a "jail-bar" pattern. This has been characterized as a multiplicative effect and can be corrected by scaling the affected readout columns with a median of the other three unaffected readouts. This is only possible of course if the background doesn't show complex structure. The second effect is when the saturating source is bright enough to leave "dark" latents in many subsequent images. They appear dark because the slopes are fitted to saturated ramps, which turn out to be lower than average on the array. Given a sufficient number of frames with no complex structure, dark latents can be corrected by creating time-ordered sequences of self-calibration flats and dividing these into the BCDs.

Overall, the behavior of the 24μm array since launch can be described as excellent. Unless bright saturating sources are inadvertently observed, very few instrumental residuals are present. Future work will focus on reducing the residuals just described.

Acknowledgments. This work was carried out at the *Spitzer* Science Center, with funding from NASA under contract 1407 to the California Institute of Technology and the Jet Propulsion Laboratory.

References

Gordon, K. D., Rieke, G. H., Engelbracht, C. W., et al. 2004, PASP, (in press)

Laher, R., & Rector, J. 2005, in ASP Conf. Ser., Vol. 347, ADASS XIV, ed. P. L. Shopbell, M. C. Britton, & R. Ebert (San Francisco: ASP), 39

Lee, W., Laher, R., Fowler, J., Masci, F., & Moshir, M. 2005 in ASP Conf. Ser., Vol. 347, ADASS XIV, ed. P. L. Shopbell, M. C. Britton, & R. Ebert (San Francisco: ASP), 594

Makovoz, D., & Khan, I. 2005, in ASP Conf. Ser., Vol. 347, ADASS XIV, ed. P. L. Shopbell, M. C. Britton, & R. Ebert (San Francisco: ASP), 81

Masci, F. J., Makovoz, D., & Moshir, M. 2004, PASP, 116, 842

Moshir, M. 2001, in ASP Conf. Ser., Vol. 281, ADASS XI, ed. D. A. Bohlender, D. Durand, & T. H. Handley (San Francisco: ASP), 336

Moshir, M., et al. 2003, in ASP Conf. Ser., Vol. 295, ADASS XII, ed. H. E. Payne, R. I. Jedrzejewski, & R. N. Hook (San Francisco: ASP), 181

Rieke, G. H., Young, E. T., Engelbracht, C. W., et al. 2004, ApJS, 154, 25

ASTRONOMICAL DATA ANALYSIS SOFTWARE AND SYSTEMS XIV
ASP Conference Series, Vol. 347, 2005
P. L. Shopbell, M. C. Britton, and R. Ebert, eds.

The CIAO Website Unveiled

Elizabeth C. Galle, D. J. Burke, Christopher Stawarz, and Antonella Fruscione

Harvard-Smithsonian Center for Astrophysics, 60 Garden Street Cambridge, MA 02138

Abstract. The Chandra Interactive Analysis of Observations (CIAO) website is the primary resource for users of the CIAO software package. Several hundred pages of content are written in XML, allowing for a baseline set of text from which many types of documentation may be created. Development and production (i.e., public) HTML versions of the site are generated from the XML via conversion scripts and XSL stylesheets. We present an overview of the back-end of the CIAO website, including custom markup tags, stylesheets, and CSS. The success of the project led to the use of this system in maintaining five websites at the Chandra X-ray Center (CXC).

1. Introduction

The CIAO software package[1] evolves relatively quickly, with an average of one major release and several smaller "patch" releases per year. The documentation needs to be updated at least as quickly as the software, and often on a much shorter timescale. By maintaining all of the documentation online, we are able to keep it up to date in real time and make modifications available to the user community immediately.

The result of the CIAO documentation effort is a set of websites that contain several hundred pages of current content, as well as archived information for previous software versions.

Maintaining this information led us to create a customized content management system which would allow multiple authors to write documents, review them in a closed development environment, and then post them on the live site. Here we present an overview of the back-end of the publishing system and discuss some of the benefits from its use.

2. Architecture

The publishing system (hereafter referred to as "CIAODOC") used to create the CIAO website consists of several interrelated components which control content, layout, and style. The following process is currently used to maintain five different sites (including one internal) at the CXC.

[1] `http://cxc.harvard.edu/ciao/`

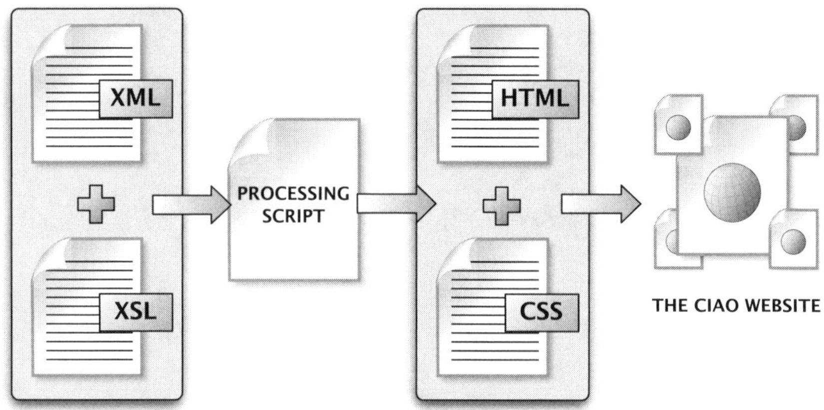

Figure 1. The CIAODOC processing system.

As shown in Figure 1, CIAODOC takes the XML source code, transforms it with XSL stylesheets (via a processing script), and produces HTML pages. CSS is used to apply color and other styles, completing the CIAO website.

2.1. XML Source Code

All of the CIAO webpages are written in XML (Extensible Markup Language)[2], a versatile text format from which many types of documentation may be created. The customized XML syntax is similar to (X)HTML and has many tags in common with it: paragraphs (`<p>`), `<table>`s, `<div>`s. While the basic structural elements are the same, a number of new tags were defined to simplify the writing, notably for the creation of links.

CIAODOC uses over twenty different custom link tags in place of the HTML anchor (`<a>`) tag. The convenience of these link tags is that the author is not required to give a complete path, just the name of the target page. CIAODOC determines the path to that page from the current location and creates the link accordingly. A few examples of link tags used in the XML source and the corresponding HTML destination are shown in Table 1.

Another customization is the creation of different page types. Each XML file begins with a tag that specifies the page type, such as `thread` or `dictionary`. The tag is required in order for CIAODOC to apply the correct transformations during the processing (see Section 2.2.).

The files are stored in a directory structure separate from the web pages, so that they are not visible to the user. SCCS (Source Code Control System) is used for version control, a vital feature in a multiple author environment. SCCS prevents more than one person from editing an XML file at a time, and maintains a history file for each XML file. The history file tracks the edits to the XML and makes it possible to retrieve any previous version of the document.

[2] `http://www.w3.org/XML/`

Table 1. Sample Link Tags.

XML tag	HTML link
`<aguide>`	http://cxc.harvard.edu/ciao/guides/
`<ahelppage>`	http://cxc.harvard.edu/ciao/ahelp/
`<threadpage>`	http://cxc.harvard.edu/ciao/threads/
`<why>`	http://cxc.harvard.edu/ciao/why/

2.2. XSL Transformations

XSLT (Extensible Stylesheet Language Transformations)[3] is a language used to transform XML into an output format, whether it be HTML, text, or new XML. The CIAODOC XSLT consists of three dozen stylesheets used to process the XML into HTML. There are specific stylesheets to handle different page types, as well as many all-purpose stylesheets that are used in processing every type of page.

For example, when processing an analysis thread[4], CIAODOC uses one stylesheet which creates thread-specific structure (overview, table of contents, section blocks, summary), while another stylesheet is employed to process all the links in the document. Depending on how complex the content is, additional stylesheets could be called to perform other transformations.

Some error handling is also built directly into the XSLT to prevent authors from omitting required information, such as link content.

2.3. Processing Script

A Perl script which calls xsltproc[5] is used to process the XML files. The script serves two main functions:

1. **HTML creation:** the XML source code is converted to HTML by applying XSLT stylesheets (see Section 2.2.)
2. **File management:** the newly generated files are copied to the designated output directory.

This process is known as "publishing" the pages. By default, the script publishes the page to a development server so that it may be reviewed for accuracy. Once it is ready to be made public, a command-line switch is used to publish it to the live server.

The script may also be run on non-XML files (such as images, text files, or postscript files); in this case the files are simply copied to the correct directory.

---

[3] `http://www.w3.org/TR/xslt`

[4] `http://cxc.harvard.edu/ciao/threads/all.html`

[5] `http://xmlsoft.org/XSLT/xsltproc2.html`

2.4. CSS

Once CIAODOC has created the HTML version of the pages, Cascading Style Sheets (CSS)[6] are applied. CSS is primarily used to control the look of the navigation bar (e.g., the signature blue background), as well as the behavior of links when they are selected. It also adds highlighting in other areas of the site, and handles some minor positional information. The CSS file for the CIAO website can be viewed online:

`http://cxc.harvard.edu/ciao/ciao.css`

CSS is not used more extensively in part because Netscape 4.7x - a browser still favored by a number of our users - has substandard support for CSS.

2.5. Creating Hardcopies

Since the CIAO documentation is exclusively released online, we provide two convenient methods of obtaining hardcopies of the webpages.

- **PDF files:** after the HTML pages are generated, CIAODOC creates PDF files of them in Letter and A4 formats. Links to the PDF files are provided at the top and bottom left of every webpage for download.
- **Print-specific CSS:** just as stylesheets are used to format the pages for display in your browser, we use a stylesheet that relays print formatting instructions to modern browsers. This stylesheet is applied when you send a page to the printer or use the "print to file" option.

In both cases, some minor modifications are made to the page content before creating the hardcopy. The URL of the page is included at the top of the document, and details that aren't useful in print, such as the navigation bar, are omitted.

3. Validation

The XML source code is checked before publication to make sure that it is both well-formed and valid. A "well-formed" document is structurally correct (e.g., all tags are closed and properly nested), while a "valid" one contains all the required elements for that particular document type. Both of these criteria must be met in order for CIAODOC to handle the code properly. A RelaxNG[7] schema which defines the structure of the document, similar to a document type definition (DTD) for an HTML page, is used to examine the XML file prior to processing.

CIAODOC is designed to create HTML that meets the "4.01 Transitional (no frames)" DTD. The live HTML pages are periodically validated via the W3C Markup Validation Service[8] to ensure we are achieving this goal.

---

[6] `http://www.w3.org/Style/CSS/`

[7] `http://www.relaxng.org/`

[8] `http://validator.w3.org/`

4. Conclusion

Originally developed with only the CIAO website in mind, the CIAODOC publishing system is now used to maintain four public CXC websites: CIAO, *Sherpa*[9], ChaRT[10], and CalDB[11].

Since all of these sites are built by CIAODOC, it is easy to apply the same layout and color schemes (i.e. "branding") across the board. After a user is introduced to one of the sites, he will be able to recognize instantly that he is browsing one of the CXC software webpages on subsequent visits.

Along with the consistent user experience afforded by this system, there are a number of benefits to both the developers and the users:

- Any page structure beyond basic organization is determined by the publishing scripts rather than being hardcoded into the site. This allows us to make site-wide changes easily, instead of editing every individual XML file.
- Another benefit of separating the content and presentation is the ability to maintain a consistent look and feel in a multiple-author environment. An author can control the basic organization of the page content, but the layout is determined by the XSL stylesheets.
- Forward compatibility: producing valid code helps ensure that the website will continue to work correctly as newer browsers are developed.

Acknowledgments. This work was supported by the Chandra X-ray Center under NASA grant NAS8-03060.

[9] `http://cxc.harvard.edu/sherpa/`

[10] `http://cxc.harvard.edu/chart/`

[11] `http://cxc.harvard.edu/caldb/`

ASTRONOMICAL DATA ANALYSIS SOFTWARE AND SYSTEMS XIV
ASP Conference Series, Vol. 347, 2005
P. L. Shopbell, M. C. Britton, and R. Ebert, eds.

The Impact of the ACIS Readout Streak and Pileup on Chandra Source Detection

Michael L. McCollough and Arnold H. Rots

Harvard-Smithsonian Center for Astrophysics, Cambridge, MA 02138

Abstract. As part of the development of the Chandra Level 3 (L3) data products (Chandra Source catalog) the impacts of the ACIS *Readout Streak* and Pileup have been examined. A method has been developed which allows us to determine, for a given observation, which columns of the ACIS CCDs are impacted by *Readout Streaks*, resulting in the detection of false sources. A discussion is given of how to use this information to prevent false detections and how to identify and determine the characteristics of real sources located in the *Readout Streak*. The implications of this for the creation of L3 products are discussed.

1. Introduction

X-Ray observations made with the Chandra Advanced CCD Imaging Spectrometer (ACIS) containing bright X-ray sources often contain a *Readout Streak*. This steak is the result of the finite amount of time (40 μsec per row) that it takes to read out the CCD after an integration period. As a result all pixels along a given readout channel are exposed to all points on the sky that lie along that readout channel. The result is a streak along channels which contain a bright source. This streak can cause problems when one is trying to detect, and determine the properties of, the X-ray sources in the data.

Currently under development is the Chandra Level 3 (L3) data products (Chandra Source Catalog). This project seeks to do source detection on all public Chandra data ($\sim$ 4200 observations) and create data products for all sources which are found. Due to the number of observations it is important to automate the source detection and minimize the number of spurious sources. A strong *Readout Streak* can easily cause on the order of 100 false sources being found.

Figure 1a is an image of the S3 chip of an observation of M81 (OBSID: 735). Apparent is the *Readout Streak* of the central X-ray source and to a lesser extent the *Readout Streak* of the ultra-luminous object X6 (to the left edge of the image). Overlayed are the sources detected from a run of WAVDETECT (see Freeman et al. 2002). It is apparent that a large number of false sources detected are caused by the *Readout Streak* with 60 of the 158 sources along the *Readout Streak*. We describe a simple algorithm to address this problem.

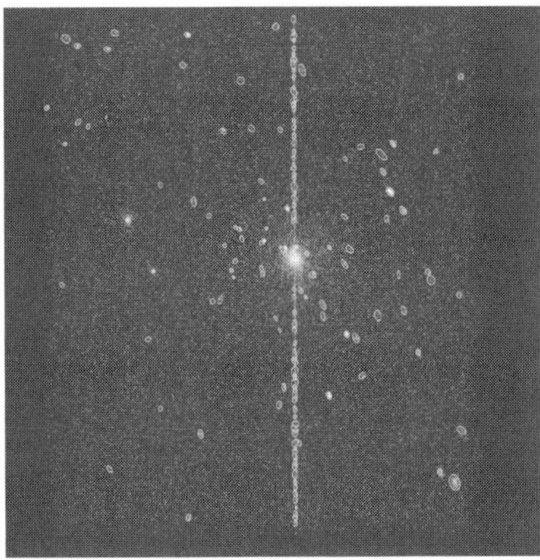

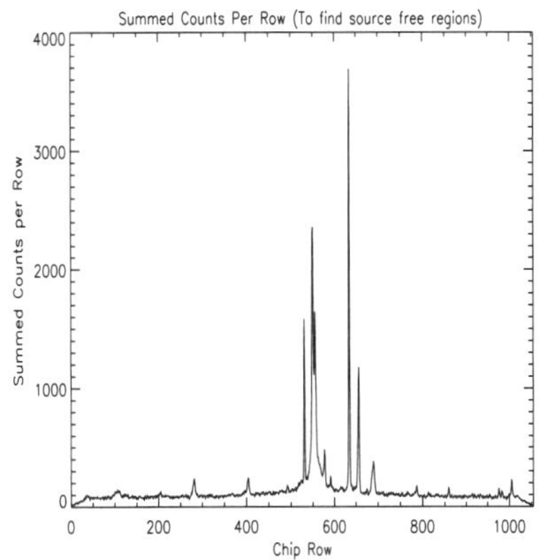

Figure 1. *a (left)*: Image of the S3 chip of an ACIS observation of M81 with WAVDETECT source locations overlayed. *b (right)*: Plot of the counts summed across readout channels vs. chip row. This is used to identify chips rows which do not contain counts from X-ray sources.

2. Background + Readout Streak

To address the problem of the *Readout Streak* in source detections we outline a technique in which from the data we create a background image containing the *Readout Streak*. The resulting background image can either be used to determine which sources are contained in the *Readout Streak*, or it can be input into a source detection algorithm (such as WAVDETECT) as a background map.

- We first create an image for each chip. The resulting background images can be later merged to create an image to match the image being used by WAVDETECT.
- We rotate the image such that the X-axis will be along the chip rows (across readout channels) and Y-axis will be along the direction of the readout channels.
- We first sum along the X-axis (across readout channels) to make a plot of summed counts per row (see Figure 1b) in order to determine which rows on the chip are source free.
- We find the lowest summed value (excluding off-chip and dither regions). This is represented by the dotted line in Figure 2a. The square root of this value is used as a σ (standard deviation).
- We use as an upper bound the lowest value + n $\times$ σ (dashed line in Figure 2a). The rows whose summed values fall between the minimum value and this upper bound are considered the source free region. Note that n is a parameter that one can vary; we have used a value of 4.
- We then take the rows which are source free and sum along the direction of the readout channels and divide by the total number of rows. The result is a plot of counts/pixel vs. readout channel (see Figure 2b). This represents the average background + *Readout Streak* contribution per readout channel.

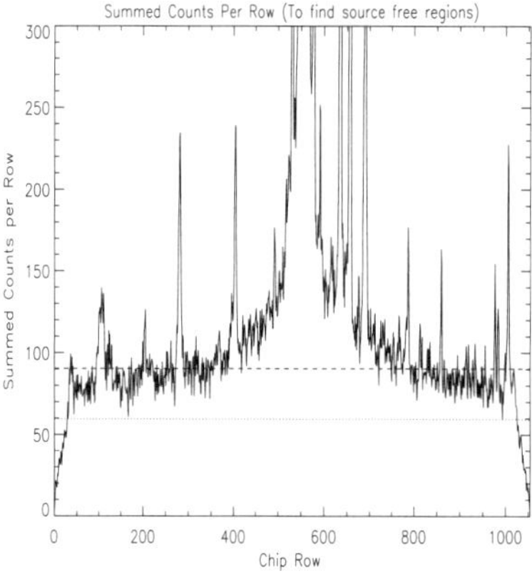

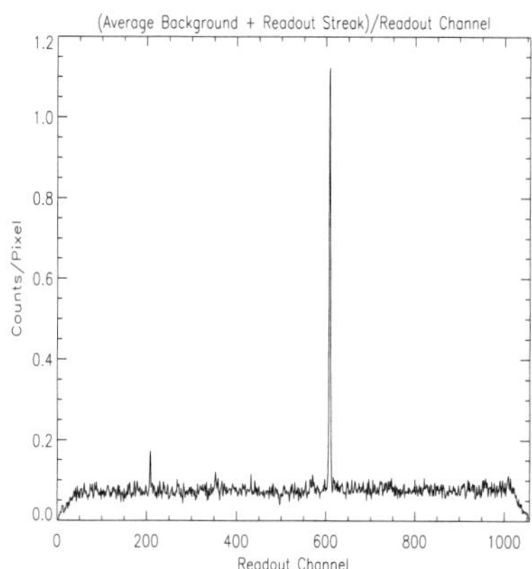

Figure 2. *a (left)*: Same plot as Figure 1b with limits used to determined which chip rows do not contain an X-ray source. *b (right)*: The summed counts of the source free rows divided by the total number of source free rows vs. readout channel. This represents the background counts + *Readout Streak* counts as a function of readout channel.

- This can now be replicated (copied) for each row of the chip to create a background image which also contains the *Readout Streak* contribution (see Figure 3a). Care needs to be taken in handling the edges of the chip.
- This background + *Readout Streak* image can now be input into the source detection algorithm (WAVDETECT) as a background map. Note that the use of this type of map is important since the *Readout Streak* affects the noise statistics significantly.

3. Results of this Approach

Using the background + *Readout Streak* image created in the manner described above for the S3 chip of the M81 observation we found the following:

- One can see the suppression of false detections caused by the background + *Readout Streak* as well as improved source detection due to supplying a background. (see Figure 3b).
- There are now 95 sources found instead of 158 sources. All of the "missing" sources were those previously found in the *Readout Streak*.
- Of the sources still found in the *Readout Streak* all appear, upon closer examination, to be real sources.
- Near the edge of the chip and near the center of M81 (where there is extended emission) there are a few additional sources found (not found in the initial run of WAVDETECT).
- The *Readout Streak* from the background image can also be used to "clean" the image and remove the *Readout Streak* from the image (see Figure 4). But one should be careful of using this image for source detection since the noise statistics are significantly different for the *Readout Streak* than

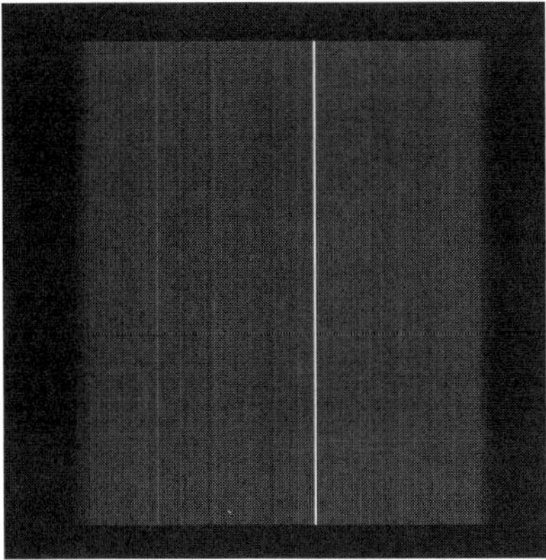

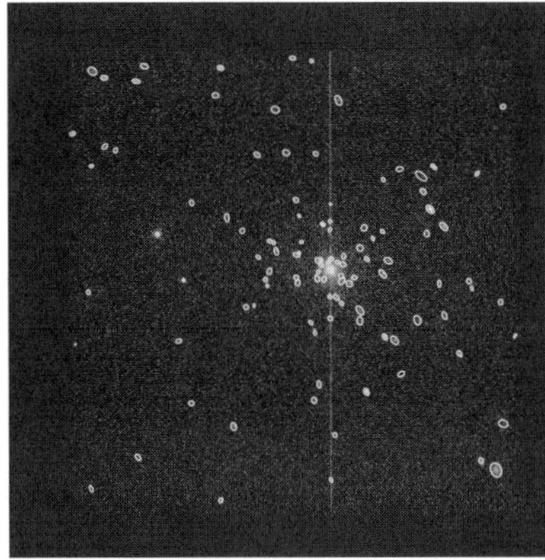

Figure 3. *a (left)*: A background + *Readout Streak* image created by taking the background + *Readout Streak* (shown in Figure 2b) and replicating it for each chip row. *b (right)*: Image of the S3 chip observation of M81 with source locations (WAVDETECT) found using the background map shown in Figure 3a. Note the lack of false sources found in the *Readout Streak*.

for the rest of the background. This a cosmetic operation that has little scientific meaning.

4. Future Work

This method holds great promise in terms of identifying and removing spurious *Readout Streak* sources. Some additional areas of study and work we are pursuing are:

- **Two-dimensional Background Images**: From the source free rows there should be additional two-dimensional information about the background. Thus, in principle we should be able to create a two-dimensional background image containing *Readout Streaks*.
- **Pileup**: In general, sources which are bright enough to have a noticeable *Readout Streak* also suffer from Pileup (multiple photons per pixel during integration). The exposure to the bright source of the pixels in the *Readout Streak* is so short that they in general should not have this problem. Thus an examination of the *Readout Streak* should give one information on how much Pileup a source is experiencing.
- **Different X-Ray Missions**: This technique can be applied to any X-ray mission which is using a CCD detector and may experience *Readout Streak* problems. The use of this method for other X-ray missions (Current: XMM, Past: ASCA, Future: ASTRO-E2) is being investigated.
- **L3 and CIAO**: We are working to make this tool able to be run in the L3 pipeline environment and also to release a version as a CIAO tool for the general user.

This work is supported by NASA contract NAS8-03060 (CXC).

Figure 4. Image of the S3 chip observation of M81 "cleaned" using the *Readout Streak* information shown in Figure 3a.

References

Freeman, P. E., Kashyap, V., Rosner, R., & Lamb, D. Q. 2002, ApJS 138, 185

ASTRONOMICAL DATA ANALYSIS SOFTWARE AND SYSTEMS XIV
ASP Conference Series, Vol. 347, 2005
P. L. Shopbell, M. C. Britton, and R. Ebert, eds.

Filtering of Signal Dependent Noise Applied to MIPS Data

M. Pesenson, M. Moshir, D. Makovoz, D. Frayer, and D. Henderson

Spitzer Science Center (SSC), California Institute of Technology

Abstract. Linear filtering and usual nonlinear median filtering are not effective for signal-dependent noise removal. We apply here an approximate decoupling of signal and noise by means of a nonlinear transform. The transform is followed by a linear filter and the corresponding inverse transform. This procedure allows us to mitigate the signal-dependent noise in the images obtained by the Multiband Imaging Photometer for *Spitzer* (MIPS), 70μm imaging band.

1. Introduction

Extraction and detection of point sources is of a paramount importance in astronomy. Thus one of the ultimate goals of processing astronomical images is to secure more reliable point source identification. In general, to achieve that it is necessary first to smooth images in order to reduce the noise. Linear filtering, or even nonlinear median filtering are not useful for signal dependent noise removal, due to the nonlinear coupling of signal and noise. Therefore, the first step would be to decouple the noise from signal, which is done by applying a nonlinear transformation. This transformation is followed by a linear filter and a nonlinear inverse transformation thus leading to the removal of the signal dependent noise (Arsenault & Denis 1983; Kotropoulos, Papas, & Pitas 2001). It is argued that since real images have different types of noise, a combination of filtering techniques should be used.

2. Homomorphic Image Processing

There are many different ways of including noise into image processing procedures, depending on how the noise is defined. The most common approach is based on the the so called additive noise model

$$I(x,y) = s(x,y) + N(x,y) \tag{1}$$

where $I(x,y)$ is the observed image, $s(x,y)$ denotes the signal and $N(x,y)$ is Gaussian white noise.

A more complicated model considers the so called multiplicative noise

$$I(x,y) = s(x,y)N(x,y) \tag{2}$$

We consider here a more general model which describes the signal-dependent noise and reduces to the aforementioned equations in more simple situations. The model is described by the following equation

$$I(x,y) = f(s) + g(s)N(x,y) \tag{3}$$

The functions f, g are nonlinear. We assume that $N(x,y)$ is uncorrelated with the signal s. Multiplicative noise is of course a special case of (3). Photoelectron noise can also be described as a special case of (3)

$$I(x,y) = \alpha s^p + \beta s^q N(x,y) \tag{4}$$

The desirable decoupling can be obtained (Arsenault & Denis 1983; Kotropoulos et al. 2001) by applying to the observed image (4) a nonlinear transformation $G(I)$ which is chosen in such a way that

$$G(I) = G[f(s) + g(s)N(x,y)] \sim V(s) + N(x,y) \tag{5}$$

Strictly speaking, the exact separation can be achieved only for multiplicative noise, when $G(s) = \gamma H(s)$, where γ is a constant. For the photoelectron noise (4) it can be done approximately by choosing $G(I)$ in the following form

$$G(I) = [(p\alpha^{-p/q} - p/q)/(\beta(p-q))]I^{(p-q)/p}, \tag{6}$$

Then the expression (5) becomes

$$G(I) \sim G(f(s)) + N(x,y) \tag{7}$$

To process the images below, the following values of the parameters were used $\alpha = \beta = 1, p = 0.75, q = 0.45$.

3. Linear Filtering of the Transformed Image and the Restoring Tranformation

The removal of the noise $N(x,y)$ here can be done by a linear filtering and the choice of the appropriate filter depends on the probability density function (PDF) of $N(x,y)$. For example, if the noise has a Gaussian distribution the simple arithmetic mean will do the job. If the noise is a mixture of positive and negative spikes, the median filters will be more effective. After linear filtering the image must be restored by the following nonlinear transform $T(G(f(s))) = s$. For the model (3) it takes the following form

$$T(z) = (z\beta(p-q)/(\alpha p))^{1/(p-q)} \tag{8}$$

4. Mosaics

A mosaic of astronomical observation request AOR3865856 is presented in Figure 1. The image was obtained by MIPS in the 70μm imaging band. The mosaic size is 0.9 deg x 0.5 deg. The effective exposure time for one AOR is 30 sec. The basic calibrated data (BCD) is from SSC pipeline version 10.0. The mosaic image is made by MOPEX (Makovoz & Khan 2005).

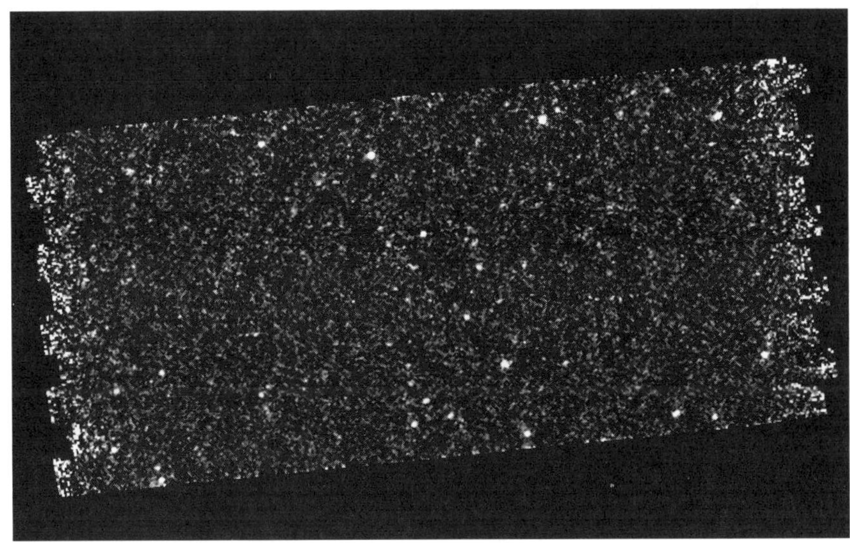

Figure 1. AOR3865856 mosaic.

Figure 2. AOR3865856 mosaic after a linear filtering.

Figure 3. AOR3865856 mosaic after the homomorphic filtering.

5. Conclusion

The signal to noise ratio (SNR) obtained by applying the homomorphic filtering is higher than SNR obtained after direct linear filtering. This leads to a more reliable point source extraction. It also indicates presence of a signal-dependent noise. In reality, the additive, multiplicative and signal-dependent noises are all present which calls for consecutive application of a few different filtering procedures. The filtering procedure applied here, together with the nonlinear denoising and segmentation methods described in Pesenson et al. (2005), are complementary to each other, which makes them especially comprehensive.

Acknowledgments. This work was carried out at the *Spitzer* Science Center, with funding from NASA under contract 1407 to the California Institute of Technology and the Jet Propulsion Laboratory.

References

Arsenault, H., & Denis, M. 1983, Image processing of signal-dependent noise, Canadian J. Phys, 61, 309

Kotropoulos, C., Papas, M., & Pitas, I. 2001, Nonlinear mean filters and their applications in image filtering and edge detection, in Nonlinear Image Processing, Ed. S. Mitra, Academic Press

Makovoz, D. & Khan, I. 2005, in ASP Conf. Ser., Vol. 347, ADASS XIV, ed. P. L. Shopbell, M. C. Britton, & R. Ebert (San Francisco: ASP), 81

Pesenson, M., Moshir, M., Makovoz, D., Frayer, D., & Henderson, D. 2005, in ASP Conf. Ser., Vol. 347, ADASS XIV, ed. P. L. Shopbell, M. C. Britton, & R. Ebert (San Francisco: ASP), 153

ASTRONOMICAL DATA ANALYSIS SOFTWARE AND SYSTEMS XIV
ASP Conference Series, Vol. 347, 2005
P. L. Shopbell, M. C. Britton, and R. Ebert, eds.

Java-Based Astronomical Software - The HERSCHEL/ PACS Common Software System as Example

Ekkehard Wieprecht

Max Planck Institut f"ur extraterrestrische Physik, Garching/Germany

R. Huygen, B. Vandenbussche, and W. de Meester

Institute of Astronomy K.U. Leuven, Leuven/Belgium

S. Guest

Rutherford Appleton Laboratory, Chilton/United Kingdom

A. de Jonge and P. Zaal

Space Research Organization Netherlands, Groningen/The Netherlands

S. Osterhage and M. Wetzstein

Max Planck Institut f"ur extraterrestrische Physik, Garching/Germany

Abstract.
ESA's Herschel Space Observatory, planned to be launched in 2007, is the first space observatory covering the full far-infrared and submillimetre wavelength range (60 - 670 microns). The Photodetector Array Camera and Spectrometer (PACS) is one of its three science instruments. It employs two Ge:Ga photoconductor arrays and two bolometer arrays to perform imaging line spectroscopy and imaging photometry in the 60 - 210 micron wavelength band. The HERSCHEL Common Software System (HCSS) forms the base of the HERSCHEL ground segment. HCSS is implemented using JAVA/Jython technology and interacts with an object oriented database. It is written in a common effort by the HERSCHEL Science Centre and the three instrument teams. The PACS Common Software System (PCSS) is based on the HCSS. Both systems were designed for a smooth transition between the different phases of the project. Instrument engineers use PCSS during instrument tests, calibration specialists for instrument characterization on ground and in orbit, and finally the observer will use it for data reduction. The design allows use of the same components for automatic processing, quick look analysis, and interactive processing. We outline some design aspects of such a highly complex system and present user experiences with PCSS, gathered during first instrument tests of PACS.

1. Introduction

The cold instrument tests of the PACS qualification model were conducted at MPE Garching/Germany. A combination of simulators and real hardware components (including a number of detectors and readout electronics) were used to qualify the functionality and complete first calibrations of the instrument.

During this test period the PACS Common Software System (based on the Herschel Common Science System - HCSS) was used to analyze the data. We describe the HCSS/PCSS components utilized, considering the technical aspects and first user experiences.

The requirements for this test period on HCSS/PCSS were rather basic :

- near real-time housekeeping data monitoring
- near real-time detector readout visualization including basic processing steps (background subtraction, etc.)
- test data management (saving, access, backup, etc.)
- offline visualization of raw and on board reduced detector data
- offline housekeeping data visualization
- interactive data manipulation and mathematical operations on housekeeping and detector data (array manipulation, functions)
- usable documentation and framework which allow instrument engineers and calibration specialists to write their own data analysis scripts in a quick and efficient way (easy scripting language, not JAVA)

Five components were used to fulfill the requirements.

- Satellite Control and Operating System (SCOS 2000)
- Quick Look Analysis (QLA)
- Housekeeping (HK) Analysis tool
- Interactive Environment inclusive scripting language and numerics
- IA Plotting and Imaging tools

Except SCOS 2000, all components are written in JAVA. Almost all applications are shared by the three instruments of HERSCHEL.

2. Satellite Control and Operating System - SCOS 2000

Although it is possible to do near real-time visualization of HK data within HCSS/PCSS, the external tool SCOS 2000 was used for commanding and HK display in near real-time. SCOS 2000 is a software system to control spacecraft and to command the instruments during ground tests. It is written in C++ and CORBA.

3. Quick Look Analysis

QLA is a GUI application for near real-time science data visualization, including basic data reduction. Ideally the QLA application is just a sequence of interactive components, working on a near real-time data stream. This concept is only partly implemented for PACS QLA as the QLA development was started before the work on the common IA Framework.

The first tests to plug in interactive JYTHON scripts in the near real-time data stream showed possible performance drawbacks. Performance was always an issue for the near real-time application. JAVA profilers, like OptimizeIt, were essential to optimise this application.

The GUI interface is simple enough for the general user. Currently only developers can contribute code to the system; the framework is complex and JAVA knowledge is required.

4. Housekeeping Analysis Tool

This tool is used to inspect the Telecommand (TC) and Telemetry (TM) Headers and Housekeeping parameters within PCSS. A complex, very flexible system, using the Mission Implementation Base, builds the fundamental base of this application (Wieprecht et al. 2003). On top of it a GUI application provides convenient access to all packets and parameters. Packet and parameter names can be browsed, parameters extracted, displayed and exported into PCSS sessions or to FITS and ASCII.

5. Interactive Development Environment and Numerics

PACS users require a command line environment for script development and data analysis. Such a scripting language must also support easy data manipulations.

Although JYTHON was used as the base for the interactive environment, basic JAVA data types like Int1d, Double3d, ... needed definition . These data types are hiding object orientation, can be stored in the database, and are conveniently usable both in JAVA programs and from the Jython command line. Mathematical functions needed implementations for all data types and dimensions.

Convenient array manipulation within HCSS/PCSS has been successfully implemented up to three dimensional data cubes. Also a library of mathematical functions and algorithms for these arrays are available. Although templates are used to generate parts of the code, the framework is very large and not easy to maintain.

Initially, the PACS user familiar with another scripting language (like IDL) had to get used to this new scripting and its data types :

- First, the possibility of using JYTHON data types and the newly defined JAVA data types caused confusion.
- Many users were not used to the case sensitivity of JYTHON.
- PACS users had to get used to the object-oriented design (data types are coming with their functions, a=b is a reference, use a=b.copy(), etc.).
- PACS users had problems using the JAVA API documentation.

But once users passed the initial problems, the easy array manipulation and mathematical functions were used successfully. Users continue to propose ways to make the system more convenient, but in general it seems to fulfill the most important requirements.

One remaining problem is the huge amount of data that PACS analysts have to handle. Especially during the early instrument tests, the instrument engineers need to analyze the raw, not reduced telemetry data. In the future it is expected that the raw data will be reduced to standard products of smaller size. The current design of the PACS interactive system assumes that all data to be analyzed are kept in memory. JAVA needs to allocate the expected virtual memory at the start of the application. Once this is done it is not possible to change the allocated virtual memory within the interactive session. In any case, if the system starts swapping, the performance drops drastically.

6. Plot and Display

To enable plotting and image visualization of HERSCHEL data, layers provide convenient access to the underlying JAVA packages: JFreeChart and JSky. Both plotting and imaging is extensively used and working well. In the beginning users had problems getting used to the object oriented syntax (Layer objects, Axis objects, etc.). Also the syntax was consider too verbose, so function wrappers had to be written.

7. Conclusions

JAVA is well suited for Interactive Analysis Frameworks. JYTHON was a good element for interactive usage of JAVA. Nevertheless a significant effort is needed to fulfill the user requirements. In principle, the IA Framework is independent of the HERSCHEL data analysis system. As the framework had to be developed first, other IA functionalities for instrument tests sometimes came too late. Therefore, the smooth transition between the phases was not completely successful.

The HCSS system is the agreed framework for all three instruments. This has many advantages for users and developers. Many users had to get used to the object orientation and new data types. Although documentation was ready in time, training and examples were essential to assist the users.

The use of external libraries saved significant development time. But the maintenance and version control of these libraries caused problems from time to time. Also there is only limited influence on the development of external packages.

The idea of re-usable components is not completely achieved due to timing between Framework and other applications. The performance of IA script components within near real-time systems has not been completely evaluated. Now the Framework is a solid base for components of other systems - like QLA.

HCSS/PCSS was successfully used for PACS instrument tests of the qualification model.

References

Wieprecht, E. et al. 2004, in ASP Conf. Ser., Vol. 314, ADASS XIII, ed. F. Ochsenbein, M. Allen, & D. Egret (San Francisco: ASP), 376

ASTRONOMICAL DATA ANALYSIS SOFTWARE AND SYSTEMS XIV
ASP Conference Series, Vol. 347, 2005
P. L. Shopbell, M. C. Britton, and R. Ebert, eds.

The SIP Convention for Representing Distortion in FITS Image Headers

David L. Shupe, Mehrdad Moshir, Jing Li,[1] David Makovoz, and Robert Narron
Spitzer Science Center, California Institute of Technology, Pasadena, CA 91125, USA

Richard N. Hook[2]
Space Telescope European Coordinating Facility, Karl-Schwarzschild-Str. 2, D-85748, Garching, Germany

Abstract. The SIP (Simple Imaging Polynomial) convention provides a convenient means for representing non-linear geometric distortion as polynomials in FITS header keywords. This scheme has been incorporated by the Spitzer Science Center (SSC) into its imaging products, and it is under consideration for wider use. The SIP keywords are read and applied by the SSC's MOPEX mosaicking software, and by several standard analysis tools, including WCSTools, SAOImage, DS9, GAIA/Starlink AST, IDL ASTROLIB, and Drizzle. We provide definitions of how these keywords are used together with standard FITS World Coordinate System (WCS) keywords to relate pixel coordinates to the sky. Some examples illustrate the application of this scheme to actual distorted data.

1. Introduction

Images from the instruments on board the *Spitzer Space Telescope* are distorted by a few percent relative to a regular sky grid. These distortions are well-represented by polynomials, and the Spitzer Science Center (SSC) has developed a convention for inserting the coefficients into the FITS headers of each Basic Calibrated Data (BCD) product. The post-BCD processing software MOPEX[3] developed by the SSC (Makovoz & Khan 2004) accounts for the distortions in regridding and coadding the images. Writing the coefficients to the FITS headers was motivated particularly by the optics of the Multiband Imaging Photometer for Spitzer (MIPS) instrument (Rieke et al. 2004)—the distortion changes with scan mirror position, and hence from one image to the next.

The development of the SIP convention proceeded in parallel with work on the World Coordinate System (WCS) standard. The first two papers in this series (Greisen & Calabretta 2002, "Paper I"; and Calabretta & Greisen 2002,

---

[1]Princeton, NJ, USA

[2]currently based at STScI, Baltimore, MD 21212, USA

[3]`http://ssc.spitzer.caltech.edu/postbcd/`

"Paper II") specifying the WCS keywords (sans distortion) have been approved by the IAU FITS Working Group. "Paper IV" will address distortion–it has been drafted[4] but is not yet final (cf. Calabretta et al. 2004). The Spitzer keywords are compliant with the first two papers, and have been influenced by discussions of Paper IV, but are distinct from the proposed keywords in Paper IV.

2. Definitions of the Distortion Keywords

The SIP convention derives its name from the four characters '-SIP' that are appended to the values of CTYPE1 and CTYPE2. These extra characters were included in early drafts of Paper IV to denote the distortion representation; however, the most recent draft has dropped this form. We chose '-SIP' to be distinct from the '-PLP' that was to be used in Paper IV for polynomials, and because it has the useful mnemonic "Simple (or Spitzer) Imaging Polynomial".

We define u, v as relative pixel coordinates with origin at CRPIX1, CRPIX2. Following Paper II, x, y are "intermediate world coordinates" in degrees, with origin at CRVAL1, CRVAL2. Let $f(u,v)$ and $g(u,v)$ be the quadratic and higher-order terms of the distortion polynomial. Then

$$\begin{pmatrix} x \\ y \end{pmatrix} = \begin{pmatrix} \texttt{CD1\_1} & \texttt{CD1\_2} \\ \texttt{CD2\_1} & \texttt{CD2\_2} \end{pmatrix} \begin{pmatrix} u + f(u,v) \\ v + g(u,v) \end{pmatrix} \tag{1}$$

We define $\texttt{A}\_p\_q$ and $\texttt{B}\_p\_q$ as the polynomial coefficients for polynomial terms $u^p v^q$. Then

$$f(u,v) = \sum_{p,q} \texttt{A}\_p\_q u^p v^q, \qquad p + q \leq \texttt{A\_ORDER}, \tag{2}$$

$$g(u,v) = \sum_{p,q} \texttt{B}\_p\_q u^p v^q, \qquad p + q \leq \texttt{B\_ORDER}. \tag{3}$$

For example, for a third-order polynomial,

$$f(u,v) = \texttt{A\_2\_0}u^2 + \texttt{A\_0\_2}v^2 + \texttt{A\_1\_1}uv + \texttt{A\_2\_1}u^2v + \texttt{A\_1\_2}uv^2 + \texttt{A\_3\_0}u^3 + \texttt{A\_0\_3}v^3$$

The $\texttt{CD}i\_j$ keywords encode skew as well as rotation and scaling. The CD-matrix values together with the higher-order distortion polynomials, as in Equations 1, 2, and 3, define a unique transformation from pixel coordinates to the plane-of-projection.

For Spitzer, we also provide polynomials for the reverse transformation, for fast inversion. Corrected pixel coordinates U, V are found from

$$\begin{pmatrix} U \\ V \end{pmatrix} = CD^{-1} \begin{pmatrix} x \\ y \end{pmatrix} \tag{4}$$

then the original pixel coordinates are computed by

$$u = U + F(U,V) = U + \sum_{p,q} \texttt{AP}\_p\_q U^p V^q, \qquad p + q \leq \texttt{AP\_ORDER}, \tag{5}$$

$$v = V + G(U,V) = V + \sum_{p,q} \texttt{BP}\_p\_q U^p V^q, \qquad p + q \leq \texttt{BP\_ORDER}. \tag{6}$$

[4] http://www.atnf.csiro.au/people/mcalabre/WCS/index.html

3. Example: Spitzer-IRAC Channel 4

We take as an example the distortion of the Spitzer Infrared Array Camera (IRAC) instrument (Fazio et al. 2004), which is characterized by quadratic coefficients. Polynomial distortion of this form (plus linear terms) was fit to data taken during In-Orbit Checkout (see Spitzer Observer's Manual). The linear terms are folded into the CDi_j. An excerpt from an actual BCD header produced by the SSC pipeline (Moshir et al. 2004) for IRAC Channel 4 is shown below.

```
CTYPE1   = 'RA---TAN-SIP'          / RA---TAN with distortion
CTYPE2   = 'DEC--TAN-SIP'          / DEC--TAN with distortion
CRVAL1   =     248.588845714283 / [deg] RA at CRPIX1,CRPIX2
CRVAL2   =    -46.7664545714299 / [deg] DEC at CRPIX1,CRPIX2
CD1_1    = 3.79787003748175E-05
CD1_2    = 0.000336011759341397
CD2_1    = 0.000337467809645458
CD2_2    = -3.6366440945003E-05
CRPIX1   =                 128.
CRPIX2   =                 128.
A_ORDER  =                    2 / polynomial order, axis 1
A_0_2    =            1.569E-05 / distortion coefficient
A_1_1    =            5.232E-05 / distortion coefficient
A_2_0    =             3.31E-05 / distortion coefficient
B_ORDER  =                    2 / polynomial order, axis 2
B_0_2    =            4.172E-05 / distortion coefficient
B_1_1    =            2.213E-05 / distortion coefficient
B_2_0    =           -9.819E-07 / distortion coefficient
AP_ORDER=                     2 / polynomial order, axis 1
AP_0_1   =            5.677E-05 / distortion coefficient
AP_0_2   =           -1.569E-05 / distortion coefficient
AP_1_0   =            5.871E-05 / distortion coefficient
AP_1_1   =           -5.231E-05 / distortion coefficient
AP_2_0   =           -3.309E-05 / distortion coefficient
BP_ORDER=                     2 / polynomial order, axis 2
BP_0_1   =            4.432E-05 / distortion coefficient
BP_0_2   =           -4.172E-05 / distortion coefficient
BP_1_0   =            2.091E-05 / distortion coefficient
BP_1_1   =           -2.213E-05 / distortion coefficient
BP_2_0   =            9.814E-07 / distortion coefficient
```

In this case, the reverse coefficients have the opposite sign and roughly the same absolute values as the corresponding forward coefficients. However, this is not true for some more distorted fields of view, so the Spitzer headers retain the reverse coefficients in general.

The SSC has developed library routines to implement this coefficient naming scheme. The functions key off the extended `CTYPEn`. The order in which the keywords are displayed in the example is the order in which the software searches for them and is the most efficient for lookups using CFITSIO.

4. Software that Reads and Applies the Coefficients

The SSC's mosaicking package MOPEX (Makovoz & Khan 2004) applies the SIP distortion coefficients in the SSC pipelines. Support has also been added to IPAC's Skyview display program[5]. Doug Mink has implemented Spitzer distortion support in his WCS routines[6]. SAOimage and DS9 use these routines and hence automatically handle the SIP distortions. The Montage software[7] (Laity et al. 2004) also uses Mink's routines and applies the coefficients. Support in the GAIA viewer has been added via David Berry's ASTLIB. Wayne Landsman has added support to the IDL ASTROLIB. The Drizzle software (Fruchter & Hook 2002) has also recently been modified to read these coefficients.

5. SIP for Hubble

Of the active cameras on board the Hubble Space Telescope, the distortion is largest by far for the Wide Field Channel (WFC) of the Advanced Camera for Surveys (ACS) where it amounts to more than fifty pixels at the corner of the image in addition to an even larger (linear) skew term.

The image distortion for Hubble cameras is currently characterized by a FITS table known as the Image Distortion Correction Table (IDCTAB) which includes information about the positioning of the instrument aperture in the telescope focal plane as well as the distortion polynomial coefficients. Software has been developed that will combine the IDCTAB information with the normal information from the telescope's pointing control software to write out a header which makes the header WCS keywords fully compatible with the table values and also populates the SIP-keywords (or at least the most important ones). An example of the resultant header is given in Table 1.

Currently the writing of these SIP keywords is an unsupported feature for Hubble data. However, it would be very easy to formally include such headers in future to provide users with a full, self-describing distortion model without the need for access to external files in non-standard formats. The software to read the coefficients and apply them to remove image distortion already exists within the standard Hubble data processing tools.

Acknowledgments. We are grateful to Mark Calabretta for significant comments and suggestions, and Jane Morrison for discussions of MIPS distortions and the CD matrix. We thank Doug Mink, Wayne Landsman, David Berry, and Booth Hartley for implementing SIP in their software. The work carried out at the SSC was funded by NASA under contract 1407 to the California Institute of Technology and the Jet Propulsion Laboratory.

---

[5] `http://www.ipac.caltech.edu/Skyview/`

[6] `http://tdc-www.harvard.edu/software/wcstools`

[7] `http://montage.ipac.caltech.edu`

Table 1. SIP coefficients for the Hubble ACS Wide Field Channel.

Keyword	Value	Keyword	Value
CTYPE1	'RA---TAN-SIP'	CTYPE2	'DEC--TAN-SIP'
CRPIX1	2048.0	CRPIX2	1024.0
CRVAL1	5.6260667398471	CRVAL2	-72.076963036772
CD1_1	-7.8481866550866E-06	CD2_1	1.1406694624771E-05
CD1_2	1.0939720432379E-05	CD2_2	8.6942510845452E-06
A_0_2	2.1634068532689E-06	B_0_2	-7.2299995118730E-06
A_1_1	-5.194753640575E-06	B_1_1	6.1778338717084E-06
A_2_0	8.543473309812E-06	B_2_0	-1.7442694174934E-06
A_0_3	1.0622437604068E-11	B_0_3	-4.2102920235938E-10
A_1_2	-5.2797808038221E-10	B_1_2	-6.7603466821178E-11
A_2_1	-4.4012735467525E-11	B_2_1	-5.1333879897858E-10
A_3_0	-4.7518233007536E-10	B_3_0	8.5722142612681E-11
A_0_4	1.4075878614807E-14	B_0_4	6.5531313110898E-16
A_1_3	-1.9317154005522E-14	B_1_3	1.3892905568706E-14
A_2_2	3.767898933666E-14	B_2_2	-2.9648166208490E-14
A_3_1	5.0860953083043E-15	B_3_1	-2.0749495718513E-15
A_4_0	2.5776347115304E-14	B_4_0	-1.812610418272E-14
A_ORDER	4	B_ORDER	4

References

Calabretta, M.R., & Greisen, E.W. 2002, A&A, 395, 1077 (Paper II)

Calabretta, M.R., Valdes, F., Greisen, E.W., & Allen, S.L. 2004, in ASP Conf. Ser., Vol. 314, ADASS XIII, ed. F. Ochsenbein, M. Allen, & D. Egret (San Francisco: ASP), 551

Fazio, G., et al. 2004, ApJS, 154, 10

Fruchter, A.S. & Hook, R.N. 2002, PASP, 114, 144

Greisen, E.W., & Calabretta, M.R. 2002, A&A, 395, 1061 (Paper I)

Laity, A.C., Anagnostou, N., Berriman, B., Good, J.C., Jacob, J.C., & Katz, D.S. 2004, in ASP Conf. Ser., Vol. 347, ADASS XIV, ed. P. L. Shopbell, M. C. Britton, & R. Ebert (San Francisco: ASP), 34

Makovoz, D., & Khan, I. 2004, in ASP Conf. Ser., Vol. 347, ADASS XIV, ed. P. L. Shopbell, M. C. Britton, & R. Ebert (San Francisco: ASP), 81

Moshir, M. 2004, in ASP Conf. Ser., Vol. 347, ADASS XIV, ed. P. L. Shopbell, M. C. Britton, & R. Ebert (San Francisco: ASP), 434

Rieke, G., et al. 2004, ApJS, 154, 25

Part IX
Ground-based Observatories

ASTRONOMICAL DATA ANALYSIS SOFTWARE AND SYSTEMS XIV
ASP Conference Series, Vol. 347, 2005
P. L. Shopbell, M. C. Britton, and R. Ebert, eds.

FITS World Coordinates for DEIMOS Slitmasks

Steven L. Allen

UCO/Lick Observatory, University of California, Santa Cruz, CA 95064

Abstract. Efforts are underway to supply more World Coordinate System meta-data in the FITS files produced by DEIMOS. The goal is to improve the utility of the images, both for the sake of real-time display to the astronomers and for archival purposes. This is a progress report on the acquisition and reduction of data aimed at providing metrology for the instrument.

1. Introduction

The DEep Imaging Multi-Object Spectrograph (DEIMOS) installed at the Keck II telescope began producing direct images and multi-slit spectra in 2002 (Faber et al. 2003). Pixel data from the mosaic of 8 CCDs are stored in multi-extension FITS (MEF) files compatible with the NOAO/IRAF mosaic (Valdes 1997). Each IMAGE extension contains keywords for multiple World Coordinate Systems (WCSs) that conform to the recently adopted FITS WCS standards (Greisen & Calabretta 2002; Calabretta & Greisen 2002). Allen (2003) provided an overview of the content of the DEIMOS FITS files and their world coordinate systems.

Currently the WCS keywords in DEIMOS images only describe coordinate systems related to the pixel layout of the mosaic detector. There have been no WCS keywords for finding coordinates on the metal of the multi-object slit masks or on the sky. Consequently, the real-time display cannot indicate the identity of the object(s) within, nor the wavelengths along, the spectrum produced by any given slitlet. The MEF files from DEIMOS contain the slit and object information in FITS tables which accompany the images. Those tables are used during data reduction, but it would be desirable for the information to be available instantly as part of the real-time display for DEIMOS.

2. The WCS Calibration Mask: Creation and Use

The mapping from the sky to the metal of DEIMOS slitmasks was necessarily verified during installation of the instrument. The FITS WCS keywords, however, are couched in terms of the mapping from the pixels in the CCD mosaic to the sky and to the slitmask.

The illuminated area of DEIMOS slitmasks is 733 mm (16.7 arcmin) by 223 mm (5.3 arcmin). Typical slitmasks for DEIMOS contain up to 130 slitlets. Typical slitlets are 5 mm (7 arcsec) long by 0.5 mm (0.7 arcsec) wide.

Figure 1 shows the design of the calibration slitmask. The illuminated area of the mask has a semi-regular grid of about 700 holes. The irregularity permits unique identification of each hole. No special arrangements were made for the

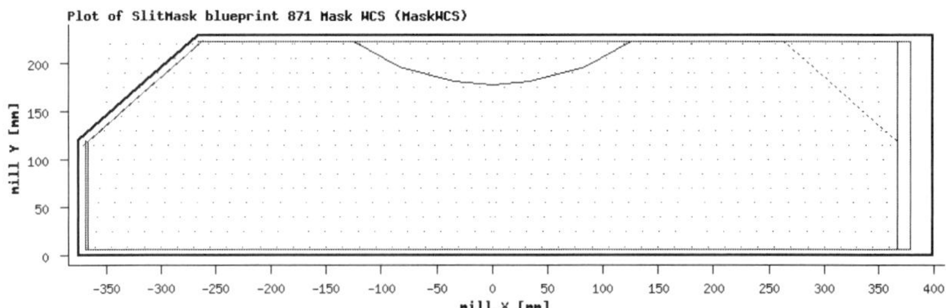

Figure 1. The WCS Calibration Mask

```
CRPIX1M =      5229.18103595138 / ref pixel along FITS axis j
CRPIX2M =      1805.80497303895 / ref pixel along FITS axis j
CRVAL2M =      69.5160901692975 / coord val on WCS axis i @ refpix
CRVAL1M =      86.5703178606477 / coord val on WCS axis i @ refpix
CD2_1M  =   1.40745050410589E-4 / CTM i_j from pixel to WCS
CD2_2M  =   -0.0863028991914294 / CTM i_j from pixel to WCS
CD1_1M  =    0.0859505852594443 / CTM i_j from pixel to WCS
CD1_2M  =   1.41321968438173E-4 / CTM i_j from pixel to WCS
CTYPE2M = 'MASK_Y  '            / coord/proj type for WCS axis i
CTYPE1M = 'MASK_X  '            / coord/proj type for WCS axis i
CUNIT1M = 'mm      '            / physical unit for WCS axis i
CUNIT2M = 'mm      '            / physical unit for WCS axis i
WCSNAMEM= 'mask    '            / DEIMOS slitmask coordinates
CRDER1M =                   0.2 / random error for WCS axis i
CRDER2M =                   0.2 / random error for WCS axis i
CSYER1M =                   0.5 / systematic error WCS axis i
CSYER2M =                   0.5 / systematic error WCS axis i
```

Figure 2. typical FITS header cards for a mask WCS

production of this WCS mask. The mask was cut using the 0.015 inch diameter milling tool that is used for almost all other slitmask manufacturing.

DEIMOS is located at a Nasmyth focus; the entire instrument rotates around a horizontal axis. We loaded the calibration slitmask into DEIMOS and observed at a variety of different rotation angles in order to determine the amount of any systematic shift caused by flexure. We obtained both direct images and spectrally dispersed images using single-amplifier and dual-amplifier readout modes for the CCDs in the mosaic.

3. Initial results of the WCS Fitting

We processed the MEF files from DEIMOS using various tasks from the `mscred` external package for IRAF (Valdes 1998). This provided WCS keywords which map from FITS pixels to the milling coordinate system on the metal of the slit-

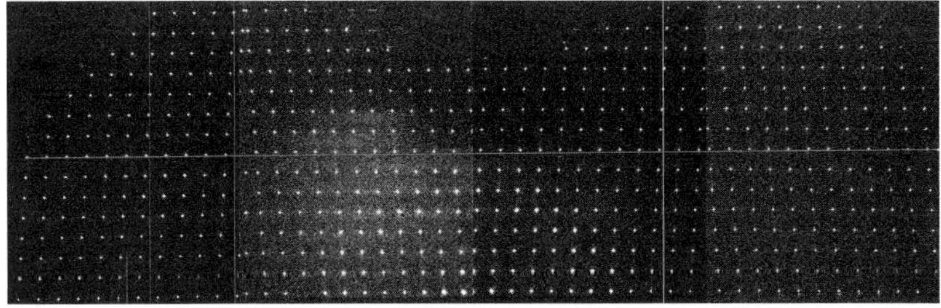

Figure 3. 4-amplifier direct image in ds9 using PANE coordinates

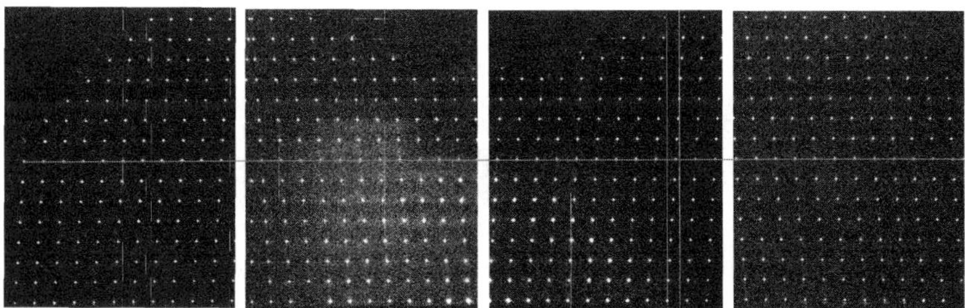

Figure 4. 4-amplifier direct image in ds9 using MASK coordinates

mask. In addition to the results produced by `mscred`, other standard keywords were inserted into the final mask WCSs. Figure 2 shows a mask WCS extracted from a typical `IMAGE` extension header.

In this example the WCS from FITS pixels to mm on the mask is a purely linear transformation; there is no attempt to map the distortions in the DEIMOS optical system. Nevertheless, the RMS residual error over any single CCD is less than 0.2 mm. The errors within the central CCDs are much smaller. Therefore a linear fit is sufficient for unambiguous identification of each slitlet.

Figure 3 shows the calibration mask using the default layout of the real-time display for a DEIMOS image. The default layout is optimized for speed; no attempt is made to depict the subtle aspects of the geometry of the CCD mosaic.

Figure 4 shows the calibration mask as displayed by ds9 using the WCS results from the fit by `mscred`. In this case the horizontal rows of holes in the mask are straight and parallel.

4. Spectra through the Calibration Mask

Figure 5 shows a portion of the mosaic with a spectral exposure. Each of the holes in the WCS calibration mask produces one spectrum. In this case the spectra show continuum and lines from comparison lamps. One goal of this effort is to identify the object and wavelength in real-time as the cursor moves over such images.

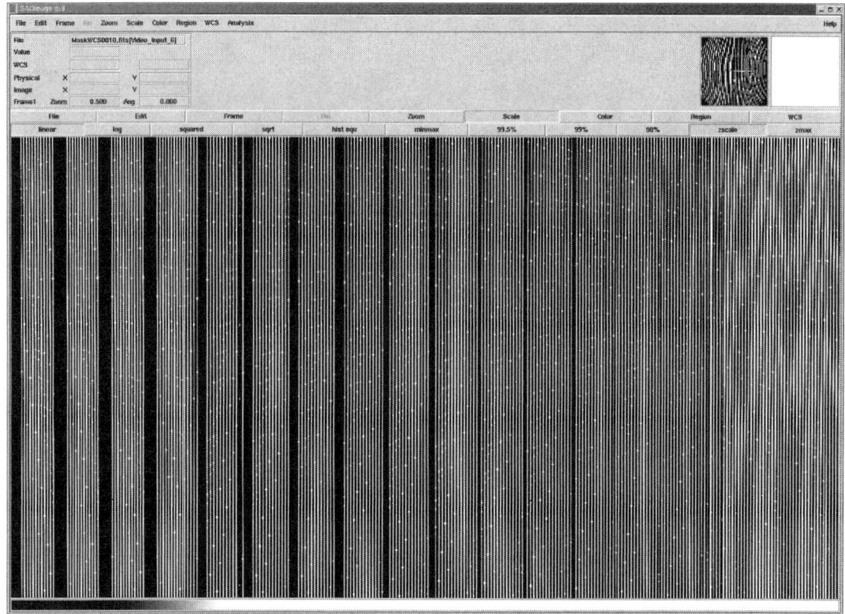

Figure 5. 8-amplifier spectral image in ds9 using PANE coordinates

5. Ongoing Effort

It is straightforward to modify the DEIMOS data acquisition system to produce FITS files with additional world coordinate systems. After obtaining a final version of the WCS information from the exposures described here we intend to add the slit mask coordinate system to the DEIMOS images.

The current draft of FITS WCS Paper III (Representations of spectral coordinates in FITS) is available via the web site of its principal author, Greisen[1]. The text of Paper III describes FITS WCS for cases where there is only a single spectrum. In order to provide the desired quick-look capabilities we expect to append additional FITS tables to the DEIMOS files. Within the additional tables we will use the vocabulary of FITS WCS Paper III to define coordinates for each of the slitlets in a DEIMOS slitmask. We envision making further enhancements to ds9, as described by Clarke (2005), that will ascertain which slitlet is under the cursor and present the object and coordinate information to the observers.

We will also investigate how the non-linearities in the DEIMOS optical system might be handled by the algorithms described in the current draft of WCS Paper IV on distortions in FITS coordinates. The current draft FITS WCS Paper

---

[1] `http://www.aoc.nrao.edu/~egreisen/`

IV (Representations of distortions in FITS) is available via the web site of its principal author, Calabretta[2].

Acknowledgments. Robert Kibrick of UCO/Lick Observatory and Greg Wirth of Keck Observatory provided support for obtaining the observations through the WCS calibration mask. Bill Joye enhanced the SAOImage ds9 display program in response to the mask WCSs in these MEF files.

Further information about DEIMOS is available online from UCO/Lick Observatory[3] and Keck Observatory[4].

The content of the poster for this paper presented at ADASS XIV is online at `http://spg.ucolick.org/Docs/ADASS/2004/Allen/`.

References

Allen, S.L., et al. 2003, in ASP Conf. Ser., Vol. 295, ADASS XII, ed. H. E. Payne, R. I. Jedrzejewski, & R. N. Hook (San Francisco: ASP), 295

Calabretta, M.R. & Greisen, E.W. 2002, A&A, 375, 1077

Clarke, D.A., et al. 2005, in ASP Conf. Ser., Vol. 347, ADASS XIV, ed. P. L. Shopbell, M. C. Britton, & R. Ebert (San Francisco: ASP), 114

Faber, S.M., et al. 2003, Proceedings of the SPIE, 4841, 1657

Greisen, E.W. & Calabretta, M.R. 2002, A&A, 375, 1061

Valdes, F. 1997, in ASP Conf. Ser., Vol. 125, ADASS VI, ed. G. Hunt & H. E. Payne (San Francisco: ASP), 459

Valdes, F.G. 1998, in ASP Conf. Ser., Vol. 145, ADASS VII, ed. R. Albrecht, R. N. Hook, & H. A. Bushouse (San Francisco: ASP), 53

[2]`http://www.atnf.csiro.au/people/mcalabre/`

[3]`http://deimos.ucolick.org/`

[4]`http://www2.keck.hawaii.edu/inst/deimos/`

ASTRONOMICAL DATA ANALYSIS SOFTWARE AND SYSTEMS XIV
ASP Conference Series, Vol. 347, 2005
P. L. Shopbell, M. C. Britton, and R. Ebert, eds.

Architecture of the WFCAM/eSTAR Transient Object Detection Agent

Brad Cavanagh

Joint Astronomy Centre, 660 N. A'ohōkū Place, University Park, Hilo, HI 96720, USA

Alasdair Allan

School of Physics, University of Exeter, Stocker Road, Exeter EX4 4QL, U.K.

Tim Jenness, Frossie Economou, Paul Hirst, and Andy Adamson

Joint Astronomy Centre

Tim Naylor

University of Exeter

Abstract. The eSTAR Project and the Joint Astronomy Centre are developing an agent for automated transient and moving object detection, identification, and alerts. The agent is being developed for the United Kingdom Infrared Telescope's Wide Field Camera, and will use output from a data reduction pipeline to cross-correlate results with known objects, taken from pre-existing survey databases and a running database of objects previously observed with WFCAM. The agent will be able to identify transient and moving objects and alert astronomers depending on their specific interests. This paper will describe the architecture of the agent and its integration with the eSTAR Project, systems at the JAC, and external resources.

1. Introduction to WFCAM

In February 2005 the United Kingdom Infrared Telescope (UKIRT) will begin scientific observations using the Wide Field Camera, or WFCAM for short (Henry et al. 2003). This near-infrared instrument has four 2048×2048 arrays spaced by 94% in the focal plane such that four separately pointed observations can be tiled together to cover a filled square of sky covering 0.75 degrees. To compensate for the large 0.4-arcsecond pixel scale and fully sample the sub-arcsecond seeing typically observed at UKIRT, a microstepping mode will be used to achieve an interleaved image of 0.2-arcsecond or 0.13-arcsecond pixel scale. This microstepping mode creates four or nine separate observations at each jitter position that are then interleaved together to form a higher-resolution image.

WFCAM has been specially designed for large-scale survey observations, with the UKIRT Infrared Deep Sky Survey (UKIDSS for short, Warren 2003) taking the majority of WFCAM's observing time. Approximately 250 GB of data will be produced per night and reduced using a data reduction pipeline for data quality assessment (Cavanagh et al. 2003; Lewis et al. 2004). The raw data will be transferred to the Cambridge Astronomical Survey Unit in the UK for full off-line processing. The results from the off-line processing will be transferred to the WFCAM Science Archive operated by the Wide Field Astronomy Unit in Edinburgh (Read et al. 2004). The WFCAM Science Archive will be the primary data source for both UKIDSS and regular PATT observing results.

2. Introduction to eSTAR

The eSTAR Project[1] is a program to develop an intelligent robotic telescope network (Allan et al. 2004). Intelligent agents reside on a user's local machine and can request observations from telescopes on the eSTAR network. These user agents can also receive results of observations on the network to analyse and potentially follow up on interesting results. The user agents will communicate with discovery nodes on the network via Grid middleware. These discovery nodes are a collection of sub-systems that can receive observation requests from an intelligent agent, and through a series of interactions produce some astronomical data which the agent requested. A discovery node is typically a telescope, however, it may also be a middleware wrapper around an online astronomical database or other data-serving entity. The transient object detection agent described in this paper is one example of a discovery node.

3. Transient Object Detection Agent Overview

Since science results will only be accessible via the WFCAM Science Archive, and data can take some time to ship to the UK from Hawaii for reduction, there is the potential that time-sensitive scientific results would be lost. For this reason a real-time data mining transient object detection agent was proposed. This agent will examine data from each of the four summit data reduction pipelines (one for each WFCAM camera), identify transient and moving objects, and alert astronomers depending on their specific interests.

The overall architecture of the system, shown in Figure 1, has several processes. The correlation daemons correlate data from each of the four data reduction pipelines and the database of previously observed objects. The database web service handles interactions between the correlation daemons, the data mining agent, and the alert agent. The data mining agent determines what type of transient object was found, correlating previous data through interaction with the Grid. The alert agent alerts astronomers depending on the type of object found and that astronomer's specific interests.

[1]`http://www.estar.org.uk`

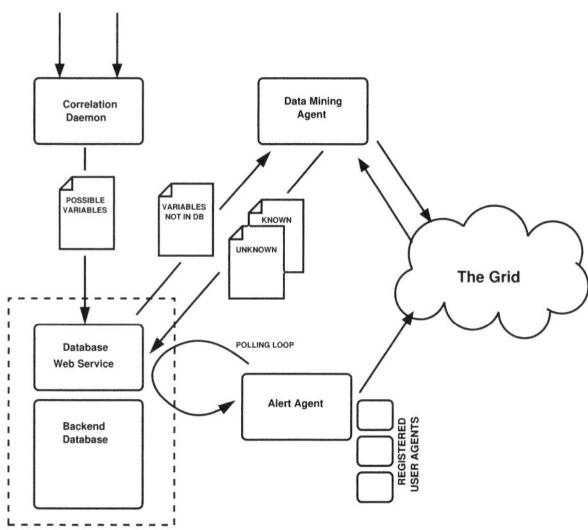

Figure 1. Data flow diagram for the WFCAM/eSTAR Transient Object Detection Agent described in this paper.

4. Agent Details

4.1. Correlation Daemon

The correlation daemon correlates results from multiple sources, including 2MASS, a database of objects previously observed with WFCAM, and the microstepped observations described in Section 1.

WFCAM is a more sensitive instrument than those used for 2MASS, and as such the overlap of good 2MASS detections which are not saturated in WFCAM will be small, so the computational overhead of this correlation is likely to be minimal. Therefore this correlation may not be necessary for the agent. Infrastructure for putting this into place exists using our catalogue infrastructure, so the overhead for implementing 2MASS correlation will be small.

WFCAM will be performing two deep surveys, the Deep Extra-Galactic Survey (DXS) and the Ultra-Deep Survey (UDS). On the first occasion a deep survey field is imaged a reference point-source catalogue is generated. On subsequent observations the daemon will correlate the current point source catalogue with the reference catalogue.

As WFCAM takes multiple frames of the same field for interleaving, the corresponding frame catalogues can be correlated to find fast-moving or quickly-varying objects. These frames are of particular interest as only the interleaved frames will be used for scientific study at CASU, so information from the individual microstepped frames will be lost.

4.2. Database and Web Services

The backend database is the core component of the WFCAM object detection agent, as all of the other components exist to populate or query the database.

Web services sit on top of the backend database and handle messages to and from the other components in the system. These messages are passed via RPC SOAP calls which are authenticated using HTTP cookies. An eSTAR mechanism exists to standardise this procedure throughout the agent.

After receiving the SOAP message from the correlation daemon the DB web service generates a unique ID for that list of candidate variables. This context stays with the targets and is used in asynchronous calls to the data mining agent, which forms the basis of the contextual web services.

The database web service adds all sources not previously known to the database. Sources that are known have the observation data attached to the existing source to build a lightcurve of that source.

The DB web service will also pass the list of objects not found in the database to the data mining agent for identification. The SOAP message passes the contextual information, host and port to which the data mining agent should respond, along with a list of the objects as a wrapped VOTable document. In this case only a simple list of coordinates is required.

4.3. Data Mining Agent

The data mining agent is responsible for locating objects passed to it by querying pre-existing databases available through the Grid. Examples of databases and data sources used by the data mining agent are SIMBAD and the Minor Planet Center's Minor Planet Checker.

Candidates are identified as known sources if that candidate is recognised by one of these data sources. The tolerance radius will be determined by the accuracy of SIMBAD and MPC positions and by the astrometric accuracy of the WFCAM survey.

The data mining agent will return two serialised VOTable documents, one for known sources and the other for unknown sources, along with the contextual ID string. Both known and unknown sources are flagged as such in the backend database.

4.4. Alert Agent

The alert agent is responsible for alerting astronomers about new results. The astronomer's user agent will register interest in certain types of objects with the alert agent. The alert agent polls the database web service for objects of these types and notifies the appropriate user agents when new observations of these objects are found. The user agent, operating under partial-plan autonomy, may independently initiate follow-up observations or contact the astronomer, depending on the wishes of the astronomer and the user agent's goals.

4.5. Database Portal

The backend database facilitates queries and presents a browsable interface. The portal will allow direct access to useful information, such as a lightcurve of the object built up from multiple WFCAM passes, a thumbnail finding chart, and additional information about the object obtained from SIMBAD and other sources.

5. Early Results

During early commissioning of WFCAM in November 2004, one field in the DXS was imaged. In this field the MPC's Minor Planet Checker predicted the presence of two minor planets. One of these, 2000 SB277, a minor planet with V magnitude 19.3, was easily found by eye in Y, J, H, and K filters[2], demonstrating that WFCAM can be used for discovering or recovering minor planets. With the WFCAM/eSTAR agent infrastructure in place an astronomer would be notified of this recovery minutes after the observations were taken.

6. Summary

The WFCAM/eSTAR transient object detection agent takes advantage of the latest generation of infrared survey instruments with WFCAM. Variable and transient objects are first detected through cross-correlation. These objects are then stored in the backend database, where web services deliver unknown variables to the data mining agent, which queries pre-existing databases on the Grid. Known and unknown variables are returned by the data mining agent and updated in the backend database. An alert agent periodically polls the backend database for objects that are of interest to specific astronomers, and if any of those objects are found follow-up observations can be requested and the astronomer can be notified.

References

Allan, A., Naylor, T., Steele, I., Jenness, T., Cavanagh, B., Saunders, E., Adamson, A. J., Etherton, J., & Mottram, C. 2004, in Proceedings of the SPIE, in press

Cavanagh, B., Hirst, P., Jenness, T., Economou, F., Currie, M. J., Todd, S. D., & Ryder, S. D. 2003, in ASP Conf. Ser., Vol. 295, ADASS XII, ed. H. E. Payne, R. I. Jedrzejewski, & R. N. Hook (San Francisco: ASP), 237

Henry, D. M., Casali, M. M., Mongomery, D., Burch, K., Laidlaw, K., Ives, D. J., Vick, A. J. A., Bridger, A., Lunney, D., Adamson, A. J., Rees, N. P., Chylek, T., & Chuter, T. C. 2003, in Proceedings of the SPIE, Vol. 4841, ed. Iye, M. & Moorwood, A. F. M., 63

Lewis, J. R., Irwin, M. J., Hodgkin, S. T., Bunclark, P. S., Evans, D. W., & McMahon, R. G. 2005, in ASP Conf. Ser., Vol. 347, ADASS XIV, ed. P. L. Shopbell, M. C. Britton, & R. Ebert (San Francisco: ASP), 599

Read, M., Hambly, N., Sutorius, E., Mann, B., & Bond, I. 2005, in ASP Conf. Ser., Vol. 347, ADASS XIV, ed. P. L. Shopbell, M. C. Britton, & R. Ebert (San Francisco: ASP), 669

Warren, S. 2002, in Proceedings of the SPIE, Vol. 4836, ed. Tyson, J. A.. & Wolff, S., 313

[2]See http://www.jach.hawaii.edu/software/estar/2000SB277.gif for an animated GIF of the asteroid in the DXS field.

ASTRONOMICAL DATA ANALYSIS SOFTWARE AND SYSTEMS XIV
ASP Conference Series, Vol. 347, 2005
P. L. Shopbell, M. C. Britton, and R. Ebert, eds.

The Visible–Light Magnetograph at the Big Bear Solar Observatory: Hardware and Software

Sergiy Shumko, Volodymyr Abramenko, Carsten Denker, Philip Goode, Alexandra Tritschler, and John Varsik

Big Bear Solar Observatory, New Jersey Institute of Technology, 40386 North Shore Ln, Big Bear City, CA 92314, USA

Abstract. In this paper we report about the current status of the control and acquisition software package developed to control the visible-light imaging magnetograph (VIM) system at the Big Bear Solar Observatory (BBSO). The instrument is designed to perform high-spatial and high-temporal observations of the solar photosphere and chromosphere utilizing the remodeled Coudé-feed of the 65 cm vacuum telescope.

1. Introduction

The visible-light magnetograph (VIM) is a Fabry-Pérot based filtergraph instrument in a telecentric configuration, it will expand the capabilities for measuring solar magnetic fields at BBSO (Denker et al. 2002, Shumko et al. 2002). VIM is designed to work with a combination of a narrow-band prefilter and a single Fabry-Pérot interferometer (FPI). Standard observations will be performed in the Zeeman sensitive neutral Iron lines at 630.10 nm and 630.25 nm respectively. Modifications of the optical setup will also allow for observations in other spectral lines in order to probe the solar atmosphere from the photosphere to the chromosphere. The two-dimensional filtergraph will replace the Digital Vector Magnetograph (Spirock et al. 2001), attached to the 25 cm telescope, after the New Solar Telescope is operational. VIM is currently operating in a spectrometer mode. The integration of an adequate polarization optics (calibration unit, modulation package, analyzer) and the development of the corresponding control mechanisms is a future task.

2. Hardware

The key parts of the instrument are formed by (a) the SIFAM (formerly Queensgate) servo-stabilized FPI system and (b) a high-speed DALSA CCD camera used for image acquisition. Both parts, the FPI system and the camera are controlled and synchronized via a single processor PC (2.8 GHz INTEL CPU) operated under Windows XP. The user operates the whole system via a graphical user interface (GUI). Three serial ports (RS-232) are used to control the external hardware. Two parallel ports, currently not in use, are available for additional hardware. Figure 1 visualizes schematically the VIM remote control system.

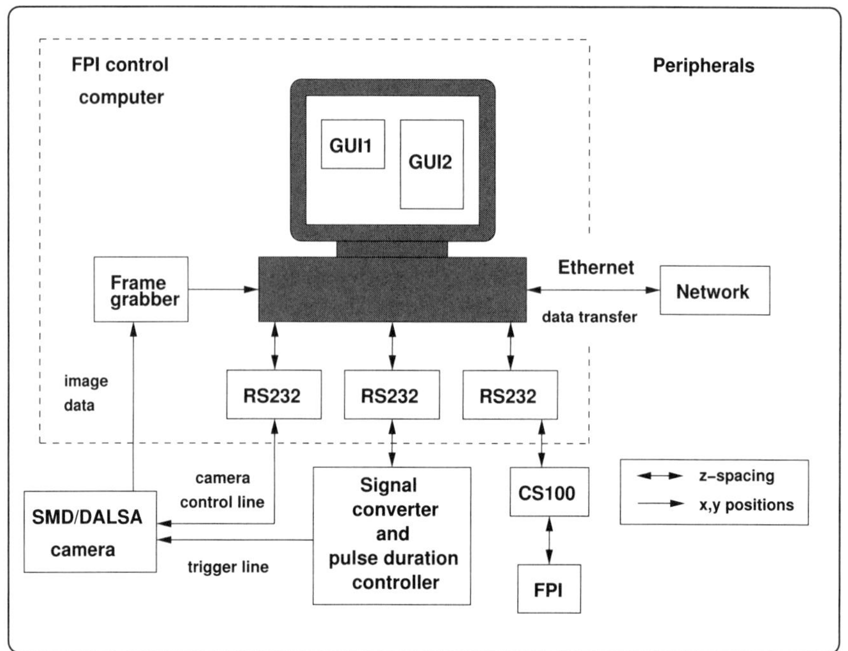

Figure 1. Schematic control and acquisition structure of the instrument.

The detector is a 1024×1024 pixel high–speed DALSA camera with a 12 bit digitization. The pixel size is 14×14 μm. In the full resolution mode (1k×1k) the maximum frame rate is 60 fps or 110 fps in the 2×2 binning mode. Images are acquired via an IC-PCI frame grabber manufactured by Coreco Imaging. Due to limitations of the frame grabber 12 bit are not supported in the full resolution mode. Instead, the 2×2 binning mode is used to obtain 12 bit images. Camera parameters, such as binning, gain, exposure time (four factory preset values), frame rate (five factory preset values), and the synchronization mode are set through a serial port. The camera features also a TTL *trigger-in* synchronization input which is used to control the exposure time and to synchronize the camera with the etalon.

Due to an inability of the camera to set a custom exposure time and the instability of the computer produced pulses, which can be up to 5 ms off we had to develop an external controller, the *signal converter*. The signal converter communicates with the computer via a serial port. The control computer can set exposure times between 3 ms and 850 ms with a step increase of approximately 3 ms and a precision in the range of 1 μs. After the exposure time is set, the computer starts the integration by sending a signal to the signal converter which sets a TTL signal for the camera. When the preset exposure time is ended, the converter drops the TTL signal to the camera to finish the exposure and sends a signal back to the computer. If the exposure time exceeds 850 ms, the computer is used to produce an exposure pulse and the signal converter just converts the pulse, i.e. the converter does not change duration of the pulse, into a TTL pulse and resends it to the camera.

The SIFAM (former Queensgate) FPI system comprises an ET-series II fused silica, ambient temperature etalon and the CS100 control unit. The CS100 controller operates the FPI, stabilizes the etalon spacing and parallelism and serves as the interface to the control software. The CS100 is a three channel controller using capacitance micrometers and Piezo actuators (incorporated into the etalon) to monitor and correct plate spacing and parallelism. The controller has both manual and remote controls via the RS-232 serial port. The remote control has some limitations, e.g. only the z-spacing value, i.e., the plate separation, can be read back. The plate parallelism (x- and y-channel) can be set remotely, but the current values cannot be obtained. At the moment, we only set the z-spacing remotely.

3. The Software

The control software has to perform and synchronize different tasks: the communication with the Telescope Control Computer (TCC), the acquisition and storage of data, setting the cavity spacing of the Fabry-Pérot etalon and controlling the polarization optics. All tasks can be performed in a stand-alone mode (one task at a time) or in three different observing modes: a 2D-spectrometer mode (only Stokes I), a longitudinal mode (only Stokes I and V), and a full 2D-spectropolarimetric mode (Stokes I,Q,U, and V). With regard to these rather event-oriented requirements the GUI and the program modules are written in Microsoft Visual C++ and compiled with Microsoft Visual C++ version 6. To control the frame grabber the program uses the Coreco C routines library ITEX. The images are stored taking advantage of the public-domain CFITSIO library provided by NASA.

The code consists of a couple a dozen C++ classes which can be separated in three groups: the GUI group, the hardware control group and the acquisition classes. A separate class enables communication with the TCC via the TCP/IP protocol and sockets. Currently this class only retrieves the telescope pointing information and active region numbers. The GUI classes provide an user interface that allows the observer to set numerous camera and FPI parameters.

The hardware control classes control various parts of the instrument. For each part of VIM (with exception of the signal converter) there is one class which provides the methods to control it. The acquisition classes group consists of three classes: one class controls various aspects of the acquisition process and synchronizes the tasks; another one keeps track of the settings and is also responsible for their saving and retrieval. The third class, `CImage`, provides methods to acquire images from the camera.

3.1. The CImage class

The implementation of the `CImage` class and its corresponding hardware control class `CCameraControl` is an example of our idea to hide hardware details from a high-level programmer. With this approach it is easy to replace the current camera (and/or frame grabber) with a different one. We only need to rewrite methods of the underlying hardware class `CCameraControl`, keeping names of

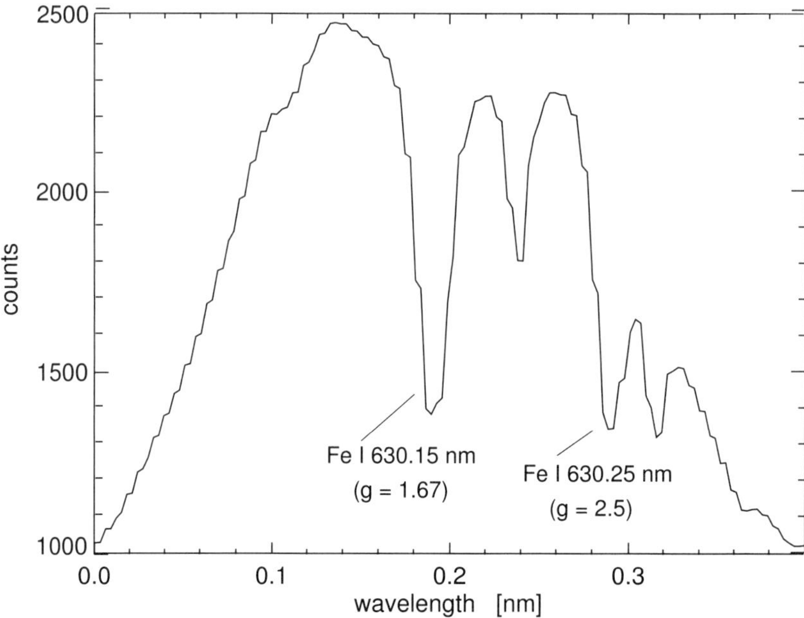

Figure 2. The spectrum of the pre-filter range.

those methods and their parameters intact. All the other parts of the code will remain unchanged.

To perform image acquisition tasks, all that the programmer needs is to create an object of the `CImage` class and then use one of the *Acquire* set of functions to get an image from the camera. Then, with the image data already in the computer memory, the programmer can display the image with the help of the *Display* function and save it as a FITS file using the *Save* function.

4. First–light Measurements

First sun-light observations with VIM were taken on February 6, 2004. The data set covers a full scan (801 narrow-band filtergrams) of quiet solar granulation using the entire FPI scanning range of about 1.2 nm. A 0.3 nm wide (FWHM) interference filter was used to preselect the wavelength band around the Zeeman sensitive Fe I 630.25 nm ($g = 2.5$) spectral line. With a step size of 3 pm the whole scan consists of 801 narrow-band filtergrams. The exposure time was set to 360 ms for each individual filtergram, which resulted in total scan time of 4.8 min. Figure 2 displays the spatially averaged (over the VIM field of view) spectrum of the pre-filter range, corresponding to 133 wavelength points.

5. Conclusions

We continue to work on both, hardware and software improvements. The GUI will be replaced with an upgraded and more userfriendly version. We want to point out that the object-oriented approach has proven to be useful during the GUI alterations. Furthermore, we plan to add another CCD detector to the project to allow for simultaneous broad-band observations.

References

Denker, C., Ma, J., Wang, J., Didkovsky, L., Varsik, J., Wang, H., Goode, P.R., 2002, Proc. SPIE, 4853, 223

Shumko, S., Denker, C., Vasik, J., Didkovsky, L., Marquette, W., Goode, P.R., 2002, Proc. SPIE, 4848, 483

Spirock, T., Denker, C., Chen, H., Chae, J., Qui, J., Varsik, J., Wang, H., Goode, P.R., Marquette, W. 2001, in ASP Conf. Ser., Vol. 236, Advanced Solar Polarimetry, ed. M.Sigwarth (San Francsisco: ASP), 65

ASTRONOMICAL DATA ANALYSIS SOFTWARE AND SYSTEMS XIV
ASP Conference Series, Vol. 347, 2005
P. L. Shopbell, M. C. Britton, and R. Ebert, eds.

IRAF Package for GNIRS Data Reduction — A Product of the Gemini/NOAO Collaboration

Andrew Cooke

Cerro Tololo Inter-American Observatory

Bernadette Rodgers

Gemini Observatory

Abstract. In August 2003 NOAO and Gemini began working together to improve the IRAF based data reduction packages used at the two telescopes. Here we report on the GNIRS (Gemini Near Infra-Red Spectrograph) Gemini/IRAF package - one of the first fruits of the collaboration.

The package helps the user generate high quality, calibrated spectra from the raw observations. It supports longslit, cross-dispersed and, experimentally, IFU (Integral Field Unit) observing modes - the processing of different data types is handled automatically, with the user following a similar reduction procedure in all cases. It can also be used as part of an automated pipeline and is generic enough to support other instruments with a similar data format.

This work is based on the earlier NIRI (Near Infra-Red Imager) spectroscopy routines, written by Joe Jensen (Gemini). All the tasks have been rewritten and, although they remain CL scripts, they now call new, compiled tasks to implement common functionality. This gives cleaner, more reliable code and an improved user response.

In this paper we illustrate the flow of data through the package, outlining how GNIRS observations are reduced.

This software was released as part of the Gemini IRAF Package, version 1.7, on October 27th 2004.

1. Introduction

Gemini and NOAO have been collaborating since August 2003 to improve the IRAF based software used for reducing data from the twin telescopes. The project is making progress along two related fronts – extending core IRAF support for commonly required functionality and improving the Gemini–specific packages with script refactoring and compiled tasks. On the NOAO side, Mike Fitzpatrick is technical lead, while Inger Jorgensen, from Gemini, is scientific lead.

Although many of the improvements are general, they are driven by deadlines for specific instruments. A set of tasks to support the Gemini Near Infra-Red Spectrograph (GNIRS), which is being commissioned during 2003-2004, was therefore chosen as the first development target. One of us (Bernadette Rodgers, the GNIRS Instrument Scientist) was responsible for the package design, the other (Andrew Cooke, Scientific Programmer) for the implementation.

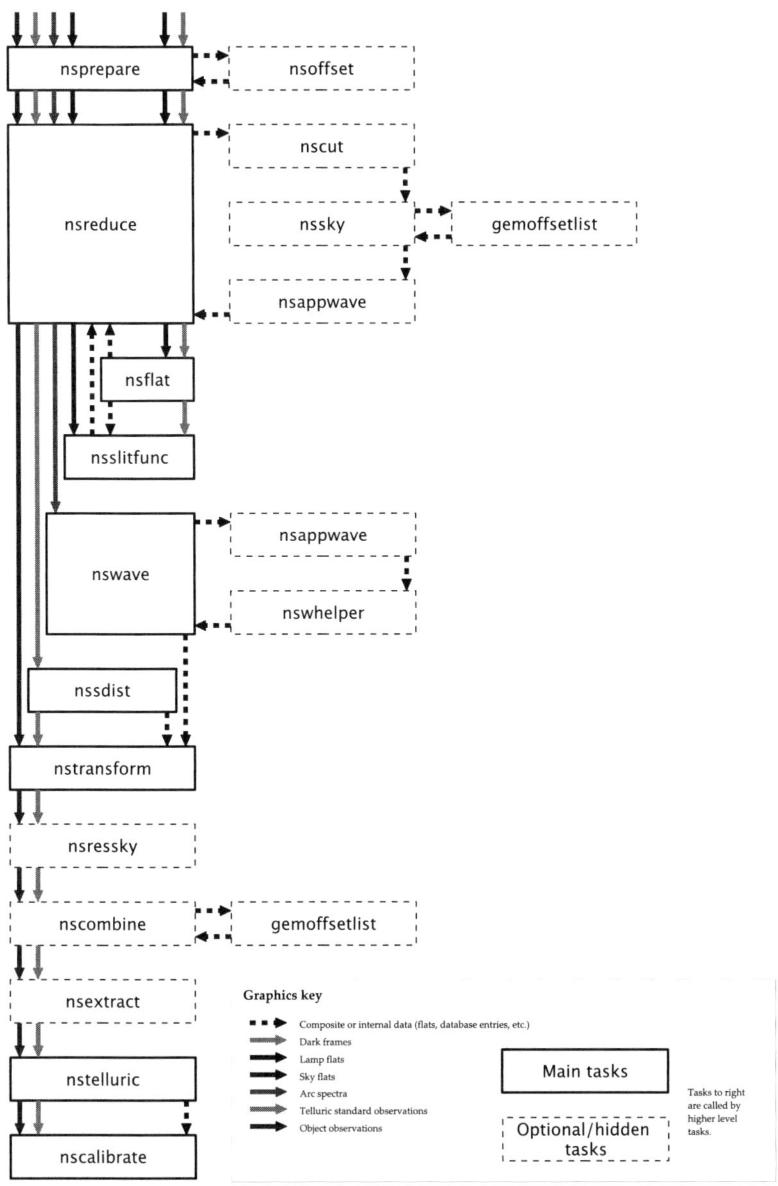

Figure 1. Data flow through the main package tasks. The different types of data (distinct coloured arrows) are typically processed in order, from right to left.

GNIRS was designed to offer a wide variety of spectroscopic capabilities including long-slit (single order) spectroscopy within the 1-5.5 μm range at 2-pixel spectral resolutions of approx 1700, 5900 and 18000 with a choice of 2 pixel scales (0.15″/pix and 0.05″/pix); cross-dispersed spectroscopy giving simultaneous coverage of 1-2.5 μm at R $\sim$ 1700 and 5900; integral field unit (IFU) spectroscopy from 1-5.5 μm over a 5 arcsec field with a pixel scale of 0.15″/pix; and a Wollaston prism for spectral polarimetry with the Gemini polarization unit GPOL.

The first release of the GNIRS package, which is based on tasks written for the spectroscopy mode of the Near Infra-Red Imager (NIRI), will process long-slit and cross-dispersed spectroscopy, with limited support for IFU data. The same tasks work with all these different modes through the careful separation of instrument specific information and multi–extension FITS files.

Section 2. describes the main package tasks. Section 3. gives background implementation details.

2. Data Flow

Figure 1 illustrates the data flow through the main package tasks. This section describes the flow from top to bottom.

nsprepare Preparation corrects data for low noise reads, the detector offset, and non-linear response. It also modifies the noise model to compensate for digital averages and co-adding and, optionally, calculates variance and data quality frames.

The detector offset is a constant value that must be measured from a dark area of the data and then subtracted. This must be done before non-linearity corrections and noise estimates are made. It has a separate source from dark current, but any measurement (and consequent correction) will include both offset and dark levels. This is unfortunate, as dark current *does* contribute to the noise in the data. For most data we believe that the offset will be more significant than the dark, so by default **nsprepare** will subtract the offset, but it is possible to override this behaviour and remove the dark level within **nsreduce** (ie. after noise estimates have been calculated).

Header values in the observation determine the observation mode and so identify a particular Mask Definition File (MDF), which is associated with the data (see section 3.). The MDF describes the illumination pattern on the detector — information used later by **nscut**.

Once the data have been processed by **nsprepare** they are in a reasonably instrument–independent format. Similar preparation tasks can be written for other instruments, which will then be able to take advantage of the remaining tasks in the GNIRS package. Indeed, such tasks already exists for Michelle and the Thermal-Region Camera Spectrograph (T-ReCS).

nsreduce Reduction removes the instrument signature (flat fielding and dark subtraction), identifies and subtracts sky frames, provides an approximate wavelength solution, and separates the observation into separate spectral regions.

All these steps are optional and the task is typically called several times to process the different observation types. For example, data that will be used to generate the flat field may be processed by **nsreduce** without applying the (still to be generated) flat field.

The **nscut** task isolates the useful sections of each frame. For cross-dispersed and IFU data each section is placed in a separate MEF extension.

nsflat, nsslitfunction A flat can be constructed from lamp data alone (**nsflat**), or with sky flat observations to correct for variations in response along the slit (**nsslitfunction**). Flat field data are cut to size before processing and so pass through **nsreduce** first. The bad pixel map generated with the flat can be used to flag other data during preparation (**nsprepare**).

nswavelength Wavelength calibration uses the core IRAF identify/reidentify tasks. These have been improved (Frank Valdes, NOAO), but still require careful selection of parameter values, so the help page includes step-by-step troubleshooting information. 3-5μm and mid-IR data can use skylines rather than arcs.

nssdist The spatial alignment of spectra on the detector is measured with nssdist using standard star or pinhole observations.

nstransform The measurements of dispersion and spatial variation (the output from nssdist and nswavelength) are used by **nstransform** to produce rectified data on a uniformly sampled wavelength scale.

nsressky Optionally, residual sky emission can be subtracted by fitting to the data.

nscombine This task will shift and combine 2D frames based on the the output from **gemoffsetlist**, which measures relative offsets using WCS header data.

nsextract If desired, 1D spectra can be extracted using **nsextract**.

nstelluric This removes the atmospheric signature by dividing by a 1D telluric standard spectrum. This can be applied to 1 or 2 dimensional data.

nscalibrate Final flux calibration is calculated from a suitable standard with nscalibrate.

3. Implementation

The Gemini packages make extensive use of the Multi-Extension FITS support (MEF) in IRAF. MEF allows a single FITS file to contain many different data sets, called extensions. These are typically identified by name and version. The most important extensions in Gemini data are called SCI (the data), VAR (variance) and DQ (data quality). Versions correspond to different spectral orders (cross-dispersed data) or sub-apertures (IFU data).

Storing all the data for a single observation within one file simplifies the user interface. It is only necessary for the user to supply the file name to the tasks within the package; within each task each order or aperture is processed in turn, automatically.

Variance and data quality values are also carried forward through the tasks (as much as is possible, given the current IRAF core and simplifying assumptions about correlated statistics) to help the user assess the significance of features in the data.

Two other extensions are also important. The Primary Header Unit (PHU) contains header information applicable to all extensions (e.g., the initial World Coordinate System, or WCS, which reflects the telescope pointing and orientation). The Mask Definition File (MDF) describes how the SCI, VAR and DQ extensions are extracted from the original data.

To examine the extensions present in a MEF file, use the **gemextn** task (in the **gemtools** package). This is a new, compiled (SPP) task, developed to help simplify the task scripts and provide an easier interface to the MEF files.

4. Conclusion

We hope that the GNIRS package will provide a useful, flexible and efficient tool for people working with data from this instrument. Revised help pages (wavelength calibration in particular, which now contains detailed step–by–step recipes for solving common problems) should make the tasks more useful; flexibility is already evident in the processing of Michelle and T-ReCS spectral data; efficiency should already be improved by the compiled **gemextn** task for file handling, with further advances expected in the next release as tasks dedicated to calculations (**gemarith**, Craig Allen, Gemini), logging (**gemlog**, Kathleen Labrie, Gemini) and header handling (NOAO) are more tightly integrated into the package.

Acknowledgments. Many thanks to all collaboration members (too numerous to include as authors); Kathleen Labrie's fastidious quality control was particularly appreciated.

ASTRONOMICAL DATA ANALYSIS SOFTWARE AND SYSTEMS XIV
ASP Conference Series, Vol. 347, 2005
P. L. Shopbell, M. C. Britton, and R. Ebert, eds.

An Overview of the Mount Stromlo Observatory Telescope Control System

Mark A. Jarnyk and Gary R. Hovey

Research School of Astronomy and Astrophysics, Australian National University, Canberra ACT 2611, Australia

Abstract. A new telescope control system, MSOTCS, is being written to support the remote control and automation functionality of the Siding Spring 2.3m telescope. Key features of MSOTCS include: the use of a proprietary telescope pointing/astrometry kernel called TCSpk; QNX6 as the operating system; the use of a central shared database for all astrometric, configuration and control system data; and the use of a publish and subscribe database for the telescope data required by multiple remote clients. MSOTCS takes the form of an embedded controller which communicates with external user interfaces via RPC.

1. Overview of the System

The Mount Stromlo Observatory Telescope Control System (MSOTCS), is software written in C++ running on the real-time operating system QNX6. It is being written to accommodate new remote observing software, TAROS (Wilson et al. 2005). MSOTCS is responsible for the control and safety of the hardware which constitutes the optical support structure of the telescope and of associated plant which controls the observing environment such as ventilation, air-conditioning, building/dome rotation, shutters, windscreens, safety interlocks, etc. Most importantly, the TCS implements or controls the servo-mechanisms which effect telescope axis motion, instrument rotation, focusing and secondary mirror positioning.

2. Architecture

MSOTCS (Figure 1) is a suite of cooperating autonomous processes operating around a central shared memory database. External processes communicate with the TCS using RPC, UDP and TCP sockets. Within the TCS interprocess communication is either through shared memory or via a resource manager (a QNX resource manager is a process which creates a device, such as `/dev/cmd`). In QNX this communication mechanism is fast and efficient: indeed, the QNX operating system itself is just a set of cooperating processes communicating by this message passing mechanism.

System startup is effected by launching the watchdog process. Watchdog reads a process configuration file to determine which processes are to be started, what action to take when a process dies, and the privilege level at which they are scheduled. This makes MSOTCS configurable for any number of different pro-

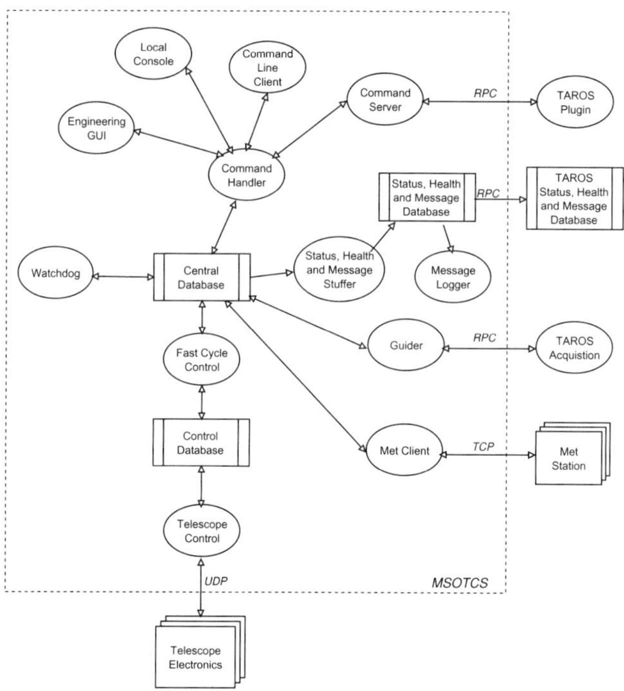

Figure 1. MSOTCS architecture showing processes and shared-memory databases

cesses; e.g., different external interfaces could be run to accommodate different observing software.

3. External Interfaces

MSOTCS runs on an industrial x86 machine as an embedded controller. Observers do not log on to the MSOTCS machine, but interact with the telescope using other software (such as TAROS) on other machines. Two interfaces to this external software are provided: command handling and published data.

3.1. Commands

MSOTCS provides a command server, which is an RPC server process. The command server forwards commands to the command handler, and returns replies from the command handler. This channel is normally only used to direct actions of the telescope; for obtaining information the alternative channel (published data) is preferred. (Command handling is described in more detail in Section 4.)

3.2. Published Data

Telescope parameters are usually communicated using the Status, Health and Message (SHAM) databases which contain parameters of interest to client software. This mechanism automatically establishes a mirror of this database on the client-side. Values updated in the database on the TCS automatically update values in the mirror.

While the component data may be changing at the control frequency (20 Hz), the database stuffer publishes this at the slower rate of 1 Hz, aligned with the UTC second.

The SHAM has been implemented specifically for TAROS but MSOTCS is not reliant upon the SHAM. It is just one way of extracting data of interest to clients and could, in another environment, be replaced by a different process.

4. Command Handling

Commands from a multiplicity of sources are dealt with by a central command handler implemented as a QNX6 resource manager providing the interface `/dev/cmd` to the TCS. Any process may write to this device, sending a command and awaiting a response. The multi-threaded command handler manages the interleaving of commands from any number of sources: observer commands, local sources including an engineering GUI, a command line client and pushbutton control from a local hard-wired console.

A command is a structure consisting of an arbitrary string and two housekeeping parameters. The first token in the string is understood to be the command verb. Whenever the command handler receives a command, a separate thread is spawned. The command handler implements the command pattern using a singleton factory class (Alexandrescu 2001) . The factory class examines the command verb and if it finds a match in its table of commands, creates the appropriate command object. There is one C++ class corresponding to each command, as shown in Figure 2. Since each thread has its own instance of the command class, concurrency issues are reduced. It also means that commands can be added simply by adding the command verb to the command list (an STL map), and implementing a concrete implementation of the Base_Cmd abstract base class. The command list supports synonyms for commands.

The command-response protocol is used, i.e., each client that sends a command waits upon a response. A response is an informative string and an integer code indicating the success or failure of the command. The response is normally immediate. However, if the keyword "wait" appears in the command string, a response is not given until the command action completes. These mechanisms allow operation of the TCS to be scripted.

5. Fast Cycle Control

The time-critical tasks of encoder and time capture, reference generation, hardware control and servo implementation are performed by the "fast cycle control" process. Each logical hardware component (windscreen, focuser, etc.) has an independent control object which implements an interface called `Base_Controller`.

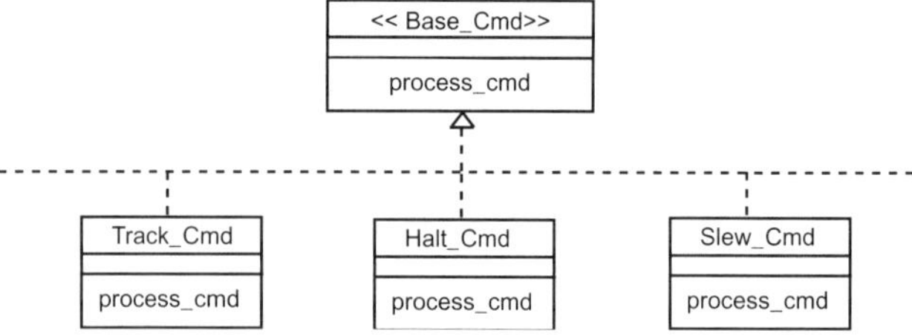

Figure 2. Each command implements the `Base_Cmd` abstract base class

The `Base_Controller` interface consists of several functions, the most important being the `process()` function. The `process()` function of each controller object is called consecutively in the main loop of the fast cycle control process. A control object need not run on each fast cycle: in this way, control processes may operate at some subharmonic of the control frequency. Note the logical nature of these controller objects: a logical hardware component may encapsulate several real hardware components; the `Pointing_Controller` manages the altitude, azimuth and rotator servos and encoders.

The `Pointing_Controller` is the most complex of the controller objects. This is where the virtual telescope loops which generate the tracking position references for the telescope axes and the instrument rotator are implemented. The context in which the fast cycle control process runs (i.e., the set of data needed for the computations in these virtual loops) must always be internally consistent. It may change slowly (e.g., changing barometric pressure will alter the refraction corrections) or may need to be recalculated completely (e.g., when an observer selects a new celestial object).

The number, control-frequency and type of controller objects are established at run-time by a configuration file. For the Siding Spring 2.3m telescope these objects read feedback data from, and write control values to, a control database. A separate process ("Telescope Control" in Figure 1) is run consecutively with the fast cycle control process to convey the data between the control database and the telescope hardware interface.

6. Pointing Control

Central to MSOTCS is the concept of a virtual telescope, widely used since the design of the Anglo-Australian Telescope (Straede & Wallace 1976) in the 1970's. The virtual telescope concept allows an ideal telescope to be abstracted from the specific telescope mount and control plant and facilitates efficient and rigorous control of the main axes, secondary mirror, autoguiding and other complex functions. Similar schemes have been used on the Keck and Gemini telescopes and in PTCS, a telescope control system developed by Jeremy Bailey (Bailey & Prestage 1997)

Rigorous calculation of the astrometry and geometry of the telescope axis pointing, mirror control and instrument rotation functions is necessary to realize the full benefits of the virtual telescope approach. Yet speed and computational robustness is equally necessary in the real-time control environment of a TCS, particularly if multiple virtual telescope loops are run in order to implement autoguiders and other features. For MSOTCS we chose to employ a proprietary astrometric kernel called TCSpk (Wallace 2002) because (i) it is computationally rigorous but reasonably efficient, (ii) it shares code with the TPOINT telescope pointing analysis package ensuring consistency of pointing correction implementation, (iii) it is linked with the widely used SLALIB astrometry and conversion routines, and finally (iv) because it is a library of well-documented C-language routines it offers a kernel which can be cleanly incorporated into our chosen system architecture.

7. Simulation

Individual hardware components may be simulated. The simulation takes place at the point where data is written to, or read from the control database. This means that virtually all of the software which would normally control the component is still exercised. Simulation can be enabled or disabled, and the selection of simulated vs activated hardware changed without having to restart the system.

8. Conclusion

MSOTCS is a versatile and scalable telescope control system. It consists of a suite of cooperating processes identified at run-time by a configuration file. Different processes would be used for different telescopes. Commands are defined at compile time, but each command is managed as an independent object. The main pointing loop (comprising reference generation and control) is largely independent of the target hardware and specific mount characteristics.

Acknowledgments. We gratefully acknowledge the generous support of a QNX-in-education grant.

References

Alexandrescu, A. 2001, "Modern C++ Design: Generic Programming and Design Patterns Applied" (Addison-Wesley)

Bailey, J. & Prestage, R. 1997, Proc. SPIE, 3112, 124

Straede, J.O., & Wallace, P.T. 1976, PASP, 88, 792

Wallace, P. 2002, Proc. SPIE, 4848, 125

Wilson, G., Czezowski, A., Hovey, G., Jarnyk, M., Nielsen J., Roberts, W., Sebo, K., Smith, D., Vaccarella, A., & Young, P. 2005, in ASP Conf. Ser., Vol. 347, ADASS XIV, ed. P. L. Shopbell, M. C. Britton, & R. Ebert (San Francisco: ASP), 563

ASTRONOMICAL DATA ANALYSIS SOFTWARE AND SYSTEMS XIV
ASP Conference Series, Vol. 347, 2005
P. L. Shopbell, M. C. Britton, and R. Ebert, eds.

NBodyLab Simulation Experiments with GRAPE-6a and MD-GRAPE2 Acceleration

Vicki Johnson

Interconnect Technologies Corporation, Claremont, CA

Alper Ates

Physics Department, Pomona College, Claremont, CA

Abstract. NbodyLab is an astrophysical N-body simulation testbed for student research. It is accessible via a web interface and runs as a backend framework under Linux. NbodyLab can generate data models or perform star catalog lookups, transform input data sets, perform direct summation gravitational force calculations using a variety of integration schemes, and produce analysis and visualization output products. NEMO (Teuben 1994), a popular stellar dynamics toolbox, is used for some functions. NbodyLab integrators can optionally utilize two types of low-cost desktop supercomputer accelerators, the newly available GRAPE-6a (125 Gflops peak) and the MD-GRAPE2 (64-128 Gflops peak). The initial version of NBodyLab was presented at ADASS 2002. This paper summarizes software enhancements developed subsequently, focusing on GRAPE-6a related enhancements, and gives examples of computational experiments and astrophysical research, including star cluster and solar system studies, that can be conducted with the new testbed functionality.

1. GRAPE Hardware Acceleration

The design objectives and basic functionality of NBodyLab[1] were described by Johnson (2002). Interfaces have been developed for two types of Japanese GRAPE (Gravity Pipeline) desktop supercomputer PCI boards.

Applications using large arrays of GRAPE chips have six times won the annual international Gordon Bell competition for top supercomputer performance. At least two multi-year, well-funded Japanese research programs are developing next-generation GRAPE systems aimed at dominating high-performance astrophysical research and bio-computing. This promising future is further motivation to expose students to GRAPE technology and for researchers and software engineers to gain experience with the capabilities and software interfaces to the current generation of GRAPE's.

NBodyLab development began by developing interfaces to an MD-GRAPE2 accelerator board (Molecular Dynamics - Gravity Pipeline) running at 64 Gflops peak (128 Gflops peak with optional booster card), handling up to 500K particles for a cost of less than $12K. The MD-GRAPE2 was designed for molecular dynamics (MD) simulations and calculates a general force between all pairs of

[1] `http://www.nbodylab.com`

particles in an N-body particle simulation, supporting Coulomb, van der Waals and user-defined forces. NBodyLab utilizes the MD-GRAPE2 only as a basic Newtonian accelerator for gravitational forces.

In late 2003 the GRAPE-6a, developed at Tokyo University, became available with a PCI interface. Generations of GRAPE chips have been developed specifically for astrophysical research. GRAPE designers Makino and Taiji (1998) describe GRAPE simulations of planetary formation, star clusters, galactic nuclei, galaxy interactions and galaxy formation. (The MD-GRAPE R&D was spun-off into a Japanese research institute and is now aimed at biosciences such protein modeling, headed by Dr. Taiji). A GRAPE-6a board runs at 128 Gflops peak, can handle 64K particles, and costs less than $9K.

GRAPE cards facilitate quick NBodyLab runs by students using a web browser. A 10,000 particle simulation with a stable and accurate integrator takes 10 seconds using the GRAPE-6a; without GRAPE-6a acceleration, the same run takes 10 minutes (10K Plummer sphere; t=0 to 1; Hermite individual time step integrator; 3 GHz P4 host).

2. Integration Schemes

The focus of development of integration options for NBodyLab has been direct integration methods because the GRAPE cards implement direct summation of forces in hardware and direct summation is a straightforward implementation of Newton's Law of Gravitation and easier for undergraduates to understand than, for example, tree algorithms. Hardware-accelerated direct summation is preferable for many classes of astrophysics problems. (An option to use the classic tree algorithm, from NEMO, is available for comparison runs.)

The MD-GRAPE2 force call is simple: *m2_calculate_forces (mu, x, npart, a)* which returns the computed accelerations for each particle. In the initial version of NBodyLab, 1st, 2nd and 4th order (e.g., Runge-Kutta RK4) constant timestep integrators were implemented for the MD-GRAPE2, following basic numerical analysis texts. Astrophysics and MD researchers explained to co-author and developer Vicki Johnson that better schemes were available. The theory, analysis, performance and physical interpretation of numerical integration methods has long been an challenging area of research (Aarseth 2003).

For example, Prof. Makino, lead designer of the astrophysics GRAPE family, has discussed in personal communications with Johnson some considerations for MD and astrophysics applications: In MD, simulations typically follow the vibration of atoms for very long periods. The typical period of oscillation is on the order of 10 femtoseconds and an MD simulation evolves the system for pico, nano, or microseconds, so energy conservation must be very good. Constant time steps can be used and symplectic (or time-symmetric) integration schemes work well in these cases. By contrast, RK4 gives better accuracy per orbit, but it can cause long-term linear drift of energy. Symplectic schemes do not cause linear drift. For collision-less problems in astrophysics, it is not necessary to follow a large number of orbits. The high accuracy of RK4 is not needed per orbit, so computational speed may be gained by using a 2nd-order symplectic (also known as leapfrog or Verlet) algorithm, which requires less memory to compute. For collisional problems, astrophysical simulations need to use variable and individual timesteps, and leapfrog algorithms do not suffice. High-order schemes are needed, and Hermite has become the method of choice.

The name Hermite is used in numerical analysis to denote a polynomial based on the function and its derivatives. A 4th-order Hermite integrator is implemented by the GRAPE-6a. In the Hermite individual timestep scheme, each particle has its own timestep and integrations proceed as a global time step is incremented by the particle i with the smallest timestep. The positions of all particles at the global time are estimated and the gravitational force on particle i is calculated and its position, velocity, time, timestep, etc. are updated. The 4th-order Hermite scheme is well-proven for accuracy, efficiency, and algorithmic simplicity. The GRAPE-6a hardware performs the expensive calculation of the derivative of acceleration, which is not computed by the MD-GRAPE2.

Prof. Makino provided Interconnect with leapfrog, 4th and 6th order symplectic schemes with constant timestep for incorporation into NBodyLab. Individual and blocked timestep Hermite test codes provided with the GRAPE-6a were adapted to NBodyLab. An interface to a 2nd-order leapfrog method for the GRAPE-6a was developed as well. Versions of all the NBodyLab integration options have been developed for hosts without GRAPE support. Although the MD-GRAPE2 offers software emulation of the hardware (which has its own numerical precision and scaling characteristics), these were not used, in favor of faster direct summation calculated by the PC host processor.

3. Computational Experiments

Undergraduate students not exposed to numerical analysis can be surprised by the results of varying, sometimes only slightly, the parameters or inputs to different integration options. Experiments with small data sets can demonstrate to students the importance of carefully and skeptically interpreting the results of larger n-body simulations, and confirm the value of robust schemes like the Hermite.

NBodyLab's web interface makes it easy to perform a variety of computational physics experiments with different input data sets, transformations, integrators and parameters like softening. The advantages of NBodyLab's newly developed GRAPE-6a support and Hermite options are evident in runs with an n=100 Plummer model – see Makino and Taiji for an analysis of interesting system behaviors that can be studied with this small model.

With the new GRAPE-6a related features, students can achieve better performance, stability and accuracy for studies of small systems, such as binary, triple and binary-binary encounters, and larger stellar dynamics simulations. They can compare conservation of energy, momenta, etc., for GRAPE-6a versus MD-GRAPE accelerated evolutions and use the more computationally expensive Hermite for large numbers of particles or long time scales. Comparisons of individual timestep algorithms with variable blocked timestep or constant fixed-step schemes are now possible.

4. Astrophysical Investigations

Ates (2004) used NBodyLab with the MD-GRAPE2 to model the geometric evolution of nearby stars and an open star cluster Collinder 70. This research utilized a new NBodyLab tool developed to search Hipparcos and SKY2000 star catalogs, find stars in a user-specified volume, extract values, and convert and scale their units and estimate masses from their luminosity. NBodyLab simulated the geometric evolution of the selected regions. Ates analyzed errors and investigated scaling by simulating two 1 solar mass stars freely falling on each other; the time to reach 1/2 distance and the velocity at that point was calculated classically. Time and velocity values simulated with NBodyLab with RK4 and symplectic integrators were compared with calculated theoretical values. Ates (2003) also used NbodyLab to study star clusters, using catalog data, and he compared the simulations to telescope observations he made using Pomona College observatories. His most recent work concentrates on the Solar System. Recently simulations from Ates' work have been run with the new NBodyLab GRAPE-6a and Hermite options. Small systems studied by Ates using NBodyLab have included:

- *Ursa Major* – Using data obtained from SKY2000 catalog, the well-known 7 members of Big Dipper were modeled. This model was created to check the accuracy of Cartesian coordinate conversion algorithms. An OpenGL 3D visualizer revealed the well-known Big Dipper shape. Since Big Dipper is just 7 members of a larger cluster this work concentrated on positional accuracy, not the time evolution.
- *Star cluster Hyades* – The flatness of the cluster parallel to the galactic plane was observed and convergent motion was successfully simulated.
- *Collinder 70* – Very accurate astrometric data for this cluster is available, but surprisingly no prediction on the future geometric evolution of this cluster had been made. Using the NBodyLab catalog extraction, 46 members of the cluster were retrieved and simulated for 75,000 years and the future shape of the cluster predicted.
- *Solar System* – Since the heliocentric coordinates and mass for each planet are well-known, the study focused on accurate scaling and achieving long term stability and accuracy of the orbits. Software to convert heliocentric coordinates was developed and a scaling scenario designed involving the Sun and the Jupiter, where Jupiter is left free to fall onto Sun and the time it takes to reach half distance and the velocity at that point classically calculated. Once scaling was done the Heliocentric positions of the planets obtained from VSOP routines for any given date were converted into Cartesian units, and the orbital velocities calculated and scaled for the NBodyLab simulation. Resultant orbits show good eccentricities and are stable over 200 years. Hermite schemes generally have lower energy errors for planetary N-body simulations than other NBodyLab integrators, as confirmed with GRAPE-6a runs.
- *Halley's comet* – The simulation shows 76 year periodic orbits. Surprisingly the comet is seen to be approaching from z= 300 to z=270 in 4 revolutions. The fast change in the inclination angle deserves future investigation of errors and calculation of initial conditions.

Wertheimer (2003) used NBodyLab with the MD-GRAPE2. He started with a single Plummer sphere to simulate a globular cluster and let it evolve in a constant potential. He imposed a tidal radius to simulate the tidal force from the galaxy. Periodically the cluster was shocked by letting it pass through the disk of a simulated Milky Way derived from GalactICS. Runs took 10-20 hours with RK4. Running this simulation with the GRAPE-6a Hermite option will likely yield interesting differences in energy and radial distributions.

Acknowledgments. Peta Computing and Prof. Jun Makino of Tokyo University generously loaned GRAPE hardware to Interconnect Technologies for NBodyLab development and demonstrations during the period fall 2002 to the present, during which Interconnect funded and developed enhancements for GRAPE-6a support, symplectic and Hermite integration options, and star catalog retrieval with HR diagram generation. Prof. Makino and Prof. Milan Mijic provided valuable scientific assistance. Pomona College provided access to computing facilities and partially funded development of software for Wertheimer's thesis and solar system simulations. Pomona College funded development of the initial version of NBodyLab, in 2002.

References

Aarseth, S. 2003, "Gravitational N-Body Simulations", Cambridge University Press

Ates, A., et al. 2003, N-body simulations of Star Clusters and Observational Comparisons, AAS Jan 2003 poster

Ates, A. 2004, N-body simulations and HR Diagrams of Nearby Stars, California State University, Los Angeles, Dept. of Physics and Astronomy, Masters Thesis

Johnson, V. L., et al. 2002, in ASP Conf. Ser., Vol. 295, ADASS XII, ed. H. E. Payne, R. I. Jedrzejewski, & R. N. Hook (San Francisco: ASP), 365

Makino, J., & Taiji, M. 1998, "Scientific Simulations with Special-Purpose Computers–the GRAPE System", Wiley

Teuben, P.J. 1994, in ASP Conf. Ser., Vol. 77, ADASS IV, ed. R. A. Shaw, H. E. Payne, & J. J. E. Hayes (San Francisco: ASP), 398

Wertheimer, J. 2003, Tidal Shocking of Globular Clusters, Pomona College Senior Thesis

ASTRONOMICAL DATA ANALYSIS SOFTWARE AND SYSTEMS XIV
ASP Conference Series, Vol. 347, 2005
P. L. Shopbell, M. C. Britton, and R. Ebert, eds.

Optimizing Observing Sequence Design for Periodic and Non-periodic Phenomena: A Bayesian Approach

Mark D. Johnston and Russell Knight

Jet Propulsion Laboratory, California Institute of Technology
4800 Oak Grove Drive, M/S 126-347, Pasadena, CA 91109-8099
email: {firstname.lastname}@jpl.nasa.gov

Abstract. The problem of designing observing sequences to detect and characterize periodic phenomena occurs regularly in astronomical investigations. Examples of current interest include Cepheid variable searches in external galaxies (with Hubble Space Telescope), and future high accuracy astrometric observations of nearby stars with the Space Interferometry Mission (SIM) satellite to search for planetary companions. Various sampling strategies have been proposed to obtain good phase coverage over an interesting range of periods. Recently, Loredo and Chernoff have proposed the use of "Bayesian adaptive exploration", a model-based Bayesian method that exploits observations made to date to determine the best future observation times according to a maximum information criterion. While this method makes the best possible use of any results already obtained, it does not address the "bootstrap" problem of scheduling in advance of any data collection. It also is highly compute-intensive, which is especially problematic when an integrated observing schedule for hundreds of targets is required, taking into account all of the various other constraints and preferences that come into play. In this paper we report on our progress on addressing these issues. We have developed an approximate expression for the uniformity of phase coverage that can be used when scheduling to assess candidate sample times. We describe the results obtained using this estimator, and compare them with detailed simulations. We describe our progress and plans for integrating optimizing criteria for both periodic and non-periodic observations into a single observation sequence.

1. Introduction

Data analysis techniques for periodic phenomena have become very sophisticated (see e.g., Schwarzenberg-Czerny 1999, Bretthorst 2001), providing powerful techniques for the estimation of periods and other source parameters. From the perspective of planning to *acquire* data, the question naturally arises of how best to schedule data collection, in order to optimize post-observation data analysis. Information-theoretic approaches have been proposed, such as described by Sebastiani and Wynn (2000), motivated by the criterion that it is preferable to make observations which provide the greatest additional information, given all that is currently known.

This approach has been elaborated in detail by Loredo and Chernoff (2003) who propose a technique called *Bayesian Adaptive Exploration* (BAE). Their approach is based on a Bayesian model of the phenomena, from which is derived an expression for the expected information gain as a function of (a) results from observations to date and (b) the derived model probability distribution. Such an approach is extremely appealing in that it makes maximum use of all available information, and is specific to the detailed form of the model for the pheonmena under investigation. However, there are some practical drawbacks to the method: it is computationally very expensive (multiple Markov Chain Monte Carlo simulations may be required to numerically evaluate the model), and it requires data from past observations to estimates times for future ones. In large scale planning and scheduling applications, where many future observation possibilities must be considered in the context of interacting constraints and preferences, it is infeasible to use this kind of approach.

In the following we describe an approximation, motivated and based on the Bayesian modeling approach, which allows us to estimate the uniformity of phase coverage as a function of a set of observing times, past and future. We first describe the approach, then provide several examples of how our approximation can be used in practice. We conclude with some comments on future research and application areas.

2. Approach

Our approach is based on the results of Bretthorst (2001) who analyzed in detail the case of a noisy sinusoidal signal with non-uniform samples. He derived an expression for the Bayesian posterior probability distribution for the frequency f, given a set of observations $\mathbf{d}_i(t) = \mathbf{A}\exp(-2\pi i f t_i) + \mathbf{n}(t_i)$, where $\mathbf{d}_i$ are the data samples at times t_i, $\mathbf{A}$ is the signal amplitude, and $\mathbf{n}$ the noise level. If we examine the expression for $P(f)$, the posterior probability of frequency f, we note that the denominator of Bretthorst's $\overline{h^2}$ statistic is a function only of t_i (i.e., is independent of the data values $\mathbf{d}_i$) and provides a measure of the variation in the measured probability that arises *solely* from the choice of sampling times t_i. This expression has the form:

$$V(f) = \left(\sum_i \cos^2(2\pi f t_i)\right)\left(\sum_i \sin^2(2\pi f t_i)\right) - \left(\sum_i \cos(2\pi f t_i)\sin(2\pi f t_i)\right)^2$$

We use the normalized mean square deviation of the quantity $V(f)$, over a frequency range of interest $[f_{min}, f_{max}]$, as a measure of *phase coverage nonuniformity* — the quantity to minimize. We have compared this quantity to the results of simulation runs with noisy data analyzed to determine the signal frequency: we find that this approximation tracks quite well the likelihood that the true frequency will be determined.

3. Examples

1. Sample Time Selection — Consider the case of selecting in advance a set of observation times with the goal of detecting a periodic signal of frequency in the range $[f_{min}, f_{max}]$. This situation arises in Cepheid searches (e.g., Freedman et al. 1994) and other variable star observations. A comparison of power law and geometric series sampling strategies is shown in Figure 1 for a range of sample

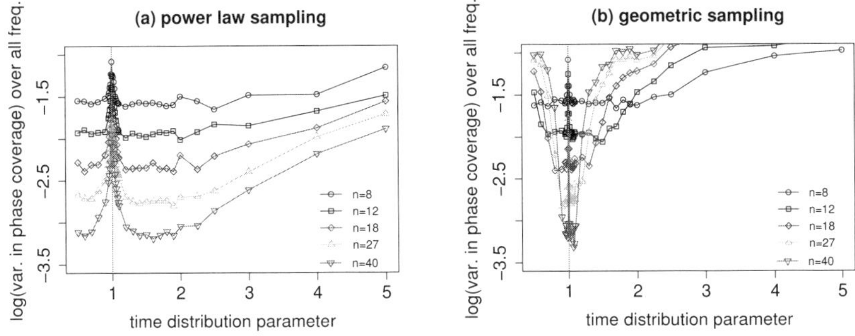

Figure 1. Comparison of (a) power law and (b) geometric series sampling for a range of sample sizes.

sizes (time span of 100d, period range 2-100d). The conclusion that power law sampling is preferred is consistent with previous investigations.

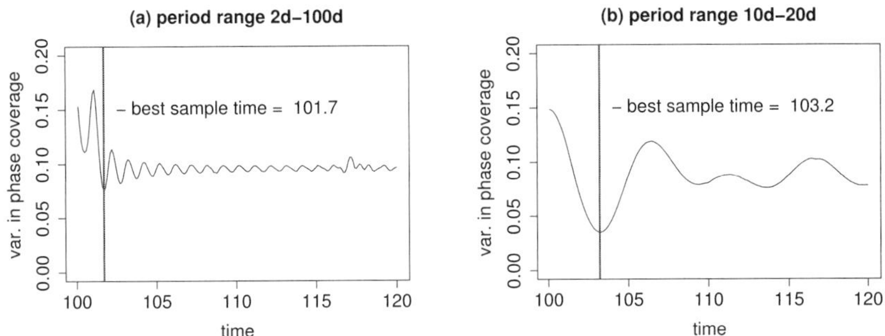

Figure 2. Variation in phase coverage as a function of scheduled time of observation $N + 1$. The results depend on the frequency range of interest: in (a) the period range is 2d to 100d, while in (b) the range is 10d-20d.

2. Incremental Scheduling — Suppose we have scheduled N observations and seek the best time to schedule observation $N + 1$. This is the situation addressed by BAE when there exists both data and a model. When we have no data, we can use our approximation to find a time of minimal variation in phase coverage given a frequency range of interest. An example where $N = 4$ is shown in Figure 2, where the best sample time depends on what one expects to be the period range of interest.

3. Eliminating Redundant Schedule Times — In some cases we have given a set of candidate schedule times that is larger than can be actually accomplished – a situation which arises in some satellite mission scheduling problems such as SIM (Schwartz and Wehrle 2004). The problem is to *reduce* the set to one of manageable size. The approximation formula can be used to assess candidate subsets. We have conducted numerical experiments that show that a guided selection of a good subset can improve by factors of several the variation in phase coverage. Such an improvement can make a dramatic difference in the detectability of periodicities at the sensitivity limit.

4. Future Directions

There are several directions for future research and application of this work:

- Incorporate information gained from past observations
- Broaden to non-sinusoidal variability (if analysis shows this useful)
- Automate the analysis of redundant candidate schedule times
- Investigate integration with schedulers such as ASPEN (Chien et al. 2000) as a specialized heuristic

Acknowledgments. The research described in this paper was carried out at the Jet Propulsion Laboratory, California Institute of Technology, under a contract with the National Aeronautics and Space Administration. The authors wish to recognize and thank the SIM team at JPL, and Arnie Schwartz in particular, for very useful discussions related to this work.

References

Bretthorst, G. L. 2001, "Nonuniform Sampling: Bandwidth and Aliasing", in Maximum Entropy and Bayesian Methods in Science and Engineering, ed. J. Rychert, G. Erickson and C. R. Smith, AIP: 1-28

Chien, S., G., Rabideau, et al. 2000, "ASPEN – Automating Space Mission Operations using Automated Planning and Scheduling", SpaceOps 2000, Toulouse, France

Freedman, W. L., Hughes, S. M., et al. 1994, ApJ, 427, 628

Loredo, T. J., & Chernoff, D. F. 2003, "Bayesian Adaptive Exploration", in Statistical Challenges in Modern Astronomy III, ed. E. D. Feigelson and G. J. Babu (Springer-Verlag, New York) 57

Schwartz, A. A., & Wehrle, A. E. 2004, "Formulating an Observation Plan for SIM", AAS 203rd Meeting

Schwarzenberg-Czerny, A. 1999, ApJ, 516, 315

Sebastiani, P., & Wynn, H. P. 2000, J. R. Statist. Soc. B, 62, 145

ASTRONOMICAL DATA ANALYSIS SOFTWARE AND SYSTEMS XIV
ASP Conference Series, Vol. 347, 2005
P. L. Shopbell, M. C. Britton, and R. Ebert, eds.

The Remote Control System for the ASTE Telescope

Takeshi Kamazaki

Institute of Astronomy, University of Tokyo, 2-21-1 Osawa, Mitaka, Tokyo 181-0015, Japan, Email: kamazaki@ioa.s.u-tokyo.ac.jp

Hajime Ezawa and Ken'ichi Tatematsu

National Astronomical Observatory of Japan, 2-21-1 Osawa, Mitaka, Tokyo 181-0015, Japan

Nobuyuki Yamaguchi, Nario Kuno, and Koh-ichiro Morita

Nobeyama Radio Observatory, 462-2 Minamimaki, Minamisaku, Nagano 384-1305, Japan

Kiyohiko Yanagisawa and Osamu Horigome

FUJITSU LIMITED, 1415 Tsurugamidoricho, Nagano, Nagano 380-0813, Japan

Jun Maekawa

MAEKAWA Co., Ltd., 8240-6078 Nishi-Ide, Oizumi, Hokuto, Yamanashi 409-1501, JAPAN

Abstract. The Atacama Submillimeter Telescope Experiment (ASTE) is a project to operate a 10-m submillimeter telescope at a high altitude site (4,800 m) in the Atacama desert in northern Chile. The key to successful telescope operation under the severe environment of the observing site is to realize a stable remote control system. The remote control system for ASTE consists of a newly developed operating software capable of remote observation and a satellite network facility (56-64 kbps), which connects the telescope site to the outside world including the operation base in San Pedro de Atacama (altitude 2,400 m) or institutes in Japan. The control software was developed based on the existing COSMOS3 system, which has been used for the 45-m telescope and the Millimeter Array (NMA) of the Nobeyama Radio Observatory (NRO) in Japan. The telescope operation with N-COSMOS3 in Chile started in April 2002. We have succeeded in the first remote observation from the base in San Pedro de Atacama in July 2002, followed by the first successful remote observation from Japan in October 2003.

1. Introduction - the ASTE project

The Atacama Submillimeter Telescope Experiment (ASTE; Figure 1) is a project to operate a 10-m submillimeter telescope at a high altitude site (4,800 m) in the Atacama desert in northern Chile. The aims of this project are to observe the southern sky through submillimeter waves, and to develop and establish a high

Figure 1. The ASTE telescope and its infrastructure at Pampa la Bola.

performance submillimeter telescope (see also Ezawa et al. 2004). The ASTE is a telescope of the National Astronomical Observatory of Japan (NAOJ), while the telescope is currently operated collaboratively by the ASTE team including NAOJ, University of Tokyo, Nagoya University, Osaka Prefecture University, and Universidad de Chile.

2. The ASTE control system

The telescope site is far from Japan and under a severe environment. Therefore, the key to successful telescope operation is to realize a stable remote control system. The control system for ASTE consists of a newly developed operating software capable of remote observation (N-COSMOS3) and a satellite network facility (56-64 kbps).

2.1. The ASTE network

The satellite network connects the telescope site to the outside world including the operation base in San Pedro de Atacama (altitude 2,400 m) or institutes in Japan through INTERNET2 collaborated with AccessNova. There are currently five operation sites as shown in Figure 2. Two sites (the telescope site and the base at San Pedro de Atacama) are in Chile, and the other three sites (NAOJ, NRO and University of Tokyo) are located in Japan.

2.2. N-COSMOS3

The new control software was developed based on the existing COSMOS3 system (Morita et al. 2003), which has been used for the NRO 45-m telescope and the NMA in Japan. This new system for ASTE employs the following features:

- **Remote operation through narrow-band network connection (56-64 kbps).**
  The original COSMOS3 was expanded with new functions as shown in Figure 3. Distinct features are information servers and traffic shaping.
- **Robust system to survive with sudden network connection failure.**

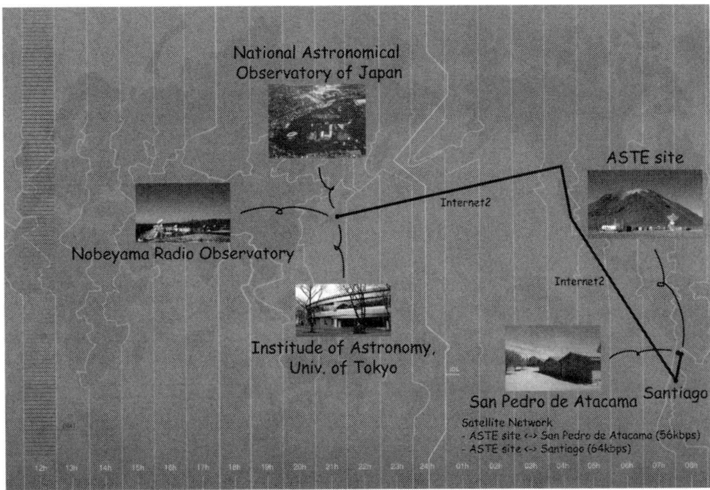

Figure 2. The current operation sites and the network routing. The satellite network connects the ASTE site to the base in San Pedro de Atacama and Universidad de Chile in Santiago. Internet2 lines collaborated with Access-Nova are used between NAOJ in Japan and Universidad de Chile in Santiago.

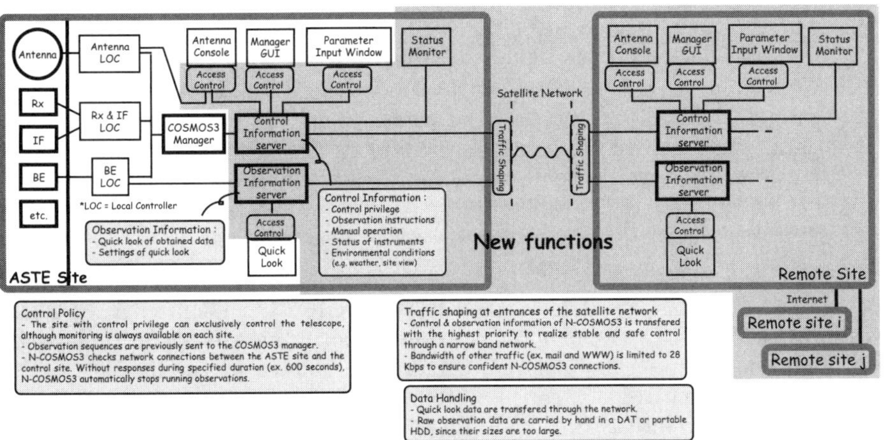

Figure 3. A diagram of N-COSMOS3. The gray zone indicates the expanded parts with new functions. All of the operational data are transfered via control and observation information servers, using a server-client method.

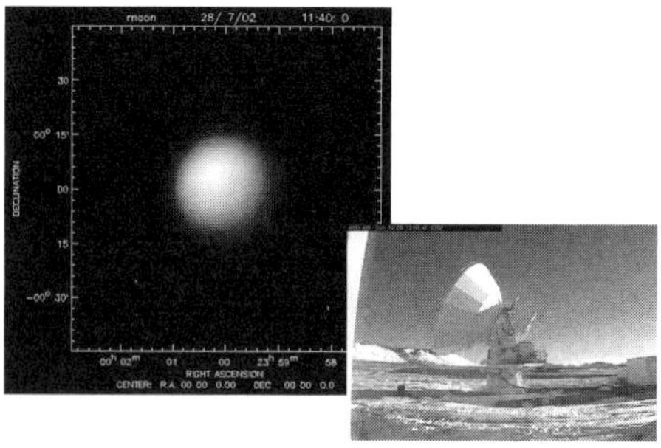

Figure 4. The first remote observations from the operation base in San Pedro de Atacama (Moon@345GHz & Web-camera image).

Observation tables are loaded into the COSMOS3 manager, and N-COSMOS3 at the ASTE site operates the telescope under the tables. N-COSMOS3 also checks network connections between the ASTE site and the control site. Without a response during a specified duration (e.g., 600 seconds), N-COSMOS3 automatically stops running observations.

- **Traffic shaping to guarantee necessary communication for telescope control.**
  Control and observation information of N-COSMOS3 is transfered with the highest priority to realize stable and safe control through a narrow bandwidth network. In addition, the bandwidth of other traffic (e.g., mail and WWW) is currently limited to 28 kbps to ensure confident N-COSMOS3 connections.
- **Multiple remote operating sites, including access control and easy change of remote sites.**
  The site with control privilege can exclusively control the telescope, although monitoring is always available at each site. Control and observation information such as control privilege and quick look is transferred through information servers via a server-client method. Raw observation data (maximum 15 GB/day) are too large to transfer through the satellite network; they are carried by hand on DATs or portable HDDs.

3. Status

We started operating the telescope with N-COSMOS3 in Chile in April 2002. We have succeeded in the first remote observation from the operation base in San Pedro de Atacama in July 2002 (Figure 4), followed by the first successful remote observation from Japan in October 2003. Now, most of observations and measurements are carried out from remote sites including the Japanese sites (Figure 5). Figure 6 shows astronomical results taken with such remote operations.

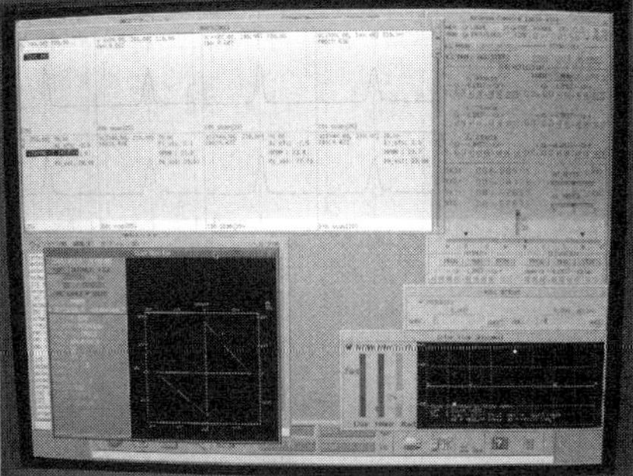

Figure 5. Remote observations in San Pedro de Atacama and a screen of N-COSMOS3.

Figure 6. Astronomical observations of ^{12}CO(J=3-2) emission toward the Carina region taken with remote operations (Yamaguchi et al. 2003).

References

Ezawa et al. 2004, Proc SPIE 5489, 763

Morita et al. 2003, in ASP Conf. Ser., Vol. 295, ADASS XII, ed. H. E. Payne, R. I. Jedrzejewski, & R. N. Hook (San Francisco: ASP), 166

Yamaguchi et al. 2003, IAUS221, 236

ASTRONOMICAL DATA ANALYSIS SOFTWARE AND SYSTEMS XIV
ASP Conference Series, Vol. 347, 2005
P. L. Shopbell, M. C. Britton, and R. Ebert, eds.

On the Use of IDL for Instrument Control

S. Mazuk and C. Venturini

The Aerospace Corporation P.O. Box 92957 Mail Code M2-266 Los Angeles, CA 90009-2957

Abstract. The Aerospace Corporation has developed a near-infrared and visible spectrograph that is used for astronomical observations at Lick observatory's 3 meter telescope. This paper describes the instrument control and data handling system, which employs the Interactive Data Language (IDL) for both the user interface and instrument control. The system employs IDL in a client-server design to control all aspects of data acquisition, and has been operational for several years. The use of IDL has simplified the system design and allowed for extensive modifications.

1. Introduction

The Aerospace Near-Infrared Imaging Spectrograph (NIRIS) is a long-slit spectrograph that covers the wavelength range from 0.38-2.5 micrometers. It uses two 1024x512 HcCdTe focal- plane arrays (FPAs) for the Near-IR (NIR), and one 1024x256 deep depleted Si FPA for the visible channel. Each FPA is fed by a separate camera, collimator and grating. A common slit and field lens ensures that all three channels of the spectrograph view the same field simultaneously, with beam splitters to separate the wavelengths to each array. The slit jaws are reflective, allowing a CCD camera to image the light not passing into the spectrograph for guiding, and provides an exact view of what is observed by the spectrograph. Figure 1 shows an image of Saturn on the slit, while Figure 2 shows the associated near infra-red spectra obtained. Figure 3 shows the spectra converted to reflectance units.

2. System Design

The control system for the instrument had several requirements. The system is planned to be used for several years, spanning multiple operating system and computer upgrades. The system was to be hardware independent, e.g. requiring no plug-in boards. The instrument design allowed for upgrades to the hardware, so the software architecture needed to be extensible as well. And the software was to be accessible and modifiable by the instrument Principal Investigator (PI) .

A previous version of the control software was written for the Microsoft Windows 3.1 operating system using the C language. This required expertise in using both the C language and the Windows Software Development Kit (SDK). With support for that OS ending there was a need to move to the Windows NT/2000 environment, which required a complete rewrite of the control software.

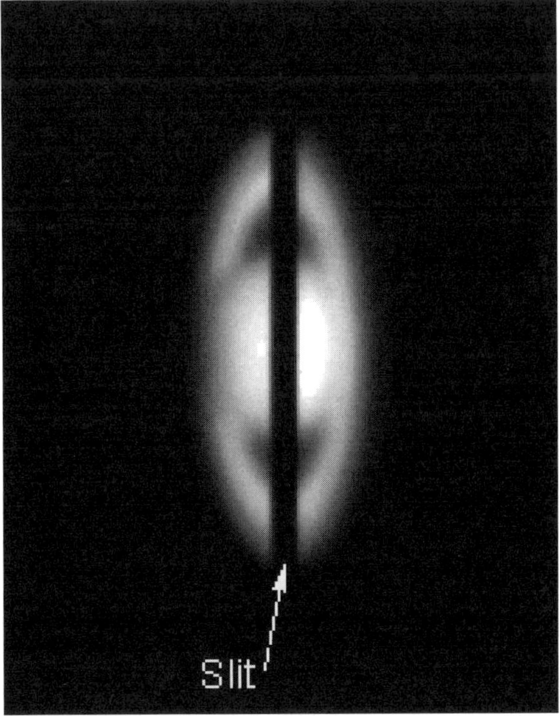

Figure 1. NIRIS entrance slit image showing the planet Saturn, taken using the Shane 3 meter telescope at Lick observatory.

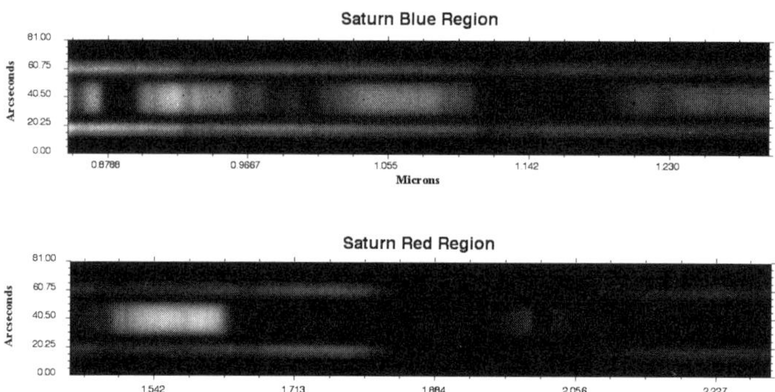

Figure 2. The near infrared spectra obtained from the regions of Saturn in the slit.

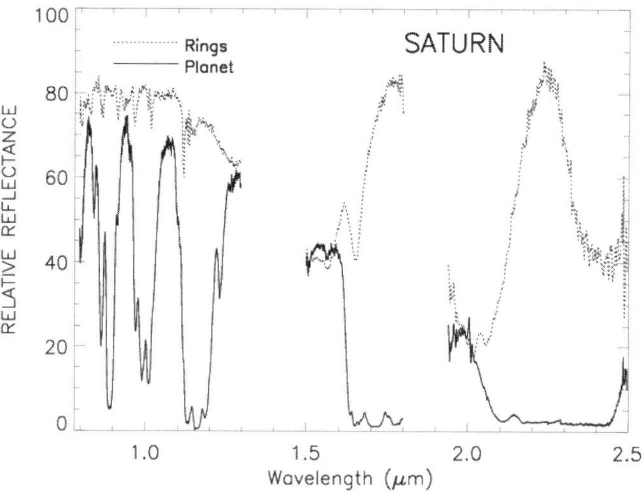

Figure 3. The spectra of Saturn reduced and converted to reflectance units.

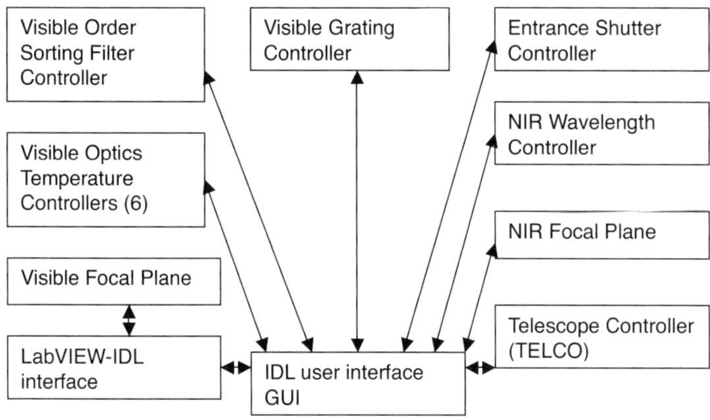

Figure 4. Functional block diagram of the NIRIS control system.

Since the data analysis software for the instrument was entirely written in IDL, and all the team members were facile with the language, we decided to use IDL to implement the control system as well. IDL implements the Unix socket communications protocol, which enables network communications with no additional hardware. This is a lossless bi- directional protocol, required for a control application, with IDL implementing client-side connections only. Since IDL is running as an application environment on top of the operating system, this insulates us somewhat from changes in the underlying OS.

The system design implication of going to a socket communication architecture is that all devices must connect to the local area network (LAN). Devices can be direct connected or can use interface conversion boxes, e.g. Ethernet-to-serial. The communications protocols must be clearly documented as well, since the socket standard only defines the interface and not the protocol. The protocols can be defined by the manufacturer for commercial off the shelf (COTS) products, or developed internally for lab-fabricated controllers.

The control system functions shown schematically in Figure 4 is implemented as an IDL widget graphical user interface (GUI). The visible and near infra-red focal planes are separate systems, and the focal planes are operated independently and asynchronously. Timing events generated by the GUI drive the system timing during both acquisition and idle times, using a state machine internal organization. During idle times the system monitors the temperatures of lenses in the visible channel. These temperatures are actively controlled to compensate for changes in the refractive index of the lenses and differential expansion of materials in the optical chain.

The commercial CCD camera used as the visible focal plane did not have a public communications protocol available, however a driver for the LabVIEW environment was available. We developed a socket communication protocol to communicate between IDL and LabVIEW to deal with this problem. The LabVIEW application handles all the camera control functions, and communicates with the IDL-based control program via the socket connection. One advantage of this approach was that we were able to tailor the communications protocol to meet our specific operational needs. Both the LabVIEW and IDL applications run on the same computer system at present, but they could be readily moved to separate machines as performance needs change.

3. Summary

This paper has described the control system used for the Aerospace NIRIS instrument. The control system is implemented in IDL to allow for rapid development and ease of modification. The use of socket communications simplifies system design so no plug-in cards are required in the controlling PCs. This results in an all-software system that can be easily transferred to another machine, eliminating a single point of failure. An IDL-LabVIEW socket communication application has been developed for control of a visible CCD focal plan. Implementing an IDL-LabVIEW interface allows control of equipment that may not have a public communications protocol available.

Acknowledgments. This work was supported under the Aerospace Corporation's Mission Oriented Investigation and Experimentation program, funded by the U.S. Air Force Space and Missile Systems Center under Contract No. FA8802-04-C-0001.

ASTRONOMICAL DATA ANALYSIS SOFTWARE AND SYSTEMS XIV
ASP Conference Series, Vol. 347, 2005
P. L. Shopbell, M. C. Britton, and R. Ebert, eds.

Database of Instrumental Characteristics in a Real Observatory

A. Micol[1] and D. Sforna[1]

ESA/ESO Space Telescope European Coordinating Facility, Karl Schwarzschild Str.2, D-85748 Garching, Germany

Abstract. The build of a central repository of the various characteristics of the science instrument contingent of HST and ESO Observatories is here illustrated. This repository, along with associated smart user interfaces, could become a fundamental tool for many and equally important reasons: for proposal preparation, or for archive research, and to help the various groups within the Observatory Operations (from the User Support to the Quality Control, down to the Archive) in their daily activities. The repository is central to many of the VO services that a data provider must develop (Access Protocols, Product Characterisation, etc.). While discussing the various problems a data provider has to solve on the way, an illustration of the first implementation is here shown. Indications on how this tool could be beneficial to the efficiency of the Observatory operations are given. Last but not least, applications in the Virtual Observatory context are suggested, especially with reference to Sforna et al. (2005).

1. Introduction

A repository of all the primary characteristics of the Hubble Space Telescope astronomical instruments is currently under development at the Space Telescope European Coordinating Facility. The possibility of extending it to the European Southern Observatory instruments is also contemplated.

The idea is to store in a Relational Database all the main characteristics (e.g., the field of view, the pixel scales, the transmission curves) for each of the available configurations of a given instrument,[2] with the ability to track changes due to human intervention or to unavoidable instrument degradation.

The aim is to have at hand a system and accompanying tools which can help both the entire Observatory to describe and monitor, and the Astronomers to understand, the performance of each instrument, or even more, of each instrumental configuration.

If properly constructed, and by taking into account that the users of such tools are not necessarily knowledgeable of all the internal intricacies of the Observatory Data Flow System, this tool can be used both by Astronomers during proposal preparation, or for archive research, and by the various groups within

[1]Affiliated with the Space Telescope Division of the European Space Agency, ESTEC, Noordwijk, Netherlands

[2]Or at least for the most common and supported instrument modes.

the Observatory Operations (from the User Support to the Quality Control, down to the Archive) for their daily activities.

While many are the possible uses of such repository, our main intent in building such system is to associate to each archived observation a snapshot of the actual instrument setup at the time of the observation.[3]

2. Current Implementation

Designing a database to collect the main characteristics of a single, simple instrument is an easy exercise; quite different is the design of a repository for a multitude of both simple and complex instruments, some for imaging (e.g., HST/WFPC2), some for spectroscopic investigations (e.g., HST/FOS, VLT/GIRAFFE), some for both (e.g., HST/STIS, VLT/VIMOS). Having to learn on the way the various idiosyncrasies of the more complex instruments, but willing to build some simple tools immediately useful for the simpler instruments, it has been decided to move on into a prototyping phase.

Our current prototype covers the aspects shown in Table 1.

Table 1. Sketch of the current database implementation

Aspects	Characteristics
Detector Geometry	Field of view, pixel scales, position on the focal plane, etc.
Filters	Central wavelength, min and max wavelengths, etc.
Dispersive Elements	Central wavelength, typical resolution, etc.
Transmission Curves	Overall transmission curves for combination of filter, detectors and optics
Dispersion Curves	Dispersion curves for gratings, prisms, grisms.

These aspects cover mostly what in the Virtual Observatory Data Modelling effort is called the Coverage and Sampling of an observation.

Apart from storing the physical characteristics, of which some were listed in the previous table, the database design contemplates other attributes:

- Instrument and telescope identifier, and instrumental modes
- Date ranges within which a certain characteristic was known to have a certain value
- A flag ("current") to indicate whether a given piece of information reproduces our current knowledge, or whether it has been declared obsolete

It is via the instrumental mode that a relation amongst various tables can be established. (In other words, it is the instrumental mode that specifies which filters

[3]We firmly believe in the reproducibility of a scientific measurement. Hence we claim it is a mandatory task for a Science Archive to store and provide easy access to the actual setup of the instrument that carried out a scientific experiment.

ESO & HST Detectors by Characteristics Query Form

Search Reset query Help | Home

Instrumental Bandpass

Match a Standard Bandpass johnson-morgan.V

and/or specify your criteria: Type Broad Medium Narrow

Central Wavelength.......... : nanometers

FWHM (nm)................... : nanometers

Detectors Pixel Scale

Pixel Scale X (")........... : <0.3

Pixel Scale Y (")........... :

Detectors Field of View

Fov (')..................... : >15 operator number (e.g., > 3

Num Pixels X................ :

Num Pixels Y................ :

Fov X (')................... :

Fov Y (')................... :

Other Choices

Telescope................... : HST 2P2 Instrument ACS WFPC2 WFI

Detector.................... :

Filter Name................. :

ObsMode..................... :

Current..................... :

Figure 1. Instruments by Characteristics Query Form

and which detectors can be used together.) Versioning is another important concept not only to know which information is obsolete, but to also associate with an old archived observation the correct knowledge of the instrument characteristics, that is, the characteristics as they were at the time of the observation. To clarify with and example, think to the overall transmission curve of an imaging instrument; the actual transmission might change because of a changed optical element, or because some element's transmissivity degrades. The database can keep track of such changes, and can hence be used both for trend analysis and for describing the status of an instrument using the best knowledge currently available for that particular moment in time (which could be different than the best knowledge available at the time of the observation).

Ancillary information are also stored in the database, like for example a table of characteristics of standard bandpasses, with references to the publications where the information was extracted from.

3. An Example

We built some tools to demonstrate the capabilities of the current prototype. In particular, a web form allows the Astronomers to pose questions like: *which HST or ESO instruments are capable of observing in a certain spectral band, with a resolution better than X and with a field of view bigger than Y?* (Figure 1).

Using the geometric and spectral coverage tables, an astronomer can easily discover which instrument is more appropriate for his/her scientific case. It is to be stressed that such a question can be posed using a terminology that any astronomer can understand. That is, the astronomer does not need to know any of the many specific terms and conventions adopted by the engineers that built the instrument or any other aspect of the observatory data flow. (A typical example of instrument-specific conventions are the names of the filters: e.g., the Wide Field Imager's filter $BB\#Ic/lwp\_ESO845$ resembles - it is not identical to - the standard I band passband.) All the concepts are described using numbers and units the astronomer is used to. For example, using the mentioned table of standard bandpass characteristics, a similarity criterion is applied to identify which instrument filters resemble best the standard bandpass specified by the astronomer.

Description of other examples, and of the actual technologies used, can be found in Sforna et al. (2005).

4. Benefits

Our main intent is to provide better Archive services and to publish HST (and possibly ESO) data to the Virtual Observatory. Nevertheless, the instrumental characteristics database can be used in many different contexts:

- by the Astronomers, to easily compare characteristics of different instruments and to identify the best instrument mode for a given scientific project, e.g., during the proposal preparation or archive exploitation;
- by the User Support group, to provide detailed information to the observers and to other astronomers -e.g., non-optical astronomers- in a standard way;
- by the Exposure Time Calculator team (Ballester et al. 2000), to provide estimates of the exposure time required to reach a given signal to noise ratio; versioning allows to revert such computation to provide an estimate of the limiting flux of archived data (Voisin et al. 2003);
- by the Quality Control Instrument Scientists, for trend analysis allowing the inspection of the changes of any instrument characteristic;
- by the Archive team, to build more scientific interfaces (e.g. a la Querator, queries by wavelength instead of by obscure filter names) and to link each observation to the actual throughput at the time of the observation, for better calibration and data exploitation;
- by the Virtual Observatory, to gather detailed instrumental capability of an Observatory, to build enhanced services, to implement access protocols (like IDHA, SIAP/SSAP), web services (Sforna et al. 2005), etc..

5. Status and Challenges

The database design is still in its infancy; more tables will be required to store many more parameters and to cope with the variety of ESO and HST instruments, and with the hundreds of different observing modes. The database design is a challenge in itself. A very mind-boggling problem which could benefit from (and be beneficial to) the work of the IVOA Observation Data Modeling group.

Collecting all the necessary information to populate such database is almost straightforward in the case of HST. A lot of work is daily done by the many Instrument Scientists at STScI; various databases or calibration reference files already contain the required parameters, and tools exist to extract them. For ground-based observatories this is not yet the case. Dealing with changes in a geographically dispersed operational environment (the Mountains in Chile, the Main centre in Europe) is not as easy. A proper communication channel must be established.

Maintenance of the database is an essential part of the problem. When an instrument characteristic changes, it could be because our knowledge has improved, or because the instrument has changed or degraded. The effect in the two cases is quite different. Versioning and Validity ranges are important concepts to track down which piece of information was used when, to make sure that a scientific experiment is reproducible.

In the HST case, the maintenance is partially based on manual intervention and partially based on scheduled scripts ("cronjobs") which automatically detect any difference in the reference files associated with the various instruments, and propagate the information to the database changing te previous valid "current" record to a "non-current" value, and resetting the involved validity ranges appropriately.

Acknowledgments. We thank Pascal Ballester for the useful discussions we had on the topic, and for having provided the transmission curves of the Wide Field Imager.

References

Ballester, P., et al. 2000, in ASP Conf. Ser., Vol. 216, ADASS IX, ed. N. Manset, C. Veillet, & D. Crabtree (San Francisco: ASP), 331

Sforna, D. et al. 2005, in ASP Conf. Ser., Vol. 347, ADASS XIV, ed. P. L. Shopbell, M. C. Britton, & R. Ebert (San Francisco: ASP), 257

Voisin, B., et al. 2004, in ASP Conf. Ser., Vol. 314, ADASS XIII, ed. F. Ochsenbein, M. Allen, & D. Egret (San Francisco: ASP), 125

ASTRONOMICAL DATA ANALYSIS SOFTWARE AND SYSTEMS XIV
ASP Conference Series, Vol. 347, 2005
P. L. Shopbell, M. C. Britton, and R. Ebert, eds.

Building a General Purpose Beowulf Cluster for Astrophysics Research

Matthew W. L. Phelps

Harvard - Smithsonian Center for Astrophysics

Abstract. The challenges of designing and deploying a high performance, Linux based, Beowulf cluster for use by many departments and projects are covered. Considerations include hardware, infrastructure (space, cooling, networking, etc.), and software; particularly scheduling systems.

1. Background

The Harvard - Smithsonian Center for Astrophysics (CfA) is a large institution with a wide variety of computational requirements. Today, some 300 Smithsonian and Harvard scientists and several hundred more support staff cooperate in broad programs of astrophysical research supported by Federal appropriations and University funds as well as contracts and grants from government agencies. These scientific investigations, touching on almost all major topics in astronomy, are organized into seven divisions. The Computation Facility (or CF) manages (among other things) the scientific computing resources of all divisions except High Energy Astrophysics.

Deploying a High Performance Computing cluster for a wide range of needs, using a wide range of funding sources, provided a considerable challenge.

2. Requirements

- Cost to be spread over the several groups participating.
- Limited personnel resources for hardware and software installation and management.
- Compatibility with existing environment of Linux and SPARC Solaris platforms.
- 64-bit addressing for large data sets in a single process.
- Resource allocation for each group based on size of contribution (i.e. number of machines purchased) to the cluster, however "public" access to available resources should be provided to the CfA community at large.

3. Hardware Solution

- AMD Opteron based SunFire V20z nodes from Sun Microsystems
  - Dual processor, 64-bit capable, 1U rack-mountable servers
  - Runs 64-bit Linux with compatibility for 32-bit applications

- Existing Sun - CfA relationship made Educational Promotional discount
- Sun Compute Grid Rack System
  - Pre-assembled 32 node cluster with gigabit network switch all in one cabinet
  - Just add power (a lot!) and a gigabit connection to the public network

4. Software Solution

- ROCKS Cluster Distribution from http://rocksclusters.org
  - Based on RedHat Enterprise Linux Release 3
  - Supports Opteron Hardware in 64-bit mode
  - Simple install of frontend node, followed by automatic Kickstart install of compute nodes via PXE boot
- Sun Grid Engine Enterprise Edition
  - Extremely flexible resource queuing system
  - Included in ROCKS distribution
  - Queues based on projects, Unix groups, or individual users
  - Resource allocation of a percentage of all cluster resources based on Projects

5. Problems

- Weight: Cabinet is 1800 pounds (815 kg), machine room tiles need to be reinforced.
- Size: Cabinet needed professional movers to go from loading dock to machine room.
- Heat: AC in machine room may need boosting
- Power: 60 A 220 V power source needed (4 big plugs [NEMA L6-20P])
- User accounts: ROCKS uses 411 instead of NIS for user account (and other) information which doesn't integrate with our existing environment. Separate accounts (with matching UIDs and GIDs) are necessary.
- Software installs: ROCKS can install any RPM based software on all nodes, but non-RPM based software (e.g. IRAF and IDL) is problematic. An NFS mounted directory on an existing Network Appliance disk server was set up for this. It is a sub-optimal solution because of the network traffic.

6. Web Resources

- Sun Fire V20z http://www.sun.com/servers/entry/v20z/
- Sun Compute Grid Rack System http://www.sun.com/servers/computegrid
- ROCKS High Performance Linux Cluster Solution http://rocksclusters.org

ASTRONOMICAL DATA ANALYSIS SOFTWARE AND SYSTEMS XIV
ASP Conference Series, Vol. 347, 2005
P. L. Shopbell, M. C. Britton, and R. Ebert, eds.

A New Data Flow System for the Nordic Optical Telescope

Saskia Prins and Peter M. Sørensen

Nordic Optical Telescope, Apartado 474, E-38700 Santa Cruz de La Palma, Canarias, Spain

Abstract. To remain scientifically interesting for the Nordic community in the 8-10m era, the NOT focuses its resources on flexible instrumentation and near real-time data assessment to accommodate quick-response observations. To this end we are currently implementing a new data flow system which includes new FITS formats and headers, quality control tools and science data reduction pipelines, metadata databases and several upgrades to the observing system.

1. Introduction

The Nordic Optical Telescope[1] (NOT) is a 2.6 m telescope for observations at optical and near-infrared wavelengths. It is operated jointly by Denmark, Finland, Iceland, Norway, and Sweden on La Palma, Canary Islands, Spain.

In view of the expected change of focus of resources to 8+ m telescopes, the NOT must be operated in a way that is both cost-efficient and scientifically interesting. To remain scientifically interesting for the Nordic community, the NOT focuses its resources on flexible instrumentation and near real-time data assessment to accommodate quick-response observations. The NOT data flow system will therefore have to adjust to a likely future situation of increasing data volumes and complexity with decreasing manpower to handle these data.

To this end we are currently implementing a new data flow system which includes new FITS formats and headers, quality control tools and science data reduction pipelines, metadata databases and several (mostly low-level) upgrades to the observing system. Figure 1 illustrates the development of this data flow system.

2. An Overview of the NOT Data Flow System

- **Handling of incoming proposals** Proposals are submitted by e-mail and processed by an e-mail parser. Proposals and schedules are kept in relational databases.
- **Collection of astronomical data and various metadata** The observing system is undergoing major mostly low-level upgrades; the new TCS and ICS using SOAP for subsystem access (Verkkoniemi et al. 2004) will become operational in 2005. At the same time the network speed on the

[1] `http://www.not.iac.es/`

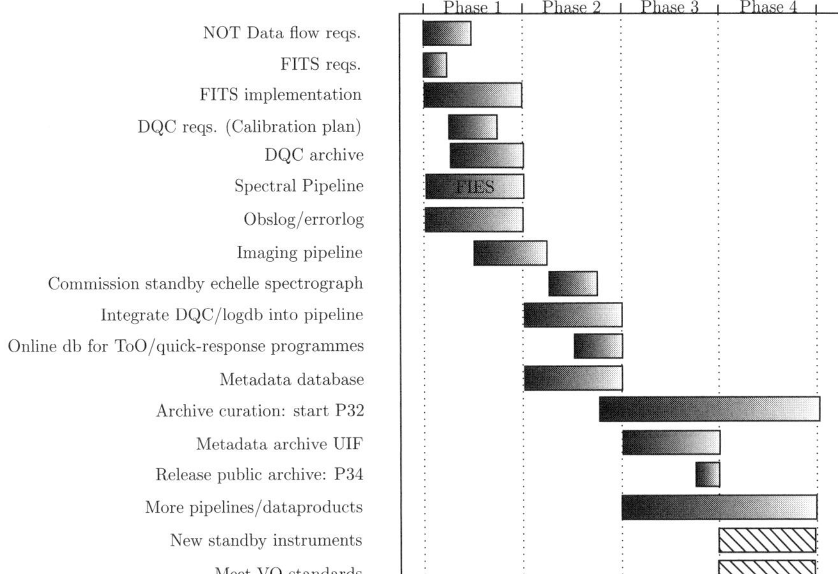

Figure 1. Phased development of the NOT data flow system. Phase 1: preparations for automated postprocessing. Phase 2: implementation of automated postprocessing. Phase 3: implementation of a searchable online archive. Phase 4: preparations for next generation stand-by instruments, meet VO standards.

mountain will be upgraded with a gigabit switch. For online storage of data and metadata on site a RAID-array is used.

- **Near real-time data assessment** Quicklook tools, quality control tools for detector characteristics and instrument performance monitoring, spectroscopic and imaging data reduction pipelines are currently being developed. Quality control curation is performed at planned intervals.
- **Archiving and distributing of raw data and data products** Data copying onto CD or DAT (depending on data rates) for visiting observers and the archive will be replaced by a robotic DVD burner placed in a controlled-humidity environment.
- **Public science archive** Archive requirements for a searchable online database for metadata of raw data and pipeline products, and an off-line system for handling of data set requests are being defined.

2.1. Enhancements to NOT FITS Files

As an important step towards pipeline data reduction, automated observing logs, quality control tools, and a searchable data archive, the data structures and header information in the NOT FITS files are being revised. Mostly for historic reasons there is no uniform data structure in use at the NOT for multi-channel, multi-window or multi-readout data. These different formats (separate files, single tiled image, data cubes) are all being changed into MEF.

Secondly, the FITS header information is being extended, rearranged into classes and properly documented in a FITS dictionary. The new classes include:

- FITS mandatory keywords
- WCS
- Data description (including PI and programme info)
- Observation
- Object information
- Site information
- Meteo (measurements from external weather station)
- TCS (including dome and internal measurements)
- Adapter (CCDprobe, atmospheric dispersion corrector, other pieces of foreoptics, like a polarisation unit)
- Autoguider
- Calibration unit(s)
- FASU (Filters and shutter)
- Instrument
- Detector
- Pipeline/postprocessing/Quality Control

In Memoriam

On October 19, 2004 our dear friend and colleague Markku Verkkoniemi died in an accident while driving up to the observatory. In times of ever-increasing virtual observatories, we must not forget that most ground-based observations still require observatory staff to travel to remote mountain sites.

References

Verkkoniemi, M., Clasen., J. W., & Sørensen, P. M. 2004 in ASP Conf. Ser., Vol. 314, ADASS XIII, ed. F. Ochsenbein, M. Allen, & D. Egret (San Francisco: ASP), 744

ASTRONOMICAL DATA ANALYSIS SOFTWARE AND SYSTEMS XIV
ASP Conference Series, Vol. 347, 2005
P. L. Shopbell, M. C. Britton, and R. Ebert, eds.

The CARMA Correlator

Kevin P. Rauch
Department of Astronomy, University of Maryland, College Park, MD 20742-2421

Rick Hobbs and David W. Hawkins
Caltech-OVRO, 100 Leighton Lane, Big Pine, CA 93513

Abstract. We describe the software development effort for the first-light correlator of the Combined Array for Research in Millimeter-wave Astronomy (CARMA[1]), the merger of the Owens Valley Radio Observatory (OVRO[2]) and Berkeley-Illinois-Maryland Association (BIMA[3]) millimeter arrays that is expected to reach first-light in 2005. The digital hardware relies on programmable logic devices (FPGAs), signal processors (DSPs), and Linux-based host computers acting in concert to produce baseline visibility data. We summarize the hardware configuration, the distribution of computational tasks among components, and the tools and techniques used to program these devices.

1. Introduction

The CARMA millimeter-wave radio array, currently under construction at a site in the mountains above Big Pine, CA, is scheduled to reach first-light in Fall 2005. CARMA will initially be a 15-station interferometer consisting of 9 six meter dishes (formerly part of the BIMA array) and 6 ten meter dishes (the former OVRO array), eventually growing to 23 stations with the addition of the 8 three meter Sunyaev-Zeldovich Array (SZA) dishes upon completion of SZA survey operations (currently underway on the Owens Valley floor). At first-light a total observable bandwidth of 4 GHz (in each of two sidebands) is anticipated, distributed among eight individually-configurable 500 MHz correlator bands. Individual bands can be reconfigured for high-resolution spectral operation, with a selectable bandwidth ranging from 2 MHz to 250 MHz.

The relatively wide bandwidth and large number of baselines for CARMA, combined with the dynamic reconfiguration required of individual correlator bands, demand that the CARMA correlator be a flexible instrument capable of cross-correlation with high throughput. These objectives are met through the use of reprogrammable digital logic devices (field-programmable gate arrays, or FPGAs) working together with generic embedded signal processors (DSP or

---

[1] `http://www.mmarray.org/`

[2] `http://www.ovro.caltech.edu/`

[3] `http://bima.astro.umd.edu/`

CPU devices) to deliver baseline visibility data to a Linux host computer. The first-light correlator will be a hybrid system containing three bands (1.5 GHz total bandwidth) of reprogrammed digital hardware originally used in the Caltech Owens Valley Broadband Reconfigurable Array (COBRA) correlator, the remaining bands consisting of newly revised and upgraded digital components. The estimated spectral resolution of the revised correlator bands is shown in Table 1.

Table 1. Revised CARMA Correlator Resolution Estimates

Bandwidth (MHz)	Channels (per sideband)	δV[3 mm] (km/s)	$V_{\rm tot}$[3 mm] (km/s)	δV[1 mm] (km/s)	$V_{\rm tot}$[1 mm] (km/s)
500	100	15	1500	5	500
250	175	4.3	750	1.4	250
125	250	1.5	375	0.5	125
62	300	0.63	188	0.21	62
31	350	0.27	94	0.09	31
8	400	0.06	23	0.02	8
2	400	0.015	6	0.005	2

The digital hardware consists of a set of compact PCI (cPCI) digitizer, correlator, and Intel x86 CPU cards housed in cPCI crates with custom backplanes. The following sections summarize the digital signal processing tasks for which each component is responsible, and the mechanisms used to transfer data between them and ultimately up into the high-level software pipeline.

2. Digitizer Board Processing

Antenna signal digitization and spectral band shaping is performed by a set of digitizer boards; each contains two digitizer modules and processes two independent antenna signals. A COBRA digitizer board is shown in Figure 1. Each band can be configured to process a total bandwidth ranging from 2 MHz to 500 MHz, in incremental factors of two to four (cf. Table 1). Spectral band shaping is accomplished using digital filtering techniques inside the digitizer board FPGAs. Digitized input signals first pass through a digital delay line and fractional sample delay filter to correct for geometric and other offsets in the analog domain. The fractional delay filter is implemented as an asymmetric finite-impulse response (FIR) digital filter, for which numerous design techniques exist (see Laakso, Valimaki, & Karjalainen 1996 for a review).

The 2-bit quantized, 1 GHz sampled (500 MHz bandwidth) data streams—one per antenna, two antennas processed per digitizer board—then pass through a series of digital filtering and decimation stages until the Nyquist bandwidth is reduced from 500 MHz to that of the spectral mode requested by the observer. The design of the decimator is based on multi-stage, multi-rate techniques (Crochiere & Rabiner 1983); the modularity of each stage allows multiple bandwidth con-

Figure 1. COBRA digitizer board, including: two 1GHz 2-bit digitizers, 4 Altera FLEX10K FPGAs, a TI DSP, a PCI interface, and 4 LVDS front panel connectors.

figurations to be supported with a high degree of code reuse. The FPGA logic devices are programmed using the hardware description language VHDL (e.g., Ashenden 2002), an open industry standard whose syntax is reminiscent of Pascal. Finally, the decimated data signals are re-quantized to 2-bit samples for transmission to the correlator boards via a set of 32-bit LVDS (UltraSCSI) outputs. In wideband mode (500 MHz), this involves time-demultiplexing the data by a factor of 8 to reduce the absolute transmission frequency from 1 GHz to 125 MHz.

3. Correlator Board Processing

CARMA implements a lag (i.e., XF) correlator. A correlator board is responsible for calculating the cross-correlation lag spectrum of a portion of the $N(N-1)/2$ possible antenna baselines, where N is the number of antennas in the array; in the COBRA design (Figure 2), 10 baselines are processed per board, one per FPGA. As the cross-correlation logic in these devices is limited to a maximum operating frequency of approximately 125 MHz, all spectral modes wider than 62 MHz bandwidth require the use of parallel logic structures to support the required data throughput; this significantly decreases the spectral resolution in these modes (cf. Table 1). COBRA hardware is limited to 2-bit digitizer samples and computes lag values using a specially-weighted 2-bit by 2-bit cross-correlation scheme (the "modified two-bit" scheme of Hagen & Farley 1973) with an efficiency of 0.87 relative to an ideal multi-bit correlation. The revised CARMA digitizers will support input samples up to eight bits wide, which may allow higher efficiency cross-correlation schemes to be implemented, subject to data transmission and spectral resolution constraints.

The correlation logic accumulates lag values for 16 ms (determined by a hardware phase-switch period); the lags are then dumped to on-chip SRAM. A DSP integrated onto the board reads and processes these lags for all FPGAs on the

Figure 2. COBRA correlator board, including: 10 Altera FLEX10K FPGAs, a TI DSP, a PCI interface, and 4 LVDS front panel connectors.

board while the FPGAs continue with the next integration cycle. The DSP is responsible for computing the lag spectra via an FFT and then further accumulating the lag spectra for 500 ms, the time interval on which the visibility data is transferred to the Linux host. DSP functionality is implemented using the MicroC OS real-time kernel (Labrosse 2002).

4. Host CPU Card Processing

Each correlator band contains one Intel x86 CPU card. It is a diskless PXE client and runs the Linux operating system, booted via a network file server. The Linux host is responsible for aggregating the baseline visibilities from all correlator boards and making them available on the network to the high-level software processing pipeline. New data is published every 500 ms as a CORBA distributed object using the CORBA Notification Service. Data transfer from the correlator boards to the Linux host occurs over the PCI backplane via a custom Linux kernel driver (an interrupt is triggered on the Linux host by the correlator board DSP to alert the host to new data). Network communication is through a Gigabit ethernet switch. The visibility data from all bands is further aggregated and processed by the software pipeline computer, for eventual storage in an off-site archive.

5. Conclusion

The extensive use of programmable logic in the CARMA correlator results in a highly flexible instrument in which spectral resolution, bandwidth, and observing efficiency can be balanced according to the needs of the individual observer. The continuation of Moore's Law will allow future hardware revisions to further increase overall capability while maintaining compatibility with the existing VHDL codebase. This approach minimizes hardware development time so that technological advances can be incorporated into each generation in a timely fashion.

Acknowledgments. This overview was written on behalf of the CARMA consortium, which includes the University of California (Berkeley), the University of Maryland (College Park), the University of Illinois (Urbana-Champaign), the California Institute of Technology Owens Valley Radio Observatory and the University of Chicago. This work was supported by the National Science Foundation through grant NSF AST-0028963 to CARMA.

References

Ashenden, P. 2002, The Designer's Guide to VHDL (Morgan Kaufmann)

Crochiere, R. & Rabiner, L. 1983, Multirate Digital Signal Processing (Englewood Cliffs: Prentice-Hall)

Hagen, J.B. & Farley, D.T. 1973, Radio Science, 8, 775

Laakso, T.I., Valimaki, V. & Karjalainen, M. 1996, IEEE Signal Processing Magazine, 13, 30

Labrosse, J. 2002, MicroC OS II: The Real Time Kernel (CMP Books)

ASTRONOMICAL DATA ANALYSIS SOFTWARE AND SYSTEMS XIV
ASP Conference Series, Vol. 347, 2005
P. L. Shopbell, M. C. Britton, and R. Ebert, eds.

PESO - The Python Based Control System of the Ondřejov 2m Telescope

Petr Škoda, Jan Fuchs, and Jaroslav Honsa

Astronomical Institute, Academy of Sciences of the Czech Republic, 25165, Ondřejov

Abstract. Python has been gaining a good reputation and respectability in many areas of software development. We have chosen Python after getting the new CCD detector for the coudé spectrograph of Ondřejov observatory 2m telescope. The VersArray detector from Roper Scientific came only with the closed source library PVCAM of low-level camera control functions for Linux, so we had to write the whole astronomical data acquisition system from scratch and integrate it with the current spectrograph and telescope control systems. The final result of our effort, PESO (Python Exposure System for Ondřejov) is a highly comfortable GUI-based environment allowing the observer to change the spectrograph configuration, choose the detector acquisition mode, select the exposure parameters, and monitor the exposure progress. All of the relevant information from the control computers is written into the FITS headers by the PyFITS module, and the acquired CCD frame is immediately displayed in an SAO DS9 window using XPA calls. The GTK-based front end design was drawn in the Glade visual development tool, giving the shape and position of all widgets in single XML file, which is used in Python by a simple call of the PyGlade module. We describe our experience with the design and implementation of PESO, stressing the easiness of quick changes of the GUI, together with the capability of separate testing of every module using the Python debugger, IPython.

1. Introduction

At the end of 2003 a new CCD camera system for the Ondřejov coudé spectrograph was bought. It is an LN2-cooled VersArray 2048B with and EEV 2kx2k thinned chip from Roper Scientific. The provided data acquisition software (for MS Windows only) is, however, not suitable for astronomical purposes (targeted rather to laboratory spectroscopy). As we needed to incorporate the detector in a structure of several Linux based computers, called by the control system of another CCD camera as well, we decided to write an entirely new data acquisition program from scratch. After a short evaluation of requirements, we have chosen Python as the main development language. The fully functional system was ready in three months (including the GUI), designed by one programmer (who just started to learn Python) and one astronomer.

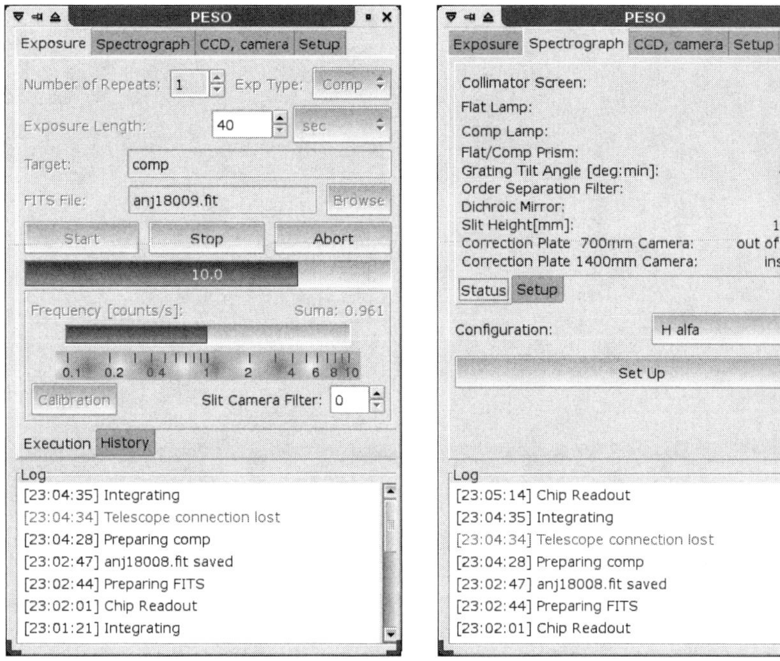

Figure 1. The Exposure (left) and Spectrograph (right) window tabs

2. PESO Overview

PESO is a multi-threaded Python script providing a comfortable GUI interface for controlling all subsystems involved in CCD spectra acquisition. It sends commands to spectrograph micro-controller units, creates FITS files from CCD data acquired, including the information gathered into FITS headers from the telescope and spectrograph control computers, monitors progress of exposures, maintains observing logs, and sends the final image for display to DS9 using XPA calls. It speaks to the CCD controller using a binary library provided by Roper Scientific in a Linux version as well. The PVCAM library provides only low-level functions for setting the readout parameters (speed, geometry of sub-region, binning, shutter mode), getting the status (temperature, integration mode), and starting the exposure (including opening the shutter). All higher-level work is done in Python. Our program requires the observer to do things in a chronological order and complains if something important was left out (e.g., observation of a star requires first its name to be given as well as the names of the observers). On the other hand, it works with a number of reasonable predefined values (either built-in or read from configuration files) and can be extended easily. It reconfigures the widgets on-line according to current state of the exposure (e.g., disables certain buttons, creates logs and history).

The GUI of PESO consists of four main window tabs (see Figures 1 and 2) containing widgets for controlling the main areas of the CCD acquisition process.

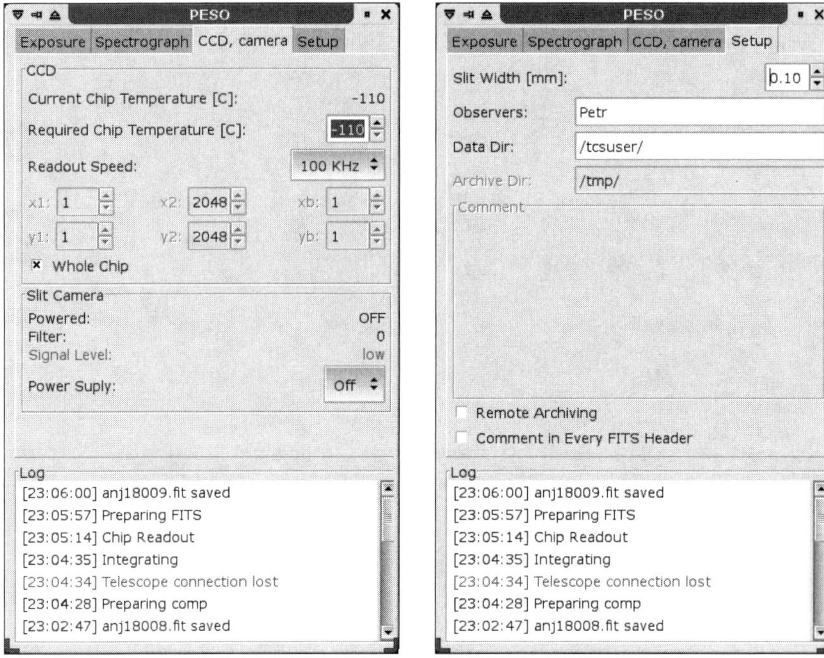

Figure 2. The CCD and Camera (left) and global Setup (right) tabs

Exposure tab: This is the main window, where the exposure is started and its progress monitored (sub-tab *Execution*) or all previous frames may be displayed together with the most important header items (sub-tab *History*). The type of exposure (flat, zero, comp, dark, target) is selected. If it is the target, the name of the object is requested, or a list of all objects exposed in the current session is shown. At the same time their last exposure time or count value is preset. For calibrations (flat, comp,) a recommended exposure time is preset according to the spectrograph setup. Each exposure may be repeated several times.

An exposure may be finished either after the given time elapses or after accumulating a given number of counts from an exposure meter (option button [`sec`]/[`Mcounts`]). The FITS name is auto-incremented according to a strict scheme (first two letters represent year and month) but may be changed. Once an exposure is running, the progress bar shows the elapsed time and on star the count frequency and total sum of counts. The exposure may be prematurely *Stop*ped (with immediate readout) or *Abort*ed (without chip readout). The *Calibration* button re-calibrates the bar to unity if the star gets fainter (e.g., cirrus clouds).

The function of changing the slit camera filter is put here, as the filter has to be changed for different stars.

Every action is logged to a log file and shown in the log window - warnings and errors are highlighted, and a warning sound is played. The success sound is played at the end of each exposure, and another if the signal of the exposure meter falls too rapidly, i.e., the star goes out of the slit. Every FITS file after its creation is sent by an XPA interface to DS9 for preview, including the x and y slices.

Spectrograph tab: Required spectrograph configuration (grating tilt, filters, ...) can be set from configuration files (e.g., for Hβ) or accorded values from the last observation of the same target. Recommended integration times for flats and arcs are set and displayed in the Exposure tab as well. All values may be further modified in the *Setup* sub-tab or watched in the *Status* sub-tab.

CCD and Camera tab: The chip readout parameters (speed, sub-frame size, binning) and chip temperature are set here. In addition, the slit-viewing camera is switched on from this tab.

Setup tab: Several global values are required to be set before the start of observation. The current version expects the names of the observer(s), the slit width, and the default directory for primary FITS storage. If the *Remote Archiving* mark is checked, a copy of the same file is sent to a (preconfigured) archive directory.

3. Some Highlights of PESO Development

3.1. The GUI Development

One of the most time-consuming tasks is the GUI design. We have done it quickly using the GTK library together with the Glade-2 development environment. All widgets are simply prepared by their visual placement and changing their properties, and the final GUI is written in a single XML file `PESO.glade`. This file is used by a Python module `gtk.glade` to create the objects for all of the widgets. Thus the GUI is built after PESO startup in memory. The change of widget objects properties is done interactively. An example of the part of `PESO.glade` follows. It shows an (abridged) definition of a simple file section menu with an OK button, and the activation of this menu from Python.

```
<widget class="GtkFileSelection" id="file_FS">
  <property name="title" translatable="yes">File Selection</property>
  <property name="window_position">GTK_WIN_POS_MOUSE</property>
   ...
  <child internal-child="ok_button">
    <widget class="GtkButton" id="ok_button1">
      <property name="visible">True</property>
      <property name="focus_on_click">True</property>
    </widget>
  </child>
</widget>

  def file(self, file_BT):
    self.xml_file_FS = gtk.glade.XML('PESO.glade', 'file_FS')
    self.xml_file_FS.signal_autoconnect(self)
    self.file_FS = self.xml_file_FS.get_widget("file_FS")
    result = self.file_FS.run()
```

3.2. Calls of PVCAM Functions

The detector control is done by calling internal functions hidden in the PVCAM library from their C wrappers. The C wrapper is a C function returning a pointer to a Python object and accepting parameters in a special structure. Such a library of C functions is then recognized as a Python module. All PVCAM functions are finally called as Python functions of this module – even from a debugger window. Here is an example of a C wrapper for calling a PVCAM library function `pl_exp_start_seq` from Python as a `pvcam` module function `pvcam.exp_start_seq()`:

```
PyObject* py_exp_start_seq(PyObject* self, PyObject* args){
  if (!pl_exp_start_seq(cam, buffer))
    pl_error_message(pl_error_code(), msg);
  return Py_BuildValue("s", msg); }
PyMethodDef pvcam_methods[] = {
  {"pl_exp_start_seq", (PyCFunction)py_exp_start_seq, METH_NOARGS, NULL},
...  };

void initpvcam(void) { Py_InitModule("pvcam", pvcam_methods); }
```

3.3. Threads

In PESO there run several threads all of the time (including GTK event threads), but some are spawned upon request of a particular function (e.g., timer of exposure integration). To allow this behaviour, the Python `threading` module must be used. If some changes in widgets are needed, `gtk.threads_enter` must be called on entry and `gtk.threads_leave` after leaving this function. An example of hiding the progress bar at the end of an exposure follows:

```
gtk.threads_enter()
self.countdown_PB.set_child_visible(gtk.FALSE)
gtk.threads_leave()
```

3.4. Debugging

The `IPython` debugger is a module which simply spawns another thread attached to the terminal window, from which a running process may be interactively controlled. The functions may be called with custom parameters, the values of variables may be examined, and the properties of GUI widgets changed.

4. Conclusions

We have shown that Python is well-suited for the rapid development of instrument control systems. It allows the quick design of a GUI interface using the XML-based Glade-2 development environment, easy calling of external library functions (even closed-source) by encapsulating C calls in Python modules, and can provide complex process control using a multi-threads module. Its usage for astronomical data acquisition is eased by the PyFITS module. The debugging of such a complex system is very simple using `IPython`.

Acknowledgments. This work was supported by the grant 102/02/1000 GA ČR. The Astronomical Institute Ondřejov is supported by projects K2043105 and Z1003909.

ASTRONOMICAL DATA ANALYSIS SOFTWARE AND SYSTEMS XIV
ASP Conference Series, Vol. 347, 2005
P. L. Shopbell, M. C. Britton, and R. Ebert, eds.

Telescope Automation and Remote Observing System (TAROS)

Greg Wilson, Adam Czezowski, Gary R. Hovey, Mark A. Jarnyk, Jon Nielsen, Bill Roberts, Kim Sebo, Dione Smith, Annino Vaccarella, and Peter Young

Research School of Astronomy and Astrophysics, Mt Stromlo Observatory, Australian National University, Canberra, Australia

Abstract. TAROS is a system that will allow for the Australian National University telescopes at a remote location to be operated automatically or interactively with authenticated control via the internet. TAROS is operated by a Java front-end GUI and employs the use of several Java technologies - such as Java Message Service (JMS) for communication between the telescope and the remote observer, Java Native Interface to integrate existing data acquisition software written in C++ (CICADA) with new Java programs and the JSky collection of Java GUI components for parts of the remote observer client. In this poster the design and implementation of TAROS is described.

1. Introduction

TAROS is a remote observing system currently being implemented at the Mt Stromlo Observatory. The approach has been to use proven technologies and to aim for platform independence. TAROS will allow for telescopes at a remote location to be operated in fully automated, remote interactive, assisted remote interactive or local interactive observing modes.

- Automated observing is an observing mode in which the telescope and instruments are fully controlled by computer and observations are scheduled by the computer. Monitoring will be possible from SSO and remote sites.
- Remote observing is an observing mode in which an observer at a remote location has full control of the telescope and instrument, with no assistance from the staff.
- Assisted observing is an observing mode in which an observer, usually at a remote location, has full control of the telescope and instrument, with limited assistance provided by an observing assistant at the telescope site.
- Local observing is an observing mode in which an observer at the telescope site has full control of the telescope and instrument.

TAROS will provide for the integrated control of the telescopes independently of their mode of operation. From the observer's point of view there is no difference between local and remote observing. TAROS will also control the flow and archiving of data obtained from instruments attached to the telescopes. It will provide fully automated environmental monitoring to ensure that telescopes are only operated when appropriate. In what follows we provide an overview of the overall TAROS architecture and then focus on two of the more mature components of TAROS - the status and message server and the displayer sub-system.

2. Architecture

2.1. Technologies

TAROS is implemented using the programming languages C, C++ and Java. It is a distributed, multi-threaded and multi-process system. Processes communicate with each other using standard and reliable protocols. Technologies that are employed in TAROS include:

Java Web Start - for distribution and launching of the observer's GUI.

JMS - Java Message Service, underpins a significant proportion of inter-process communication within TAROS, both point-to-point and broadcast;

JNI - Java Native Interface, allows Java and non-Java code to work together;

JSky - is a collection of reusable Java components for use in astronomy;

RPC - Remote Procedure Calls, provide a means for a process on one machine to call a process on another such that the programmer can treat both local and remote procedure calls in a like manner;

XDR - External Data Representation Standard, a standard used for the description and encoding of data across multiple hardware platforms;

XML - Extensible Markup Language, acts as the "specification language" for observation blocks.

2.2. Core Processes

The overall architecture for TAROS is shown in Figure 1. The main components are briefly described below:

- **Control** - this process starts the system and ensures that processes are still running.
- **Monitor** - this process monitors the health of sub-systems.
- **Gateway** - this process authenticates TAROS users and relays TAROS commands from the user to the other TAROS components.
- **Status, Health and Message Server** - provides a central repository of system status and health information which will be available to all other TAROS components. Both "message queue" and "subscriber-provider" models of message/data distribution are available.
- **Scheduler** - accepts requests for observation from external schedulers, including the interactive GUI. It interacts with the various TAROS sub-systems to adjust observation priorities and to initiate observations.

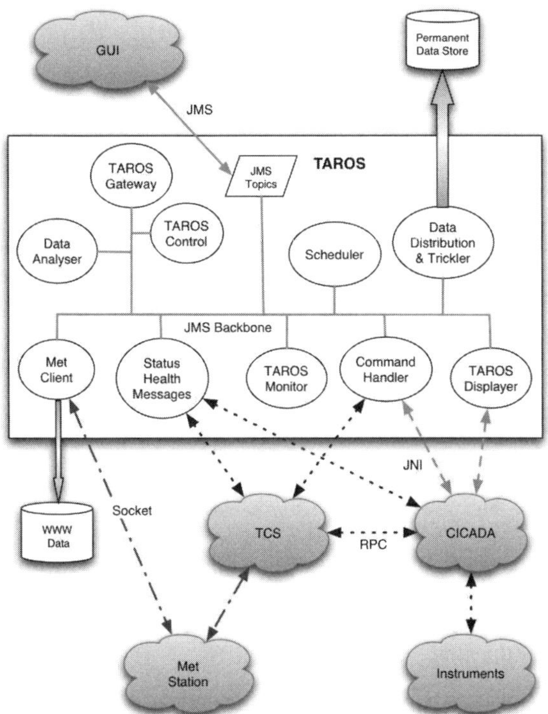

Figure 1. TAROS Architecture

- **GUI** - Java based GUI that can be used from anywhere on the Internet.
- **Displayer** - supplies images taken by the data acquisition system to the observer's image display tool. Images may be compressed to a level determined by the observer.
- **Telescope Control System** - The TCS will have its own internal mechanisms for startup and error recovery. It is responsible for all telescope hardware, handling telescope commands and maintaining status information such as interlocks, position and motion status (see Jarnyk & Hovey 2004 for more details).
- **Meteorological Station** - is also an independent subsystem that provides meteorological data. This information will be obtained by TAROS subsystems via the Met Client.
- **Data Distributor** - performs data "trickling" to remote observers and data archiving on various media types.
- **CICADA** - CICADA (Young et al. 1997, Young et al. 1999) is the current data acquisition software used at the ANU telescopes. TAROS has been

designed so that this software can be reused with a few modifications as possible.

3. Status, Health and Messages

In implementing a remote observing system it is necessary for the remote client software (and observer) to be informed of changes in telescope parameters as well as messages from the telescope control system and data acquisition software. TAROS provides this functionality. Sets of system parameters are defined using a subset of the Remote Procedure Call Language (RPCL - we do not include pointers or programs). This allows for parameters to be of any standard RPCL types such as arrays, structures, unions, and enumerated types. A shared memory database is used to store parameters in machine independent XDR format. Parameters are converted from XDR format to native machine format as they are retrieved.

The system requires a master database to be configured on one machine. Slave databases running on other machines can send a copy of the entire contents of their database or any structure within their database as an atomic update to the master database through the transfer of a block of XDR formatted memory. In this way it is ensured that telescope control system parameters are always kept as a self-consistent block in the TAROS master database.

The underlying database implementation is in C++, using RPC as a means of transferring parameter updates from slave databases to the master database. A Java Native Interface is also provided to allow full access to the database from Java applications running on the same machine as the master database. To facilitate access from remote Java clients (such as the GUI) a server application has been developed in Java. It receives incoming commands via a JMS queue and provides parameter and message updates via JMS topics. Parameter values can be explicitly requested from the server or the client application can elect to receive updates automatically whenever any relevant parameters change in the database.

4. Image Compression and Display

Real time image display can be a difficult problem when setting up a remote observing system. Depending on bandwidth and image size it can be impractical to download every image before the next exposure (for instance, an image of 140 MB[1], would take over an hour to download using a 256Kb/s broadband connection). TAROS will use compression to enable users to view some/all of their images. The H-Compress algorithm (White & Percival 1994) is used to perform the compression. While this form of compression is lossy for compression ratios of around 20 and greater there is no significant effect on the astrometric and photometric properties of the uncompressed image. This enables observers to use the compressed image to measure seeing and line widths etc. It should

---

[1]140 MB is the size of the largest image that our instruments generate currently, but future instruments will generate larger images.

be noted however that the images saved by the displayer client only represent quick look data. The observer receives the uncompressed images via download some time after the observation run.

The TAROS displayer is composed of two main components - an image server and a client:

The display server The server is mostly written in Java. It interfaces with the CICADA data acquisition system via JNI. The JNI interface receives the images in chunks as the image is being read out. Each chunk is compressed to the desired level and then forwarded to the client via JMS. Due to the nature of JMS this allows more than one client to receive the real time display images just by subscribing to the real time display topic.

The display client The displayer GUI is built from the JSky Java components. It is closely based on the JSkyCat tool. Essentially the client subscribes to the displayer JMS topic and receives the image chunks as they are published. The client then decompresses the image, updates the display and writes a copy to disk.

As well as the JSky Java package we have made use of several existing Java packages to simplify the implementation of the displayer code. To perform the H-Compression TAROS uses the Java H-Compression package provided by Jim Dowler[2]. The Java FITS package written by Tom McGlynn[3] is also used.

5. Final Remarks

Currently the system is fully designed but only partially implemented. The full system is due to be completed in early 2006 in time for the new SkyMapper telescope. When completed TAROS will provide a reliable, maintainable and easy-to-use automatic and remote observing system.

References

Jarnyk, M. & Hovey G. 2004, in ASP Conf. Ser., Vol. 347, ADASS XIV, ed. P. L. Shopbell, M. C. Britton, & R. Ebert (San Francisco: ASP), 519.

White, R. L. & Percival, Jeffre. W. 1994, Proc. SPIE Advanced Technology Optical Telescopes V, 2199, 703

Young, P. J., Brooks, M., Meatheringham, S. J., & Roberts W. H. 1997, in ASP Conf. Ser., Vol. 125, ADASS VI, ed. G. Hunt & H. E. Payne (San Francisco: ASP), 385

Young, P. J., Roberts, W. H. & Sebo K. M. 1999, in ASP Conf. Ser., Vol. 172, ADASS VIII, ed. D. M. Mehringer, R. L. Plante, & D. A. Roberts (San Francisco: ASP), 115

[2]http://cadcwww.dao.nrc.ca/software/hcompress/

[3]http://heasarc.gsfc.nasa/gov/docs/heasarc/fits/java

ASTRONOMICAL DATA ANALYSIS SOFTWARE AND SYSTEMS XIV
ASP Conference Series, Vol. 347, 2005
P. L. Shopbell, M. C. Britton, and R. Ebert, eds.

Optical Camera with High Temporal Resolution to Search for Transients in a Wide Field

I. Zolotukhin and A. Biryukov

Sternberg Astronomical Institute of Moscow State University, Universitetskij pr. 13, Moscow, Russia, 117415

G. Beskin and S. Karpov

SAO RAS, Nizhnij Arkhyz, Karachai-Cherkessia, Russia, 369167

S. Bondar, E. Ivanov, and E. Katkova

Research Institute for Precision Instrumentation, Moscow, Russia

K. Hurley

Space Science Laboratory, University of California, Berkeley, USA

A. Pozanenko

Space Research Institute, Profsojuznaja 84/32, Moscow, Russia, 117997

Abstract. For successful observation of optical emission from gamma-ray bursts (GRBs), one should monitor the sky continuously to be independent from space-borne telescopes. According to this idea we have developed a Wide Field Optical Camera with high temporal resolution and software for real-time detection of fast optical transients in a wide field.

1. Introduction

Gamma-ray bursts (GRBs) are one of the most powerful transient events in the Universe and are probably related to the compact relativistic objects. The structure and properties of their central engine define the fine time structure of GRB emission.

At the same time a number of models predict the generation of considerable optical flux synchronous with the GRB event which can achieve $10 - 12^m$ for 0.2 sec (Eichler & Beskin 2000) or even $8 - 9^m$ for $0.1 - 10$ sec (Liang et al. 1999). In the model of early afterglows in wind shells, the optical flash is estimated to be as bright as $9 - 10^m$ in $0.2 - 0.5$ sec with $0.5 - 60$ sec lag from proper GRB event (Wu et al. 2003).

Thus, a search and study with high temporal resolution of optical transients (OTs) accompanying GRBs can provide statistically reliable information about the nature of these phenomena. To be successful, such observations have to be carried out independently of alerts received from space-borne gamma-ray telescopes and use optical instruments with a wide field of view. As a side effect this method is also able to detect and investigate short stochastic flares from different variable objects - SNs, flare stars, CVs, X-ray binaries, and NEOs, natural and artificial.

2. Hardware

To realize the approach described above, a wide field optical camera with high temporal resolution has been developed (see Tables 1 and 2 for main parameters). The main objective (15 cm diameter, F/1.2) of the camera projects 21×16 degrees of sky area onto the image intensifier photocathode with 90 mm diameter (quantum efficiency - 10%, gain - 150, scaling factor - 0.18). Special transmission optics transfers the image from the intensifier's output to the VS-CTT285-2001 TV-CCD camera (1280×1024 pixels with size of 6.5 microns) with aframe frequency of 7.5 Hz (0.13 sec exposure time).

The camera is placed in the Northern Caucasus (close to the 6-m optical telescope BTA) at a height of 2030 m above the sea level.

Table 1. Main parameters of the Camera.

Main objective		Intensifier	
Diameter	150 mm	Photocathode	S 25
Focus	180 mm	Photocathode diameter	90 mm
D/F	1/1.2	Gain	150
Field of view	21×16 deg	Scaling coefficient	5.5 / 1
		Quantum efficiency	10%

3. Software

In order to process in real time the 13 Mb/sec data stream from the camera, one cannot use any standard reduction routines usually applied for field photometry and source extraction. For this reason special software for the detection and investigation of OTs has been created.

The software is installed on three PCs operated under WINDOWS and LINUX. The incoming information is a sample of 1280×1024 pixel CCD frames with an exposure time of 0.13 sec. The software performs the following tasks (see Figure 1):

- data transfer in real time from TV-CCD to LAN
- accumulation of initial data with volume up to 0.5 Tb per night on the RAID array
- data reduction in real time - detection and classification of OTs, determination of their equatorial coordinates and magnitudes, their possible identification with known objects, and the transfer of information about OTs (alerts) to the local and global networks.

Table 2. TV-CCD camera parameters.

CCD Camera	
Dimensions	1280×1024 pix
Image scale	50″ / pix
Exposures available	0.13-10 sec
Pixel size	6.5 micron
Data stream	400 Gb / night

The OT detection algorithm is based on the comparison of the current frame with one averaged over the 10-100 previous frames and consists of the following steps:

1. Extraction from the current frame of all pixels with intensity deviating from the mean by a given fraction of the RMS noise.
2. Location of any continuous regions of such pixels on the frame and determination of their parameters - coordinates and fluxes. All these regions are considered as optical transients (OTs) if observed on at least 3 successive frames.
3. Analysis of the OTs' shape (on a single frame) and motion (on three successive frames), their classification as meteors, satellites, or stationary transients, and determination of their parameters (trajectory, light curve, etc.).
4. For the two latter cases - comparison of object parameters with known objects from star and satellite catalogs.
5. For stationary transients with an absence of catalog identifications, information on their parameters are sent to robotic telescopes or global networks.

Thus, any transient may be classified if it is seen on at least 3 successive frames (in 0.4 sec). The faintest detectable object has 11.5 magnitude (close to V band). This limit increases to 12.5 - 14 magnitude after simultaneous analysis of samples from summarized frames (over 7.5 - 750).

4. Two-Telescope Complex

The Wide Field Optical Camera has been designed for the independent detection of OTs and hardly can be used for detailed investigation of the events. Followup should be done with larger telescopes and more sensitive detectors. Thus, our aim is to build FAVOR (FAst Variability Optical Registration) – a complex of two instruments: a rapid wide-field camera and a robotic telescope for more precise measurement of the OTs.

The Camera is already operative and able to detect OTs, determine their coordinates and brightness, classify them and send these data to the robotic telescope.

The 60-cm robotic telescope must be equipped with a Fast Universal Spectropolarimeter (FUSP) which consists of: an input unit with three observing

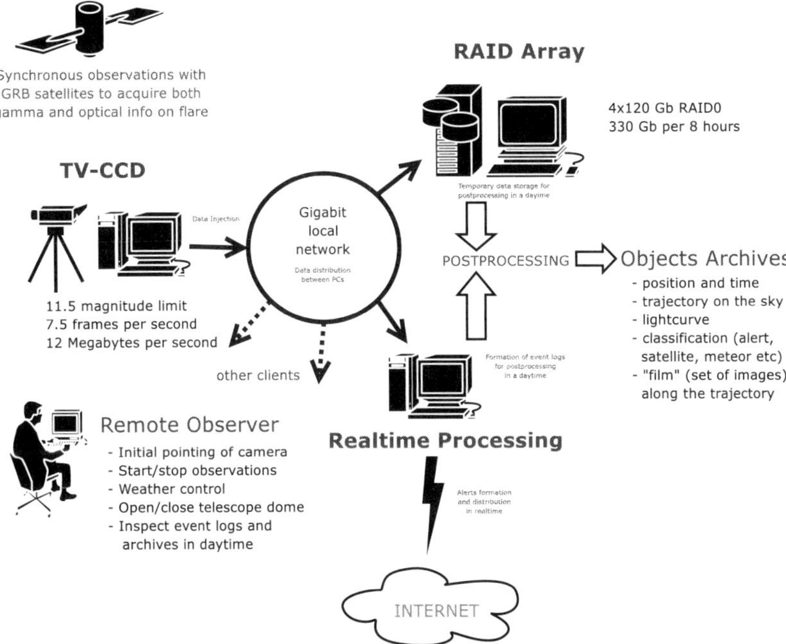

Figure 1. The observational data is transmitted to the local PC which broadcasts it through the LAN to the storage computer equipped with a RAID array and to the PC for real-time processing. The chosen network configuration allows us to keep raw data with a volume up to 0.5 Tb (partially available at http://rokos.sao.ru/history/history.php) obtained during an 8-hour observing set with 0.13 s resolution and detect on the fly both stationary and moving OTs up to $V < 11^{\mathrm{m}}.5$. The detection limit is $13^{\mathrm{m}}.5$ for a 13 s timescale (sum of 100 frames).

modes (photopolarimetry, slitless, and slit spectropolarimetry); a polarization unit (double Wollaston prism); a dispersion element; a position-sensitive detector (PSD) with time resolution of 1 microsecond; and an acquisition system (Debur et al. 2003). The FUSP records photons of four images or four spectra with different polarization plane orientations for each object in a field of $5'$ or for an object and comparison star in the slit (with size $2'' \times 4'$). The spectral range is 4000-6500Å and the resolution is 200-250Å. Linear spectropolarimetry with signal-to-noise ratio of 10 can be performed for objects with V $\sim 11^{\mathrm{m}}$ – 12^{m} in 1 s. A telescope slewing rate of 5 deg/s allows one to point at the OT withqin 1-2 seconds after its detection, thus FAVOR will give high time resolution photometric, spectroscopic, and polarimetric information about the transient a few seconds after the trigger. The robotic telescope will be placed in the vicinity of the Wide Field Camera.

The complex will be completed with the commissioning of the robotic telescope. Its tests are expected to start in 2005 when the SWIFT satellite will be fully operative.

Also possible is the use of the REM (Rapid Eye Mount) telescope as a robotic telescope. In this configuration, a similar wide field camera should be placed at La Silla Observatory at the REM telescope mount.

5. Preliminary Results

Since November 2003 we have accumulated more than 150 nights of observations with the Wide Field Optical Camera. Having an average observation time per night of 6 hours duration, we get $150 * 6 = 900$ hours of observations, that in a 16×21 degree field gives 0.6 of a BATSE GRB. We observe approximately 100 satellites and 200 meteors per night. Expected numbers of GRBs hitting in the field of view of the camera which will be registered simultaneously with WXM (HETE-2) and BAT (SWIFT) are 1-2 and 4 per year, respectively.

A more detailed description of the software and the current status of the Camera (including recently occurred unidentified events) is available at `http://rokos.sao.ru/favor/`

Acknowledgments. This work was supported by grants of CRDF (No. RP1-2394-MO-02) and RFBR (No. 04-02-17555).

I.Z. is grateful to the Local Organizing Committee of ADASS XIV and Thomas Handley personally for financial support of my attendance of the conference. G.B. thanks cordially the Cariplo Foundation for their fellowship and Merate Observatory for their hospitality.

References

Debur, V., et al. 2003, Nucl. Instrum. Meth., A513, 127

Eichler, D., & Beskin, G. 2000, Phys.Rev.Lett, 85, 13

Liang et al. 1999, ApJ, 519, L21

Wu, et al., 2003 MNRAS, 342, 1131

Part X
Data Pipelines

ASTRONOMICAL DATA ANALYSIS SOFTWARE AND SYSTEMS XIV
ASP Conference Series, Vol. 347, 2005
P. L. Shopbell, M. C. Britton, and R. Ebert, eds.

Using Perl in Basic Science and Calibration Pipelines for Spitzer Infrared Array Camera Data

H. Brandenburg, P. Lowrance, R. Laher, J. Surace, and M. Moshir

Infrared Processing and Analysis Center, Caltech, Pasadena, CA 91125, Email: heidi@ipac.caltech.edu

Abstract. Object oriented Perl language pipelines generate calibration products and basic calibrated data from raw images taken by the Infrared Array Camera (IRAC) onboard NASA's Spitzer Space Telescope. The pipelines gather input data and control files, initiate database interactions, and manage data flow through C, C++, and Fortran component programs. The compiled component programs perform instrumental signature correction, calibration, and data characterization.

Core pipeline functionality is provided by two compact Perl object hierarchies - one for pipelines and another for images. The objects allowed flexible and agile response to change during Spitzer's first year of operations, which offset the cost-impact of utilizing interpreted Perl on production data processing.

1. Execution Environment

Data processing takes place on a cluster of low-end Solaris servers ("pipeline drones"); the drones may run multiple, possibly unrelated, jobs concurrently. The Pipelines write products to a disk on their host, then copy the final products to a network drive at the end of processing.

Job preparation and control software (written in C), under the control of an executive, prepares directories, seeds a FITS file, and then initiates and waits for the pipeline.

Most intraprogram communication occurs via environment variables. Pipelines communicate with the science operations database primarily through a Perl API. Calibration files and control data files (CDFs, or resource files) are written directly to a pipeline's working directory.

2. Classes & Methods

Two class hierarchies provide most of the required pipeline functionality (Figure 1). Each hierarchy is rooted in an _Initializable container class which constructs a new hash-based object and blesses it into its derived class, and then calls an _init() method on that class to force object initialization to begin in the derived class and bubble up through parent classes.

2.1. Image

By SSC convention all images are assumed FITS formatted. The Image parent class provides read-only header keyword access.

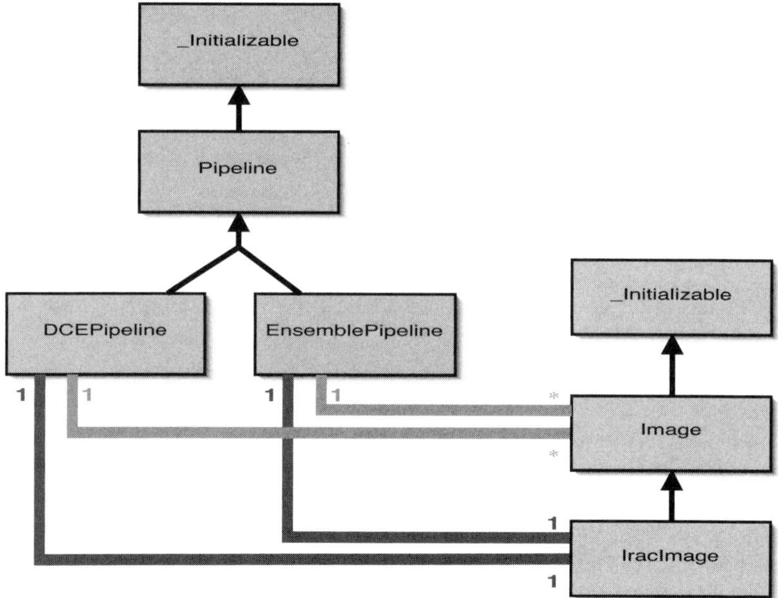

Figure 1. Class Relationship and Inheritance

2.2. Image::IracImage

Image::IracImage contains data and methods specific to IRAC observing. Its private methods include image integrity and telemetry checks. Public methods provide access to observing parameters (frametime, fowler number, wait periods, etc.), assembly of calibration transfer parameters, and pack/unpack functionality for sub-array data.

2.3. Pipeline

The Pipeline class contains a hash holding the current pipeline state, including the current input file name, the next output file name, and where to find calibration data and other control file resources. The class contains private methods for environment checks, parsing control files, and database interactions. It also contains public methods to allow the pipeline scripts to access and mutate the state hash, load control files, execute programs, and end the pipeline.

2.4. Pipeline::DCEPipeline

The DCEPipeline class contains functionality specific to single image pipelines. It also implements public methods for single DCE related database interactions.

2.5. Pipeline::EnsemblePipeline

The EnsemblePipeline class contains functionality specific to processing sets of images together. It extends the basic pipeline state hash with ensemble-related items and provides accessor and mutator methods for them. The class also provides additional public methods for ensemble-related database interactions.

Pipeline objects require an image object for instantiation. During pipeline processing, messages may be passed from Pipeline objects to Image objects.

3. Pipeline Scripts

The driving design aesthetic in the launch era IRAC pipeline was to create a simple code base with no redundancy in order to minimize labor and risk during change. Extant pipelines for calibration product processing and basic science reduction total around 2000 source lines. Each of eight pipeline scripts contain between 100 and 550 lines of executable perl source in a single file.

All pipelines begin by instantiating an IracImage object with the FITS file seeded by the job preparation and control program. The image object is then used to instantiate a pipeline object (the image becomes part of a pipeline's internal state). The pipeline script calls a public method on the pipeline object to load control data (resource) files.

Pipeline scripts then execute compiled component programs for instrumental signature correction, data characterization, and data calibration (Figure 2). When numerical processing is finished, the pipeline registers products with the science operations database and moves them into a network file system.

Detailed pipeline and module behavior is changed by configuring a control data file. For example, all calibration observations are preprocessed by the same pipeline script, but darks and flats are processed with parameter sets appropriate to each kind of observation (e.g., Figure 3).

4. In Operations

There is an intrinsic cost to utilizing an interpreted language like Perl for pipeline processing. Although we have not yet quantified the percent of pipeline time spent by the interpreter or in process bookkeeping for the modular infrastructure (anecdotal evidence from very early phases of development indicated that the overhead was in the few percent range), we have met the project IRAC processing throughput requirements -using nearly outdated Netra-class hardware in operations. In particular all data are processed at least 5 times faster than the rate of data acquisition.

Nearly as important, the Perl code allows the easiest response to change. To build and deploy a compiled patch, change management procedures require a new release, increasing workload and turnaround time. Since SSC Perl scripts and modules do not require a build step for configuration and install on SSC systems, Spitzer staff were able to support a larger number of changes during in-orbit checkout than would have been possible with compiled code.

Acknowledgments. This work was carried out at the Spitzer Science Center, with funding from NASA under contract to the California Institute of Technology and the Jet Propulsion Laboratory.

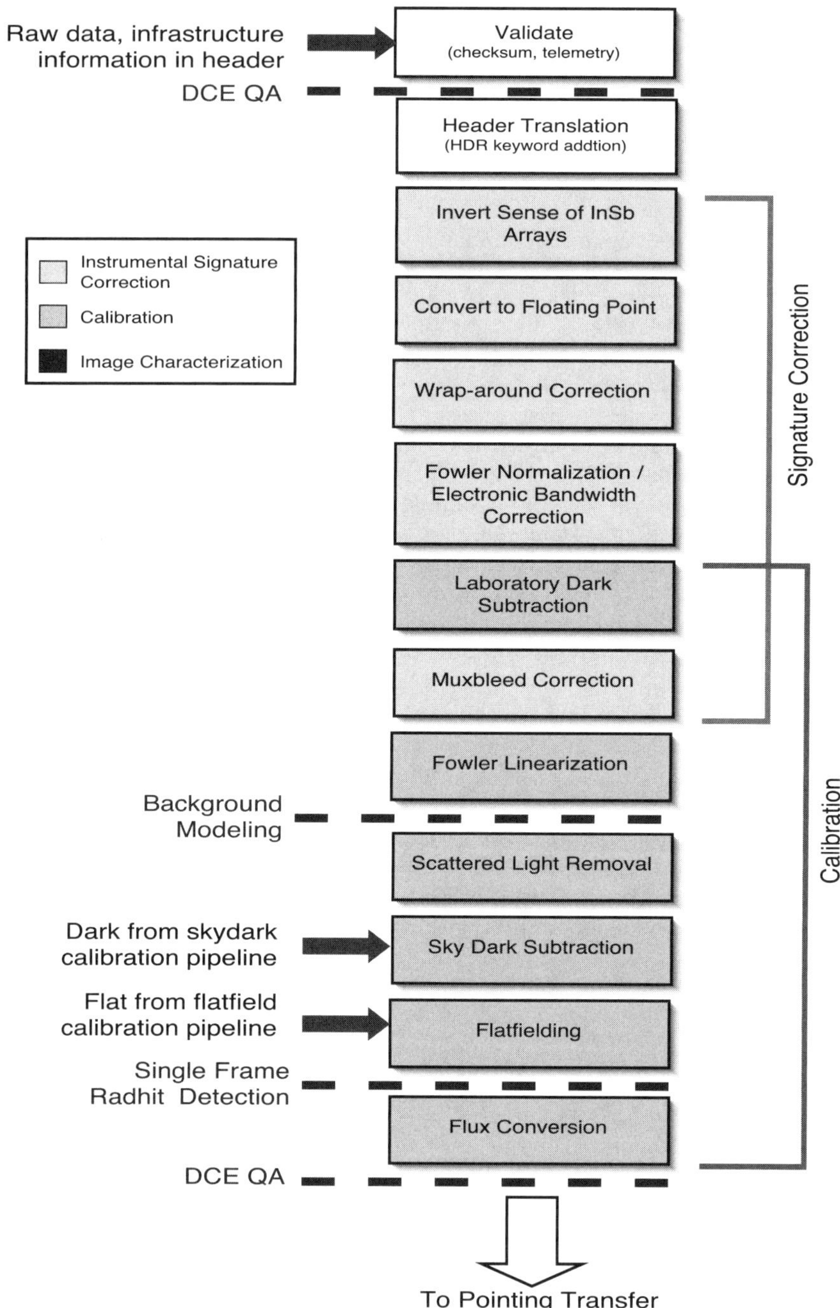

Figure 2. Science (Basic Calibrated Data) Pipeline

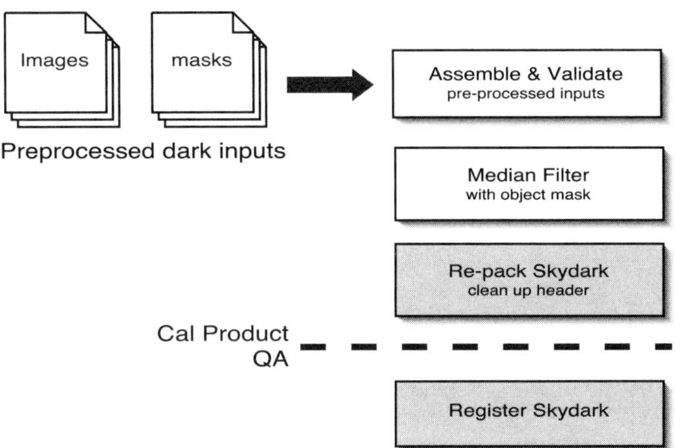

Figure 3. Skydark Creation Ensemble Pipeline

ASTRONOMICAL DATA ANALYSIS SOFTWARE AND SYSTEMS XIV
ASP Conference Series, Vol. 347, 2005
P. L. Shopbell, M. C. Britton, and R. Ebert, eds.

A Photometric Comparison of ECLIPSE Jitter and the ORAC-DR Equivalent Recipe for ISAAC

Malcolm J. Currie
Starlink Project, Rutherford Appleton Laboratory, Chilton, Didcot, Oxfordshire OX11 0QX, United Kingdom

Abstract. Motivated by a request from astronomers demanding accurate and consistent infrared photometry, I compare the photometry and quality of mosaics generated by the ECLIPSE *jitter* task and the ORAC-DR *JITTER_SELF_FLAT* recipe in two fields. The current (v4.9.0) ECLIPSE produces photometry a few percent fainter than ORAC-DR; the systematic trend with magnitude seen in v4.4.1 is now removed. Random errors arising from poor flat-fielding are not resolved. ECLIPSE generates noisier mosaics; ORAC-DR has poorer bias removal in crowded fields and defaults to larger mosaics. ORAC-DR runs a few times slower than ECLIPSE, but its recipe development is measured in weeks, not years.

1. Introduction

ORAC-DR is an intelligent, data-driven, tailorable pipeline system, developed at the Joint Astronomy Centre Hawaii. Object-oriented Perl data-reduction recipes use Starlink applications to process the bulk data (Currie et al. 1999). Amongst the ORAC-DR pipelines are several for infrared instruments, including ISAAC at ESO.

In the spring of 2003, two UK astronomers contacted me to see if my ORAC-DR pipeline could provide consistent photometry for their gamma-ray burster (GRB) afterglow observations. They had analysed mosaics reduced by the ECLIPSE *jitter* command (Devillard 1999) and had found large random photometric errors. Comparative aperture photometry of two reference stars over a few nights are shown in Table 1. ORAC-DR's *JITTER_SELF_FLAT* recipe was consistent with its quoted errors, typical for infrared photometry. ECLIPSE v4.4.1 photometry exhibited a much larger scatter in K_s and H bands.

Table 1. Variations of the magnitude difference of two comparison stars for the *JITTER_SELF_FLAT* recipe and the v4.4.1 *jitter*.

Band	Error in mean		Standard deviation of mean		Maximum deviation of mean	
	ECLIPSE	ORAC-DR	ECLIPSE	ORAC-DR	ECLIPSE	ORAC-DR
J	0.007	0.004	0.016	0.010	0.038	0.031
K_s	0.014	0.004	0.032	0.009	0.084	0.025
H	0.022	0.004	0.054	0.011	0.149	0.027

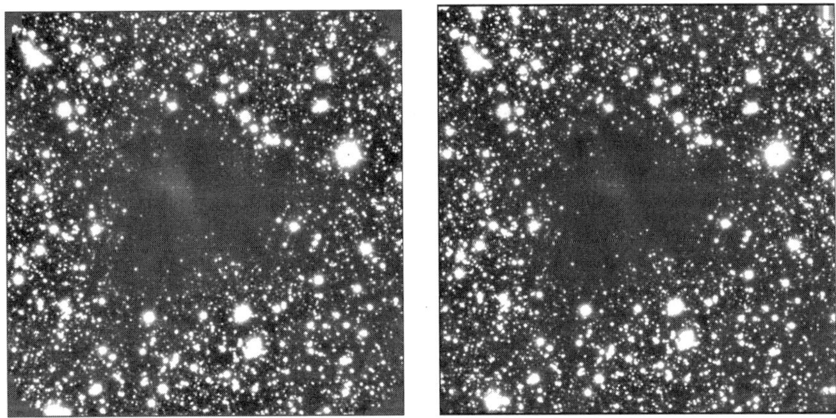

Figure 1. ECLIPSE (left) and ORAC-DR (right) plotted to the same 20, 80-percentile scaling limits around the sky.

A comparison of the mosaics produced by both packages revealed that ECLIPSE leaves shadow artefacts from the sky subtraction and has more noise. The presence of shadows was surprising, since I had discussed this problem and its solution with Nic Devillard in 1998 at ADASS VIII.

These findings were discussed informally at the 2003 ADASS with Michele Peron and others from ESO. We agreed that further comparisons be made.

2. Crowded-field Comparison

ESO provided test raw data and ECLIPSE-reduced mosaics and ancillary files. By delivery time ECLIPSE had progressed to 4.9.0, and had allegedly cured its photometric problems, including the systematic variations (Iovino 2001).

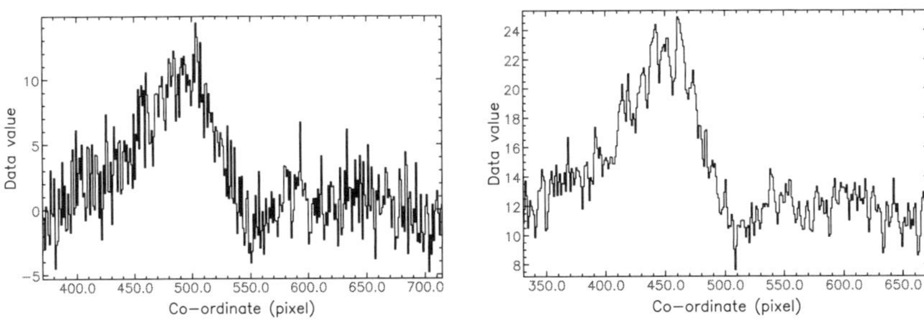

Figure 2. Matching x slices through the nebula, ECLIPSE (left) and ORAC-DR (right).

The first comparison was of a crowded field containing a dark nebula (Figure 1) based upon thirty dithered integrations. The two mosaics are quite similar. The shadows are gone. The ECLIPSE background subtraction seems to be biased by the dark nebula, as this has a near zero signal, and the background drops in the outer regions. That is not a problem provided the photometry is measured against the local background. The ORAC-DR bias correction is poorer, revealed by some weak residual banding. The PSFs and registrations agree.

Background profiles, after first masking the majority of the signal from the stars and nebula, indicate that ECLIPSE shows a greater variation. Note the ORAC-DR recipe used does not sky subtract, but forms a flat field from the dithered observation frames themselves, and thus some naturally occurring patterns might be dampened. The self-flat approach does well even in the crowded field, where there are no shadow artefacts visible. The ECLIPSE mosaic exhibits 50% greater noise, for example see Figure 2, affecting faint-source detection.

Figure 3 shows the magnitude differences of matched sources. The mean difference is -0.021 ± 0.004 for all 1407 values within a 500-pixel radius to mitigate the mismatching from distortion. This becomes -0.037 ± 0.002 for $H < 17.5$, and parity around $H = 19.5$. One interpretation is that ECLIPSE is still subtracting object signal as part of the sky.

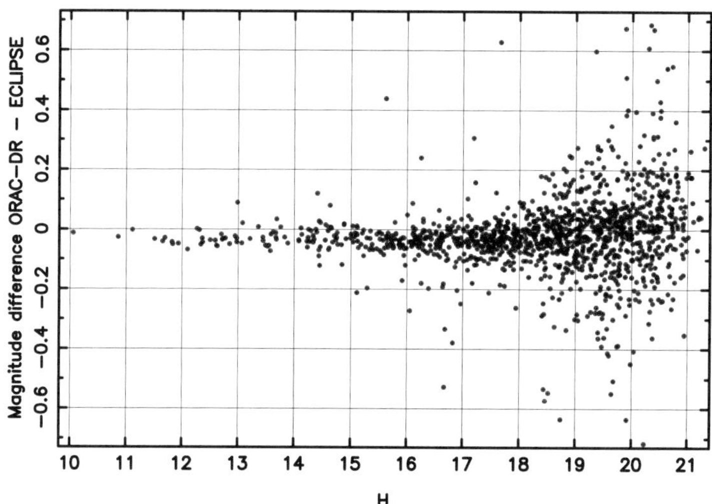

Figure 3. The difference in magnitudes ORAC-DR minus ECLIPSE for the crowded field from SEXTRACTOR isophotal photometry using default settings.

Some outliers are due to the imperfect matching of sources caused by distortion correction being applied only in ORAC-DR or objects near the masked regions. While not a problem in this field, the lack of spatial-distortion correction would adversely affect dithers with larger offsets for a wide-field panorama.

Unfortunately, ECLIPSE does not generate files containing the individual sky-subtracted frames forming the mosaic. Hence I was unable to investigate the photometric errors for stars dithered to different detector positions, and check the quality of the flat field. Although using the ESO-supplied flat in ORAC-DR produces a ~3.5% varying background as seen in Figure 4. This non-uniformity may be present but invisible in the low counts of the sky-subtracted ECLIPSE mosaic; however, it would increase the photometric scatter.

Figure 4. As Figure 1 (right), but using the ESO-supplied twilight flat.

3. Second Comparison

The second dataset for the comparison comprises five frames in a medium-density field (see Figure 5). In contrast to the first dataset the dither offsets are an appreciable fraction of the field size.

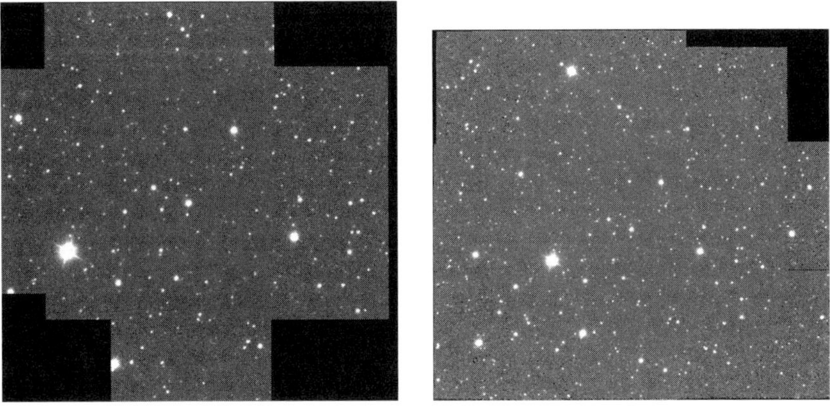

Figure 5. ECLIPSE (left) and ORAC-DR (right) scaled to the same 2, 98-percentile limits around the sky. Note the different spatial scales.

Again the noise is larger in the ECLIPSE mosaic, but only by 10%; and the coverage is narrower. The background is more uniform in the ORAC-DR where smoothed contours show a range of 0.3 counts in the frame-intersection area, whereas the level varies by 0.8 counts for ECLIPSE. Looking at the photometry (Figure 6), there is a steeper trend to positive differences for $K_s > 16$. Brightward of that, the mean difference is -0.09 ± 0.01, but overall it is -0.005 ± 0.011.

4. Conclusions

ECLIPSE now produces reliable photometry. However, possible flat-fielding uncertainties have not been resolved in this poster, and they require further work. For instance, I am attempting to produce a revised Table 1 with v4.9.0.

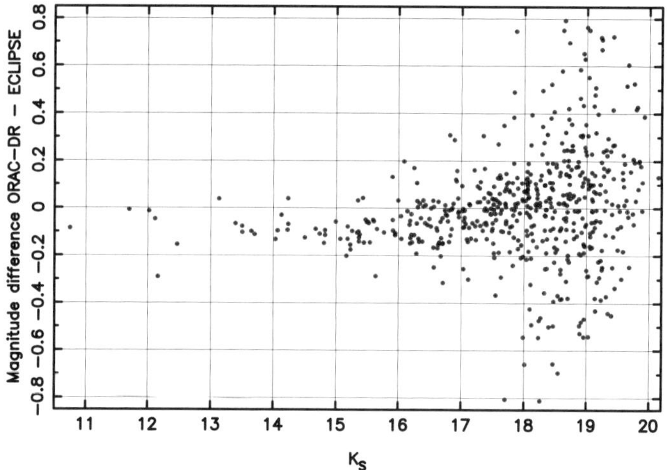

Figure 6. As Figure 3, but for the second test dataset.

ORAC-DR magnitudes are a few percent brighter than ECLIPSE's. ECLIPSE runs much faster than ORAC-DR—over three times for similar operations and up to nine times where the latter does more, such as flat-field creation, and distortion correction. Much of the difference arises because of the large quantities of file I/O and task invocations. The former does have some advantage in that intermediate frames can assist in the diagnosis of pathological problems, and permits frame-to-frame comparisons.

ORAC-DR through code reuse is faster to develop, and its recipes are easily tweaked or augmented, such as adding a correction for electronic ghosting and bias for ISAAC. This development took weeks for ORAC-DR with Starlink applications, whereas it took five years before ECLIPSE generated accurate photometry.

ECLIPSE generates noisier mosaics, possibly because it combines fewer frames; ORAC-DR defaults to larger mosaics, and has poorer bias removal in crowded fields, because it uses the observations themselves to measure the bias.

Acknowledgments. The data were kindly supplied by Wolfgang Hummel, Nial Tanvir, and Shariq Pak. Mark Taylor answered questions concerning TOPCAT cross-matching.

References

Currie, M. J., Wright, G. S., Bridger, A., & Economou, F. 1999, in ASP Conf. Ser., Vol. 172, ADASS VIII, ed. D. M. Mehringer, R. L. Plante, & D. A. Roberts (San Francisco: ASP), 175

Devillard, N. 1999, in ASP Conf. Ser., Vol. 172, ADASS VIII, ed. D. M. Mehringer, R. L. Plante, & D. A. Roberts (San Francisco: ASP), 333

Iovino, A. 2001, `ftp://ftp.hq.eso.org/pub/eclipse/jitter-photometry/iovino.ps.gz`

ASTRONOMICAL DATA ANALYSIS SOFTWARE AND SYSTEMS XIV
ASP Conference Series, Vol. 347, 2005
P. L. Shopbell, M. C. Britton, and R. Ebert, eds.

Design of the SCUBA-2 Quick Look Display and Data Reduction Pipeline

A. G. Gibb and D. Scott

Dept of Physics and Astronomy, University of British Columbia, Vancouver, BC, V6T 1Z1, Canada

T. Jenness and F. Economou

Joint Astronomy Centre, 660 N. A'ohōkū Place, University Park, Hilo, HI 96720, USA

B. D. Kelly and W. S. Holland

UK Astronomy Technology Centre, Royal Observatory Edinburgh, Blackford Hill, Edinburgh, EH9 3HJ

Abstract. SCUBA-2 is planned to be the largest submillimeter bolometer array in the world and is scheduled for delivery to the 15-m JCMT in May 2006. With over 10,000 pixels at two wavelengths and a maximum sampling rate of 200 Hz, the data rate is expected to be up to 0.5 TB per observing night. Furthermore the data will be reduced concurrently with observing to permit rapid performance and scientific assessments. This paper presents an overview of the instrument and the design of the pipeline, and discusses some of the hurdles which must be overcome in order to cope with the data flow.

1. Introduction: SCUBA-2 and submillimeter astronomy

SCUBA-2 is a second generation submillimeter bolometer camera for the 15-m James Clerk Maxwell Telescope (JCMT) at the 4000-m summit of Mauna Kea, Hawaii (Holland et al. 2003). Like its predecessor, SCUBA, it will operate simultaneously at 850 μm and 450 μm, but the similarity ends there. SCUBA-2 has a field of view 8 arcminutes in diameter which is fully-sampled at 850 μm, removing the need to "jiggle" the secondary mirror.

Each array has over 5000 pixels in four subarrays (each 40×32 pixels), which is small by current optical and infrared standards but unprecedented in the submillimeter. However, SCUBA-2 will take data at a rate of 200 Hz yielding approximately 0.5 TB (uncompressed) per 16-hour observing night. The key challenge then becomes dealing with the data in a timely manner.

With its larger, more sensitive, and better-sampled arrays SCUBA-2 will be capable of mapping the sky up to 1000 times faster than SCUBA, allowing sensitive and truly wide-field submillimeter maps covering many square degrees to be made. The first submillimeter all-sky survey (with 7-14 arcsec resolution) looks to be a distinct reality with SCUBA-2.

2. Observing Methods and Operational Modes

While the SCUBA-2 arrays have a CCD-like appearance, in practice the bolometers behave very differently, while the properties of the earth's atmosphere at submillimeter wavelengths means that specialized observing techniques are required. At submillimeter wavelengths, the sky is never "dark": the atmosphere is the dominant source of emission picked up by detectors. Traditionally, submillimeter observing has used chopping to cancel out this sky signal, which can be highly variable on short timescales. Unfortunately chopping also removes source structure on scales of the chop throw, as well as increasing the confusion limit due to continuously subtracting two images of the sky. SCUBA-2 will not use chopping and so will yield images sensitive to structure on all possible scales (limited only by the size of the map or size of the array), as well as lowering the confusion limit by a factor of $\sqrt{2}$.

To remove the sky in a DC-coupled mode requires rapid sampling and SCUBA-2 will dump frames at 200 Hz. However, observing in this manner places stringent limits on the accuracy and stability of the flat-field correction. Furthermore, the detectors themselves are limited in their stability by $1/f$ noise. A cold shutter will take "dark" frames to compensate for instrumental drifts, while it is anticipated that observational techniques will alleviate the flat-field requirement (see Jenness et al. 2004).

From the observer's point of view, SCUBA-2 will have two primary observing modes: 1) Imaging of areas similar to the array size, and 2) Scanning for making large (degree-scale) maps. Polarimetry and spectroscopy modes (using a Fourier transform spectrometer, or FTS) will also be available as supersets of Imaging. These modes are outlined in Jenness et al. (2004) and only updates will be mentioned here.

2.1. Imaging: Stare and DREAM

Imaging will be used for making images comparable with the size of the field of view (~8 arcmin) or mosaicking several such fields together (to fill in the gaps between the subarrays and produce fully-sampled images at 450 μm). In this mode, the 200-Hz data will be combined into frames which are then written to disk at a rate of no more than 1 Hz, with a lower limit imposed by sky rotation (since SCUBA-2 will be located at one of the Nasmyth foci of the JCMT but without an image rotator).

2.2. Scan-mapping

In scan-map mode the telescope slews at up to 600 arcsec s^{-1} while the electronics read out continuously at the full 200 Hz data rate. The scanning pattern is not yet decided but options include a traditional raster (scan and fly-back), "ploughed-field" scan (boustrophedon) or a billiard-ball (box-scan) pattern. The optimum scanning technique will include a way to observe sky positions at least twice over a scan period in such a way as to avoid periodic (fixed frequency) duplications while maximizing sampling and areal coverage and minimizing astrometric and photometric errors.

3. Data Flow and Pipeline Infrastructure

The maximum sampling rate for SCUBA-2 is 200 Hz, which translates to a maximum data rate of approximately 1 MB of uncompressed data per subarray per second. For the eight subarrays (four per wavelength ×2 wavelengths) this is then 8 $\mathrm{MB\,s^{-1}}$, or approximately 0.5 TB for a 16-hour observing night. Data compression will reduce the storage requirements by a factor of ~2.

Each of the four subarrays at each wavelength will have its own dedicated data acquisition (DA) computer which will write the raw frames to a local disk. In imaging mode, a file will be written to disk every 1 – 5 seconds; in scanning mode a file will be written approximately every minute.

The SCUBA-2 data processing pipeline will use the established ORAC-DR (Economou et al. 1999) pipeline infrastructure. ORAC-DR is a proven pipeline technology employed on UKIRT, JCMT and the AAT and will be able to handle data from SCUBA-2 (see Jenness et al. 2004 for further references). In the context of ORAC-DR, the SCUBA-2 data reduction (DR) project is responsible for writing new Recipes, Primitives, and Algorithm engines where necessary. We will make use of existing algorithms (such as those already developed for SCUBA) and Starlink (Draper et al. 2005) applications wherever applicable, but new applications (algorithm engines) will have to be written to support the particular requirements and observing modes for SCUBA-2.

4. Pipeline Processing

As far as possible, the DR pipeline must yield well-calibrated and scientifically-meaningful images, free from imaging artefacts (e.g., cosmic rays, bolometer drifts). There will be up to four pipelines running simultaneously at the telescope, two for each wavelength (see Figure 1 for an example of the data flow for one wavelength). One of these is the Quick-Look (QL) pipeline, designed to provide rapid feedback to observers, and the other is a full reduction (DR) pipeline designed to produce scientifically-meaningful (i.e., calibrated, astrometrically correct) images. The QL pipeline is similar to the full pipeline, differing only in the recipes applied to the incoming data. Simple despiking, coadding and regridding onto a preferred frame (e.g., RA/Dec) will enable the observer to make online assessments of the data.

The pipelines will be run on separate computers, reading the raw data from the remote disks and processing them locally to minimize network-induced lag. The pipeline will produce high quality images in so-called "near real time." The practical definition of real-time is to reduce 16 hours' data in no more than 16 hours. For the QL pipeline, this means that the data must be processed and displayed before the next observation is complete. Images generated in DREAM/Stare mode will be processed and displayed as they come in, on timescales of order 1 second. Preliminary testing of a QL pipeline shows that modest hardware (e.g., 1.6 GHz AMD Opteron) is comfortably capable of keeping up with the data rate for DREAM/Stare mode. Scan-map data taken at 200 Hz will be processed and displayed before the next scan is completed. While the DA system will be writing data at a rate of 200 Hz, the DR and QL pipelines will not have to reduce

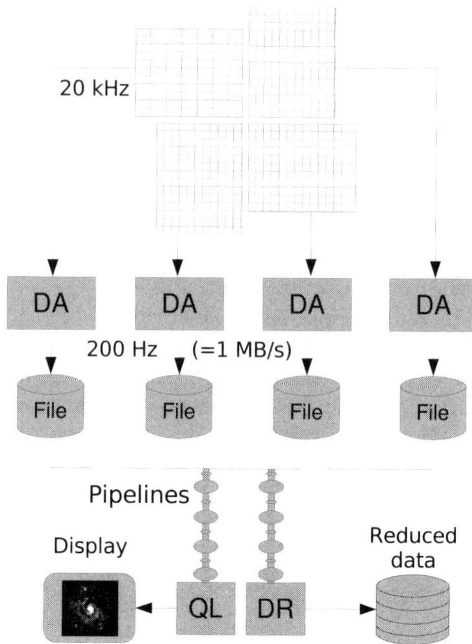

Figure 1. Schematic view of the data flow from one of the SCUBA-2 bolometer arrays showing the two pipelines running concurrently. The data from each subarray is handled by a single DA computer, from which the pipelines read to produce images.

200 files each second. Rather, the 200 Hz data will be written into a single file which will be finalized after about 1 minute.

The pipeline will have an offline mode which will make use of calibration data derived from the whole night (rather than using the most recent) as well as allowing for some flexibility in control by users. The offline mode will also allow data from multiple nights to be combined (e.g., Jenness et al. 2002) and users will have full access to the 200 Hz data in all modes.

4.1. Reduction of Submillimeter Bolometer Data

Another important difference between CCD data and those from the bolometer arrays is that data from bolometer arrays are analyzed in the time domain (e.g., Borys, Chapman, & Scott 1999). The reason for this is that it is easier to deal with the sky noise variations, and that (in general) the telescope is scanning across the sky so the position of each pixel on the sky is constantly changing. This has implications for the pipelines, especially the QL, since a complete reduction making full use of the redundancy in the scan pattern can only take place after a scan has completely finished, which may take longer than the file-dump interval.

5. Monitoring and Feedback

The pipeline will also monitor the incoming data to provide diagnostic information for assessing instrument performance, observing conditions, and data quality. The QL pipeline will provide direct visual feedback when bright sources are observed, while other measures of performance will also be stored and made available to a Stripchart plotter. The Stripchart tool will be configurable to monitor various parameters and pass updates to one or more plots for tracking the variation in, say, the flux conversion factor, seeing, or any other parameter generated as by the pipeline. The high sensitivity of SCUBA-2 and the production of fully-sampled images on timescales of a second will allow for significant improvements in the calibration accuracy of the pipeline. For example, real-time updates to the telescope pointing may be derived through a shift-and-add approach on a bright point source. Centroid measurements on a compact source will yield estimates of the submillimeter seeing or checks on the telescope focus.

6. Atmospheric Science

The unprecedented size of the SCUBA-2 array and the high brightness of the atmosphere mean that it will be possible to conduct experiments in atmospheric physics simply as part of observing, in particular testing models of the structure of the atmosphere (e.g., the wind-blown screen model) and the distribution of water vapour. The 2-D imaging capability will allow distinct atmospheric features (clouds) to be tracked across the array and correlated with the signal from the water-vapour monitor (Wiedner et al. 2001). Improved understanding of the atmosphere will lead to enhancements in calibration effectiveness and image quality at submillimeter wavelengths (Archibald et al. 2002).

7. Timeline

SCUBA-2 is scheduled for delivery to the JCMT in May 2006, and will be available to the community in time for the 07A observing semester.

References

Archibald, E. N., et al. 2002, MNRAS, 336, 1

Borys, C., Chapman, S. C., & Scott, D. 1999, MNRAS, 308, 527

Draper, P. W., et al. 2005, in ASP Conf. Ser., Vol. 347, ADASS XIV, ed. P. L. Shopbell, M. C. Britton, & R. Ebert (San Francisco: ASP), 22

Economou, F., et al. 1999, in ASP Conf. Ser., Vol. 172, ADASS VIII, ed. D. M. Mehringer, R. L. Plante, & D. A. Roberts (San Francisco: ASP), 11

Holland, W. S., et al. 2003, Proc. SPIE, 4855, 1

Jenness, T., et al. 2002, MNRAS, 336, 14

Jenness, T., et al. 2004, in ASP Conf. Ser., Vol. 314, ADASS XIII, ed. F. Ochsenbein, M. Allen, & D. Egret (San Francisco: ASP), 428

Wiedner, M. C., Hills, R. E., Carlstrom, J. E., & Lay, O. P. 2001, ApJ, 553, 1036

ASTRONOMICAL DATA ANALYSIS SOFTWARE AND SYSTEMS XIV
ASP Conference Series, Vol. 347, 2005
P. L. Shopbell, M. C. Britton, and R. Ebert, eds.

Spitzer Space Telescope MIPS Germanium Pipeline

David Henderson, David Frayer, Ted D. Hesselroth, and Meyer Pesenson

Spitzer Science Center, MS 314-6, California Institute of Technology,Pasadena, CA 91125

Abstract. The MIPS Germanium data reduction pipelines present challenges to remove a wide variety of detector artifacts and still operate efficiently in a loosely coupled multiprocessor environment. The system scheduling architecture is designed to sequentially execute four stages of pipelines. Each pipeline stage is built around perl scripts that can invoke Fortran/C/C++ modules or Informix database stored procedures. All inter-pipeline communication is via the database.

The pipeline stages are the elimination of nonlinear and radiation artifacts in the flux measurement, the calibration of the fluxes with both onboard and stellar calibration sources, applying post-facto pointing information, and assembling individual exposures into mosaics.

1. Germanium Detector Description

The MIPS germanium detectors arrays are a 32 × 32 gallium-doped germanium (Ge:Ga) array for 70 microns and a 5 arc-minute field; and a 2 × 20 Ge:Ga array, mechanically stressed to extend its photoconductive response to 200 microns, with a field of 0.5 × 5 arcminutes. On-board calibrators are provided for each array. Additionally, it has a scan mirror to provide mapping with a very efficient use of telescope time. The three arrays (two germanium and a 24 micron silicon array), calibrators, scan mirror, and optics compose the cryogenic part of the MIPS. That is, these components are mounted, as in the pictures below, inside the SIRTF cryostat and cooled by superfluid liquid helium to a temperature of about 1.5K, 1.5 degrees above absolute zero, or -457 degrees Fahrenheit.

2. Detector operation and deriving photometric flux

The detectors generate a current ideally proportional to the incident flux. The detector electronics integrates this current in a charge well for each pixel. The accumulated charge is sampled 8 times per second for each pixel. Typical integration times range from 3 to 10 seconds.

1. Several nonlinearity, hysteresis, and environmental effects must be compensated for by the pipeline software.
2. Saturation - the charge well has a finite capacity and the A/D converter has a maximum value of 65535.
3. Nonlinearity - the voltage/charge curve has a predictable, nonlinear, and reproducible relationship.

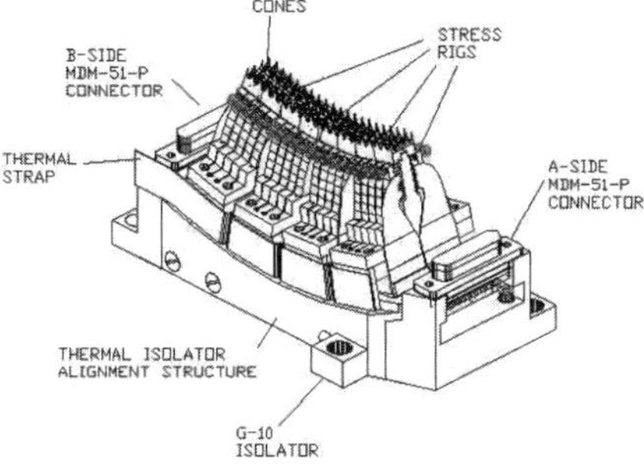

Figure 1. 70 micron and 160 micron MIPS detectors.

4. Responsivity variations - the instantaneous responsivity of the detectors varies over time due to various factors (accumulated radiation dose since anneal, hysteresis over several time scales).
5. Single Event Radiation Hits - charged particles (cosmic rays) introduce discontinuities in the stored well charge.
6. Dark Current - each pixel has a dark current present even in the absence of photometric illumination.
7. Illumination Correction - each pixel has an optical path to the sky characterized by a coefficient for each pixel.

3. Dataprocessing pipeline organization

The data processing activity is broken into four major pipeline threads for scheduling purposes. Each thread may execute multiple instances of a processing script on a particular data set. Within a given thread, the script execution on behalf of any one instance is independent of all other instances. This enables loosely coupled multiprocessing with as many as forty networked servers, which offers many performance advantages.

This pipeline operates on a dataset where each pixel is sampled many times during the integration interval (e.g., 32×32×80 for a 10 second exposure). It produces a single array representing the current presented to the well during the integration time.

There are four threads which are invoked to complete the data processing.

1. Flux calibration thread - this pipeline operates on many different exposures and generates the binary calibrated data for each exposure. This pipeline takes out the detector signatures of varying responsivity, dark current, and illumination correction. The source of the responsivity variation is removed using the internal stim calibration source. Another responsivity variation relates to the average background pixel illumination and is removed by a hipass filter to enhance data from sources of small angular extent on the sky.

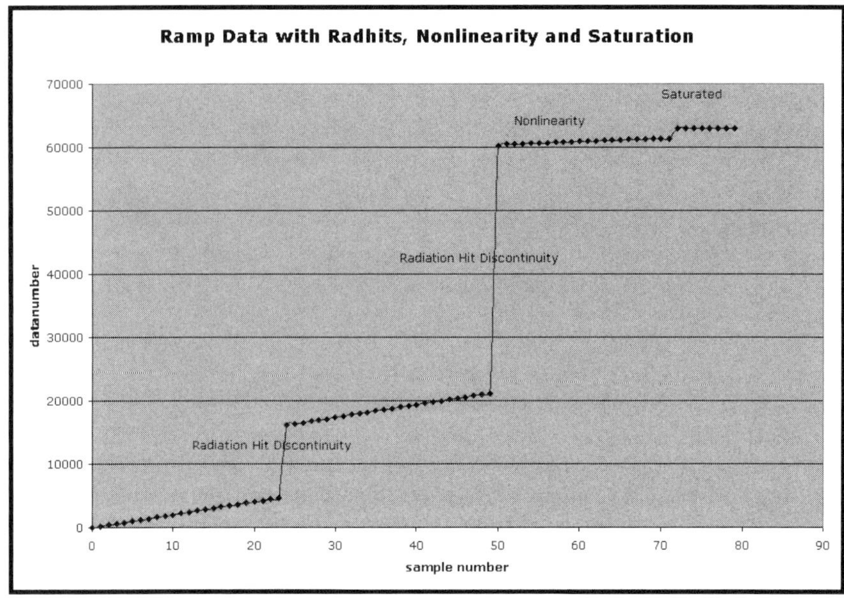

Figure 2. A generic charge ramp.

2. Pointing refinement thread - this inserts the observed coordinates for an observation, replacing the predicted coordinates available up to this point.
3. Image mosaic thread- this thread takes many exposures and uses the redundant information to generate an aggregate image.

3.1. Current Estimation Thread

Here are the functional steps for the Current Estimation Thread

1. Saturation detection
2. Electronic nonlinearity
3. Detect radiation hits and segment
4. Segment slope fusion yielding current and uncertainty

3.2. Flux Calibration Thread

Here are the functional steps for the Flux Calibration Thread.

1. Interpolate stimflash exposures (internal calibration source)
2. Normalize all exposures by interpolated stimflash
3. Subtract the dark current (calibration file)
4. Divide by the illumination correction
5. Mark all observations too close to stims
6. Output raw image and uncertainty
7. Perform hipass filtering operation
8. Output filtered image and uncertainty

Acknowledgments. This work was carried out at the Spitzer Science Center, with funding from NASA under contract 1407 to the California Institute of Technology and the Jet Propulsion Laboratory.

References

Hesselroth, T. D., Ha, E. C., Pesenson, M., Kelly, D. M., Rivlis, G., & Engelbracht, C. W. 2000, Proc. SPIE Int. Soc. Opt. Eng., 4131, 26

ASTRONOMICAL DATA ANALYSIS SOFTWARE AND SYSTEMS XIV
ASP Conference Series, Vol. 347, 2005
P. L. Shopbell, M. C. Britton, and R. Ebert, eds.

Caltrans Keeps the Spitzer Pipelines Moving

Wen Lee, Russ Laher, John W. Fowler, Frank J. Masci, and Mehrdad Moshir

Spitzer Science Center, MS 314-6, California Institute of Technology, Pasadena, CA 91125[1]

Abstract. The computer pipelines used to process digital infrared astronomical images from NASA's Spitzer Space Telescope require various input calibration-data files for characterizing the attributes and behaviors of the onboard focal-plane-arrays and their detector pixels, such as operability, dark-current offset, linearity, non-uniformity, muxbleed, droop, and point-response functions. The telescope has three very different science instruments, each with three or four spectral-band-pass channels, depending on the instrument. Moreover, each instrument has various operating modes (e.g., full array or sub-array in one case) and parameters (e.g., integration time). Calibration data that depend on these considerations are needed by pipelines for generating both science products (production pipelines) and higher-level calibration products (calibration pipelines). The calibration files are created in various formats either "off-line" or by the aforementioned calibration pipelines, depending on the above configuration details. Also, the calibration files are generally applicable to a certain time period and therefore must be selected accordingly for a given raw input image to be correctly processed. All of this complexity in selecting and retrieving calibration files for pipeline processing is handled by a procedural software program called "caltrans". This software, which is implemented in C and interacts with an Informix database, was developed at the Spitzer Science Center (SSC) and is now deployed in SSC daily operations. The software is rule-based, very flexible, and, for efficiency, capable of retrieving multiple calibration files with a single software-execution command.

1. Introduction

The provision of the correct calibrations for reducing the raw data measured by a scientific instrument can be a challenge because the properties of the instrument generally change over time and, therefore, so do the calibration data. Pipeline data-processing of digital infrared astronomical images from NASA's Spitzer Space Telescope is even more complex because there are three uniquely different, multi-channel science instruments onboard (IRAC, MIPS, and IRS), and there are correspondingly many different kinds of calibrations required (e.g., Masci et al. 2005). The Spitzer pipeline-processing is done in parallel on about 50 workstations, and multiple parallel processes running on each workstation can access common calibration files stored on a local hard disk. For this system,

[1]Author e-mail addresses, respectively: wplee@ipac.caltech.edu, laher@ipac.caltech.edu, jwf@ipac.caltech.edu, fmasci@ipac.caltech.edu, and mmm@ipac.caltech.edu.

fast software for precise calibration-file retrieval, utilizing data management in a database, is essential for determining the best set of calibration files for a given process and then copying them to the target workstation.

This paper describes the calibration-file-retrieval software and associated database schema that we have developed and implemented for the Spitzer mission. The software is called "caltrans", which is short for calibration transfer. It has been successfully used in operations at the Spitzer Science Center (SSC) since the telescope's launch in August 2003.

2. High-Level Requirements

A primary requirement levied on caltrans, perhaps the most important of all, is that the complications involved in retrieving calibration data should be decoupled from the design and development of the other pipeline software programs. These programs were difficult enough to develop without the added complexity of every one of them having to have similar logic about which calibration files to use, where to get them, etc.

Other important requirements of caltrans include fast performance and minimal impact on the project's Informix database. Also, to ensure that a complete set of calibration files will always be retrieved by caltrans, a requirement was made that "fallback" calibration files will always be available as substitutes if caltrans database queries fail to find the requested calibration files.

3. Database Schema

Caltrans queries several database tables for calibration files (see Figure 1). The caltransControls table associates a unique database index (calId) with a given Spitzer instrument, channel number (each instrument has three or four channels, depending on the instrument), and the specific database table names where meta-data about the corresponding calibration files are stored. Calibration files are classified as either "fallback" or "non-fallback" in the database.

In order to keep the number of caltrans database tables manageable, there is a channel-number field (chnlNum) in the database tables for IRAC and IRS, which is permissible because of the parallel nature of their respective data channels. On the other hand, because the attributes of three MIPS channels are so different, it was necessary to design separate database tables for them.

Meta-data about the *fallback* calibration files are stored in five database tables: iracFallback, irsFallback, mips1Fallback, mips2Fallback, and mips3Fallback. Each of these tables has a primary key (fbId) for fast access to the meta-data. Records in these database tables are populated manually via a script calling database stored procedures.

Meta-data about the more generally used *non-fallback* calibration files, which are generated by automated calibration pipelines in Spitzer operations, are also stored in five database tables, separately from the fallbacks: iracMetaData, irsMetaData, mips1MetaData, mips2MetaData, and mips3MetaData. Since these calibration files are registered by calibration pipelines in the ensembleProducts database table, where each file is assigned a unique database ID called epId,

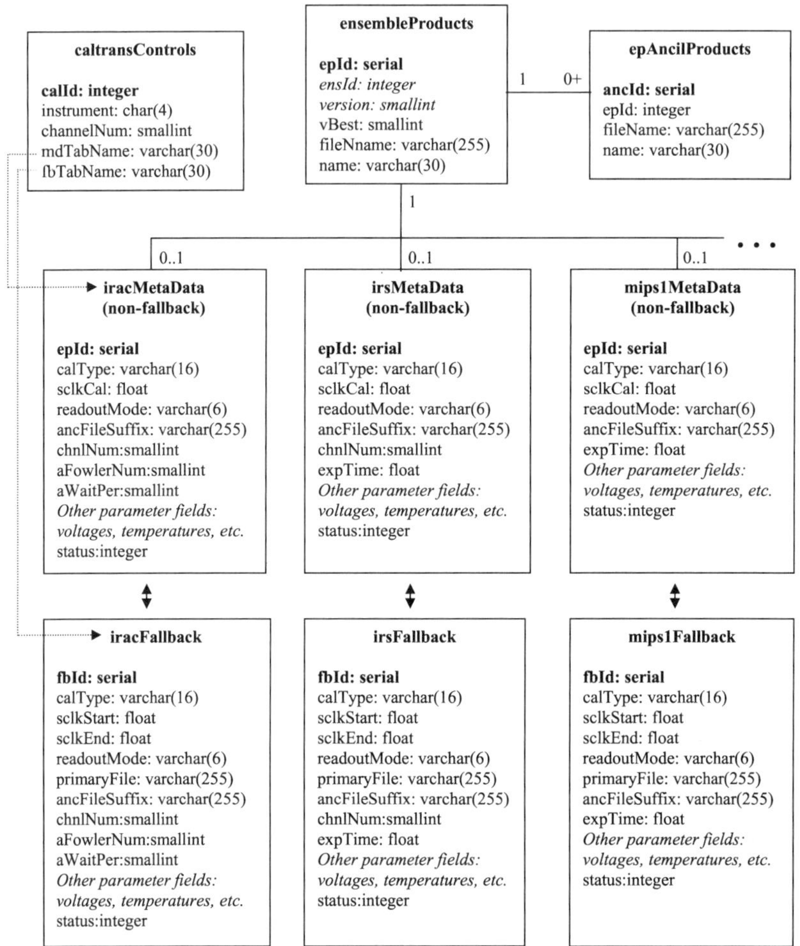

Figure 1. Schema for the database tables used by caltrans. Not shown are the database tables for MIPS channels 2 and 3.

the calibration non-fallback database tables use epId as a primary key, which is distinctly different from the calibration fallback database tables.

Records in the calibration fallback and non-fallback database tables contain a number of useful fields for distinguishing a given calibration file from another. The calType field allows various broad categories of calibration types to be defined, such as “labdark” or “skyflat”. Each record is applicable to a primary calibration file (e.g., a dark-offset correction image) and zero or more ancillary calibration files (e.g., associated mask image, uncertainty image, etc.).

For the calibration fallback database tables, the complete path/filename of the primary calibration file is stored in the primaryFile field and ancillary-file suffixes are stored in the ancFileSuffix field as a comma-separated list. It is assumed for fallbacks that the primary and ancillary files are stored together in the same directory and the ancillary filenames are basically derivable by inserting the ancillary-filename suffixes in the primary filename before the extension.

For the calibration non-fallback database tables, the complete path/filename of the primary file is stored in the filename field of the ensembleProducts database table, and so a table join is necessary in the caltrans database query. The ancillary filenames are derived as described above, and for calibration files that have been archived (and renamed to have archival filenames), the archival locations are looked up in the corresponding record in the epAncilProducts database table.

A given calibration file is generally applicable to a certain range or point in time. The sclkStart and sclkEnd (sclk stands for Spitzer spacecraft time) fields in the calibration fallback database tables allow a time range to be specified for the calibration file. The sclkCal field in the calibration non-fallback database tables is for storing the average time associated with the calibration file as computed by a calibration pipeline.

4. Software

Caltrans is a stand-alone procedural software program that was implemented in C for the fastest performance possible. Its architecture includes embedded SQL functionality for querying the calibration database tables. The software runs under the Solaris operating system in Spitzer operations; it has also been shown to compile/build successfully under Linux.

Caltrans is normally executed as an early pipeline step. A single caltrans command is capable of retrieving all the calibration files that the pipeline may require. After a database connection has been established, multiple database queries are performed in order to find all the requested files, and then the database connection is closed. Making maximal use of the open database connection fulfills our requirement for minimally impacting the database.

Caltrans has available several rules for retrieving calibration files (Table 1 give the basic ones). A different rule can be applied for each individual calibration file in the set of files to be retrieved. Note that a variant of rule 400, which employs a random-walk method of interpolating uncertainties between sample times (see Moshir et al. 2003), is currently under development.

Input configuration/control parameters are read from either a specially formatted input file (“namelist”), the command line, or a combination of both.

Table 1. Basic calibration-file retrieval rules implemented in caltrans.

Rule	Description
0	Get fallback-only calibration file
100	Get nearest-in-time-before calibration file
200	Get nearest-in-time-after calibration file
300	Get nearest-in-time (before or after) calibration file
400	Create calibration file via linear interpolation in time

Command-line parameters, by convention, override any identical ones that may be present in the namelist. There are options for tailoring the database query; e.g., chnlNum=0, ASLDTmpA=6.21±0.01 (IRS Short-Low-Detector Temperature A), etc.

The reference time of the calibration files must be input to caltrans, in units of Spitzer spacecraft clock, for querying the calibration database tables. The reference time can either be read from the caltrans namelist, command line, or header of a specified FITS image (SCLK_OBS keyword).

Only those calibration files with the status field in the fallback and non-fallback database tables set to values ranging from 1024 to 8191 are retrievable by caltrans; a value outside of this range causes the associated calibration file to be effectively removed from the system. Additionally, only non-fallback calibration files with the vBest field in the ensembleProducts database table set to a value greater than 0 are retrievable by caltrans.

A local-disk "cache" directory may be specified as caltrans input. Caltrans will check the cache to see if the calibration file already exists there from a prior execution before retrieving the file from the networked file system. This reduces network traffic, which is important for multi-workstation parallel computing.

Acknowledgments. This work was performed at the Spitzer Science Center as part of a mission/project managed by Jet Propulsion Laboratory, California Institute of Technology, under a contract with the National Aeronautics and Space Administration.

References

Masci, F., Laher, R., Fang, F., Fowler, J., Lee, W., Stolovy, S., Padgett, D., & Moshir, M. 2005, in ASP Conf. Ser., Vol. 347, ADASS XIV, ed. P. L. Shopbell, M. C. Britton, & R. Ebert (San Francisco: ASP), 468

Moshir, M., Fowler, J., & Henderson, D. 2003, in ASP Conf. Ser., Vol. 295, ADASS XII, ed. H. E. Payne, R. I. Jedrzejewski, & R. N. Hook (San Francisco: ASP), 181

ASTRONOMICAL DATA ANALYSIS SOFTWARE AND SYSTEMS XIV
ASP Conference Series, Vol. 347, 2005
P. L. Shopbell, M. C. Britton, and R. Ebert, eds.

VISTA Data Flow System: Pipeline Processing for WFCAM and VISTA

James R. Lewis, M. J. Irwin, S. T. Hodgkin, Peter S. Bunclark, and D. W. Evans

Cambridge Astronomy Survey Unit, Institute of Astronomy, Madingley Road, Cambridge CB3 7XT, UK

R. G. McMahon

Institute of Astronomy, Madingley Road, Cambridge CB3 7XT, UK

Abstract. The UKIRT Wide Field Camera (WFCAM) on Mauna Kea and the VISTA IR mosaic camera at ESO, Paranal, with respectively 4 Rockwell 2k $\times$ 2k and 16 Raytheon 2k $\times$ 2k IR arrays on 4m-class telescopes, represent an enormous leap in deep IR survey capability. However with an expected data rate of an image of the sky every 5-30s and combined nightly data-rates of typically 1 TB, automated pipeline processing and data management requirements are paramount.

Pipeline processing of IR data is far more technically challenging than for optical data. IR detectors are inherently more unstable, while the sky emission is over 100 times brighter than most objects of interest, and varies in a complex spatial and temporal manner.

The pipelines are designed around a selectable modular scheme, driven by processing recipes for maximum flexibility. Our general philosophy is that all fundamental data products are in multi-extension FITS files with headers describing the data taking protocols in sufficient detail to trigger the appropriate pipeline processing components. All derived information, DQC, photometric and astrometric calibration and processing details are also incorporated into the FITS headers. Generated catalogues are stored in FITS binary tables. The headers provide a basis for ingest into databases for archiving, real time monitoring of survey progress and survey planning.

To reduce the data storage I/O overheads and transport requirements, we intend to use, as much as possible, the lossless Rice tile compression scheme as used transparently, for example, in CFITSIO. For this type of data (32 bit integer) the algorithm typically gives a factor of 3-4 compression

1. WFCAM Pipelines

There will be essentially three separate pipelines for WFCAM. A summit pipeline will assess data quality on-the-fly, provide feedback to survey planning and progress tools and provide first pass science products. Figure 1 schematically shows the corresponding data flow operations at the summit on Mauna Kea.

A standard calibration pipeline to do full instrumental signature removal, astrometric and photometric calibration and catalogue generation will be run in Cambridge. Data compression will be absolutely vital for these projects as, with compression, we anticipate the yearly storage requirements for WFCAM

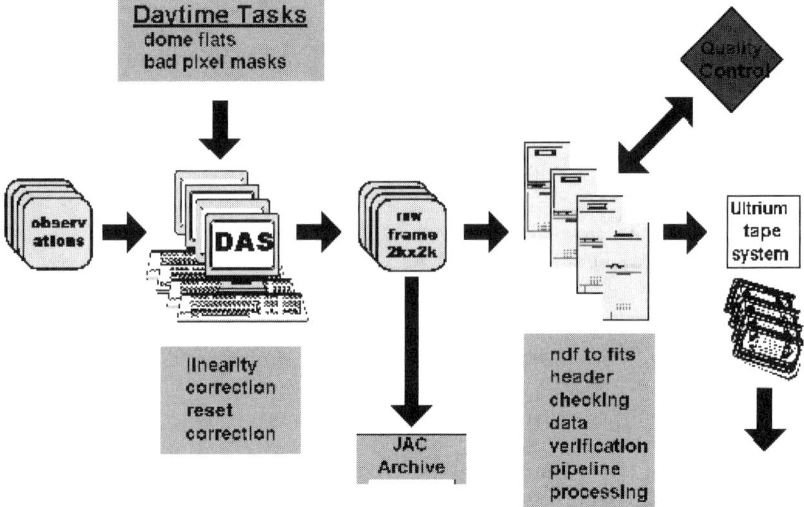

Figure 1. An overview of the summit processing setup for WFCAM. Each detector feeds a separate data acquisition PC, which then forwards the images in parallel to the summit data processing PCs. Data transport to the UK for science processing is via Ultrium LTO-I tapes

to be ≈10 Tbytes per year and for VISTA ≈50 Tbytes per year. Transfer of processed data from Cambridge to the WFCAM and VISTA Science Archives in Edinburgh will be via the Internet. The Cambridge pipeline setup is illustrated in Figure 2.

A third pipeline to generate advanced data products (e.g., deep stacking) will be run in Cambridge and Edinburgh.

2. Image Processing

The first stage of image processing will be done in the DAS before the data is written out. This includes combining individual reads into a single exposure and removing the reset (bias) frame. By and large data obtained from NIR arrays are strongly non-linear. This effect will be removed in the DAS for WFCAM but not for VISTA.

The main reduction steps and their approximate location in the pipeline are:

- Linearity correction: if it has not been done in the DAS.
- Dark and reset anomaly correction: removes two additive effects caused by dark current and residual bias structure
- Flat field correction: to remove pixel-pixel sensitivity differences, gain differences between data channels and between detectors.
- Confidence map generation: we define a confidence map as a normalised inverse variance weight map denoting the "confidence" associated with the flux value in each pixel.

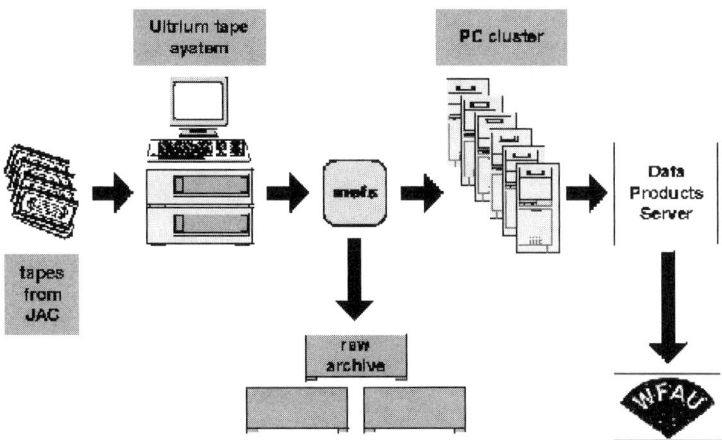

Figure 2. A schematic of the data processing setup for WFCAM in Cambridge. Incoming tapes are ingested, the data verified, converted to MEF format and fed to the processing cluster. A raw data archive is held on-line, whilst processed products are automatically transferred via the internet to the Wide Field Astronomy Unit (WFAU) in Edinburgh.

- Defringing: If the fringe spatial pattern is stable, it is possible to decouple sky correction from fringe subtraction.
- Sky subtraction: This can be done using the dither sequences themselves or with observations of offset sky regions.
- Image Persistence and Detector Crosstalk: These are electronic effects which can be modeled and removed.
- Interleave: To recover lost resolution when observations are undersampled (as will be the case with WFCAM and, in best seeing conditions, VISTA) observations may be "microstepped" into a series of exposures with subpixel shifts. The resulting image pixels are interwoven to form a single output image.
- Dithering: Shift and average all the exposures in a dither sequence into a single output image.
- Catalogue generation: Objects are extracted from the images and catalogued to FITS binary tables. Standard object descriptors include assorted aperture flux measures, intensity weighted centroid estimates, and shape information, such as intensity-weighted 2nd moments to encode the equivalent elliptical Gaussian light distribution.
- Astrometric calibration: Catalogued objects are matched to astrometric standards to define a WCS for each image/catalogue.

Figure 3. The result of processing and stacking 1760 individual 120 K_s band images (44 observing blocks)

- Photometric zeropoint: for the purposes of quality control a photometric zeropoint will be determined for each observation by direct comparison of instrumental magnitudes with 2MASS. The complete night's observations will allow for a more accurate value to be determined later.

3. Trial Commissioning

We have tested the WFCAM and VISTA pipeline modules on infrared data from several existing cameras. This includes the NIR ISAAC camera on the VLT. As a test of our pipeline and commissioning procedures, we downloaded the raw data for the Faint InfraRed Extragalactic Survey (FIRES) from the ESO archive. These data have already been reduced and presented by Labbé et al. (2003) and therefore make an excellent benchmark against which to test our reduction procedures. A result of this testing is shown in Figure 3.

References

Labbé, I., et al. 2003, AJ, 125, 1107

ASTRONOMICAL DATA ANALYSIS SOFTWARE AND SYSTEMS XIV
ASP Conference Series, Vol. 347, 2005
P. L. Shopbell, M. C. Britton, and R. Ebert, eds.

Time Domain Explorations with Digital Sky Surveys

Ashish A. Mahabal, S. G. Djorgovski, M. J. Graham, Priya Kollipara, Benjamin Granett, Elisabeth Krause, Roy D. Williams, and M. Bogosavljevic

California Institute of Technology, Pasadena, CA 91125

C. Baltay, D. Rabinowitz, A. Bauer, P. Andrews, N. Ellman, S. Duffau, and J. Jerke

Yale University, New Haven, CT 06520

A. Rengstorf and R. J. Brunner

NCSA/UIUC, Champaign, IL 61820

J. Musser, S. Mufson, and M. Gebhard

Indiana University, Bloomington, IN 47405

Abstract. One of the new frontiers of astronomical research is the exploration of time variability on the sky at different wavelengths and flux levels. We have carried out a pilot project using DPOSS data to study strong variables and transients, and are now extending it to the new Palomar-QUEST synoptic sky survey. We report on our early findings and outline the methodology to be implemented in preparation for a real-time transient detection pipeline. In addition to large numbers of known types of highly variable sources (e.g., SNe, CVs, OVV QSOs, etc.), we expect to find numerous transients whose nature may be established by a rapid follow-up. Whereas we will make all detected variables publicly available through the web, we anticipate that email alerts would be issued in the real time for a subset of events deemed to be the most interesting. This real-time process entails many challenges, in an effort to maintain a high completeness while keeping the contamination low. We will utilize distributed Grid services developed by the GRIST project, and implement a variety of advanced statistical and machine learning techniques.

1. Introduction

Systematic exploration of previously poorly covered regions of the observable parameter space is a major source of discoveries in astronomy (Djorgovski et al. 2001a, 2001b). In particular, exploration of the time variability on the sky over a broad range of flux levels and wavelengths is rapidly becoming a new frontier of astronomical research (Paczynski 2000; Diercks 2001). All manner of variable stars, stellar explosions such as SNe and GRBs, variable AGN, pulsars, microlensing events, etc. are some of the examples of currently known exciting time domain astrophysical phenomena. Many more are certain to be discovered over time as planned ambitious synoptic sky surveys (Tyson 2002; Kaiser et al. 2002) join the existing ones involving the time domain (see, e.g., Paczynski 2001, http://www.astro.princeton.edu/~bp/ for listings). The important factors in such programs are (1) the area covered, (2) the depth of coverage, (3) number of wavelengths used, and (4) the baseline(s) in time.

We briefly describe here a pilot project we carried out with the plate overlap regions of the Digital Palomar Observatory Sky Survey (hereafter DPOSS). We then outline ongoing programs with the Palomar-QUEST sky survey (hereafter PQ) and describe a realtime transient detection system and its challenges.

2. Variability with DPOSS overlap regions

DPOSS covers the entire Northern Hemisphere in three filters viz. J, F and N plates (calibrated to Gunn g, r, i). Each plate is 6.5 degrees wide with adjacent plates overlapping by strips that are 1.5 degrees wide. This results in at least 40% of the sky being imaged at least twice in each of the three filters. We conducted an exploratory search for highly variable objects (Granett et al. 2005) and optical transients (Mahabal et al. 2005) using $\sim$ 8000 deg^2 in the NGP and SGP areas of these plate overlap regions. The effective depth of these searches was $r_{max} \approx 19$ mag for the "high" states, with a plate limits $r_{max} \approx 21$ mag. Time baselines ranged from days to years, with $\sim$ 2 yrs being typical. After eliminating various artifacts and contaminants, and applying well defined statistical criteria for selection, we identified a large number of highly variable objects, and followed up spectroscopically a subset of them at the Palomar 200-inch telescope. They turned out to be a heterogeneous collection of flaring M-dwarfs, OVV QSOs and BL Lacs, CVs (including a rare magnetic one), and some otherwise non-descript stars. Approximately a third to a half of all highly variable objects down to this magnitude level, at moderate and high Galactic latitudes appear to be associated with AGN.

We also found a number of optical transients (operationally defined as high-S/N, PSF-like objects, detected only once). We estimate that a single-epoch "snapshot" down to this flux level contains up to $\sim$ 1000 transients/sky. Their nature remains unknown, but in at least 2 cases deep follow-up imaging revealed apparent faint host galaxies, which now await spectroscopy.

This pilot study gave us some hints as to what may be expected in a dedicated, wide-field, synoptic sky survey at comparable magnitudes. The faint variable sky has a very rich and diverse phenomenology.

3. The driftscans of Palomar-QUEST

The Palomar-Quest synoptic sky survey (Djorgovski et al. 2004, 2005; Baltay et al. 2005), a collaborative project between Yale, Caltech, and NCSA (some other groups are also involved in more specific collaborations) is a new major digital sky survey conducted at the Samuel Oschin 48-inch Schmidt telescope at Palomar. The survey uses a special 112-CCD, 162-Megapixel camera built especially for this purpose. Some of the salient features of the survey are: (1) Data taking in the Point-and Stare (PS) mode, covering $\sim$ 9.2 deg^2 per exposure, or in a Drift Scan (DS) mode, in strips 4.6° wide, with a typical coverage of $\sim$ 500 deg^2/night, (2) Near simultaneous observations in one of two filter sets in the DS mode: Johnson-Cousins *UBRI* or SDSS $r'i'z'z'$, (3) In good conditions, typical limiting magnitudes for point sources: $R_{lim} \approx 22$ mag, $I_{lim} \approx 21$ mag, (4) In the DS mode, useful Declination range $-25^{\circ} < \delta < +30^{\circ}$, for a total anticipated survey area of $\sim 15,000$ deg^2, (5) Multiple-pass coverage, with at least 4 passes per year at each covered location, (6) Time baselines for repeats ranging from days to months, anticipated to extend to multi-year time scales over the next 3 to 5 years or beyond, (7) NVO standards, protocols, and connections built in from the start.

The survey has started producing a steady stream of science-grade data, from the summer of 2003. In the DS mode, it typically generates ~ 1 TB of raw image data per month (assuming ~ 14 clear nights). This unprecedented amount of data makes this the largest synoptic survey of its kind both in terms of area covered and depth. A broad range of science is envisioned for the survey, but exploration of the time domain will be one of the main focal areas.

PQ coverage as of Oct. 2004 is as follows: ~ 11500 deg^2 have been covered in *UBRI* (~ 8900 deg^2 at least twice) and ~ 12100 deg^2 in $r'i'z'z'$ (~ 9300 deg^2 at least twice). We have been testing our transient detection techiniues on areas that have been observed a large number of times. These techniques are being perfected in preparation of a realtime pipeline described in the next section. Besides the more exciting transient detections, other science involves objects that simply vary in interesting ways. A large number of epochs with a range of baselines for several tens of thousands of quasars will provide good limits to differentiate between the models and further lead to better estimates of quasar lifetimes. This is possible by combining PQ data with earlier surveys like SDSS, DPOSS, DSS etc. Figure 1 shows B-band structure function for 500 quasars with $1 < z < 4$, a starting point for such a study.

4. Detecting Transients in Real Time

The existing *PQ* pipeline is geared to complete processing of a night's data by the next day. In a matter of hours catalogs become available in the four filters used and can be compared with other PQ epochs, or with other sky surveys and pointings for the area covered that night. This is sufficient for most *PQ* projects, including those involving variable objects, e.g. SNe. However, for the follow-up of rapidly fading sources and transients a faster turn-around time is needed. Thus, we have started work on a real-time processing pipeline which will enable detections of such sources within minutes or less. The following steps will be part of the process: (1) Compare nightly catalogs with older catalogs from *PQ* itself, (2) Compare the catalogs with other surveys and archives, using NVO infrastructure and methodology, (3) Compare positions of possible transients with those of known variable sources, known asteroids, etc. (see Figure 2 for an example), (4) A source classification engine will use the catalog data and the image data to catgorize the possible transients to determine the likely types, (5) A decision engine will determine the significance of the object: should it be broadcast to the email subscribers list so that it can be followed up immediately, or should it go to the website to be announced as a lower priority but interesting transient, or just flagged as an interesting object to be looked at again in future epochs (in addition to other catagories like possible variable, orphan GRB etc.).

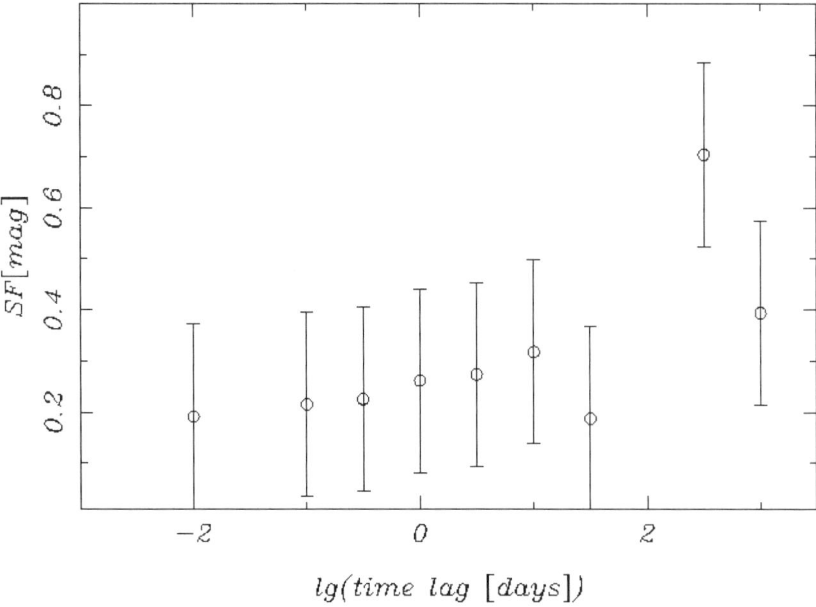

Figure 1. B-band structure function of ~ 500 quasars for $1 < z < 4$ using PQ and SDSS data. We are extending this to tens of thousands of quasars.

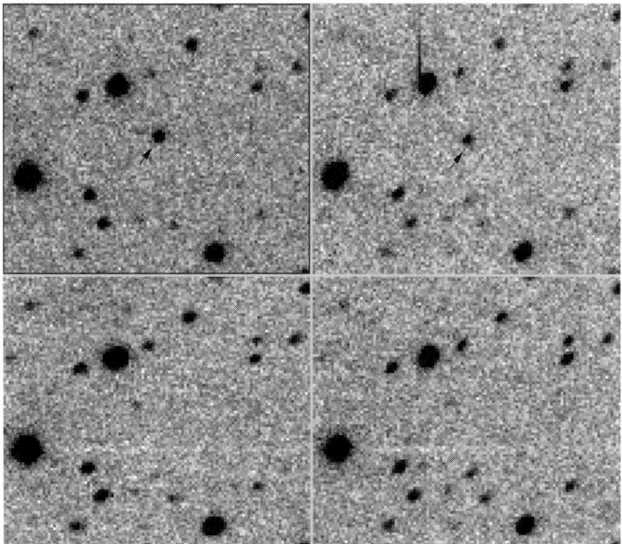

Figure 2. Example of an asteroid from R and I images from August and June 2004. The object easily passed as a star. Only after using the known asteroid database could it be identified with one. Faster asteroids are often easier to track as they form an elongated image.

Given the large area coverage of *PQ* and the results from our exploratory DPOSS project noted above, we estimate that we will be detecting up to several tens of highly variable or transient sources per night. Points (4) and (5) above summarize the key challenge of dealing with this abundance of data in an effective manner – maintaining a high completeness in terms of the interesting variable and transient sources discovered, while maintaining a low contamination rate by spurious or uninteresting sources – and doing it in real time. We will utilize distributed Grid based services developed by the GRIST project (Jacob et al. 2005) along with a variety of advanced statistical (Mahabal et al. 2004; Graham et al. 2005) and Machine Learning techniques towards this end.

Acknowledgements: This work was supported in part by the NSF grants AST-0326524, AST-0407448 and the NASA contract NAG5-9482. PK and EK were supported in part by SURF Fellowships at Caltech. SGD acknowledges a partial support from the Ajax Foundation.

References

Baltay, C., et al. 2005, in preparation

Diercks, A., 2001, in: Virtual Observatories of the Future, ed. R. Brunner, S.G. Djorgovski, & A. Szalay, ASPCS, 225, 52

Djorgovski, S. G., et al. 2001a, in: Mining the Sky, ed. A. Banday et al., Berlin: Springer Verlag, p. 305

Djorgovski, S. G. et al. 2001b, in: Astronomical Data Analysis, ed. J.-L. Starck & F. Murtagh, Proc. SPIE, 4477, 43

Djorgovski, S. G. et al. (The PQ Team) 2004, BAAS, 36, 805

Djorgovski, S. G. et al. 2005, in preparation

Graham, M. J., et al. 2005, in ASP Conf. Ser., Vol. 347, ADASS XIV, ed. P. L. Shopbell, M. C. Britton, & R. Ebert (San Francisco: ASP), 394

Granett, B., et al. 2005, in preparation

Jacob, J., et al. 2005, in ASP Conf. Ser., Vol. 347, ADASS XIV, ed. P. L. Shopbell, M. C. Britton, & R. Ebert (San Francisco: ASP), 306

Kaiser, N., et al. 2002, in: Survey and Other Telescope Technologies and Discoveries, ed. A. Tyson & S. Wolff, Proc. SPIE, 4836, 154

Mahabal, A.,et al. (The VOStat Team) 2004, BAAS, 36, 806

Mahabal, A.,et al. 2005, in preparation

Paczynski, B. 2000, PASP, 112, 1281

Paczynski, B. 2001, in: Mining the Sky, ed. A. Banday et al., Berlin: Springer Verlag, p. 481

Tyson, A. 2002, in: Survey and Other Telescope Technologies and Discoveries, ed. A. Tyson & S. Wolff, Proc. SPIE, 4836, 10

ASTRONOMICAL DATA ANALYSIS SOFTWARE AND SYSTEMS XIV
ASP Conference Series, Vol. 347, 2005
P. L. Shopbell, M. C. Britton, and R. Ebert, eds.

Planck/LFI Pipeline - The Demonstration Model

F. Pasian, A. Zacchei, and C. Vuerli

INAF-Osservatorio Astronomico di Trieste

D. Maino

University of Milan, Department of Physics

C. Baccigalupi

SISSA - International School for Advanced Studies, Trieste

on behalf of the DPC Development Team

Abstract. LFI is one of the two instruments installed on board Planck, the M3 mission of the ESA Horizon 2000+ program, the main goal of which is to produce maps of the CMB (Cosmic Microwave Background) with an unprecedented combination of angular resolution, sensitivity, spatial coverage, and frequency range. The Demonstration Model (DM) is the second version of the LFI DPC pipeline, built on top of the BBM (Bread-Board Model), released in July 2002. The DM has been integrated at DPC using the Process Coordinator and FITS files for standard data exchange; in the future, instead of using FITS files, the pipeline will be compliant with the Planck DMC (Data Management Component) standard. The DM goal is to demonstrate that the current pipeline is able to handle the foreseen LFI data flow and to perform an end-to-end processing of the data, from telemetry to the production of the scientific results. The DM development was mainly concentrated on understanding and removing systematic effects; at the moment the pipeline is able to detect and remove all systematic effects which are understood and modeled within the Planck simulation pipeline, such as thermal fluctuations, 1/f noise, the side-lobe effects, beam distortion, and so on.

1. Introduction

A specific development scheme has been envisaged for the DPC of Planck/LFI; it foresees the creation of three models: the Bread-Board Model (BBM), the Demonstration Model (DM), and the Operations Model (OM). This paper is centered on the DM whose implementation has been just completed and whose verification/test phase is currently in progress.

The data processing at the two Planck DPCs is organized in four different levels ranging from *level 1* through *level 4*; an additional *level S* deals with the preparation and running of simulation software whose goals are: 1) check the robustness and reliability of those pipeline modules that will be used during the operative phase of the mission and 2) process data coming from the ground-based instrument test/calibration phase. Figure 1 gives an overview of how the data processing is spread over the four levels of the DPC.

2. Planck/LFI DPC Models Overview

The BBM was completed in July 2002. The tasks provided by the BBM may be summarized as follows:

1. Implement level 1 facilities to a fairly complete degree, so as to be able to support instrument ground testing activities (ILTs and ISTs).
2. Implement the other levels (in particular 2 and 3) to such a degree that a simplified but complete processing chain can be performed on simulated data.

The DM implementation has now ended. The verification/test phase is currently in progress. The major improvements introduced in the DM are listed below:

1. Work with full time order data size (1 minute rings) instead of scan circles averaged over 1-hour periods (one re-pointing period)
2. Include a more realistic treatment of systematic effects and their proper handling in the data analysis phase.
3. Usage of the Process Coordinator (ProC) for creation/handling and running of pipelines.

Items 1 and 2 of the above list are significant in terms of the data-size that the LFI DPC can handle and of the knowledge of the instrument that is expected to improve as soon as ground tests will start. The usage of the ProC (Process Coordinator) represents the first step of integration of the LFI DPC with the IDIS components. It allows a great flexibility in creating a pipeline where different S/W modules could be placed in a different order of processing. This allows us to properly test/validate pipelines and assess quality of the produced data. A huge amount of knowledge of real H/W behaviour will come out from the ground tests, so we expect to build new S/W modules specifically tailored to the flight H/W, which will allow better control and removal of systematic effects.

The OM (to be implemented) will be ready 1 year before launch. It will be able to support the end-to-end tests on the ground including the full ground segment chain.

3. The Bread-Board Model

In this section the main aspects characterizing the BBM are exposed, describing what for this model is developed for each of the five levels (including the level S) in which the Planck data processing is articulated at the two DPCs.

Level S

- Quantization process of the simulated Time Ordered Data (TOD) (LFI DPC). Given the quantization scheme, this step returns the quantized TOD.
- Creation of Scientific TM packets from simulated TOD (LFI DPC).

- Creation of H/K Telemetry packets. An editor capable of generating H/K TM has been created. The TOOTSIE application, furthermore, is capable of reading the output of the SCI TM simulator and generating complete streams of LFI TM ingested by SCOS-2000, the ESA multi-mission tool to handle telemetry and telecommands.

Level 1

- The TM data acquired is captured by the SCOS-2000 environment and stored therein. RTA runs on the acquired TM directly within SCOS-2000.
- Telemetry Quick Look (TQL) performs similar functions on scientific (SCI) telemetry, but, given the current limitation of SCOS-2000 in handling scientific packets, it works outside of it.

Level 2

- Extraction of point sources from TOD (both calibrated and un-calibrated) using optimal pseudo filters and wavelets and production of cleaned TOD together with a point source catalogue.
- TOD calibration using the CMB dipole as a main calibration. The output is a calibrated TOD as well as a calibration constant as a function of time. Useful for future trend analysis and monitoring of the health of the instrument.
- Use of simple destriping technique to remove systematic effects like 1/f noise artefacts and output fluctuations induced by sorption cooler fluctuations. The output consists of maps stored in the Frequency Maps Archive.
- Creation of frequency and channel maps (Map-making) from cleaned (from point source and systematic effects) and calibrated TOD.

Level 3

- Component Separation by using different approaches, the most important of which are fastICA (a blind component separation technique) and Wiener Filtering (classical approach).
- C(l) extraction according to the MASTER approach developed by E. Hivon and for which different implementations are available.

Level 4

- Access to the simulated data through a prototype level 4 user interface written in Java, and operating on the IDIS Data Management Component.

4. The Demonstration Model

This section gives a detailed description of what has been done for the simulations level (level S) and for levels 1 through 3 in the framework of the DM.

Level S. All the Level S S/W modules extracted from the CVS repository in ESTEC are used. This is S/W developed in common between LFI and HFI and has the main goal of producing simulated TOD for each Planck feed horn on the focal plane. Level S contains also common libraries (e.g., FITSIO, FFTW, and others) that are used by modules belonging to other DPC levels.

Level 1. DM activities have involved mainly the update of the Instrument Database, the tailoring of SCOS-2000 tables for the implementation of the RTA (Real Time Analysis), the development of the TQL (Telemetry Quick Look) systems, and the implementation of the TMH (Telemetry Handler) system, i.e., TM2TOI converter.

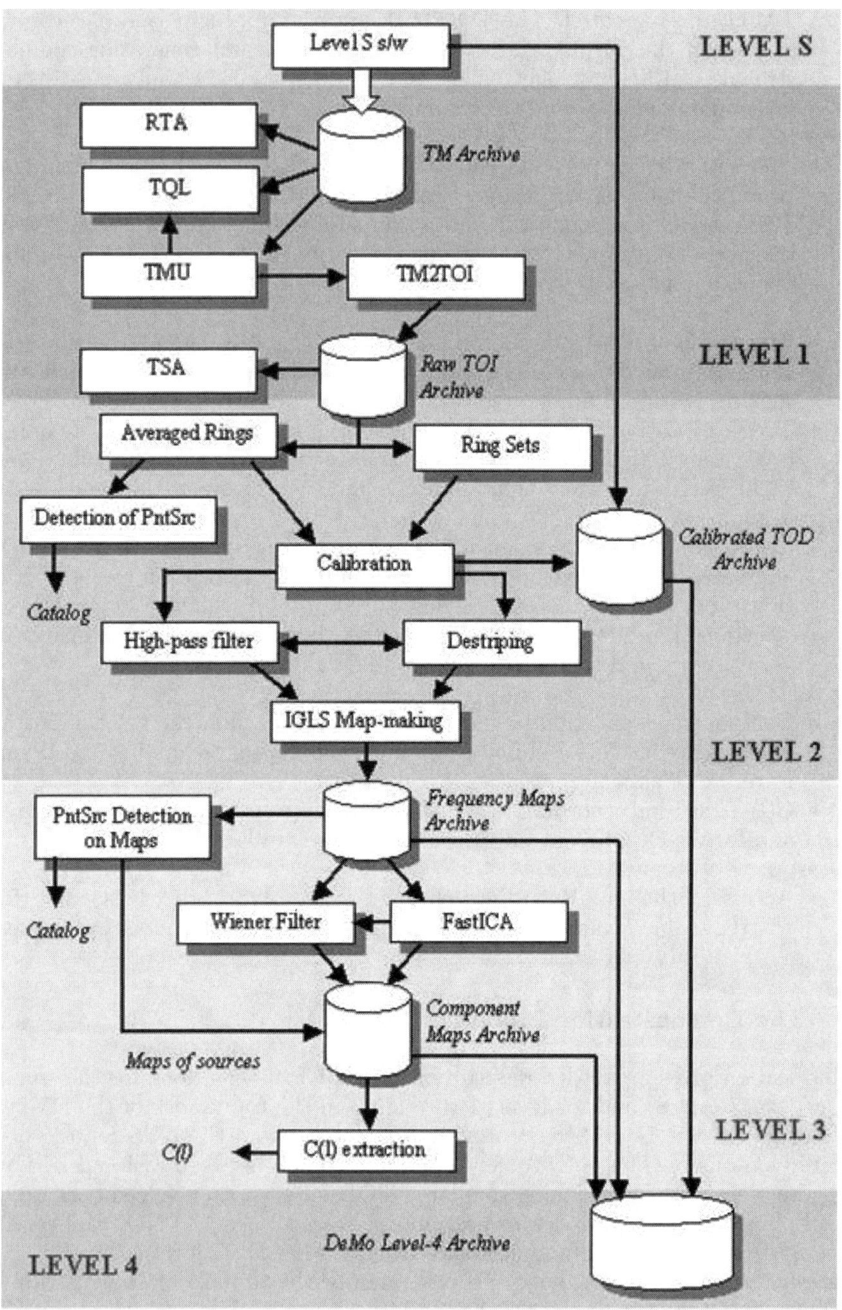

Figure 1. Data flow and processing scheme for the DPC Demonstration Model

- Update of the LFI Instrument DB created for the BBM and extension of the RTA features by tailoring other tables, within the SCOS-2000 environment (OATs). The L1 software is now table-driven: TQL and TMU (Telemetry Unscrambler) use directly the SCOS-2000 MIB tables to interpret the TM packets.
- DM deals also with decompression. The algorithm used on board as part of the REBA at IAC, is based on arithmetical compression (IAC). The decompression module is integrated with L1 modules (TQL and TMU).
- The TQL (Geneva Planck Team) is performed in the DM using the same mechanism as in the BBM: SCI TM extraction from the SCOS-2000 and TM stream creation; decompression and visualization (TQL); TM unscrambling and UTM storage.
- Creation of time-ordered streams from UTM through TM2TOI.

Level 2. Level 2 mainly performs calibration on TOD and produces frequency maps from TOD of each feed horn. The software modules are:

- Removal of Systematic Effects. Beside classical destriping of the BBM, DM includes a generalized destriping in which it is possible to fit a given systematic effect with base functions other than a simple baseline, such as Fourier modes and Legendre polynomials. A high-pass filter code used to study the removal of period effects (such as those from the sorption cooler) is also included.
- Map-Making algorithms. Besides the Rome group IGLS map-making algorithm, we added MADcap. A major feature of this code is the ability to perform a final overall calibration using the proper S/C motion along its orbit.

Level 3

- Component Separation. FastICA has been updated with major changes in the I/O handling and a more user-friendly parameter interface. It deals with total intensity as well as polarisation data with two approaches: a) independent pipeline on Q and U Stokes parameters; b) a unique data vector represented by Q and U joined. The code accepts priors on signal pattern of foreground emission, if that is known at an arbitrary frequency.
- Wiener filter. The code accepts priors on the signal to recover represented by their frequency scalings and angular power spectra. It works in total intensity as well as polarisation. It is capable of a full separation pipeline, but it is better exploited to clean residual noise in the output of the FastICA separation process.
- C(l) estimation. The MASTER like approach is exploited due to its capability to control the effects of incomplete sky coverage/masks on the reconstructed C(l) power spectra. It may be exploited by performing massive cuts of the Galactic plane as well as on the nearly all sky outputs of the FastICA component separation procedure. The results of the two approaches are compared in terms of faithfulness of the C(l) reconstruction.
- Systematics toolkit. This is made of the modules developed in order to control systematics and estimate/restore the component separation capability in the presence of systematic effects. Monte Carlo chains, sky weights, and masks are examples of modules contained in this package.

ASTRONOMICAL DATA ANALYSIS SOFTWARE AND SYSTEMS XIV
ASP Conference Series, Vol. 347, 2005
P. L. Shopbell, M. C. Britton, and R. Ebert, eds.

FLIPPER: a FLexIble PiPeline framEwoRk

Francesco Pierfederici

National Optical Astronomy Observatory, Tucson, AZ 85719

Abstract. FLIPPER is a lightweight pipeline framework able to handle blackboard-based pipelines. Its ease of configuration and operation and the fact the it is extremely lightweight make FLIPPER the ideal solution for both small and large pipeline systems. FLIPPER is currently being used for the NOAO Mosaic Pipeline and has been chosen as a pipeline framework for the NOAO NEWFIRM pipeline system. The present paper describes the architecture, features, and design choices of FLIPPER in detail.

1. Overview

FLIPPER is a lightweight pipeline framework able to handle blackboard-based pipelines. Some of its main features are:

- Only one polling process per pipeline.
- Very few software dependencies (just Python and the XML parser).
- All the major Operating Systems are supported (Linux, Mac OS X, Windows, and most unices).
- Pipelines are fully described in XML files (one file per pipeline).
- Pipeline modules are activated when trigger conditions are met.
- The most popular trigger types are supported (time, file and event based).
- Pipeline modules can define pre and post-processing actions.
- Pipeline modules can be written in any language.

2. Architecture

FLIPPER's Data Model (Figure 1) is centered around the concepts of Pipeline and Pipeline System. A Pipeline System is a collection of Pipelines and dependency rules. Similarly, Pipelines are collections of Modules (atomic pieces of code) and dependency rules between them.

Each Pipeline System is handled by a Resource Manager (Figure 2). Its main role is to make sure that the dependencies between Pipelines are satisfied. Its other function is to act as both a load balancer between processing machines and as a registry of the resources available to the Pipeline System.

Each Pipeline has a Pipeline Manager associated to it. The Pipeline Manager keeps track of the instantaneous state (stored in two separate Blackboard objects) of the corresponding Pipeline.

Pipelines are sets of Modules. Each Module is controlled by a Module Manager, which, querying the parent Pipeline Manager, ensures that all the relevant

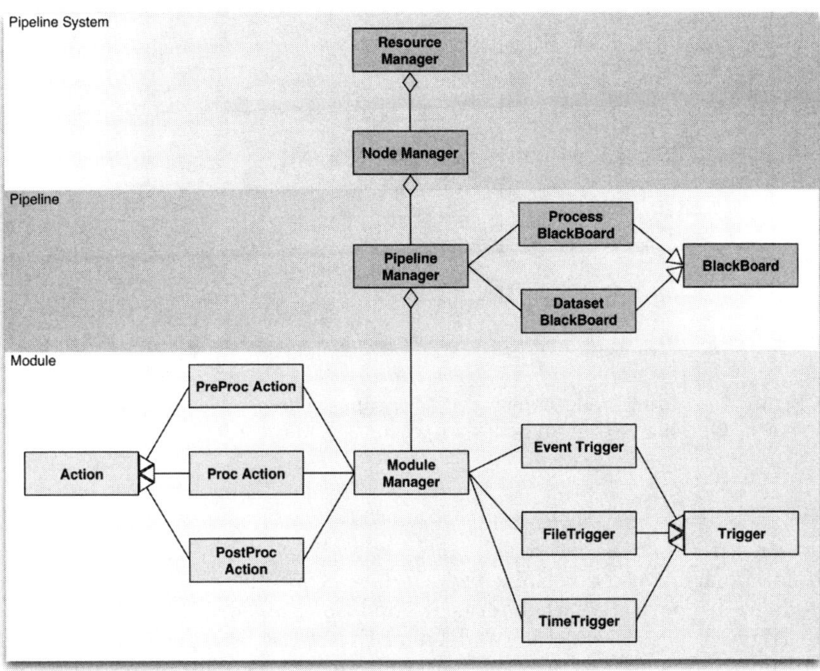

Figure 1. Class diagram for FLIPPER.

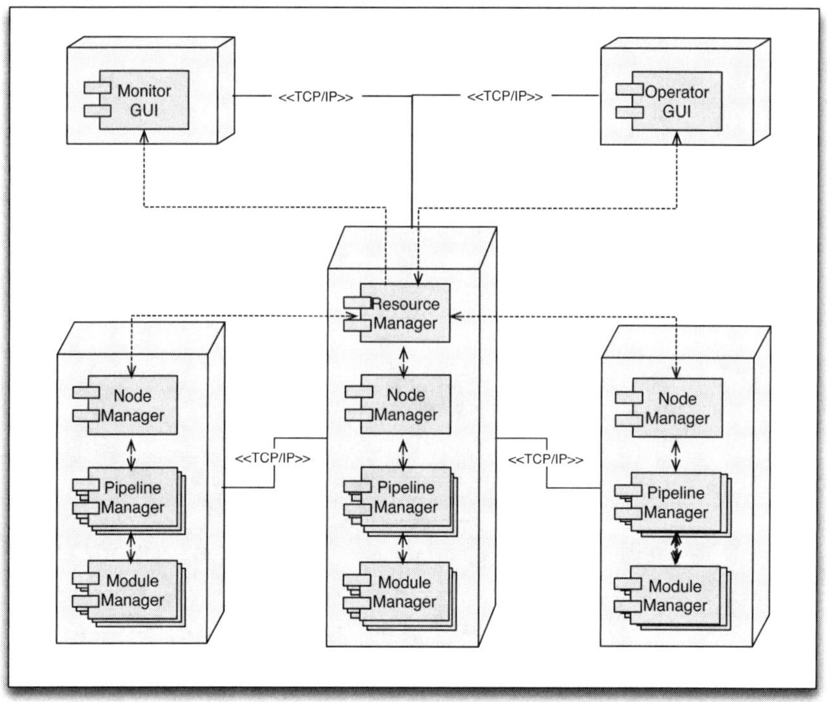

Figure 2. Multi-node deployment diagram. Three processing modes are shown. Optional GUIs can be used to monitor and control the whole processing network.

dependency rules are satisfied. The interaction between Pipeline Managers and Module Managers also determines which dataset Modules need to operate on.

Modules are activated when Trigger conditions are met. FLIPPER defines three Trigger types: Event, File, and Time. Event Triggers specify which dataset processing state activates any given Module (e.g., bias subtraction must occur before the flat field can be applied). File Triggers activate Modules when a file matching a given name pattern appears in a given directory. Time Triggers define Modules that need to execute at a given point in time and/or at a specified interval.

FLIPPER offers the possibility to define actions to be performed (i.e., executables to be run) just before each Module is activated and just after it has finished execution. It is possible to have different post-processing actions depending on the exit code of the corresponding Module.

Each processing machine is controlled by a Node Manager (Figures 2 and 3). The Node Manager monitors the machine load, available disk space and, communicating with the Resource Manager, makes its machine (and Pipelines) available to the Pipeline System.

Operators can monitor and interact with the System by using optional GUIs (that communicate directly with the Resource Manager). Alternatively, it is

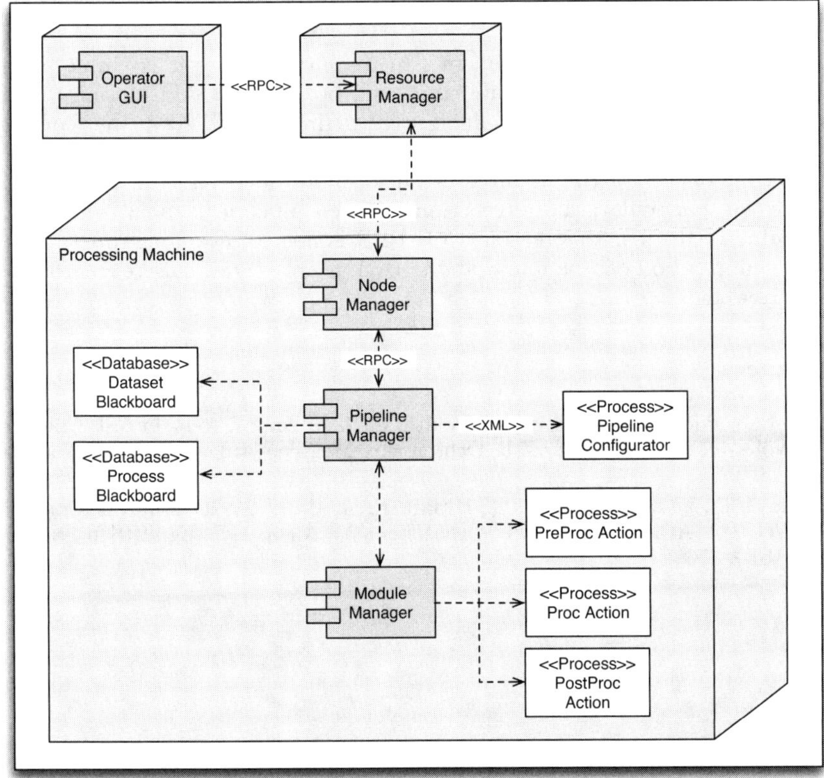

Figure 3. Deployment diagram for a single processing node. Resource Manager and Operator GUIs are shown for completeness.

possible to log into any machine and use host-callable commands to query the status of the different components.

3. Configuration

Each Pipeline is fully described in a single XML file. This pipeline description uses the objects defined in the Data Model (Figure 1). A separate XML file is used to describe the full system (and how different Pipelines interact).

This approach not only simplifies the overall setup and configuration of the whole system, but also allows for Pipelines to be build up dynamically at run-time. It then becomes easy to envision a system in which, say, archive users are able to build customized Pipelines on the fly.

4. Status and Availability

FLIPPER is still under development. In its current state it is stable enough to be used in operational environments (e.g., both the NOAO Mosaic Pipeline System and the NEWFIRM Pipeline System; Hiriart et al. 2004; Pierfederici et al. 2004).

Two components of the architecture need more development: the Resource Manager (in particular, the interaction between pipelines) and the Operator/ Monitoring GUIs. For this reason, FLIPPER is not yet available to the public.

More information can be found at the URL `http://iraf.noao.edu/projects/mosaicpipeline/`.

References

Hiriart, R., Valdes, F., Pierfederici, F., Smith, C., & Miller, M. 2004, in ASP Conf. Ser., Vol. 314, ADASS XIII, ed. F. Ochsenbein, M. Allen, & D. Egret (San Francisco: ASP), 74

Pierfederici, F., Valdes, F., Smith, C., Hiriart, R., & Miller, M. 2004, in ASP Conf. Ser., Vol. 314, ADASS XIII, ed. F. Ochsenbein, M. Allen, & D. Egret (San Francisco: ASP), 476

ASTRONOMICAL DATA ANALYSIS SOFTWARE AND SYSTEMS XIV
ASP Conference Series, Vol. 347, 2005
P. L. Shopbell, M. C. Britton, and R. Ebert, eds.

Wavelength Calibration in Physical Model Based Pipelines

M. Fiorentino, P. Bristow, F. Kerber, and M. Rosa

Space Telescope European Co-ordinating Facility, ESO, Karl-Schwarzschild-Str 2, D85748, Garching bei Muenchen

Abstract. We show how the wavelength calibration of complex 2D-spectrograms, such as echelle and long slit spectra, can be achieved in a fully self-consistent manner by a physical model based pipeline. The ST-ECF is currently upgrading the HST STIS data calibration pipeline in a step-by-step manner. In a first step the Calibration Enhanced CE-calstis pipeline will be able to generate the geometric transformations during wavelength calibration and spectral extraction from a physical (ray trace) model of the STIS optics. The configuration reference data are obtained using a special tools package developed to optimize the model predictions on the routine long exposure wavelength calibration data sets. In parallel the actual HST STIS implementation is going to be re-used in a synergetic manner for the ESO VLT infrared high resolution spectrograph CRIRES. The analysis of wavelength calibration data across the life cycle of STIS (1997 - 2004) shows that the complex 2D spectrograms of the STIS high resolution Echelle Modes can be predicted with very high accuracy for each central wavelength by varying just two parameters, namely the tilt angles of the cross dispersers. All the other parameters describing the optics remain stable throughout the entire STIS lifetime (1997-2004).

1. Introduction

Wavelength calibration of 2D echellograms and spectral extraction of the orders is usually considered a challenging task, requiring software very specialized for each individual echelle spectrograph. At the basis of the canonical approach is an attempt to find two-dimensional polynomial solutions for the geometry in each of the many modes (central wavelength settings) such instruments can be used. We have chosen instead to solve the wavelength/geometry projection problem with a ray optics physical description of the image forming process in multi-grating spectrographs. The configuration is thus described by a number of engineering parameters which are mainly the tilt angles, grating constants and focal lengths of optical elements, and the physical dimensions and orientation of the detector array. As in reality, changes from one mode to another - e.g., central wavelength settings or insertion of different cross dispersers - are achieved by varying only the relevant configuration data.

2. The Model

A generic model of a spectrograph consists of a complete geometric description of its optical elements (i.e., tilt angles, focal length, grating constants, etc.). The accurate knowledge of these parameters allows the user to "predict" which part of the detector will be hit by the incoming rays. Without the spectrograph description an empirical solution is needed which forces the users to revise it whenever an angle of an optical element or an environmental condition has been modified. Instead, by using the model and by knowing the configuration, a predictive calibration is not only possible but operationally more efficient to use and more reliable. Our model of the spectrograph on board HST (STIS) consists of 35 parameters (17 for the optical elements and 18 to support a "not yet physical" polynomial description of small scale (< 1 pix) optical distortions or detector irregularities) which allow an accurate description of the 2D echelle spectrograms (see Ballester & Rosa 1997, for a generic description of ray-optics based spectrograph models). Moreover, whenever configuration parameters are unknown - the "Instrument Scientist Question Case" - it is possible to use the model in a boot-strap manner on the estimated positions of a few brilliant lines only in order to infer the unknown displacements and update the current spectrograph configuration. The spectrograph description so obtained can be easily generalized. The IPMG group is currently working also on a port of the STIS model to the ESO/VLT CRIRES spectrograph.

3. The Reference File Production

The spectrograph model, if bound into an operational pipeline, requires supporting tools. While the model itself is an elegant description of the structure of the spectrograph, it becomes operationally useful only once the configuration specific parameters can be obtained with sufficient precision from real data obtained with the instrument in a very robust manner. To estimate the basic parameters - those which describe the rigid, or only slowly varying structure of the instrument - with great accuracy a very detailed line list and several very deep calibration exposures from wavelength standard lamps are needed. The line list is used together with the first estimate of the model parameters to obtain a reasonable set of line positions (ExtractSubCatalog). This set of line measurements will then be used in a bootstrap manner (non-interactive) to accurately identify the observed lines on the exposure (Fitlines). At the end of this two-stage process a set of observed and predicted line positions is available which can be used to improve the model configuration by using an optimization technique known as simulated annealing (Kirkpatrick et al. 1983; see `www.stecf.org/poa/gendocs/ADA3.ppt` for more details). Optimizing several wavecal exposures with slightly varying grating settings (differing central wavelengths), it is possible to gain information about which parameters are really mode relevant, and which describe the (hopefully very rigid) optical bench of the system once and forever. In the case of STIS, analysis of data from 1997 to 2004 demonstrates that the model is able to describe the different central wavelengths of the mode E140H by just varying two parameters, namely the tilt angles of the cross dispersers (see Figure 1).

4. Calstis Modification

In order to use the STIS model the existing calstis pipeline needs to be modified in the modules Calstis 4 and Calstis 6. Calstis 4 is actually calculating the offset

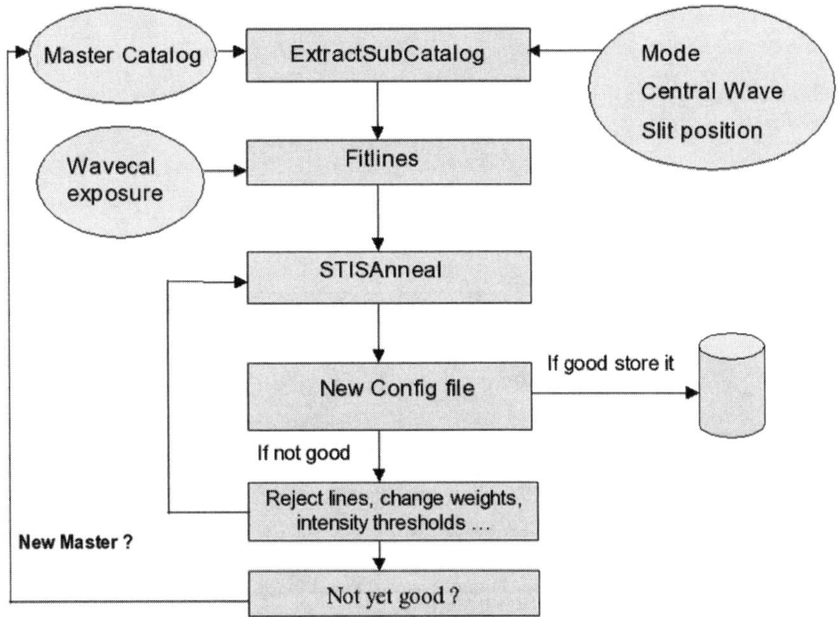

Figure 1. Reference file production dataflow.

(SHIFTA1 and SHIFTA2 keywords) by cross-correlating a collapsed spectrum along the spatial direction with a template. In our implementation Calstis 4 will calculate deltas of the major tilt angles of the cross-disperser and then update the model configuration accordingly for use in Calstis 6. In Calstis 6 the model will be used in conjunction with the bisection method to replace the empirical wavelength calibration.

5. Results

The STIS model has been optimized and tested against a very long wavecal exposure in order to infer a very accurate estimate of the model parameters. The configuration so obtained is able to match nearly all the lines with better than one pixel accuracy (see Figure 2).This configuration is now the baseline for all of the other central wavelengths for the same mode. The analysis of several other images showed that this configuration is able to match all the other images by varying just one tilt angle of the cross disperser (see Figure 3). Among others there is the direct advantage that even central wavelengths now not supported could be easily managed by just deducing the corresponding tilt angle at the cross-disperser from a very linear relationship such as the one shown in Figure 3. In the traditional calibration scheme a complete set of new reference files will have to be produced painstakingly in order to cope with the use of a non-calibrated mode.

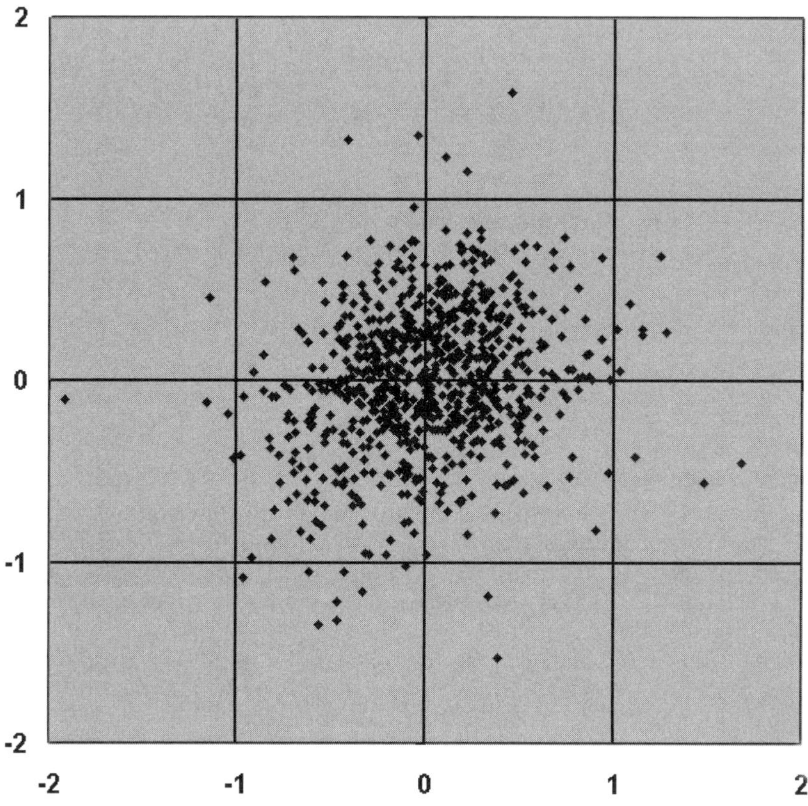

Figure 2. Scatter between the observed and predicted lines.

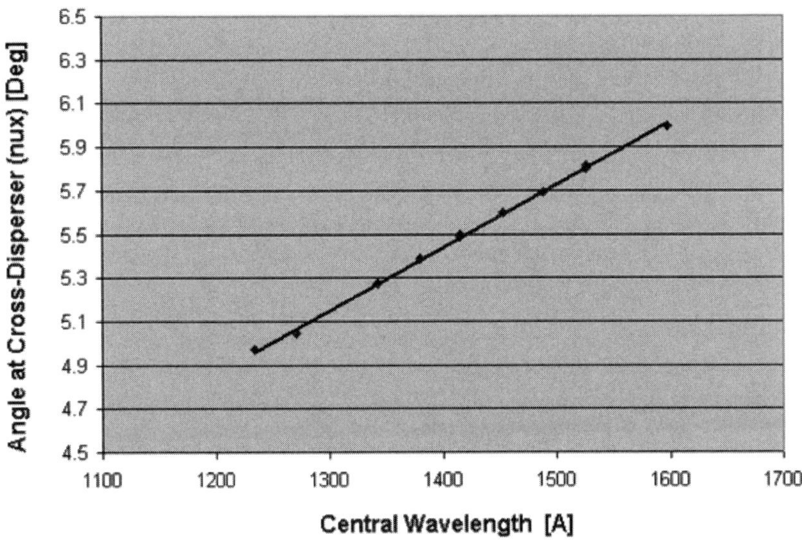

Figure 3. Central wavelength vs. Cross-Disperser tilt angle (nux)

6. Conclusions and Future Goals

We have demonstrated that modeling the STIS Echelle modes is a concrete and valuable alternative to the empirical calibration. The pipelines of the ESO UVES and CASPEC spectrographs were equipped with early study versions of ray-optic models (See Ballester and Rosa 2004). The IPMG group of the ST-ECF has already implemented more advanced models for the FOS of HST. It is therefore natural to seek an extension into an even wider variety of spectrographic equipment in space and on the ground. The tools developed alongside this process, mainly concerning optimization of configuration data and housekeeping of the databases, are deliberately kept very generic to allow cross-project utilization.

Acknowledgments. We are very grateful to Phil Hodge from STSci for the useful discussions on the STIS model insertion into the calstis pipeline.

References

Ballester, P., & Rosa, M.R. 1997, A&AS, 126, 563

Ballester, P., & Rosa, M.R. 2004, in ASP Conf. Ser., Vol. 314, ADASS XIII, ed. F. Ochsenbein, M. Allen, & D. Egret (San Francisco: ASP), 481

Kirkpatrick, S., Gelatt Jr., C.D., & Vecchi, M.P. 1983, Science, 220, 4598, 671

Part XI
Data Archives

ASTRONOMICAL DATA ANALYSIS SOFTWARE AND SYSTEMS XIV
ASP Conference Series, Vol. 347, 2005
P. L. Shopbell, M. C. Britton, and R. Ebert, eds.

The Design of the W. M. Keck Observatory Archive

G. B. Berriman

Michelson Science Center and Infrared Processing and Analysis Center, California Institute of Technology, Pasadena, CA 91125, Email: gbb@ipac.caltech.edu

D. R. Ciardi

Michelson Science Center, California Institute of Technology, Pasadena, CA 91125

A. C. Laity and N. D. Tahir-Kheli

Infrared Processing and Analysis Center, California Institute of Technology, Pasadena, CA 91125

A. Conrad, J. Mader, and H. Tran

W. K. Keck Observatory, Kamuela, HI 96743

T. Bida

Lowell Observatory, Flagstaff, AZ 86001

Abstract.
The Michelson Science Center(MSC) and the W. M. Keck Observatory are building an archive that will serve data obtained at the Keck Observatory. The archive has begun operations and is ingesting Level 0 (uncalibrated) observations made with the recently upgraded High Resolution Echelle Spectrometer (HIRES); these observations will be publicly accessible after expiration of a proprietary period. Observatory staff have begun using the archived data to determine the long-term performance of the HIRES instrument. The archive is housed at the Michelson Science Center (MSC) and employs a modular design with the following components: (1) Data Evaluation and Preparation: images from the telescope are evaluated and native FITS headers are converted to metadata that will support archiving; (2) Trans Pacific Data Transfer: metadata are sent daily by e-mail and ingested into the archive in a highly fault tolerant fashion, and FITS images are written to DVDs and sent to MSC each week; (3) Science Information System: inherited from the NASA/IPAC Infrared Science Archive, it provides all the functionality needed to support database inquiries and processing of requests; and a Web-based (4) User Interface, a thin layer above the information system that accepts user requests and returns results. The design offers two major cost-saving benefits: it overcomes the geographical separation between the telescope and the archive and enables development at Keck and at MSC to proceed independently; and it permits direct inheritance of the IRSA architecture.

1. Introduction

The W. M. Keck Observatory Archive (KOA) (http://msc.caltech.edu/koa.html, and http://www2.keck.hawaii.edu/realpublic/koa) is a collaboration between the Michelson Science Center (MSC) and the W. M. Keck Observatory (WMKO). It aims to:

1. Promote the National Aeronautics and Space Administration (NASA) Navigator Program goal of searching for extra-solar planets
2. Curate and disseminate observations made on Keck Single Aperture instruments to maximize science return from the observatory
3. Enable long-term instrument performance studies that will benefit development of observing programs

The KOA entered operations on August 18, 2004, when it began ingesting level 0 (uncalibrated) data obtained with a major upgrade to the High Resolution Echelle Spectrometer (HIRES). The upgrade replaced a single CCD chip with a mosaic of three 2048 x 4096 15μm pixel MIT/Lincoln Labs CCDs. The HIRES supports a broad range of astrophysical research, but is most celebrated for its role in the discovery of extra-solar planets.

By November 3, 2004, the KOA had ingested metadata from 54 nights of observations, totaling 6923 observations, and 43 nights (118 GB) of data in Flexible Image Transport System (FITS) format. Currently, access to these datasets is limited to HIRES Principal Investigators (PIs), who are able to query, subset and download their own datasets. The archive contents will be made public after expiration of a proprietary period that is under negotiation between the California Association for Research in Astronomy (CARA) and NASA.

2. Design of the KOA

The science and calibration data, written in FITS format, along with ancillary data, such as summaries of weather conditions and observing logs, are transferred from the observatory to the MSC, where they are curated and served. The KOA employs a modular design consisting of four uncoupled components, as follows:

1. Data Evaluation and Preparation (performed at the Observatory).
2. Trans-Pacific Data Transfer from the Observatory to MSC.
3. Science Information System (maintained at MSC).
4. User Interface (maintained at MSC).

The relationships between these components are shown in Figure 1.

This modular approach to design offers three substantial cost-saving benefits to the archive, as itemized below:

1. It overcomes the geographical separation between telescope and archive.
2. It reduces maintenance costs, as one component can be upgraded independently of the others.
3. It enables inheritance of existing operational software architecture that underpins the NASA/IPAC Infrared Science Archive (IRSA).

The following subsections describe the functions of these components in more detail, identifying where software was reused.

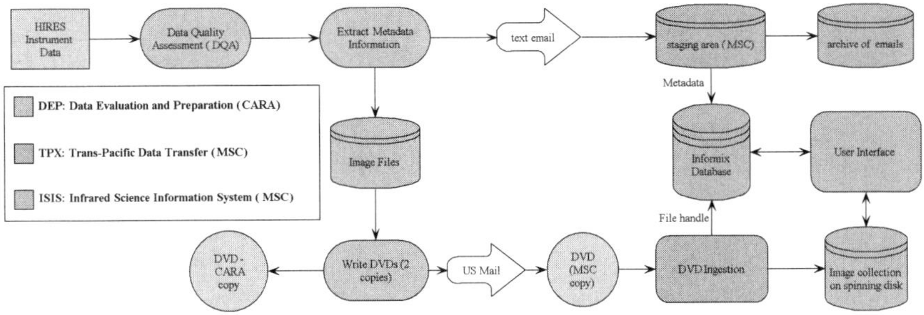

Figure 1. The Design Components of the Keck Observatory Archive.

2.1. Data Evaluation and Preparation (DEP)

This component validates the integrity of the headers and science content of the FITS files created at the telescope. It adds content to the FITS headers to support archiving, as follows:

1. A unique file identifier.
2. An image type identifier that distinguishes program objects from calibration files and identifies the type of calibration file.
3. Data integrity keywords that describe the status of the instrument and its configuration.

Following validation of data, DEP writes all the keywords into a metadata table, with one record for each FITS file (in column delimited format, suitable for ingestion into the database at MSC); places the FITS files and ancillary files in a local disk farm, and copies them to DVDs: one DVD resides at the Observatory and the other resides at MSC.

2.2. Trans-Pacific Data Transfer (TPX)

The metadata are ingested into the archive within 24 hours of observation to support short-term instrument performance analysis by KOA staff at the Observatory. The metadata are transferred via text email to a staging area at MSC. Following automated validation of the content of the metadata, they are ingested into an Informix database within 24 hours of observation, and the emails themselves are then archived.

The FITS files and ancillary data are ingested within 30 days of the observations. The data are written on DVDs that are sent to MSC via U.S. mail, where the contents are copied to a staging area and validated in two ways: by comparison of the MD5 checksums with values of checksums included on the DVDs, and by comparison of the datatypes and values of the metadata in the FITS files with those in the database. The files are then copied to the configured archive disk farm, and file handles are attached to the metadata entries in the database.

The TPX processes are essentially wrappers that automate ingestion tools developed by IRSA that have up to now been used for ingestion of static datasets. The content of the FITS keywords and the structure of the data on the DVDs are managed through an interface control document.

2.3. Science Information System and User Interface (UI)

The KOA has inherited the component-based architecture used that underpins the NASA/IPAC Infrared Science Archive (IRSA[1]), which has been operational since spring of 1999. The architecture has been designed to archive and serve the types of data used by astronomers, such as FITS files and ASCII tables. Its applicability is unrestricted by wavelength, despite being developed specifically for an infrared archive, because it consists of a collection of tools or libraries, each of which performs a specific task. The architecture is optimized for two-dimensional astronomical spatial searches by applying a Hierarchical Triangular Mesh[2] as a spatial indexing scheme.

User services are developed by linking existing modules through a simple executive program. New functionality is added as required. While invariably driven by the needs of a specific data provider, this functionality is written for generality, to maximize the flexibility of the archive services. There are two such augmentations made to support the KOA:

1. *Secure PI Access.* The KOA leverages the Request Object Management Environment, ROME (Good, Kong and Berriman 2004), built at IPAC to support the National Virtual Observatory. ROME is middleware designed expressly to manage user access and time intensive jobs, but the KOA takes advantage only of the user access capabilities. ROME is written as Enterprise Java Bean servelets running under the JBoss Open Source Application Server. It uses Java Data Base Connectivity (JDBC) to provide an interface to an Informix database, which contains a permanent store of authorized users, their passwords and a handle to a permanent workspace that acts as a staging area for retrieved data.
2. *Calibration File Association* The KOA has implemented simple algorithms for associating calibration files with science files. Once science files satisfying a query have been retrieved, the algorithms find the calibration files that have the same instrument and CCD configuration parameters, within specified tolerances.

3. KOA Services to Users

3.1. Secure Access for Keck Principal Investigators

In a given semester, KOA assigns and stores each PI with a user name and password that is stored by ROME and passed on to PIs after the first night of a run. PIs are automatically notified when their data have been ingested into the archive.

3.2. User Interface

Through a simple web form, the KOA supports queries for science and/or calibration files by target name, position (in all common coordinate systems) and radius, date and time, observation parameters (e.g., exposure time, wavelength

---

[1]`http://irsa.ipac.caltech.edu`

[2]`http://www.sdss.jhu.edu/htm`

coverage), and program information (e.g., PI, program title). The current interface is an advanced prototype that is expected to evolve according to usage patterns and comments from PIs.

The interface offers three options: return science files, calibration files, or the science files and associated calibration files that share instrument and CCD configuration parameters with the science files, as described above. Ancillary data are not yet accessible through the user interface.

The KOA also supports a Structured Query Language (SQL) interface that allows completely general queries of all metadata fields that are stored in the database. Currently, this interface is accessible only by MSC and WMKO archive staff, who are exploiting it to begin investigations of the long term performance of the HIRES instrument.

3.3. Query Results Page and File Download

The web interface returns an HTML page that lists in separate tables all science and calibration files that satisfy the input query criteria. They are available for download one at a time, or through a packaging mechanism that stages a tarball of selected files on a URL-accessible staging area and informs the user by email that the file is ready for download. The package contains, as applicable, science and calibration files, and tabulations of associated calibration files.

4. Planned Functionality

The archive plans to implement the following functionality:

1. More efficient calibration grouping algorithms that overcome the I/O limitations of the current implementation
2. Queries at multiple input positions
3. Access to ancillary data
4. Access to quick-look products
5. Compliance with National Virtual Observatory Standards
6. Support for data tagging initiatives currently being developed by NASA data centers and astronomical journals

Acknowledgments. The W. M. Keck Observatory Archive is funded by the Navigator Program of the National Aeronautics and Space Administration (NASA).

References

Kong, M., Good, J . C. and Berriman, G. B., 2004, in ASP Conf. Ser., Vol. 347, ADASS XIV, ed. P. L. Shopbell, M. C. Britton, & R. Ebert (San Francisco: ASP), 213

ASTRONOMICAL DATA ANALYSIS SOFTWARE AND SYSTEMS XIV
ASP Conference Series, Vol. 347, 2005
P. L. Shopbell, M. C. Britton, and R. Ebert, eds.

Software Architecture of the Spitzer Space Telescope Uplink/Archive

Joe Chavez, Tatiana Goldina, Annie Hoac, William Roby, and Xiuqin Wu

Spitzer Science Center, California Institute of Technology

Abstract. The Spitzer Science Center (SSC) provides a set of science user tools to support planning and archive access via the Internet. We will present the software architecture and design principles that underlie the Uplink/Archive subsystem of the SSC. Included in the discussion will be a review of the original Uplink architecture as presented in P1-59 ADASS 1999 and the evolutionary changes for the current deployment. The Archive subsystem is based on the same set of core components used in the Uplink subsystem but is based on Web services technology to allow open access to the Archive. Web services technology provides a basis for searching the archive and retrieving data products.

1. Introduction

- SSC Uplink is responsible for the development of software and tools to support the science community with planning and accessing observations with the Spitzer Space Telescope.
- To support the geographically diverse science community the Java 2 Standard Edition platform is the basis for the development of platform independent client software (Spot).
- To support the Spot client software the Java 2 Enterprise Edition is the basis for the development of robust scalable services.
- Also part of the system are C programming language libraries that support for the resource estimation and astronomical visibility services using the Java Native Interface (JNI).
- The software architecture of the SSC Uplink/Archive is presented in the 4+1 views format.

2. Architecture Views

The 4+1 software architecture view divides the architecture into 5 distinct areas of concern (Figure 1). The "+1" view is the Use Case View which contains the set of use cases that define the system behavior. The Logical View represents the logical system structure without regard to implementation specific choices. The Implementation View is the realization of the Logical View in the selected programming languages and tools. The Deployment View shows the physical mapping of components to the hardware nodes in the system. The Process View defines the development processes that are used to construct the system.

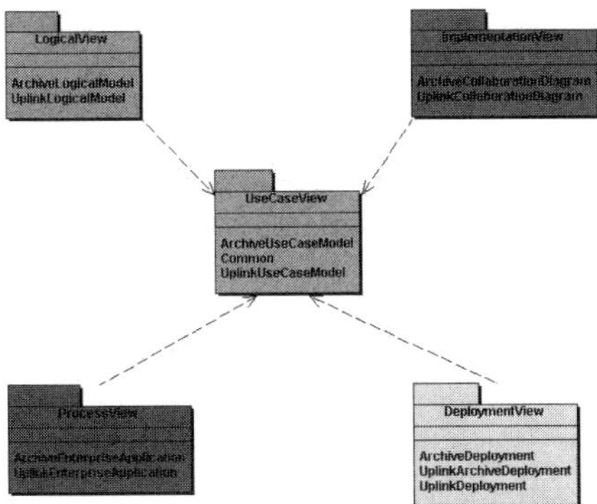

Figure 1. 4+1 Architecture Views

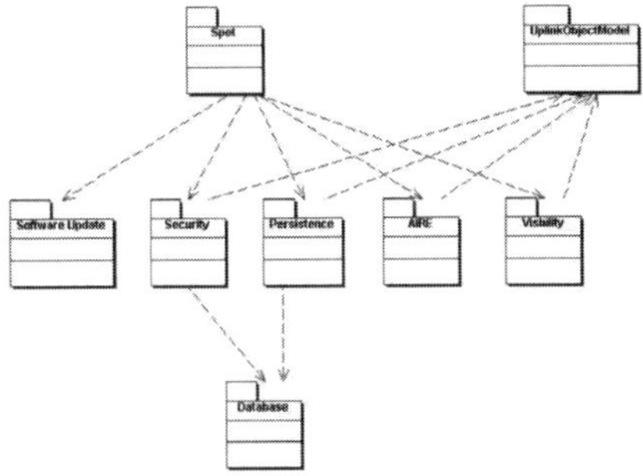

Figure 2. Uplink Component Model

3. Use Case View

The Use Case View contains Use Cases that capture the key functions of the system at a level of detail from which the first iteration of the system architecture can be derived. In the Use Case view there are two main actors: the Astronomer and Operator. The Astronomer represents the class of users of the Spot software. An Operator represents the class of users that operate the Spitzer Science Center.

4. Logical View

This Logical View organizes the component of a system in a vertical stack with increasing order of dependency flowing from top to bottom.

The Uplink component model (Figure 2) consists of the Spot client and the Uplink services. The SoftwareUpdate update component provides a means to update the Spot software at the installed location. This component is split into the client access in Spot and the server management in the J2EE application server. The Security component provides authentication and authorization services to the Spot client. The Persistence component is responsible for managing the persistence of the UplinkComonentModel in the Database. The AIRE com-

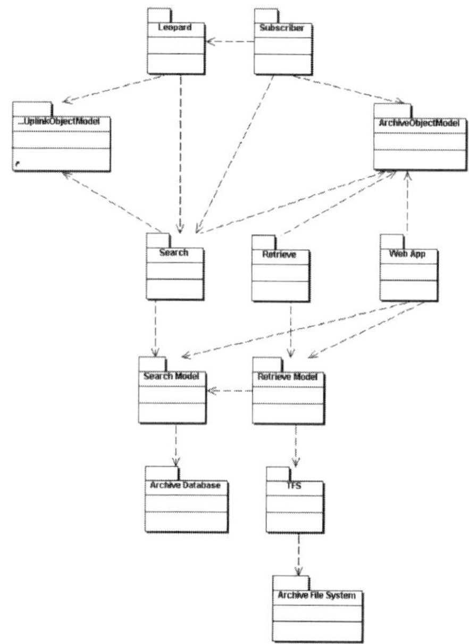

Figure 3. Archive Component Model

ponent is used by Spot to perform resource usage estimation when planning observations for the Spitzer Space Telescope. The Visibility component provides information relating to the ability to view positions at given time with the telescope as well as other estimation services.

The Archive component model (Figure 3) consists of the Leopard and Subscriber clients. The services components are Search and Retrieve. The WebApp component resides on the server and represents the Web browser interface to the Spitzer Archive. The SearchModel provides the data content to the Search service. The RetrieveModel provides the retrieve data to the RetrieveService. The ArchiveDatabase contains the set of Meta-Data for the files contained in the archive. While the SearchModel encapsulates access to the ArchiveDatabase. The RetrieveModel uses the TFS component to access the ArchiveFileSystem via a proxy service provided by the Spitzer Science Data Management group.

5. Implementation View

The Implementation View adds behaviors and interactions to the components defined in the Logical View. In Figure 4, the Uplink component collaborations are shown. This diagram is a representative view of the set of collaborations among the Uplink components. The diagram contains association links between Spot and the various service handler component that provide the implementation. Additional properties of the relationships between the components. The AIREServerHandler component The Archive component collaborations are shown in Figure 5. In this collaboration diagram the primary application flow

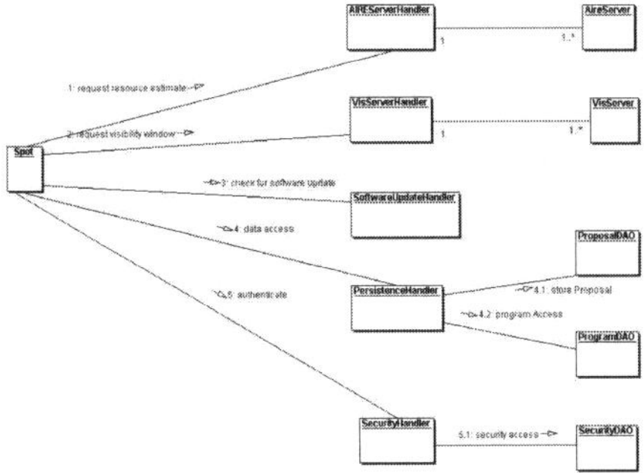

Figure 4. Uplink Collaboration Diagram

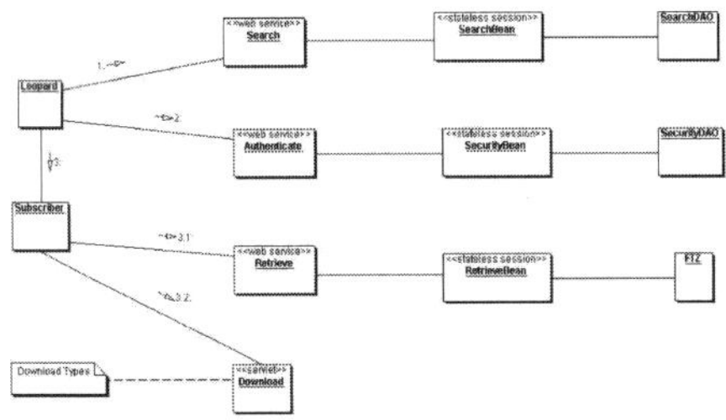

Figure 5. Archive Collaboration Diagram

is depicted.

6. Process View

The Process View encompasses the build and deployment of the system. In the case of the Services Tier the components are packaged into distributable enterprise archive (EAR) files. The EAR files are deployed into a J2EE compliant container. For client software deployment is managed using java archive (JAR) files and native OS installation programs.

7. Deployment View

The deployment diagram shows the combined Uplink/Archive component set as deployed in each tier of the architecture. Each of the components is colored to represent the generation in which the software architecture was designed. The blue colored components represent the original Uplink design circa 1999 and the yellow colored components represent the combined Uplink/Archive design of 2003. The combined Uplink/Archive Deployment model is shown in Figure 6 Some of the key characteristics of this software architecture are:

- Shared Object Model - The Uplink and Archive object model components contain a consistent view of the data that is used across the tiers in the system. In additional to a single view of the data the object model also

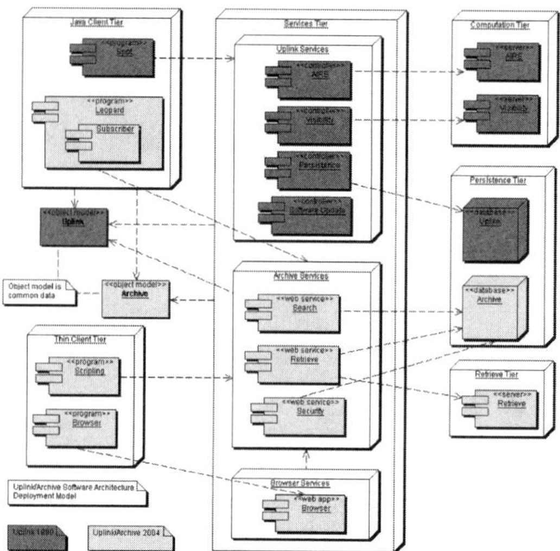

Figure 6. Uplink/Archive Deployment Diagram

provides a set of operations for manipulating the data that helps to maintain data integrity. Internet Protocols - Where appropriate established Internet protocols (HTTP/S) are used for communications between the tiers. In addition, the Archive interface employs Web services via SOAP over HTTP to provide services to any SOAP compatible client.

- Multiple Client Types - The Services Tier exposes multiple interface types to allow many types of clients to access the system services. The types of clients include Java applications, Web browsers and scripting languages (support SOAP).
- Multiple Client Platforms - The Java Client tier allows for the deployment to many client platforms including Unix, Windows, Linux and Mac. The software update service facilitates the update of remote software installations with a minimum of user interaction.
- Distributed Processing - A distributed processing architecture is key to performance and scalability. Where available existing protocols and resource pool management mechanisms are used for inter-process communication. The backend tiers Computation and Retrieve can all scale via server based pool management and clustering. Access to the shared Persistence tier is controlled via JDBC connection pools.

ASTRONOMICAL DATA ANALYSIS SOFTWARE AND SYSTEMS XIV
ASP Conference Series, Vol. 347, 2005
P. L. Shopbell, M. C. Britton, and R. Ebert, eds.

Data Processing Discovery Agents in the Gemini Science Archive

Adrian Damian, Norm Hill, Patrick Dowler, David Bohlender, Sharon Goliath, Severin Gaudet, and Geoffrey Melnychuk

Canadian Astronomy Data Centre

Colin Aspin

Gemini Observatory

Abstract. Data Processing Discovery Agents (Discovery Agents or DAs in short) represent one of the key design concepts of the newly released Gemini Science Archive (GSA). It is a simple but promising solution for decomposing the functionality of the increasingly complex astronomical data archive systems. This paper presents the rationale behind this concept and discusses its main characteristics with examples from the GSA.

1. Introduction

The Gemini Science Archive is a new archive developed by the Canadian Astronomy Data Centre (CADC) to provide the scientific community with tools for effective on-line access to data collected by the Gemini telescopes (Bohlender et al. 2005; Damian et al. 2005). The main objective of the archive is to ensure that data collected by the two state of the art telescopes are fully exploited and that maximum scientific value is returned to the communities. Although started as a basic archive, it is expected that advanced exploration and data mining tools will be gradually added to the system in order to meet this goal.

This gradual increase in complexity and size of the archive system poses a number of design challenges. Performance penalties, increased development and integration times or time consuming regression testing are just some of the main problems that have to be overcome and a monolithic design might not be up for the job. A more appropriate approach is to decompose the overall functionality of the system into smaller tasks that can be implemented by a number of independent applications. Discovery Agents are such applications that perform predefined computational tasks on the metadata of the archive. The basic GSA already incorporates a number of DAs and it is expected that future planned features will further make use of this concept.

2. What Exactly is a Discovery Agent?

A Data Processing Discovery Agent is a standalone software component designed to discover patterns of input data in an archive catalogue and perform relevant tasks, called recipes, without the user's intervention. A DA can be characterized as:

- **Idempotent:** Mathematically, function $f : D \rightarrow D$ is idempotent if $f(fx) = fx$ for all x in D. Applied to DA, this means that unless the input data changes, repeated invocations of the DA have the same effect as one. This property greatly simplifies scheduling the execution of DAs.
- **Autonomous:** The execution of a DA does not require any user's assistance. Therefore, a DA can be triggered by external events, other DAs or can be manually invoked by the user. With no dependencies on the presence of the user, it can automatically perform its task and exit.
- **Independent of other components:** Because Discovery Agents are standalone applications there is little or no coupling between them. DAs use the catalogue data as the input source for their tasks, so changes to one DA will have no direct impact on the other DAs. This is in sharp contrast with the decomposition at the class or library level approach where changes in the interface, for example, need to be propagated into the client code.

3. Discovery Agents Advantages

The DA concept is appealing for the development of astronomical archives for a number of reasons:

- **"Divide and conquer" approach.** The ever-increasing complexity and size of the future archives becomes hard to handle with a monolithic system. Breaking down the functionality of the system into well-defined tasks that can be carried out independently by Discovery Agents greatly simplifies the design of the system.
- **Arbitrary execution order.** Because DAs are idempotent they can be invoked multiple times without modifying the result when the input is unchanged. Therefore, they can start processing even if the input data are incomplete. Subsequent invocations will pick up the changes in the input space but will not affect the results of the data already processed. As a result, depending on the importance of the task, DAs can be scheduled to run periodically, in preset or random order or can be triggered by certain events.
- **Inherently suitable for parallelization.** The DA properties make them prime candidates for parallelization. This allows DAs to exploit the advantages of a distributed computing environment such as time-sharing and load balancing. DAs, for example, can be invoked on different hosts and at different times (e.g., idle time). They might have priorities associated with them so that highly important tasks take precedence over unimportant ones, which can be executed later on when resources are available.
- **Increased robustness.** A large monolithic application has more interactions, which tends to lead to more bugs and makes it more prone to system failures. Discovery Agents, on the other hand, are independent of each other. The whole system continues to function when parts of it (DAs) are down. Fixes or changes in a recipe automatically propagate in the system through the changes in the catalogue data.

- **Easily expandable system.** Another advantage of this decomposition of the functionality is that new tasks can be easily added to the system. Implementing the new functionality in DAs that can be easily integrated with the existing systems can greatly reduce the deployment efforts.

4. Discovery Agents Challenges

One of the main disadvantages of using the DA approach when working with astronomical archives is the size of the input catalogue data. There is a steep decrease in performance if, at every run, the DA has to look at and process all the records in the catalogue in spite of the fact that only a small number have changed since the last execution. Mechanisms should instead be put in place to enable the DA to quickly skip the execution of unchanged input datasets between two consecutive executions. Time-stamping the records in the input data or using other means to flag changed data without the need to look at and analyze it are just two examples of such techniques.

Another potential drawback is the DA's dependency on the format of the input catalogue data. Changes to the database schema need to be propagated to all DAs that use it and this often requires modifying the source code of the agent. It is therefore desirable to have a general, stable database schema for the catalogues or have the DAs use a configuration file for accessing the catalogues.

5. Examples of CADC Discovery Agents

The design of the Gemini Science Archive benefited from the knowledge and experience of other astronomical archives developed and hosted by CADC. The DA concept has been used before in CADC. The preview generation feature present in some archives relies on this mechanism to regularly investigate the catalogue data in order to generate preview images associated with the datasets.

WFPC2 Associations (`http://cadcwww.dao.nrc.ca/wfpc2/`) is another classical example of a DA. Using specific algorithms, agents search the catalogues for associated datasets. The associated datasets are then processed and new entries are created in the catalogue. The produced association has a greater scientific value than the individual components and can replace them, thus saving space in the archive.

6. GSA Discovery Agents

With the development of the GSA, the DA concepts are further formalized. As depicted in Figure 1, a common interface has been defined for the Java implementation of the DAs, and specialized classes implement generic functionality for different categories of DAs. These classes are described in the following sections.

6.1. DiscoveryAgent Class

At the top of the hierarchy is the **DiscoveryAgent** class that defines the common interface for every DA. The majority of the methods are abstract and need

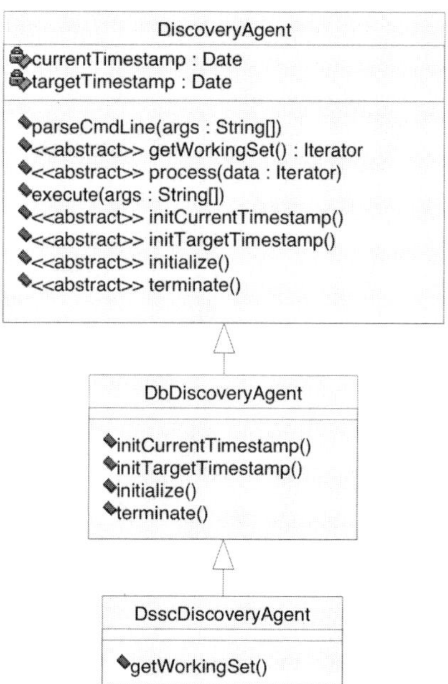

Figure 1. Discovery Agents Class Diagram.

to be implemented by the derived classes. There is however a notable exception: the execute method implements the workflow of the DAs by defining the main steps of the execution. It does so by calling each of the abstract methods in a predefined order (see Template Method design pattern; Gamma et al. 1995).

If the DA is used in standalone mode, this class parses the command line arguments, opens and initializes the log file. The subclasses are responsible for implementing the code for initializing and terminating the execution, for calculating the time window of the input datasets (when they are time-stamped) and for implementing the recipe in the process method.

6.2. DbDiscoveryAgent Class

The **DbDiscoveryAgent** class is the parent class of all DAs that work with catalogs of data stored in databases. This class implements ***initCurrentTimestamp*** and ***initTargetTimestamp*** that allows the DA to look only at data that has changed within the time window of two consecutive executions. It relies on the datasets having a timestamp to indicate when the last change occurred. This class persists the time of the last execution, which is compared with the timestamps of the datasets to quickly filter out the unchanged data.

6.3. DsscDiscoveryAgent Class

The **DsscDiscoveryAgent** class was designed for DAs that use the GSA catalogue. It uses the specific database schema of this catalogue to read the input data in getWorkingSet method. So far, all the DAs in GSA are derived from this class.

7. GSA Discovery Agent Instances

The basic GSA contains three DAs:

- **gsaIngest:** ingests data from the metadata database to the catalogues.
- **gsaFieldGen:** enhances the catalogue by making derived attributes available for querying
- **gsaScience:** populates a table in the catalogue with the science datasets. This enables faster searches on the science datasets in the catalogue.

All these applications are implemented in Java. They can be invoked from the command line as standalone applications or from another Java application as a package. They have two working modes:

- incremental: gives optimal execution performances. In this mode the DAs use the timestamps of the catalogue records to filter out the datasets that have not been modified since the last execution.
- full: processes all the input datasets. This mode, while not optimal, can be used in case the DA is required to process all datasets. This might be the case when the processing algorithm (the recipe) has changed. Another option in this mode is to specify a range of datasets to process.

The GSA DAs can be invoked periodically or only when new metadata are ingested in the catalogues. They can also be invoked to process a specific dataset only. In addition, because they are lightweight processes, they can be called one after the other when ingesting new dataset into the catalogues without incurring delays. If a DA implements a heavier processing task, it can be scheduled to run when resources are available.

These are only the first DAs in GSA. New ones are planned for the advanced capabilities features to be released in future versions of the GSA. They will perform tasks such as: preview generation, data mining and association identification. With the flexibility given by the DA approach, the new tasks can be easily integrated in the existing GSA system.

8. Conclusion

The Discovery Agent approach offers a flexible, modular way for implementing and carrying out the tasks associated with a science archive. They are suitable to automated activities such as: new data arrival, catalogues maintenance, data mining, etc. Discovery agents are used in the first release of the GSA and are expected to play an important role in the development of the advanced capabilities of future GSA releases.

References

Bohlender, D., et al. 2005, in ASP Conf. Ser., Vol. 347, ADASS XIV, ed. P. L. Shopbell, M. C. Britton, & R. Ebert (San Francisco: ASP), 642

Damian, A., et al. 2005, in ASP Conf. Ser., Vol. 347, ADASS XIV, ed. P. L. Shopbell, M. C. Britton, & R. Ebert (San Francisco: ASP), 655

Gamma, E., Helm, R., Johnson, R., & Vlissides, J. 1995, in Design Patterns: Elements of Reusable Object-Oriented Software (Addison-Wesley)

ASTRONOMICAL DATA ANALYSIS SOFTWARE AND SYSTEMS XIV
ASP Conference Series, Vol. 347, 2005
P. L. Shopbell, M. C. Britton, and R. Ebert, eds.

The Gemini Science Archive: Current Status and Future Prospects

David Bohlender, Adrian Damian, Severin Gaudet, Sharon Goliath, Norm Hill, and Geoffrey Melnychuk

Canadian Astronomy Data Centre

Colin Aspin

Gemini Observatory

Abstract. The most recent addition to the services provided by the Canadian Astronomy Data Centre (CADC) is the Gemini Science Archive (GSA). Developed as a collaborative effort with the Gemini Observatory and released in September 2004, the GSA provides the scientific community with the tools required for effective online access to data collected by both Gemini telescopes. This paper provides a summary of the current capabilities of the GSA and discusses some of the advanced capabilities to be developed in the near future.

1. Introduction

Well before the Gemini telescopes were ready for scientific operation, the GSA was foreseen to play an integral part of the planning, observation, calibration, data reduction, and data distribution systems in operation at the two Gemini sites. This foresight allowed for the implementation of a number of novel approaches to data handling and processing in the CADC's development of the GSA. These include

- support for data produced by the telescopes' instruments as well as related metadata such as weather images and observing logs, etc.
- use of a database schema that enables separate storage of metadata and catalogues
- an XML data dictionary used to specify data origins, destinations, formats, ranges, etc.
- rapid availability of data due to electronic transfer of all data and metadata from the telescopes
- an advanced user interface that allows the user to make full use of these advanced design elements.

Details of many of these items can be found in the companion papers in this volume (Damian et al. 2005a,b; Goliath et al. 2005; Melnychuk et al. 2005). This paper's role is to present a summary of the current scientific capabilities of the GSA and discuss some of the developments GSA users can look forward to in future releases.

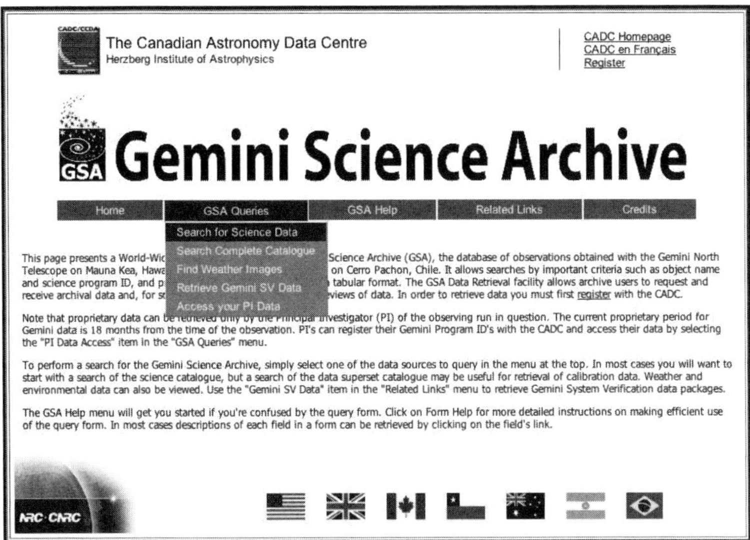

Figure 1. The GSA home page, `http://cadcwww.hia.nrc.ca/gemini`.

2. Overview of the GSA

The GSA is located at `http://cadcwww.hia.nrc.ca/gemini`. The user is presented with several menus of which the most frequently used will be the "GSA Queries" entry. We describe some of the querying options available below.

2.1. Science Queries

Most GSA users will first want to use the "Search for Science Data" menu entry as shown in Figure 1. In addition to the standard object or coordinate searches, users can constrain their selection by Gemini site, instrument, exposure time, or other scientifically interesting catalogue metadata produced by GSA discovery agents (see Damian et al. 2005a) such as pixel scale, wavelength bounds or spectral resolution (Figure 2).

2.2. Full Archive Search

To search for calibration files to accompany archival science observations the user selects the "Search Complete Catalogue" query. The user is presented with a selection of instruments for which archival data are available and if, for example, NIRI is selected he/she is presented with a complete list of all of the potentially interesting NIRI metadata. After selecting the parameters which are of interest, a query form containing these entries is built on the fly and the user can then enter constraints and execute a search (Figure 3).

2.3. PI and Gemini Staff Access

Additional links provided in the GSA web site enable PI's to access their own proprietary data. Users must first register with the CADC and then for each

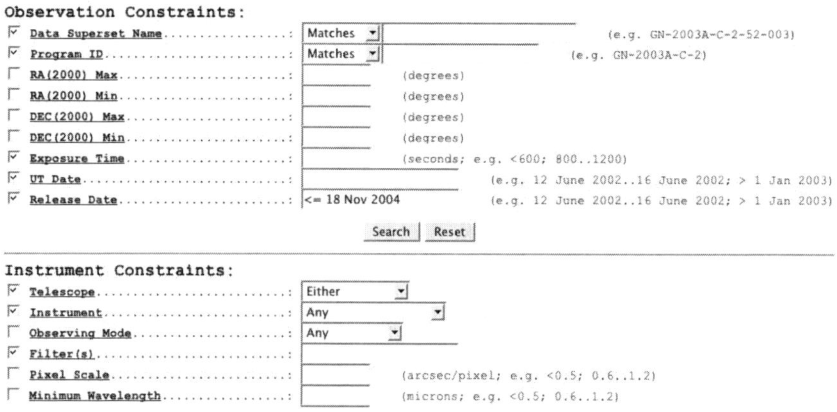

Observation Constraints:

- [x] Data Superset Name: Matches (e.g. GN-2003A-C-2-52-003)
- [x] Program ID: Matches (e.g. GN-2003A-C-2)
- [ ] RA(2000) Max: (degrees)
- [ ] RA(2000) Min: (degrees)
- [ ] DEC(2000) Max: (degrees)
- [ ] DEC(2000) Min: (degrees)
- [x] Exposure Time: (seconds; e.g. <600; 800..1200)
- [x] UT Date: (e.g. 12 June 2002..16 June 2002; > 1 Jan 2003)
- [x] Release Date: <= 18 Nov 2004 (e.g. 12 June 2002..16 June 2002; > 1 Jan 2003)

Search Reset

Instrument Constraints:

- [x] Telescope: Either
- [x] Instrument: Any
- [ ] Observing Mode: Any
- [x] Filter(s):
- [ ] Pixel Scale: (arcsec/pixel; e.g. <0.5; 0.6..1.2)
- [ ] Minimum Wavelength: (microns; e.g. <0.5; 0.6..1.2)

Figure 2. A portion of the GSA science query form.

of their Gemini observing programs they can obtain authorization for retrieval of their proprietary data by entering their Gemini Phase II password. Figure 4 shows an example of a result set from such a proprietary query. Proprietary data which the PI has access to are highlighted in the "Mark" column.

Authorized Gemini staff have unrestricted access to all GSA data through a separate password-protected link under the "Related Links" menu.

2.4. Other Data and Metadata

Other GSA menus let users search for and view additional supporting data for their archive research. This includes satellite images, weather and seeing conditions, all-sky camera images, etc. (see Figure 5). A link to Gemini System Verification data, well-defined packages of images or spectra obtained with various Gemini instruments but not available individually in the GSA, is also provided.

3. Future Developments

The release of the current basic GSA represents the end of Phase II of the GSA development. We are currently beginning GSA Phase III activities, or advanced capabilities. These will be incremental additions to the basic archive which will enhance both Gemini operations and PI and archive researcher use of the GSA. Some of the expected advanced capabilities include:

- automated electronic distribution of science and calibration observations, and relevant additional metadata to PI's
- archiving of processed data
- preview generation
- on-the-fly reprocessing of Gemini data for facility instruments
- generation of VO-enabled data products
- advanced catalogues, including
  - advanced search capabilities based on object class

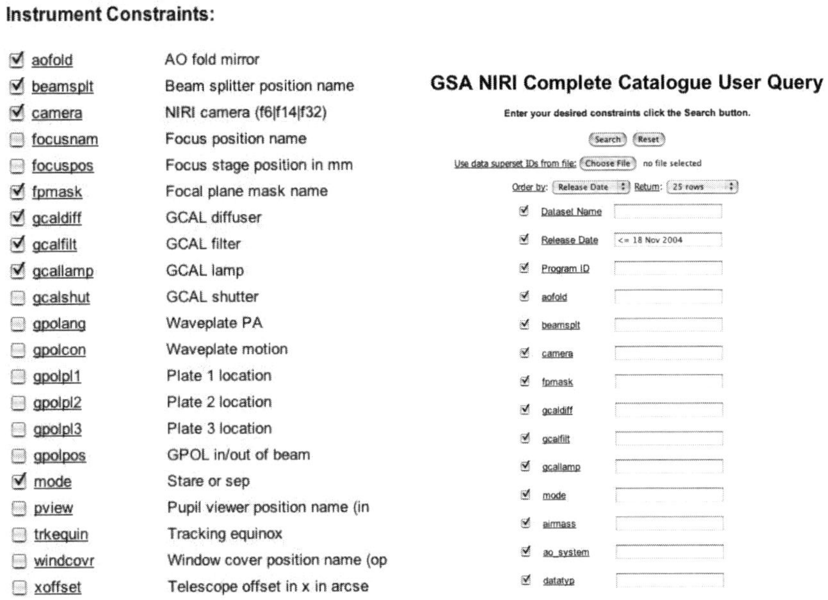

Figure 3. Full archive search constraint selection and query form.

  - advanced dataset association
  - advanced source cataloging
- advanced dataset presentation/visualization via web tools
- association of Gemini datasets with published results

4. Conclusion

Despite its very recent release, the basic GSA already contains a rich collection of superb astronomical data with more than 103,000 science datasets, of which almost 61,000 are public, at the time of writing. Already one of the most powerful archives for such general purpose telescopes, the ongoing development of advanced capabilities will ensure that the GSA continues to lead the way in ground-based astronomical archives.

References

Damian, A., et al. 2005a, in ASP Conf. Ser., Vol. 347, ADASS XIV, ed. P. L. Shopbell, M. C. Britton, & R. Ebert (San Francisco: ASP), 637

Damian, A., et al. 2005b, in ASP Conf. Ser., Vol. 347, ADASS XIV, ed. P. L. Shopbell, M. C. Britton, & R. Ebert (San Francisco: ASP), 655

Goliath, S., et al. 2005, in ASP Conf. Ser., Vol. 347, ADASS XIV, ed. P. L. Shopbell, M. C. Britton, & R. Ebert (San Francisco: ASP), 660

Melnychuk, G., et al. 2005, in ASP Conf. Ser., Vol. 347, ADASS XIV, ed. P. L. Shopbell, M. C. Britton, & R. Ebert (San Francisco: ASP), 647

Request Marked Datasets | Save Marked Datasets | MarkAll | UnMarkAll

Mark	Target Name	RA (J2000)	DEC (J2000)	Data Superset Name	Release Date	All Program Obs.	Calibration Files	Instrument
N/A	eta_car	10 45 03.63	-59 41 04.5	GS-2003B-Q-48-2-0145	Jun 10 2005 12:00AM	List	Cal.	PHOENIX
N/A	eta_car	10 45 03.63	-59 41 04.5	GS-2003B-Q-48-2-0144	Jun 10 2005 12:00AM	List	Cal.	PHOENIX
N/A	eta_car	10 45 03.63	-59 41 04.5	GS-2003B-Q-48-2-0143	Jun 10 2005 12:00AM	List	Cal.	PHOENIX
N/A	eta_car	10 45 03.66	-59 41 04.5	GS-2003B-Q-48-2-0142	Jun 10 2005 12:00AM	List	Cal.	PHOENIX
N/A	eta_car	10 45 03.68	-59 41 04.5	GS-2003B-Q-48-2-0141	Jun 10 2005 12:00AM	List	Cal.	PHOENIX
N/A	eta_car	10 45 03.68	-59 41 04.5	GS-2003B-Q-48-2-0140	Jun 10 2005 12:00AM	List	Cal.	PHOENIX
	HD66522	08 01 35.47	-50 36 19.6	GS-2003B-Q-19-3-0098	Jun 10 2005 12:00AM	List	Cal.	PHOENIX
	HD66522	08 01 34.95	-50 36 19.6	GS-2003B-Q-19-3-0097	Jun 10 2005 12:00AM	List	Cal.	PHOENIX
	HD66522	08 01 34.95	-50 36 19.6	GS-2003B-Q-19-3-0096	Jun 10 2005 12:00AM	List	Cal.	PHOENIX
	HD66522	08 01 35.48	-50 36 19.6	GS-2003B-Q-19-3-0095	Jun 10 2005 12:00AM	List	Cal.	PHOENIX
	HD66522	08 01 35.22	-50 36 19.6	GS-2003B-Q-19-3-0094	Jun 10 2005 12:00AM	List	Cal.	PHOENIX
	HD66522	08 01 35.22	-50 36 19.7	GS-2003B-Q-19-3-0093	Jun 10 2005 12:00AM	List	Cal.	PHOENIX
	HD66522	08 01 35.22	-50 36 20.2	GS-2003B-Q-19-3-0092	Jun 10 2005 12:00AM	List	Cal.	PHOENIX
	HD66522	08 01 35.22	-50 36 20.2	GS-2003B-Q-19-3-0091	Jun 10 2005 12:00AM	List	Cal.	PHOENIX
	HD66522	08 01 35.13	-50 36 20.2	GS-2003B-Q-19-3-0090	Jun 10 2005 12:00AM	List	Cal.	PHOENIX
N/A	lmc3	05 23 46.66	-70 14 22.5	GS-2003B-Q-49-32-0065	Jun 10 2005 12:00AM	List	Cal.	PHOENIX
N/A	lmc3	05 23 45.89	-70 14 22.5	GS-2003B-Q-49-32-0064	Jun 10 2005 12:00AM	List	Cal.	PHOENIX
N/A	lmc3	05 23 45.12	-70 14 22.5	GS-2003B-Q-49-32-0063	Jun 10 2005 12:00AM	List	Cal.	PHOENIX

Figure 4. An example of a result set from a proprietary access query.

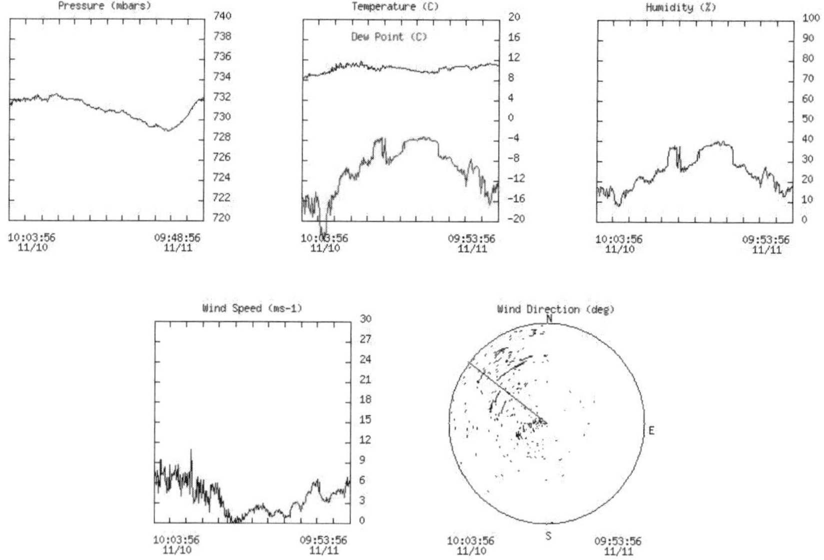

Figure 5. A sample of GSA weather metadata.

ASTRONOMICAL DATA ANALYSIS SOFTWARE AND SYSTEMS XIV
ASP Conference Series, Vol. 347, 2005
P. L. Shopbell, M. C. Britton, and R. Ebert, eds.

Gemini Science Archive Automatic Data Transfer

Geoffrey Melnychuk, Adrian Damian, Sharon Goliath, Ling Shao, Norm Hill, Severin Gaudet, and David Bohlender

Canadian Astronomy Data Centre

Colin Aspin

Gemini Observatory

Abstract. The Canadian Astronomy Data Centre (CADC) has developed an automatic system for the transfer of data from the Gemini telescopes to the Gemini Science Archive (GSA). Each step in the process between the data validation at the telescope base facilities and the data becoming available to archive users is automated, including the file transfer, the ingestion of catalogue metadata, and the insertion of the file into the archive tracking system. File names, content, and metadata are checked for conformance with GSA expectations before data transfer and ingestion take place. The CADC has been using this system for transfer of GSA data since 2004 January, and has adapted it for use in the Canada-France-Hawaii Telescope Legacy Survey.

1. Introduction

Faced with a requirement to make Gemini GMOS-South UDF observations available to the entire Gemini community in a timely fashion, the Gemini Observatory asked the CADC to investigate the feasibility of rapid, accurate electronic transfer of Gemini data and metadata from the telescope base facilities to the GSA.

This paper presents details of the CADC's resulting prototype electronic file transfer system currently in operation for the GSA (Damian et al. 2005; Bohlender et al. 2005) as well as for the Canada-France-Hawaii Telescope Legacy Survey.

2. The Electronic Transfer System

The electronic transfer system is designed as a finite-state machine, driven by the data files it is processing. Possible states were derived from the steps required to add a file to the archive. As the system is implemented, these states are represented by directories and any data file in that directory is in that state. The initial states, "new" and "replace", are directories on remote servers at each of the Gemini North and Gemini South sites. Files are placed in one of these directories by Gemini operations staff, according to whether or not the file has been archived before. This allows files in the archive to be easily replaced with better or corrected versions. State transitions are implemented as lightweight Perl scripts; each state has one script to handle its transitions.

3. File State Transitions

Figure 1 gives a schematic presentation of the file state transitions that occur during electronic transfer. Each of these transitions is described in the sections below.

3.1. File Checking

Each file destined for the GSA (FITS image, weather image, science program, etc.) is first tested for validity before transfer to the archive. These tests include:

- use FITSverify to ensure that FITS files meet current standards
- use gsaNameCheck to ensure that files conform to the GSA's file naming conventions
- check to see if the file is already in the archive (i.e., supposedly "new" files that are already in the archive, or "replacement" files that are not already in the archive indicate a potential discrepancy between which data are thought to be archived and which data actually are)
- make other sanity checks to ensure that the files that get to the archive are actually meant to be there (make sure files called ".xml" actually contain XML, for example)

3.2. File Transfer

Once checking is complete the transfer script moves files from the Gemini sites to the CADC. Files are transferred using the new bbftp transfer protocol which is optimized for the transfer of large files through the use of multiple TCP streams. A crc check of each file is made both before and after transfer to ensure that no errors have been introduced.

3.3. File Insertion

Once a file has been transferred to the CADC, the file metadata are checked against the specifications of the GSA data dictionary (Goliath et al. 2005) to make sure that the file is ready for archiving. This is done after the file is transferred instead of in the above "Check" step so that the GSA data dictionary does not have to be synchronized between sites. The CADC infrastructure ensures integrity and availability of data through periodic checksumming, redundant online storage, and offsite nearline backups, etc.

3.4. Metadata Ingestion

After a file has been inserted into the CADC archive infrastructure the gsaDataIngest package is used to put the file metadata into the GSA system for catalogue use, again using the GSA data dictionary to determine sources and destinations for the metadata.

3.5. File Removal

After successful transfer and ingestion of archive files and metadata the final script removes files from staging areas, writes a log file to indicate which files have traversed the system successfully, or in the case of errors generates a record of the errors that occurred.

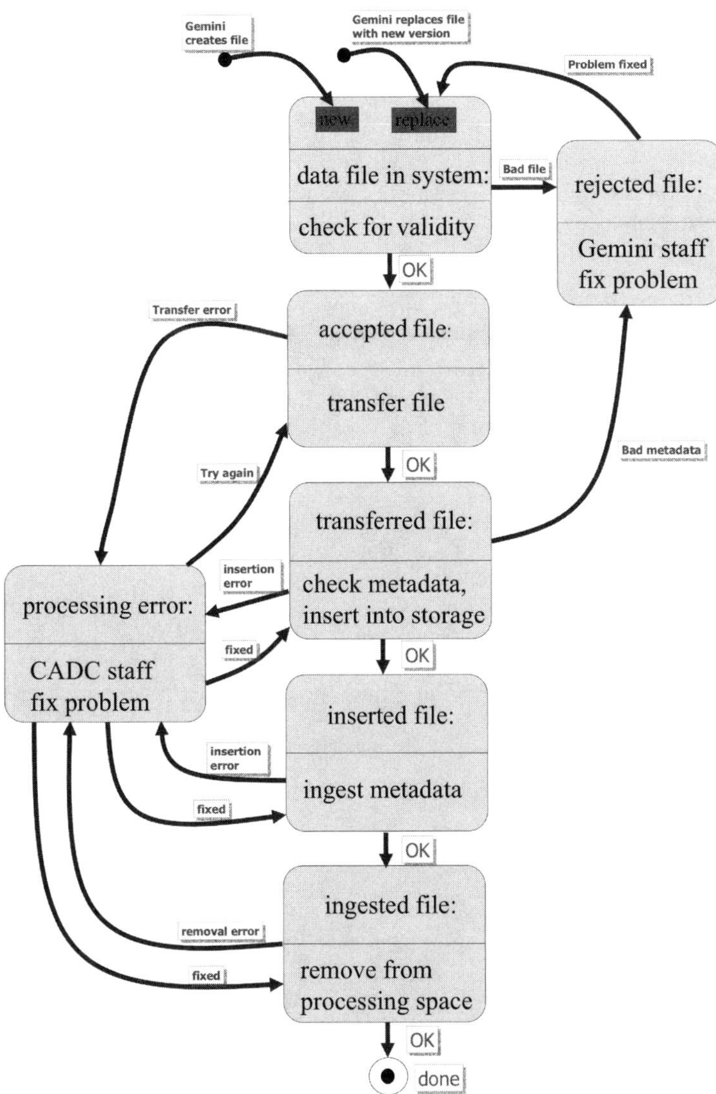

Figure 1. Schematic of the file state transitions that can occur during the electronic transfer of Gemini data from the base facilities to the Gemini Science Archive at the CADC in Victoria, BC.

4. Conclusion

The CADC's development of a system for the electronic transfer of data and metadata from the Gemini telescopes to the GSA has greatly reduced the load on Gemini Observatory operations staff and reduced the time between Gemini data acquisition and subsequent availability in the archive.

Previously, after observations were made, data acquired, and initial processing was done, data waited at the telescope operations centres to be written to storage media, boxed, and sent to the archive site. After traveling to the archive site, the media were read and ingested into the archive. The latency between data acquisition and archival storage was often on the order of weeks or months. In cases where magnetic media were being transported, errors were often introduced into the data, requiring it to be shipped again.

With automated transfer, this latency and shipping cost can be eliminated, and updating and maintaining the archive is made easier. Instead of waiting at the operations centre for media to be written, data can be transferred as soon as they are ready. In current operational use, the time it takes for a file to become accessible to archive users has been reduced to only a few minutes from the time it is made available by Gemini staff.

References

Bohlender, D., et al. 2005, in ASP Conf. Ser., Vol. 347, ADASS XIV, ed. P. L. Shopbell, M. C. Britton, & R. Ebert (San Francisco: ASP), 642

Damian, A., et al. 2005, in ASP Conf. Ser., Vol. 347, ADASS XIV, ed. P. L. Shopbell, M. C. Britton, & R. Ebert (San Francisco: ASP), 655

Goliath, S., et al. 2005, in ASP Conf. Ser., Vol. 347, ADASS XIV, ed. P. L. Shopbell, M. C. Britton, & R. Ebert (San Francisco: ASP), 660

ASTRONOMICAL DATA ANALYSIS SOFTWARE AND SYSTEMS XIV
ASP Conference Series, Vol. 347, 2005
P. L. Shopbell, M. C. Britton, and R. Ebert, eds.

The D4A Digitiser

Jean-Pierre De Cuyper

Royal Observatory of Belgium, Ringlaan 3, B-1180 Ukkel, Belgium

Lars Winter

Roehrigstrasse 40, D-22763 Hamburg, Germany

Abstract. The D4A (Digital Access to Aerial- and Astro-photographic Archives) project aims to acquire the necessary know-how, hardware and software to digitise the astro-photographic collections of the Royal Observatory of Belgium and the aerial-photographic collections of the National Geographic Institute and the Royal Museum of Central Africa in collaboration with AGFA-Gevaert. The D4A digitiser under construction consists of a granite based Aerotech ABL 3600 open frame air bearing XY positioning system, with custom build hardware to mount glass plates, film sheets and film rolls. The optical subsystem consists of a C-Cam Technologies CMOS camera mounted to a Schneider telecentric Xenoplan lens and will be illuminated by a computer controlled LED. The maximum scanning area is 350mm × 350mm with a speed of 6 plates (240mm × 240mm) per hour. A first benchmark of a prototype ABL 3600 was done and the proposed illumination system tested as detailed below.

1. Introduction

The D4A Digitiser project, as described in J.-P. De Cuyper, L. Winter & J. Vanommeslaeghe (2004), is progressing toward completion. The Aerotech ABL 3600 XY table including supporting subsystems was ordered in September 2004. We are finalising the drawings right now and expect the machine to arrive in Brussel around May 2005. The airconditioned clean room is being prepared at the Royal Observatory of Belgium in Brussel and will be ready by the time the machine arrives.

We finished most of the preliminary tests of the CMOS camera and illumination. We are building a thermoelectrically cooled version of the BCi4 CMOS camera in cooperation with C-Cam Technologies to improve the signal to noise ratio of the camera still further as well as the geometric stability.

A geometric benchmark was carried out in Pittsburgh in the Aerotech plant to check the behaviour of the ABL 3600 XY positioning table. To test the illumination subsystem of the Digitiser in connection with the telecentric lens and the CMOS camera, a test assembly of all necessary components was done in Hamburg and a variety of light sources checked.

2. Description of the digitiser

The D4A digitiser will consist of a granite based Aerotech ABL 3600 open frame air bearing XY positioning system, with an automatic plate holder assembly suited for mounting glass plates and film sheets up to 350mm x 350mm. Extended with an automatic film roll transport system and a plate tray handling and storage assembly with plate rotator, the Digitiser will be able to process nearly all known transparent photographic material.

A cooled CMOS camera from C-Cam Technologies (using a 12bit ADC) is mounted to a Schneider Xenoplan telecentric 1:1 lens. This optical subsystem will be illuminated by a light source with a very bright LED (lifetime min. 50,000h) that is computer controlled through a precision power supply for adjusting the exposure of each individual subimage. The D4A digitiser is intended to measure astronomical (up to 350mm x 350mm) as well as aerial (up to 240mm $\times$ 240mm) photographs at a rate of at least 6 plates per hour.

Special software is being developed for handling the 18 GByte/h data rate with online processing and storage of all images. The goal is to provide astrometrically and photometrically calibrated digital images with overlayed identified stars. The extracted information will be stored in a database. The digital images will be compressed lossless in order to reduce the storage size and time.

3. Benchmark

The geometric benchmark carried out in Pittsburgh at the Aerotech plant was intended to measure the stability of the ABL 3600 air bearing XY table. This open frame system performed very well and we decided to use it as the base for our Digitiser.

The ABL 3600 was fitted with a cross beam above the moving stages. A small Z-stage was mounted to this beam to allow focusing of the CCD camera. The Z-stage carried the geometric benchmark assembly, which consisted of a microscope objective and an Electrim EDC2000S CCD camera. Thus focusing the camera was done by changing the distance of the microscope objective to the plate mounted on the inner frame of the ABL 3600.

To carry out the benchmark, we used a glass plate with a grid of tiny chromium dots on its upper surface (i.e., the surface facing the microscope objective). This geometric grid glass plate was mounted in a sturdy aluminum plate holder in the centre of the inner frame of the ABL 3600. The X and Y frames of the ABL 3600 are open and the granite table they are moving on also has a central opening so that the light of the illumination can pass through it (open frame construction). To get flat illumination, we used a diffuser between the halogen lamp and the geometric grid glass plate.

The ABL3600 uses Heidenhain ZERODUR glass scales for the positioning of the XY table. These determine the positioning accuracy of the Digitiser, while the centering error of the chromium dots in the field of view of the digital camera is dominated by the poison noise of the bright background illumination, as we are measuring black dots on a white background. It was possible to do the centering to a precision below 0.1 μm and the repeatability found for this benchmark was below 0.09 μm. Due to the temperature change in the room we could only verify that the ABL3600 prototype would meet our requirements, but were unable to determine a reliable measuring accuracy. As the D4A Digitiser will be located in a temperature (18 °C ±0.1 K) and humidity (50 % RH ±1 % RH) stabilised clean room build by Becker Reinraumtechnik, the final accuracy of the XY table will be better than 0.1 μm. This corresponds to the proposed submicron accuracy for digitising astrometric plates (Zacharias et al. 2004)and for reassembling the individual overlapping foot prints into one (aerial) image.

4. Illumination Experiments

To determine the optical quality of the telecentric lens, the CMOS (Complementary Metal Oxide Semiconductor) camera and the light source, an existing test assembly (Winter, 1999) was modified. By illuminating the photographic plate or film with a highly diffuse light source a good mapping of the plates density scale onto the CMOS cameras intensities is obtained. As the measurement itself is carried out in transmittance, the intensity of the measured light is proportional to the transmittance. Hence the density of the plate has to be calculated out of this. For the whole optical system a photometric benchmark was designed, so that the machine can be checked at all times for its correct photometric behaviour.

As it is well known that halogen lamps are not very stable and have a very moderate mean time before failure (MTBF, typical 2,000h to 4,000h) and in order to avoid problems that are caused by changing the light "bulb", we choose to illuminate the photographic plate with a high power light emitting diode (LED) that has a very stable light output depending only on the stability of the supplied current and that has a MTBF of approximately 50,000 hours.

The influence the illumination has on the digitised image is shown in Figures 1 and 2. The first picture (Figure 1) was taken with a very diffuse illumination, while the second picture (Figure 2) was taken with light that had a high percentage of parallel light. Note that the scratches on the emulsion show up only in the latter case, while all the details of the astronomical image are the same in both. A more detailed analysis shows that the grain noise in Figure 2 is doubled as compared to Figure 1. Obviously we need a very diffuse illumination for our Digitiser.

References

De Cuyper, J.-P., Winter, L., & Vanommeslaeghe, J. 2004, in ASP Conf. Ser., Vol. 314, ADASS XIII, ed. F. Ochsenbein, M. Allen, & D. Egret (San Francisco: ASP), 77

Winter, L. 1999, PhD Thesis

Zacharias, N., & Urban, S. 2004, PDPP Newsletter Scan-It, 2, 24

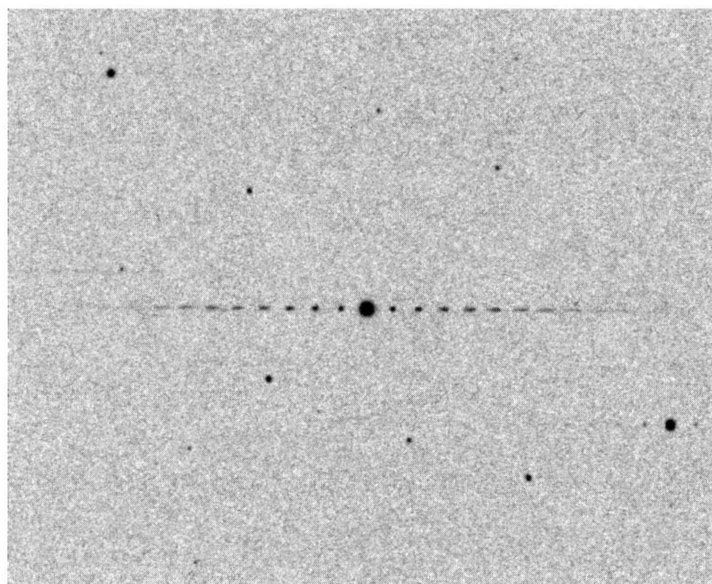

Figure 1. Diffuse illumination. Note the upper left corner underneath the medium bright star shows an empty field.

Figure 2. Diffuse and parallel illumination. Note the scratch in the upper left corner that is uncovered by the parallel light. A lot of starlike images appear everywhere, these are caused by dust particles on the surface of the photographic emulsion.

ASTRONOMICAL DATA ANALYSIS SOFTWARE AND SYSTEMS XIV
ASP Conference Series, Vol. 347, 2005
P. L. Shopbell, M. C. Britton, and R. Ebert, eds.

Overview of the Gemini Science Archive

Adrian Damian, Sharon Goliath, Severin Gaudet, David Bohlender, Norm Hill, and Geoffrey Melnychuk

Canadian Astronomy Data Centre

Colin Aspin

Gemini Observatory

Abstract. The Gemini Science Archive (GSA) is a new archive developed by the Canadian Astronomy Data Centre (CADC) with the aim to provide the scientific community with tools for effective on-line access to data collected by the Gemini telescopes. The first version of the GSA basic archive was released in September 2004, with subsequent releases containing advanced capabilities to follow. This paper depicts the overall architecture of the GSA basic archive, the product of a fruitful collaboration between the data provider (Gemini Observatories) and the development team (CADC).

1. Introduction

The Gemini Science Archive plays an important role in the full exploitation of the unique characteristics of the scientific data collected by the two state of the art Gemini telescopes. As such, it is considered an important part of the capabilities of the Gemini observatories and it has been designed with this goal in mind. The basic GSA was released in September 2004 and contains a number of innovative approaches briefly described in the following sections of this paper. More details about these features can be found in Melnychuk et al. (2005), Goliath et al. (2005), and Damian et al. (2005). Bohlender et al. (2005) describe the GSA from the archive user's point of view.

2. GSA Requirements

The design of the GSA benefited heavily from the experience of past archives developed and hosted by the CADC. Besides the typical requirements of a basic science archive, the main features of the design also include:

- **Prompt and reliable availability of the data.** It became increasingly apparent that the PIs and their collaborators make up an important category of users of the GSA. In order to make the archive appealing to this category of users, their datasets should be made available for retrieval promptly.
- **Easy characterization of the data.** Most archives provide finite products or the raw data together with calibration files. The GSA should go

beyond this, offering other complementary data such as weather information, observing logs and information about the science programs.

- **Easy maintenance of the archive.** It is expected that an average of two new instruments will be added each year to the Gemini observatories. While most of the descriptors in the metadata are common to all of these instruments, instrument specific descriptors should be accommodated without changes to the software or the database schema.
- **Extensibility of the system.** The design of the archive should be flexible enough to allow for the integration with future features to be implemented for the GSA advanced capabilities.

3. Overall Design

Figure 1 depicts the software architecture of the GSA system with its main components and their interaction. The modules of the system are implemented in C/C++, Java and Perl and use a Sybase database. The archived data are stored on magnetic disks.

The software modules in the figure above support the three main phases of data archiving in GSA: Data Acquisition, Catalogue Creation and Data Access and each of these phases has some features worth mentioning.

3.1. Data Acquisition (Electronic Transfer and Bulk Storage Modules)

Data is transferred from the Gemini observatories to CADC through electronic transfer, which became feasible with the advent of the faster network links between CADC and Gemini Observatory sites (see Melnychuk et al. (2005) for details). Files that are ready for archiving at the Gemini sites are placed in specific directories. The Electronic Transfer module checks the integrity of the data before transferring and archiving it at the CADC. All the data are reliably transferred and archived on magnetic disks and become immediately accessible to GSA users without an operator's assistance.

3.2. Catalogue Creation (Data Ingest, Catalogues and Data Processing Modules

Each GSA dataset is characterized by its own metadata, and querying the metadata represents the main method for searching through the archive. The Data Ingest module extracts metadata from each recognized file type (currently FITS, weather, log and science programs are the supported file types) and stores them in the Metadata Store (MS) database. Metadata are typically extracted from the dataset file but sometimes they are contained in the file name, such as in the case of weather files. Next, a subset of the metadata is copied to the searchable Catalogue Store (CS) database. In the GSA, this subset makes up the set of attributes that are available for querying the archive. To limit the size of the CS database and improve its performances, instrument specific descriptors and descriptors that have limited usefulness to the general user are not transferred from the MS database. In order to assist the user in searching for the data, the catalogue is instead populated with derived descriptors that are generated by the Data Processing module from the existing metadata. Examples of such derived

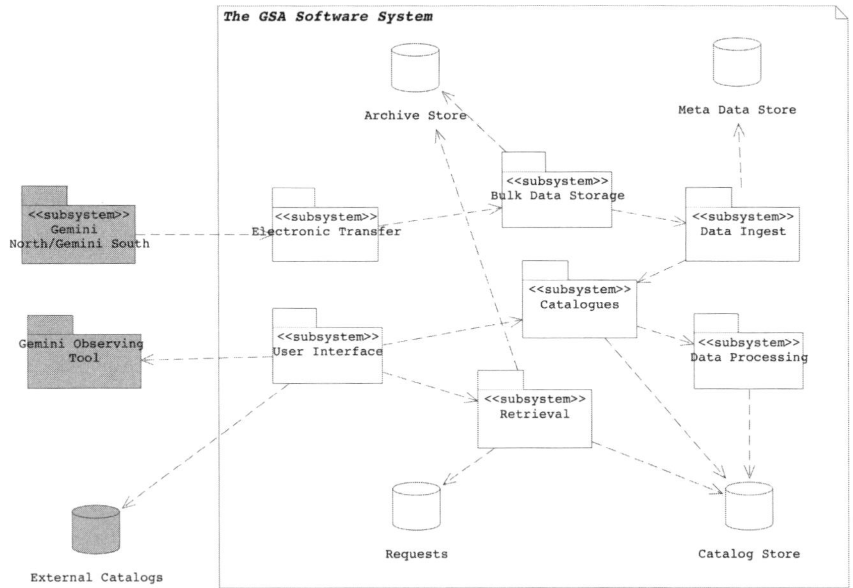

Figure 1. Schematic of the software architecture of the GSA.

descriptors include: wavelength information, pixel scale, spectral resolution and min/max coordinates.

Using the Discovery Agents (see Damian et al. 2005) approach, the whole functionality of the system is based on small individual entities that can run independently. Therefore, new Discovery Agents can be incrementally integrated into the system to perform other activities required by the advanced capabilities.

Data Dictionary The Data Dictionary (DD) is an XML document that specifies the characteristics of the metadata – the metadata of the GSA metadata (Goliath et al. 2005). Through it, the operator must:

- Determine the categories of datasets. All the input datasets are categorized in the DD based on the value of some specified descriptors. Each category contains a complete set of descriptors pertaining to the member datasets.
- For each category, specify all of the metadata descriptors with their source (file type), destinations (CS or just MS) and intermediate points (e.g., MS, CS) in the archive system. The names of the descriptors in each of these points are specified, as well as their types, transformations and ranges.

The Data Dictionary is one of the key concepts of the GSA archive that is used by the majority of the modules in the system. In data acquisition, the DD is used to check the files candidate for archiving and reject them if they don't comply with a required format, previously agreed upon with the data provider. The DD is also used in ingesting the metadata into MS and in the creation of the catalogues.

It is expected that the use of the DD together with an improved database schema will significantly enhance the maintainability of the system. Changes to accommodate different formats or support new metadata from new instruments are now limited to changing the DD without any modification of the database or source code. With the DD the archive operator has a superior view of the whole metadata and can easily detect inconsistencies. Furthermore, the electronic transfer allows problems to be promptly reported to the Gemini observatories and remedies worked out. It is the common understanding of both the data supplier and the archive operator that "bad data" must not be archived as it has limited usefulness for the user and degrades the overall quality of the archive.

3.3. Data Access (User Interface and Retrieval Modules)

The Data Access portion of the system is responsible for offering the user a seamless experience in utilizing the archive. The User Interface, for example, offers a variety of queries for different types of tasks:

- science query – tuned for fast access to the science data through the most common descriptors
- complete catalogue query – designed for querying all the descriptors for a specific instrument available in the catalogue. The user is prompted to specify the instrument and the query form is tailored to display only the descriptors pertaining to the selected instrument. This form is also used for obtaining the appropriate calibration files.
- weather query – allows for querying the database for the available weather images.
- Gemini System Verification data – makes the system verification data available to the user.
- PI access – password protected mechanism that allows PIs to access their data before they become public.

Taking advantages of the Web interface, hyperlinks are used to enable seamless navigation between related datasets such as observations, calibration, weather, science program information etc. Other features allow the user to save the queried metadata in a file, manipulate it and load it back into the User Interface to retrieve the selected data.

While the metadata becomes public as soon as it is available in the CS database, the actual data file cannot be accessed during the standard 18 month Gemini proprietary period. The only exception is the PI access mentioned earlier which allows PIs to access their own data. The Retrieval module enforces this rule when data are requested. Since data are archived on magnetic disks and hence online at all times, the delay in the data retrieval is kept to a minimum. Data are made available through ftp or http proxies.

4. Future Work

While the basic GSA archive has just been released, planning for the next version with more advanced capabilities is under way. Distribution of data to PIs is one of the main features that are planned to be incrementally added. This feature will monitor the data arrival and notify the interested parties when the complete data set for an observation program becomes available. It will ensure that PIs have access to their data in a timely, secure and reliable fashion from everywhere via the Internet. Currently, only raw data are being archived; however, calibrated data will be produced and archived as well. Once the calibrated data are archived, new modules can process it and generate previews that the user can access to have a quick look at the data before requesting it. Interfaces and services to make the archive accessible through the Virtual Observatory initiative are also planned for the future releases. They will be implemented once the Virtual Observatory infrastructure becomes more mature and stable. Data mining is another potential advanced capability that will allow cross correlation and sophisticated analysis on an integrated view of the data. This becomes more important as the archive grows and the brute force navigation of the data is not sufficient.

5. Conclusions

Benefiting from the knowledge and experience with other telescope archives built and operated by CADC, the GSA was designed to be one of the first scientifically effective archives of ground-based operations. As such, from the very beginning the GSA was considered an integral part of the planning, observation, calibration, data reduction and data distribution processes of the Gemini telescopes. This fundamental shift in data archiving allowed for the implementation of some novel approaches to the data handling and processing:

- Flexible database schema that does not require changes when the ingested metadata changes
- Easy to edit data dictionary to specify the flow and format of the metadata
- Rapid availability of data as the result of the electronic transfer of data from the Gemini telescopes
- Generation of derived metadata for easy access by the user or for compatibility with other telescopes and instruments
- Support for observation data produced directly by telescopes' instruments as well as other related data (weather, logs etc)
- Improved Graphical User Interface

The GSA archive became operational in September 2004 and new data from existing and new instruments are added daily. The archive is going to be enhanced over time with new features that will give the user new means for seeking and exploring Gemini data.

References

Bohlender, D., et al. 2005, in ASP Conf. Ser., Vol. 347, ADASS XIV, ed. P. L. Shopbell, M. C. Britton, & R. Ebert (San Francisco: ASP), 642

Damian, A., et al. 2005, in ASP Conf. Ser., Vol. 347, ADASS XIV, ed. P. L. Shopbell, M. C. Britton, & R. Ebert (San Francisco: ASP), 637

Goliath, S., et al. 2005, in ASP Conf. Ser., Vol. 347, ADASS XIV, ed. P. L. Shopbell, M. C. Britton, & R. Ebert (San Francisco: ASP), 660

Melnychuk, G., et al. 2005, in ASP Conf. Ser., Vol. 347, ADASS XIV, ed. P. L. Shopbell, M. C. Britton, & R. Ebert (San Francisco: ASP), 647

ASTRONOMICAL DATA ANALYSIS SOFTWARE AND SYSTEMS XIV
ASP Conference Series, Vol. 347, 2005
P. L. Shopbell, M. C. Britton, and R. Ebert, eds.

The Gemini Science Archive Data Dictionary

Sharon Goliath, Adrian Damian, Severin Gaudet, Norm Hill, David Bohlender, and Geoffrey Melnychuk

Canadian Astronomy Data Centre

Colin Aspin

Gemini Observatory

Abstract. The Gemini Science Archive (GSA) is a new science archive developed by the Canadian Astronomy Data Centre (CADC) to provide the scientific community with tools for effective on-line access to data collected by the Gemini telescopes. CADC has developed a MetaData Data Dictionary (MDDD) for the GSA to meet the goal of providing dependable and consistent archive data. This MDDD defines the rules that govern data manipulation on meta-data obtained from a variety of data sources. The MDDD is specified in XML, and so provides flexibility while reducing source code maintenance. The paper will discuss the rules that may be specified for the data sources and individual meta-data attributes, the role of the data dictionary in populating the GSA, verification of the consistency of the dictionary itself, and possible future improvements in the data dictionary.

1. Introduction

The GSA, as described in the overviews of Bohlender et al. (2005) and Damian et al. (2005), was released on 20 September 2004. The GSA content consists of pixel data, as well as meta-data such as telescope and instrument configuration information, observation logs, weather information, calibration data, and science programs. The continuing success of the GSA within the user community is dependent on ensuring all data are consistent in form and quality. The success of GSA within CADC is dependent on ensuring this data consistency in a timely and efficient fashion.

Data consistency in the GSA is maintained by evaluating all input sources against common criteria. The MetaData Data Dictionary (MDDD) was developed to specify these criteria in a configurable manner, and is used to maintain GSA content. The description of these criteria, how they are captured as configuration information, and the use of the criteria follows.

2. The Criteria

The MDDD encapsulates two types of criteria: those that address the data consistency needs of the archive users, and those that address the efficiency needs of the archive developers. All criteria are expressed at the level of the individual attributes that make up the meta-data from a data source. The data consistency criteria may affect the archiving of an entire data source, or they may only affect the archiving of individual attributes. The data consistency criteria are:

- all required attributes from an input source must be present in an input source,
- no unexpected attributes are present in a data source,
- attribute values must conform to expected primitive and complex data types and formats, and
- attribute values are within optional minimum and maximum boundaries.

The efficiency criteria facilitate the success of the GSA processing path for the data, because they guarantee the presence, type, and format of attributes for all data sources. These criteria are considered only for individual attributes. The efficiency criteria make it possible to:

- specify the transformation of attribute values between data types and different representations of the same data type,
- identify whether an attribute should be catalogued, and if so, identify the destination database tables for that attribute,
- rename an attribute in the destination database table, and
- provide a default value if an attribute is missing from a data source.

3. Making the Criteria Configurable

The CADC invested in a prototyping effort to evaluate the feasibility of quantifying the identified criteria. The prototype data dictionary was stored in a relational database table, and limited the expected data sources to the keywords from FITS files.

This effort was insufficient to process the wide selection of data sources maintained in the GSA, and thus the production MDDD expanded to handle metadata from FITS files, science programs, weather and log files. With this expansion of input sources came the support for multiple destinations. The input and output sources supported by the MDDD are illustrated by Figure 1.

The prototyping effort found that the most complex data consistency criteria to make configurable are the ones that identify allowed and required keywords by data source. The prototype explicitly named instrument, facility, and instrument mode keywords, as well as date keywords for range checking in an effort to define valid keyword combinations. The MDDD has generalized this naming so that any attribute from any data source may now serve as the criterion for identifying expected attribute combinations.

An additional aspect highlighted by the prototype was the difficulty in communicating the correct keyword combinations for FITS files using relational database tables. XML was chosen to represent the production data dictionary in part because of its ability to organize information hierarchically, and thus better visually communicate expected attribute combinations. A hierarchical representation also lends itself to reducing the duplication of information that occurred in the prototype data dictionary.

The MDDD abandoned the prototype's use of the database as the storage location for the data dictionary content, because the searchability benefits pro-

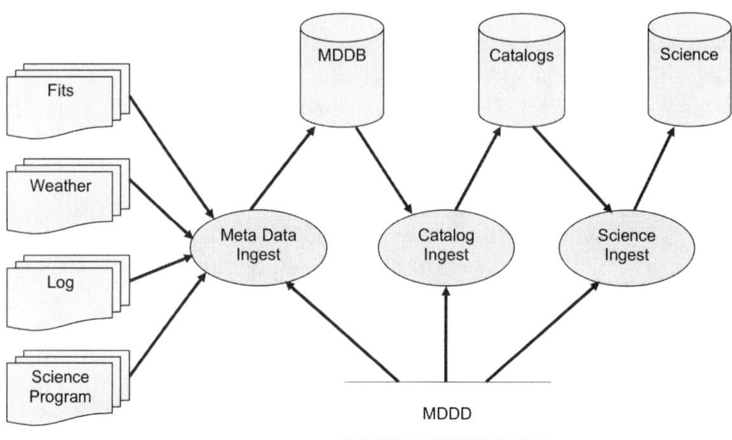

Figure 1. The GSA processing path dependence on the MDDD.

vided by the database were outweighed by the communication and maintenance benefits of XML.

The MDDD encapsulates the data correctness and efficiency criteria in two ways. Attribute-level rules specify the criteria that apply to the individual meta-data attributes of a data source, such as data type, data format, and destination table name. Data source level rules identify the exists/doesn't exist data consistency criteria for a data source as a whole, such as which attributes are mandatory in a data source.

A descriptor is used to define the attribute-level rules, and is the unit of organization for these rules in the MDDD. A descriptor can be thought of as defining the characteristics of meta-data as it arrives from is data source, and the characteristics of meta-data as it will be stored in its destination. Each descriptor defines several data sources and destinations (i.e., this attribute is expected from a FITS file, and must end up in the MetaData DataBase (MDDB) and the Data SuperSet Catalogue (DSSC), but not the Science tables). Descriptors specify if an attribute must be available from the source, the default value if the attribute doesn't exist in the source, the minimum and maximum if the value should be range checked, and how to transform the value from one data type and/or representation to another.

Categories are how the data source level rules are defined. They are also the MDDD unit of organization for descriptors, and tend to cluster descriptors by data source characteristics. Only certain categories, and thus certain descriptors, are valid for an individual data source. Categories limit the set of descriptors used as the source of the individual attribute rules. A descriptor that refers to the same attribute may exist in more than one category.

Each category is selected by a set of conditionals. If these conditionals are met by a data source the category is selected, and the descriptors within the category become the source of the attribute rules to which the meta-data must conform.

Categories may be nested within categories, and the conditionals for the nested category includes the conditionals of the parent category. The MDDD currently has categories that differentiate between source types, images, image extensions, instrument type, observation types, calibration files, mask definition files, and observation dates.

The MDDD makes provision for three types of conditionals: existence, value and range. Existence requires that a particular attribute be present in a data source. Value requires that a particular attribute be present in the data source, and it must have the specified value. Range requires that a particular attribute be present in the data source, and it must have a value that falls between the specified limits.

If an attribute does not conform to its descriptor rules, the attribute does not move to the next destination on the GSA processing path. If a data source does not conform to its data source level rules, the data source and all attributes originating from it does not move to the next destination on the GSA processing path.

Categories, conditionals, and descriptors are specified by an astronomer familiar with telescope instrumentation. The need for shepherding software through design, development, and testing to capture the specific rules has been removed.

4. The Criteria in Operation

Figure 1 illustrates GSA processing which is dependent on the MDDD.

Meta Data Ingest is the entry point for all meta-data into the GSA. It is responsible for checking each attribute against the MDDD descriptor and data source rules. Additionally, Meta Data Ingest uses the MDDD to identify the name of the database table where the attribute will be stored. The MDDB stores all meta-data, whether or not it conforms to MDDD rules. Data sources which do not conform to the rules are brought to the attention of GSA Operations Staff.

Catalog Ingest separates the meta-data into distinct catalogs as data is moved along the GSA archive path from the MDDB. It checks each attribute against the MDDD descriptor and data source rules. Additionally, Catalog Ingest uses the MDDD to identify the catalog and the database table where the attribute will be stored. The catalogs store only data that conforms to MDDD rules.

Science Ingest is responsible for determining the science usability of the datasets in the catalogs. The MDDD provides Science Ingest with the list of attributes in the Science database table. The Science table stores attributes supporting more elaborate queries.

The MDDD is verified for consistency against the set of descriptors the database tables can accept – no descriptors may appear in the MDDD that are not known to the tables, and no descriptors may appear in the tables that are not known to the MDDD.

5. Future Possibilities

The following items have been suggested for addition to the capabilities of the MDDD:

- provide attribute value replacement by specifying hard-coded values for descriptors,
- support more sophisticated data transformations than the already existing modification of data type and representation by the use of tags to identify sets of descriptors as input and output,
- provide support for the construction of FITS files, and
- provide more internal consistency checks. There are currently 2296 descriptors in the GSA MDDD, and more are added for each new instrument, so automated consistency checking is helpful.

6. Conclusion

The MDDD has succeeded in specifying in a configurable manner the data consistency and efficiency rules. GSA data is more consistent because all meta-data have their descriptor and data source level rules specified. GSA processing software maintenance costs are lower because all data-dependent knowledge has been restricted to the data dictionary, and modification of the MDDD does not require the maintenance expense of source code.

Experience has proven these claims. When Gemini Thermal-Region Camera Spectrograph (TReCS) data began arriving, the MDDD was modified to handle its keyword peculiarities, and FITS files produced by this instrument traveled the GSA processing path successfully.

References

Bohlender, D., et al. 2005, in ASP Conf. Ser., Vol. 347, ADASS XIV, ed. P. L. Shopbell, M. C. Britton, & R. Ebert (San Francisco: ASP), 642

Damian, A., et al. 2005, in ASP Conf. Ser., Vol. 347, ADASS XIV, ed. P. L. Shopbell, M. C. Britton, & R. Ebert (San Francisco: ASP), 655

ASTRONOMICAL DATA ANALYSIS SOFTWARE AND SYSTEMS XIV
ASP Conference Series, Vol. 347, 2005
P. L. Shopbell, M. C. Britton, and R. Ebert, eds.

FITS Image Cutout Web Services

Vivek Haridas, William O'Mullane, Tamás Budavári, Maria A. Nieto-Santisteban, Alexander S. Szalay, Samuel Carliles, and Ani R. Thakar

Center for Astrophysical Sciences, JHU, 3701, San Martin Dr., Baltimore, MD 21218-2695

Abstract. The Sloan Digital Sky Survey is on its way to build a very large map of the universe. We look at implementing web services that help in getting image cutouts from the FITS files in the SDSS archive. These cutouts would be provided in the traditional FITS format. We present the interface, the algorithm and usage of the FITS Cutout Web Service.

1. Introduction

The Fits Cutout is a XML Web service that provides methods to create the best available rectangular cutouts of a requested size at a given point in the sky. The cutouts are extracted from the FITS files stored in the SDSS Data Archive System.

It also provides some related facilities like conversions between Pixel and World Coordinate System (WCS) locations and reading contents of the header of a FITS files on the Web.

The web service is available at `http://voservices.net/FitsCutOut/`.

2. The Extract cutouts Method

This is one of the main methods provided by the Web service. It extracts rectangular sections and returns them as DIME attachments in the form of FITS and JPEG images.

A rectangular section is defined by a center point, width and height in pixels and the band 'u','i','r','g' or 'z' of observation. The image center point has to be provided in right ascension (Ra) and declination (Dec) in terms of J2000 degrees.

The possible outcomes of a cutout operation are listed below.

2.1. Ok

This is a case where the entire rectangular section existed completely in a single field of observation. The entire section will be returned in FITS and JPEG formats for all the requested bands as dime attachment.

2.2. Incomplete

This is a case where the rectangular section exists only partially in any single field of observation. The best available incomplete rectangular section will be returned in FITS and JPEG formats for all the requested bands as dime attachments. The best available field is the one which has maximum overlapping area with the cutout section.

2.3. Not found

This is the case where no observations are available for the given point in the sky. No attachments are returned.

3. The Extract cutouts Algorithm

The steps of the algorithm are listed below.

3.1. Fields

Get all fields of observation that have the given point. This is done using the SdssFields Web Service[1]. This web service finds fields in the Sloan Digital Sky Survey around the given point with a given search radius.

3.2. Best Fit

Find the field with the best fit for the requested rectangular section. This field has information on the FITS file which has the processed image of the observation, the WCS and all the required information.The steps involved are:-

a. Read the WCS values from the field table.

b. Calculate the pixel values (x,y) local to the image for the point, from the World Coordinate Values (Ra,Dec). This is done using a C-Sharp implementation of the JSky package.

c. Find the first field which contains all the corners of the rectangular section within the image. If no field exists with the complete section inside it, then find the field which covers maximum area of the rectangular section.

3.3. Individual Band operations

For each of the requested bands:

a. Upload the FITS file corresponding to the band.

b. Cut Out a rectangular section into a new FITS image. The steps involved are:

(i) Read the WCS values from the FITS image.

(ii) Calculate the pixel values (x,y) local to the image for the center point of the section, from the World Coordinate Values (Ra,Dec). This is done using a C-Sharp implementation of the JSky package.

(iii) Now calculate the starting and ending points of the rectangular section.

(iv) Create a new FITS image for the cut out and copy the rectangular section from the original FITS image.

---

[1]`http://skyserver.sdss.org/vo/dr2fields/sdssfields.asmx`

(v) Recalculate the WCS values for the new FITS image according to the new local pixel values of the original reference point of the World Coordinates System.

(vi) Copy the relevant key word values from the original FITS Image and the new WCS values into the new FITS image. All the operations on the FITS images are performed using the FITSLIB library[2] along with the CFITSIO package.

c. Convert the FITS image to a JPEG image

d. Attach the FITS and JPEG images as dime attachments

3.4. Result

Return the result of the operation (Ok / Incomplete / Not Found)

4. Other Methods

Here are the other methods:

4.1. World Coordinate - Pixel conversions

Given WCS values or a valid FITS file on the web with WCS headers, this interface provides conversions between World Coordinate values and Pixel values for a given point in the image.

4.2. FITS Header Reader

Given a valid FITS file on the web, this interface reads the keyword values of the primary FITS Header.

4.3. FITS file cutout

Provides a cutout of a given FITS file on the web.

4.4. FITS cutout image pixel data extraction

This is helpful in cases where the a entirely new FITS file is not needed, but rather the need is to know only the pixel values.

4.5. Get Best Field

This provides the best field information available for the cut out from the SDSS-Dr2 database.

5. Command Line Tool

A sample command line tool has been developed to use the Fits Cutout Web service. The source code and binaries are available on the Fits Cutout web page[3].

Here are some commands of the tool:

---

[2]`http://skyservice.pha.jhu.edu/develop/FitsLib/`

[3]`http://voservices.net/FitsCutOut/`

5.1. Extract CutOuts

This method finds the best cutout available in the archive for the requested bands at the provided location in the sky.

The usage is ExtractCutOuts [Ra] [Dec] [Width] [Height] [Bands]

5.2. Cut out Section

This method cuts out a rectangular section from a given FITS file of requested size at the provided location.

The usage is CutOutSection [fitsUrl] [Ra] [Dec] [Width] [Height]

5.3. Pixel to World coordinate conversion

This method converts a given pixel location into a WCS value using the information available in the provided FITS file.

The usage is Pix2World [fitsUrl] [X] [Y]

5.4. World to Pixel coordinate conversion

This method converts a given World coordinate location into a pixel location value using the information available in the provided FITS file.

The usage is World2Pix [fitsUrl] [Ra] [Dec]

5.5. Listing Header

This method lists the header values of the provided FITS file.

The usage is ListHeader [fitsUrl]

6. Useful Links

HEASARC, CFITSIO User's Reference Guide,
`http://heasarc.gsfc.nasa.gov/docs/software/fitsio/c/c_user/cfitsio.html`

NASA/GSFC, A User's Guide for the Flexible Image Transport System,
`http://archive.stsci.edu/fits/users_guide/`

Microsoft Corporation, Microsoft Developer Network,
`http://msdn.microsoft.com/`

JSky, Java Components for Astronomy,
`http://archive.eso.org/JSky/`

ASTRONOMICAL DATA ANALYSIS SOFTWARE AND SYSTEMS XIV
ASP Conference Series, Vol. 347, 2005
P. L. Shopbell, M. C. Britton, and R. Ebert, eds.

The WFCAM Science Archive and Other WFAU Work

M. A. Read, N. C. Hamby, E. T. W. Sutorius, R. G. Mann, I. A. Bond, and P. M. Williams

Institute for Astronomy, University of Edinburgh, Royal Observatory, Blackford Hill, Edinburgh EH9 3HJ, UK, Email: mar@roe.ac.uk

Abstract. The Wide Field Astronomy Unit (WFAU) of the Institute for Astronomy, University of Edinburgh, will host the data taken with WFCAM, the new wide-field infrared imager on UKIRT. This, the WFCAM Science Archive, is the final stage in the data flow from telescope to user.

Starting in 2005, WFCAM will carry out five main surveys over a 7 year period. Observations will typically result in 100Gb of data per night and generate a total archive size of tens of terabytes.

In addition to holding data, science archives should provide for fast and highly flexible querying and science driven data products that are significantly enhanced with respect to the ingest data. With this in mind, and following an agreed set of requirements laid out in a science requirements analysis document, WFAU designed the WFCAM Science Archive (WSA) described here. Areas covered include hardware choices, software architecture, curation tasks and user access.

Archive scalability is also a key issue as the WFCAM Science Archive forms the first phase VISTA Data Flow System (VDFS) science archive. WFAU's involvement with the VDFS and other developments are briefly discussed.

1. Hardware

Balancing performance with cost, the WSA hardware is based around 32-bit PC technology. The modular design allows for straightforward expansion as the archive grows.

A schematic diagram of the WSA network is shown in Figure 1; features include:

- the WSA network is private in that it is isolated from the general site ethernet and network links. This ensures the archive has maximum bandwidth for all tasks, namely ingest, curation and querying;
- interconnected at 1 Gbit s^{-1} the archive hardware has a 1 Gbit s^{-1} internet connection direct to a JANET backbone access router straight onto the 10 Gbit s^{-1} JANET backbone;
- RAID5 arrays of low cost SATA disks form a mass pixel storage area with good access speed;
- the catalogue servers employ Ultra320 SCSI RAID arrays which provide very high aggregate disk IO rates;
- there are two catalogue servers providing redundant storage and allowing the isolation of the day-to-day ingest and curation tasks (performed on the load server) from user queries accessing periodically released data products (carried out on the public catalogue server);

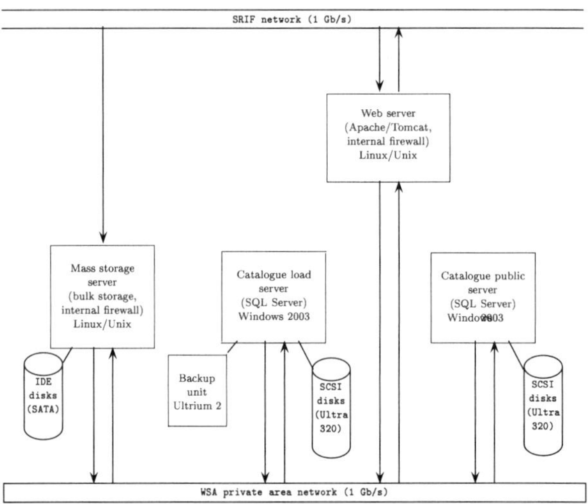

Figure 1. Schematic diagram of the WSA private area network.

- Ultrium-II LTO tape is the catalogue backup medium. (A copy of the pixel data is kept in Cambridge where the pipeline reductions are performed.)

The configuration of the Ultra320 SCSI disks utilized in the catalogue servers is the result of substantial investigation and experimentation. The current set-up, which offers good read performance and stability with an acceptable write performance penalty, employs multi-channel PCI-X interface adapters to support "spanned arrays" (RAID50).

2. DBMS

A database management system (DBMS) is required to house the archive and to facilitate flexible user access. The factors affecting our choice of DBMS include

- simplicity of application interfaces;
- scalability to large database volumes;
- availability of up-to-date comprehensive documentation and technical support;
- ease of administration and maintenance;
- cost.

These factors led us to consider only off-the-shelf relational DBMS solutions that come with industry standard interfaces and built-in management tools. The scalability issue narrowed the field to three systems: DB2, Oracle and Microsoft SQL Server. The Skyserver archive for the SDSS employs SQL Server. Their

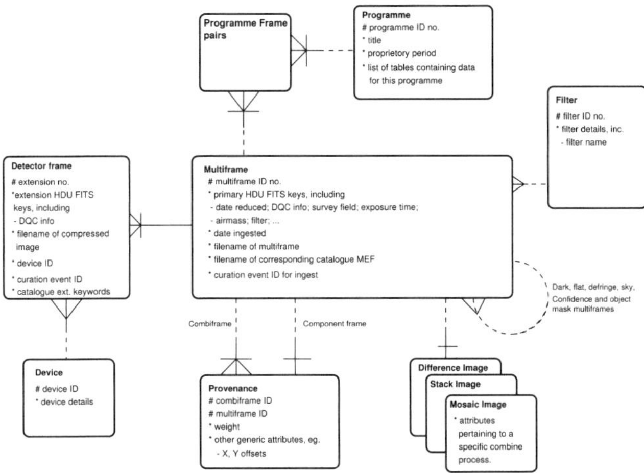

Figure 2. Data model for all of the image data in the WSA.

experience coupled with the astronomy-specific developments they have made for that system and the assistance they have given us have led us to choose SQL Server as the WSA DBMS.

3. Schema

A schematic model of the WSA was derived using a set of generalised entity-relationship (ER) models. These models describe the object catalogue data, image data, calibration metadata and database-driven functions (ingest, curation and generation of database products). Some of the modeling proved quite complex as it must describe the different WFCAM surveys and handle issues such as re-calibration. Figure 2 shows the ER model for all image data in the WSA. This model describes the smallest, basic unit of image data (a single device detector image) through calibration, multi-device camera "paw prints", stacks and mosaics to the most heavily processed database driven products like tiled and stacked contiguous images.

SQL schema script files were created to implement the tables, their attributes and relationships described in the models. These scripts are in turn used to drive the data ingest and generate the schema documentation.

4. Software - curation operations

A set of curation use cases were drawn up during the design process. The use cases range from the daily transfer of the pipeline-reduced data from Cambridge to our mass storage server, through the upload and ingest of data (e.g. image metadata, merged source catalogues) onto our catalogue load server, to the release of data products on the public server.

These curation uses have driven the software development. In designing and writing this software, and with ease of maintenance and scalability in mind, a modular approach with "clean" programming interfaces was adopted. The software architecture developed includes the following features:

- high level application code written in Python;
- large scale intensive operations (e.g. source merging, bulk ingest) coded in efficient C/C++ bound to Python scripts and operated outside the DBMS;
- use of off-the-shelf middleware and an abstracted, layered approach that compartmentalises the curation software;
- client/server architecture between the curation driving server and Microsoft SQL Servers employing standard, layered interfaces (e.g. ODBC, XML–RPC).

5. User access

User access to the WSA will initially be via a set of web based forms and consist of:

- proximity searches - radial or rectangular searches of the main WFCAM survey tables;
- menu driven queries - a series of forms guide the user in constructing simple SQL queries (Figure 3);
- freeform SQL - users can access the full power of SQL to mine the WSA;
- cross-match - uploaded source lists can be paired against the WFCAM survey tables,
- image extraction - given a position, a small area pixel image (up to the size of a single detector) can be extracted.

In addition to WFCAM data, the archive DBMS will also house external datasets e.g. SDSS and 2MASS. Joint queries of these datasets and relevant WFCAM surveys will be possible via pre-existing neighbour tables.

The forms action Java servlets, running on a Tomcat server, which execute the parsed SQL query via a JDBC connection. Extracted data will be returned in a variety of user-selected standard data formats (e.g. FITS, VOTable).

Comprehensive online documentation including a database schema browser and SQL cookbook will guide the user in exploiting the full flexibility of the archive.

Much of the functionality described above has been implemented in the SuperCOSMOS Science Archive which was built by WFAU and has served as a prototype for the WSA.

Further enhancements to the user interface are planned (e.g. extraction of image mosaics spanning several detector frames). In addition, users will be able to access the WSA via the AstroGrid (the UK's VO project) infrastructure.

6. Other WFAU work

As mentioned previously the WSA is the first phase archive in the Vista Data Flow System. There are many similarities between the data produced by the WFCAM and VISTA surveys, the main difference will be in the volume and rate

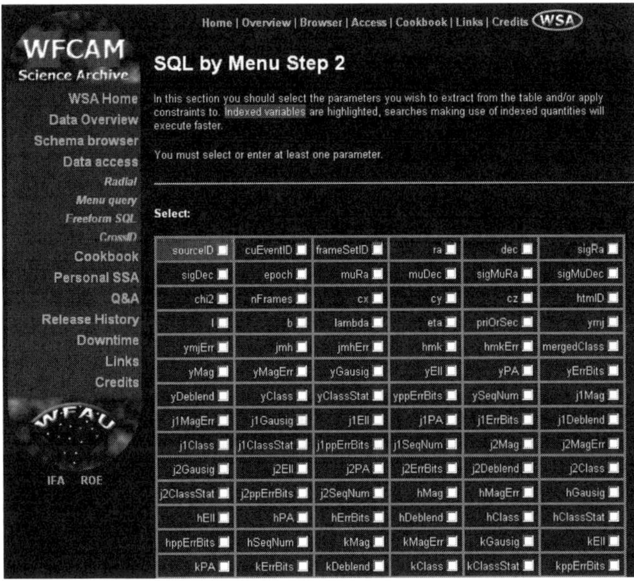

Figure 3. Screen-shot of the menu driven interface.

of data. As due consideration was given to the issue of scalability most of the design work and software developed for the WSA will be readily applied to the Vista Science archive.

WFAU hosts the SuperCOSMOS Sky Surveys (includes the southern Galactic plane H-alpha survey). Object catalogues and image data for the POSS-II Bj and R surveys were added in mid-2004 giving full sky coverage. The southern sky UK Schmidt survey catalogues are also available via the SuperCOSMOS Science Archive.

The archive for the 6dF Galaxy Redshift Survey is also designed and hosted by the WFAU.

7. Links

WFCAM/VISTA Science Archive Development: `http://www.roe.ac.uk/~nch/wfcam/index.html`
WFAU pages: `http://www.roe.ac.uk/ifa/wfau/`
SuperCOSMOS Science Archive: `http://surveys.roe.ac.uk/ssa/`
SuperCOSMOS Sky Surveys: `http://www-wfau.roe.ac.uk/sss/`
6dF Galaxy Redshift Survey: `http://www-wfau.roe.ac.uk/6dFGS/`
UKIDSS pages: `http://www.ukidss.org/`

ASTRONOMICAL DATA ANALYSIS SOFTWARE AND SYSTEMS XIV
ASP Conference Series, Vol. 347, 2005
P. L. Shopbell, M. C. Britton, and R. Ebert, eds.

ESO Archive Services

Nathalie Rossat, Nausicaa Delmotte, Nathalie Fourniol, Benoit Pirenne, and Adam Dobrzycki

ESO, European Southern Observatory, Karl-Schwarzschild-Str. 2, D-85748 Garching bei München, Germany

Abstract. The ESO/ST-ECF Science Archive Facility (SAF) is a joint collaboration of the European Southern Observatory (ESO) and the Hubble Space Telescope-European Coordinating Facility (ST-ECF). The archive provides access to data from both the Hubble Space Telescope (HST) and the European Southern Observatory (ESO).

The ESO Science Archive offers a number of on-line information services relevant to ESO users and to the community at large. In this paper, we will go through the different steps that enable a user to retrieve raw and processed data, from the query form to the actual processing. The concept of "Header On The Fly (HOTF)" that "repairs" possibly wrong data headers with meta-data from a database is also presented, together with the fact that calibrated data from ESO instruments will shortly be made available.

The emphasis in this paper will be on how the meta-data describing each ESO observation is kept in a relational database in instrument specific tables, and how this information is used to correct data upon distribution. The avenues for making ESO science products available to the Virtual Observatory (VO)[1] will also be explored.

1. Introduction

The European Southern Observatory (ESO) overseas two major observatories with state–of–the–art observing facilities located on Paranal and La Silla, Chile. Those observatories produce huge amounts of data which are stored in the ESO Archive. This paper describes the different on-line services that are provided to the ESO users and to the astronomical community at large, from proposal information to the distribution of data.

2. Data Distribution System

For about 13 years, ESO and HST archive data have been distributed to the astronomical community by the ESO Science Archive Facility. Requests of public data can be submitted through the web (see Sect. 3.) and are serviced using *Request Handler*. It is a set of software entities that processes asynchronous user requests for data stored on archive storage media. Request handling can be decomposed into a number of steps :

---

[1] `http://www.euro-vo.org`

- Submission of requests by users.
- Retrieval of the files from their archive support media to temporary data area identified by a request number. The currently supported archive data systems include CDs/DVDs, offline or in jukeboxes, the Next Generation Archive System (NGAS) and HTTP file access.
- Optionally, execution of a number of transformations on the files retrieved from storage.
- Sending a formatted message to the requestor notifying him/her of the service completion.

The Request Handler uses a database management system to continuously keep its status up-to-date. The core control database has about 5 different tables, and few more tables are need for the management of. The disk space, user and proprietary right management uses a few more. Database transaction security controls the parallelism of the task.

2.1. Submission of the Requests

Before retrieving data, one must register as a user under the menu "User On-Line Registration" of the main webpage of the SAF[2]. A complete set of information (like postal addresses, telephone number and password) is desirable in case media have to be sent to the user. At present we have 3585 users of the Archive. Once the registration is done, the user can request a collection of data from the Archive by providing his/her unique user_id and password.

The submission consists of entering the request, constituted in a body and a header, in some database tables. The body contains a selection of one or more data products from the ones available in the Archive. The header contains the user identification (user_id), together with site/archive specific options (on-the-fly recalibration, write to DVD, etc).

Once a request is submitted, it is handled by the first of the three main asynchronous components constituting the Request Handler: the File Retrieval.

2.2. File Retrieval (RH_GET)

Once entered in the database, the request can be processed. Its availability is "discovered" by a program called RH_GET which periodically looks up the database tables for files to be retrieved. In the case where all the data is on-line (NGAS, HTTP) or quasi on-line (DVDs in jukeboxes) or off-line, in which case the operator is prompted to mount the relevant volume in a suitable drive, the procedure can run unattended.

Once files are available for retrieval, the RH_GET program takes them one by one and tries to either mount the medium on which they are to be visible in an available reader (jukebox in case of DVDs), or to copy them directly from hard disks (NGAS) or from other machines on a remote site (HTTP). The status of the data changes and data are ready to be processed.

---

[2] `http://archive.eso.org`

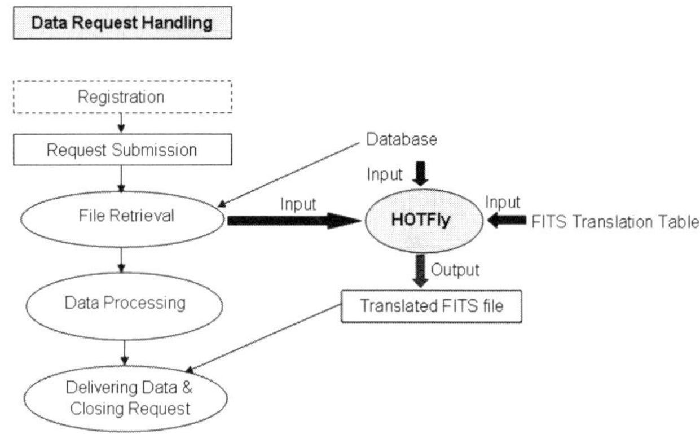

Figure 1. ESO Archive Data retrieval

2.3. Data Processing (RH_PROC and Header On The Fly)

Similarly to RH_GET, the program RH_PROC will periodically review the database queue for files/requests ready to be processed. Processing can be either a simple decompression, a sophisticated on-the-fly re-calibration or on-the-fly reconstruction of headers. A FITS file is composed of a header, containing observation meta-data, and a data part. Most important keywords from the FITS file header are stored in the relational database to keep track of the information related to the observation. The values of keywords stored in the database are sometimes updated, following the information obtained from the ESO Quality Control scientists or from the personnel from the observatories in Chile. The user may request either raw files, as observed at the telescope, or files updated with the most recent information. This is accomplished with the help of a tool called HOTFly (Headers On The Fly). This tool takes the original FITS file, searches for updates in the `observations` database, then generates a translated or corrected file on-the-fly and delivers it to the user.

The corrections are always applied to the keywords inside the header of the FITS file. The corrected values are stored in the database, which shall overwrite the erroneous ones in the original headers. In that sense was implemented a Data Base FITS Update and Correction Utility (DBFUCU). DBFUCU takes as input a configuration file, which is the main information source for carrying out the corrections from the Database to the original fits file.

A schematic view of the process is given in Fig. 1.

HOTFly can be used whenever ESO needs to distribute data to the users. HOTFly can be applied in two different ways: either through the web when the user wish to display the header of a file he selected, or by the Request Handler which retrieves files from the Archive when the user submits a request.

Once the files have been processed, their status change again together with the request status in the database and this is the signal for the Request Handler that the request is ready to be closed.

2.4. Request Closure (RH_CLOSE)

Once all pending files in the service queue have been successfully retrieved from their storage media and copied to the staging areas, and once all retrieved files have successfully undergone the various possible processing steps for a given request (which means the request and its files have the adequate status in the database), a request is eligible for closure.

Like for RH_GET and RH_PROC, the RH_CLOSE procedure periodically looks up the database for any request satisfying the above conditions. If such request is found, RH_CLOSE sends an acknowledgment message to the user. It updates the request status in the database to the final status, indicating that the request is completed and set the completion date to present.

3. User Web Interfaces to the ESO Archive

Requesting ESO Archive data is done through on-line query forms based on Web to database Interface written in Perl. This toolset provides an easy mechanism for crossing databases, by allowing data retrieved to be converted into hypertext links. In this section, we will review the major news and updates regarding the ESO Archive query forms.

3.1. Observing Program Information and Scheduling

Starting Oct 3, 2004 and following the decision of the Science and Technology Committee (STC), the abstracts of ESO proposals are publicly available once the proprietary period of the associated data is over. Access is enabled through the regular ESO Observation Schedule interface:

`http://archive.eso.org/wdb/wdb/eso/sched_rep_arc/form`

The result pages of this query form now include links to raw and, where available, science products associated with Observing Programmes. Also included are links to ESO data-based publications recorded in the ESO Telescope Bibliography database.

3.2. Public Datasets and Packages

Starting from the end of the Science Verification period, all data obtained with an instrument are public once the proprietary period is over. In addition, some data from the Commissioning and Science Verification phases of an instrument may be made public to the astronomical community. In such cases, those data are released through packages, as public datasets:

`http://archive.eso.org/archive/public_datasets.html`

ESO also provides advanced data products from surveys like EIS.

3.3. General and Instrument Specific Query Forms

The ESO Archive plans to offer instrument specific query forms for each instrument present in the Archive. The aim of an instrument specific form is to quickly retrieve information and identify frames related to the relevant instrument, based on specific keywords, like filter, grism, chip, etc. Some of them are already available:

FORS1	`http://archive.eso.org/wdb/wdb/eso/fors1/form`
FORS2	`http://archive.eso.org/wdb/wdb/eso/fors2/form`
UVES	`http://archive.eso.org/wdb/wdb/eso/uves/form`
ISAAC	`http://archive.eso.org/wdb/wdb/eso/isaac/form`

Those forms serve a complementary role to the generic observation form, which is based on principal information such as time or position:

VLT `http://archive.eso.org/wdb/wdb/eso/vlt-obs/form`

3.4. ESO Data-based Publication Query Form

This web interface links scientific papers based on ESO observations with underlying observing proposals and archival data. It is accessible from:

`http://archive.eso.org/wdb/wdb/library/publications/form`

It offers search options for bibliographic details of individual papers as well as for queries by observatory site, telescope, instrument and proposal ID. This ESO Telescope Bibliography can also be queried through a filter implemented at the ADS Abstract Service[3], with soon-to-come active links to the ESO Archive. These services are prerequisites to the astronomical Virtual Observatory as they allow the user to keep track of the entire lifetime of a scientific proposal: from scheduling, to observations and publications.

4. Toward a Science VO Compliant Archive

Reduced ESO Science Products from UVES Service Mode are currently added to the Archive on a regular basis as data packages. It will be extended to all instruments and Large Programs (Pirenne & Quinn 2004).

Internal FITS data integrity for HOTFly–1 concerns mainly generic keywords, thus a new release of HOTFly is on-going. Indeed, one of the difference between HOTFly–1 and HOTFly–2 will be the extension to all the instrument specific keywords, so the full instrument support.

Benefits of these initiatives (reduced data, high quality meta-data) are large for the future VO users of the Archive.

References

Pirenne, B., & Quinn P. 2004, The Messenger, 116, 48

[3]`http://esoads.eso.org`

ASTRONOMICAL DATA ANALYSIS SOFTWARE AND SYSTEMS XIV
ASP Conference Series, Vol. 347, 2005
P. L. Shopbell, M. C. Britton, and R. Ebert, eds.

The NOAO Data Cache Initiative - Building a Distributed Online Datastore

Rob Seaman, Irene Barg, and Nelson Zárate

National Optical Astronomy Observatory, Data Products Program, Tucson, AZ 85726

Chris Smith and Nelson Saavedra

National Optical Astronomy Observatory, Cerro Tololo Inter-American Observatory, La Serena, Chile

Abstract. The Data Cache Initiative (DCI) of the NOAO Data Products Program is a prototype Data Transport System for NOAO and affiliate facilities. DCI provides pre-tested solutions for conveying data from our large suite of instrumentation to a central mountain data cache. The heart of DCI is an extension of the Save-the-Bits safestore, running for more than a decade (more than 4 million images saved, comprising more than 40 Tbytes). The iSTB server has been simplified by the removal of STB's media handling functionality, and iSTB has been enhanced to remediate each incoming header with information from a database of NOAO instrumentation and an interface to the NOAO proposal database. Each mountain data cache has been implemented on commodity hardware running Redhat 9.0. Software RAID 1 runs over hardware RAID 5 to provide maximum storage reliability for each copy of the data. Each image is transferred from Kitt Peak or Cerro Tololo to the corresponding datastore at the Tucson or La Serena data centers using an rsync-based queue adopted from NCSA. From each data center, the files are transported to the other NOAO data center and also to NCSA for off-site storage using the Storage Resource Broker (SRB) of the San Diego Supercomputer Center. Thus we have three copies of each file on spinning disks or near-online. Major institutional users will be given access to the datastores.

1. Overview

The Data Cache Initiative (DCI) of the NOAO Data Products Program serves as a prototype Data Transport System for data originating from NOAO telescopes as well as from affiliate facilities such as WIYN, SMARTS, and SOAR (Figure 1). The intent is to provide simple pre-tested solutions to the challenges of conveying raw data products from our large and growing suite of instrumentation to a central online datastore. At the heart of the system is an extension of the Save-the-Bits safestore which has been running at NOAO for more than a decade. This new “iSTB” system benefits from the easy configuration and lightweight footprint that have always been the hallmark of the STB concept on the client data acquisition hosts. The iSTB server has been greatly simplified by the excision of STB's entensive tape and CD handling functionality. On the other hand, iSTB has been greatly enhanced with new functionality to remediate each incoming header with information from a newly compiled database of NOAO instrumentation and a new interface to the NOAO proposal database.

DCI has been commissioned with data from telescopes on Kitt Peak in Arizona, and from telescopes on Cerro Tololo and Cerro Pachon in Chile. Data from instruments at a particular site flow into a mountainwide data cache - one on Kitt Peak and one on Cerro Tololo that is shared by the SOAR telescope on Cerro Pachon via high-speed link. The iSTB server runs on the KP and CT data cache machines, receiving connections from two or three dozen supported instrument configurations at about a dozen supported telescopes. As each image is processed from the queue, keywords are added to each header ranging from the observing proposal ID to a public release date. A precise directory location within the random access datastore is generated given the telescope, date, proposal ID and so forth. The FITS Checksum is computed, the file is compressed, and then the MD5 checksum is computed for good measure. Each remediated and compressed file is transferred from the mountain cache to the corresponding datastore at the Tucson or La Serena data centers using an rsync-based queuing mechanism extended from one used at the National Center for Supercomputing Applications (NCSA). From each hemisphere's data center, the compressed images are transmitted across the equator to the other NOAO data center and also to NCSA for off-site storage using The Storage Resource Broker (SRB) of the San Diego Supercomputer Center. This provides us with three copies of each image on spinning disks or near-online systems. A limited set of institutional users will be allowed to access the datastores, providing a basic data distribution system for these users. DCI's underlying distributed data transport architecture will inform future decisions in the NOAO Science Archive. We expect many of the specific subsystems developed for DCI to be useful indefinitely.

2. Telescopes to Mountain Cache

As with other tape and CD based versions of Save-the-bits, iSTB relies on the BSD Unix line printer daemon, lpd, to provide queued network data transfers. This has been implemented under LPRng on Redhat servers. For each incoming file, the lpd daemon spawns a "bitf" filter that receives the contents of the file on its standard input and writes a validated and remediated copy to a unique location /cache/calendar date/telescope/proposal/site+serial.fits.gz within the datastore. Each resulting file is FITS checksum validated after being written to disk. After each file is compressed, its MD5 checksum is calculated and the pathname and checksum are passed to the script which will trigger the next step in the file transport. Keywords are added to each header such as proposal ID and principle investigator (see Figure 2).

As each file is processed, its header information is stripped off and is available for directly populating downstream databases. Archive ingest does not require reading the original files from the datastore. Data and header metadata are crossreferenced via a separately maintained index, allowing either datastore or metadata storage to be recast without affecting the other.

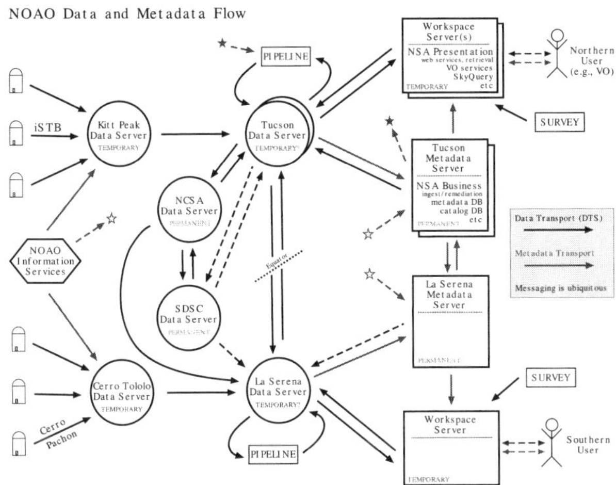

Figure 1. NOAO Data Flow. Proposed data and metadata flow for NOAO Science Archive and related facilities such as pipelines and data transport. The DCI project corresponds roughly to the left hand side of this diagram.

```
DTSITE  = 'kp                  '  /  observatory location
DTTELESC= 'kp4m                '  /  telescope identifier
DTINSTRU= 'flamingos           '  /  instrument identifier
DTCALDAT= '2004-10-18          '  /  calendar date from observing schedule
DTPUBDAT= '2006-04-19          '  /  calendar date of public release
DTOBSERV= 'NOAO                '  /  scheduling institution
DTPROPID= '2004B-0909          '  /  observing proposal ID
DTPI    = 'Nick Raines         '  /  Principal Investigator
DTPIAFFL= 'National Optical Astronomy Observatory'  /  PI affiliation
DTTITLE = 'Test and characterize new image slicer capability for FLAMINGOS IR s'
DTACQUIS= 'flmn-4m-1a          '  /  host name of data acquisition computer
DTACCOUN= '4mguest             '  /  observing account name
DTACQNAM= 'dark.sp.oct19.75s.8rd.0009.fits'  /  file name supplied at telescope
DTNSANAM= 'kp077017.fits       '  /  file name in NOAO Science Archive
DTCOPYRI= 'AURA                '  /  copyright holder of data
RECNO   =                77017  /  NOAO Science Archive sequence number
CHECKSUM= 'UJeAVGZ1UGd8UGZ8'     /  ASCII 1's complement checksum
DATASUM = '3132862946'           /  checksum of data records
```

Figure 2. Example Headers. Example keywords that are added to each image header. The data acquisition hostname specifies the telescope and instrument. The calendar date is generated from each headers DATE-OBS and other keywords using a heuristic with a noon pivot and knowledge of the timezone. The publication date is this date plus one day more than the proprietary period. The date and telescope are used to retrieve proposal and investigator information from the observatory schedule database. The RECNO is the primary key for locating a file and the associated metadata.

DCI Components and Deployment, Southern Data Flow

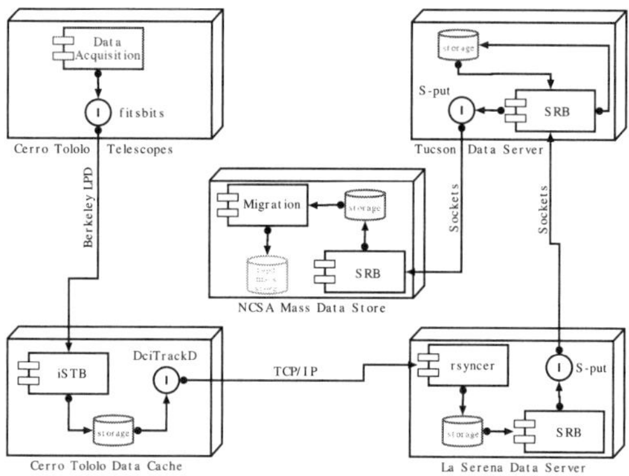

Figure 3. DCI Components - La Serena. Deployment diagram representing the flow of raw data products from Cerro Tololo (and soon, the SOAR telescope on Cerro Pachon). Note the balance between the use of COTS hardware and software (rsyncer from NCSA and SRB from SDSC) and in-house software (NOAO data acquisition/handling and iSTB).

3. Mountain to Downtown Segment of DCI

The transfer of files between the Mountain and Downtown Cache systems is accomplished with a set of Perl scripts contributed by David Mehringer at NCSA, and modified for NOAO use. The two core elements of this suite of code are briefly described as follows:

- **DciTrackD** - A Perl implementation of a TCP server daemon. It manages the list of files waiting to be transferred from the mountain.
- **rsyncer** - The data transferrer code wrapped around *rsync.*

iSTB sends file transfer requests to the mountain **DciTrackD** daemon. On the downtown system, **rsyncer** queries the **DciTrackD** daemon every 10 minutes, for a list of files waiting to be transfered, then transfers the files using *rsync.* Upon a successful transfer, the file is removed from the **DciTrackD** queue. The segment 2 code suite relationship to the other DCI segments (iSTB and SRB) is shown in Figure 3. At noon each day, another program called **xferStatus**, is run as a cron job. **xferStatus** provides a summary of transfers within the last 24 hours, and reports any errors.

4. Internationally and from NOAO to NCSA

To distribute the data from Tucson to La Serena and NCSA as well as from La Serena to Tucson and NCSA we use the Storage Resource Broker system from the San Diego Supercomputing Center. To transfer files from Tucson to La Serena the following steps are taken:

1. An input list files is continuously updated every 5 minutes from files coming from KPNO.
2. A Perl script reads this list and looks at new files being appended, which are then transfered to LS with an SRB command of the form:

   ```
   Sput -v -m -R 5 -S la_serena -P path input_file >& logfile
   ```

   The output from the command is captured on the logfile.
3. An entry is appended on a local list with the file just processed. On the next iteration of the script, both the input list and this local list are compared to look for new entries.
4. If no new entries are found, the script waits for 5 minutes and repeats.

The advantages of using SRB are the ability to break the output stream into several threads speeding the transmission of large files over a 4 Mbits per second network bandwith to La Serena; the handling of metadata information from the SBR database; a uniform interface to all the data in different nodes and many tools to access files and resource information.

5. Data Access

In addition to serving as a prototype for archive data transport and storage, a goal of DCI is to provide institutional users such as the SMARTS consortium and the SOAR and WIYN Observatories with simple password-protected access to their data. This is achieved by relying on the NOAO observing schedule/proposal database to characterize the telescope/calendar date ownership of each dataset. All data are owned by the user "cache", but group ownerships are assigned to each file corresponding to the appropriate institution. World read access is denied to all holdings of the DCI datastore.

References

Mehringer, D. 2001, "The BIMA Data Archive Real Time Transfer System", http://bimaarch.ncsa.uiuc.edu/RTT.html

San Diego Supercomputer Center - Storage Resource Broker, http://www.sdsc.edu/DICE/SRB

Seaman, Zarate, Lauer, and Warner 2003, in ASP Conf. Ser., Vol. 295, ADASS XII, ed. H. E. Payne, R. I. Jedrzejewski, & R. N. Hook (San Francisco: ASP), 100

Seaman, R. 2000, in ASP Conf. Ser., Vol. 216, ADASS IX, ed. N. Manset, C. Veillet, & D. Crabtree (San Francisco: ASP), 133

Seaman, R. 1997, in ASP Conf. Ser., Vol. 125, ADASS VI, ed. G. Hunt & H. E. Payne (San Francisco: ASP), 306

Seaman, R. 1995, in ASP Conf. Ser., Vol. 77, ADASS IV, ed. R. A. Shaw, H. E. Payne, & J. J. E. Hayes (San Francisco: ASP), 247

Seaman, R. 1994, in ASP Conf. Ser., Vol. 61, ADASS III, ed. D. R. Crabtree, R. J. Hanisch, & J. Barnes (San Francisco: ASP), 119

ASTRONOMICAL DATA ANALYSIS SOFTWARE AND SYSTEMS XIV
ASP Conference Series, Vol. 347, 2005
P. L. Shopbell, M. C. Britton, and R. Ebert, eds.

New Features for VO-Enabled Data Intensive Science with the SDSS Data Release 3

Ani R. Thakar, Alexander S. Szalay, William O'Mullane, Tamás Budavári, María A. Nieto-Santisteban, George Fekete, Nolan Li, and Sam Carliles

Center for Astrophysical Sciences, JHU, thakar@pha.jhu.edu

Jim Gray

Microsoft Research

Robert Lupton

Department of Astrophysical Sciences, Princeton University

Abstract. We describe new and enhanced features for data access with the third data release (DR3) of the SDSS, particularly in the context of data intensive science and the VO. These include several enhancements to the CasJobs batch query workbench system, improved JPEG color images in the Visual Tools, extensive usage and traffic logging with harvesting of multiple remote site logs, enhancements to the HTM spatial library and a more versatile object crossid facility. We briefly describe these features and list future enhancements anticipated with SQLServer Yukon.

1. Introduction

With the third data release (DR3) of the Sloan Digital Sky Survey (Abazajian et al. 2005), the SDSS Catalog Archive Server (CAS) serves up 3 Terabytes of catalog data for nearly 142 million celestial objects via the SkyServer web pages[1]. Several changes and new features in DR3 facilitate data-intensive science and advance VO standards and technologies.

2. CasJobs and MyDB - A SQL Workbench

The CasJobs batch query system was introduced at ADASS XIII (O'Mullane et al. 2003). CasJobs provides a queue-based SQL query interface and a personal workspace database called MyDB. New features for DR3 include:

1. The ability to create SQL functions and stored procedures in the user's workspace **MyDB**. This enables users to *bring their program to the data* rather than the other way around, which is far more efficient for data-intensive applications. Users can now perform complex tasks (see MaxBCG, P1.1.16) within CasJobs and share results with collaborators.

[1]`http://skyserver.sdss.org/`

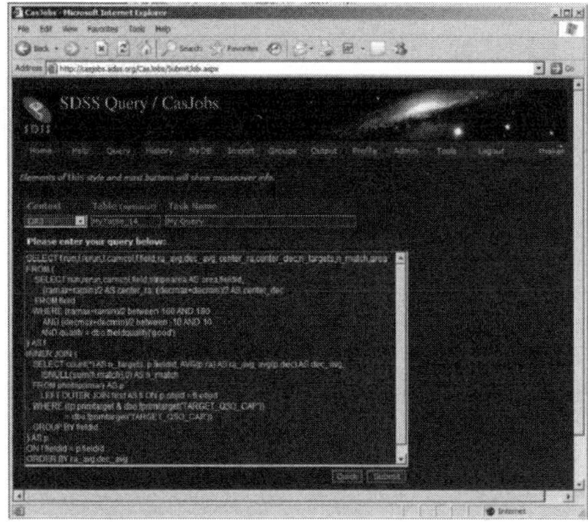

Figure 1. CasJobs query input screen.

2. A simplified queue system, with only 2 queues:
   - a quick (one minute) queue, and
   - a submit (long, currently 8 hours) queue.
3. We replicate user account details between installations to allow distributed CasJobs configurations. This form of replication raises issues that are crucial to the VO community. We are also attempting to secure the system using X509 certificates through WS-Security. This ties in with the distributed storage effort within the VO.
4. Prefix syntax has been both simplified and made more powerful. You can now use any context name as a prefix; in any situation that is appropriate in normal SQL. This allows users to reference many different catalogs in a single query.
5. User guide has been extensively updated and reflects these changes
6. For the user determined to extract massive amounts of data in a single scipt, a Java **command-line tool** has been made available.
7. Import form has been updated to accept more data formats and includes file upload as well as cut-and-paste.
8. "Table" field on submit page will automatically insert results into a table, without users having to write a "select into" clause.

Figure 1 shows the query input screen in CasJobs, and Figure 2 is a plot of CasJobs monthly usage statistics logged in the logging system (see below).

3. VO-Enabled Mirage

We have enhanced the Mirage Java visualization application developed at Bell Labs (Ho) to make it VO-compatible. This was presented last year at ADASS (Carliles et al. 2003). New features added since then include:

1. The imaging module has been separated out and written in pure Java so that JAI is no longer required for the application to run. The module can now be easily incorporated into other applications.

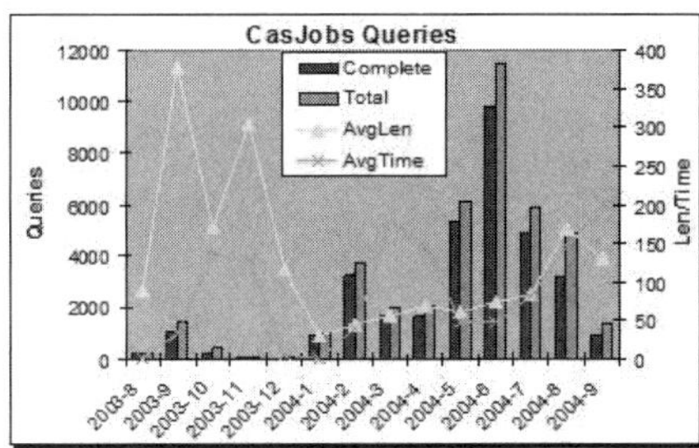

Figure 2. CasJobs usage from logging system.

2. Integration of the new IVOA FITS Package into the new astronomical imaging module, and in turn into Mirage for FITS viewing. The IVOA FITS Package uses nom.tam.fits to load image data from a FITS file, then creates a Java BufferedImage object so that any Java app can display the image. The imaging module also uses the asinh scale described below for JPEG images.
3. Integration of the IVOA Client Package to provide IVOA Registry access and subsequent access to registered Cone and SIAP servers, with data retrieval into Mirage. Then the retrieved VOTable data is parsed using JAVOT or SAVOT, and fed to Mirage in its native data format (this is all done seamlessly without the user noticing).

4. Enhanced Color JPEGs

Traditionally, in scaling image data for color images, log stretches or even linear or square-root stretches are used, but CCDs tend to saturate at the high end of the scale. The **asinh** stretch (Lupton et al. 2004) is particularly suited to CCD images. The asinh method is two pronged:

- it allows for a linear-to-log scale switch, which allows for faint objects to show up better;
- saturated objects are found and filled in with an appropriate surrounding color to prevent all bright objects looking simply white.

The new images are deeper and more feature-rich. This scaling is used both in SkyServer and in the Mirage package[2].

5. Faster HTM

Several speed and interface enhancements have been applied to the Hierarchical Triangular Mesh spatial indexing library.

- The HTM2 Spatial Toolkit (Fekete et al. 2003) has been rewritten under the .NET framework in C#. This new reference platform has built-in

[2]`http://voservices.org/`

regression and unit testing harnesses, and is the basis for other language bindings, namely C++ and Java.
- New releases of the development tools that are 64-bit ready provide even further speed-ups for HTM lookup and cover operations.
- Topological operations, such as neighbor finding and morphologicaloperations, such as trixel region growing is now performed by pure symbolic means on the HTM addresses directly.

With SQL Server 2005 (Yukon), C# code can be written right in the database just like SQL, avoiding the performance penalty due to the C++-to-SQL interface (DLL). We are currently incorporating a C# version of HTM into Yukon.

6. Object CrossID

The imaging CrossID page in the DR3 SkyServer (Tools→CrossID→Upload) lets you search for SDSS matches to your own list of positions within a given search radius. It has been enhanced for DR3 in the following ways:

- You can now combine your crossid list with a SQL query. This lets you customize your output as well as modify your search to include other criteria, including searching for spectra matching your imaging objects.
- You can prepend your own columns to the position columns so that they are included in the resulting output.
- You can search for primary as well as non-primary objects
- You can also search for objects matching the 5-part SDSS ID (run-rerun-camcol-field-object).

The spectro CrossID page (Tools→CrossID→Speclist), already has the option to include your own query with the spectro search.

7. Usage and Traffic Logging

In order to monitor and maintain performance for data-intensive science, it is critical that the system usage and traffic be extensively logged and regularly analyzed. The SkyServer has had such a traffic logging system since its inception, and we now harvest logs from multiple mirror sites at JHU. There are several benefits of such a system:

- Usage/traffic profiles great management resource:
    - Invaluable for funding proposals and reviews
    - Monitor impact of press articles
- Use data for load balancing:
    - Find crawlers and "inconsiderate" users
    - Track server performance
- Guide schema and interface design:
    - How often are SDSS photo flags being used?
    - Are people getting clean photometry?
    - Are they filtering out invalid data values?
    - How comfortable are people with SQL/complex queries?
- Help prospective mirror sites to budget hardware

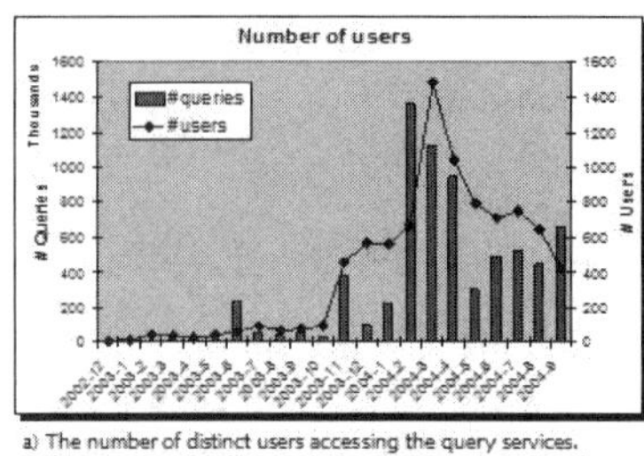

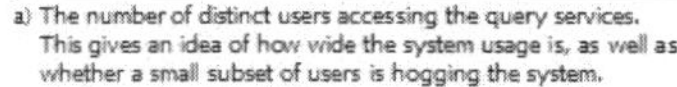
a) The number of distinct users accessing the query services. This gives an idea of how wide the system usage is, as well as whether a small subset of users is hogging the system.

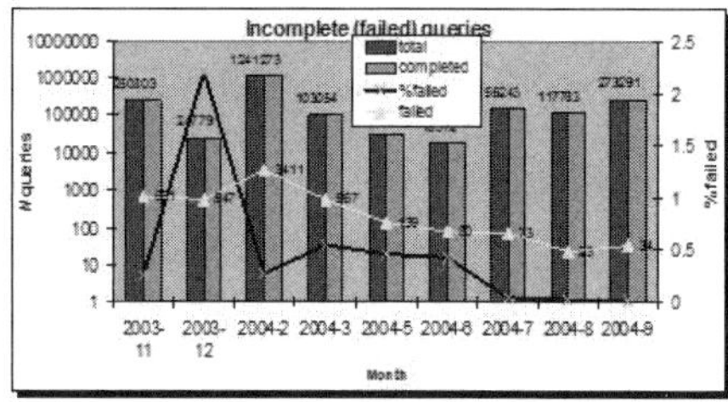

b) Analyzing error and failure rates is a very important application of the logging system. Here we count the queries that did not complete in real numbers as well as a percentage of the total number of queries submitted per month.

Figure 3. Examples of logging data analysis.

Figure 3 shows examples of the kind of data that can be obtained from the logs. The Support Services specification proposed for IVOA includes a logging section modeled on this approach. We are working with the VO community to develop and finalize the logging data model.

8. Future of Data Intensive Science with SDSS

As mentioned above, the next release of SQL Server (SQL Server 2005 or Yukon) will enable us to have a much faster version of HTM. There are other performance enhancements with Yukon that we are looking forward to, such as fatter indices (currently limited to 16 columns) and considerably enhanced query optimization (including elimination of the bookmark lookup problem).

We have also begun to investigate partitioning of the SDSS data across a cluster of database nodes using a zone-based partitioning approach. This will provide fast parallel access to the data for data intensive applications like all-sky cross-matches between SDSS and 2MASS, for example. We are developing a parallel cross-match engine that will enavle such cross-matches with our OpenSkyQuery federated query service. For the longer term, we are deploying a 100-TB "data-wulf" cluster at JHU for the analysis of multi-TB turbulence simulation datasets, as part of a data intensive science laboratory for the analysis of prodigious amounts of data expected from future archives like LSST.

References

Abazajian, K. et al. 2005, AJ, in press

Carliles, S., Ho, T. K., & O'Mullane, W. 2004, in ASP Conf. Ser., Vol. 314, ADASS XIII, ed. F. Ochsenbein, M. Allen, & D. Egret (San Francisco: ASP), 300

Fekete, G., Szalay, A., & Gray, J. 2003, in ASP Conf. Ser., Vol. 314, ADASS XIII, ed. F. Ochsenbein, M. Allen, & D. Egret (San Francisco: ASP), 289

Lupton, R. H., Blanton M. R., Fekete, G., Hogg, D., O'Mullane, W., Szalay, A.S., & Wherry, N. 2004, PASP, 116, 133

O'Mullane, W., Gray, J., Li, N., Budavári, T., Nieto-Santisteban, M. A., & Szalay, A. 2004, in ASP Conf. Ser., Vol. 314, ADASS XIII, ed. F. Ochsenbein, M. Allen, & D. Egret (San Francisco: ASP), 372

ASTRONOMICAL DATA ANALYSIS SOFTWARE AND SYSTEMS XIV
ASP Conference Series, Vol. 347, 2005
P. L. Shopbell, M. C. Britton, and R. Ebert, eds.

A Conceptual Domain Model for the NOAO Science Archive

Phillip Warner, Rafael Hiriart, Frank Valdes, Tod Lauer, and Sean Points

National Optical Astronomy Observatory, 950 North Cherry Ave., Tucson, AZ, 85719

Abstract. A domain model for the NSA has been developed using classes and concepts devised by Valdes (2000). The NSA domain model incorporates and generalizes these classes to provide a generic model for data acquisition. Other classes representing data products derived by pipelines, e.g., calibration, reduction, and analysis pipelines, are also included. Concepts modeled by the generalization of these classes include collections of elements involved in data acquisition, collections of element states (where relevant), high-level data classification, and internal and external resources.

1. Introduction

The NOAO Science Archive (NSA) team is working on a design for the next generation science archive. The goal is to expand the current NSA data holdings[1] (Seaman et al. 2003) by providing access to all raw data produced by observatories that are under the direction of NOAO and partner organizations, and to NOAO Pipeline-processed data products.

The NSA Domain Model, represented in part by the diagrams below, captures many of the elements from the observation and data processing domains. Each element and its relationship to other elements assist in defining the appropriate software components that will be used in the NSA. A few generalizations have been made that lead the effort into the development of a design model. This effort is a work-in-progress, i.e., not all elements of the domain have yet been included or discovered.

For further information, the full model can be retrieved from the IVOA Wiki[2].

2. Domain Model

The Domain model, which represents portions of the astronomical domain, as well as the domain of the NSA, includes three sections, i.e., the Observational domain, the Data Product domain, and a preliminary view of the Pipeline domain. Elements of this domain are used in defining the metadata elements that may be implemented in the NSA system.

[1] `http://archive.noao.edu/nsa/holdings.html`

[2] `http://ivoa.net/internal/IVOA/IvoaDataModel/NSA-DM-v0.4.0.pdf`

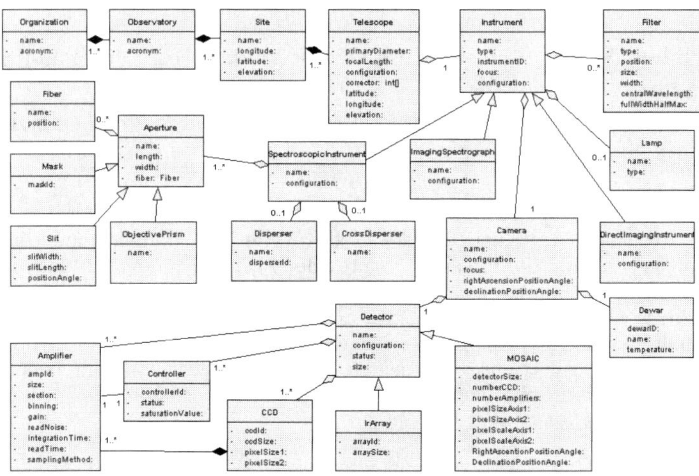

Figure 1. Observational Domain Model

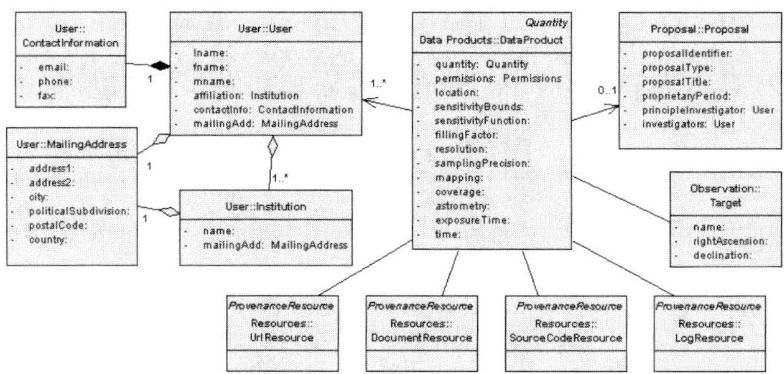

Figure 2. Domain Model - Data Product

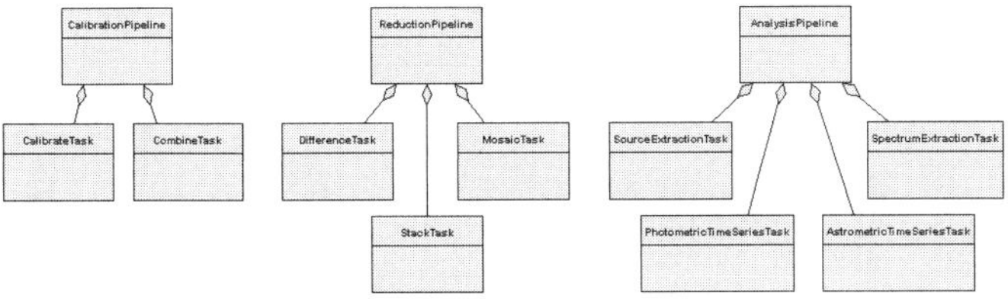

Figure 3. Domain Model - Pipelines

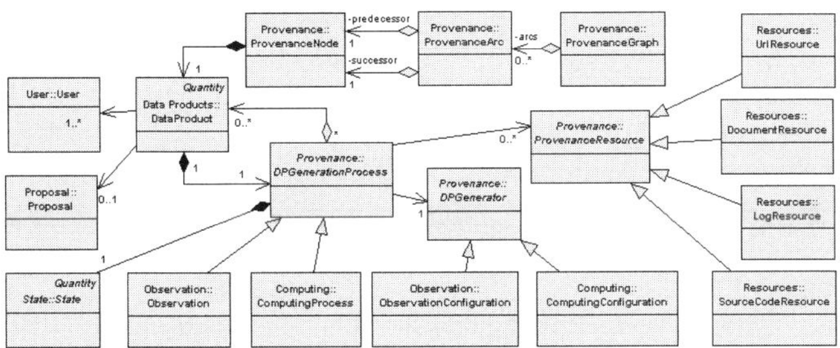

Figure 4. Provenance Model

Figure 1 shows a view of the Observational domain for optical astronomy. The observational model was derived from the classes and concepts devised by Valdes (2000). Although the relationships between observation elements can be complex, they will likely not hold in all cases. Hence, the capture of metadata from these elements, in the implementation of the NSA data model, is handled by a composite object, as seen below (§3.).

The Data Product model (Figure 2) includes the deliverable that is visible by the NSA user. Classes associated with the `DataProduct` class provide insight and references to information such as why, and by whom, the `DataProduct` was generated.

Pipeline facilities (Figure 3), include calibration, reduction, and analysis processes. Although elements of the Pipeline domain are currently fairly high level, they are intended to capture information about the generation of data products from raw, calibrated, or reduced data products, such as mosaics, stacks, time series, and catalogs.

3. Provenance

Provenance elements, including generalizations of domain elements, are shown in Figure 4. These elements represent metadata capture associated with the generation of a `DataProduct`. The high-level elements that were derived are: `DPGenerationProcess`, `DPGenerator`, `ProvenanceResource`, and `State`.

The top-level element in this model is the `DataProduct`, which aggregates the `DPGenerationProcess`. More specifically, the elements containing the metadata produced by the domain elements that generate the `DataProduct`, are the

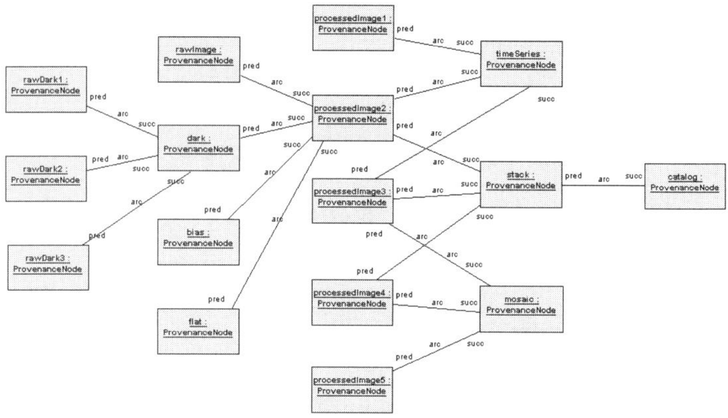

Figure 5. Provenance Model Example

`Observation` and the `ComputingProcess`. These elements are also responsible for containing references to other elements of the domain, which are associated with the `DataProduct`, i.e., the `DPGenerator` classes. Although the relationships between the domain elements are defined in Figures 1, 2, and 3, a `DPGenerator` should only capture the elements as, e.g., a list; hence the `DPGenerator` contains these elements as a composite class, remaining a simple as possible.

`ProvenanceResource` contains information associated with the data acquisition, such as logs, references, source code, manuals, etc. `State`, also a composite class (see §5.), captures status, configuration, and other dynamic metadata coming from the various elements that emit state information. Other elements of the provenance model include three classes that represent a historical view of a given `DataProduct`, i.e., `ProvenanceGraph`, `ProvenanceNode`, and `ProvenanceArc`. As a side note, as of publication these three elements are superfluous, as the provenance is implicit in the relationship between the `DataProduct` and the `DPGenerationProcess`; these element remain in the model to provide a basis for the the example described in the next section.

4. Provenance Example

An example of the provenance hierarchy is shown in Figure 5. Each `DataProduct` is a node in the graph, where a given `DPGenerationProcess` is represented by the association (the `ProvenanceArc`, or "arc") between each `ProvenanceNode`. The graph is read from left to right, where the text "pred" indicates the predecessor end of the arc, while "succ" indicates the successor end of the arc. Objects that are not associated with a parent object represent raw data (provided by the acquisition system, i.e., from an observation). Objects associated with parent objects are products that were derived from a processing system, e.g., a pipeline. Possible operations of a pipeline are calibration; reduction, i.e., post-calibration processing; and analysis, which produces, e.g., image stacks and mosaics, time series, and catalogs.

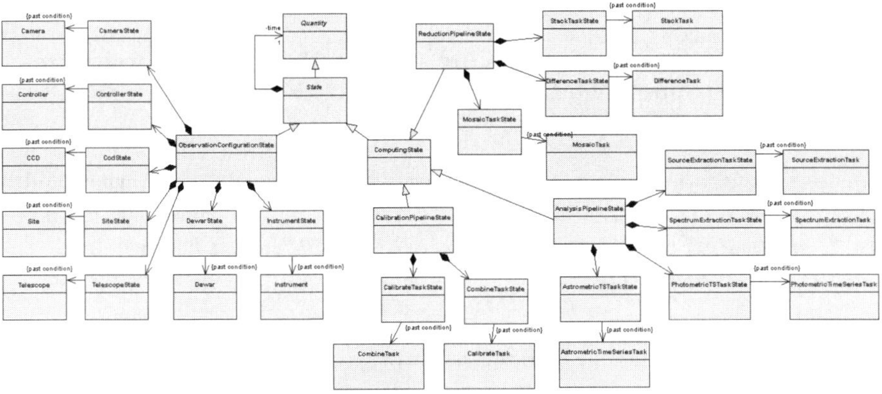

Figure 6. State Model

5. States

The State model (Figure 6) represents the state of various domain elements, i.e., state information. The state classes contain only dynamic state information and a reference to the appropriate class that represents the domain element for which the state information was generated. The top-level element, Quantity, is a class, or set of classes, as devised by the Data Model[3] group of the IVOA; hence, a State is a Quantity, and also contains other States (or Quantities), as a composite class. The decomposition of state classes may or may not be part of the implementation; further analysis is required to understand the common attributes and associations.

Acknowledgments. The National Optical Astronomy Observatory (NOAO) is operated by the Association of Universities for Research in Astronomy, Inc. (AURA) under cooperative agreement with the National Science Foundation.

References

Seaman, R., Zárate, N., Lauer, T., & Warner, P. B. 2003, in ASP Conf. Ser., Vol. 295, ADASS XII, ed. H. E. Payne, R. I. Jedrzejewski, & R. N. Hook (San Francisco: ASP), 100

Valdes, F., *Classes Describing Astronomical Observations*, available at `http://www.noao.edu/projects/ccdmosaic/imagedef/classes.html`

[3]`http://ivoa.net/twiki/bin/view/IVOA/IvoaDataModel`

ASTRONOMICAL DATA ANALYSIS SOFTWARE AND SYSTEMS XIV
ASP Conference Series, Vol. 347, 2005
P. L. Shopbell, M. C. Britton, and R. Ebert, eds.

A Case Study of Applying Object-Relational Persistence in Astronomy Data Archiving

Shilong Stanley Yao, Rafael Hiriart,[1] Irene Barg, Phillip Warner, and Dave Gasson

Data Products Program, NOAO/AURA, 950 North Cherry Ave., Tucson, AZ 85719

Abstract.
The NOAO Science Archive (NSA) team is developing a comprehensive domain model to capture the science data in the archive. Java and an object model derived from the domain model weil address the application layer of the archive system. However, since RDBMS is the best proven technology for data management, the challenge is the paradigm mismatch between the object and the relational models.

Transparent object-relational mapping (ORM) persistence is a successful solution to this challenge. In the data modeling and persistence implementation of NSA, we are using Hibernate, a well-accepted ORM tool, to bridge the object model in the business tier and the relational model in the database tier. Thus, the database is isolated from the Java application. The application queries directly on objects using a DBMS-independent object-oriented query API, which frees the application developers from the low level JDBC and SQL so that they can focus on the domain logic. We present the detailed design of the NSA R3 (Release 3) data model and object-relational persistence, including mapping, retrieving and caching. Persistence layer optimization and performance tuning will be analyzed. The system is being built on J2EE, so the integration of Hibernate into the EJB container and the transaction management are also explored.

1. Introduction

Astronomy data archiving manipulates large volumes of semantic-rich data. Meta-data, as the portal to the massive data products, is of special interests, because it contains numerous science parameters describing a large number of domain objects with complex structures and interactions. An effective modeling of meta-data is essential to the archive design. The NSA (NOAO Science Archive; Seaman et al. 2003) is developing a comprehensive domain model to better capture the science data. Therefore a powerful server side development platform is desired to realize this complex model.

The NSA release 2 was successfully developed majorly based on MySQL and PHP. As the domain model evolves, this platform is no longer effective enough due to its limit on inter-operation, integration and scalability. Application servers emerged recently to provide standardized, scalable and flexible platforms to develop and deploy enterprise systems. J2EE is among the pop-

[1]NOAO/CTIO, Casilla 603, La Serena, Chile

Figure 1. Hibernate in the Big Picture

ular and proved standards for the application servers. It is backed with Java technology and provides cross-platform, easy integration across multiple tiers, and high scalability. As a standard, it also guarantees the compatibility across J2EE compliant application servers. Due to the various benefits, for the next generation NSA R3, we will migrate onto the J2EE technology and Java (e.g., Servlet, JSP, EJB) would be the major server-side programming language.

J2EE enterprise applications normally consist of 4 tiers, namely client, presentation, business and EIS (enterprise information systems, or database). RDBMS (Relational Database Management System) is the best proved technology for EIS. However, there is a major challenge in this architecture, which is the paradigm mis-match between the Java object oriented model and the RDBMS relational model. Many mapping solutions have been developed to sew these two pieces together, such as Hibernate, Entity Beans, JDO. After a fully comparison in terms of robustness, maturity, features and performance, we chose Hibernate to be our ORM solution.

2. Architecture

Figure 1 shows how the Hibernate fits in the overall architecture of NSA. Hibernate locates between the business and database tiers of the application and abstracts the database details to provides the business tier with a common interface to persistent objects, connection pool, transactions, and so on, no matter which specific underlying DBMS it is and if the system is running in a managed container on some application server or directly through JDBC.

3. Relational Model

According to the object-oriented development process, the NSA team produced an object model derived from the domain model. This is usually a fine-grained model. It is trivial to do a one-to-one mapping between classes in the object model and tables in the RDBMS, but it will bring poor performance. For a better performance, we did normalization and de-normalization on the relational model. Generally, we try to keep the tables at 3NF (3rd Normal Form). However, when business keys are not obvious, we use auto-generated surrogate keys; some fine-grained per-object tables are merged. These downgrade some tables from 3NF, but improve the performance.

4. Object-Relational Mapping

4.1. Mapping

Mapping between objects and tables is the core part of the O/R persistence. We will address a few most important issues here. Inheritance hierarchy mapping is important to support efficient polymorphic query on the objects. We have two major mapping strategies depending on different situations. When subclasses tend to have more common attributes, we use table-per-hierarchy mapping, where the whole class hierarchy is mapped onto a single table, with a special column to discriminate different classes in the hierarchy. With this mapping, no join is required during a polymorphic query. When subclasses tend to have more uncommon attributes, table-per-hierarchy won't work well due to too many null values in the table. Instead, we use table-per-subclass mapping, in which each subclass has a separate table storing attributes only owned by itself. Common attributes inherited by subclasses, are mapped to the super class's table. This mapping requires table joins during queries, but also reduces the number of null values, which actually improves the overall performance.

Hibernate provides mapping of most conventional Java types, like primitive types, wrapper classes (e.g., Long), etc. However, astronomy data has many flexible types like Dimension, vectors, arrays. So mapping non-conventional java types efficiently is critical. A simple default solution is to map those types to binary data type in RDBMS (e.g., Blob). But this will create difficulty to the query performance, because WHERE clause involving any part of the blob require reading the whole binary block. We use Hibernate API for customized persistent user types instead, which can map each part of a complex data type to a separate column in the table. This greatly optimizes the database operations.

In the object model, some small objects could form a logical group, for example, astronomer's home address, contact (e.g., email, phone), and NSA account (e.g., ID, role). Having one table for each of the 3 objects and joining them during the query is inefficient. In the reality, there will be very few astronomers sharing the same home address, phone number, or account ID. So mapping all three objects into a single table would reduce the number of tables and the cost of joins. We do this by mapping objects as components.

Other issues like mapping of class associations, auto-generated object ID, etc. are implemented in standard ways with hints from the object and relational model, which we won't discuss in detail due to space limit.

4.2. Retrieving

The object model really presents us a "graph", in which classes are nodes and the associations are edges. Querying is done on this graph, while the database tables are joined and rows are transformed into resulting objects. Hibernate does this by issuing SQL statements to the database. So the key issue of database performance is to minimize the SQL round-trips and the data bandwidth for each query. And this is the challenge of object retrieval.

Hibernate provides "lazy" retrieval, which means that objects are not retrieved until the application actually visits them. For example, an astronomy observation object has a long list of facility objects as its members describing the details of the telescope, camera, filter and so on, with which the observation was taken. If the application is answering a user's question regarding to the date of the observation, it doesn't make sense to load all the member facility objects. With "lazy" retrieval, this overhead is eliminated.

But in some cases, we need "eager", the opposite of "lazy". Continuing the above example, if a user want to see the full details of the facility configuration of the observation, it's better to join the tables of all facility objects and the observation table and finish the query with a single SQL statement. So in the NSA, we configure the mapping to use "lazy" retrieval as the default and explicitly override it with "eager join" on some queries, which we are sure all members will get loaded anyway.

4.3. Caching and Concurrency Control

Hot-spot data is frequently accessed. For example, some telescopes are extensively used by many recent observations. It is best to cache them in the memory rather than loading them every time they are referenced. Hibernate provides two levels of such cache with different scopes, namely level-1 per-session cache and level-2 per-process cache. Level-1 cache is required, while level-2 is optional for caching process-wide frequently read but less frequently modified objects. We specify objects, like Institute, Permission, to be level-2 cached.

Hibernate caching feature provides more flexible concurrency control mechanism. Hibernate provides optimistic locking (based on versioning and level-1 caches) to support "repeatable reads" isolation. This improves performance in most cases in the NSA, where concurrent read-write and write-write conflicts tend to be rare. Versioning info (a sequence number) is a special attribute added to the object. In this case, more expensive DBMS level pessimistic locks are not acquired. Instead, the version number is incremented after each modification and during committing the version number of the object is checked against the old version number when we read it in. If no other transaction modified this object on our back, the two version numbers should match and we could safely commit. Otherwise, we abort.

Although optimistic locking works well for us most of the time, there are cases where conflicts are highly possible, and we still want to use the DBMS pessimistic locking. Multiple instances of running ingestion services is an example. In this case, we explicitly specify, during the queries, that Hibernate locking is bypassed and a certain type of pessimistic lock is acquired at the DBMS level.

4.4. Transaction and Session

In concept, we have two kinds of transactions. Application transaction is a coarse-grained unit of work from the point of view of the users. Database transaction is the transaction from the RDBMS's point of view. Nested transactions are not well supported by EJB, Hibernate or JTA, so we use flat CMT (Container Managed Transaction). Hibernate supports CMT, because its transaction API is only an abstraction of the underlying JTA. In case of session bean calling session beans in the same transaction, the callee joins the caller's transaction instead of starting a new one.

Hibernate session dictates the level-1 cache and object identity scope. For a long application transaction (e.g., astronomers read some data, modify it, and store it), we use session-per-request with detached objects. This reduces resources tied up during the modification and minimize the stale data.

5. Implementation

We build the NSA on JBoss application server. Hibernate is used at the ORM layer. With the aid of ANT build tool and xDoclet code generation engine, most code and script generation is automated. Hibernate mapping files, which specify the details of ORM, are generated from the tags we embed in the Java source code. Although, Hibernate API is already very concise, we built utility wrappers around it to further simplify the persistence code in the business tier. Hibernate resources, including library files and mapping files, are automatically packaged into a service deployment unit for the JBoss server, which is utilized by the application through JNDI.

6. Performance Tuning

For retrieving, we will adjust the lazy fetching and the global maximum depth of eager fetching. Both consulting with the astronomers and collecting production system performance hints will be used to study the data access pattern, such as which ones among the many observation parameters are most interesting. The less frequent accessed data will be tagged for lazy fetching.

In a polymorphic query, the number of SQL interactions or the number of tables joined is roughly a linear function of the hierarchy depth. So the shallower hierarchy depth is more desired. For a good balance of the performance and soundness of the object model, an iterative approach should help, which suggests a revisit to the object modeling with the hints from Hibernate implementation.

The level-1 cache of the Hibernate provides opportunities for optimizations. Besides the optimistic locking, immutable objects is another example. Immutable objects will never be flushed. Facility objects (e.g., filter, CCD) are good candidates of immutable objects. From the data access pattern, we will adjust the list of immutable objects. However, another tradeoff related to this is the performance improvement and the portability of immutable objects. We learned that not all ORM middle-ware provide this facility.

References

Seaman, R., Zarate, N., Lauer, T., & Warner, P. B. 2003, in ASP Conf. Ser., Vol. 295, ADASS XII, ed. H. E. Payne, R. I. Jedrzejewski, & R. N. Hook (San Francisco: ASP), 100

Author Index

Subject Index

ASTRONOMICAL SOCIETY OF THE PACIFIC
(ASP)

An international, nonprofit, scientific and educational organization founded in 1889, established the

ASP CONFERENCE SERIES

in 1988, to publish books on recent developments in astronomy and astrophysics.

A LIST OF VOLUMES PUBLISHED SINCE 2000 ARE LISTED NEXT

For additional volumes available
please see our web site at:
http://www.astrosociety.org

For electronic versions of volumes available in color
please see our e-book site at:
http://www.aspbooks.org

ASP CONFERENCE SERIES VOLUMES

Published by the Astronomical Society of the Pacific

PUBLISHED: 2000 (* asterisk means OUT OF PRINT)

Vol. CS-195 IMAGING THE UNIVERSE IN THREE DIMENSIONS:
Astrophysics with Advanced Multi-Wavelength Imaging Devices
eds. W. van Breugel and J. Bland-Hawthorn
ISBN 1-58381-022-6

Vol. CS-196 THERMAL EMISSION SPECTROSCOPY AND ANALYSIS OF DUST, DISKS, AND REGOLITHS
eds. Michael L. Sitko, Ann L. Sprague and David K. Lynch
ISBN: 1-58381-023-4

Vol. CS-197 XVth IAP MEETING DYNAMICS OF GALAXIES:
FROM THE EARLY UNIVERSE TO THE PRESENT
eds. F. Combes, G. A. Mamon and V. Charmandaris
ISBN: 1-58381-24-2

Vol. CS-198 EUROCONFERENCE ON “STELLAR CLUSTERS AND ASSOCIATIONS: CONVECTION, ROTATION, AND DYNAMOS”
eds. R. Pallavicini, G. Micela and S. Sciortino
ISBN: 1-58381-25-0

Vol. CS-199 ASYMMETRICAL PLANETARY NEBULAE II:
FROM ORIGINS TO MICROSTRUCTURES
eds. J. H. Kastner, N. Soker and S. Rappaport
ISBN: 1-58381-026-9

Vol. CS-200 CLUSTERING AT HIGH REDSHIFT
eds. A. Mazure, O. Le Fèvre and V. Le Brun
ISBN: 1-58381-027-7

Vol. CS-201 COSMIC FLOWS 1999: TOWARDS AN UNDERSTANDING OF LARGE-SCALE STRUCTURES
eds. Stéphane Courteau, Michael A. Strauss and Jeffrey A. Willick
ISBN: 1-58381-028-5

Vol. CS-202 * PULSAR ASTRONOMY – 2000 AND BEYOND, IAU Colloquium 177
eds. M. Kramer, N. Wex and R. Wielebinski
ISBN: 1-58381-029-3

Vol. CS-203 THE IMPACT OF LARGE-SCALE SURVEYS ON PULSATING STAR RESEARCH, IAU Colloquium 176
eds. L. Szabados and D. W. Kurtz
ISBN: 1-58381-030-7

Vol. CS-204 THERMAL AND IONIZATION ASPECTS OF FLOWS FROM HOT STARS: OBSERVATIONS AND THEORY
eds. Henny J. G. L. M. Lamers and Arved Sapar
ISBN: 1-58381-031-5

Vol. CS-205 THE LAST TOTAL SOLAR ECLIPSE OF THE MILLENNIUM IN TURKEY
eds. W. C. Livingston and A. Özgüç
ISBN: 1-58381-032-3

Vol. CS-206 HIGH ENERGY SOLAR PHYSICS – *ANTICIPATING HESSI*
eds. Reuven Ramaty and Natalie Mandzhavidze
ISBN: 1-58381-033-1

Vol. CS-207 NGST SCIENCE AND TECHNOLOGY EXPOSITION
eds. Eric P. Smith and Knox S. Long
ISBN: 1-58381-036-6

ASP CONFERENCE SERIES VOLUMES

Published by the Astronomical Society of the Pacific

PUBLISHED: 2000 (* asterisk means OUT OF PRINT)

ATLAS	VISIBLE AND NEAR INFRARED ATLAS OF THE ARCTURUS SPECTRUM 3727–9300 Å eds. Kenneth Hinkle, Lloyd Wallace, Jeff Valenti and Dianne Harmer ISBN: 1-58381-037-4
Vol. CS-208	POLAR MOTION: HISTORICAL AND SCIENTIFIC PROBLEMS, IAU Colloquium 178 eds. Steven Dick, Dennis McCarthy and Brian Luzum ISBN: 1-58381-039-0
Vol. CS-209	SMALL GALAXY GROUPS, IAU Colloquium 174 eds. Mauri J. Valtonen and Chris Flynn ISBN: 1-58381-040-4
Vol. CS-210	DELTA SCUTI AND RELATED STARS: Reference Handbook and Proceedings of the 6th Vienna Workshop in Astrophysics eds. Michel Breger and Michael Houston Montgomery ISBN: 1-58381-043-9
Vol. CS-211	MASSIVE STELLAR CLUSTERS eds. Ariane Lançon and Christian M. Boily ISBN: 1-58381-042-0
Vol. CS-212	FROM GIANT PLANETS TO COOL STARS eds. Caitlin A. Griffith and Mark S. Marley ISBN: 1-58381-041-2
Vol. CS-213	BIOASTRONOMY `99: A NEW ERA IN BIOASTRONOMY eds. Guillermo A. Lemarchand and Karen J. Meech ISBN: 1-58381-044-7
Vol. CS-214	THE Be PHENOMENON IN EARLY-TYPE STARS, IAU Colloquium 175 eds. Myron A. Smith, Huib F. Henrichs and Juan Fabregat ISBN: 1-58381-045-5
Vol. CS-215	COSMIC EVOLUTION AND GALAXY FORMATION: STRUCTURE, INTERACTIONS AND FEEDBACK The 3rd Guillermo Haro Astrophysics Conference eds. José Franco, Elena Terlevich, Omar López-Cruz and Itziar Aretxaga ISBN: 1-58381-046-3
Vol. CS-216	ASTRONOMICAL DATA ANALYSIS SOFTWARE AND SYSTEMS (ADASS) IX eds. Nadine Manset, Christian Veillet and Dennis Crabtree ISBN: 1-58381-047-1 ISSN: 1080-7926
Vol. CS-217	IMAGING AT RADIO THROUGH SUBMILLIMETER WAVELENGTHS eds. Jeffrey G. Mangum and Simon J. E. Radford ISBN: 1-58381-049-8
Vol. CS-218	MAPPING THE HIDDEN UNIVERSE: THE UNIVERSE BEHIND THE MILKY WAY – THE UNIVERSE IN HI eds. Renée C. Kraan-Korteweg, Patricia A. Henning and Heinz Andernach ISBN: 1-58381-050-1
Vol. CS-219	DISKS, PLANETESIMALS, AND PLANETS eds. F. Garzón, C. Eiroa, D. de Winter and T. J. Mahoney ISBN: 1-58381-051-X

ASP CONFERENCE SERIES VOLUMES

Published by the Astronomical Society of the Pacific

PUBLISHED: 2000 (* asterisk means OUT OF PRINT)

Vol. CS-220 AMATEUR – PROFESSIONAL PARTNERSHIPS IN ASTRONOMY: The 111th Annual Meeting of the ASP
eds. John R. Percy and Joseph B. Wilson
ISBN: 1-58381-052-8

Vol. CS-221 STARS, GAS AND DUST IN GALAXIES: EXPLORING THE LINKS
eds. Danielle Alloin, Knut Olsen and Gaspar Galaz
ISBN: 1-58381-053-6

PUBLISHED: 2001

Vol. CS-222 THE PHYSICS OF GALAXY FORMATION
eds. M. Umemura and H. Susa
ISBN: 1-58381-054-4

Vol. CS-223 COOL STARS, STELLAR SYSTEMS AND THE SUN: Eleventh Cambridge Workshop
eds. Ramón J. García López, Rafael Rebolo and María Zapatero Osorio
ISBN: 1-58381-056-0

Vol. CS-224 PROBING THE PHYSICS OF ACTIVE GALACTIC NUCLEI BY MULTIWAVELENGTH MONITORING
eds. Bradley M. Peterson, Ronald S. Polidan and Richard W. Pogge
ISBN: 1-58381-055-2

Vol. CS-225 VIRTUAL OBSERVATORIES OF THE FUTURE
eds. Robert J. Brunner, S. George Djorgovski and Alex S. Szalay
ISBN: 1-58381-057-9

Vol. CS-226 12th EUROPEAN CONFERENCE ON WHITE DWARFS
eds. J. L. Provencal, H. L. Shipman, J. MacDonald and S. Goodchild
ISBN: 1-58381-058-7

Vol. CS-227 BLAZAR DEMOGRAPHICS AND PHYSICS
eds. Paolo Padovani and C. Megan Urry
ISBN: 1-58381-059-5

Vol. CS-228 DYNAMICS OF STAR CLUSTERS AND THE MILKY WAY
eds. S. Deiters, B. Fuchs, A. Just, R. Spurzem and R. Wielen
ISBN: 1-58381-060-9

Vol. CS-229 EVOLUTION OF BINARY AND MULTIPLE STAR SYSTEMS
A Meeting in Celebration of Peter Eggleton's 60th Birthday
eds. Ph. Podsiadlowski, S. Rappaport, A. R. King, F. D'Antona and L. Burderi
IBSN: 1-58381-061-7

Vol. CS-230 GALAXY DISKS AND DISK GALAXIES
eds. Jose G. Funes, S. J. and Enrico Maria Corsini
ISBN: 1-58381-063-3

Vol. CS-231 TETONS 4: GALACTIC STRUCTURE, STARS, AND THE INTERSTELLAR MEDIUM
eds. Charles E. Woodward, Michael D. Bicay and J. Michael Shull
ISBN: 1-58381-064-1

Vol. CS-232 THE NEW ERA OF WIDE FIELD ASTRONOMY
eds. Roger Clowes, Andrew Adamson and Gordon Bromage
ISBN: 1-58381-065-X

ASP CONFERENCE SERIES VOLUMES

Published by the Astronomical Society of the Pacific

PUBLISHED: 2001 (* asterisk means OUT OF PRINT)

Vol. CS-233 P CYGNI 2000: 400 YEARS OF PROGRESS
eds. Mart de Groot and Christiaan Sterken
ISBN: 1-58381-070-6

Vol. CS-234 X-RAY ASTRONOMY 2000
eds. R. Giacconi, S. Serio and L. Stella
ISBN: 1-58381-071-4

Vol. CS-235 SCIENCE WITH THE ATACAMA LARGE MILLIMETER ARRAY (ALMA)
ed. Alwyn Wootten
ISBN: 1-58381-072-2

Vol. CS-236 ADVANCED SOLAR POLARIMETRY: THEORY, OBSERVATION, AND INSTRUMENTATION, The 20th Sacramento Peak Summer Workshop
ed. M. Sigwarth
ISBN: 1-58381-073-0

Vol. CS-237 GRAVITATIONAL LENSING: RECENT PROGRESS AND FUTURE GOALS
eds. Tereasa G. Brainerd and Christopher S. Kochanek
ISBN: 1-58381-074-9

Vol. CS-238 ASTRONOMICAL DATA ANALYSIS SOFTWARE AND SYSTEMS (ADASS) X
eds. F. R. Harnden, Jr., Francis A. Primini and Harry E. Payne
ISBN: 1-58381-075-7

Vol. CS-239 MICROLENSING 2000: A NEW ERA OF MICROLENSING ASTROPHYSICS
eds. John Menzies and Penny D. Sackett
ISBN: 1-58381-076-5

Vol. CS-240 GAS AND GALAXY EVOLUTION,
A Conference in Honor of the 20th Anniversary of the VLA
eds. J. E. Hibbard, M. P. Rupen and J. H. van Gorkom
ISBN: 1-58381-077-3

Vol. CS-241 THE 7TH TAIPEI ASTROPHYSICS WORKSHOP ON COSMIC RAYS IN THE UNIVERSE
ed. Chung-Ming Ko
ISBN: 1-58381-079-X

Vol. CS-242 ETA CARINAE AND OTHER MYSTERIOUS STARS:
THE HIDDEN OPPORTUNITIES OF EMISSION SPECTROSCOPY
eds. Theodore R. Gull, Sveneric Johannson and Kris Davidson
ISBN: 1-58381-080-3

Vol. CS-243 FROM DARKNESS TO LIGHT:
ORIGIN AND EVOLUTION OF YOUNG STELLAR CLUSTERS
eds. Thierry Montmerle and Philippe André
ISBN: 1-58381-081-1

Vol. CS-244 YOUNG STARS NEAR EARTH: PROGRESS AND PROSPECTS
eds. Ray Jayawardhana and Thomas P. Greene
ISBN: 1-58381-082-X

Vol. CS-245 ASTROPHYSICAL AGES AND TIME SCALES
eds. Ted von Hippel, Chris Simpson and Nadine Manset
ISBN: 1-58381-083-8

ASP CONFERENCE SERIES VOLUMES

Published by the Astronomical Society of the Pacific

PUBLISHED: 2001 (* asterisk means OUT OF PRINT)

Vol. CS-246 SMALL TELESCOPE ASTRONOMY ON GLOBAL SCALES, IAU Colloquium 183
eds. Wen-Ping Chen, Claudia Lemme and Bohdan Paczyński
ISBN: 1-58381-084-6

Vol. CS-247 SPECTROSCOPIC CHALLENGES OF PHOTOIONIZED PLASMAS
eds. Gary Ferland and Daniel Wolf Savin
ISBN: 1-58381-085-4

Vol. CS-248 MAGNETIC FIELDS ACROSS THE HERTZSPRUNG-RUSSELL DIAGRAM
eds. G. Mathys, S. K. Solanki and D. T. Wickramasinghe
ISBN: 1-58381-088-9

Vol. CS-249 THE CENTRAL KILOPARSEC OF STARBURSTS AND AGN: THE LA PALMA CONNECTION
eds. J. H. Knapen, J. E. Beckman, I. Shlosman and T. J. Mahoney
ISBN: 1-58381-089-7

Vol. CS-250 PARTICLES AND FIELDS IN RADIO GALAXIES CONFERENCE
eds. Robert A. Laing and Katherine M. Blundell
ISBN: 1-58381-090-0

Vol. CS-251 NEW CENTURY OF X-RAY ASTRONOMY
eds. H. Inoue and H. Kunieda
ISBN: 1-58381-091-9

Vol. CS-252 HISTORICAL DEVELOPMENT OF MODERN COSMOLOGY
eds. Vicent J. Martínez, Virginia Trimble and María Jesús Pons-Bordería
ISBN: 1-58381-092-7

PUBLISHED: 2002

Vol. CS-253 CHEMICAL ENRICHMENT OF INTRACLUSTER AND INTERGALACTIC MEDIUM
eds. Roberto Fusco-Femiano and Francesca Matteucci
ISBN: 1-58381-093-5

Vol. CS-254 EXTRAGALACTIC GAS AT LOW REDSHIFT
eds. John S. Mulchaey and John T. Stocke
ISBN: 1-58381-094-3

Vol. CS-255 MASS OUTFLOW IN ACTIVE GALACTIC NUCLEI: NEW PERSPECTIVES
eds. D. M. Crenshaw, S. B. Kraemer and I. M. George
ISBN: 1-58381-095-1

Vol. CS-256 OBSERVATIONAL ASPECTS OF PULSATING B AND A STARS
eds. Christiaan Sterken and Donald W. Kurtz
ISBN: 1-58381-096-X

Vol. CS-257 AMiBA 2001: HIGH-Z CLUSTERS, MISSING BARYONS, AND CMB POLARIZATION
eds. Lin-Wen Chen, Chung-Pei Ma, Kin-Wang Ng and Ue-Li Pen
ISBN: 1-58381-097-8

Vol. CS-258 ISSUES IN UNIFICATION OF ACTIVE GALACTIC NUCLEI
eds. Roberto Maiolino, Alessandro Marconi and Neil Nagar
ISBN: 1-58381-098-6

Vol. CS-259 RADIAL AND NONRADIAL PULSATIONS AS PROBES OF STELLAR PHYSICS, IAU Colloquium 185
eds. Conny Aerts, Timothy R. Bedding and Jørgen Christensen-Dalsgaard
ISBN: 1-58381-099-4

ASP CONFERENCE SERIES VOLUMES

Published by the Astronomical Society of the Pacific

PUBLISHED: 2002 (* asterisk means OUT OF PRINT)

Vol. CS-260 INTERACTING WINDS FROM MASSIVE STARS
eds. Anthony F. J. Moffat and Nicole St-Louis
ISBN: 1-58381-100-1

Vol. CS-261 THE PHYSICS OF CATACLYSMIC VARIABLES AND RELATED OBJECTS
eds. B. T. Gänsicke, K. Beuermann and K. Reinsch
ISBN: 1-58381-101-X

Vol. CS-262 THE HIGH ENERGY UNIVERSE AT SHARP FOCUS: CHANDRA SCIENCE, held in conjunction with the 113th Annual Meeting of the ASP
eds. Eric M. Schlegel and Saeqa Dil Vrtilek
ISBN: 1-58381-102-8

Vol. CS-263 STELLAR COLLISIONS, MERGERS AND THEIR CONSEQUENCES
ed. Michael M. Shara
ISBN: 1-58381-103-6

Vol. CS-264 CONTINUING THE CHALLENGE OF EUV ASTRONOMY: CURRENT ANALYSIS AND PROSPECTS FOR THE FUTURE
eds. Steve B. Howell, Jean Dupuis, Daniel Golombek, Frederick M. Walter and Jennifer Cullison
ISBN: 1-58381-104-4

Vol. CS-265 ω CENTAURI, A UNIQUE WINDOW INTO ASTROPHYSICS
eds. Floor van Leeuwen, Joanne D. Hughes and Giampaolo Piotto
ISBN: 1-58381-105-2

Vol. CS-266 ASTRONOMICAL SITE EVALUATION IN THE VISIBLE AND RADIO RANGE, IAU Technical Workshop
eds. J. Vernin, Z. Benkhaldoun and C. Muñoz-Tuñón
ISBN: 1-58381-106-0

Vol. CS-267* HOT STAR WORKSHOP III: THE EARLIEST STAGES OF MASSIVE STAR BIRTH
ed. Paul A. Crowther
ISBN: 1-58381-107-9

Vol. CS-268 TRACING COSMIC EVOLUTION WITH GALAXY CLUSTERS
eds. Stefano Borgani, Marino Mezzetti and Riccardo Valdarnini
ISBN: 1-58381-108-7

Vol. CS-269 THE EVOLVING SUN AND ITS INFLUENCE ON PLANETARY ENVIRONMENTS
eds. Benjamín Montesinos, Álvaro Giménez and Edward F. Guinan
ISBN: 1-58381-109-5

Vol. CS-270 ASTRONOMICAL INSTRUMENTATION AND THE BIRTH AND GROWTH OF ASTROPHYSICS: A Symposium held in honor of Robert G. Tull
eds. Frank N. Bash and Christopher Sneden
ISBN: 1-58381-110-9

Vol. CS-271 NEUTRON STARS IN SUPERNOVA REMNANTS
eds. Patrick O. Slane and Bryan M. Gaensler
ISBN: 1-58381-111-7

Vol. CS-272 THE FUTURE OF SOLAR SYSTEM EXPLORATION, 2003–2013 Community Contributions to the NRC Solar System Exploration Decadal Survey
ed. Mark V. Sykes
ISBN: 1-58381-113-3

ASP CONFERENCE SERIES VOLUMES

Published by the Astronomical Society of the Pacific

PUBLISHED: 2002 (* asterisk means OUT OF PRINT)

Vol. CS-273 THE DYNAMICS, STRUCTURE AND HISTORY OF GALAXIES
eds. G. S. Da Costa and H. Jerjen
ISBN: 1-58381-114-1

Vol. CS-274 OBSERVED HR DIAGRAMS AND STELLAR EVOLUTION
eds. Thibault Lejeune and João Fernandes
ISBN: 1-58381-116-8

Vol. CS-275 DISKS OF GALAXIES: KINEMATICS, DYNAMICS AND PERTURBATIONS
eds. E. Athanassoula, A. Bosma and R. Mujica
ISBN: 1-58381-117-6

Vol. CS-276 SEEING THROUGH THE DUST: THE DETECTION OF HI AND THE EXPLORATION OF THE ISM IN GALAXIES
eds. A. R. Taylor, T. L. Landecker and A. G. Willis
ISBN: 1-58381-118-4

Vol. CS 277 STELLAR CORONAE IN THE CHANDRA AND XMM-NEWTON ERA
eds. Fabio Favata and Jeremy J. Drake
ISBN: 1-58381-119-2

Vol. CS 278 NAIC–NRAO SCHOOL ON SINGLE-DISH ASTRONOMY: TECHNIQUES AND APPLICATIONS
eds. Snezana Stanimirovic, Daniel Altschuler, Paul Goldsmith and Chris Salter
ISBN: 1-58381-120-6

Vol. CS 279 EXOTIC STARS AS CHALLENGES TO EVOLUTION, IAU Colloquium 187
eds. Christopher A. Tout and Walter Van Hamme
ISBN: 1-58381-122-2

Vol. CS 280 NEXT GENERATION WIDE-FIELD MULTI-OBJECT SPECTROSCOPY
eds. Michael J. I. Brown and Arjun Dey
ISBN: 1-58381-123-0

Vol. CS 281 ASTRONOMICAL DATA ANALYSIS SOFTWARE AND SYSTEM (ADASS) XI
eds. David A. Bohlender, Daniel Durand and Thomas H. Handley
ISBN: 1-58381-124-9 ISSN: 1080-7926

Vol. CS 282 GALAXIES: THE THIRD DIMENSION
eds. Margarita Rosado, Luc Binette and Lorena Arias
ISBN: 1-58381-125-7

Vol. CS 283 A NEW ERA IN COSMOLOGY
eds. Nigel Metcalfe and Tom Shanks
ISBN: 1-58381-126-5

Vol. CS 284 AGN SURVEYS
eds. R. F. Green, E. Ye. Khachikian and D. B. Sanders
ISBN: 1-58381-127-3

Vol. CS 285 MODES OF STAR FORMATION AND THE ORIGIN OF FIELD POPULATIONS
eds. Eva K. Grebel and Walfgang Brandner
ISBN: 1-58381-128-1

PUBLISHED: 2003

Vol. CS 286 CURRENT THEORETICAL MODELS AND HIGH RESOLUTION SOLAR OBSERVATIONS: PREPARING FOR ATST
eds. Alexei A. Pevtsov and Han Uitenbroek
ISBN: 1-58381-129-X

ASP CONFERENCE SERIES VOLUMES

Published by the Astronomical Society of the Pacific

PUBLISHED: 2003 (* asterisk means OUT OF PRINT)

Vol. CS 287 GALACTIC STAR FORMATION ACROSS THE STELLAR MASS SPECTRUM
eds. J. M. De Buizer and N. S. van der Bliek
ISBN:1-58381-130-3

Vol. CS 288 STELLAR ATMOSPHERE MODELING
eds. I. Hubeny, D. Mihalas and K. Werner
ISBN: 1-58381-131-1

Vol. CS 289 THE PROCEEDINGS OF THE IAU 8TH ASIAN-PACIFIC REGIONAL MEETING, VOLUME 1
eds. Satoru Ikeuchi, John Hearnshaw and Tomoyuki Hanawa
ISBN: 1-58381-134-6

Vol. CS 290 ACTIVE GALACTIC NUCLEI: FROM CENTRAL ENGINE TO HOST GALAXY
eds. S. Collin, F. Combes and I. Shlosman
ISBN: 1-58381-135-4

Vol. CS-291 HUBBLE'S SCIENCE LEGACY:
FUTURE OPTICAL/ULTRAVIOLET ASTRONOMY FROM SPACE
eds. Kenneth R. Sembach, J. Chris Blades, Garth D. Illingworth and Robert C. Kennicutt, Jr.
ISBN: 1-58381-136-2

Vol. CS-292 INTERPLAY OF PERIODIC, CYCLIC AND STOCHASTIC VARIABILITY IN SELECTED AREAS OF THE H-R DIAGRAM
ed. Christiaan Sterken
ISBN: 1-58381-138-9

Vol. CS-293 3D STELLAR EVOLUTION
eds. S. Turcotte, S. C. Keller and R. M. Cavallo
ISBN: 1-58381-140-0

Vol. CS-294 SCIENTIFIC FRONTIERS IN RESEARCH ON EXTRASOLAR PLANETS
eds. Drake Deming and Sara Seager
ISBN: 1-58381-141-9

Vol. CS-295 ASTRONOMICAL DATA ANALYSIS SOFTWARE AND SYSTEMS (ADASS) XII
eds. Harry E. Payne, Robert I. Jedrzejewski and Richard N. Hook
ISBN: 1-58381-142-7

Vol. CS-296 NEW HORIZONS IN GLOBULAR CLUSTER ASTRONOMY
eds. Giampaolo Piotto, Georges Meylan, S. George Djorgovski and Marco Riello
ISBN: 1-58381-143-5

Vol. CS-297 STAR FORMATION THROUGH TIME,
A Conference to Honour Robert J. Terlevich
eds. Enrique Pérez, Rosa M. González Delgado and Guillermo Tenorio-Tagle
ISBN: 1-58381-144-3

Vol. CS-298 GAIA SPECTROSCOPY: SCIENCE AND TECHNOLOGY
ed. Ulisse Munari
ISBN: 1-58381-145-1

Vol. CS-299 HIGH ENERGY BLAZAR ASTRONOMY, An International Conference held to Celebrate the 50th Anniversary of Tuorla Observatory
eds. Leo O. Takalo and Esko Valtaoja
ISBN: 1-58381-146-X

ASP CONFERENCE SERIES VOLUMES

Published by the Astronomical Society of the Pacific

PUBLISHED: 2003 (* asterisk means OUT OF PRINT)

Vol. CS-300 RADIO ASTRONOMY AT THE FRINGE, A Conference held in honor of Kenneth I. Kellermann, on the occasion of his 65th Birthday
eds. J. Anton Zensus, Marshall H. Cohen and Eduardo Ros
ISBN: 1-58381-147-8

Vol. CS-301 MATTER AND ENERGY IN CLUSTERS OF GALAXIES
eds. Stuart Bowyer and Chorng-Yuan Hwang
ISBN: 1-58381-149-4

Vol. CS-302 RADIO PULSARS, In celebration of the contributions of Andrew Lyne, Dick Manchester and Joe Taylor – A Festschrift honoring their 60th Birthdays
eds. Matthew Bailes, David J. Nice and Stephen E. Thorsett
ISBN: 1-58381-151-6

Vol. CS-303 SYMBIOTIC STARS PROBING STELLAR EVOLUTION
eds. R. L. M. Corradi, J. Mikołajewska and T. J. Mahoney
ISBN: 1-58381-152-4

Vol. CS-304 CNO IN THE UNIVERSE
eds. Corinne Charbonnel, Daniel Schaerer and Georges Meynet
ISBN: 1-58381-153-2

Vol. CS-305 International Conference on MAGNETIC FIELDS IN O, B AND A STARS: ORIGIN AND CONNECTION TO PULSATION, ROTATION AND MASS LOSS
eds. Luis A. Balona, Huib F. Henrichs and Rodney Medupe
ISBN: 1-58381-154-0

Vol. CS-306 NEW TECHNOLOGIES IN VLBI
ed. Y. C. Minh
ISBN: 1-58381-155-9

Vol. CS-307 SOLAR POLARIZATION 3
eds. Javier Trujillo Bueno and Jorge Sanchez Almeida
ISBN: 1-58381-156-7

Vol. CS-308 FROM X-RAY BINARIES TO GAMMA-RAY BURSTS
eds. Edward P. J. van den Heuvel, Lex Kaper, Evert Rol and Ralph A. M. J. Wijers
ISBN: 1-58381-158-3

PUBLISHED: 2004

Vol. CS-309 ASTROPHYSICS OF DUST
eds. Adolf N. Witt, Geoffrey C. Clayton and Bruce T. Draine
ISBN: 1-58381-159-1

Vol. CS-310 VARIABLE STARS IN THE LOCAL GROUP, IAU Colloquium 193
eds. Donald W. Kurtz and Karen R. Pollard
ISBN: 1-58381-162-1

Vol. CS-311 AGN PHYSICS WITH THE SLOAN DIGITAL SKY SURVEY
eds. Gordon T. Richards and Patrick B. Hall
ISBN: 1-58381-164-8

Vol. CS-312 Third Rome Workshop on GAMMA-RAY BURSTS IN THE AFTERGLOW ERA
eds. Marco Feroci, Filippo Frontera, Nicola Masetti and Luigi Piro
ISBN: 1-58381-165-6

ASP CONFERENCE SERIES VOLUMES

Published by the Astronomical Society of the Pacific

PUBLISHED: 2004 (* asterisk means OUT OF PRINT)

Vol. CS-313 ASYMMETRICAL PLANETARY NEBULAE III: WINDS, STRUCTURE AND THE THUNDERBIRD
eds. Margaret Meixner, Joel H. Kastner, Bruce Balick and Noam Soker
ISBN: 1-58381-168-0

Vol. CS 314 ASTRONOMICAL DATA ANALYSIS SOFTWARE AND SYSTEMS (ADASS) XIII
eds. Francois Ochsenbein, Mark G. Allen and Daniel Egret
ISBN: 1-58381-169-9 ISSN: 1080-7926

Vol. CS 315 MAGNETIC CATACLYSMIC VARIABLES, IAU Colloquium 190
eds. Sonja Vrielmann and Mark Cropper
ISBN: 1-58381-170-2

Vol. CS 316 ORDER AND CHAOS IN STELLAR AND PLANETARY SYSTEMS
eds. Gene G. Byrd, Konstantin V. Kholshevnikov, Aleksandr A. Mylläri, Igor' I. Nikiforov and Victor V. Orlov
ISBN: 1-58381-172-9

Vol. CS 317 MILKY WAY SURVEYS: THE STRUCTURE AND EVOLUTION OF OUR GALAXY, The 5th Boston University Astrophysics Conference
eds: Dan Clemens, Ronak Shah and Tereasa Brainerd
ISBN: 1-58381-177-X

Vol. CS 318 SPECTROSCOPICALLY AND SPATIALLY RESOLVING THE COMPONENTS OF CLOSE BINARY STARS
eds. Ronald W. Hilditch, Herman Hensberge and Krešimir Pavlovski
ISBN: 1-58381-179-6

Vol. CS-319 NASA OFFICE OF SPACE SCIENCE EDUCATION AND PUBLIC OUTREACH CONFERENCE
eds. Carolyn Narasimhan, Bernhard Beck-Winchatz, Isabel Hawkins and Cassandra Runyon
ISBN: 1-58381-181-8

Vol. CS-320 THE NEUTRAL ISM OF STARBURST GALAXIES
eds. Susanne Aalto, Susanne Hüttemeister and Alan Pedlar
ISBN: 1-58381-182-6

Vol. CS-321 EXTRASOLAR PLANETS: TODAY AND TOMORROW
eds. J. P. Beaulieu, A. Lecavelier des Etangs and C. Terquem
ISBN: 1-58381-183-4

Vol. CS-322 THE FORMATION AND EVOLUTION OF MASSIVE YOUNG STAR CLUSTERS
eds. Henny J. G. L. M. Lamers, Linda J. Smith and Antonella Nota
ISBN: 1-58381-184-2

Vol. CS-323 STAR FORMATION IN THE INTERSTELLAR MEDIUM: In Honor of David Hollenbach, Chris McKee and Frank Shu
eds. D. Johnstone, F. C. Adams, D. N. C. Lin, D. A. Neufeld and E. C. Ostriker
ISBN: 1-58381-185-0

Vol. CS-324 DEBRIS DISKS AND THE FORMATION OF PLANETS: A Symposium in Memory of Fred Gillett
eds. Larry Caroff, L. Juleen Moon, Dana Backman and Elizabeth Praton
ISBN: 1-58381-186-9

ASP CONFERENCE SERIES VOLUMES

Published by the Astronomical Society of the Pacific

PUBLISHED: 2004 (* asterisk means OUT OF PRINT)

Vol. CS-325 THE SOLAR-B MISSION AND THE FOREFRONT OF SOLAR PHYSICS, The Fifth Solar-B Science Meeting
eds. Takashi Sakurai and Takashi Sekii
ISBN: 1-58381-187-7

Vol. CS-326 GRAVITATIONAL LENSING: A UNIQUE TOOL FOR COSMOLOGY
eds. David Valls-Gabaud and Jean-Paul Kneib
ISBN: 1-58381-188-5

Vol. CS-327 SATELLITES AND TIDAL STREAMS
eds. F. Prada, D. Martínez Delgado and T. J. Mahoney
ISBN: 1-58381-190-7

PUBLISHED: 2005

Vol. CS-328 BINARY RADIO PULSARS
eds. F. A. Rasio and I. H. Stairs
ISBN: 1-58381-191-5

Vol. CS-329 NEARBY LARGE-SCALE STRUCTURES AND THE ZONE OF AVOIDANCE
eds. A. P. Fairall and P. A. Woudt
ISBN: 1-58381-192-3

Vol. CS-330 THE ASTROPHYSICS OF CATACLYSMIC VARIABLES AND RELATED OBJECTS
eds. J.-M. Hameury and J.-P. Lasota
ISBN: 1-58381-193-1

Vol. CS-331 EXTRA-PLANAR GAS
ed. Robert Braun
ISBN: 1-58381-194-X

Vol. CS-332 THE FATE OF THE MOST MASSIVE STARS
eds. Roberta M. Humphreys and Krzysztof Z. Stanek
ISBN: 1-58381-195-8

Vol. CS-333 TIDAL EVOLUTION AND OSCILLATIONS IN BINARY STARS: THIRD GRANADA WORKSHOP ON STELLAR STRUCTURE
eds. Antonio Claret, Alvaro Giménez and Jean-Paul Zahn
ISBN: 1-58381-196-6

Vol. CS-334 14TH EUROPEAN WORKSHOP ON WHITE DWARFS
eds. D. Koester and S. Moehler
ISBN: 1-58381-197-4

Vol. CS-335 THE LIGHT-TIME EFFECT IN ASTROPHYSICS
Causes and Cures of the $O - C$ Diagram
ed. Christiaan Sterken
ISBN: 1-58381-200-8

Vol. CS-336 COSMIC ABUNDANCES as Records of Stellar Evolution and Nucleosynthesis, in honor of Dr. David Lambert
eds. Thomas G. Barnes, III and Frank N. Bash
ISBN: 1-58381-201-6

Vol. CS-337 THE NATURE AND EVOLUTION OF DISKS AROUND HOT STARS
eds. Richard Ignace and Kenneth G. Gayley
ISBN: 1-58381-203-2

ASP CONFERENCE SERIES VOLUMES

Published by the Astronomical Society of the Pacific

PUBLISHED: 2005 (* asterisk means OUT OF PRINT)

Vol. CS-338 ASTROMETRY IN THE AGE OF THE NEXT GENERATION OF LARGE TELESCOPES
eds. P. Kenneth Seidelmann and Alice K. B. Monet
ISBN: 1-58381-205-9

Vol. CS-339 OBSERVING DARK ENERGY
eds. Sidney C. Wolff and Tod R. Lauer
ISBN: 1-58381-206-7

Vol. CS-340 FUTURE DIRECTIONS IN HIGH RESOLUTION ASTRONOMY: A Celebration of the 10th Anniversary of the VLBA
eds. Jonathan D. Romney and Mark J. Reid
ISBN: 1-58381-207-5

Vol. CS-341 CHONDRITES AND THE PROTOPLANETARY DISK
eds. Alexander N. Krot, Edward R. D. Scott and Bo Reipurth
ISBN: 1-58381-208-3

Vol. CS-342 1604–2004: SUPERNOVAE AS COSMOLOGICAL LIGHTHOUSES
eds. M. Turatto, S. Benetti, L. Zampieri and W. Shea
ISBN: 1-58381-209-1

Vol. CS-343 ASTRONOMICAL POLARIMETRY: CURRENT STATUS AND FUTURE DIRECTIONS
eds. Andy Adamson, Colin Aspin, Chris J. Davis and Takuya Fujiyoshi
ISBN: 1-58381-210-5

Vol. CS-344 THE COOL UNIVERSE: OBSERVING COSMIC DAWN
eds. C. Lidman and D. Alloin
ISBN: 1-58381-211-3

Vol. CS-345 FROM CLARK LAKE TO THE LONG WAVELENGTH ARRAY: Bill Erickson's Radio Science
eds. Namir E. Kassim, Mario R. Pérez, William Junor and Patricia A. Henning
ISBN: 1-58381-213-X

Vol. CS-346 LARGE SCALE STRUCTURES AND THEIR ROLE IN SOLAR ACTIVITY
eds: K. Sankarasubramanian, M. Penn and A. Pevtsov
ISBN: 1-58381-214-8

Vol. CS-347 ASTRONOMICAL DATA ANALYSIS SOFTWARE AND SYSTEMS (ADASS) XIV
ed: Patrick L. Shopbell, Matthew C. Britton and Rick Ebert
ISBN: 1-58381-215-6

A listing of the IAU volumes published by the ASP follows on the next page

INTERNATIONAL ASTRONOMICAL UNION VOLUMES

Published by the Astronomical Society of the Pacific

PUBLISHED: 1999 (* asterisk means OUT OF PRINT)

Vol. No. 190 NEW VIEWS OF THE MAGELLANIC CLOUDS
eds. You-Hua Chu, Nicholas B. Suntzeff, James E. Hesser and David A. Bohlender
ISBN: 1-58381-021-8

Vol. No. 191 ASYMPTOTIC GIANT BRANCH STARS
eds. T. Le Bertre, A. Lèbre and C. Waelkens
ISBN: 1-886733-90-2

Vol. No. 192 THE STELLAR CONTENT OF LOCAL GROUP GALAXIES
eds. Patricia Whitelock and Russell Cannon
ISBN: 1-886733-82-1

Vol. No. 193 WOLF-RAYET PHENOMENA IN MASSIVE STARS AND STARBURST GALAXIES
eds. Karel A. van der Hucht, Gloria Koenigsberger and Philippe R. J. Eenens
ISBN: 1-58381-004-8

Vol. No. 194 ACTIVE GALACTIC NUCLEI AND RELATED PHENOMENA
eds. Yervant Terzian, Daniel Weedman and Edward Khachikian
ISBN: 1-58381-008-0

PUBLISHED: 2000

Vol. XXIVA TRANSACTIONS OF THE INTERNATIONAL ASTRONOMICAL UNION REPORTS ON ASTRONOMY 1996–1999
ed. Johannes Andersen
ISBN: 1-58381-035-8

Vol. No. 195 HIGHLY ENERGETIC PHYSICAL PROCESSES AND MECHANISMS FOR EMISSION FROM ASTROPHYSICAL PLASMAS
eds. P. C. H. Martens, S. Tsuruta, and M. A. Weber
ISBN: 1-58381-038-2

Vol. No. 197 * ASTROCHEMISTRY: FROM MOLECULAR CLOUDS TO PLANETARY SYSTEMS
eds. Y. C. Minh and E. F. van Dishoeck
ISBN: 1-58381-034-X

Vol. No. 198 THE LIGHT ELEMENTS AND THEIR EVOLUTION
eds. L. da Silva, M. Spite and J. R. de Medeiros
ISBN: 1-58381-048-X

PUBLISHED: 2001

IAU SPS ASTRONOMY FOR DEVELOPING COUNTRIES
Special Session of the XXIV General Assembly of the IAU
ed. Alan H. Batten
ISBN: 1-58381-067-6

Vol. No. 196 PRESERVING THE ASTRONOMICAL SKY
eds. R. J. Cohen and W. T. Sullivan, III
ISBN: 1-58381-078-1

Vol. No. 200 * THE FORMATION OF BINARY STARS
eds. Hans Zinnecker and Robert D. Mathieu
ISBN: 1-58381-068-4

INTERNATIONAL ASTRONOMICAL UNION VOLUMES

Published by the Astronomical Society of the Pacific

PUBLISHED: 2001 (* asterisk means OUT OF PRINT)

Vol. No. 203 RECENT INSIGHTS INTO THE PHYSICS OF THE SUN AND HELIOSPHERE: HIGHLIGHTS FROM SOHO AND OTHER SPACE MISSIONS
eds. Pål Brekke, Bernhard Fleck and Joseph B. Gurman
ISBN: 1-58381-069-2

Vol. No. 204 THE EXTRAGALACTIC INFRARED BACKGROUND AND ITS COSMOLOGICAL IMPLICATIONS
eds. Martin Harwit and Michael G. Hauser
ISBN: 1-58381-062-5

Vol. No. 205 GALAXIES AND THEIR CONSTITUENTS AT THE HIGHEST ANGULAR RESOLUTIONS
eds. Richard T. Schilizzi, Stuart N. Vogel, Francesco Paresce and Martin S. Elvis
ISBN: 1-58381-066-8

Vol. XXIVB TRANSACTIONS OF THE INTERNATIONAL ASTRONOMICAL UNION REPORTS ON ASTRONOMY
ed. Hans Rickman
ISBN: 1-58381-087-0

PUBLISHED: 2002

Vol. No. 12 HIGHLIGHTS OF ASTRONOMY
ed. Hans Rickman
ISBN: 1-58381-086-2

Vol. No. 199 THE UNIVERSE AT LOW RADIO FREQUENCIES
eds. A. Pramesh Rao, G. Swarup and Gopal-Krishna
ISBN: 58381-121-4

Vol. No. 206 COSMIC MASERS: FROM PROTOSTARS TO BLACKHOLES
eds. Victor Migenes and Mark J. Reid
ISBN: 1-58381-112-5

Vol. No. 207 EXTRAGALACTIC STAR CLUSTERS
eds. Doug Geisler, Eva K. Grebel and Dante Minniti
ISBN: 1-58381-115-X

PUBLISHED: 2003

Vol. XXVA TRANSACTIONS OF THE INTERNATIONAL ASTRONOMICAL UNION REPORTS ON ASTRONOMY 1999–2002
ed. Hans Rickman
ISBN: 1-58381-137-0

Vol. No. 208 ASTROPHYSICAL SUPERCOMPUTING USING PARTICLE SIMULATIONS
eds. Junichiro Makino and Piet Hut
ISBN: 1-58381-139-7

Vol. No. 209 PLANETARY NEBULAE: THEIR EVOLUTION AND ROLE IN THE UNIVERSE
eds. Sun Kwok, Michael Dopita and Ralph Sutherland
ISBN: 1-58381-148-6

Vol. No. 210 MODELLING OF STELLAR ATMOSPHERES
eds. N. Piskunov, W. W. Weiss and D. F. Gray
ISBN: 1-58381-160-5

INTERNATIONAL ASTRONOMICAL UNION VOLUMES

Published by the Astronomical Society of the Pacific

PUBLISHED: 2003 (* asterisk means OUT OF PRINT)

Vol. No. 211 BROWN DWARFS
ed. Eduardo Martín
ISBN: 1-58381-132-X

Vol. No. 212 A MASSIVE STAR ODYSSEY: FROM MAIN SEQUENCE TO SUPERNOVA
eds. Karel A. van der Hucht, Artemio Herrero and César Esteban
ISBN: 1-58381-133-8

Vol. No. 214 HIGH ENERGY PROCESSES AND PHENOMENA IN ASTROPHYSICS
eds. X. D. Li, V. Trimble and Z. R. Wang
ISBN: 1-58381-157-5

PUBLISHED: 2004

Vol. No. 202 PLANETARY SYSTEMS IN THE UNIVERSE: OBSERVATION, FORMATION AND EVOLUTION
eds. Alan Penny, Pawel Artymowicz, Anne-Marie LaGrange and Sara Russell
ISBN: 1-58381-176-1

Vol. No. 213 BIOASTRONOMY 2002: LIFE AMONG THE STARS
eds. Ray P. Norris and Frank H. Stootman
ISBN: 1-58381-171-0

Vol. Nol 215 STELLAR ROTATION
eds. André Maeder and Philippe Eenens
ISBN: 1-58381-180-X

Vol. No. 217 RECYCLING INTERGALACTIC AND INTERSTELLAR MATTER
eds. Pierre-Alain Duc, Jonathan Braine and Elias Brinks
ISBN: 1-58381-166-4

Vol. No. 218 YOUNG NEUTRON STARS AND THEIR ENVIRONMENTS
eds. Fernando Camilo and Bryan M. Gaensler
ISBN: 1-58381-178-8

Vol. No. 219 STARS AS SUNS: ACTIVITY, EVOLUTION AND PLANETS
eds. A. K. Dupree and A. O. Benz
ISBN: 1-58381-163-X

Vol. No. 220 DARK MATTER IN GALAXIES
eds. S. D. Ryder, D. J. Pisano, M. A. Walker and K. C. Freeman
ISBN: 1-58381-167-2

Vol. No. 221 STAR FORMATION AT HIGH ANGULAR RESOLUTION
eds. Michael Burton, Ray Jayawardhana and Tyler Bourke
ISBN: 1-58381-161-3

PUBLISHED: 2005

Vol. No. 13 HIGHLIGHTS OF ASTRONOMY
ed. Oddbjorn Engvold
ISBN: 1-58381-189-3

Vol. No. 201 NEW COSMOLOGICAL DATA AND THE VALUES OF THE FUNDAMENTAL PARAMETERS
eds. Anthony Lasenby and Althea Wilkinson
ISBN: 1-58381-212-1

Vol. No. 216 MAPS OF THE COSMOS
eds. Matthew Colless, Lister Staveley-Smith and Raylee Stathakis
ISBN: 1-58381-202-4